Biology

Marcus Barbor
Mike Boyle
Mike Cassidy
Kathryn Senior

Withdrawn

Biology

Marcus Barbor
Mike Boyle
Mike Cassidy
Kathryn Senior

Collins Educational

An imprint of HarperCollins*Publishers*

Published by Collins Educational
An imprint of HarperCollins*Publishers* Ltd
77–85 Fulham Palace Road
London W6 8JB

© Marcus Barbor, Mike Boyle, Mike Cassidy and Kathryn Senior 1997

First published 1997
ISBN 000 322 3272

Marcus Barbor, Mike Boyle, Mike Cassidy and Kathryn Senior assert the
moral right to be identified as the authors of this work.

British Library Cataloguing in Publication Data:
A catalogue record for this book is available from the British Library.

Editor: Kathryn Senior
Publishing coordinator: Pat Winter
Designer: Glynis Edwards
Cover design: Michael Faulkner
Illustrations: Barking Dog Art, Tom Cross, Jerry Fowler, Lois Hague,
Peter Harper, Illustrated Arts, Pantek Arts, TTP International.
Picture research: Marilyn Rawlings
Indexer: Julie Rimington

Printed and bound by Scotprint LTD, Musselburgh.

CONTENTS

About this book

Biology: the science of life

DURING THE LAST couple of centuries, our knowledge of biology has expanded at a staggering rate. Two hundred years ago the average person in the UK had a life expectancy of about 45. Tuberculosis, smallpox, cholera, plague and typhoid all took their toll. People didn't know anything about bacteria or viruses; there was no effective sewage treatment and clean water was a rare commodity. Malnutrition was common: many people survived by growing their own crops but they knew nothing about photosynthesis or plant physiology. And as for surgery – with no anaesthetics or antiseptics – you can probably imagine the horrors involved.

This child has smallpox; the last recorded case of this viral disease was in 1977, in Somalia

At the turn of the millennium, we take for granted many of the greatest advances in biology and medicine. Relatively few people in the more developed countries now die from diseases caused by other organisms. The average life expectancy of someone born in the UK is currently about 80. Many killer diseases are either very rare or have been eradicated completely – as in the case of smallpox. Heart disease and cancer are now the biggest killers in the western world but progress in the treatment of both is advancing rapidly. In the next century, it might even become possible to control some aspects of the genetic causes of ageing, extending life expectancy beyond 100.

As the human population continues to grow and impact on the natural world, our understanding of

plant biology and our increasing awareness of the complexities of ecology is helping us to cope. Application of the latest biological techniques to agriculture, for example, is allowing us to develop new crop plants to increase the efficiency of food production and to grow crops with a higher resistance to drought and disease.

Genetic engineering of crop plants is a controversial development but it is one which could solve the problem of food shortage in many parts of the world

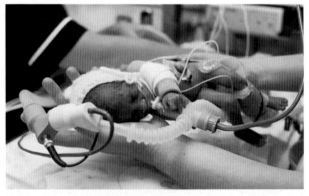

Fifty years ago, a baby born prematurely would have had little chance of survival. Today, advances in technology can save babies born as early as 23 weeks gestation

Of course, any advance in knowledge brings its problems, and the biological revolution is no exception. As people live longer, an increasing number of older people now look forward to a retirement longer than their working life. As medical technology improves, so does the demand for its benefits. And then there are the moral and ethical dilemmas which surround, for example, genetic engineering of crop plants and infertility treatment.

So what will the next century bring? We are certain to see a continuation of the genetic revolution started by Crick and Watson when they worked out the structure of DNA. The Human Genome Project – a quest to map out all of the human DNA sequences – is already underway. Its completion will be a major milestone: in the future we will need to find out exactly what individual genes do, how they interact to switch each other on and off, and how they control growth and development.

To the student

We have written this book to give you a thorough introduction to Biology at advanced level – this may be A-level, Advanced GNVQ, Scottish Higher or International Baccalaureate. We also hope that other students such as nurses and first-year undergraduates will find it a valuable introduction to study at a higher level. The book contains the material which is part of any core syllabus, as well as many of the common options, such as physiology, ecology, diversity and biotechnology.

But we have aimed to do more than just give you the facts that you need in order to cope well with your exam. We hope we have also been able to:

- support and reinforce your understanding of the subject rather than just bombard you with facts.

- convey our enthusiasm and enjoyment of this fascinating subject.

- de-mystify the jargon-filled areas of biology.

- emphasise the 'So what?' factor. We try to stress the applications of what you are learning so that you can relate the facts to the living world.

A recent survey showed that the average A-level biology syllabus contains more new words than a GCSE language course. It often seems that scientists want to keep their subject to themselves by describing their work in Greek and Latin names. In this book, we do, of course, introduce you to new biological terms, but they are always explained carefully. We have kept the non-biological language very clear and straightforward and it has been written to give you a taste of the drama of the subject.

The 'guided tour'

Before you start using this book, it is a good idea to become familiar with its major features.

First, the sections. The book is divided into sections which cover six themes:

- Section 1: The basic unit of life

- Section 2: Energetics

- Section 3: The variety of living organisms

- Section 4: Life processes

- Section 5: Genetics

- Section 6: The environment

Each section begins with a theme opener which introduces the chapters in a section.

Next, the individual chapters. Many teachers start covering a subject at A level assuming that students are familiar with the 'basics'. This can make some areas of the syllabus (such as respiration!) difficult and stressful. In this book we have tried to assume as little as possible and we have written five special introductory chapters designed to be a gentle scene-setter to give an overview of major areas covered in sections 2, 3, 4, 5 and 6.

The other chapters in each section have the following features in common.

The opener

Each chapter begins with a short piece of text that focuses on a practical application of the science covered later - allergies, using rainforest plants to make medicines to treat AIDS, the ageing process, the origins of life, studying plant growth in space, genetic disease, diving, the sex life of wierd animals and many more. Their aim is to emphasise the vital 'So what?' factor.

12 The animal kingdom

IN DECEMBER 1995, a tabloid newspaper introduced a completely new animal to its unsuspecting readers. *Symbion pandora*, a small jelly-like blob less than a millimetre in length, lives on the 'lips' of the Norwegian lobster. You may know this lobster by its more common name: scampi.

So why is this tiny creature of so much interest – to zoologists in particular? Strangely, *Symbion pandora* does not belong to any of the recognised animal phyla – it is definitely not a vertebrate, an arthropod or mollusc, or any other type of animal that we know about. And its structure, behaviour and life-style make other stories that you read in tabloid newspapers seem distinctly tame.

It has no head, but can have two mouths at some stages of its life cycle. These have circular mouthparts and are sited next to the creature's anus – an unusual design feature by anyone's standards. Its sex life is peculiar, too. The female reproduces asexually most of the time, budding off male 'babies' from the surface of her body. Males have two penises and cling on to the first female they come across. Female 'babies' are also produced: these grow inside the female's body. Some of these babies have their own babies while still there, and then two generations of females burst out of the mother's/grandmother's body.

When the scampi sheds its skin, *Symbion* is shed with it. The female *Symbion* then needs to find a mate to reproduce sexually. She literally dissolves and reforms inside her 'skin', and becomes a new form that is more independent and can move about in search of a mate.

Symbion pandora – it may not be pretty, but it can turn heads in the world of zoology

The text

The main text introduces ideas from scratch. Key words are highlighted in bold and explained. Throughout the chapter the text is supported by full colour diagrams and photos with explanatory labels, annotations and captions.

Marginal reminders

Marginal reminders are small boxes headed by a tick. They summarise an essential point in the main text or give you an instant reminder of information contained in other chapters. Some are useful exam hints.

Lipid breakdown

The fat stores in the body can be used wh supply. Stored triglycerides are broken **fatty acids**. The latter are further br 2-carbon **acetyl fragments** by a prod These fragments are combined with co Krebs cycle and, eventually, the electro is converted into **dihydroxyacetone p** used as a fuel in glycolysis.

> ✓ Lipids can fuel aerobic exercise but not glycolysis. Aerobic exercise is therefore a good way to burn excess fat (see Assignment on page 146).

Self-test questions

Each chapter contains several self-test questions which students find valuable. Try to answer them as you work through the text: they test your progress and make you think. (The answers, thankfully, are at the back of the book – but no cheating – think first!)

Using the respiratory quotient

If we measure the volumes of oxygen a exchanged by an organism and find tha that the organism is respiring main substrates will produce different RQ val

- Carbohydrate respiration gives an RQ
- Protein respiration gives an RQ of 0.
- Lipid respiration gives an RQ of 0.7.
- Fermentation gives an RQ of infinity

The problem with trying to work ou respired by looking at RQ values is that

> ? **J (a)** Why is there a range of values for the respiratory quotient of proteins?
> **(b)** Why is the respiratory quotient for fermentation infinity?

Feature boxes

The feature boxes contain topics of special interest – historical features, practical techniques or modern applications of biology. Often, the information is not specifically related to the syllabus, but we have included it because we think it's interesting. Premature babies, chlorophyll, plankton and global warming, hormone replacement therapy, using plant hormones as herbicides, cartilage and footballers' knees, hangovers - it's all good stuff.

acids are fed into the Krebs cycle and are respired (see Fig 8.12).

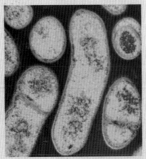

THE ORIGINS OF RESPIRATION: SOME SPECULATION

EVEN THE SIMPLEST life form needs to release the energy locked in organic molecules and it is thought that glycolysis was one of the first metabolic pathways to evolve.

Since there was very little free oxygen at the time, aerobic respiration was impossible. It seems reasonable to assume that the process of glycolysis, as practised by early, bacteria-like organisms (Fig 8.15), used up much of the available food, giving the organisms a major problem: that of energy supply. Solving this problem proved to be a massive evolutionary step: the development of photosynthesis. Not only did photosynthesis provide a new supply, but it also released free oxygen as a by-product. The oxygen in the atmosphere is due to photosynthesis.

The build-up of oxygen in the atmosphere kick-started another evolutionary leap: the appearance of organisms that could respire aerobically. This probably happened fairly quickly because of the massive survival advantage gained by organisms that respire aerobically to produce the much larger amounts of ATP.

It is at this stage that prokaryotic organisms, which could respire aerobically, may have developed a close association with large cells, so forming primitive eukaryotic cells. Find out more about the **endosymbiont theory**, on page xxx.

Fig 8.15 *Clostridium botulinum*, the anaerobic bacterium that causes botulism (serious food poisoning), is found in food such as meat and fish that have been badly canned. Anaerobic bacteria like these are believed to have been the only organisms that could have survived in the oxygen-poor environment 3500 million years ago

Extension boxes

These cover ideas which go beyond the normal A-level syllabus requirements. They are there to give you more in-depth knowledge to help you to better understand a concept that is central to the syllabus.

chlorophyll

The pigments involved in photosynthesis

If you grind up a leaf in a solvent and then pass the mixture through a filter, you end up with a coloured solution. This contains a variety of pigments which can be separated by chromatography.

Sunlight (white light) contains all the wavelengths of the light spectrum that we see as colours in a rainbow. Most pigments absorb light of only certain wavelengths. Other wavelengths are reflected. Green plants are green because the chlorophyll they contain reflects green light. Leaves contain two very similar forms of chlorophyll, **chlorophyll A** and **chlorophyll B**. Some also have **accessory pigments** such as **carotene**. Carotene can trap energy from wavelengths of light which chlorophyll cannot, and so it helps the plant to make more efficient use of sunlight.

It is possible to measure how much of each wavelength is absorbed by a particular pigment. The graph this produces is called an **absorption spectrum** (Fig 9.5). Each pigment has a unique absorption spectrum which is a sort of 'fingerprint'.

Measuring the photosynthetic activity of a plant under light of different wavelengths gives a graph called an **action spectrum**. By taking a pigment and comparing its action spectrum of photosynthesis with its absorption spectrum, you can see whether or not the pigment takes part in the overall process.

The dark red pigment **anthocyanin**, sometimes found outside the chloroplasts in the cytosol, has an absorption spectrum which is completely different from the action spectrum of photosynthesis, showing that anthocyanin does not capture light for photosynthesis.

Fig 9.5 **Chlorophyll** pigments in a green leaf are found to have an action spectrum for photosynthesis which is very similar to their absorption spectrum

Examples

Worked Examples take you through some of the more difficult exam questions. Examples often include calculations, and are designed to show you that using maths in biology is easier than you thought.

136 ■ 8 Cell respiration

EXAMPLE

Q How many ATP molecules are produced, per molecule of glucose, in aerobic cell respiration?

A ATP can be made in one of two ways, by **substrate level phosphorylation** and by **oxidative phosphorylation** (see page 131). We shall take each process in turn.

Substrate level phosphorylation

Glycolysis and the Krebs cycle produce ATP in this way. We can complete the following table in this way:

Stage	ATP produced
Glycolysis	2 (2 used but 4 made)
Pyruvate oxidation	0
The Krebs cycle	2 (1 per turn)
Total	4

Oxidative phosphorylation

This is the production of ATP by the electron transport chain. It relies on a continuous supply of 'high energy' electrons from the preceding processes. Before we can calculate the final ATP total, we need to calculate the number of electrons being provided by NADH and FADH$_2$.

We can draw up a second table, as follows. Remember, pyruvate oxidation and the Krebs cycle occur twice for every glucose molecule put into the system.

Stage	NADH produced	FADH$_2$ produced
Glycolysis	2	0
Pyruvate oxidation	2	0
The Krebs cycle	6 (3 per turn)	2 (1 per turn)
Total per glucose	10	2

We know that;

3 ATP molecules are produced for every molecule of NADH fed into the chain,

and that;

2 ATP molecules are produced for every molecule of FADH$_2$ fed into the chain.

We can now estimate the final ATP total:

10 NAD molecules produce 3 ATPs, each giving a total of	30
2 FADH molecules produce 2 ATPs, each giving a total of	4

Therefore, the total number of ATP molecules produced by oxidative phosphorylation is **34**

The total number of ATP molecules produced by substrate level phosphorylation is **4**

The grand total per glucose molecule is therefore 38 molecules of ATP, most of which are produced by the electron transport chain.

In practice, the figure is usually 36 molecules of ATP. Because the mitochondrial membrane is impermeable to NADH, the 2 molecules of NADH made during glycolysis cannot carry their electrons directly into the electron transport chain. To get over this, most cells have a shuttle system in which the electrons are released by the NADH and passed across the mitochondrial membrane where they are picked up by FADH$_2$. As we saw earlier, FADH$_2$ only produces 2 ATPs compared with the 3 made by NADH, thus reducing the final total by 2.

This is the ATP total under *ideal* conditions. Several other factors can also reduce ATP production even further, but these are beyond the scope of this book

3 OTHER FUELS IN RESPIRATION

So far, we have studied the respiration of glucose, but many organic molecules can be fed in the same central pathway to create ATP. Look back at Fig 8.12 and then read on to see how different substrates can be fed into the central pathway.

Lipid breakdown

The fat stores in the body can be used when carbohydrate is in short supply. Stored triglycerides are broken down into **glycerol** and **fatty acids**. The latter are further broken down to give several 2-carbon **acetyl fragments** by a process called **beta-oxidation**. These fragments are combined with coenzyme A and so enter the Krebs cycle and, eventually, the electron transport chain. Glycerol is converted into **dihydroxyacetone phosphate** so that it can be used as a fuel in glycolysis.

> ✓ Lipids can fuel aerobic exercise but not glycolysis. Aerobic exercise is therefore a good way to burn excess fat (see Assignment on page 146).

144 ■ 8 Cell respiration

SUMMARY

By the end of this chapter, you should know and understand the following:

■ Cell respiration is the release of energy from food. The energy released is transferred to ATP, a chemical that can provide the cell with energy in small, instant, controllable amounts.

■ Cell respiration consists of many reactions. Each reaction is controlled by a different enzyme. Most of the reactions occur inside the mitochondrion.

■ Complete aerobic respiration of glucose has four stages: glycolysis, pyruvate oxidation, the reactions of the Krebs cycle and the reactions of the electron transport chain.

■ Glycolysis: Glucose is broken down into 2 molecules of pyruvate. Only 2 molecules of ATP are made. This happens in the cytosol of cells.

■ Pyruvate oxidation (the 'link' reaction): In the presence of oxygen, pyruvate enters the mitochondrion and is converted into acetyl coenzyme A.

■ The Krebs cycle: A series of reactions, fuelled by acetyl coenzyme A, which produces only 2 molecules of ATP but many electrons and hydrogen ions.

■ The electron transport chain: A series of redox reactions, fuelled by the electrons and protons from the preceding 3 processes, as supplied by the coenzymes NAD and FAD. Electron transport releases enough energy to synthesise most of the ATP produced by aerobic respiration.

■ Most food chemicals – carbohydrates, lipids and proteins – can be respired to provide energy.

■ The rate of respiration can be measured as a respiratory quotient, the ratio of CO_2 evolved to O_2 absorbed. It can be calculated from measurements made using a respirometer.

■ Most organisms respire aerobically (in the presence of oxygen). The substrate (eg glucose) is broken down completely to produce carbon dioxide, water and a lot of ATP.

■ Some organisms and some tissues respire anaerobically. In the absence of oxygen, substrates are only partially broken down, leading to waste products such as lactate (in animals and bacteria) or ethanol and carbon dioxide (in plants and yeast).

QUESTIONS

1 The diagram below shows a calorimeter used to determine the energy content of a sample of food.

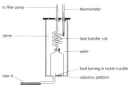

to filter pump
thermometer
stirrer
heat transfer coil
water
food burning in nickel crucible
tube A
asbestos platform

a) State the function of each of the following parts of the apparatus: **(i)** the filter pump, **(ii)** the heat transfer coil, **(iii)** the stirrer.
b) **(i)** Name the gas which enters the apparatus through tube A.
(ii) What is the function of this gas?
c) Suggest *two* sources of error which may arise when using this apparatus to determine the energy content of food.

[ULEAC 1996 Biology/Human Biology: Specimen Paper 4, q.4]

2
a) Define the term *Basal Metabolic Rate*.
b) Assuming a low Physical Activity Level (PAL = 1.4) at work and leisure, the average requirements for energy for adults are given in the table below.

Age (years)	Average energy requirement/MJ day⁻¹	
	Men	Women
19–49	10.60	8.10
50–59	10.60	8.00
60–64	9.93	7.99
65–74	9.71	7.96
75+	8.77	7.61

(i) What **two** general conclusions can you draw from this table?
(ii) Average Energy Requirement (AER) can be calculated from the Basal Metabolic Rate (BMR) and the Physical Activity Level (PAL) using the formula:
AER = BMR × PAL
Using data given in the table, calculate the BMR for a male adult under 60 years of age with a PAL of 1.4.
c) Sugars, complex carbohydrates, proteins and lipids are all eaten by adults in the UK. Although each type of food provides energy, they may be eaten in

Summaries

The main text of each chapter ends with a Summary. This lists the main points of the text. Excellent for revision, or to gain an overview of the topics covered in the chapter.

Questions

Most chapters contain a selection of recent examination questions from the major examining boards. Research has shown that one of the biggest factors in examination success is practice of past paper questions.

References to specific questions are included at appropriate points in the text.

Do I need advanced chemistry and maths?

Generally – no. Recent A-level syllabuses have deliberately reduced their mathematical and chemical content to make the subject more accessible.

In this book we reflect this. The chemical details are kept to a minimum. The mathematical aspects covered centre around the application of numeracy – a basic **key skill**. So, no calculus or complex algebra, but there are exercises in which you work out percentages or the scale of a diagram. The Examples take you through more complex maths where necessary. We look at the chemistry of water and carbon in a bit more detail in the two Appendices at the end of the book.

Assignments

Assignments appear at the end of most chapters. They focus on an application of the subject matter and contain questions which give the opportunity to practise key skills.

160 ■ 9 Photosynthesis

Assignment

ARE PLANTS GOOD AT PHOTOSYNTHESIS?

One way to measure photosynthetic efficiency is to look at the proportion of light that plants can convert into stored energy in food.

Let us start with the energy in a glucose molecule. We can base our calculations on a mole of glucose molecules. A mole is the amount of a substance equal to its relative molecular mass in grams. A mole contains 6×10^{23} particles. When a mole of glucose is burned it releases about 2820 kJ.

1 How much does a mole of glucose weigh?

In the first half of this century, scientists believed that 24 photons of light were needed to synthesise one mole of glucose. Let us see if they were right. You have met the following summary reaction for photosynthesis:

$$6CO_2 + 6H_2O \rightarrow C_6H_{12}O_6 + 6O_2$$

Look back at Fig 9.7 and the text which describes it.

2
a) **(i)** How many electrons are needed to release one molecule of oxygen at PSII?
(ii) How many electrons are needed to release 6 molecules of oxygen?
b) If one photon of light is needed for each electron excited, then how many photons would be needed to generate 6 molecules of oxygen at PSII?

But there is a complication to consider. The chloroplast has two photosystems which do not act independently of one another. Each electron released from PSII ends up at PSI, where another photon is absorbed.

3 Taking into account both photosystems, what is the minimum number of photons required to produce one molecule of glucose?

The latest research shows that the figure is closer to 54 photons per glucose molecule, but perhaps the difference you find may be caused by problems in making accurate measurements.

Scientists have compared the amount of energy that would be trapped in a glucose molecule with the energy in the photons needed to produce it. The actual efficiency, the **yield**, for most crop plants is below 1 per cent.

4 List as many factors as you can which could affect the efficiency of photosynthesis.

Crop researchers have estimated the energy in joules of light falling on a crop and compared it with the yield observed at harvest time. The UK receives sunlight at 100 joules per metre squared per second (100 J m⁻² s⁻¹) in a 24-hour period.

5
a) Calculate the average energy input per hectare per year. (1 hectare = 10 000 m²)
b) Assume the energy value of the crop of potatoes shown in Fig 9.A1 is 17.85 kilojoules per gram dry weight (17.85 kJ g⁻¹). If light were converted to chemical energy with an efficiency of 5 per cent, then what is the maximum crop (dry weight) you could expect in tonnes per hectare?
(1 tonne = 1000 kg)

Fig 9.A1 **Potato crop in Norfolk**

6 Copy and complete the following table:

Yield	% conversion of light to chemical energy	Total dry weight /tonnes hectare⁻¹
Possible maximum	5	
Record UK crop		22
Average		3

7 The values of the record and average conversion efficiencies are very low. Find out the strategies that scientists and farmers are using to improve on them.

Practical work

Practical work is a central feature of any biology course. Biological research is practically based, and students at A level should aim to be competent practical biologists. This book is not a practical manual and so does not contain the details for carrying out experiments, but we do feature experimental biology because of its central importance in research. The Environment section includes notes which you should find useful when you are out on a field trip.

THE CELL: BASIC UNIT OF LIFE

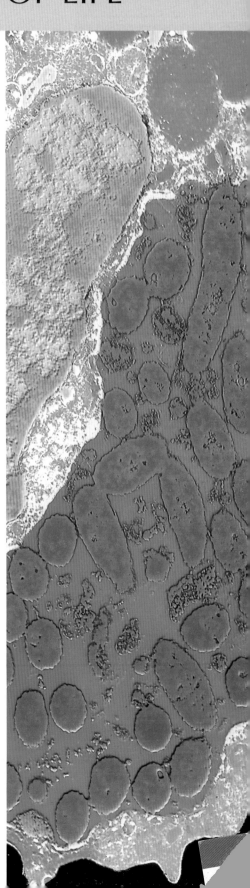

BIOLOGY IS ABOUT understanding how living organisms work. As we have learned more about the world around us, we have zoomed in from whole organisms to look more closely at the individual cells and molecules that make them tick. To study A-level biology today, you need to know a lot about the way cells and molecules interact in the major processes of life. You can't, for example, even start to understand how the lungs work without knowing how molecules diffuse across membranes. This itself is difficult until you know something about the detailed structure of a membrane.

So, while later in the book we look in detail at what many students consider is the really exciting stuff - the major processes which define animals and plants as living organisms - we start by looking at the basic unit of life: the cell.

As you start your course, you probably have only a vague idea about what is in a cell, and so this is the subject of Chapter 1. In Chapter 2 we look at how cells are organised into whole organisms; in Chapters 3 and 4 we find out about the basic chemicals of life and their role in the structure of cells; in Chapter 5 we focus on membranes and how they control the materials which pass in and out of cells; in Chapter 6 we look at how cells divide and how this allows organisms to grow and reproduce. This chapter is a vital introduction to genetics, a major section later in the book.

However, just because cell biology is a necessity which leads on to more advanced study of biological systems, don't get the idea that it's boring. A quick scan of the latest issue of *New Scientist* or *Scientific American* will reassure you that the study of molecules and cells is at the very forefront of modern biological and medical research into AIDS, cancer, genetic disease, heart disease, the way the brain works... and a million more major projects world-wide.

1 What's in a cell?

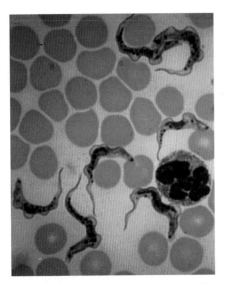

Amongst the blood cells are trypanosomes, the parasites that cause sleeping sickness. Each is a single-celled organism (a protoctist), but you should not think of it as simple. This one cell can perform all the normal processes of life – feeding, excretion, reproduction, movement, growth, cell respiration and the ability to respond to the environment

LIVING CELLS are too small for the naked eye to see but they are still remarkably complex. Individual cells can be thought of as compartments of complex chemical organisation, separated from their surroundings by membranes. Cells can absorb substances from their environment and can use them in many different life processes.

It is difficult to imagine how the complex systems inside a cell could have developed, but several landmark experiments, carried out in the 1950s, have given us some important clues. Scientists recreated the conditions that existed on Earth 3.5 million years ago, at the time when life first started to appear. Then, the atmosphere contained no free oxygen. It is likely to have been a mixture of methane, ammonia, hydrogen sulphide, carbon dioxide and hydrogen. There was no ozone layer, so ultraviolet rays from the Sun bombarded the surface of the Earth. Electrical storms were also frequent.

When scientists in the laboratory bombarded a similar mixture of gases with ultraviolet radiation, and created electrical storm conditions, the chemical reactions produced several biologically important organic molecules – sugars, amino acids and bases. This finding was significant because it is these molecules that are the building blocks of nucleic acids and proteins, the chemicals on which life is based.

1 THE COMPLEXITY OF CELLS

Every human body contains about 50 million million cells. Every day we make new skin cells to replace those that wear away, new red blood cells to replace those that die and new cells to replenish the lining of our digestive system. Not all cell types have such a high turnover, but it is still easy to think of individual cells as small, simple and disposable. And, when you look at cheek cells under a light microscope, what you can see looks fairly uninspiring.

But these impressions are deceptive. As Fig 1.1 shows, individual **animal cells** are extremely complex in both their structure and function. So are **plant cells** (see page 20). Even something like a **bacterium**, a primitive single-celled organism, is far more complex than the fluid that surrounds it (see Figs 1.3 to 1.5).

2 THE CELL THEORY

The observations of early microscopists led to the development of the **cell theory**, a general acceptance that all living things are made of cells. Modern cell theory has three central ideas:

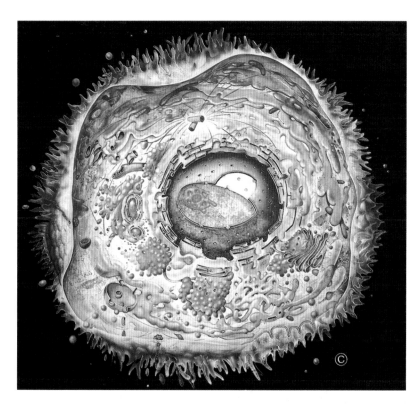

Fig 1.1 **This painting shows the amazing complexity of an animal cell. It was commissioned by the Science Museum on behalf of the Biochemical Society to provide the centrepiece of a permanent exhibition opened in 1987**

- The cell is the smallest independent unit of life.
- The cell is the basic living unit of all organisms: all living things are made up of one or more cells.
- Cells arise from other cells by division. They cannot arise spontaneously.

You will discover as you study biology in more detail that there are exceptions to every rule. In the case of cell theory, that exception is the **virus**. Viruses do not have a cellular structure or organisation, and whether they are actually *living* organisms is a subject of debate. Find out more about them in Chapter 11.

3 USING CELL STRUCTURE TO CLASSIFY ORGANISMS

Organisms can be classified on the basis of the internal organisation of their individual cells. With the exception of viruses, all organisms are either **prokaryotic** or **eukaryotic**. The main differences between prokaryotes and eukaryotes are shown in Table 1.1.

?

A 'Spontaneous generation' was the belief that living things could be created from non-living ones. A recipe for the spontaneous production of mice was reported in the 1600s by the chemist J.B. van Helmont: 'If you press a piece of underwear, soiled with sweat, together with some wheat in an open-mouthed jar, after about 21 days the odour changes and the ferment, coming out of the underwear and penetrating through the husks of wheat, changes the wheat into mice.'

Suggest what was really happening.

	Prokaryotes	Eukaryotes
Organisms	bacteria, including blue-green bacteria	plants, animals, fungi, protoctists
Size of cells	0.1–10 μm diameter	10–100 μm. Much greater volume
Site of nuclear material	DNA in cytoplasm	DNA inside distinct nucleus
Organisation of nuclear material	DNA is circular; no histones; DNA does not condense at cell division	DNA is linear; attached to histones; condenses into visible chromosomes before cell division
Internal structure	few organelles	complex membrane systems; many organelles
Cell walls	always present	present in plants and fungi and some protoctists
Flagella	have simple flagella	have modified cilia called undulipodia which consist of microtubules in a distinctive '9 + 2' arrangement

Table 1.1 **Differences between prokaryotes and eukaryotes**

Blue-green bacteria are prokaryotic organisms that belong to the Kingdom Prokaryotae (see page 177). They were previously called **blue-green algae**; some texts still refer to them under this name and others call them **cyanobacteria**.

Prokaryotes are relatively simple single-celled organisms such as bacteria and blue-green bacteria whose cells have no separate **nucleus** and show little organisation. Eukaryotes – animals, plants, fungi and protoctists – are made up from one or more larger cells with a membrane-bound nucleus. These cells have many membrane systems which form **organelles** and so show high levels of internal organisation.

4 PROKARYOTES – THE FIRST ORGANISMS?

A few billion years ago, the first living organisms to evolve on Earth were probably prokaryotes. The term 'prokaryote' literally means 'before the nucleus'; the genetic material (DNA) of these organisms is not enclosed by a nuclear envelope.

It is tempting to think of prokaryotes as inferior to eukaryotes, but is some ways they have achieved greater success. They have been on Earth more than twice as long as eukaryotes, they are present in greater numbers (there are more bacteria living on your skin than there are people on Earth) and they occupy an enormous number of different habitats. Some bacteria, for example, are able to live in volcanic springs at temperatures as high as 90 °C.

Classification of prokaryotes is covered in Chapter 11.

B (a) Estimate the size (in micrometres) of a full stop on this page.

(b) How many nanometres are there in 1 millimetre?

How small is small?

Cells and the molecules they contain are small. When we study them, we must think about measurements that are minute beyond our imagination. It is easy, for example, to develop a mental block when confronted by the statement 'The nanometre is 10^{-9} m.'

The two units commonly used to describe microscopic objects are the **micrometre** (μm) and the **nanometre** (nm). Starting with a familiar unit, the millimetre (mm), one thousandth of one millimetre is known as a micrometre (μm). The micrometre is used to describe cells and organelles.

An average animal cell is 30 to 50 μm across; the nucleus has a diameter of about 10 μm. Plant cells can reach 150 μm or more in length.

When describing small cellular components and molecules, the useful unit is the nanometre (nm). A nanometre is one thousandth of a micrometre. As a rough guide, the light microscope reveals structures that can be measured in micrometres, but you need an electron microscope to see objects measured in nanometres.

You may also come across another unit, the **Ångström** (Å). An Ångström is one tenth of a nanometre and has been used to describe distances within atoms and molecules.

Fig 1.2 **Units of measurement; how they are used and how they relate to each other**

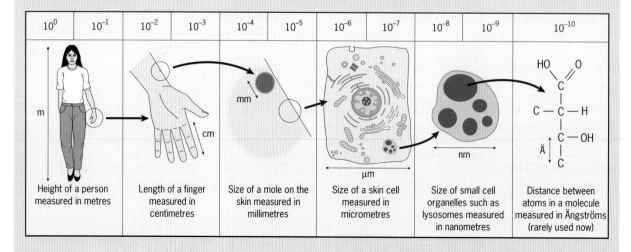

10^0	10^{-1}	10^{-2}	10^{-3}	10^{-4}	10^{-5}	10^{-6}	10^{-7}	10^{-8}	10^{-9}	10^{-10}

Height of a person measured in metres	Length of a finger measured in centimetres	Size of a mole on the skin measured in millimetres	Size of a skin cell measured in micrometres	Size of small cell organelles such as lysosomes measured in nanometres	Distance between atoms in a molecule measured in Ångströms (rarely used now)

Bacteria

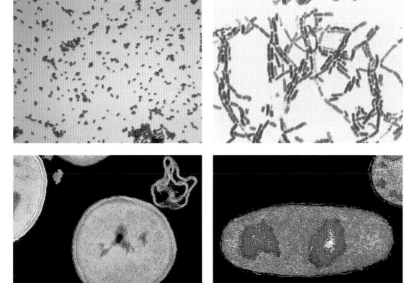

Fig 1.3 **Looking at bacteria with a light microscope reveals that there are two main forms: the spherical organisms are called** cocci (singular coccus), **and the rod-shaped forms are called** bacilli (**singular** bacillus)

Fig 1.4 **Although bacteria look much simpler than animal or plant cells, their membrane – blue in cocci (left) and fine red outline in bacillus (right) – forms a compartment in which complex chemical processes take place**

Bacteria are spherical or rod-shaped cells, usually several micrometres long (see Figs 1.3 and 1.4). Most possess a rigid protective **cell wall** made of **peptidoglycan**, a substance unique to bacteria. Beneath the cell wall the **cell surface membrane**, similar in structure to the membranes of eukaryotic cells, completely encloses the contents of the cell. Some bacteria, such as *Neisseria meningitidis* which can cause meningitis, also have a **capsule**, a sticky coat outside the cell wall. This protects them from drying out, or from digestion by intestinal enzymes, or from attack by a host's immune system.

Fig 1.5 shows the internal structure of a rod-shaped bacterium, *Escherichia coli* (often abbreviated to *E. coli.*)

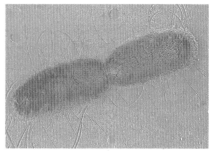

Fig 1.5 *E. coli* **as seen in an electronmicrograph above, and drawn in section and in 3D below.** *E. coli* **is a bacillus found in the human colon. The presence of** *E coli* **in water may indicate contamination by human sewage**

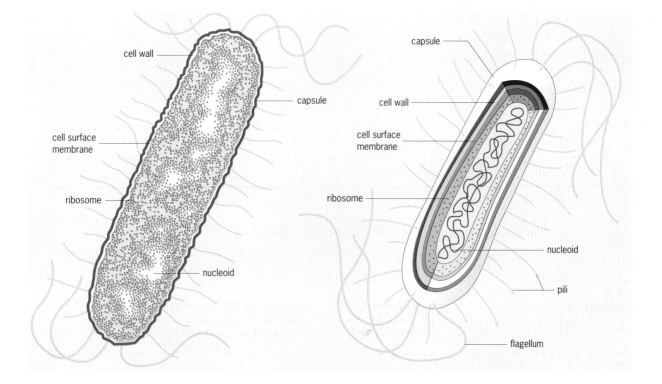

cell wall

capsule

cell surface
membrane

ribosome

nucleoid

capsule

cell wall

cell surface
membrane

ribosome

nucleoid

pili

flagellum

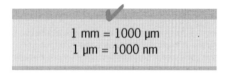

✓

1 mm = 1000 μm
1 μm = 1000 nm

?

C Suggest uses of bacteria based on the substances that some of them can digest.

Inside the cell surface membrane, the bacterial cell is a single cytoplasmic compartment which contains DNA, RNA, proteins and small molecules. Bacteria have a circular piece of DNA in a region of the cytoplasm known as the **nucleoid**. There are smaller rings of DNA, known as **plasmids**, elsewhere in the cytoplasm. Plasmids are of great interest to biologists because they often contain genes for antibiotic resistance, and can be used to carry genes between bacterial cells in genetic engineering (see Chapter 34).

Bacteria feed by **extracellular digestion**. They release enzymes into the surrounding medium and absorb the soluble products. Bacteria can synthesise a wide variety of enzymes and some species can digest unlikely substances such as oil and plastic. Protein are synthesised on **ribosomes** (page 14), and cell respiration (the breakdown of food to release energy) occurs on **mesosomes**, inner extensions of the cell surface membrane.

Some bacteria are **motile**: they can swim. They have thin fibres called **flagella** (singular; flagellum) which are corkscrew-shaped and rotate, propelling the bacteria in different directions.

The roles of bacteria in disease and recycling are discussed in Chapters 26 and 37 respectively.

Blue-green bacteria

The other main group of prokaryotes are the blue-green bacteria (Fig 1.6). These organisms photosynthesise: they have pigments that allow them to use sunlight to synthesise food molecules.

Fig 1.6 **Blue-green bacteria are the most simple photosynthetic organisms**

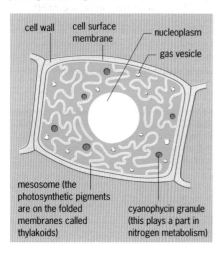

cell wall   cell surface membrane

nucleoplasm

gas vesicle

mesosome (the photosynthetic pigments are on the folded membranes called thylakoids)

cyanophycin granule (this plays a part in nitrogen metabolism)

5 EUKARYOTES

The term eukaryote means 'true nucleus', since the DNA of eukaryotic cells is confined to a definite area inside the cell enclosed by a nuclear envelope.

Eukaryotic cells also have other **organelles** which form compartments. By being in a compartment the chemicals of a particular process, such as respiration or photosynthesis, are kept separate from the rest of the cytoplasm. This allows the chemical reactions of the process to take place quickly and efficiently. This high degree of internal organisation is one of the reasons why eukaryotic cells are larger than prokaryotic cells. The fluid that occupies the space between organelles is the **cytosol**, a solution containing a complex mixture of enzymes, food and waste materials.

All species of animals, plants, fungi and protoctists are composed of eukaryotic cells. We shall now look in detail at the structure of animal and plant cells. The detailed structure of cells of organisms in the Kingdom Fungi (see page 181) and the Kingdom Protoctista (see page 179) are beyond the scope of this book.

Inside an animal cell

If you look at an **animal cell** under a **light microscope** you can see some of its internal structure (see Fig 1.8). With the right staining and illumination techniques and a good quality microscope, the **nucleus, nucleolus, chromosomes** (in a dividing cell), **Golgi complex, mitochondria** and food storage particles such as **glycogen granules**, all show up quite well. But to see far more of the intricate detail in a cell, you really need an **electron microscope**.

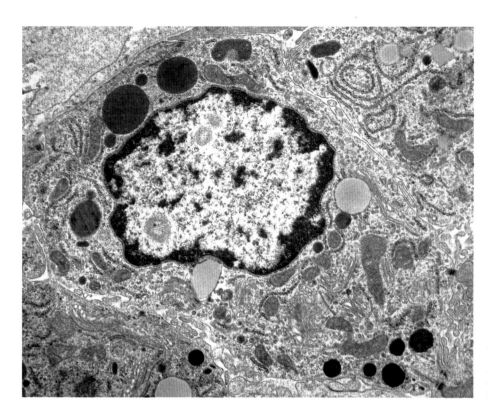

Fig 1.7 **A single whole epithelial cell seen with an electron microscope. It shows that the cytoplasm of animals cells contains a complex system of membrane-bound organelles**

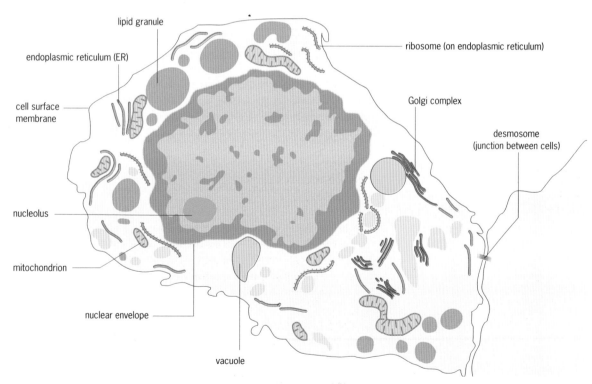

lipid granule

endoplasmic reticulum (ER)

cell surface membrane

nucleolus

mitochondrion

nuclear envelope

vacuole

ribosome (on endoplasmic reticulum)

Golgi complex

desmosome (junction between cells)

The micrograph in Fig 1.7 shows the internal structure of the animal cell. Cell components include the **nucleus, endoplasmic reticulum, mitochondria, cell surface membrane, lysosomes, Golgi complex, ribosomes** and **cytoskeleton.** All these are found in all eukaryotic cells – in plant cells and animal cells. The organelles and other components are labelled. We shall look at each one to investigate its structure more fully and to find out about how each contributes to the function of the cell as a whole.

Using microscopes

The light microscope

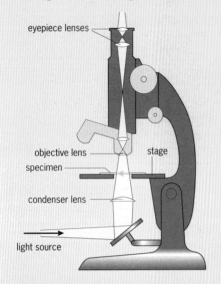

eyepiece lenses

objective lens

stage

specimen

condenser lens

light source

Fig 1.8
The standard compound microscope

Table 1.2 **Structures normally visible in animal and plant cells with the light microscope**

Organelle	Function	Animal cell	Plant cells
Nucleus	control of cell activities	✓	✓
Vacuole	storage and support	✗	✓
Chloroplasts	photosynthesis	✗	✓
Cell wall	support	✗	✓

Fig 1.8 shows the structure of a basic **light microscope** commonly used for teaching purposes. This instrument is also known as the **compound microscope** because two lenses, the **eyepiece** lens and the **objective** lens, are combined to produce a much greater **magnification** than is possible with a single lens. The total magnification is calculated by multiplying the magnification of the two lenses together. For example, if the eyepiece lens has a magnification of ×10 and the objective lens is ×50, the total magnification is ×500.

The light microscope has powers of magnification of up to ×1500, good enough to see cells, larger organelles and individual bacteria, but not powerful enough to reveal smaller structures such as cell membranes, viruses or individual molecules. Fig 1.9 shows how an animal cells appears when viewed with a light microscope and Table 1.2 shows which organelles can be seen in animal and plant cells.

An important feature of a microscope is its **resolving power**, which should not be confused with the magnification. Two objects close together may appear as one single image when viewed under the light microscope, and increasing the magnification will only show the objects as *one* larger image; the microscope is unable to *resolve* the two objects into separate images. Resolving power is as important as magnification when investigating structural details.

The limitation of the light microscope is due to the nature of light itself. The wavelength of light determines the maximum magnification and the resolving power. The wavelength of visible light is around 500–650 nm and the resolving power of the light microscope is 200 nm (0.2 μm), so two objects separated by less than 200 nm will appear as one object.

The electron microscope (EM)

Fig 1.10 shows the essential features of the **electron microscope (EM)**. Invented in the 1930s, the present day version of the EM can magnify up to 500 000 times and has a resolution of 1 nm. In other words, the EM can resolve two objects that are only 1 nm apart. As many biological molecules are larger than 1 nm, the EM can be used to study the arrangement of individual molecules that make up structures within the cell.

The development of the EM has had a huge impact on biology. A magnification of half a million means that an object the size of a full stop becomes over 200 m in diameter – the length of two football pitches. Organelles that are only blurred images when viewed with a light microscope can now be studied in great detail, and many new structures have been discovered using electron microscopy.

While the light microscope uses lenses to focus a beam of *light*, the electron microscope uses electro-

Fig 1.9 **The light microscope can reveal the basic features of animal cells**

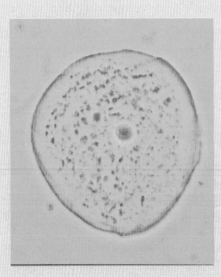

Fig 1.10 **The basic features of the transmission electron microscope. The beam of electrons is focused by electromagnets before passing through the specimen, and the image appears on a fluorescent screen**

cathode (electron gun)

anode

electron beam

condenser lens (electromagnetic 'lens')

vacuum

specimen

objective lens

projector lens

black and white image

fluorescent screen

camera chamber

Table 1.3 **Comparing light and electron microscopes**

	Light microscope	Electron microscope
Illumination	light	electrons
Focused by	lenses	magnets
Maximum magnification	×1500	×500 000
Resolving power	200 nm (0.2 µm)	1 nm
Specimens	living or dead	dead
Preparation of specimens	often simple	more complex
Cost of equipment	relatively cheap	very expensive

Scanning and transmission electron microscopes

There are two main types of electron microscope – the **transmission electron microscope** and the **scanning electron microscope**.

Transmission electron microscopes (TEMs) work on the principle that the beam of electrons is transmitted *through* the specimen. The specimen must be thin and stained using electron-dense substances such as heavy metal salts. These substances deflect electrons in the beam and a pattern is produced which is converted into an image. The electron micrograph in Fig 1.7 was produced using the TEM.

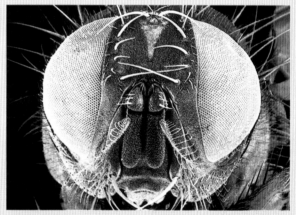

Fig 1.11 **Scanning electron micrograph of the head of a housefly, showing detail of the compound eye**

Scanning electron microscopes (SEMs) can be used to study relatively large three-dimensional objects. Thin sections are not required because the SEM records the electrons that are reflected off the surface of the object rather than passing through it. Fig 1.11 shows the sort of detailed three-dimensional images that can be produced. Although the SEM does not have the resolving power of the TEM, it is more versatile and can be used to observe many kinds of intact structures. The person operating the SEM can move the specimen about, to look at the surface of the structure from a variety of angles – a bit like taking aerial views of the countryside!

magnets to focus a beam of *electrons*. The wavelength of the electrons is much smaller than the wavelength of light, and the resolving power is increased accordingly.

The main disadvantage of the EM is that the electron beam must travel in a vacuum because, being so small, electrons are scattered when they hit air molecules. Specimens for the EM must therefore be prepared (killed, dehydrated and fixed) so that they retain their structure inside a vacuum. Such harsh preparation methods can disrupt cells, and cause **artefacts** – features that do not exist in the living cell – to appear. For example, **microsomes**, tiny vesicles surrounded by ribosomes, were seen when animal cells were examined using the electron microscope. At first, cell biologists thought that these were organelles they had not noticed before, but they later realised that microsomes were fragments of endoplasmic reticulum (see page 13) produced by the fixing process.

The main differences between electron microscopy and light microscopy are shown in Table 1.3.

D If a light microscope had an eyepiece lens of ×25 and an objective lens of ×40, what would the total magnification be?

Those organelles and structures that are found only in plant cells are described on pages 20–22. Table 1.4 summarises the functions of the major organelles.

Table 1.4 **Summary of the functions of major eukaryotic cell organelles**

Organelle	Occurrence	Size	Function
Nucleus	usually one per cell	10 µm	site of the nuclear material – the DNA
Nucleolus	inside nucleus	1–2 µm	manufacture of ribosomes
Mitochondria	numerous in cytoplasm; up to 1000 per cell	1–10 µm	aerobic respiration
Rough endoplasmic reticulum	continuous throughout cytoplasm	extensive membrane network	isolation and transport of newly synthesised proteins
Smooth endoplasmic reticulum	usually small patches in cytoplasm	variable	synthesis of some lipids and steroids
Ribosomes	free in cytoplasm or attached to rough ER	20 nm	site of protein synthesis
Golgi body	free in cytoplasm	variable	modification and synthesis of cellular chemicals
Lysosomes	free in cytoplasm	100 nm	digestion of material
Chloroplasts	cytoplasm of some plant cells, eg mesophyll	4–10 µm × 2–3 µm	site of photosynthesis
Vacuoles	usually large, single fluid-filled space in plant; smaller and more numerous in animals	up to 90 per cent of volume of whole plant cell	storage of salts, sugars and pigments;creates turgor pressure by interaction with cell wall
Cell wall	surrounds all plant cells	thickness varies	provides rigidity and strength

As you read about the different cell components, imagine that you are small enough to move around inside the animal cell, exploring its internal structure.

The cell membrane

To get into the cell, you would need to pass through the **cell surface membrane**. Also known as the **plasma membrane** or **plasmalemma**, this is the boundary between the cell and its environment. The membrane has virtually no mechanical strength but it has a vital role to play in controlling which materials pass in and out of the cell.

Although basically a double layer of phospholipid molecules, arranged tail to tail, the cell surface membrane is a complex structure, studded with protein **pores** (see Fig 1.12). The structure and function of the cell surface membrane is covered in Chapter 5.

Fig 1.12 **The cell surface membrane is a complex organelle but its structure can be represented by this simplified diagram**

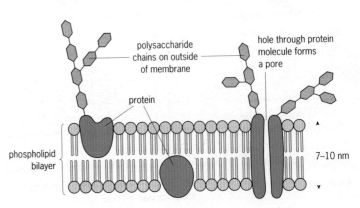

polysaccharide chains on outside of membrane

hole through protein molecule forms a pore

protein

phospholipid bilayer

7–10 nm

The nucleus

The largest and most prominent organelle in the cell is the **nucleus**. Almost all eukaryote cells possess a nucleus – red blood cells in mammals and phloem sieve tubes in plants are exceptions. Every nucleus is surrounded by the **nuclear envelope**. As Fig 1.13 shows, this is a double membrane formed by two lipid bilayers separated by a gap of 20 to 40 nm, known as the **perinuclear space**.

The nucleus is usually spherical and about 10 μm in diameter. It contains the cell's DNA, which carries information that acts as a blueprint, allowing the cell to divide and carry out all its cellular processes. From the micrographs in Fig 1.14 you can see that, in a dividing cell, the DNA is highly condensed into thread-like structures called **chromosomes** but, at other times, it is spread throughout the nucleus as **chromatin**. Nuclei also have one or more **nucleoli** (Fig 1.15). These dark-staining, spherical structures are ribosome-producing machines: they synthesise ribosomal RNA and package it with ribosomal proteins to make ribosomes.

Fig 1.13 **As the electron micrograph and the diagram show, the nucleus is bounded by the nuclear envelope, a double membrane which contains many pores, each one being about 100 nm in diameter. The** nuclear pores **represent about 15 per cent of the total surface area of the nuclear envelope, indicating the heavy traffic of materials in and out of the nucleus. The nucleus contains the cell's genetic material which exists as chromatin, loosely packed DNA attached to proteins called** histones. **The** nucleolus **can clearly be seen**

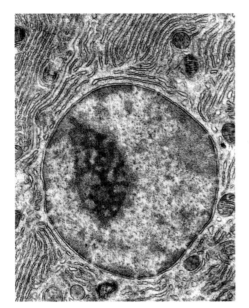
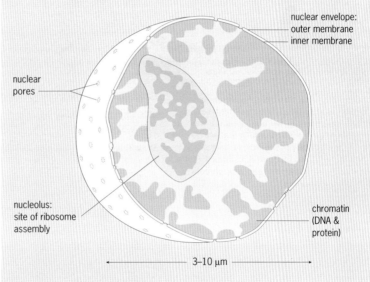

nuclear envelope:
outer membrane
inner membrane

nuclear pores

nucleolus:
site of ribosome
assembly

chromatin
(DNA &
protein)

3–10 μm

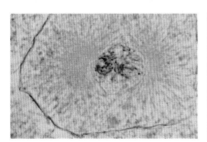

1 DNA condenses into chromosomes

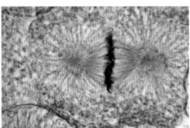

2 Chromosomes move to the centre of the cell

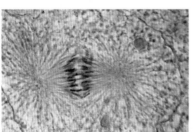

3 Chromosomes are pulled apart

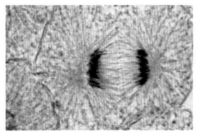

4 Each new cell receives an identical set of chromosomes

Fig 1.14 **Before a cell divides, its DNA condenses into visible chromosomes. This figure shows cells at various stages of cell division. For further information about cell division, see Chapter 6**

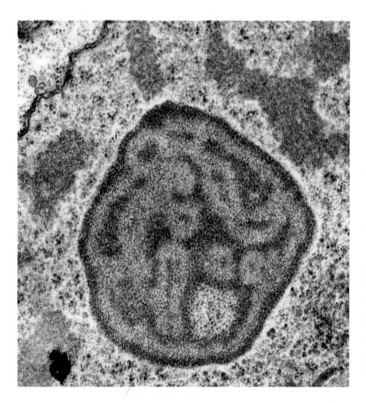

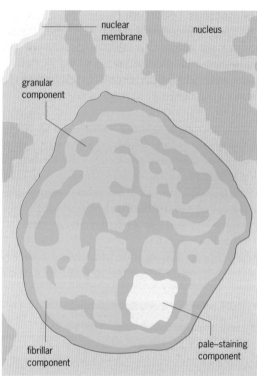

Fig 1.15 **The nucleolus has a highly organised structure. The** pale-staining component **contains DNA from the part of the cell's genome which is being 'read' to produce the RNA that will be used to make new ribosomes. The** fibrillar component **contains fibres of ribosomal proteins, and the** granular component **contains partially assembled ribosomes**

A dividing cell condenses and packages its DNA into chromosomes, a form in which it can be more easily transported inside the cell. The chromosomes must be moved to opposite ends of the cell so that, when the cell splits, each daughter cell receives the correct amount of DNA.

Most normal cells contain two copies of each chromosome (they are said to be **diploid**). **Mitosis**, the more common type of cell division, produces two daughter cells that are genetically identical to the parent cell. The parent cell duplicates its DNA, condenses it into chromosomes and then splits to form *two* new cells. Each must receive a copy of each chromosome pair. **Meiosis**, the other type of cell division, produces **gametes** (sex cells). A cell dividing by meiosis duplicates its DNA, condenses it into chromosomes, as in mitosis, but then usually splits to form *four* new cells. Each of these must contain only *one* copy of each chromosome, half the number of the parent cell (they are said to be **haploid**).

The processes of cell division are covered more fully in Chapter 6.

In a cell that is not dividing, close examination of the chromatin reveals two different levels of density. Dark-staining chromatin, consisting of tightly packed DNA, is called **heterochromatin**, while the lighter, more loosely packed material is known as **euchromatin**. In this more loosely packed state, the DNA is accessible to the proteins which 'read' genes to produce molecules of messenger RNA. Messenger RNA passes into the cytoplasm where it is used by ribosomes as a **template** to build proteins from amino acids.

Individual segments of DNA called **genes** contain the information necessary to make individual proteins, including the enzymes that control most of the cell's activities. In fact, a central concept in biology that is true for all cells, prokaryotes and eukaryotes, is that:

Genes make enzymes which, in turn, control the activities of the cell.

The process of protein synthesis is covered in detail in Chapter 31.

Endoplasmic reticulum (ER)

The nucleus is connected to the complex tunnels of the **endoplasmic reticulum** (Fig 1.16). Endoplasmic reticulum means 'network of the inner fluid'. We will refer to it as ER. All cells contain ER, which has a membrane that is continuous with the outer membrane of the nuclear envelope.

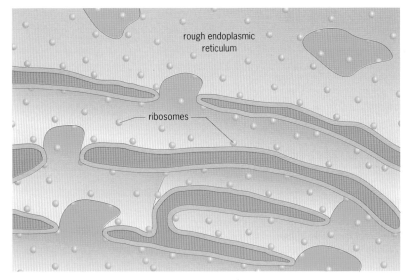

Fig 1.16 **Electronmicrograph (above) and diagram to interpret the photo (left) showing rough ER. The ER is a large sheet of membrane which is folded over on itself many times, forming stacks called** cisternae. **The space inside the cisternae, the** ER lumen, **forms an extensive transport system throughout the cytoplasm**

On much of the outside surface of the ER in a eukaryotic cell are the sites of attachment for **ribosomes**. This gives it a grainy appearance (Fig 1.17) and its name, **rough ER**. RER is folded over itself many times, forming flattened stacks called **cisternae**.

The main function of rough ER is to keep together and transport the proteins made on the ribosomes. Instead of simply diffusing away into the cytoplasm, newly made proteins are threaded through pores in the membrane and accumulate in the rough ER lumen (Fig 1.17) where they are free to fold into their normal three-dimensional shape. Not surprisingly, a mature cell which makes and secretes large amounts of protein, has rough ER that occupies as much as 90 per cent of the total volume of the cytoplasm. The rough ER is also a storage unit for enzymes and other proteins.

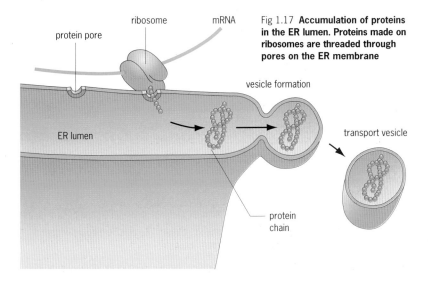

Fig 1.17 **Accumulation of proteins in the ER lumen. Proteins made on ribosomes are threaded through pores on the ER membrane**

?

E What might the cells of the liver, gut and glands have in common?

?

F Was ER discovered before or after the invention of the electron microscope?

Small vesicles containing newly synthesised proteins pinch off from the ends of the rough ER and either fuse with the Golgi complex or pass directly to the cell surface membrane.

ER with no ribosomes attached is known as **smooth ER**. Smooth ER tends to occur in small areas that are not continuous with the nuclear envelope, as in Fig 1.18. Smooth ER is not involved in protein synthesis but is the site of steroid production (many hormones are steroids). It also contains enzymes that **detoxify**, or make harmless, a wide variety of organic molecules, and it acts as a storage site for calcium in skeletal muscle cells.

Fig 1.18 **Smooth ER has no attached ribosomes. It is usually not as abundant as rough ER, although it is common in the cells of the liver, gut and some glands**

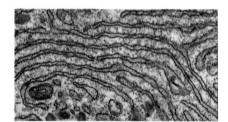

Ribosomes

Ribosomes are small, dense organelles, about 20 nm in diameter, present in great numbers in the cell. Most are attached to the surface of rough ER but they can occur free in the cytoplasm, as in Fig 1.19, which shows their distinctive shape. Ribosomes are made from a combination of ribosomal RNA and protein (65 per cent RNA: 35 per cent protein).

Ribosomes are the point of anchorage for all the chemicals involved in **protein synthesis**. We say that ribosomes are the site of **translation**. Translation is a process in which the code on messenger RNA (mRNA) is used by the ribosome to assemble

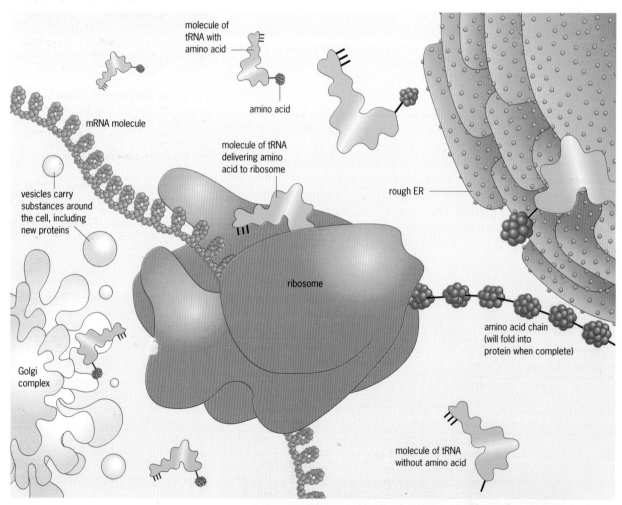

molecule of tRNA with amino acid

amino acid

mRNA molecule

molecule of tRNA delivering amino acid to ribosome

rough ER

vesicles carry substances around the cell, including new proteins

ribosome

amino acid chain (will fold into protein when complete)

Golgi complex

molecule of tRNA without amino acid

Fig 1.19 **An artist's impression of protein synthesis. A ribosome can be thought of as a giant enzyme on which a protein is assembled.** Transfer RNA **molecules (tRNA) bring specific amino acids to the ribosome where they are added to the growing amino acid chain according to the code on the mRNA which is a mobile copy of a gene (not to scale)**

amino acids in the correct order to form a specific protein. This is another central concept in biology:

> **A gene is a piece of DNA that codes for a particular protein. A copy of the gene in the form of messenger RNA passes out of the nucleus and travels to the ribosome where it controls protein synthesis.**

Further details of protein synthesis are described in Chapter 31.

Generally, proteins that are to be used inside the cell are made on the free ribosomes, while those that are to be secreted out of the cell are made on the ribosomes bound to the ER membranes.

The Golgi complex

The **Golgi complex**, also called the **Golgi body** or **Golgi apparatus**, is a tightly packed group of flattened cavities or vesicles (Fig 1.20). The whole organelle is a shifting, flexible structure; vesicles are constantly being added at one side and lost from the other. Generally, vesicles fuse with the **forming face** (nearest to the nucleus) and leave from the **maturing face** (nearest to the cell surface membrane).

The relationship between the rough ER and the Golgi complex in secretion is summarised in Fig 1.21.

The Golgi complex appears to be involved with the synthesis and modification of proteins, lipids and carbohydrates. Studies have shown that proteins made on the ribosomes are packaged into **vesicles** by the ER. Some of the vesicles join with the Golgi complex and the proteins they contain are modified before they are secreted out of the cell.

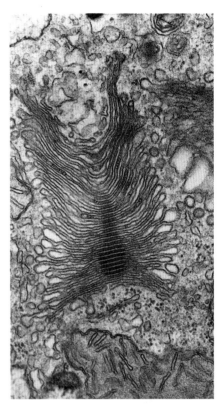

Fig 1.20 **The Golgi complex with a very large number of vesicles. The forming face is at the foot and secretory vesicles are at the top**

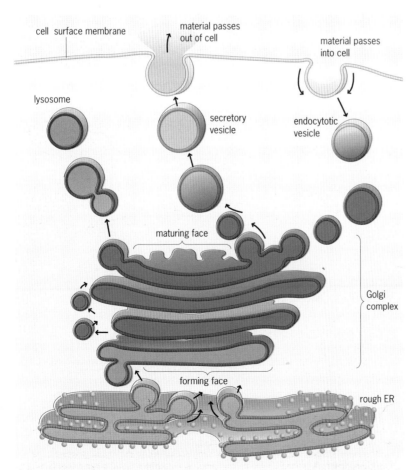

Fig 1.21 **The various functions of the Golgi complex are summarised in this diagram. Vesicle from the ER or from outside the cell bring material *into* the ER. After processing, material passes out of the cell or enters lysosomes to be taken to other organelles**

A **vesicle** is a small spherical organelle, bounded by a single membrane, which is used to store and transport substances around the cell.

$$1 \text{ mm} = 1000 \text{ μm}$$
$$1 \text{ μm} = 1000 \text{ nm}$$

Cell fractionation

See questions 3, 4 and 5. ■ In order to study the function of a particular organelle it is often helpful to isolate it from the rest of the cell. This can be done by **cell fractionation** which is outlined in Fig 1.22.

Fig 1.22 **Cell fractionation: how to isolate particular types of organelle**

1 Tissue (eg liver) is placed in an ice-cold isotonic buffer. When the cell is broken up, membranes are ruptured and many chemicals which do not normally mix are brought together. The ice-cold isotonic buffer minimises 'unusual' reactions, including self–digestion by lytic enzymes.

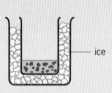 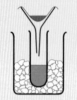 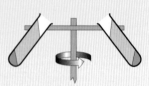

— ice

2 Tissue cut into small pieces
3 Tissue put into blender/homogeniser to break up whole cell
4 Mixture filtered to remove debris
5 Filtrate spun in centrifuge

Principle: The organelles will separate out in a particular order, according to their density and shape
After a time, the sediment containing a particular type of organelle can be separated from the supernatant (the liquid which contains the remaining organelles). The exact times vary from tissue to tissue.

	Organelle		Centrifuge setting (g)	Time
First to separate out	Nuclei		800–1000	5–10 mins
	Mitochondria		10 000–20 000	15–20 mins
	Lysosomes			
	Rough ER		50 000–80 000	30–50 mins
	Plasma membranes		80 000–100 000	60 mins
	Smooth ER			
Last to separate out	Free ribosomes		150 000–300 000	> 60 mins

Lysosomes

Lysosomes, often rather dramatically called 'suicide bags', are small vesicles (0.2–0.5 μm diameter) that contain a mixture of powerful **lytic** (digestive or, literally, breaking) enzymes (Fig 1.23). It is important that the membrane of lysosomes remains intact because if the enzymes leak out they could destroy the whole cell.

Cells need such potentially lethal structures:

● To supply the enzymes to destroy old or surplus organelles.

● To digest material taken into the cell. After a white cell has engulfed a bacterium, lysosomes discharge enzymes into the vacuole (Fig 1.24) and digest the organism. This process is called **phagocytosis**.

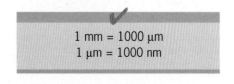

1 mm = 1000 μm
1 μm = 1000 nm

- To destroy whole cells and tissues. Parts of tissues and organs often need to be removed after they have performed their function. In mammals the muscle of the uterus is reduced after giving birth, and milk-producing tissue is destroyed after weaning.

Lysosomes are also involved in the development of organisms. For example, the tail of the tadpole is destroyed during the **metamorphosis** that produces an adult amphibian (Fig 1.25).

Lysosomes are similar in structure to many other vesicles in the cell and are thought to be made in the same way: inactive digestive enzymes from the rough ER pass through the Golgi complex where they are activated and packaged into vesicles.

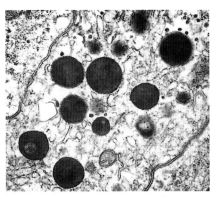

Fig 1.23 **Lysosomes are simply bags of digestive enzymes. They can be distinguished from other vesicles in the cell only by using a stain specific for the chemicals inside the lysosomes**

?

G What would happen to an organism if it lost control of its lysosomes?

?

H If we could control lysosome activity in specific tissues, how could this be used in the treatment of cancer?

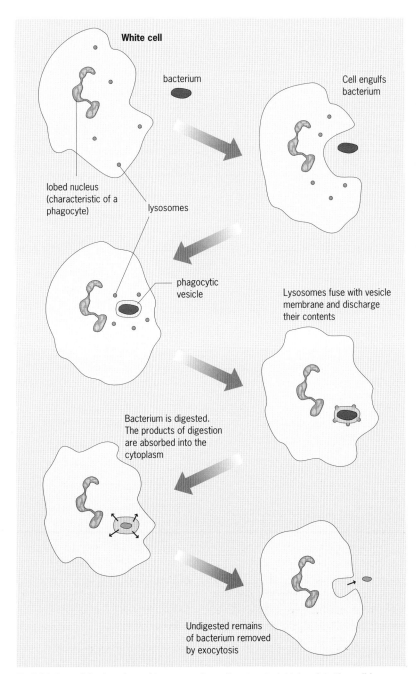

White cell

bacterium

Cell engulfs bacterium

lobed nucleus (characteristic of a phagocyte)

lysosomes

phagocytic vesicle

Lysosomes fuse with vesicle membrane and discharge their contents

Bacterium is digested. The products of digestion are absorbed into the cytoplasm

Undigested remains of bacterium removed by exocytosis

Fig 1.24 **One of the functions of lysosomes is to digest material taken into the cell by the process of phagocytosis (see page 98). Here, a white cell, known as a phagocyte, is ingesting a bacterium. Lysosomes discharge their enzymes into the temporary vacuole and the bacteria are digested**

Fig 1.25 **Lysosomes have been activated in the shortening tail of this frog, causing the tissues to break down and to be reabsorbed into the body of the animal**

?

I Why are digestive enzymes synthesised in an inactive form?

Fig 1.26 **Mitochondria are the sites of aerobic respiration within cells. The inner membrane is folded into cristae, which give a large surface area for attachment of some of the enzymes involved in cell respiration**

Mitochondria

Mitochondria are relatively large, individual organelles, usually spherical or elongated (sausage-shaped), that occur in large numbers in most cells. They are 0.5–1.5 μm wide and 3–10 μm long. Their function is to make ATP via the process of **aerobic respiration**. ATP is a molecule that diffuses around the cell and provides instant chemical energy to the processes that require it.

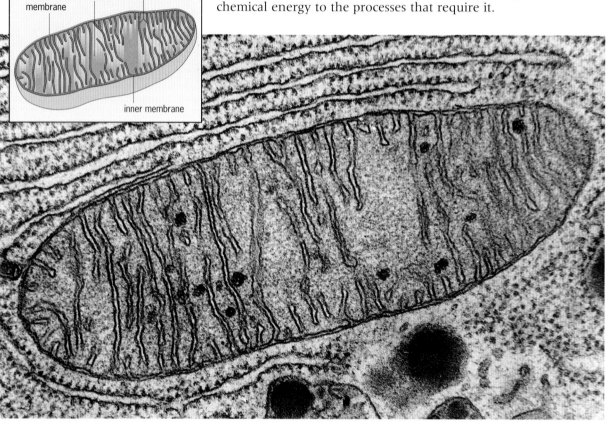

AUTORADIOGRAPHY

ELECTRON MICROSCOPES are useful tools for investigating the structure of cells, but they do not reveal much about the *processes* within a living cell. After the Second World War, as a by-product of work on the atomic bomb, chemicals called radioisotopes became available. These are splendid tools for investigating metabolism because, in the quantities in which they are used, they are unlikely to harm the cell, and their movements through the cell can be traced.

After cells are exposed to a radioactive label, such as carbon-14, they are washed to remove excess label and then fixed, embedded, sectioned and mounted on microscope slides. The slides are then taken to a dark room and are coated with photographic emulsion, heated so that it melts. After coating, the slides are cooled and kept in the dark for various time periods from a few days to several weeks. During this time, radioactive decay causes a chemical change to occur in the film of photographic emulsion. When this is developed using standard photographic techniques, silver grains can be seen in the areas exposed by the radioactivity.

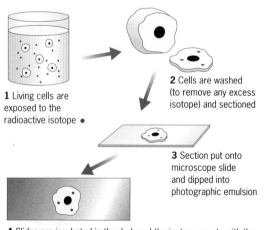

1 Living cells are exposed to the radioactive isotope •

2 Cells are washed (to remove any excess isotope) and sectioned

3 Section put onto microscope slide and dipped into photographic emulsion

4 Slides are incubated in the dark and the isotope reacts with the light-sensitive chemicals. The emulsion is then developed like a normal photograph and, when examined under the microscope, the position of the isotope in the cell can be seen

Fig 1.27 **The basic process of autoradiography**

As Fig 1.26 shows, mitochondria have a double membrane; the outer membrane is smooth while the inner one is folded. This arrangement gives a large internal surface area on which the complex reactions of aerobic respiration can take place. Mitochondria are particularly abundant in metabolically active cells and tissues such as muscle and those involved in **active transport** (see Chapter 5). The function of mitochondria is covered in more detail in Chapter 8.

Centrioles and the cytoskeleton

You might think that travelling inside a cell would be like flying through open spaces. But, from what we know of the cell's skeleton, it might be more like moving through tangles of cobwebs, packed with large obstacles, the organelles. Between the organelles of all eukaryotic cells is a complex network of fine fibres. These fibres are described as **microtubules**, **intermediate filaments** and **microfilaments**, depending on their size and chemical composition.

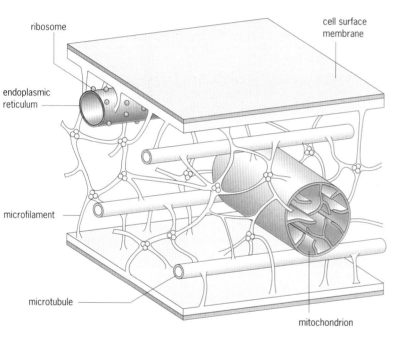

ribosome

cell surface membrane

endoplasmic reticulum

microfilament

microtubule

mitochondrion

Fig 1.28 **Inside the cell is a network of tubules which forms the cytoskeleton.**

Microtubules are fine filaments made from the protein **tubulin**. Intermediate filaments are made from one of several different proteins. In epithelial cells, for example, they contain **keratin**. Microfilaments are usually formed from strings of the protein **actin**. All three types occur throughout the cell and, as Fig 1.28 shows, they help to form a supportive scaffolding known as the **cytoskeleton** ('cell skeleton') which:

● supports the whole cell,

● maintains cell shape, as in red blood cells,

● organises and moves organelles,

● forms the spindle during cell division,

● moves the whole cell.

This last function indicates that the cytoskeleton is not a rigid structure; it can be assembled or dismantled in seconds – as happens in a moving white blood cell. As the foot, or **pseudopod**, extends outwards, the cytoskeleton forms an 'extension' to fill the pseudopod and cytoplasm follows it, flowing between the mesh.

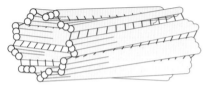

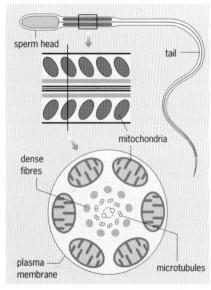

Fig 1.29 **The centrioles are two bundles of fibres that form the cytoskeleton of the cell and also give rise to the spindle during cell division**

Centrioles are short bundles of microfilaments, at right angles to each other (Fig 1.29), which are found in a clear area of cytoplasm known as the **centrosome**. Centrioles occur in all animal cells but are absent from the cells of higher plants. For a long time their function was thought to be the formation of the **spindle** – the cradle of fibres that guide chromosomes during cell division. Recent studies have shown that, in addition to spindle formation, the centrioles can act as the centre of formation for the whole cytoskeleton and therefore have become known as **microtubule organising centres**.

Microtubules make up more complex external cell structures called **cilia**. These tiny hair-like projections are used for cell movement. The tail of a sperm is, for example, a modified cilium, often called a **flagellum**, but which more strictly should be described as an **undulipodium**. (Only prokaryotic motile structures can be called flagella.) All undulipodia in animal and other eukaryotic cells contain a bundle of microtubules in a '9 + 2' arrangement, as shown in Fig 1.30.

Plant cells

A selection of plant cells seen with the light microscope is shown in Fig 1.31.

Plant cells have several features that are not found in animal cells (see Table 1.5). The plant cell is surrounded by a cell wall made of **cellulose**. A large proportion of the inside is taken up with a fluid-filled compartment known as the **vacuole**. Together, the wall and vacuole maintain the shape of the whole cell.

As you can see from the electron micrographs in Fig 1.32, plant cells have specialised organelles, the **chloroplasts**, which enable them to make their own food by **photosynthesis**. Chloroplasts bring together all the chemicals involved in photosynthesis, and provide sites for the chlorophyll molecules so that they can absorb the maximum amount of light.

Fig 1.30 **A cross-section of the tail of a human sperm showing the '9 + 2' arrangement of microtubules in its centre**

sperm head

tail

mitochondria

dense fibres

plasma membrane

microtubules

?

J How many of the following cells would fit into a line 1 cm long?

(a) Animal cells of 20 μm.

(b) Plant cells of 50 μm.

(c) Bacterial cell of 2 μm.

See question 6.

Fig 1.31 **The light microscope shows some of the basic features of plant cells. Note: vacuoles are clearly seen in (c) and (d)**

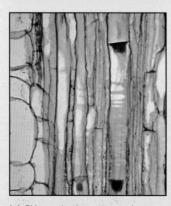

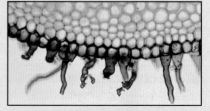

(a) Guard cells of a stoma, a pore which allows the transfer of gases in and out of leaves

(b) Chloroplasts in palisade cells of leaves. Photosynthesis occurs in chloroplasts

(c) Phloem, the tissue that carries dissolved materials

(d) Hair cells on a root. Water and soluble nutrients pass through them from the soil

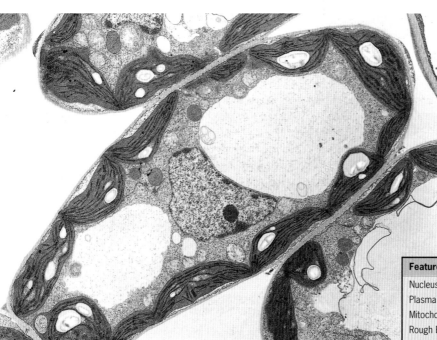

Fig 1.32 **A cell from the palisade layer of a soya bean leaf**

Golgi complex
lysosome
cell wall
cell membrane
vacuole
nucleus
vacuole
nuclear membrane
nucleolus
mitochondrion
rough endoplasmic reticulum
ribosome (on endoplasmic reticulum)
chloroplast   grana

Table 1.5 **Comparing plant cells and animal cells (*centrioles not present in higher plants)**

Feature	Cell: Animal	Plant
Nucleus	✓	✓
Plasma membrane	✓	✓
Mitochondria	✓	✓
Rough ER	✓	✓
Smooth ER	✓	✓
Golgi bodies	✓	✓
Lysosomes	✓	✓
Cell wall	✗	✓
Plastids, eg chloroplasts	✗	✓
Vacuoles	✗	✓
Centrioles*	✓	✓*

?

K Which part of a plant would be unlikely to contain chloroplasts, and why?

The cellulose cell wall

Unlike animal cells, plant cells are enclosed by a **cell wall** which consists of many cellulose fibres cemented together by a mixture of other organic substances.

Cellulose is a polysaccharide (a sugar polymer) which consists of long straight chains of glucose molecules bonded to adjacent molecules (details on page 44). In the cell wall, around 2000 parallel cellulose molecules are packed together to form **microfibrils**. These in turn are bundled together to form **fibrils** which are apparent in Fig 1.33. Like fibreglass, the cell wall has great strength because of the many strong fibres and the 'glue' that holds them together.

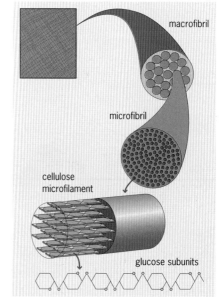

macrofibril
microfibril
cellulose microfilament
glucose subunits

Fig 1.33 **The cellulose cell wall is a composite material.** Fibrils **containing thousands of cellulose molecules are cemented together by a 'glue', a complex mixture of compounds that include** pectins **and** hemicelluloses

All plant cells start by having a **primary cell wall** which is flexible and grows with the cell. The fibrils in the primary wall run in all directions. Most plant cells develop a thicker **secondary** wall when they have reached full size. Many additional layers of fibrils are deposited outside the primary wall. In each layer of the secondary wall the fibrils run mainly in the same direction, and later layers are laid down at different angles. The overall structure therefore has great strength, and prevents any further increase in size.

In certain plant tissues, **xylem** for example, the secondary wall becomes even further strengthened by the addition of **lignin**. This greatly increases the strength of the supporting tissues of trees and shrubs. Wood consists mainly of **xylem cells** which have been thickened and strengthened by cellulose and lignin.

The cell wall has several functions:

- It provides rigidity and strength to the cell. Cell walls need to be strong enough to resist expansion to allow the cell to become turgid.

- It allows communication between cells. Cytoplasmic connections (**plasmodesmata**) run through holes (**pits**) in the cell wall.

- It forces the cell to grow in a certain way. A particular arrangement of fibrils causes the cell to assume a particular shape – for example a long, thin tube.

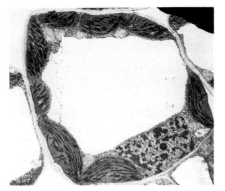

Fig 1.34 **In a mature plant cell, the vacuole becomes so large that the cytoplasm is reduced to a thin layer around the cell**

Vacuoles

Plant cells also have a **vacuole**, a large, fluid-filled cavity bounded by a single membrane, the **tonoplast**. In mature cells the vacuole can occupy over 80 per cent of the cell volume, and is filled with a fluid called **cell sap** (see Fig 1.34). The sap consists of a complex mixture of sugars, salts, pigments and waste products in water.

Vacuoles have several functions:

- They create turgor pressure. The vacuole tends to absorb water by **osmosis** and therefore swells, pushing the cytoplasm against the cell wall. In this state, a plant cell is said to be **turgid**; see pages 92–93.

- They store food substances such as sugars and mineral salts.

- They store pigments that give colour to plant structures such as petals.

- They may accumulate waste products and by-products of metabolism. In some cases, these chemicals may be toxic or have an unpleasant taste and are used by the plant to make it less palatable to herbivores.

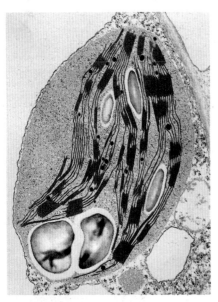

Chloroplasts and other plastids

Chloroplasts are one of a group of plant cell organelles known as **plastids**. They are surrounded by a double membrane and contain an elaborate internal membrane system that houses the photosynthetic apparatus (Fig 1.35). The structure and function of chloroplasts is covered in detail on page 150.

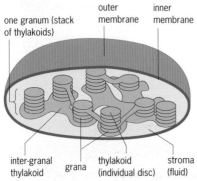

one granum (stack of thylakoids) outer membrane inner membrane

inter-granal thylakoid grana thylakoid (individual disc) stroma (fluid)

Fig 1.35 **Chloroplasts are organelles in which all the chemicals associated with photosynthesis are brought together. Chlorophyll molecules are housed on membranes called** thylakoids **to maximise light absorption. In a flowering plant, most chloroplasts are found in the palisade cells of the leaves**

There are several other sorts of plastid. One type, **chromoplasts**, found in the cells of petals and fruit, also produce special pigments. But these are reds, yellows and blues instead of the green produced by chloroplasts. Another type, **amyloplasts**, contain stores of starch and are found in seeds and in tubers such as potatoes.

The origin of mitochondria and chloroplasts: the endosymbiont theory

Scientists continue to speculate about the intriguing theory that mitochondria and chloroplasts did not originate from inside eukaryotic cells. Many researchers think that they were originally free-living prokaryotes, similar to bacteria and blue-green bacteria, which, at some point, became incorporated into larger cells.

Let's look at the evidence. Mitochondria and chloroplasts are similar to prokaryotes in several important ways:

- They are similar in size to prokaryotes.

- They have their own DNA, as prokaryotes do, suggesting an ability to reproduce independently.

- The DNA of both organelles is circular, like that of bacteria.

- Both organelles have their own ribosomes, smaller than those in eukaryotic cells but identical to those in prokaryotes.

- All reproduce by binary fission.

- The biochemistry of mitochondria is closer to that of aerobic bacteria than it is to that of the eukaryotic cell; there is an even closer similarity between chloroplasts and cyanobacteria.

This theory was first suggested by a team led by Lyn Margulis at Boston University, USA. Assuming the theory is true, how could these organisms have become incorporated into a larger cell without being destroyed by lysosomes?

Imagine a heterotrophic, anaerobic organism – similar to an amoeba in structure and movement – that feeds by phagocytosis and respires without oxygen, though oxygen is present in its environment. If such a cell were to ingest aerobic bacteria that it did not digest, the bacteria could respire aerobically, and in doing so could provide the host organism with far more ATP than the host could produce on its own. The host cell in turn would give the bacteria material to respire – food.

Similarly chloroplasts, which may originally have been free-living blue-green bacteria, could also have become incorporated into larger cells. If these were ingested – but not digested – they would continue to photosynthesise within the host and so provide it with organic food and oxygen – an obvious advantage. The host, in turn, could provide the blue-green bacterium with carbon dioxide and nitrogen.

Associations in which an autotroph lives inside a heterotroph, or an aerobe lives inside an anaerobe, are both very convenient arrangements. One organism produces what the other needs. This is the key to the success of several other associations, for example, lichens (algae and fungi) and corals (algae inside animals). Both host and 'passenger' gain a survival advantage: the host gets food and the smaller organism is protected from predators.

The idea of incorporation by ingestion could also explain why both organelles have a double membrane. The inner membrane would represent the original prokaryotic membrane while the outer one could represent the food vacuole into which the organism was taken. Although this theory may seem to be based on 'ifs and buts', there is some strong evidence for it, not least the existence of organisms like *Mixotricha* (Fig 1.36), a living example of an endosymbiotic relationship. Other organisms that show endosymbiotic relationships are covered in Chapter 11.

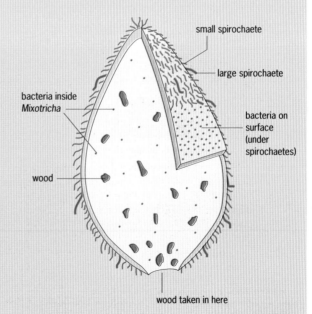

Fig 1.36 *Mixotricha* is a protoctist that lives in the gut of Australian termites. The termites eat wood but cannot digest it without the help of organisms like *Mixotricha*. The *Mixotricha* itself also forms symbiotic or mutualistic relationships with spirochaetes and other bacteria that live on and inside it. The ones on the outside allow it to move about (they behave a bit like cilia) and the ones on the inside act like mitochondria, allowing it to respire aerobically

An **autotroph** is an organism that can make its own food: a green plant is an autotroph that makes its own food by photosynthesis. A **heterotroph** is an organism that must obtain ready-made food: a cow is a heterotroph that obtains its food from grass.

SUMMARY

When you have finished this chapter you should know and understand the following:

■ All living things (except viruses) are composed of either prokaryotic or eukaryotic cells.

■ Bacteria and blue-green bacteria are prokaryotes; plants, animals, protoctists and fungi are eukaryotes.

■ Prokaryotic cells are relatively small, with very little internal membrane organisation.

■ Eukaryotic cells are relatively large, with a high degree of internal organisation.

■ Eukaryotic cells have a true, membrane-bound nucleus and many organelles – membrane-bound compartments in which particular chemical processes take place.

■ The nucleus, ribosomes, rough ER, vesicles and Golgi complex in eukaryotic cells all take part in different stages in the making and packaging of chemical products.

■ Mitochondria are the site of production of most of the eukaryotic cell's ATP, which provides the energy for other cellular processes.

■ Eukaryotic cells have a cytoskeleton, a mesh of tubules that support and shape the cell.

■ Plant cells contain structures not found in animal cells: chloroplasts, a cell wall and a vacuole.

■ We can study the cell using various experimental techniques: the electron microscope reveals the structure of the cell, cell fractionation isolates individual organelles, and radioisotopes and autoradiography allow us to follow the passage of materials through the cell.

QUESTIONS

1 The photograph is an electron-micrograph of part of a cell from the pancreas of a mammal.

a)
 (i) Name features labelled **A**, **B** and **C**.
 (ii) Briefly describe the part played by features **A** and **B** in the production of protein granules in the cytoplasm.
 (iii) Some of the features labelled **C** are elongated; others are more or less circular. Explain this difference in shape.

b)
 (i) The magnification of this photograph is 2350 times. Calculate the actual maximum diameter of the labelled protein granule **D**. (Show your working.)
 (ii) Give **one** way, apart from size and shape, by which the protein granules might be distinguished from mitochondria.
 (iii) Name **one** protein that is secreted by pancreatic cells.

[AEB 1996 Biology AS: Specimen Paper 1, q.12]

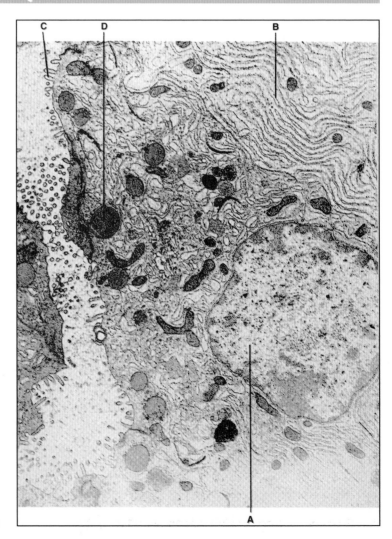

(Animal Physiology Unit, Babraham, Cambridge)

2 The arrows in the diagram show the path followed by a protein produced in a secretory cell.

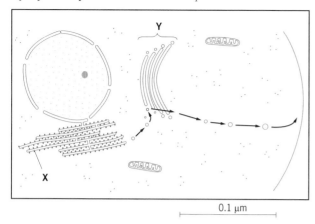

0.1 μm

a) Calculate the magnification. Show your working.

b) Identify organelle **X**.

c) With reference to the protein being produced:
   (i) identify **one** function of organelle **Y**;
   (ii) explain how the protein reaches the outside of the cell from organelle **Y**.

[AEB: 1995 Human Biology: Paper 1, q.2]

3 Liver cells were ground to produce an homogenate. The flow chart shows how centrifugation was used to separate organelles from liver cells.

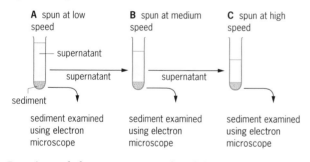

Drawings of electron micrographs of three organelles separated by the centrifugation are shown below. The drawings are **not** to the same scale.

 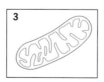

a) Complete the table below.

Electron micrograph	Name of organelle	Centrifuge tube in which the organelle would be the main constituent of the sediment
1		
2		
3		

b) Explain why it is possible to separate the organelles in this way.

[NEAB Feb 1995 Biology: Modular Life Processes Test, q.2]

4 A sample of animal tissue was treated in order to separate the cell components. A chemical analysis of the pure fractions shown in the table was carried out.

Cell component	DNA	RNA	Protein	Phospholipid
Cell surface membrane				
Rough endoplasmic reticulum				
Mitochondria				
Nuclei				

a) Complete the table to show the results of the analysis. Mark the box with a tick if you think that the chemical was present or with a cross if the chemical was absent.

b) Describe concisely how the cell components were separated.

[AEB 1994 Human Biology: Paper 1, q.2]

5 Actively secreting cells growing in tissue culture were provided with a dose of radioactive amino acids for a very short period. At various times, samples of the cells were taken and homogenised and fractionated so that different parts of the cells could be measured for their radioactivity. The following table shows the results of this experiment.

	Radioactivity present in the following cell organelles		
Time after radioactive aminoacids given (minutes)	Rough endoplasmic reticulum	Golgi apparatus	Secretory vesicles
1	123	21	7
20	84	42	7
40	39	84	7
60	28	77	7
90	27	49	28
120	24	38	56
180	28	21	63
240	18	11	20

a) Using the data provided, draw a graph to show the changes in radioactivity in the different organelles in the cells. Put all three lines on one graph.

b) Describe the role of the three organelles mentioned in the production of material to be secreted by the cells

[National Extension College Flexible Learning Pack: Biology. Assignment 1, Part 2, q.1]

Turn to the next page for question 6.

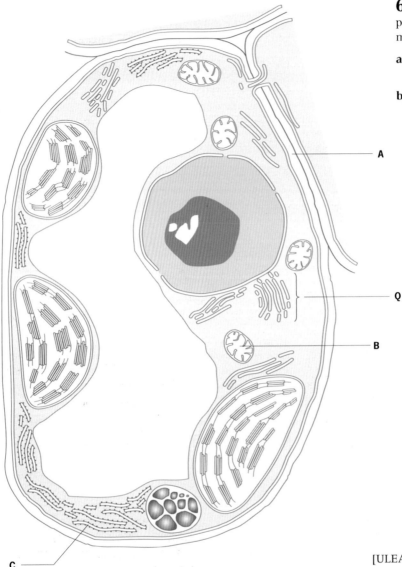

6 The drawing below shows a leaf palisade cell as revealed by an electron microscope.

a) Name the parts labelled **A**, **B** and **C**.

b)   **(i)** On the diagram, label the nuclear envelope.

     (ii) Give **one** function of the nuclear envelope.

     (iii) Identify Q and describe **one** function.

[ULEAC 1994 Human Biology: Paper 1, q.4]

Assignment

GETTING A SENSE OF PROPORTION

You have studied the separate components of cells and now this assignment should help you to gain an overview of whole cells. We shall look at:

A1 An animal cell from the epithelium (lining) of the mammalian small intestine which enables digested food to be absorbed from the intestine into the blood.

A2 A plant cell from the palisade layer of a leaf which makes food by photosynthesis.

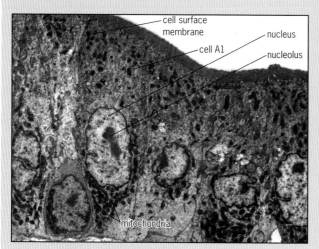

Fig 1.A1 **Electronmicrograph showing epithelial cells from the lining of a mammalian small intestine (× 1500)**

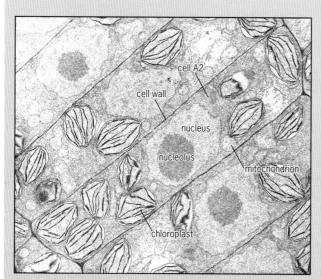

Fig 1.A2 **Electronmicrograph showing a palisade cell from a flowering plant (× 1600)**

1 Use the scales given to work out the actual size of the cells in µm.

2 Copy and complete the table below.

Organelle	Size on paper	Actual size in the cell

Cell A1:
  Nucleus
  Nucleolus
  Mitochondrion		

Cell A2:
  Nucleus
  Nucleolus
  Mitochondrion
  Chloroplast		

3 Why do you think that the endoplasmic reticulum and the cells surface membrane have been left out of the above table?

4 Imagine that the picture of the plant cell was 10 metres in diameter – the size of a large room.

a) What magnification would this represent?

b) State how big the following would be:
   (i) a mitochondrion (length and width),
   (ii) a nucleus (diameter),
   (iii) a ribosome (diameter).

The role of the cell inside the organism

As we shall see in Chapter 2, cells inside an organism are specialised and perform particular functions. The structure of cells usually reflects their specialisation.

5 The **intestinal epithelial cell** allows rapid absorption of digested food into the bloodstream. The two mechanisms for the absorption of food molecules are diffusion and active transport. (You can look at digestion in more detail in Chapter 16.) Suggest reasons why this cell has large numbers of microvilli and mitochondria.

6 The **palisade cell** is on the upper surface of a leaf, the best location for photosynthesis. Remember that photosynthesis requires light, carbon dioxide and water and produces glucose and oxygen.

Suggest how the following features help the palisade cell to perform its function:

a) the large numbers of chloroplasts,

b) the observation that chloroplasts circulate within the cell,

c) the deep, cylindrical shape of the cell.

You will gain a more complete picture of plant function when you study the structure of the whole leaf in Chapter 17.

2 Levels of organisation

IF HEART DISEASE and cancer, the biggest causes of death in the developed world, were to be eliminated tomorrow, could we expect to live forever? The answer is definitely no. Some scientists think that it would only extend life expectancy from 80 years to about 90. Even when no disease is present, different parts of our bodies begin to let us down after a certain age.

The effects of ageing range from the annoying but harmless wrinkles which form in the skin as its elastic tissue becomes less elastic, to the more serious effects found in major body systems which lead to problems with blood circulation, memory, breathing and mobility.

The process of ageing is not really understood, but we do know that it happens at the level of individual cells. One theory is that repair mechanisms which correct random mutations in DNA become much less efficient with each division. Mutations build up and the cell eventually becomes unable to function. Another theory is that damage within cells is caused by free radicals, such as negatively charged oxygen molecules. We know that mitochondria produce these during respiration. As the cell gets older, its enzymes cannot break down free radicals fast enough and they start to attack cell structures, causing cell damage and death.

Charlie Chaplin: The effects of ageing are inevitable and unstoppable

1 THE COMPLEXITY OF LIVING THINGS

All single-celled organisms are relatively small and most are microscopic. There are very few large single cells. Egg cells are an exception: some are large because they need to carry enormous food stores (Fig 2.1).

Large organisms are always multicellular. Their body systems consist of specialised organs and tissues (Fig 2.2) which interact and cooperate to keep all the cells of the body in the best possible environment. Different groups of cells carry out different tasks and functions: they have **differentiated** and become **specialised**. Different types of cell develop to carry out specific functions which contribute to the success of the organism as a whole. This is known as **division of labour**.

Fig 2.1 **The egg of an ostrich is probably the largest single cell which exists today, although some dinosaur eggs would have been larger. Much of the large volume is taken up with food material to supply the developing ostrich chick**

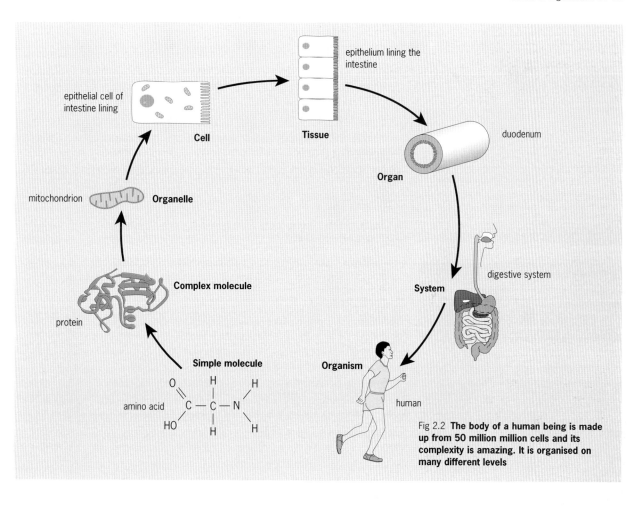

epithelial cell of intestine lining

Cell

epithelium lining the intestine

Tissue

Organ

duodenum

mitochondrion **Organelle**

Complex molecule

protein

Simple molecule

amino acid

System

digestive system

Organism

human

Fig 2.2 **The body of a human being is made up from 50 million million cells and its complexity is amazing. It is organised on many different levels**

2 TISSUES

A **tissue** is a collection of similar specialised cells which work together to achieve a particular function. Different tissues combine to form more complex structures called **organs**. An organ usually has a particular function in the organism, for example to detect light, to absorb food or to produce a hormone.

Studying the detailed structure of individual tissues can be quite dull unless you look at how the structure of a tissue relates to its function in the body of the organism. Getting familiar with normal tissues is particularly important if you later go on to a medical career. Recognising what is normal and abnormal in a particular type of tissue can mean the difference between life and death (Fig 2.3).

A **tissue** is a collection of similar specialised cells.

Histology is the study of cells, tissues and organs, using microscope techniques.

Pathology is the study of disease.

Mitosis is a type of cell division (see Chapter 6).

Oncology is the study of cancer.

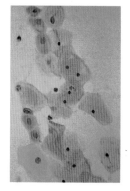
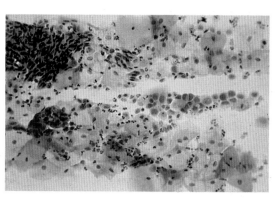

Fig 2.3 **Each of these micrographs shows a tissue sample taken from the cervix of a woman during a routine smear test.**

Left: In normal epithelial cells, the nuclei shrink as the cells get older.

Right: Cancerous cells have nuclei that take up most of the cell, as centre right; the mass of cells with dark oval nuclei at top left indicate an advanced stage of cancer

Animal tissues

Living things may contain many different organs, but the organs are all made up from the same basic types of tissue. In mammals, each organ of the body – say, the lung, kidney or eye – is constructed from a combination of the four basic tissue types:

- **Epithelial tissue** forms thin sheets of tissue which line and cover body structures. The intestines, for example, are lined by epithelial tissue.
- **Connective tissue** is usually tough and fibrous, forming structures which hold the body together. Ligaments and tendons are mainly connective tissue.
- **Muscular tissue** can contract to produce movement.
- **Nervous tissue** has the ability to conduct impulses, allowing communication between different parts of the body.

Epithelial tissue

Epithelial tissues form continuous sheets which line or cover most structures and cavities within the body. Different types of epithelia are classified according to the size and shape of their cells, and the number of layers of cells they contain (Fig 2.4).

Epithelial tissues form barriers which keep different body systems separate. They also have many other functions. For instance, the epithelial cells which line the mammalian respiratory tract are ciliated: they form a 'carpet' of hairs which trap inhaled dust and sweep it away from the lungs and out towards the throat where it is swallowed. Other epithelial tissues secrete mucus, such as the skin of a frog and the lining of the mammalian intestine.

> **?**
>
> **A** Smoking has been found to affect the lining of the respiratory system. What type of tissue is found here?

Fig 2.4 **The different types of epithelial tissues in the mammalian body. You don't need to learn all the different forms: it is more important to see how structure relates to function. For example, simple squamous epithelium is thin and is ideally suited to be an interface where substances are exchanged, such as at the surface of alveoli. Stratified squamous epithelium is thicker and can be replaced continuously, making it perfect as a protective covering to the skin**

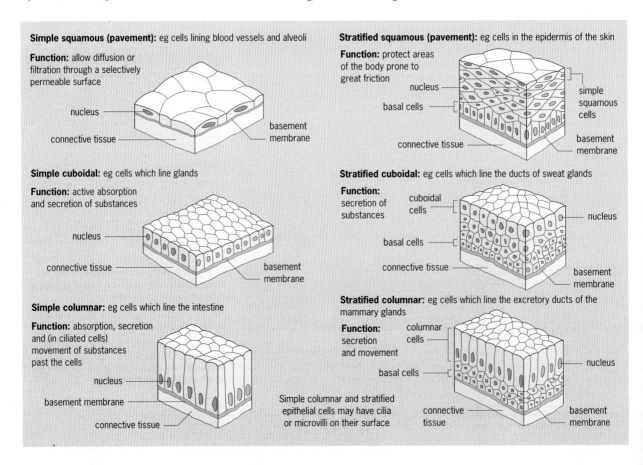

Simple squamous (pavement): eg cells lining blood vessels and alveoli

Function: allow diffusion or filtration through a selectively permeable surface

nucleus
connective tissue
basement membrane

Simple cuboidal: eg cells which line glands

Function: active absorption and secretion of substances

nucleus
connective tissue
basement membrane

Simple columnar: eg cells which line the intestine

Function: absorption, secretion and (in ciliated cells) movement of substances past the cells

nucleus
basement membrane
connective tissue

Simple columnar and stratified epithelial cells may have cilia or microvilli on their surface

Stratified squamous (pavement): eg cells in the epidermis of the skin

Function: protect areas of the body prone to great friction

nucleus
basal cells
simple squamous cells
connective tissue
basement membrane

Stratified cuboidal: eg cells which line the ducts of sweat glands

Function: secretion of substances
cuboidal cells
basal cells
nucleus
connective tissue
basement membrane

Stratified columnar: eg cells which line the excretory ducts of the mammary glands

Function: secretion and movement
columnar cells
basal cells
nucleus
connective tissue
basement membrane

Connective tissue

Connective tissues, as the name suggests, usually hold structures together. Connective tissue cells are separated by a **non-cellular matrix** (mesh) which they secrete. (This is in contrast to epithelial tissue where the cells are tightly packed.) The nature of the matrix usually accounts for the property of the tissue. In bone, for example, the matrix is a hard mineral (calcium salts) strengthened by collagen fibres.

The main types of connective tissue found in mammals are listed in Table 2.1. **Cartilage**, **tendons**, **ligaments**, **blood** and **bone** are all connective tissues. Each one is adapted to perform a particular function. All except blood have a structural role in the body. Blood is described as a fluid connective tissue because the cells are separated by a liquid, the **plasma**. Because connective tissues are mainly structural, their **metabolic demands** (their need for food and oxygen) are relatively small.

Table 2.1 **The main types of mammalian connective tissue**

Connective tissue	Function
Bone	Forms the skeleton. Protects, supports the main organs of the body. Anchors the muscles
Tendon	Attaches muscles to bones (Fig 2.5)
Ligament	Attaches bones to bones and provides support at joints
Cartilage	Smoothes surfaces at joints. Prevents collapse of trachea and bronchi
Adipose tissue	Stores fat and provides insulation
Blood	Transports substances around the body (see Chapter 18)
Areolar tissue	Protects organs, blood vessels and nerves. Gives strength to epithelial tissue. General 'packing tissue'

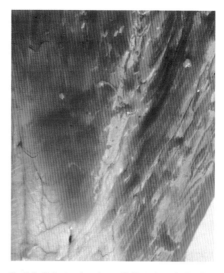

Fig 2.5 **Pale tendon tissue linking muscle (red) to bone (left). Tendons are made from fibrous connective tissue which contains many closely packed collagen fibres. In some situations, such as in the legs of horses, the tendons must be able to withstand enormous forces**

✔

Connective tissue consists of cells in a matrix with fibres.

Cartilage

Cartilage is a tough, smooth and flexible connective tissue. There are three types of cartilage:

- **Hyaline cartilage**. This is a compressible and elastic tissue found in the nose and trachea, and at joints where it covers the ends of bones to provide smooth surfaces.
- **White fibrous cartilage**. This is found between the vertebrae and acts as a shock absorber.
- **Yellow elastic cartilage**. This is a very elastic tissue found in the ears, **pharynx** (throat) and **epiglottis** (the flap which closes over the airway during swallowing).

Fibrous cartilage is a good shock absorber because it is **compressible** (it can withstand pressure without being damaged) and it can absorb the impact of a rapid jolt which would otherwise damage a more brittle structure such as a bone.

Studies into the structure of fibrous cartilage reveal how it is adapted to perform its function. The toughness, not surprisingly, is provided by **collagen fibres** (see Chapter 3). But a second type of chemical, **proteoglycan**, has proved to be especially interesting. Proteoglycan binds loosely to water molecules, but releases them when under pressure. So water seeps out of cartilage when it is compressed, and goes back again when the pressure is released, allowing cartilage to act as a very tough sponge to reduce the shock on the rest of the body.

Despite this, fibrous cartilage cannot absorb the impact of strenuous exercise. When we run and jump, for example, much of the cushioning effect is due to the tendons, ligaments and muscles. To illustrate this, think about what happens when you do a standing jump. If you land on your toes, there is no problem. The shock is absorbed by the muscles and tendons in your feet and legs. If you land on your heels (even from a few centimetres) the jarring effect is very obvious. Don't try this!

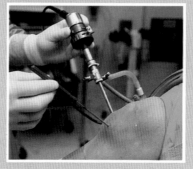

Fig 2.6 **The knee joint has to take the weight of most of the body, and in strenuous sports such as football, the twisting and turning can easily damage the cartilage. This player is having key-hole surgery to remove a fragment of cartilage which has been torn**

B Which type of muscle would you expect to have the fewest mitochondria? Explain.

Fig 2.7 **Micrographs of muscle. Top: skeletal, middle: smooth, and bottom: cardiac muscle. See** Table 2.2 **for descriptions**

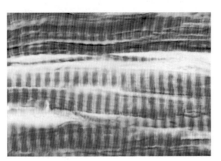

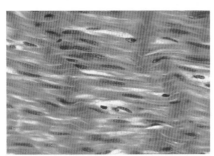

C What features of skeletal muscle make it unsuitable to form the muscles of the heart?

Muscular tissue

Muscular tissues are **contractile**: they contain protein fibres, **actin** and **myosin**, which produce movement when they slide over each other (Chapter 19). As Fig 2.7 shows, there are three types of muscle: **skeletal muscle**, **smooth muscle** and **cardiac muscle**. The properties of the three types are summarised in Table 2.2.

A large proportion of the body weight of most vertebrates is due to their muscles. As you might expect, active muscle tissue has a very high metabolic rate, and many mitochondria can be seen packed between the muscle fibres.

Table 2.2 **Comparing the three types of mammalian muscle**

Muscle:	Skeletal	Smooth	Cardiac
Other names	striped, striated, voluntary	unstriped, unstriated, involuntary	heart
Site in the body	attached to skeleton	tubular organs; gut, reproductive system, glands, bronchioles	heart
Function	movement	usually controls movement of substances along tubes	heartbeat
Control	voluntary	involuntary (autonomic), modified by autonomic nervous system	myogenic (self-generating)
Speed of contraction	rapid	slow, usually known as peristalsis	rapid
Speed of fatigue	rapid	slow	slow

Nervous tissue

Nervous tissue contains specialised cells called **neurones** which transmit electrical impulses. Nervous tissue controls and coordinates the activities of the whole organism. Changes in its internal and external environment act as **stimuli** which are detected by **receptor cells** in **sense organs**. From here, impulses travel along **sensory neurones** to the **central nervous system** (CNS). This consists of the brain and spinal cord, both made up mainly of neurones. The brain processes the information it receives and decides which response to make. Impulses are sent out along motor neurones to **effector organs**: the muscles and glands.

Control and coordination in the nervous system is covered in Chapter 21: for more information about sensory systems in animals, see Chapter 23.

The origin of tissues

To gain a more complete understanding of tissues and the role they play in an organism, we must study the development of an organism. This is a branch of science in itself, called **embryology** (study of the embryo).

All animals begin life as a single cell which multiplies to form a hollow ball of cells, a **blastula**. Soon afterwards, the ball buckles at one point of its surface and cells flood inwards. The cells makes contact with the opposite side of the blastula. (Think about what would happen if you pushed your finger into a balloon until it touched the other side.) This process is called **gastrulation** and it produces a **three-layered embryo** (Fig 2.8).

The three layers are the **ectoderm** ('outer skin'), the **mesoderm** ('middle skin') and the **endoderm** ('inner skin'). Different organs of the body develop from each layer. The process is obviously complex but overall:

- The ectoderm produces the skin, central nervous system and peripheral nerves.
- The mesoderm gives rise to the muscles, blood and blood vessels, the skeleton, kidneys and reproductive system.
- The endoderm produces the intestine and associated organs (liver and pancreas).

Some organs, the eyes for example, originate from cells of more than one layer.

Plant tissues

Like animals, plants are made up of different tissues which together form a whole functioning organism. There are, of course, basic differences between plants and animals, and plant tissues reflect these differences. Tissues of a multicellular plant are adapted for the following functions:

- photosynthesis
- transport
- support
- storage
- protection
- reproduction

Plants contain **simple tissues** (tissues made of one type of cell) and **compound tissues** (tissues which have more than one type of cell). Simple tissues include **parenchyma**, **collenchyma** and **sclerenchyma**, compound tissues include **xylem** and **phloem**.

The basic structure of a flowering plant is shown in Fig 2.9, and the tissues of a plant are summarised in Table 2.3. You may find it useful to refer back to this table when you study the physiology of plants (in Chapters 17, 18, 20 and 29).

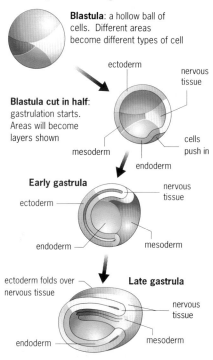

Fig 2.8 **The process of gastrulation produces a three-layered embryo which differentiates into different parts of the organism**

Fig 2.9 **The basic anatomy of a flowering plant and the position of some of its major cell types**

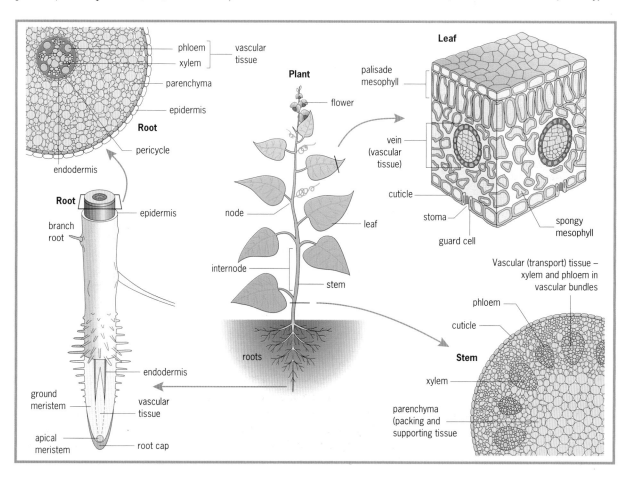

Tissue	Diagram	Distribution	Functions
Parenchyma		cortex, pith, medullary rays, vascular bundles	basic packing tissue, support and storage
Modified parenchyma:		covers all outer surfaces of plant:	
1 Epidermis		stem, leaves, roots	flattened cells protect from infection and drying out
2 Mesophyll	palisade / spongy	main part of leaf	palisade mesophyll is adapted for photosynthesis, spongy mesophyll is loosely packed to create air spaces for gas exchange
3 Endodermis	casparian strip	surrounds vascular tissue, so forms the innermost layer of cortex	selective barrier which controls movement of water and ions in and out of xylem
4 Pericycle		in roots between vascular tissue and endodermis	growth of lateral roots and secondary growth
Collenchyma		outer cortex, angle of stems and midribs	provides support
Sclerenchyma:			
1 Fibres		outer region of cortex, pericycle of stems (around vascular bundle)	support
2 Sclereids		cortex, pith, phloem, shells and stones of seeds	support/protection
Xylem*		vascular system	transport of water and dissolved ions
Phloem:			
1 Sieve tubes	companion cell	vascular bundles	translocation of food
2 Companion cells	sieve tube	vascular bundles	control of sieve tubes

Table 2.3 **Summary of the main tissues of a flowering plant**

*Note:
Xylem tissue contains sclerenchyma, parenchyma, tracheids and vessels.

3 ORGANS AND BODY SYSTEMS IN ANIMALS

An **organ** is a collection of tissues which work together to perform a particular function. The kidney, for instance, is an organ in which blood vessels (made from epithelial tissue), kidney tubules (also epithelium) and connective tissue (to hold the whole all the parts together) combine to make up an organ which can act as a filter to remove waste products from the body and can control its water balance.

Within a whole organism, organs form **systems** which carry out life processes such as digestion, excretion, reproduction. The combined systems of the body also need to create a stable internal environment to keep cells bathed in fluid which has the optimum temperature and composition.

Consider the systems of a mammal:

- The **digestive system** extracts from food the simple food molecules which cells need.

- The **excretory system** removes the cell's waste products from the body.

- The **respiratory system** provides oxygen and removes carbon dioxide.

- The **circulatory system** transports all necessary substances to the cells, and removes waste products.

- The **muscles**, **skeleton** and **nervous system** combine to ensure that the organisms obtains food and avoids danger.

?

D Which body system – not mentioned on the list on this page – does not directly contribute to the maintenance of internal conditions?

SUMMARY

When you have studied this chapter, you should understand the following ideas.

■ Multicellularity allows a **division of labour**, that is, cells can **specialise.**

■ Specialised cells form **tissues**, tissues form **organs** and organs form **organ systems**.

■ A tissue is a collection of similar specialised cells.

■ Animal tissues can be classified into four broad groups: **epithelial**, **connective**, **muscular** and **nervous**.

■ Plant tissues can be divided into **simple** (containing one type of cell) or **compound** (containing more than one).

■ The whole organism consists of all the systems working together to produce an individual which is able to control its internal conditions and reproduce.

Assignment

BURNS AND TISSUE GRAFTS

The severity of a burn can be measured on a three-point scale:

First degree burns are minor or 'partial thickness' burns (part-way through the skin), often caused by a mild scald or over-exposure to sunlight. The main symptom is redness of the skin. Though painful, they usually heal quickly and need minimal medical treatment.

Second degree burns are partial thickness burns caused by severe scalding, contact with flames or very severe sunburn. The dermis of the skin is damaged as well as the epidermis, and redness is accompanied by blistering. This type of burn is extremely painful and usually needs medical attention.

Third degree burns are severe, full thickness burns. Though they may not be painful, third degree burns completely destroy the skin and often some of the underlying tissue such as fat, muscle and bone.

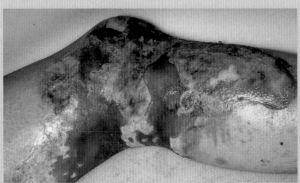

Fig 2.A1

1 Why might a patient with third degree burns often be in less pain than one with less severe burns?

Estimating the damage

The extent of a burn is expressed as a percentage of the total body area the burn covers. As a rough guide, medical staff use the rule of nines, shown in Fig 2.A2.

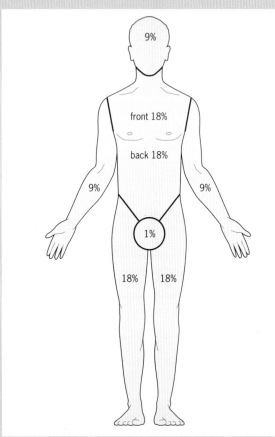

Fig 2.A2 **The extent of a burn can be estimated by using the 'rule of nines'**

2

a) A patient has burns to the whole of his back and all of both legs. What percentage of his body surface is this?

b) From your knowledge of the functions of skin, list some of the problems which will result from the loss of a large area. Refer to page 404 if you need a reminder about skin function.

Treatment

For many years, the main treatment available for burns, where the damage was not too extensive, was a **skin graft**. Skin from another region of the body was transferred onto the damaged area where it would eventually grow and provide an acceptable, if often unsightly, working replacement (Fig 2.A3). Burns covering up to 40 per cent of the skin surface can usually be treated by grafting techniques.

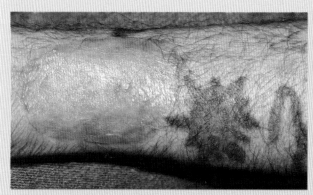
Fig 2.A3 **A skin graft on a man's forearm (to remove tattoos)**

3 Why is it desirable to use a patient's own skin rather than a graft from another person?

New techniques now allow a sample of the patient's own skin to be removed and grown outside the body. If a 2 cm² sample is kept in the right conditions, it grows into a sheet of skin which may measure as much as 10 000 times the original area – more than enough to cover a whole body. Burns of 70 per cent and more, previously thought to mean certain death, can now be treated in this way.

Tissue culture

Once removed from the body, the healthy skin sample must be kept in the correct conditions for growth. Cells must be surrounded by a culture medium which contains all the amino acids necessary for protein synthesis, and salts, glucose and several vitamins. Human cells also need an extract of human blood serum. It appears that the serum contains one or more proteins which act as growth factors, stimulating cell division and so rapid growth.

4

a) Predict the optimum temperature essential for the growth of skin cells.

b) Can you think of any problems which may arise if skin cells are provided with the ideal conditions for growth? How do you think these problems may be overcome?

Although skin grows much more rapidly in culture than it could grow in the patient, the process is not immediate. The patient therefore requires a temporary skin covering and this is usually tissue taken either from a pig or from a dead person.

5 Think of any problems that could be associated with using these two materials.

When the patient is stable and their wound is free from infection, the cultured skin is put in place. The patient must be kept sedated to give the new skin a chance to 'take', but reports from medical centres who have pioneered this work report a success rate of as high as 80 per cent.

There are problems associated with this technique: the new skin may fail to adhere (stick) to the underlying tissue, and it may be too thin and fragile. In addition, the cost is very high, anything from £650 to £5 500 for every 1 per cent of body area covered!

The future

Tissue culture techniques are advancing rapidly. In the near future it is quite likely that skin grafts will adhere reliably and function like the original tissue, even after the most extensive burns.

An interesting advance on the use of cadaver (corpse) skin has been developed in the USA. Skin consists of fibroblast cells which secrete a protein (collagen) matrix or 'scaffolding'. It is the cells themselves which stimulate the immune reaction (causing rejection), and so researchers have been experimenting with cadaver skin in which the cells have been removed, leaving only the protein matrix. Once in place, the matrix releases growth factors which stimulate growth of the victim's own fibroblasts. Thus the healing process is greatly speeded up and the patient leaves hospital earlier.

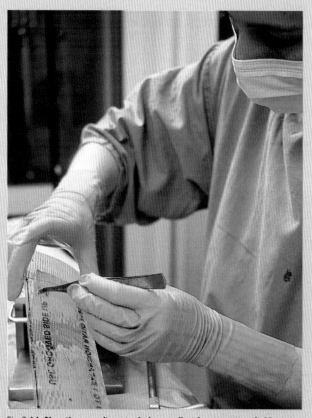
Fig 2.A4 **New tissue culture techniques allow large areas of skin to grow from a few healthy skin cells. Here, the cultured skin is sliced to stretch over three times its area**

Skin is not the only tissue type to be cultured; cartilage, bone and liver have all been successfully grow outside the body.

6 List some of the medical problems which could be treated if we could reliably replace cartilage, bone and liver.

3 The chemicals of life

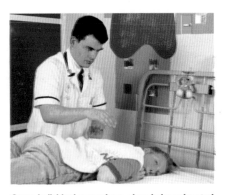

Some individual genes have already been located and sequenced – one example is the gene that underlies cystic fibrosis. Human trials to see if a healthy copy of the cystic fibrosis gene can be used to cancel out the effects of a sufferer's own damaged gene are currently under way, and the end of the decade may see the first successful genetic therapy. Meanwhile, this cystic fibrosis sufferer has regular physiotherapy to clear the mucus accumulation in his lungs

THE 1990S HAVE SEEN the most ambitious project ever to be undertaken in molecular biology: the Human Genome Project (HGP). The aim of this project is simple: to map out the sequence of DNA that makes up all the human genes. It is a phenomenal task. The sequence of chemicals involved is estimated to be 3.2 billion base pairs long.

Critics of the project look to the cost: estimates are currently running at anything between 1 and 10 US dollars per base pair. The money could be better spent, they argue, on projects that will give a much faster return on the investment. The completion date for the project is the year 2005. Despite the cost and the critics, most scientists think that the potential benefits of such knowledge are enormous. If we know where individual genes can be found, and we know the DNA sequence of those genes in a normal healthy individual, we should be better equipped to diagnose and treat genetic diseases.

1 STUDYING THE CHEMICALS OF LIFE

The study of the chemicals and reactions that take place inside living things is a science in its own right – **biochemistry**. Biochemists aim to understand how living things work by studying how the molecules within their cells interact. In this chapter we look at carbohydrates, lipids, proteins and nucleic acids, and describe the characteristics that make them the *chemicals of life*.

Some central themes in biochemistry

The chemical systems which make up living things seem to be incredibly complicated, but there are some simple underlying patterns:

- Living organisms contain a huge range of macromolecules, but they are built from a small number of simple molecules.

- Since these simple building blocks are very similar in all living organisms, we can infer that all life has a common origin.

- The characteristics of each species, and of individuals within a species, is determined by the information contained in their DNA.

- The DNA contains information that the cell can use to make proteins. Many proteins are **enzymes**, which control the physical and chemical activities of an organism.

- The chemical activities within an organism can be given the general term **metabolism**.

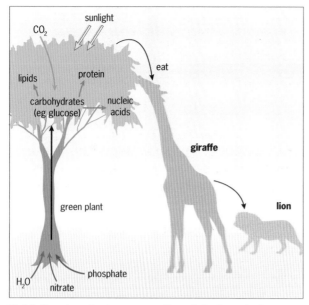

Fig 3.1 **Organic molecules are normally made in plants via the process of photosynthesis. These molecules are then available to the rest of the food chain**

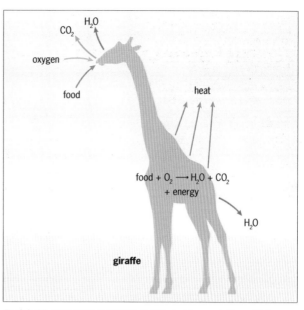

Fig 3.2 **All organisms, including plants, need energy. They obtain energy by oxidising (breaking down) organic molecules (food). This process is called respiration**

- Metabolic reactions can be divided into two general categories: **anabolic** and **catabolic**. Anabolic reactions build up large molecules from smaller ones, while catabolic reactions do the reverse, breaking down larger molecules.

- Anabolic reactions usually involve **condensation** reactions in which building-block molecules are joined together and water is produced.

- Catabolic reactions, for instance digestion, usually involve **hydrolysis** reactions in which larger molecules are split as they react with water.

- In **photosynthesis** (Fig 3.1), plants use the energy from sunlight to build up organic molecules such as sugars from simple ones such as carbon dioxide and water.

- All organisms need a supply of energy, which they obtain via **respiration** (Fig 3.2). In respiration, organic molecules are **oxidised** into simpler molecules, usually carbon dioxide and water. The resulting energy is used to fuel the many energy-requiring processes within the organism.

2 CARBOHYDRATES

Carbohydrates are organic molecules that contain three elements: carbon, hydrogen and oxygen. Carbohydrates include sugars and starches, but there are other important examples (Table 3.1 and Fig 3.3). There are three basic types of carbohydrate molecule, named according to their structure and size:

- *Mono*saccharides are *single* sugars.
- *Di*saccharides are *double* sugars (made from two monosaccharides).
- *Poly*saccharides are *multiple* sugars (polymers of many monosaccharides).

(a) **Racing cyclists have glucose-rich drinks to restore their energy**

Fig 3.3 **The term carbohydrate covers a range of chemicals that have many uses and functions in living organisms**

(b) **Flour and sugar are the main ingredients of cakes like these**

(e) **The exoskeletons of insects are made of chitin, a sugar polymer that contains nitrogen**

(c) **Cellulose fibres give strength and rigidity to plants...**

(d) **... and these properties are used in objects made of paper**

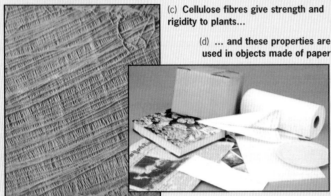

Carbohydrates are the first molecules made in photosynthesis. Lipids, proteins and nucleic acids are modified carbohydrates.

CHO type	Compound	Sub-units	Occurrence in living things
Monosaccharide	glucose		widespread
	fructose		sweet fruits
	galactose		milk
Disaccharide	maltose	2 × glucose	germinating seeds
	sucrose	glucose + fructose	fruit
	lactose	glucose + galactose	milk
Polysaccharide	starch	glucose	plants (storage)
	glycogen	glucose	animals (storage)
	cellulose	glucose	plant cell walls
	chitin	glucosamine	arthropod exoskeletons

Table 3.1 **Some common carbohydrates**

Table 3.2 **Common monosaccharides**

No. of C atoms	Name	Common examples
3	triose	glyceraldehyde (see Chapter 8)
5	pentose	ribose, deoxyribose (in RNA and DNA)
6	hexose	glucose, fructose, galactose

Monosaccharides and disaccharides

In monosaccharides, the three elements, carbon, hydrogen and oxygen are always present in the same ratio. The **empirical** (basic) formula for a monosaccharide is $(CH_2O)_n$. Monosaccharides and disaccharides are classed as sugars, and usually have names ending in *-ose*. Monosaccharides can be classified according to the number of carbon atoms they possess, 3, 5 and 6 being the most usual (Table 3.2). In glucose, for example, n is 6.

Glucose

Glucose is a useful compound with which to begin a study of carbohydrates: it is the main source of energy for many organisms. Most of the common polysaccharides are glucose polymers.

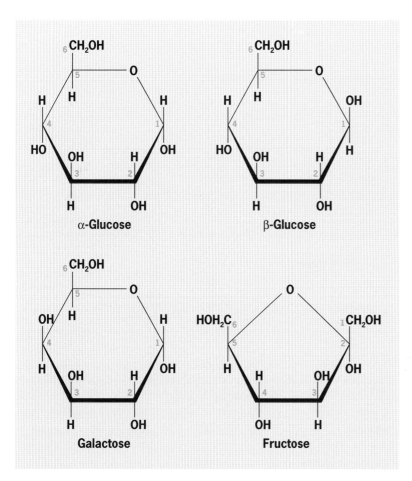

Fig 3.4 **Variations on a theme: common hexose sugars. All of these sugars share the formula $C_6H_{12}O_6$ but, as you can see, they all have a slightly different structure. They are said to be isomers. You may think that these differences are slight, but they greatly affect the properties such as taste (sweetness), digestibility and the nature of the polymers they form**

Glucose is a **hexose** (6-carbon) sugar which has the formula $C_6H_{12}O_6$. All other hexose sugars, such as **fructose** and **galactose**, have the same formula (Fig 3.4).

What is a 'reducing sugar'?

The term 'reducing' reflects the fact that some sugars have **carbonyl groups** (C=O), which can be **oxidised** to **carboxylic acids** (—COOH), **reducing** other chemicals in the process. For a full explanation of reduction–oxidation reactions (redox reactions), see page 129.

A standard test for a reducing sugar involves boiling the sample with **Benedict's solution**, a blue solution that contains copper sulphate. If a reducing sugar is present, the Cu(II) ions in copper sulphate are reduced to Cu(I) ions, resulting in an orange precipitate (Fig 3.5). Glucose, fructose, galactose, maltose and lactose are all reducing sugars but sucrose is not. However, after sucrose is boiled with dilute acid to hydrolyse (split) it into its monosaccharides, it does produce a positive result.

Monosaccharides link by means of glycosidic bonds

When two monosaccharides join together, the bond between them, a **glycosidic** link, centres around a shared oxygen atom (Fig 3.6). Two α-glucose molecules join together by an α-**1,4- glycosidic** link to make one molecule of **maltose**. Sucrose, the familiar sugar used to sweeten food and drinks, consists of one molecule of glucose and one of fructose. Lactose, the main sugar in milk (see page 43), is a

Fig 3.5 **The Benedict's test can be used to estimate the amount of reducing sugars present in foods such as fruit juice. Samples with no reducing sugars remain blue, as right; those with a low concentration produce a green suspension, those with more produce yellow and orange suspensions, and juices very rich in reducing sugars produce a brick red precipitate, as shown left**

?

A In the Benedict's test, why does sucrose give a negative result before hydrolysis but a positive result after hydrolysis?

disaccharide that contains glucose and galactose. The structure of sucrose and lactose is shown in Fig 3.7.

Fig 3.6 **Two glucose molecules join to form maltose. Like many anabolic (building-up) reactions, this is a** condensation **reaction which involves the production of water**

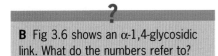

B Fig 3.6 shows an α-1,4-glycosidic link. What do the numbers refer to?

Fig 3.7 **The structure of sucrose and lactose. Sucrose is made by the condensation of glucose and fructose. Lactose is made by the condensation of glucose and galactose**

C What type of glycosidic linkages are found in sucrose and lactose?"

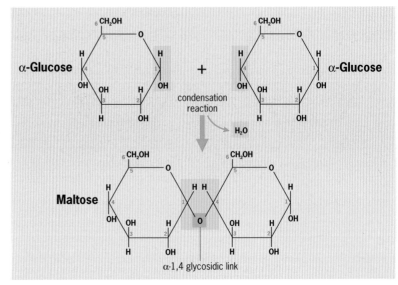

LACTOSE

ALL MAMMALS FEED their young on milk (Fig 3.8). In almost all mammals the main carbohydrate in milk is the disaccharide lactose. But why should the mother go to the trouble of combining two monosaccharides into lactose when the baby will simply have to break it down again to use it as an energy source?

The answer lies in the size of the lactose molecule. In the mother, milk is made in the mammary gland and is stored there, sometimes for several hours between feeds. Since it is a disaccharide, lactose stays in the milk, rather than diffusing into the surrounding tissue as smaller monosaccharides would tend to do. In the baby, lactose remains in the gut for the same reason and is broken down only gradually to form glucose. So there is a steady absorption of glucose into the infant's blood, rather than a sudden surge at feeding time.

Several medical conditions can arise because of a failure to digest lactose. In adult life, many species of mammal, including many cats and most non-European humans, cease to make the enzyme **lactase** and so cannot break down lactose. Undigested lactose cannot be absorbed and accumulates in the gut, encouraging the growth of bacteria which produce large amounts of carbon dioxide and lactic acid. The result is diarrhoea and wind. Individuals with this problem are said to be **lactose intolerant**.

A sort of lactose intolerance that results in far more serious problems is found in people with an inherited condition called **galactosaemia.** Affected people can break down lactose into glucose and galactose but their liver cannot convert galactose to glucose. Galactose builds up to dangerous levels and the condition can be fatal.

Fig 3.8 **Milk contains all of the food chemicals needed by a young mammal. The main carbohydrate in milk is lactose**

Polysaccharides

Starch

Starch is the most abundant storage chemical in plants (Fig 3.9) and it is the single largest provider of energy for most of the world's population .

Starch has the three properties that are necessary for a storage compound. It is:

- compact,
- insoluble,
- readily accessible when needed.

Starch is a mixture of two compounds, **amylose** and **amlyopectin**. Amylose is an unbranched polymer in which glucose monomers are joined by α-1,4-glycosidic linkages. These bonds bring the monomers together at a slight angle and, when they are repeated many times, a spiral molecule is produced. In amylose there are six glucose residues in a turn of the spiral.

The glucose chains of amylopectin have α-1,4-glycsidic linkages *and* α-1,6-glycosidic linkages. This allows branching (Fig 3.10).

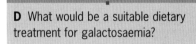

D What would be a suitable dietary treatment for galactosaemia?

E If a storage compound was soluble, what would it do to the cytoplasm of the cells in which it was stored?

■ See question 1.

Fig 3.9 **In many plants, starch is the main storage carbohydrate. Foods such as rice, pasta and potatoes all have a high starch content. When treated with iodine/ potassium iodide solution, a blue-black starch – iodide complex is formed. This reaction has been used here to test for the presence of starch in a potato**

Fig 3.10 **Amylose and amylopectin, the two different polymers that make up starch. When a plant needs to break down starch to provide glucose for respiration, it removes the terminal (end) units of amylose and amylopectin to release glucose. Since the branched amylopectin molecule has more terminal glucose units that can be removed simultaneously, it can be broken down more quickly than amylose**

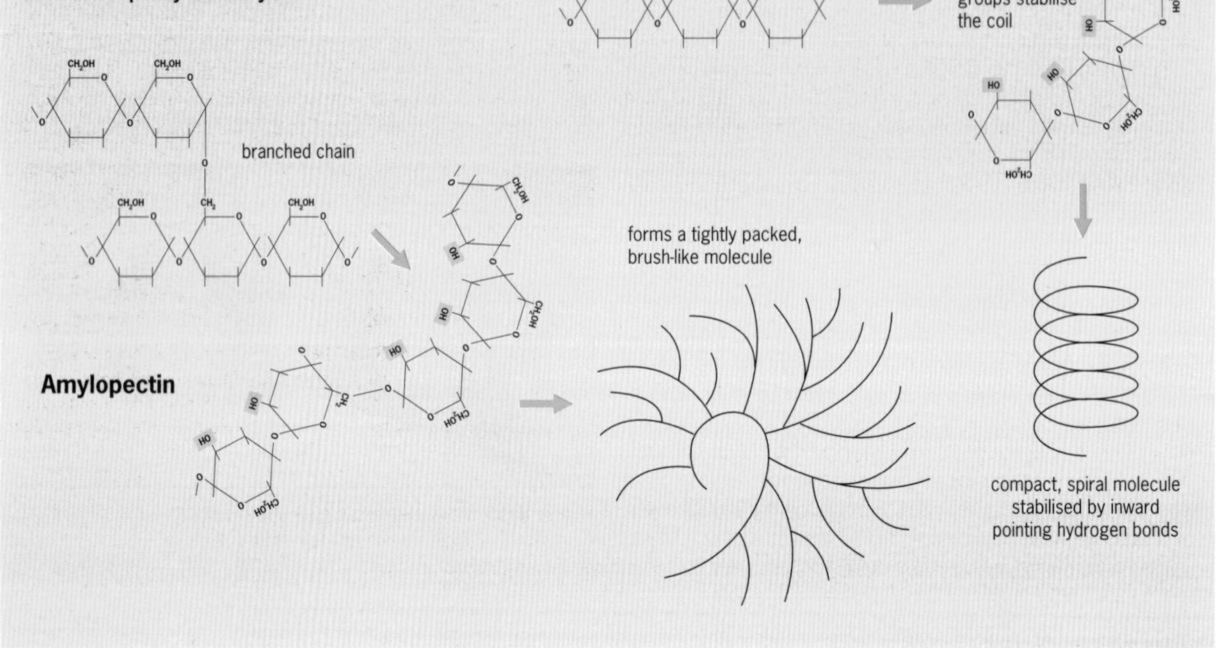

Amylose

unbranched chain produces a spiral

hydroxyl groups stabilise the coil

branched chain

Amylopectin

forms a tightly packed, brush-like molecule

compact, spiral molecule stabilised by inward pointing hydrogen bonds

microfibril

microfibril (enlarged)

cellulose chains

cellulose β-1,4 linkages

Glycogen

Glycogen is the major storage carbohydrate in animals. Its structure is similar to amylopectin, but it is even more frequently branched. In humans, glycogen is stored in significant amounts in the liver and the muscles. During prolonged exercise, when the immediate supply of glucose is used up, the body restores its supplies by breaking down glycogen. If an average person goes without food, his or her glycogen stores last for about a day, but prolonged exercise such as marathon running can use all of the body's glycogen in a few hours. When glycogen runs out, the body turns to using stored fat.

Cellulose

Cellulose is a structural polysaccharide: it gives strength and rigidity to plant cell walls (see Chapter 1). Individual cellulose molecules are long unbranched chains containing many β-1,4-glycosidic linkages (Fig 3.11). The molecules are straight and lie side by side, forming hydrogen bonds along their entire length. This results in strong **microfibrils**.

Cellulose is probably the most abundant structural chemical on Earth, but few animals can digest it. Most multicellular animals do not make the necessary enzyme, **β-1,4-glycosidase** (**cellulase**). Herbivorous animals, whose diet contains a large proportion of cellulose, are only able to deal with it because they have cellulase-producing bacteria and protoctists in their digestive system (see Chapter 15). Human beings may not be able to digest cellulose, but they make good use of it in other ways (Fig 3.12).

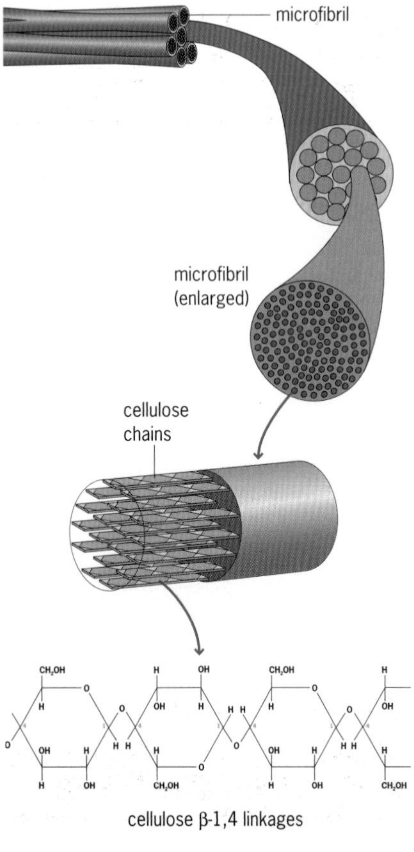

Fig 3.11 **Cellulose and the structure of the plant cell wall**

Fig 3.12 **The abundance and strength of cellulose as a material has led to its use in a variety of commercial products. These include cotton, paper, paint (until the 1980s, most car paints were based on cellulose), nail varnish, photographic film (celluloid), fibres (rayon and Lycra) and Cellophane**

?

F Why do cellulose-based materials such as cotton take a long time to rot?

Fig 3.13 **The chitin exoskeleton of an insect provides strength and flexibility without being too heavy. This may be one of the reasons why insects were the first organisms to develop flight**

Chitin

The exoskeletons of arthropods such as insects and spiders are lightweight, strong and waterproof (Fig 3.13). These properties are provided by **chitin**, a polysaccharide that contains many **glucosamine** units. A glucosamine is formed when an amino group (NH_2) is added to a glucose molecule.

3 LIPIDS

Lipids are a varied group of compounds that include the familiar **fats** and **oils**. As they are non-polar molecules, most lipids are insoluble in water but soluble in non-polar solvents such as alcohol and ether. An important exception is phospholipids, which have polar heads. The **emulsion test** for lipids, shown in Fig 3.14, is based on the solubility of lipids in alcohol.

Lipids contain the elements carbon, hydrogen, oxygen and sometimes phosphorus and/or nitrogen. They are intermediate-sized molecules that do not achieve the giant proportions of the polysaccharides, proteins and nucleic acids.

✔
Polar and ionic chemicals have particular areas of positive and/or negative charge and so will dissolve in polar solvents such as water. Lipids are non-polar (although some have polar groups) and so generally do not dissolve in water.

Fig 3.14 **The emulsion test for lipids. The food to be tested is broken up into very small pieces, mixed with pure alcohol and shaken vigorously. Any lipid present dissolves in the alcohol. This top layer is poured off, mixed with water and the mixture is shaken vigorously again. If lipid is present, the mixture turns white as an** emulsion, **a suspension of fine lipid droplets forms. If no lipid is present, the mixture remains clear**

✔
Fatty acids are organic acids. They contain a carboxyl (—COOH) group.

Lipid structure and function

(a) The fat under the skin of this elephant seal protects it from the cold

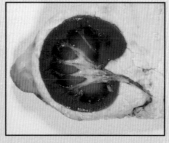

(b) Kidneys are embedded in fatty tissue

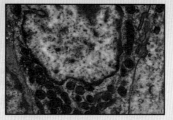

(c) The wax surface of leaves repels water, but also reduces water loss from the plant

(d) Every cell membrane contains lipids

Fig 3.15 In animals, triglycerides are used to store energy, insulate against the cold and protect organs against physical damage. Plants also use triglycerides as an energy store (in seeds, eg peanuts) but not for insulation or protection. Instead of keeping water out, plants use lipids (waxes) to keep water in

(e) Like water off a duck's back... Fatty acids on a duck's feathers repel water

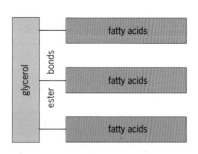

Fig 3.16 **A triglyceride is formed from a condensation reaction between one molecule of glycerol and three fatty acids. The bonds formed are called ester bonds. The diagram shows the final shape of the triglyceride**

The **triglycerides** (or **triacylglycerols**), which act mainly as energy stores in animals and plants (Fig 3.15), are a large important group of lipids. They consist of one molecule of **glycerol** and three **fatty acids**. Fig 3.16 shows how the four components join by condensation reactions to form an E-shaped molecule.

The glycerol molecule is common to all triglycerides and so the properties of different triglycerides depend on the nature of the fatty acids. Fatty acids vary in the length of their chain and in the degree of **saturation** they show.

Table 3.3 shows the chain length of some common fatty acids. Chains of about 14 to 16 carbon atoms are the average, but they range from 4 to over 28. A **saturated fatty acid** has the maximum amount of hydrogen and therefore has no double bonds. **Mono-unsaturated fatty acids** possess one C=C bond and **polyunsaturates** contain more than one.

Table 3.3 **Some common fatty acids**

Fatty acid	No. of carbon atoms	No. of double bonds	Abundant in	Melting point/°C
Palmitic acid	16	0	palm oil	63.1
Stearic acid	18	0	cocoa	69.6
Lauric acid	12	0	coconut, palm oil	44.2
Oleic acid	18	1	olive, rapeseed	13.4
Linoleic acid	18	2	sunflower, maize	−5.0

G Look at Table 3.3. Which fatty acids are unsaturated?

?

H Of the fatty acids listed in Table 3.3, which are liquid at room temperature (20 °C)? What do they have in common?

Physical and chemical characteristics of different triglycerides

Generally, triglycerides, which contain longer chain fatty acids and saturated fatty acids, form 'hard' fats such as lard and suet. Animal tissues tend to contain a higher proportion of saturated fats than plants. Plant lipids contain shorter chain, unsaturated or polyunsaturated fatty acids and so are 'light' oils that are liquid at room temperature. Unsaturation leads to a lowering of the melting point because double bonds produce kinks in the carbon chain, increasing the distance between molecules and hence the fluidity of the lipid.

HIBERNATION

SOME ANIMALS SURVIVE unfavourable seasons when food is scarce, by **hibernating** (Fig 3.17). Mammals such as dormice and bats show true hibernation: their metabolic rate is much lower than normal and their core body temperature, breathing and heartbeat rates drop significantly. Core body temperatures of around –5 °C have been recorded in hibernating bats!

Fig 3.17 **Before hibernating, dormice eat large amounts of high-energy foods to build up fat stores. During hibernation they obtain the energy they need to survive the long winter from the triglyceride stored in their adipose tissue**

Triglycerides as energy stores

Many animals store energy in the form of triglycerides: weight for weight, triglycerides yield more than twice as much energy as carbohydrates or proteins. Triglycerides are described as **highly reduced** compounds because they contain many C—H bonds which can yield energy when oxidised during respiration. Complete oxidation of a molecule of triglyceride produces a large amount of metabolic water. This is important for desert animals when water is scarce.

Phospholipids

Phospholipids are similar to triglycerides in structure but, as Fig 3.18 shows, one of the fatty acids is replaced by polar **phosphoric acid**. This gives the molecule a polar head and a non-polar tail. When placed in water, phospholipids arrange themselves with their **hydrophobic** ('water-hating') tails pointing inwards and their **hydrophilic** ('water-loving') heads pointing outwards. Fig 3.19 shows how phospholipid molecules form a surface monolayer when added to water in a beaker. Three common arrangements that result from the behaviour of phospholipids in water are the **bilayer**, the **micelle** and the **bilayered vesicle**. Phospholipid bilayers form the basis of all biological **membranes** (see Chapter 5).

Cholesterol

Many people associate **cholesterol** with heart disease, but this lipid is, in fact, a perfectly normal constituent of our bodies (Fig 3.20).

As well as eating food that contains cholesterol (Fig 3.21), we can also synthesise cholesterol in the liver. The more there is in the diet the less the liver needs to make: vegans, who eat no animal products, are easily able to make all the cholesterol they need.

Fig 3.18 **The structure of a phospholipid. The phosphoric acid gives the molecule a polar head and a non-polar tail**

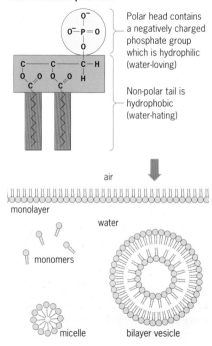

Fig 3.19 **In water, the hydrophilic polar heads face outwards while the hydrophobic tails point inwards. At the interface between water and air a phospholipid monolayer forms. Under the surface, micelles and bilayered vesicles form.**

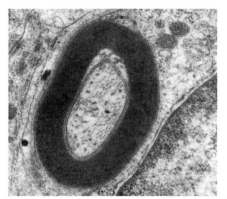

Fig 3.20 **The role of cholesterol in the cell membrane of animal cells. The membranes of Schwann cells, which form the myelin sheath round nerves, are particularly rich in cholesterol. The cholesterol 'plugs' holes in the lipid bilayer and so helps to insulate the neurone by preventing sodium and potassium ions from leaking into or out of the cell**

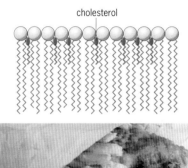

Fig 3.21 **Animal fat tends to be high in cholesterol and saturated fats. Eating low amounts of this type of dietary lipid may help to reduce our risk of suffering from coronary heart disease**

CHOLESTEROL AND HEART DISEASE

CHOLESTEROL-RELATED heart disease problems can arise in people who inherit a tendency to make large amounts of their own cholesterol, regardless of how much or little they have in their diet. This leads to a condition called **hypercholesterolaemia** (high blood cholesterol).

With high levels of cholesterol, deposits of crystalline cholesterol (and its esters) can form on the walls of arteries, a condition known as **atherosclerosis** (Fig 3.22). This narrows blood vessels and can reduce the blood supply to vital organs, particularly the heart and brain. The crisis comes when blood clots lodge in the smaller arteries, bringing blood circulation at that

point to a halt. Tissue that is deprived of blood, and therefore oxygen, dies, and a heart attack or stroke is the result.

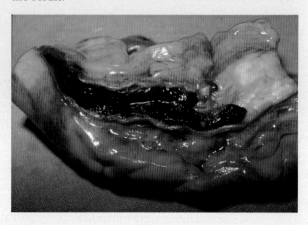

Fig 3.22 **A cholesterol deposit, known as atheroma, fills almost all the space inside this artery**

Steroid hormones

Steroid hormones have a similar structure to the cholesterol from which they are made. They include **testosterone**, **progesterone** and the **oestrogens**. The role of these hormones is discussed in Chapter 28.

Waxes

Waxes are lipids that are often used to waterproof surfaces, so preventing water loss. The cuticle of a leaf and the protective covering on an insect's body are both waxes. Waxes consist of a very long chain fatty acid joined to an alcohol molecule (not glycerol as in triglycerides). They have no nutritional value because they cannot be digested by **lipases** (lipid-digesting enzymes).

STEROIDS IN ATHLETICS

ANABOLIC STEROIDS are synthetic compounds that are similar in structure to **testosterone**, the male sex hormone. They increase anabolic (building-up) reactions, which enhance muscle size.

Athletes, notably weightlifters, began injecting testosterone in the 1950s to gain extra strength. When steroids were banned, athletes looked for other natural substances that would be undetectable in blood tests. **Human growth hormone** (**HGH**, or **somatotropin**) and also **erythropoietin** are other potential performance enhancers. HGH increases muscle size in adults and erythropoietin increases the oxygen-carrying capacity of the blood.

How anabolic steroids work

Steroid hormones are lipid soluble and so can pass through the cell membrane. Studies have shown that

anabolic steroids pass into the nucleus, where they promote the transcription of genes which code for the muscle proteins actin and myosin. As well as increasing muscle size, athletes can train for longer, and the greater aggression that they tend to develop may also give them an added competitive edge.

The use of anabolic steroids is unfair and dangerous. To achieve a muscle-enhancing effect, athletes need to inject 10 to 40 times the normal dose, and this can have irreversible side effects. Symptoms of steroid abuse include acne, impotence, sterility, diabetes, heart disease and even liver cancer. Studies of male steroid abusers have shown that their testes may decrease in size by as much as 40 per cent. Their sperm count is correspondingly lowered. Female athletes who use steroids may have fewer periods and may develop masculine features.

4 PROTEINS

Proteins play a central role in the structure and metabolism of all living organisms. Protein molecules have a huge variety of shapes and sizes (the structure of carbohydrates and lipids is relatively limited by comparison). This versatility of form and function is the key to the role of proteins in the cell.

The importance of proteins to living organisms

We can appreciate the extent to which organisms use proteins by looking at their distribution in the human body (Fig 3.23). Consider some of the needs of a living organism:

- Each cellular metabolic reaction must be catalysed by a different **enzyme**.

- Each substance which passes across a cell membrane requires a different **carrier molecule**.

- Higher animals need a different **antibody** to combat the chemicals produced by the many (and constantly changing) disease-causing organisms such as bacteria and viruses.

Protein structure allows for 'tailor made' molecules that can fulfil all these requirements.

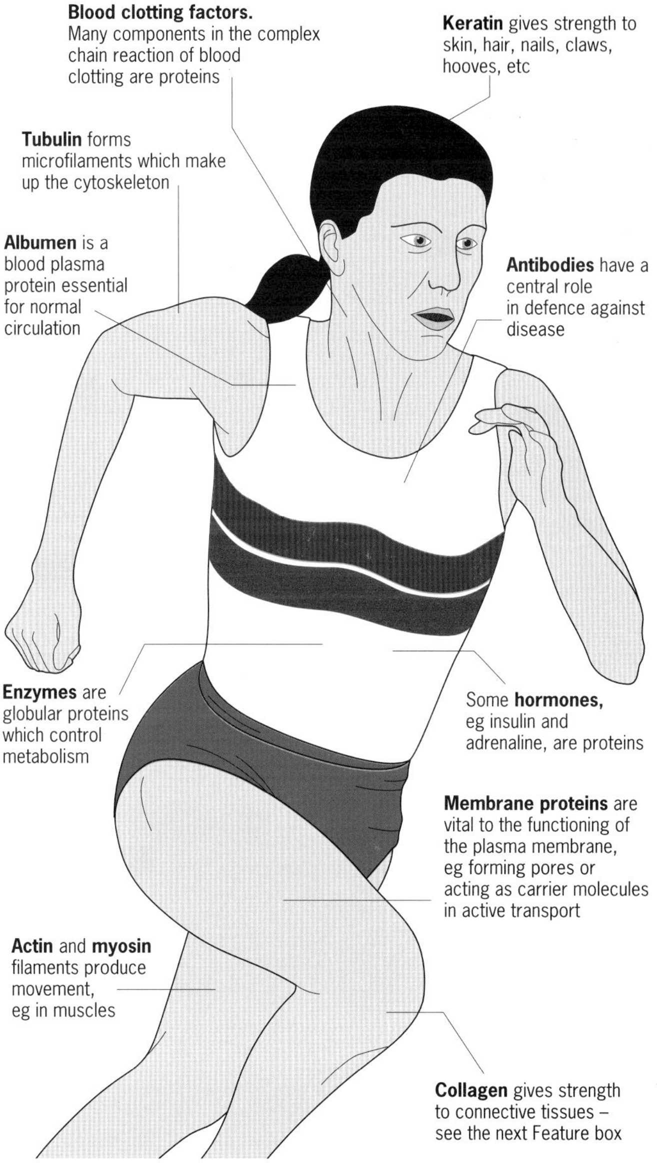

Blood clotting factors. Many components in the complex chain reaction of blood clotting are proteins

Keratin gives strength to skin, hair, nails, claws, hooves, etc

Tubulin forms microfilaments which make up the cytoskeleton

Albumen is a blood plasma protein essential for normal circulation

Antibodies have a central role in defence against disease

Enzymes are globular proteins which control metabolism

Some **hormones**, eg insulin and adrenaline, are proteins

Membrane proteins are vital to the functioning of the plasma membrane, eg forming pores or acting as carrier molecules in active transport

Actin and **myosin** filaments produce movement, eg in muscles

Collagen gives strength to connective tissues – see the next Feature box

Fig 3.23 **After water, proteins are the major constituent of our bodies**

The composition of proteins

Proteins have large relative molecular masses (Table 3.4). If a water molecule (RMM 18) was the size of a brick, proteins would be whole buildings.

In addition to the elements carbon, hydrogen and oxygen, proteins always contain nitrogen and sometimes sulphur. The building blocks of proteins are called **amino acids**. Fig 3.24 shows the basic structure of an amino acid. As the name suggests, all amino acids contain an **amino** group ($—NH_2$) and a **carboxylic acid** group ($—COOH$). Both of these groups are attached to a central carbon atom, known as the α-carbon. The 'backbone' of an amino acid are the three atoms C—C—N.

The R group varies between different amino acids. Fig 3.25 shows the different R groups in the 20 amino acids that make up all of the proteins in living organisms.

Table 3.4 **The relative molecular masses of several proteins**

Protein	Relative molecular mass
Insulin	5 700
Haemoglobin	64 500
Myoglobin	16 900
Hexokinase	102 000
Glycogen phosphorylase	370 000
Glutamine synthetase	592 000

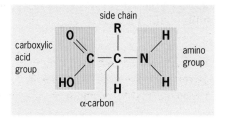

Fig 3.24 **The basic structure of an amino acid**

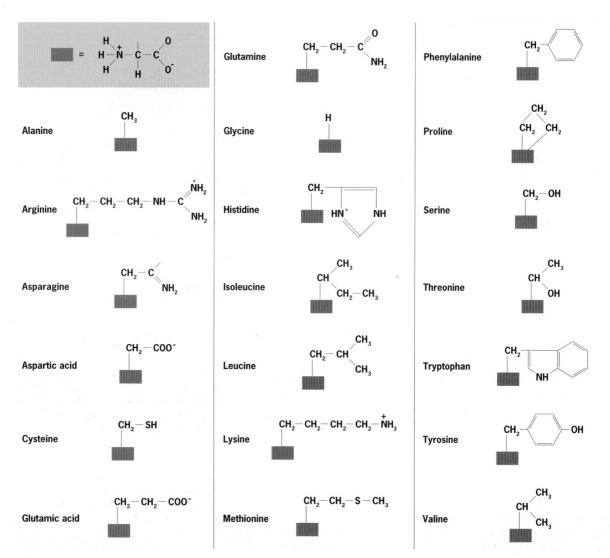

Fig 3.25 **The 20 amino acids found in living things. The R groups are highlighted in red. A further 200 or more amino acid structures are possible, but these need to be** synthesised (made artificially). They are never found in living systems. There is no need to memorise individual structures: they are for reference only

Amino acids in solution

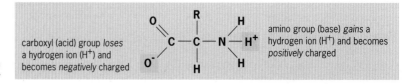

carboxyl (acid) group *loses* a hydrogen ion (H^+) and becomes *negatively* charged

amino group (base) *gains* a hydrogen ion (H^+) and becomes *positively* charged

Fig 3.26 **Ionisation of an amino acid in water**

Amino acids are soluble in water. Fig 3.26 shows what happens when an amino acid dissolves. The carboxyl group (like all acids) loses a hydrogen ion (H^+) to become COO^-. The amino group (like all bases) gains a hydrogen ion to become NH_3^+. In solution, therefore, an amino acid carries both *positive* and *negative* charges. For this reason it is known as a **zwitterion**, meaning double ion (*zwi* is German for two).

An amino acid can behave as an acid or a base. Any molecule that does this is described as **amphoteric**. Amino acids can resist a change in pH by mopping up or releasing both hydrogen ions (H^+) and hydroxyl ions (OH^-), acting as **buffers**, that is, regulators of pH. In whole proteins, the buffering effect is largely due to the side groups. Maintaining a constant pH is an important aspect of **homeostasis** (see Chapter 25).

?

I There are 20 different amino acids, and each can join with any other, so there are 400 (20 × 20) different dipeptides. How many tripeptides is it possible to make?

How amino acids form proteins

Proteins form from long chains of amino acids that are joined together by **peptide bonds**. Fig 3.27 shows how two amino acids join to form a **dipeptide**. This is another example of a condensation reaction.

Generally there are two classes of protein, **fibrous** and **globular**. Fibrous protein molecules usually bond together to form large complexes, giving great strength to structures such as tendons. Collagen and keratin are two such fibrous proteins.

In contrast, globular proteins such as enzymes are roughly spherical and soluble. They function as individual molecules. All proteins are complex molecules (Fig 3.28) and biochemists look at their structure at four different levels: primary, secondary, tertiary and quaternary.

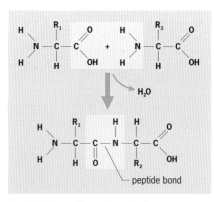

Fig 3.27 **Two amino acid molecules react together in a condensation reaction (a reaction that releases water) to produce a dipeptide**

Primary structure of proteins

The primary structure of a protein refers to the **sequence**, or order, of amino acids in that protein (Fig 3.29). A simple primary structure could be shown as:

NH_2–alanine–histidine–phenylalanine–glutamine–cysteine–COOH

A landmark in molecular biology was the discovery that the order of bases on a gene translates into the order of amino acids in a protein. The DNA sequence of the gene is copied when a messenger RNA molecule is made – this acts as a **template**. This template is used by the ribosome in **protein synthesis** (see Chapter 31).

Fig 3.28 **Using computer molecular graphics to produce a 3-D model of a protein**

Fig 3.29 **When amino acids join, there is a repeating sequence of C—C—N—C—C—N— etc. This backbone runs throughout the length of the protein**

✔ Globular proteins usually have a metabolic function in organisms, while fibrous proteins usually have a structural function.

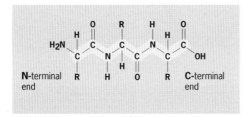

Secondary structure

When combinations of amino acids join together in a chain they tend to fold into particular shapes and patterns (such as spirals) in some places. These shapes form because the amino acids twist around to achieve the most stable arrangement of hydrogen bonds. The **secondary structure** of the protein refers to the patterns contained *within* the amino acid chain. Such patterns exist in different places in different proteins, producing an infinite variety of molecular shapes.

The main types of secondary structure in globular proteins are:

- **The α-helix,** a right-handed spiral, is the commonest type of secondary structure (Fig 3.30). Amino acid residues in the spiral twist on their axis, each residue forming hydrogen bonds with another residue four units along. These hydrogen bonds stabilise the α-helix.

- **The β-pleated sheet,** a flat structure consisting of two or more amino acid chains running parallel to each other, linked by hydrogen bonds (Fig 3.30).

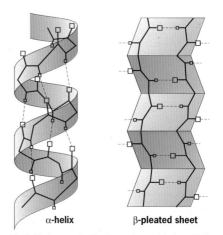

α-helix   β-pleated sheet

Fig 3.30 **In an α-helix, the polypeptide is coiled into a helix that is held in position by hydrogen bonds between different groups in the backbone. In a β-pleated sheet, the amino acid chains lie side by side, forming a sheet that is held together by hydrogen bonds between adjacent parts of the polypeptide**

The secondary structure of a protein depends on its amino acid sequence: some amino acids (or combinations of amino acids) tend to produce α helices, others usually make β sheets. A few amino acids – including proline and glycine – tend to produce a sharp bend in the chain, a vital function which allows the chain to fold back on itself. Fig 3.31 shows the secondary structures within lysozyme.

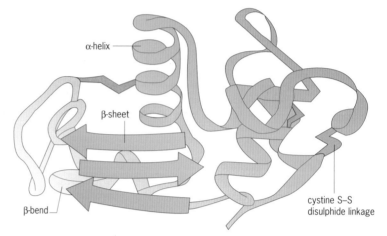

α-helix

β-sheet

β-bend

cystine S–S disulphide linkage

> A **residue** is the name given to a monomer that has become incorporated into a polymer. It is called a residue because some of the original molecule is lost in the condensation process. Thus proteins contain **amino acid residues** and starch contains **glucose residues**.

Fig 3.31 **The secondary structures within the enzyme lysozyme: the α helices and β pleated sheets are clearly visible. The secondary structure is illustrated by showing α-helices as spiral ribbons and β-pleated sheets as broad, flat arrows. This enzyme is present in secretions such as tears and sweat, where it catalyses the breakdown of some bacterial cell walls**

Tertiary structure

See question 2. ■

The **tertiary structure** of a protein is its overall three-dimensional shape and is produced as a result of:

- The sequence of amino acids which produces α-helices, β-sheets and bends at particular places along the chain.
- The hydrophobic nature of many amino acid side-chains. Globular proteins are surrounded by water and so the hydrophobic side chains tend to point inwards.

The tertiary structure is maintained by attractive forces that arise when the amino acid chain folds (Table 3.5). The strongest of these is the **disulphide bridge**. It forms when two **sulphur-containing cysteine residues** react together. Disulphide bridges occur in structural proteins (they contribute to the strength) and in extracellular proteins such as lysozyme.

If a protein consists of only one polypeptide, the tertiary structure is the final shape of the molecule. The importance of the tertiary structure cannot be overemphasised. Functional proteins, such as enzymes and antibodies, must have an exact shape – and sometimes the ability to change shape – in order to fulfil their role in the organism. Many structural proteins depend on their tertiary structure for strength: the many disulphide bridges in keratin, for example, make structures like horns very hard-wearing.

Table 3.5 **The different types of bond that maintain the secondary, tertiary and quaternary structure of a protein**

Type of bond	Formed between	Relative strength
Hydrogen bonds	H and an electronegative atom, usually O or N	weak, very common
Ionic bonds	oppositely charged ions	weak
van der Waals forces	non-specific, between nearby atoms	weak
Disulphide bridges	two —SH-containing cysteine residues<	stronger, contribute to strength of fibrous proteins

Quaternary structure

Many proteins consist of more than one polypeptide chain and sometimes also have non-protein **prosthetic** groups, often vital to the function of the protein. The quaternary structure refers to the three-dimensional structure, or **conformation**, produced when all the sub-units combine to give the final, active molecule (Fig 3.32).

Fibrous and globular proteins

The final three-dimensional structure of proteins results in two main classes of protein – **fibrous** and **globular**. Fibrous proteins contain polypeptides that associate together to form long fibres or sheets. They are physically tough and are insoluble in water (Fig 3.33).

Globular proteins contain polypeptides that are highly folded with complex tertiary and quaternary structures. They are spherical, or **globular** in shape, hence the name. Most are soluble in water and they tend to be functional proteins (Fig 3.34).

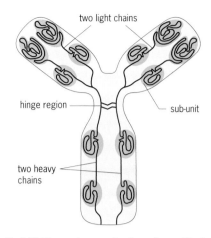

Fig 3.32 **The quaternary structure of an antibody**

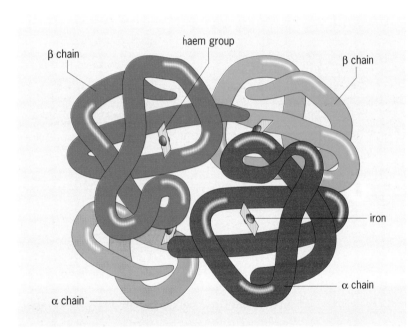

Fig 3.34 **Haemoglobin is a globular protein that picks up oxygen from the lungs and releases it to respiring tissues. It consists of four polypeptide chains held together by disulphide bonds. There is a prosthetic group, the haem group, at the centre of the each polypeptide chain. The function of haemoglobin is discussed more fully in Chapter 18**

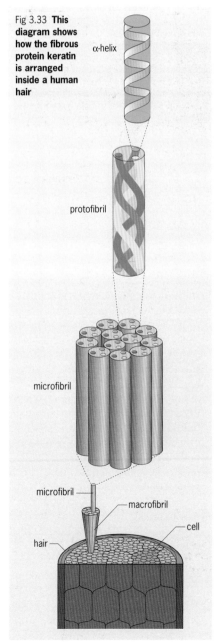

Fig 3.33 **This diagram shows how the fibrous protein keratin is arranged inside a human hair**

How stable are proteins?

As the final shape of globular proteins is maintained by relatively weak molecular interactions such as hydrogen bonds, proteins are very sensitive to increases in temperature and changes in their environment. As the temperature rises above about 40 °C, the molecular vibration increases to the point where bonds holding together the tertiary or quaternary structure break, and the shape of the molecule changes. This is known as **denaturation**. Different proteins are denatured at different temperatures, but the lethal temperature for organisms is reached when the first vital proteins become denatured, usually at around 45 °C.

Proteins can also be denatured by adverse chemical conditions. Any chemicals that affect the weak bonds tend to alter the overall structure, and even a slight change in protein shape can mean loss of function. Some proteins are particularly sensitive to changes in pH.

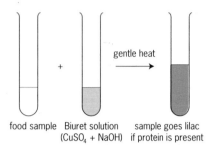

Analysing proteins

We can find out if a food contains protein using the **Biuret test** (Fig 3.35). More complex biochemical techniques can be used to tell us more about proteins in living systems. We can:

- find out which proteins are present in a mixture, such as plasma,
- work out the exact three-dimensional shape of a protein,
- find out which amino acids are present in a particular protein,
- analyse the exact amino acid sequence of a protein,
- use computers to predict the three-dimensional shape of a protein, using only its amino acid sequence.

gentle heat
+
food sample
Biuret solution
($CuSO_4$ + NaOH)
sample goes lilac
if protein is present

Fig 3.35 **The Biuret test is a simple laboratory test that detects the presence of peptide bonds**

Identifying individual proteins

The proteins present in a body fluid such as blood plasma can be separated from each other by **electrophoresis**. Each protein carries a particular overall electrical charge. Electrophoresis uses this fact to separate individual proteins, as Fig 3.36 explains.

no pattern before
electrophoresis
buffer
samples
gel
pattern after
electrophoresis
buffer
direction of
electrophoresis
Myosin (200 000)
Phosphorylase a (92 000)
Bovine serum albumin (68 000)
Ovalbumin (43 000)
Glyceraldehyde 3-phosphate
dehydrogenase (36 000)
Carbonic anhydrase (29 000)
Myoglobin (17 200)
Cytochrome c (11 700)

Fig 3.36 **Electrophoresis can be thought of as chromatography using electricity. Instead of separating chemicals according to their solubility, electrophoresis causes individual molecules to move to the positive or negative terminal according to their mass and overall charge. This is also the principle behind DNA fingerprinting – see page 62**

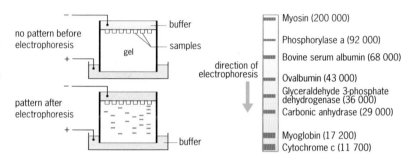

COLLAGEN

COLLAGEN IS PROBABLY the most widespread structural protein in animals. It is a fibrous protein, giving strength to tissues such as tendons, ligaments, bone and skin. Fig 3.37 shows the structure of collagen. A single collagen molecule is made from three polypeptides – each of about 1000 amino acids – intertwined to form a unique triple helix. This arrangement has great strength, mainly due to the large number of hydrogen bonds that occur along the length of the polypeptides.

Collagen is secreted in an unassembled form, **procollagen**, because the instant formation of large fibres would disable the secretory cell. Enzyme action makes procollagen into collagen: the individual polypeptides intertwine, forming very long fibres – sometimes several millimetres long. These have the tensile strength of steel and are used to strengthen bone in much the same way as metal rods reinforce concrete. The genetic disease **osteogenesis imperfecta**, 'brittle bone disease' (Fig 3.38), is due to a fault in the bonding between the collagen and the mineral component of bone.

The primary structure of collagen is very regular. It consists of a repeating sequence of glycine and two other amino acids, often proline and hydroxyproline. These amino acids do not cause the chain to gain the normal α-helix or β-sheet structure. Instead, they form long separate chains which allow the collagen triple helix to form. This is a good example of how the primary structure is ultimately responsible for the shape and properties of the whole protein.

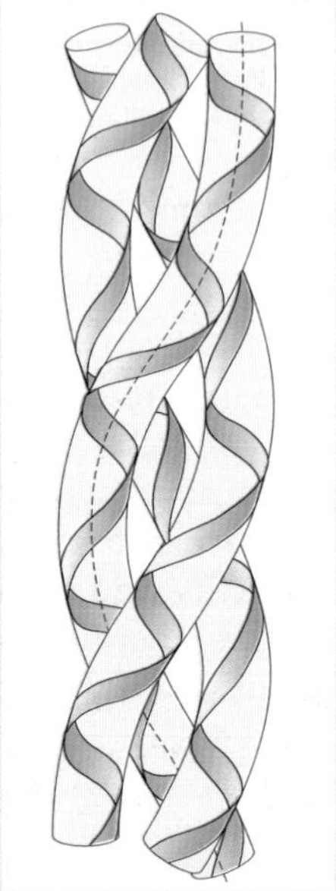

Fig 3.37 **The structure of collagen**

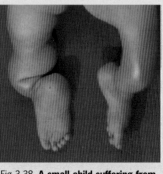

Fig 3.38 **A small child suffering from 'brittle bone' disease – an unusually severe case. Frequent fractures of the leg result in the deformities**

Working out the three-dimensional shape of a protein

X-ray diffraction is useful here. When a beam of X-rays is fired at a crystal of purified protein, the atoms in that protein **diffract** (bend) the X-rays, producing a specific pattern on a photographic plate. As Fig 3.39 shows, the shape of a protein is not immediately obvious from the information produced, but a trained scientist with a computer can produce an accurate three-dimensional model of the molecule (Fig 3.40).

Identifying which amino acids are present

The protein is digested by **protease** (protein-degrading) enzymes and then the constituent amino acids are identified using chromatography or electrophoresis. This technique does not reveal the sequence of amino acids.

Determining the amino acid sequence

The protein is digested into manageable chain lengths of amino acids and each length is analysed in turn. The polypeptide is treated with a chemical that binds to the **terminal** (end) amino acid but not to any of the others. The 'tagged' terminal amino acid is removed from the rest of the chain and identified by chromatography. Frederick Sanger, who won a Nobel prize in the 1950s, was the first scientist to sequence a complete protein (insulin). Today, the process is fully automated and has become a standard technique.

Computer modelling

As biochemists accumulate knowledge about the structure of many individual proteins, they produce computer software to help work out the shape of other complete proteins using their amino acid sequences. This is incredibly useful because of the advances in genetics. Sometimes a new gene is sequenced and no one knows its function. The DNA's sequence of triplet **codons** (described in the next section) can be given to a computer that has been programmed to work out the sequence of amino acids that would be produced if the gene was expressed in a cell. It can then 'build' the final protein and compare it to other proteins with known functions.

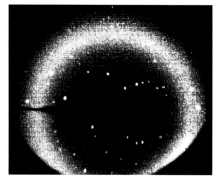

Fig 3.39 **An X-ray diffraction pattern obtained from a protein crystal**

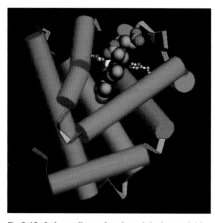

Fig 3.40 **A three-dimensional model of myoglobin, constructed as a computer graphic image with information obtained by X-ray crystallography**

See questions 3 and 4.

5 NUCLEIC ACIDS

Nucleic acids contain carbon, hydrogen, oxygen, nitrogen and phosphorus. There are two types of nucleic acid: DNA and RNA. Both are composed of sub-units called **nucleotides**.

Deoxyribonucleic acid (DNA) is the macromolecule that carries the **genetic code**, the information for making the cell's proteins. Most of the DNA in a eukaryotic cell is in the nucleus (Fig 3.41); much smaller amounts are present in mitochondria and chloroplasts.

RNA molecules are much smaller than DNA molecules. DNA may consist of over 300 000 000 nucleotides: RNA usually consists of a few hundred. RNA is also less stable. DNA molecules are the permanent store for genetic information and last for many years (there are also special enzymes that repair any damage). In contrast, RNA molecules are 'workhorses': they have a short-term function and are easily replaced. There are three forms of ribonucleic acid (RNA) in the cell:

- **Messenger RNA** (Fig 3.42(a)), can be thought of as a mobile copy of a gene. Small lengths of mRNA are assembled in the

?

J If there are 46 chromosomes in a human cell, each with an average of 5 cm uncoiled length, estimate the length of DNA contained within each cell.

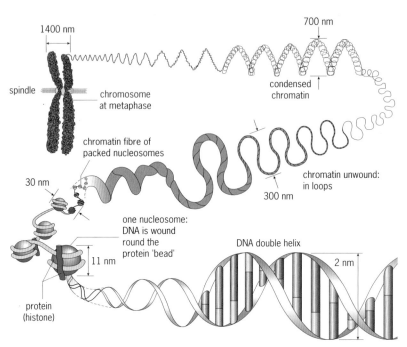

Fig 3.41 **Studies of chromosomes have shown that each one is a giant, highly coiled DNA molecule. Its shape is achieved by** supercoiling – coils within coils, condensing a huge amount of DNA into a tiny length. Each chromosome can contain 300 000 000 nucleotide units and, unravelled, the DNA within it would be between 5 and 10 *centimetres* long!

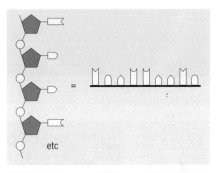

Fig 3.42(a) **Messenger RNA is a relatively delicate, short-lived molecule composed of a single strand of nucleotides. It carries the base sequence from the gene to the ribosome and provides a template for protein synthesis**

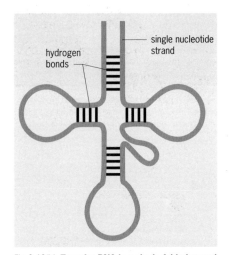

Fig 3.42(b) **Transfer RNA is a single folded strand of nucleotides that is stabilised by hydrogen bonding, producing what is often described as a 'clover leaf' structure. The tRNA brings amino acids to the ribosome so that they can be added on to the growing amino acid chain**

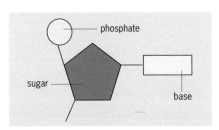

Fig 3.43 **The basic structure of a nucleotide**

?

K List the four major differences between RNA and DNA.

nucleus using a single gene within the DNA as a template. When a complete copy of the gene has been produced, the mRNA moves out of the nucleus to the ribosome, where the protein is synthesised according to the code taken from the DNA.

- **Ribosomal RNA** makes up part of the **ribosome**, a small organelle that brings together all the chemicals associated with protein synthesis.

- **Transfer RNA** (Fig 3.42 (b)), is found in the cytoplasm and is a carrier molecule, bringing amino acids to the ribosomes for assembly into a new amino acid chain, according to the order specified on the mRNA code.

The role of nucleic acids in cell division and protein synthesis are covered in more depth in Chapters 6 and 31.

The structure of nucleic acids

Nucleotides are the building blocks of nucleic acids. A nucleotide consists of three units (Fig 3.43):

- a sugar (ribose or deoxyribose),
- a **phosphate** group,
- a nitrogen-containing **base**.

As the names imply, **deoxyribo**nucleic acid has nucleotides in which the sugar is **deoxyribose**, while **ribo**nucleic acid contains **ribose**.

The nucleotides in DNA can contain any one of four nitrogenous (nitrogen containing) bases: **adenine**, **guanine**, **cytosine** or **thymine**. When DNA **replicates** (copies itself), new strands are made by adding nucleotides. These are available as free molecules in the cytoplasm. Generally, cells can synthesise their own nucleotides.

Three of the bases in RNA – **adenine**, **guanine** and **cytosine** – are the same as those in DNA. The fourth is different: RNA contains **uracil** instead of thymine.

How DNA carries the genetic code

DNA has two remarkable characteristics:

It is a store of genetic information.
It can copy itself exactly, time after time.

Looking at the structure of DNA helps us to understand how it does this.

How the bases pair
Thymine and cytosine belong to a group of chemicals called **pyrimidines** while adenine and guanine are **purines**.

Because of the shape of the two types of molecule, each purine always bonds with only one pyrimidine. So, in DNA, adenine always bonds with thymine, and cytosine with guanine. In RNA, cytosine bonds with guanine and adenine bonds with uracil:

$$\text{DNA: } A=T \quad \text{RNA: } A=U$$
$$C\equiv G \quad\quad\quad C\equiv G$$

The **base pairs** are held together by hydrogen bonds. There are two H-bonds between A and T (or U) and three between C and G.

Fig 3.44 shows how the base pairs within DNA fit together to form a double-stranded helix. The sides are formed by alternating sugar–phosphate units, while the base pairs form the cross-bridges, like the rungs of a ladder. Each base pairing causes a twist in the helix and there is a complete 360° turn every 10 base pairs.

Hint: to remember which bases are pyrimidines and which are purines, concentrate on the letter Y:

Th<u>y</u>mine and c<u>y</u>tosine are p<u>y</u>rimidines.
Adenine and guanine are purines.

L (a) If one half of a DNA strand had the base sequence ATCGTTACC, what sequence would the other strand have?

(b) What would be the same sequence for a molecule of messenger RNA that had been built on the DNA sequence in **(a)**?

How DNA stores information
The DNA that makes up an individual gene contains the information needed to build a particular protein. This information, known as the genetic code, is held in the sequence of bases.

For example, a particularly short gene may read:

DNA sequence	TAC	GGT	AGT	TGG	CGT	ATA	CCG	AAC	CTC	ATC
mRNA sequence	AUG	CCA	UCA	ACC	GCA	UAU	GGC	UUG	GAG	UAG

The second line shows the mRNA sequence that would be built on the DNA (first line) during the process of **transcription**. In DNA and mRNA, each group of three bases is a **triplet code** or **codon** which codes for a particular amino acid.

Table 3.6 shows how the mRNA codons translate into particular amino acids. If you compare the information in the table with the mRNA sequence given above, you can see that the first RNA codon, AUG, translates to the amino acid methionine. After that, each codon in the example specifies a different amino acid. Research has shown that in front of each gene there is a **promoter** sequence, which effectively says 'start transcribing here'.

See question 5.

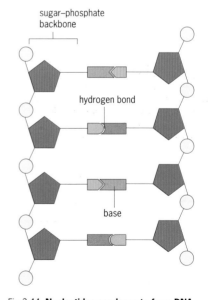

Fig 3.44 **Nucleotides condense to form DNA. There are two parallel strands of bases. Only one strand of the DNA in any gene – known as the sense strand – is used to make proteins. The other side serves to stabilise the molecule. Different genes use different sides of the molecule as the sense strand**

Why a *triplet* code?
There are only four different bases but 20 amino acids. Obviously one base can't code for one amino acid. Neither can two, since there are only 16 possible combinations (4 × 4). Three bases, however, gives plenty of different combinations (4 × 4 × 4 = 64). There are often several different triplet codes for one amino acid (see Table 3.6).

Table 3.6 **The genetic code: the base sequence in each mRNA triplet code translates to a particular amino acid. This same code is used by all organisms**

First base	G	A	C	U	Third base
G	GGG glycine	GAG glutamic acid	GCG alanine	GUG valine	G
	GGA glycine	GAA glutamic acid	GCA alanine	GUA valine	A
	GGC glycine	GAC aspartic acid	GCC alanine	GUC valine	C
	GGU glycine	GAU aspartic acid	GCU alanine	GUU valine	U
A	AGG arginine	AAG lysine	ACG threonine	AUG methionine	G
	AGA arginine	AAA lysine	ACA threonine	AUA isoleucine	A
	AGC serine	AAC asparagine	ACC threonine	AUC isoleucine	C
	AGU serine	AAU asparagine	ACU threonine	AUU isoleucine	U
C	CGG arginine	CAG glutamine	CCG proline	CUG leucine	G
	CGA arginine	CAA glutamine	CCA proline	CUA leucine	A
	CGC arginine	CAC histidine	CCC proline	CUC leucine	C
	CGU arginine	CAU histidine	CCU proline	CUU leucine	U
U	UGG tryptophan	UAG **stop**	UCG serine	UUG leucine	G
	UGA **stop**	UAA **stop**	UCA serine	UUA leucine	A
	UGC cysteine	UAC tyrosine	UCC serine	UUC phenylalanine	C
	UGU cysteine	UAU tyrosine	UCU serine	UUU phenylalanine	U

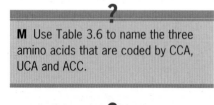

M Use Table 3.6 to name the three amino acids that are coded by CCA, UCA and ACC.

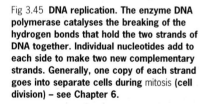

N Look at Table 3.6. What does the codon UAG code for?

How does DNA make copies of itself?

DNA copying, known as **replication**, is essential for cell division. This, in turn, is necessary for the growth and reproduction of all living organisms. Fig 3.45 shows how DNA replicates. When the molecule unwinds and separates, the two inner surfaces are exposed. Only the complementary bases can fit into the 'slots' that become available. So, the exposed single strand is remade into a double strand by adding nucleotides that are identical to those that have just been removed. The end result is two identical strands of DNA.

Mistakes do happen – but these are rare. However these **mutations** are very significant in biology. If they do not cause the death of the organism they are an important source of variation. Without variation, evolution cannot occur (see Chapter 33).

Fig 3.45 **DNA replication. The enzyme DNA polymerase catalyses the breaking of the hydrogen bonds that hold the two strands of DNA together. Individual nucleotides add to each side to make two new complementary strands. Generally, one copy of each strand goes into separate cells during** mitosis **(cell division) – see Chapter 6.**

Chapter 31 gives more detailed information about DNA replication

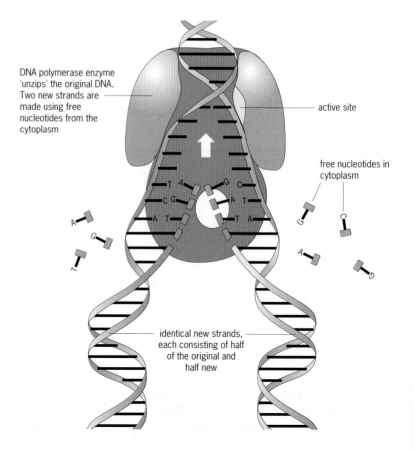

DNA polymerase enzyme 'unzips' the original DNA. Two new strands are made using free nucleotides from the cytoplasm

active site

free nucleotides in cytoplasm

identical new strands, each consisting of half of the original and half new

SUMMARY

When you have finished this chapter, you should know and understand the following:

Carbohydrates

■ Carbohydrates contain the elements carbon, hydrogen and oxygen. They are the first products made by plants in photosynthesis.

■ The term sugars describes monosaccharides and disaccharides. Their names end in the suffix **-ose**.

■ Monosaccharides include glucose, fructose and galactose. These sugars are **isomers**: they have the same formula but their atoms are arranged in different ways. Monosaccharides are linked by **glycosidic bonds**. These are formed by **condensation reactions**.

■ Disaccharides, such as maltose, sucrose (cane sugar) and lactose (milk sugar), consist of two monosaccharides linked together.

■ Most polysaccharides are polymers of glucose. Starch is used for storage in plants, glycogen is used for storage in animals and cellulose gives strength to plant cell walls.

Lipids

■ Lipids contain carbon, hydrogen, oxygen, often phosphorus and occasionally nitrogen. Most are non-polar chemicals and therefore insoluble in water.

■ Lipids are used for energy storage, protection and insulation.

■ In living things there are two main types of lipid: **triglycerides** and **phospholipids**. Triglycerides are the familiar fats and oils. Phospholipids form cell membranes.

■ Fatty acids vary in chain length, and may be **saturated** (with hydrogen) or **unsaturated**. These factors determine the properties of the triglyceride, such as its melting point and viscosity.

■ Phospholipids are similar to triglycerides, but phosphoric acid replaces one of the fatty acids. They have a polar head, allowing them to form **bilayers** (membranes) in water.

■ **Cholesterol** is a normal constituent of cell membranes.

■ **Waxes** are lipids that contain alcohols other than glycerol. They are usually indigestible and are often used for waterproofing.

Proteins

■ Proteins consist of chains of **amino acids** linked by peptide bonds.

■ There are 20 different amino acids in living things. All have a **carboxylic acid group** and an **amino group** but differ in their **R group**.

■ **Fibrous proteins** often join to form large fibrils whose function is strength or the production of movement. **Globular proteins** – including enzymes, antibodies and some hormones – are usually individual molecules with a chemical function.

■ The **primary structure** of a protein is the sequence of its amino acids.

■ The **secondary structure** refers to the patterns and shapes formed within the polypeptide chain, for example, an α-**helix**.

■ The **tertiary structure** refers to the three-dimensional shape of a polypeptide chain, which results from the interactions as the chain folds back on itself. If the protein consists of one polypeptide chain, the tertiary structure refers to the overall shape of the molecule.

■ The **quaternary structure** refers to the overall three-dimensional shape of a protein that consists of more than one polypeptide chain.

Nucleic acids

■ There are two types of nucleic acid, **deoxyribonucleic acid (DNA)** and **ribonucleic acid (RNA)**. RNA itself has three forms, **messenger RNA**, **transfer RNA** and **ribosomal RNA**.

■ DNA carries the genetic code. Its structure allows it to store information, pass information on to RNA so that proteins can be made, and to copy itself, allowing the genetic code to pass to new cells.

■ A **gene** is a length of DNA that codes for the manufacture of a particular protein

■ Messenger RNA copies the genes, to allow them to be used as templates for protein synthesis. Transfer RNA brings amino acids to the ribosome during synthesis. Ribosomal RNA is a major structural component of the ribosome.

■ DNA and RNA are composed of **nucleotides**, which themselves contain a **sugar**, a **phosphate** group and a **nitrogen-containing base**.

■ The structure of DNA, a **double helix**, indicates how it can **replicate** (copy) itself to provide new copies of the genetic code in cell division.

QUESTIONS

1 In which column are the statements correctly applied to cellulose and sucrose? (✓ = correct; ✗ = incorrect)

Property	Column A		Column B		Column C		Column D	
	Cellulose	Sucrose	Cellulose	Sucrose	Cellulose	Sucrose	Cellulose	Sucrose
It is a reducing sugar	✓	✗	✗	✓	✗	✗	✗	✗
It is a polysaccharide	✗	✓	✗	✓	✗	✓	✗	✓
It has a glycosidic link	✓	✗	✗	✗	✓	✓	✗	✓
It produces monosaccharides on complete hydrolysis	✓	✓	✗	✓	✓	✓	✓	✗

[UCLES 1996 Biology/Social Biology; Multiple Choice Specimen Paper 2, q.5]

2 The figure shows two molecules which are important components of living organisms.

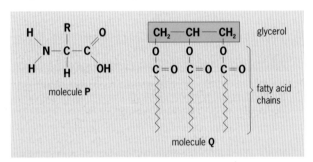

a) Name:
   (i) molecules **P** and **Q**;
   (ii) **one** element always found in molecule **P** which would not be found in a carbohydrate molecule;
   (iii) the type of molecule which would be formed by polymerisation of molecule **P**;
   (iv) the bond which is formed between molecules of **P** when it polymerises.

b) Describe a test you could carry out to investigate whether a liquid contained: **(i)** polymers of molecule **P**; **(ii)** molecules of **Q**.

[UCLES: 1995, Modular Sciences, Biology Foundation paper, q.1]

3 Electrophoresis is a technique used to separate and identify peptide residues. The peptide residues in a buffer solution are put on to a strip of filter paper wetted by the same buffer and a direct current passed along the strip. The peptide residues migrate from their starting point as a result of their overall charge.

Figs **A** and **B** show the results of a combination of electrophoretic and chromatographic techniques for the analysis under the same experimental conditions of peptide residues of human haemoglobin.

Fig **A** is for normal haemoglobin; Fig **B** is the haemoglobin from an individual suffering from sickle cell anaemia.

A

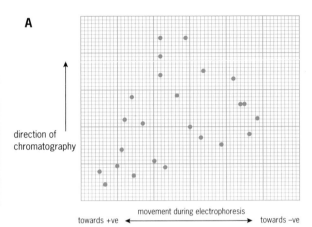

B

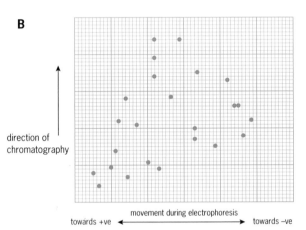

a) **(i)** Put a cross on Fig **A** to show where the peptide mixture could have been placed at the start.
   (ii) Draw a box around any **two** peptide residues which were separated by electrophoresis and not by chromatography.

b) Haemoglobin in a person with sickle cell anaemia differs from normal haemoglobin by having glutamic acid replaced by valine in part of the polypeptide chain.
   (i) Complete the formulae of the two α-amino acids below to represent the structures of the amino acids as they exist in an aqueous buffer solution of pH 7.

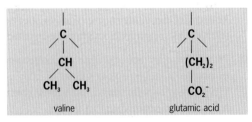

   (ii) By considering the structures of the two α-amino acids you have shown in **(b) (i)**, explain whether the compounds appear in different positions following electrophoresis.
   (iii) Draw a circle on both Figure **A** and Figure **B** to indicate the valine/glutamic acid discrepancy.

c) What is meant by describing valine as an *essential amino acid*?

d) Draw a diagram to illustrate only those atoms and bonds within the peptide linkage formed between valine when linked to leucine.

[UCLES June 1994, Biology: Biochemistry paper, q.3]

4 The diagram below represents, in two dimensions, a very small part of the structure of a naturally occurring material.

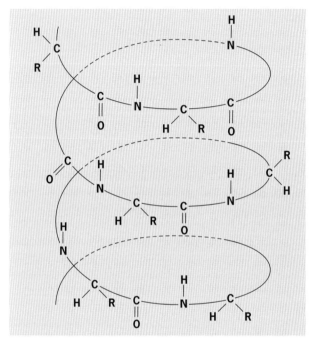

a) To what class of naturally occurring material does this substance belong?

b) What name describes the conformation into which this chain is twisted?

c) **(i)** What type of chemical bonding is involved in the cross-linking between loops within a twisted chain?
(ii) Copy and show on the diagram such a cross-linking.

d) Name and draw diagrams to show **two** other types of interaction between chains of this type.

e) **(i)** How would you attempt, in the laboratory, to break down the substance completely into its monomer units in order to investigate its structure?

(ii) Taking 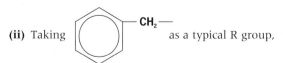 as a typical R group,

draw the structure of a monomer unit.

f) Outline the practical procedure you would use to identify the monomer units.

[UCLES December 1993 Modular Sciences: Biochemistry Module Paper, q.2]

5 The figure shows part of a DNA molecule.

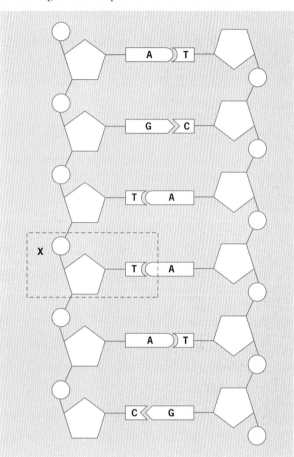

a) **(i)** Name the part of the molecule labelled **X**.
(ii) State the full names for the bases **A**, **C**, **G** and **T**.
(iii) Name the element found in each of the four bases **A**, **C**, **G** and **T** but **not** in the part of the molecule labelled **S**.
(iv) Explain how the two strands of the DNA molecule are held together.

b) Twenty-eight per cent of the bases in a DNA molecule were found to be base **A**. Calculate the percentage which would be base **G**. Show your working.

c) Outline the way in which the sequence of bases on a DNA molecule codes for protein synthesis within the cell.

d) State **three** structural differences between DNA and RNA.

[UCLES June 1994 Modular Biology: Foundation module paper, q.1]

Assignment

DNA FINGERPRINTING

In 1987, Robert Melias made legal history in the UK when he was convicted of rape based on evidence obtained by **DNA fingerprinting**. This technique – more accurately called **DNA profiling** – was developed at Leicester University by Alec Jeffreys in 1984 and is proving to be an invaluable tool in forensic medicine.

The principles behind DNA profiling

Every human body cell contains 46 chromosomes. Each one consists of a single elaborately coiled piece of DNA which, if stretched out, can be as long as 5 cm. The structure of DNA can be used to identify individuals.

Between the many genes that occur along the DNA molecule are regions which code for nothing at all. Within this non-coding DNA are **hypervariable regions**, so called because they vary enormously in length from person to person. Hypervariable regions consist of particular base sequences called **core sequences**, which are repeated again and again. Different people have different numbers of repeats and so have differently sized hypervariable regions. When these are labelled and separated according to size, a pattern is produced. Each pattern is unique to each individual person and can be used as the basis for DNA profiling (Fig 3.A1).

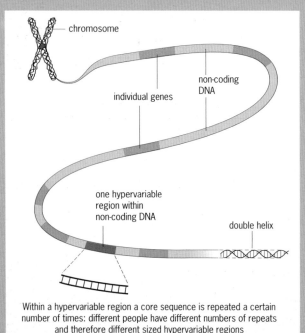

Within a hypervariable region a core sequence is repeated a certain number of times: different people have different numbers of repeats and therefore different sized hypervariable regions

Fig 3.A1 **Although each chromosome contains several thousand genes, they only account for about 10 per cent of the length: the rest is non-coding DNA. Within these 'nonsense' segments are** hypervariable regions. **These are unique to each individual and, when they are separated out, the pattern they form provides the basis of DNA profiling**

a) What is a gene?

b) What is the name given to a fault which sometimes occurs when DNA is copied?

c) Faults accumulate in the hypervariable regions more frequently than they do in the genes. Why do you think this is?

Getting a sample of DNA

All body cells contain the same DNA, and so virtually any tissue sample can be used for DNA profiling. The amount of tissue required is very small: 0.05 cm^3 of blood, 0.005 cm^3 of semen or one hair root!

a) Which blood cells contain DNA?

b) Why do forensic scientists need a larger sample of blood than of semen?

Processing the DNA and separating the fragments

Once you have a DNA sample, the next stage is to cut the molecule using **restriction enzymes**, which act rather like molecular scissors (see Chapter 34). These enzymes cut DNA at specific base sequences, giving a complex mixture of DNA fragments, some of which contain the hypervariable regions.

Electrophoresis (see page 54) separates the fragments. The fragment mixture is placed in a trough in some gel. When a current is applied, the negatively charged DNA fragments move towards the positive terminal, or **anode**. The fragments move at different speeds according to their size: small ones move faster than larger ones. The fragments separate out into bands, as shown in Fig 3.A2.

The bands of DNA are transferred from the gel onto a nylon membrane by **Southern blotting**, a process which

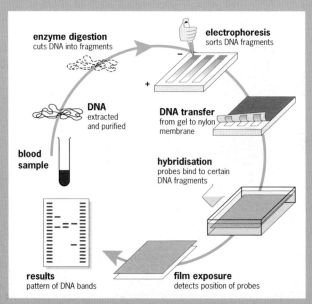

enzyme digestion
cuts DNA into fragments

electrophoresis
sorts DNA fragments

DNA
extracted and purified

DNA transfer
from gel to nylon membrane

blood sample

hybridisation
probes bind to certain DNA fragments

results
pattern of DNA bands

film exposure
detects position of probes

Fig 3.A2 **The major steps in the preparation of a DNA profile**

works by capillary action. At this stage the bands are still invisible, and must be stained so that the hypervariable regions can be seen.

Labelling the fragments

Within hypervariable regions are core sequences that are common to all humans. It is the number of times the sequences are repeated which varies from person to person.

Pieces of DNA complementary to these core sequences have been isolated and are produced in bulk for use as **genetic probes**. They are labelled with a marker chemical, commonly the enzyme **alkaline phosphatase** which fluoresces (produces light) when a particular substrate is added.

When the probes, complete with enzyme, are added to the DNA sample, they attach to the core sequences, thus marking the hypervariable regions. Excess probe is washed off, substrate for the enzyme is added, and bands which contain hypervariable regions fluoresce. If the blot is exposed to an X-ray film, dark lines appear wherever bands in the blot have emitted light, forming the familiar DNA profile.

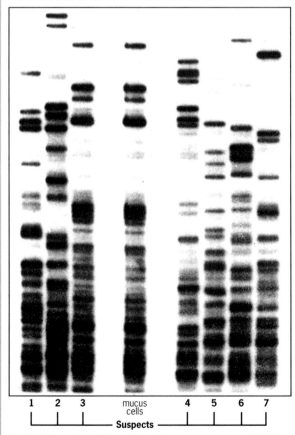

1 2 3   mucus   4 5 6 7
          cells
└──┴──┴── Suspects ──┴──┴──┴──┘

Fig 3.A3 **shows the DNA profile of six suspects compared with the sample taken from the bank**

3

a) Why do the DNA bands fluoresce?
b) Do all the DNA bands fluoresce? Explain your answer.

Case studies

You have covered the basic theory of DNA profiling: now you can put it into practice. Here are two case studies for you to interpret.

Case 1: Whodunnit?

In 1989, a man robbed a bank at gunpoint whilst suffering from a particularly heavy cold. The mucus-filled tissues found on the scene provided enough DNA for a DNA profile.

Look at Fig 3.A3.

4 On the basis of the evidence, which suspect robbed the bank?

Case 2: Whose baby?

Bands in an individuals DNA profile must have come from either the mother or the father. When a child's DNA profile is compared to the mother's, it is easy to see which bands have been inherited. Any bands in the child's profile which did not come from the mother must have come from the father. Use this information to solve the next case study.

In this case study, Mrs X is claiming that Mr Z, a famous rock musician, is the father of her 14-year-old daughter. She is seeking a large maintenance settlement. Predictably, Mr Z denies all knowledge of this and his lawyer demands that a DNA profile be carried out. Fig 3.A4 shows the results.

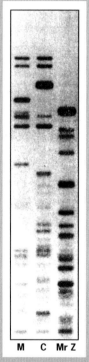

Fig 3.A4 **The DNA profiles from Case Study 2, for the mother (M), child (C) and alleged father (MrZ)**

M   C   Mr Z

5

a) Which of the daughter's DNA bands were inherited from her mother?
b) Do the remaining bands match those from Mr Z, showing that he is the father?

Questions for discussion

6 What would be the advantages and disadvantages of compulsory DNA testing for the whole UK population, so that the police have comprehensive files?

7 Would you like details of your DNA to be available to businesses, in the way that financial records are today? What problems could this cause?

4 Enzymes and metabolism

IT IS EASY to see how confectioners make chocolates with hard centres: they simply pour molten chocolate over the centre and wait for it to set. But what about the soft centres? Surely it isn't possible to pour liquid chocolate over a liquid centre and still keep the shape. So, how do they make the runny, yolk-like inside to chocolate eggs?

Chocolate lovers everywhere may be surprised to learn that the answer is: use an enzyme. To start with, the centre is solid and contains an enzyme and a polysaccharide. After the chocolate coating has set, the enzyme breaks down the long polysaccharide chains, turning the hard centre into the familiar runny filling. This is a 'fun' use of enzymes but, increasingly, enzymes are being used in medicine, industry and biotechnology.

> **Metabolism** is a general term that describes the complex and inter-related chemical reactions that take place in an organism. Metabolic reactions are controlled by **enzymes**. Enzymes, like all proteins, are built up according to the code contained within genes.

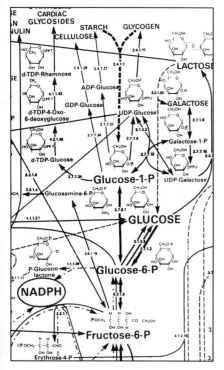

Fig 4.1 **Part of a very big metabolic chart showing some of the biochemical reactions that occur in living cells. You may notice that some reactions carry a number: this identifies the enzyme that catalyses a particular reaction**

1 WHAT ARE ENZYMES?

Enzymes are complex chemicals that control reactions in living cells. They are biochemical **catalysts**, speeding up reactions that would otherwise happen too slowly to be of any use to the organism. This definition, of course, is an over-simplification. An active enzyme may speed up a particular reaction, but living things do not need all reactions to be in full swing all of the time. The key word is control. It is more accurate to say that enzymes interact with simpler molecules to produce an ordered, stable reaction system in which the products of any reaction are made *when* they are needed, in the *amount* needed.

The role of enzymes in an organism

A metabolic pathway chart (Fig 4.1) shows some of the large number of different but interconnected chemical reactions in a living cell. The chart is purely for reference but it illustrates several important points:

- Many of the complex chemicals that living organisms need cannot be made in a single reaction. Instead, a series of simpler reactions occur, one after another, forming a **metabolic pathway**. A single pathway may have many steps in which each chemical is converted to the next. A specific enzyme controls each reaction.

- Each individual step in the chart represents one of the simple chemical reactions. At first glance, there seem to be an endless variety of reactions, but, look more closely and you will see that the same *types* of reaction occur again and again.

- Enzymes control cell metabolism by regulating how and when reactions occur. Using this very simple pathway as an example,

$$A \rightarrow B \rightarrow C \rightarrow D$$

the final product is substance D, the chemical needed by the living organism. The pathway needs three different enzymes and, when D is no longer needed, or if too much has been produced, one of the three enzymes is 'switched off'.

One enzyme, one reaction, but what a difference!

When carbon dioxide dissolves in water, a small proportion of the CO_2 molecules combine with H_2O to form carbonic acid.

$$CO_2 + H_2O \rightarrow H_2CO_3 \rightarrow H^+ + HCO_3^-$$

It is a rather unspectacular reaction, but, in the presence of the enzyme **carbonic anhydrase**, it is speeded up about 10^7 (ten million) times! One molecule of carbonic anhydrase can convert 600 000 molecules of CO_2 into carbonic acid every second. In living cells, this reaction allows other processes to occur much faster:

- In red blood cells, the enzyme speeds up the production of acid, which in turn causes haemoglobin to give up its oxygen. Without the enzyme, delivery of oxygen to the tissues would be much slower.

- In certain cells in the stomach lining, the activity of carbonic anhydrase in a pathway allows hydrochloric acid to be secreted rapidly. This creates the acidic conditions necessary for digestive enzymes in the stomach to work properly.

- In the cells of the kidney tubule, carbonic anhydrase speeds up excretion of excess acid, and so helps to maintain the pH of the body at the correct level.

?

A By what process are enzymes made within the cell?

2 THE CHEMICAL NATURE OF ENZYMES

Enzymes are globular proteins. They have a complex tertiary and often quaternary structure (see page 53) in which polypeptides are folded around each other to form a roughly spherical, or **globular** shape (Fig 4.2). The overall three-dimensional shape of an enzyme

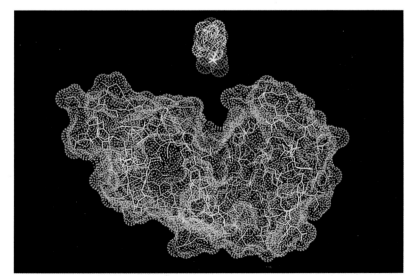

Fig 4.2 **The three-dimensional shape of ribonuclease A, an enzyme that helps break up mRNA in the cytoplasm of bacteria. The substrate at the top has been shifted away from the active site, the 'pocket' into which the substrate fits.**

Most enzymes are large molecules that are often made of two or more polypeptides. The active sites in enzymes result from the quaternary structure of the protein molecule

Fig 4.3 **This schematic diagram shows the three-dimensional structure of an enzyme molecule. Hydrogen bonds are shown in red. The shape and chemical groups on the substrate closely match that of the active site**

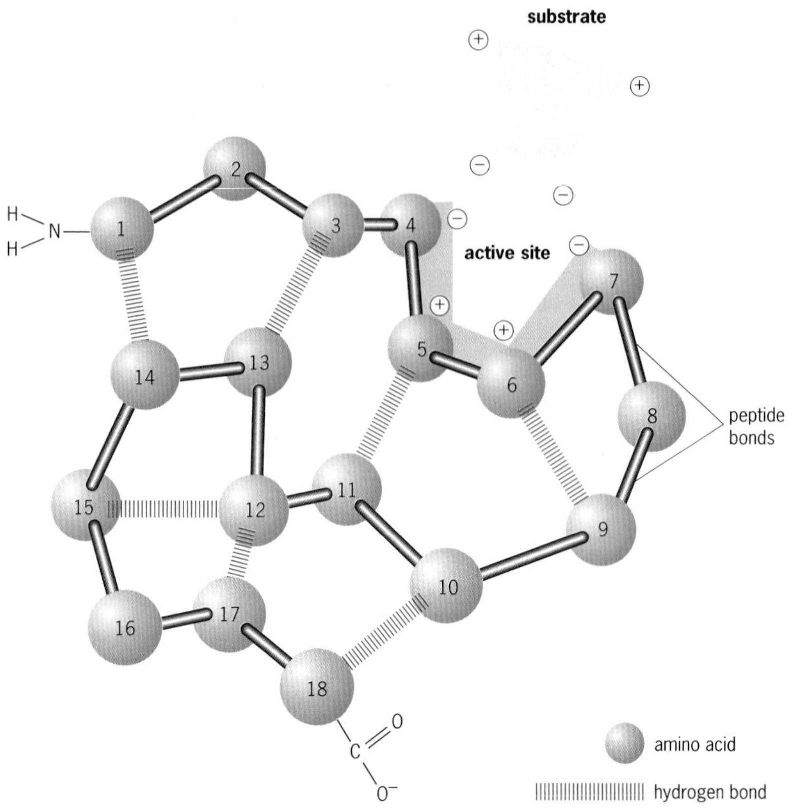

An enzyme acts on a chemical known as its **substrate**. The name of an enzyme often comes from substituting or adding -*ase* in the name of the substrate, so, for example, lactose is the substrate of the enzyme lactase.

B Predict substrates for the following digestive enzymes: protease, lipase, nucleotidase.

---

molecule is very important: if it is altered, the enzyme cannot bind to its substrate and so cannot function. Enzyme shape is maintained by hydrogen bonds and ionic forces (Fig 4.3) and their function can be affected by changes in temperature and pH.

Enzymes have several important properties:

● Enzymes are **specific**: each enzyme usually catalyses only one reaction.

● Enzymes combine with their substrates to form temporary enzyme–substrate **complexes**.

● Enzymes are not altered or used up by the reactions they catalyse, so can be used again and again.

● Enzymes work very rapidly and each has its own **turnover number** (see page 69).

● Enzymes are sensitive to temperature and pH.

● Many enzymes need other chemicals – **cofactors** – in order to function.

● Enzyme function may be slowed down or stopped by **inhibitors.**

The specificity of enzymes

Taking digestive enzymes, you might find it difficult to believe that each enzyme catalyses only one reaction. Trypsin, for example, can begin the digestion of a wide variety of foods rich in protein: eggs, pork, chicken and soya, for example. But when you look at how trypsin works at the molecular level, you can see that this enzyme *is* specific. Trypsin cuts an amino acid chain at a point between a lysine and an arginine residue, and nowhere else. Most proteins have these two amino acids next to each other at some point in their polypeptide chain, and so can be partly digested by trypsin.

Several scientists in biochemistry have devised mechanisms – often called **models** – to explain how enzymes work. In coming up with ideas, they have had to take account of enzyme specificity. Two models that may explain how enzymes work are the **lock and key hypothesis** and the **induced fit hypothesis**.

?

C If the enzyme amylase is specific, how can it catalyse the digestion of bread, potatoes and rice?

The lock and key hypothesis

This idea assumes that enzyme function depends on an area on the molecule known as the **active site**, highlighted on Figs 4.2 and 4.3. The active site is a groove or pocket in the surface of the enzyme into which the substrate molecule fits. Typically, the active site is formed by the R-groups of 3 to 12 amino acids. The size, shape and chemical nature of the active site corresponds closely with that of the substrate molecule, so they fit together like a key fits into a lock (Fig 4.4) or, perhaps more realistically, like two pieces in a three-dimensional jigsaw.

Although this model helps us to understand some of the properties of enzymes, it is now generally accepted that a modified version, known as the induced fit hypothesis, better represents what happens when an enzyme catalyses a reaction.

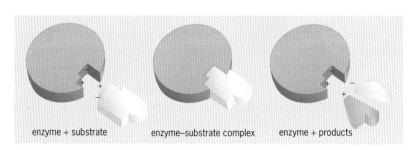

enzyme + substrate   enzyme–substrate complex   enzyme + products

Fig 4.4 **The lock and key hypothesis. The active site is a particular shape (the lock) into which only one substrate (the key) will fit. The enzyme and substrate combine for an instant to form an enzyme–substrate complex. The formation of this complex brings about the desired chemical reaction, converting substrate into product(s).**

As the drawing shows, the substrate fits the active site because it is a complementary shape and because the chemical charges (+ and –) attract each other

The induced fit hypothesis

Experimental evidence suggests that the active site in many enzymes is not exactly the same shape as the substrate, but moulds itself around the substrate as the enzyme–substrate complex is formed (Fig 4.5). Only when the substrate binds to the enzyme is the active site the correct shape to catalyse the reaction. As the products of the reaction form, they fit the active site less well and fall away from it. Without the substrate, the enzyme reverts to its 'relaxed' state, until the next substrate comes along.

Both models show why enzymes are not altered by the reactions that they catalyse: they bind to a substrate momentarily, allowing a reaction to happen, but do not themselves undergo any chemical change.

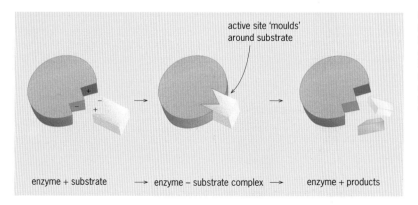

active site 'moulds' around substrate

enzyme + substrate ⟶ enzyme – substrate complex ⟶ enzyme + products

Fig 4.5 **The induced fit hypothesis. Before substrate binding, the enzyme's active site is 'relaxed'. When the substrate binds, the active site is pulled into the correct shape by molecular interactions between the two molecules, and an enzyme–substrate complex forms. As the products fall away from the active site, the molecule becomes 'relaxed' again**

The chemistry of enzyme action

In biology almost all reactions are reversible. If they were not, there would be no recycling of molecules (see Chapter 37).

A simple reversible reaction can be expressed as:

$$\underset{\text{reactants}}{A + B} \leftrightarrow \underset{\text{products}}{C + D}$$

In this reaction, the **reactants** (A and B) combine to give the **products** (C and D). If this reaction were to take place in a test-tube, a proportion of the products would react to form A and B again. Eventually, an equilibrium would be reached in which the relative proportion of reactants and products would remain the same.

In theory, an enzyme allows a reversible reaction to reach equilibrium more quickly. Depending on whether there are more reactants or products present at the start, an enzyme can speed the reaction in either direction. In living things, however, enzymes usually speed up reactions in one particular direction. This is because, in any organism, an enzyme only exists if the product of the reaction that it catalyses is needed. And, as the product is used as soon as it is made, equilibrium is rarely reached.

Enzymes work by lowering activation energy

For any chemical reaction to take place, old bonds must be broken before new ones can form. The energy needed to break these bonds, and so set the reaction in motion, is the **activation energy**.

Many reactants need a large amount of energy to push them to a state where they can take part in a reaction, so many reactions take place only at high temperature. Many substances burn, for instance, but only after the initial activation energy has been supplied, perhaps by lighting a match. In the presence of enzymes, the activation energy is greatly lowered and this allows reactions to take place at the relatively low temperatures normally found in living things. The 'half-way' point in a reaction is called the **transition state**, and this is represented by the top of the curve in Fig 4.6. The

Fig 4.6 **In order for a reaction to happen, activation energy must be supplied. Then the rest of the reaction proceeds, just as the boulder rolls down the hill once the energy has been supplied to push it to the top. Catalysts such as enzymes work by lowering the activation energy**

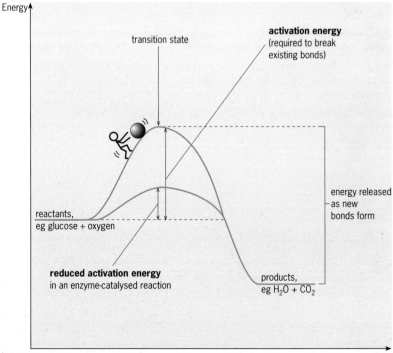

transition state represents the stage when the old bonds have been broken in order to allow new ones to form. Enzymes lower the activation energy by making it easier to achieve the transition state.

Enzymes speed up reactions in a number of ways:

- They hold the substrates close together at the correct angle – this would otherwise have to occur by chance collision.

- Any polar groups on the active site may be arranged so as to change the distribution of charge (+ or –) on the substrate, helping the reaction to occur.

- By acid or base catalysis: the active site of the enzyme behaves as an acid or a base, donating or accepting protons (hydrogen ions, H^+) from the substrate.

- Through structural flexibility: the flexible shape of the enzyme can change during catalysis. This ensures that the substrates are brought together in the correct sequence for the required reaction.

Many metabolic reactions require energy in addition to the presence of the relevant enzyme. This energy is supplied by ATP made by cell respiration (see Chapter 8). Reactions that need energy are made to happen by coupling them with ATP breakdown, and so are called **coupled reactions**.

D The enzyme lysozyme helps to split bacterial cell walls. Where would you expect this enzyme to be present in the human body?

How fast do enzymes work?

The speed at which an enzyme works is expressed its **turnover number**. This is usually defined as the number of substrate molecules turned into product in one minute by one molecule of enzyme. Values range from less than a hundred to many millions, and some specific examples are given in Table 4.1.

Naming and classifying enzymes

Older enzyme names such as **pepsin**, **catalase** and **trypsin** give no clues about the nature of the reaction they catalyse. To cope with the rapidly expanding number of new enzymes, the International Union of Biochemistry has developed a scheme for naming and classifying enzymes. Very generally, enzymes are named be adding the suffix -ase to the name of their substrate. The rest of the name attempts to indicate the nature of the reaction taking place. Alcohol dehydrogenase, for example, catalyses the removal of hydrogen from alcohol (ethanol). Further examples are given in Table 4.2.

Table 4.1 **Some turnover numbers**

Enzyme	Turnover number
carbonic anhydrase	36 000 000
catalase	5 600 000
β-galactosidase	12 000
chymotrypsin	6 000
lysozyme	60

(Source: Biochemical society guidance notes 3, *Enzymes and their role in biotechnology*.)

Table 4.2 **Some common enzymes and their substrates**

Enzyme	Substrate	Reaction catalysed
maltase	maltose	hydrolysis of maltose to glucose
amylase	starch	hydrolysis of starch to maltose
alcohol dehydrogenase	alcohol (ethanol)	removal of hydrogen from alcohol
DNA ligase	DNA	joining together two DNA strands
RNA polymerase	nucleotides that make RNA	synthesis of mRNA on a DNA molecule
glycogen synthetase	glucose	polymerisation of glucose into glycogen
ATPase	ATP	synthesis or splitting of ATP

Although there are many different enzymes, they can be put into one of six main categories according to the type of reaction they catalyse:

- **Oxidoreductases** These catalyse oxidation and reduction (redox) reactions. In aerobic respiration, most of the cells ATP is generated courtesy of redox reactions – see Chapter 8.

- **Transferases** These catalyse the transfer of a chemical group from one compound to another, such as the transfer of an **amino group** from one amino acid to another organic acid in the process of **transamination**.

- **Hydrolases** These catalyse **hydrolysis** (splitting by use of water) reactions. Most digestive enzymes are hydrolases.

- **Lyases** These catalyse the breakdown of molecules by reactions that do not involved hydrolysis.

- **Isomerases** These catalyse the transformation of one isomer into another, for instance the conversion of glucose 1,6-diphosphate into fructose 1,6-diphosphate. This is one of the first reactions in the **glycolysis**, the first stage of respiration.

- **Ligases** These catalyse the formation of bonds between compounds, often using the free energy made available from ATP hydrolysis. DNA ligase, for example, is involved in the synthesis of DNA.

> **?**
>
> **E** People with a great deal of physical stamina have been found to have a lot of **glycogen synthetase** in their muscles. Suggest what this enzyme does?

Factors affecting enzyme activity

The factors that affect enzyme activity also affect the functions of the cell and, ultimately, the organism. Enzymes are proteins and their function is therefore affected by:

- Temperature
- pH
- Substrate concentration
- Enzyme concentration
- Cofactors
- Inhibitors

> **?**
>
> **F** Suggest which type of bonds are broken when an enzyme is denatured. (Look at Fig 4.3 for help.)

Temperature

For a non-enzymatic chemical reaction, the general rule is: the *higher* the temperature, the *faster* the reaction: Fig 4.7(a). This same rule holds true for a reaction catalysed by an enzyme, but only up to about 40 to 45 °C. Above this temperature, enzyme molecules begin to vibrate so violently that the delicate bonds that maintain tertiary and quaternary structure are broken, irreversibly changing the shape of the molecule. When this happens the enzyme can no longer function and we say it is **denatured**.

The effect of temperature on a reaction can be expressed by the **temperature coefficient**, commonly known as the **Q_{10}**.

Where t is the chosen temperature, the formula for the Q_{10} is:

$$\frac{\text{rate of reaction at } t + 10°C}{\text{rate of reaction at } t}$$

Fig 4.7(b) shows how to calculate the Q_{10}. The values for living things need to fall in the range 4 to 40 °C, so we have chosen t as 20 °C.

$$Q_{10} = \frac{\text{rate at } 30°C}{\text{rate at } 20°C} = \frac{6}{3} = 2$$

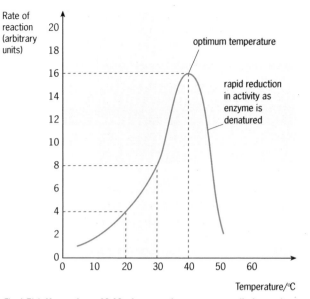

Fig 4.7(a) **Up to about 40 °C, the rate of enzyme-controlled reactions increases with temperature. The optimum temperature for enzymes is about 40 °C, although the activity of an individual enzyme may increase up to about 50 °C or beyond. However, as the temperature passes 43–44 °C, most enzymes lose their activity**

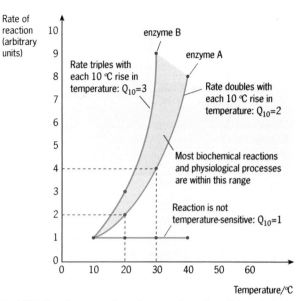

Fig 4.7(b) **Not all enzymes have the same Q_{10} values. This variation is important because enzymes, even those in the same pathway, can vary significantly in their sensitivity to temperature. Outside an organism's normal temperature range, the enzymes may begin to work at different rates. This causes a metabolic imbalance that may be lethal, and may explain why some organisms die at temperatures that would seem to be relatively mild. Some Antarctic fish, for example, live in water that remains very constant at around −2 °C, and die if placed in water above 6 °C**

In practice, most enzymes have a Q_{10} between 2 and 3. A value of 2 means that the rate of reaction doubles with a 10°C temperature rise, 3 means that it triples.

Some organisms have enzymes that are less sensitive to heat than those found in mammals. For example, certain bacteria can survive in hot volcanic springs and deep-sea hydrothermal vents (Fig 4.8) at temperatures of over 90°C, so their enzymes must be active at these extreme temperatures.

?

G Egg albumen, when heated, goes solid and white. What do you think has happened to the proteins in the egg white and why does the white not become runny again when the egg is cooled?

STABLE ENZYMES FOR WASHING POWDERS

ENZYMES ARE UNSTABLE, particularly at high temperature, so their commercial usefulness is limited. Many industrial processes need to take place in 'unnatural' environments, at high temperature and extremes of pH. But there are organisms that can thrive at high temperatures. Thermophilic bacteria have thermostable enzymes that are also more resistant to extremes of pH and other unfavourable conditions, such as organic solvents.

The genes that code for some thermostable enzymes have been transferred using recombinant DNA technology (see Chapter 34) into an appropriate organism, such as the bacterium *Bacillus subtilis*. This organism then multiplies and synthesises large amounts of the enzyme for commercial use.

An example of a thermostable enzyme is the alkaline protease **subtilisin**, the famous stain digester in biological washing powders. This enzyme is produced on a grand scale by the bacterium *Bacillus subtilis*, and is active in alkaline environments, so is compatible with the phosphates and other ingredients in washing powder. In addition, it is active at temperatures up to 60 °C, so it can be used in a wide variety of wash programmes.

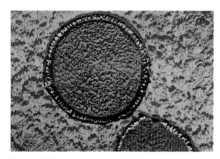

Fig 4.8 **This bacterium *Staphylothermus marinus* lives near deep sea hot-water vents in temperatures of up to 98 °C. Such thermophilic bacteria, that can survive extremely high temperatures, were at first thought to be just a curiosity. However, the thermostable enzymes in them have great potential in industry (see Feature box, left)**

?

H Why is it an advantage for mammals to have a constant body temperature of around 37 °C?

pH

Like other proteins, enzymes are stable over a limited range of pH. Outside this range, at the extremes of pH, enzymes are denatured. Free hydrogen ions (H^+) or hydroxyl ions (OH^-) affect the charges on amino acid residues, distorting the three-dimensional shape and causing an irreversible change in the protein's tertiary structure.

Enzymes are particularly sensitive to changes in pH because of the great sensitivity of their active site. Even if a slight change in pH is not enough to denature the molecule, it may upset the delicate chemical arrangement at the active site and so stop the enzyme working (Fig 4.9).

Most enzymes are **intracellular**: they work inside cells, and their optimum pH is around 7.3 to 7.4. Most organisms have buffer systems that resist changes in pH, and many are able to excrete excess acid or alkali. The control of pH is discussed in Chapter 25.

Fig 4.10 shows some enzymes that have optimums of pH that are distinctly acid or alkaline. These include digestive enzymes that normally work in the stomach or inside lysosomes. We describe enzymes that work outside cells as **extracellular**.

> **?**
>
> **I** Chymotrypsin is found in the small intestine at pH 8. What would happen to this enzyme if it found its way into the stomach?

Fig 4.9 **The effect of pH on enzyme activity**

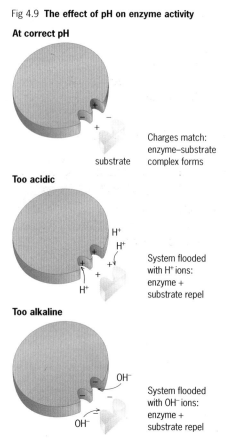

Fig 4.10 **The optimum pH of some different enzymes**

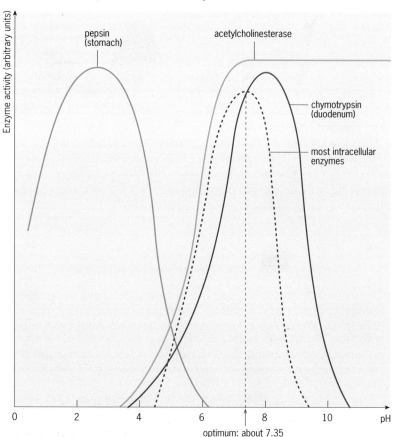

Substrate concentration

The rate of an enzyme-controlled reaction increases as the substrate concentration increases, until the enzyme is working at full capacity. At this point the enzyme molecules reach their turnover number and, assuming that all other conditions such as temperature are ideal, the only way to increase the speed of the reaction even more is to add more enzyme (Fig 4.11).

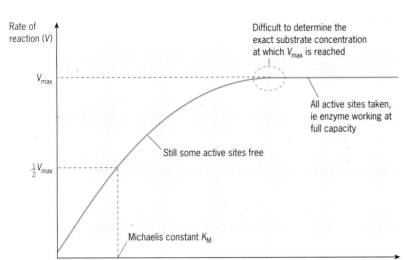

Fig 4.11 **The effect of substrate concentration on the rate of enzyme action**

The Michaelis constant

The graph in Fig 4.11, which shows the effect of increasing the substrate concentration in the presence of a fixed amount of enzyme, is called the **Michaelis curve**. The maximum reaction rate possible for a given amount of enzyme is V_{max}. Enzymes with a *high* turnover number have a *high* V_{max} because each enzyme molecule can catalyse a large number of substrate molecules per minute.

A knowledge of the V_{max} can be useful to biochemists but, as you can see from the graph, the substrate concentration at which V_{max} is reached is difficult to calculate accurately. The graph curves gradually to a plateau and it is difficult to say exactly when it becomes a straight horizontal line. The point at which the rate of reaction is half V_{max} is much easier to calculate because the graph is a straight line at that point.

The substrate concentration at which reaction rate is half V_{max} ($V_{max}/2$) is called the **Michaelis constant**, or K_M This value has several practical uses for biochemists:

- It is characteristic for an enzyme and its substrate.

- It is an indication of the affinity (degree of attraction) between enzyme and substrate.

- If an inhibitor such as a drug or a toxin is acting on the enzyme, its effect on the K_M gives some clues about its mode of action.

Enzyme concentration

In any reaction catalysed by an enzyme, the number of enzyme molecules present is very much smaller than the number of substrate molecules. If you look back at Table 4.1, which shows the turnover numbers of some enzymes, you can see that one molecule of an enzyme can convert several million substrate molecules into products every minute.

When an abundant supply of substrate is available, the rate of reaction is limited by the number of enzyme molecules present. In this situation, increasing the enzyme concentration increases the rate of reaction.

The important role of cofactors

Some enzymes work on their own, but many do not; another molecule, called a **cofactor**, needs to be present before they can function properly. Cofactors modify the enzyme complex so that it has the chemical properties necessary to catalyse a reaction. Cofactors can be classified according to their chemical nature.

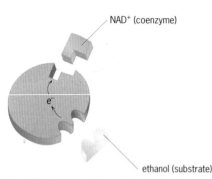

Fig 4.12 **NAD⁺ (made from vitamin B3, nicotinic acid) is the coenzyme for the enzyme alcohol dehydrogenase. The enzyme can work only when the coenzyme is in place to accept the electron that is produced by the reaction. Effectively, this enzyme has two active sites, one for the substrate and one for the coenzyme**

?

J Why are vitamins only required in the diet in tiny amounts?

?

K What is the difference between trace elements and vitamins?

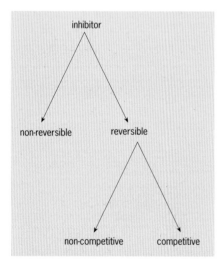

Fig 4.13 **A summary of the different types of enzyme inhibitor**

Prosthetic groups

A **prosthetic group** is an organic molecule that is permanently attached to an enzyme. The enzyme **cytochrome oxidase** catalyses one of the final steps in the electron transport chain (Chapter 8). It has a prosthetic group that joins hydrogen atoms to oxygen, forming water. Without the prosthetic group, the reaction cannot occur.

Coenzymes

Coenzymes are relatively small organic molecules that, unlike prosthetic groups, are not permanently attached to the enzyme molecule (Fig 4.12). They come in to combine with the enzyme and substrate, forming a three-membered complex. Many coenzymes are modified vitamins. Almost all of the water soluble vitamins (the many B complex compounds and vitamin C) function as coenzymes.

Metal cofactors

Metal cofactors are inorganic metal ions that also known as **enzyme activators**. They form part of the catalytic centre of the active site and produce areas of intense positive charge within the protein molecule. The protein cannot have this charge pattern without the metal cofactor. The charge pattern modifies the active site so that the reaction can take place. Cobalt, iron, zinc, magnesium and manganese are all metal cofactors: this is why we need minute amounts of these **trace elements** in our food.

Inhibitors

Inhibitors slow down or stop enzyme action. Usually, enzyme inhibition is a natural process, a means of switching enzymes on or off when necessary. Inhibition tends to be **reversible**, and the enzyme returns to full activity once the inhibitor is removed. Many 'external' chemicals, such as drugs and poisons, can inhibit particular enzymes. This type of inhibition is often **non-reversible**. Reversible inhibitors are either **competitive** or **non-competitive**.

The different types of inhibitor are summarised in Fig 4.13.

Competitive inhibitors

Competitive inhibitors compete with normal substrate molecules to occupy the active site. The inhibitor molecules must be a similar size and shape to the substrate to fit the active site, but cannot be converted into the correct product. Effectively, competitive inhibitors 'get in the way' (Fig 4.14) and reduce the number of interactions that can happen between enzyme and substrate.

Fig 4.15 shows the effect of a competitive inhibitor on the Michaelis curve: V_{max} can still be reached, but a higher substrate concentration is needed to overcome the effect of the inhibitor. K_m is lowered because the inhibitor molecules reduce the efficiency of the enzyme: the enzyme: there are fewer enzyme–substrate 'collisions'.

In the Krebs cycle (page 132) **succinate dehydrogenase** catalyses the removal of hydrogen from the substrate **succinate**. Fig 4.16 shows the structure of succinate and three competitive inhibitors.

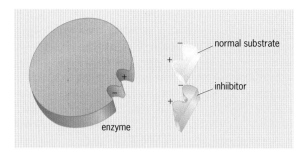

Fig 4.14 **A competitive inhibitor fits into the active site of the enzyme, preventing the real substrate from gaining access. The inhibitor cannot be converted to the products of the reaction and so the overall rate of reaction is slowed down. If an inhibitor is present in equal concentrations to the substrate, and if both types of molecule bind to the active site equally well, the enzyme can only work at half its normal rate**

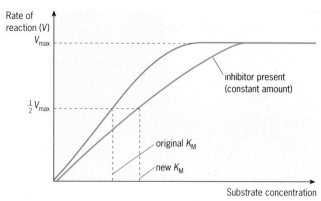

Fig 4.15 **The effect of a competitive inhibitor on the rate of reaction. Note that the effect of the inhibitor can be overcome by adding more substrate, and V_{max} can still be reached, but K_M is lowered**

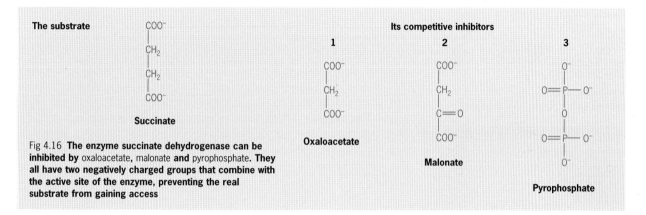

Fig 4.16 **The enzyme succinate dehydrogenase can be inhibited by** oxaloacetate, malonate **and** pyrophosphate. **They all have two negatively charged groups that combine with the active site of the enzyme, preventing the real substrate from gaining access**

Non-competitive inhibitors

Non-competitive inhibitors bind to the enzyme away from the active site but change the overall shape of the molecule, modifying the active site so that it can no longer turn substrate molecules into product (Fig 4.17). Non-competitive inhibition has this name because there is no competition for the *active site*. The presence of a non-competitive inhibitor has the same effect as lowering enzyme concentration: all inhibited molecules are taken out of action completely. Fig 4.18 shows the effect of a non-competitive inhibitor on the Michaelis curve.

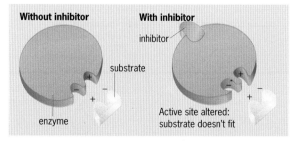

Fig 4.17 **Non-competitive inhibitors attach to enzyme molecules and alter the overall shape, so that the active site cannot function. Although the substrate may still bind, forming an** *enzyme–inhibitor substrate* **complex, the substrate cannot be turned into product. When the inhibitor molecule is removed, normal function is restored**

Fig 4.18 **Non-competitively inhibited enzymes show no activity at all, but unaffected ones work normally. Thus, V_{max} is lowered, as it would be if we simply used less enzyme, but K_m is unaltered**

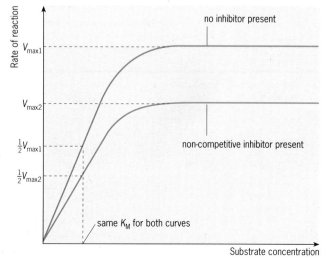

Irreversible inhibitors

Irreversible inhibitors bind permanently to the enzyme, rendering it useless. For obvious reasons organisms rarely produce this type of inhibitor for their own enzymes, but they are splendid weapons to use against other organisms. A wide variety of natural toxins are irreversible inhibitors, as are many pesticides.

Cyanide is an irreversible inhibitor of **cytochrome oxidase**, one of the enzymes involved in respiration. Organisms poisoned with cyanide die because they are deprived of ATP, their immediate energy source.

How inhibitors help to control metabolism

Many metabolic pathways are self-controlling: when a substance is needed a particular pathway is activated to produce it. When enough has been produced, the pathway is deactivated.

In many cases, metabolic pathways are controlled by switching some enzymes 'on and off'. Many enzymes consist of more than one polypeptide and therefore have a quaternary structure (page 53). The activity of these large enzymes is controlled by **effectors,** which are usually different from either the substrate or any inhibitor. Effectors work by binding to the enzyme's **allosteric site**, and they either increase or decrease enzyme activity, as in Fig 4.19.

Now comes the really clever bit: some enzymes in a metabolic pathway are inhibited by the end-product. As Fig 4.20 shows, if too much product begins to accumulate, this inhibits one of the enzymes in the pathway. When the product is again in short supply, the inhibition is lifted and the pathway becomes active again. This self-regulation is an example of **negative feedback**. This is a fundamental principle important in homeostasis (see Chapter 25).

> ✔
> The word allosteric means 'different shape'. Enzymes can have two shapes, one active and the other inactive, and metabolic control can be achieved by switching between the two.

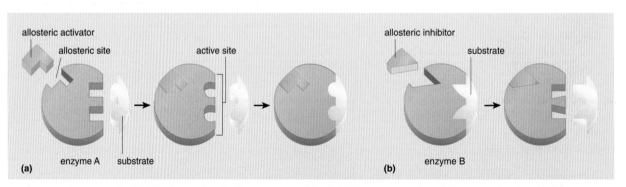

(a) allosteric activator · allosteric site · active site · enzyme A · substrate

(b) allosteric inhibitor · substrate · enzyme B

Fig 4.19 **Two ways of controlling enzyme activity. In** (a) **the enzyme works only if a particular chemical, the allosteric activator, is present. In** (b)**, the presence of another substance, the allosteric inhibitor, prevents enzyme activity. Allosteric inhibitors are non-competitive inhibitors**

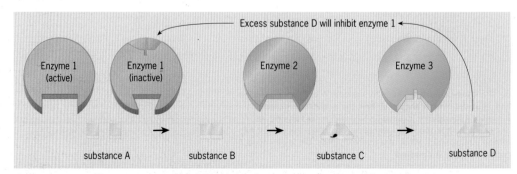

Excess substance D will inhibit enzyme 1

Enzyme 1 (active) · Enzyme 1 (inactive) · Enzyme 2 · Enzyme 3

substance A · substance B · substance C · substance D

Fig 4.20 **A metabolic pathway can be self-regulating by having the end-product act as a non-competitive inhibitor on one of the enzymes in its production pathway**

3 ENZYMES AND BIOTECHNOLOGY

The ability of enzymes to catalyse specific chemical reactions at body temperature makes them very useful tools in the commercial world. Table 4.3 outlines some of industrial applications of enzymes. This is a rapidly changing field, and new applications of enzyme technology appear all the time.

Table 4.3 **Some applications of enzymes**
(Source: Biochemical Society Guidance Notes 3, *Enzymes and their role in biotechnology*)

Enzyme	Reaction	Source of enzyme	Application
Industrial applications			
α-amylase	breaks down starch	bacteria	converts starch to glucose in the food industry
glucose isomerase	converts glucose to fructose	fungi	production of high fructose syrups
proteases	digests protein	bacteria	washing powder
rennin	clots milk protein	animal stomach linings; bacteria	cheese making
catalase	splits hydrogen peroxide into $H_2O + O_2$	bacteria; animal livers	turns latex into foam rubber by producing gas
β-galactosidase	hydrolyses lactose	fungi	in dairy industry, hydrolyses lactose in milk or whey
Medical applications			
L-asparginase	removes L-asparagine from tissues – this nutrient is needed for tumour growth	bacteria (*E. coli*)	cancer chemotherapy – particularly leukaemia
urokinase	breaks down blood clots	human urine	removes blood clots, eg in heart disease patients
Analytical applications			
glucose oxidase	oxidises glucose	fungi	used to test for blood glucose, eg in Clinistix™ diabetics
luciferase	produces light	marine bacteria; fireflies	binds to particular chemicals indicating their presence, eg used to detect bacterial contamination of food
Manipulative applications			
lysozyme	breaks 1–4 glycosidic bonds	hen egg white	disrupts bacterial cell walls
Endonucleases	breaks DNA into fragments	bacteria	used in genetic manipulation techniques, eg gene transfer, DNA fingerprinting

■ See question 1

Industrial applications of enzymes

Enzymes are both specific and sensitive: this makes them ideal for use in analysis, which often involves very small samples. One such application is in the analysis of glucose – an important technique in both medicine and industry. Some readers may have heard of 'clinistix' (Fig 4.21) – the sticks used to test for the presence of glucose in urine. Clinistix™ contain two enzymes: glucose oxidase and peroxidase.

Glucose oxidase catalyses the following reaction:

glucose + oxygen → gluconic acid + H_2O_2 (hydrogen peroxide)

In a simple, visible test, the production of peroxide is coupled to the production of a coloured dye or **chromagen**.

The second enzyme, peroxidase, catalyses the following reaction:

$$DH_2 + H_2O_2 \rightarrow 2H_2O + D$$

D stands for the chromagen, a colourless hydrogen donor. When it loses its hydrogen it becomes coloured. The intensity of colour indicates the amount of glucose present.

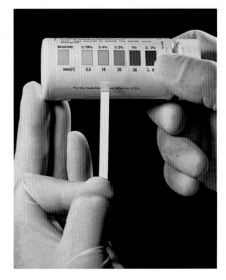

Fig 4.21 **The urine glucose test for diabetes. Glucose oxidase, peroxidase and DH₂ are fixed on a cellulose fibre pad. When this is added to a sample of urine, the colour reaction gives a quantitative measure of glucose**

SUMMARY

After reading this chapter, you should know and understand the following:

■ Enzymes are **globular** proteins with a precise, but delicate, 3D shape maintained by ionic and **hydrogen bonds**.

■ During a reaction, the substrate fits into a region on the enzyme surface called the **active site**.

■ Enzymes are **specific**: each enzyme catalyses one particular reaction.

■ Enzymes speed up reactions by lowering the **activation energy** needed to get the reaction started.

■ Between 4 °C and 40 °C, the rate of an enzyme-controlled reaction increases between two- and three-fold for every 10 °C rise in temperature. Increase in temperature beyond 40 °C usually **denatures** the enzyme. Activity is lost when its three-dimensional structure is destroyed.

■ Most enzymes have an optimum pH. For **intracellular** enzymes this is usually about 7.35. **Extracellular** enzymes such as those found in digestive juices may have optimums of extreme pH values.

■ The rate of an enzyme controlled reaction is limited by the supply of substrate, enzyme or the enzyme cofactor.

■ **Inhibitors** are substances that slow down or stop enzyme activity. They may be **reversible** or **non-reversible**. Reversible inhibitors may be **competitive** or **non-competitive**.

■ Competitive inhibitors tend to be similar in structure to the substrate and compete for the active site. Non-competitive inhibitors do not bind to the active site itself but alter the shape of the enzyme so that the active site is no longer functional.

QUESTION

1 An investigation was carried out into the effectiveness of a biological washing powder known as Bioclen in removing blood stains from a cotton fabric. Bioclen contains enzymes. The test involved soaking a blood-stained fabric in four different liquids as shown in Table 4.Q1(a) below.

Table 4.Q1(a)

Test	Liquid in which the fabric was soaked at 18 °C	Time taken to remove stain/h
A	water	not removed in 12 hours
B	1% Bioclen	5
C	1% Bioclen boiled before use then cooled	9
D	1% non-biological powder	10

a) Describe a chemical test to indicate the presence of a protein in Bioclen.

b) Which test, A, B, C or D, provides the most appropriate control for the investigation of the enzyme activity of Bioclen? Explain your answer.

c) A further experiment then investigated the effectiveness of Bioclen at different temperatures again using blood-stained cotton fabric. The results were recorded in Table 4.Q1(b) below.

Table 4.Q1(b)

Temperature/°C	20	30	40	50	60	70	80
Time taken for digestion of stain/h	5	2	0.5	0.5	2	3	3
Rate of digestion/h^{-1}	0.2						

(i) Calculate the rest of the values for the rate of digestion and enter these in Table 4.Q2(b), assuming the rate is constant at each temperature.

(ii) On 2 mm graph paper, 12 cm wide by 10 cm deep, plot a graph of the rate of digestion against temperature.

(iii) From your graph, deduce what is the optimum temperature for enzyme activity in Bioclen.

(iv) Explain why the rate of protein digestion decreases at temperatures above 50 °C.

[UCLES June 1994 Modular Sciences: Biochemistry Specimen Paper, q.2]

Assignment

ENZYMES AT WORK

Enzymes have enormous potential in the commercial world. Since they can catalyse particular reactions at relatively low temperatures, they are more versatile and much cheaper than inorganic catalysts. Once a suitable enzyme is found, it has to be made on a large scale and then purified to the level needed for the task involved.

There is a vast world-wide demand for sweeteners, mainly for confectionery and soft drinks. Traditionally, sucrose (from sugar beet or cane) has been used, but in recent years a sucrose substitute known as **high fructose syrup** has become cheaper to produce. About 40 million tonnes are produced world-wide each year, depending on the price of sugar beet. Fructose is sweeter than sucrose, and can be made from starch, a relatively abundant and cheap food-stuff.

1 What sort of carbohydrate is starch?

Fructose is produced by an enzyme-catalysed process: the starch is first broken down into glucose and then glucose is converted to fructose.

2 What type of enzyme catalyses the conversion of:
a) starch to glucose?
b) glucose to fructose?

Three enzymes are involved in the production of high fructose syrups: **bacterial β-amlyase**, **fungal amyloglucosidase** and **bacterial glucose isomerase**. In the first step of the process, the starch paste is heated to 105 °C. At this stage the β-amylase is added to reduce the viscosity (thickness) of the paste. The process results in a mixture of maltodextrins which are branched sugars.

3
a) How do we describe an enzyme that works at high temperatures?
b) What is the advantage of higher temperatures in industrial processes?
c) Why does the action of the enzyme reduce the viscosity of the paste?

The next step is the hydrolysis of the maltodextrins to glucose. The substrate is cooled to 55–60 °C and acidified to pH 4.5. The enzyme amyloglucosidase, obtained from the fungus *Aspergillis niger*, is added. This is an **exoenzyme**: it removes the terminal glucose units from the maltodextrins.

4
a) What is the difference between an endoenzyme and an exoenzyme?
b) In the hydrolysis of polymers like starch, why is it an advantage to add an endoenzyme before using an exoenzyme?
c) Sketch graphs to show the likely effect of variations in temperature and pH on the activity of amyloglucosidase.

Amyloglucosidase acts on maltodextrins to produce glucose syrup. Although not as sweet as sucrose, it still has its uses in the food industry. The production of the extra-sweet fructose needs the enzyme glucose isomerase, which is obtained from bacteria (*Bacillus* or *Streptomyces* sp.). This enzyme is fixed on rigid granules and packed into a column. The glucose syrup then flows between the granules and the enzyme converts the glucose into fructose.

5 What is the advantage of using a fixed (immobilised) enzyme?

The end product is high fructose syrup – a mixture of about 42 per cent fructose and 55 per cent glucose. In these proportions it has the same sweetening power as sucrose.

An alternative sweetener is **aspartame**, a dipeptide that is 180 times sweeter than sucrose. The commercial production of aspartame again involves enzymes – particularly **aspartase**. The ease and cost of aspartame production affects the demand for high fructose syrups.

6 What is the main dietary advantage of using aspartame instead of sucrose?

7 All of the enzymes mentioned in this assignment come from organisms such as yeast and bacteria. Why are these organisms particularly suitable for large-scale enzyme production?

8 Imagine that a scientist isolates a human gene which codes for an enzyme that catalyses a reaction giving a very valuable product. List, and perhaps discuss, the steps that would have to be taken in order to get this product into commercial production.

5 Movement in and out of cells

A kidney being prepared for a transplant operation. The pale pink colour changes to a much darker red when the new blood supply is connected

THE LIST OF PEOPLE waiting for a transplant is ever-growing. As surgical techniques and anti-rejection therapies become more sophisticated, there is a greater demand for kidneys, hearts, livers and other parts of the body.

Doctors and surgeons face major problems when transplanting organs: an organ can become available in Dundee, and the most suitable recipient may be in Norwich. Usually, therefore, the organ must be transported. However, when tissues are separated from their blood supply, cells cannot exchange materials with blood in order to carry out their normal functions. They run out of oxygen and accumulate waste and can die within a matter of minutes.

The solution to this problem is to cool the cells so that their metabolic demands are lowered. Organs packed in ice can survive a journey of several hours. During the transplant, the organ is warmed up to normal body temperature and quickly 'plumbed in' to the recipient's blood system.

1 THE CELL AND ITS ENVIRONMENT

Each living cell is a dynamic system which can exchange a volume of fluid several times bigger than its own volume every second! This can only happen if the cell is very small: there is a limiting size above which incoming nutrients could not be taken up quickly enough to satisfy demand and waste could not be expelled efficiently enough to prevent poisoning of the cell. Multicellular organisms larger than about 1 mm have developed strategies to increase the exchange of materials, to meet the needs of *all* their cells.

Material can pass into or out of cells by:

- **diffusion** (and **facilitated diffusion**),
- **osmosis**,
- **active transport**,
- **endocytosis** and **ectocytosis**.

In this chapter, we look at each of these processes in detail but, since they all involve the cell surface membrane, we first take a detailed look at this important structure.

2 THE CELL SURFACE MEMBRANE

The cell surface membrane, or the **plasma membrane** or **plasmalemma**, is the boundary between a cell and its surroundings. It has little mechanical strength, but it plays a vital role in regulating the materials that pass in and out of the cell. Many of the organelles of the eukaryotic cell are made up of membranes (Fig 5.1).

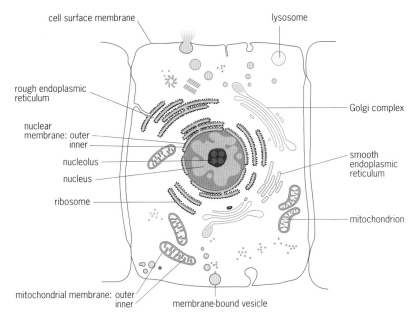

cell surface membrane
lysosome
rough endoplasmic reticulum
nuclear membrane: outer
inner
nucleolus
nucleus
ribosome
Golgi complex
smooth endoplasmic reticulum
mitochondrion
mitochondrial membrane: outer
inner
membrane-bound vesicle

Fig 5.1 **Whenever you draw a eukaryotic cell, nearly every line you draw represents a membrane. In this diagram of an animal cell, all the membranes are highlighted in red**

Fig 5.2 **The phospholipids in a membrane are arranged tail-to-tail, forming a bilayer**

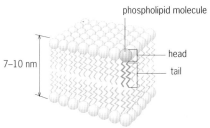

phospholipid molecule
7–10 nm
head
tail

?

A The cell surface membrane of an animal cell is about 10 nm thick. Assume that a typical animal cell is 50 μm in diameter. How thick would you make a model cell membrane if it had to fit around a cell model that was 5 metres across?

1 mm = 1000 μm
1 μm = 1000 nm

Fig 5.3(a) **The fluid mosaic model of the structure of the cell surface membrane which has been described as a collection of 'protein icebergs in a lipid sea'. Alternatively, you might find it easier to think of the lipids as a layer of ping-pong balls floating on the surface of a swimming pool, with the proteins as larger balls punctuating the lipid layer**

The structure of the cell surface membrane

As we saw in Chapter 1, the cell surface membrane is a phospholipid bilayer about 7–10 nm thick (Fig 5.2). Under a light microscope it appears as a thin line and, until the electron microscope was developed, its structure could not be studied directly. Instead, early cell biologists deduced its structure by looking at its properties.

The fluid mosaic theory

In 1972, aided by electron microscope studies and evidence from more sophisticated techniques, Singer and Nicholson put forward the **fluid mosaic theory** (Fig 5.3). Today, most scientists accept it as the model that best represents the structure of living membranes.

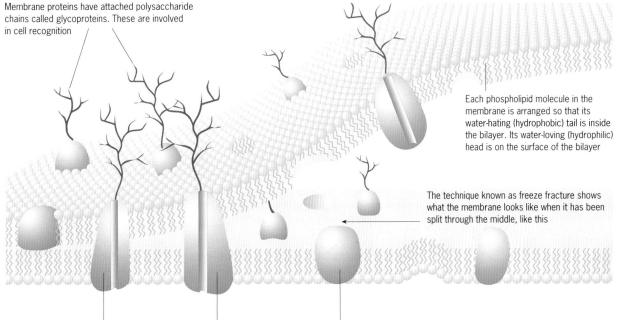

Membrane proteins have attached polysaccharide chains called glycoproteins. These are involved in cell recognition

Each phospholipid molecule in the membrane is arranged so that its water-hating (hydrophobic) tail is inside the bilayer. Its water-loving (hydrophilic) head is on the surface of the bilayer

The technique known as freeze fracture shows what the membrane looks like when it has been split through the middle, like this

Membrane proteins which span the whole width of the phospholipid bilayer form pores in the membrane which control the movement of substances into and out of the cell

Some membrane proteins do not span the whole bilayer: some are embedded in one half of the bilayer

Fig 5.3(b) **A simplified diagram of a cell surface membrane, showing the features described in the fluid mosaic theory. Most people find that this level of detail is enough to allow them to explain the essential properties and functions of the membrane. (It is also drawable for exams)**

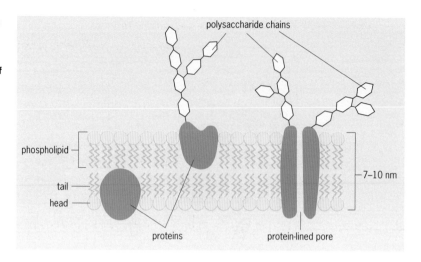

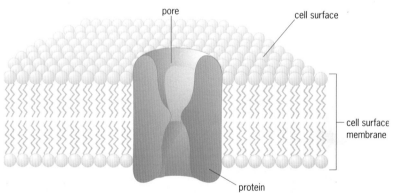

According to the fluid mosaic theory, the cell surface membrane consists of a double layer of phospholipid molecules (known as a **lipid bilayer**), studded with proteins and other molecules. The name **fluid mosaic** is used because the bilayer is a very fluid structure (the lipid molecules are in constant sideways motion) and it contains a 'mosaic' of protein molecules.

The protein molecules may be found only in the top or bottom layer of lipids or they may span the entire membrane. The cholesterol molecules fit in between the phospholipids, and the polysaccharides are attached to the membrane proteins or the lipids.

The lipid bilayer is a barrier to water and anything that is water soluble. But, the majority of the chemicals that need to pass in or out of the cell *are* water soluble. The protein molecules in the membrane act as **hydrophilic pores**, water-filled channels that allow water-soluble chemicals to pass through (Fig 5.4). Pores are usually small and highly selective: they allow only some molecules or ions through.

> **?**
>
> **B** What do you understand by the terms hydrophilic and hydrophobic? How do these terms relate to a phospholipid molecule?

> **✓**
>
> Water is a **polar molecule** (it has regions of positive and negative charge) and is a solvent for **polar substances** such as sugars, charged ions (Na^+, Cl^-, Ca^{2+}, K^+), B and C vitamins and amino acids. Polar substances do not dissolve in lipid and so can only cross a cell surface membrane by going through pores.
>
> Most fats, oils and lipids are **non-polar molecules** (they do not have charged regions) and do not dissolve in water. Other non-polar substances (such as vitamins A, D, E and K) can dissolve in lipids and so can cross a cell surface membrane without going through pores.

Fig 5.4 **Some membrane-spanning proteins form pores. These allow particles that cannot dissolve in lipid to enter and leave the cell. Membrane pores are not simply 'holes'; they can control what passes through them**

The chemical make-up of the cell surface membrane

Membranes contain **phospholipids**, **proteins**, **cholesterol** and **polysaccharides**.

Phospholipids are a major constituent of cell surface membranes. They form membranes naturally because they automatically arrange themselves into a bilayer (see Fig 5.2) that is impermeable to water and water-soluble substances.

Membrane proteins have several functions. They:

- create pores through which water and water-soluble chemicals can pass,
- act as carriers in active transport,
- form receptor sites for hormones,
- are important in cell recognition.

Cholesterol, a chemical often mentioned in relation to heart disease (Fig 5.5), is actually a vital constituent of animal cell membranes.

Polysaccharides, branched polymers of simple sugars, stick out from the outer surface of some membranes like antennae (see Fig 5.3). They attach to lipids forming **glycolipids,** or to proteins forming **glycoproteins**. Glycolipids and glycoproteins help cells to recognise each other.

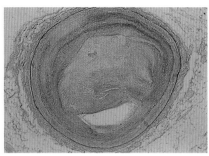

Fig 5.5 **The process that causes an artery to become 'furred up' with fatty deposits can lead to a heart attack or a stroke. Some research suggests that people with high blood cholesterol levels (cholesterol molecules that float freely in the blood) run a greater risk of developing these deposits. Although heart disease develops because of many different factors, it is probably unwise to eat a diet high in animal fat and cholesterol**

3 DIFFUSION

Diffusion involves 'molecular mixing'. Molecules and ions in a gas or liquid move continuously, bumping into each other and changing direction. In this way, particles tend to **diffuse**, or spread, so that they are spaced evenly in a gas or liquid, rather than being concentrated in one place (Fig 5.6).

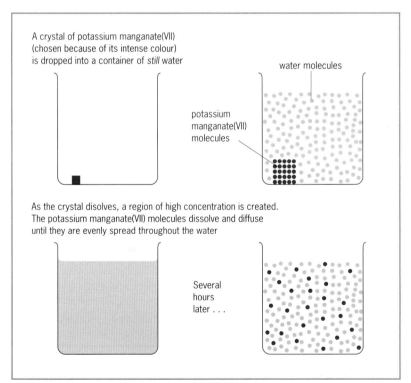

A crystal of potassium manganate(VII) (chosen because of its intense colour) is dropped into a container of *still* water

water molecules

potassium manganate(VII) molecules

As the crystal disolves, a region of high concentration is created. The potassium manganate(VII) molecules dissolve and diffuse until they are evenly spread throughout the water

Several hours later . . .

Fig 5.6 **If you drop a crystal of potassium manganate(VIII) into still water, it diffuses until evenly spread. This takes several hours (depending on temperature) but can be speeded up by stirring. Diffusion is a passive process that requires no *input* of energy; it does however depend on the *kinetic energy* of the molecules in a gas or a liquid**

A working definition of diffusion is:

The movement of particles within a gas or a liquid from a region of high concentration to a region of lower concentration until an equilibrium is reached.

Diffusion is the main process by which substances move over short distances and is a crucial process in the human body (Fig 5.7). But it is too slow to move substances efficiently over distances much greater than a few millimetres. In practice, substances usually

?

C Why can't solids diffuse?

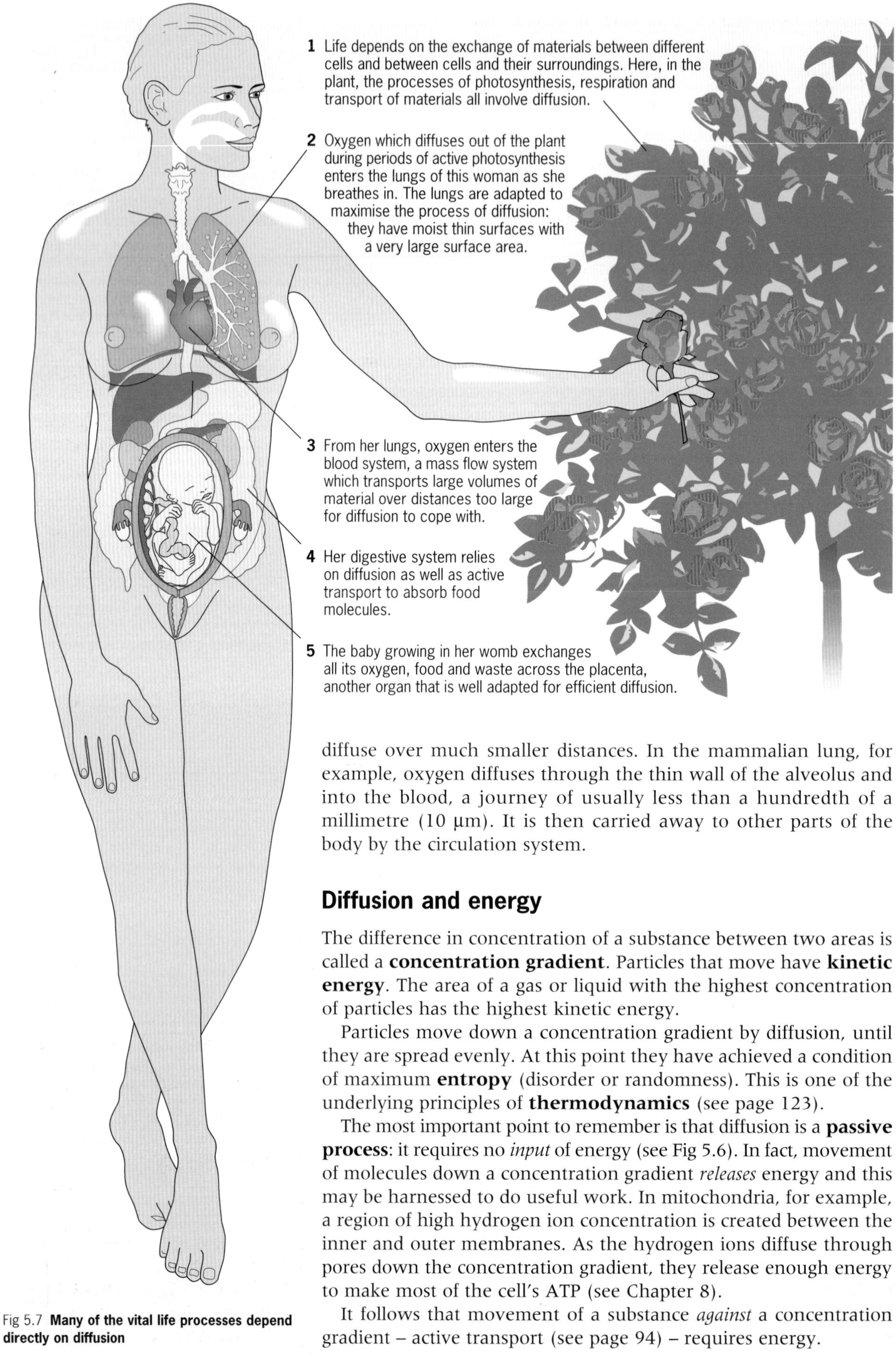

1 Life depends on the exchange of materials between different cells and between cells and their surroundings. Here, in the plant, the processes of photosynthesis, respiration and transport of materials all involve diffusion.

2 Oxygen which diffuses out of the plant during periods of active photosynthesis enters the lungs of this woman as she breathes in. The lungs are adapted to maximise the process of diffusion: they have moist thin surfaces with a very large surface area.

3 From her lungs, oxygen enters the blood system, a mass flow system which transports large volumes of material over distances too large for diffusion to cope with.

4 Her digestive system relies on diffusion as well as active transport to absorb food molecules.

5 The baby growing in her womb exchanges all its oxygen, food and waste across the placenta, another organ that is well adapted for efficient diffusion.

Fig 5.7 **Many of the vital life processes depend directly on diffusion**

diffuse over much smaller distances. In the mammalian lung, for example, oxygen diffuses through the thin wall of the alveolus and into the blood, a journey of usually less than a hundredth of a millimetre (10 μm). It is then carried away to other parts of the body by the circulation system.

Diffusion and energy

The difference in concentration of a substance between two areas is called a **concentration gradient**. Particles that move have **kinetic energy**. The area of a gas or liquid with the highest concentration of particles has the highest kinetic energy.

Particles move down a concentration gradient by diffusion, until they are spread evenly. At this point they have achieved a condition of maximum **entropy** (disorder or randomness). This is one of the underlying principles of **thermodynamics** (see page 123).

The most important point to remember is that diffusion is a **passive process**: it requires no *input* of energy (see Fig 5.6). In fact, movement of molecules down a concentration gradient *releases* energy and this may be harnessed to do useful work. In mitochondria, for example, a region of high hydrogen ion concentration is created between the inner and outer membranes. As the hydrogen ions diffuse through pores down the concentration gradient, they release enough energy to make most of the cell's ATP (see Chapter 8).

It follows that movement of a substance *against* a concentration gradient – active transport (see page 94) – requires energy.

Factors that affect the rate of diffusion

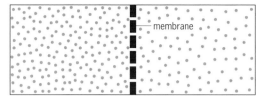

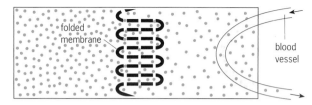

(a) Small concentration gradient and thick, unfolded membrane, so diffusion is slow.

(b) • Diffusing molecules are taken away (by blood, for example), so there are always more molecules on the left than on the right. Therefore the diffusion gradient is maintained.

• Thin membrane, so diffusion across is faster.

• Membrane folded into microvilli, creating a greater surface area for diffusion.

The rate at which molecules of a substance diffuse from one region to another has a great influence on the design of cells, organs and whole organisms (Fig 5.8). Several factors affect the rate of diffusion:

Fig 5.8 **Schematic diagram to show some of the strategies used by organisms to increase the rate of diffusion. To make the most of diffusion, organisms need to increase their surface area, maintain a diffusion gradient and keep their membranes as thin as possible. Increasing the temperature would also speed up diffusion, but this is not an option in biological systems**

● The surface area between the two regions. The greater the surface area, the greater the rate of diffusion.

● The distance over which diffusion occurs (the thickness of the membrane, for example).

● The relative concentrations of the substance in the two areas, that is, the concentration gradient. Diffusion is more efficient if the concentration gradient can be maintained: this can be achieved by transporting the substance away from the immediate area (eg by blood) once it has diffused, or by combining it with another chemical to prevent it from diffusing back.

● The size and nature of the particles. Fat-soluble substances can diffuse through the lipid bilayer of the membrane. Water-soluble substances must pass through the protein pores which tend to be small and selective. Generally, very large molecules cannot diffuse into cells at all.

● The temperature at which the process takes place. At higher temperatures, molecules in a liquid or a gas have more kinetic energy and so diffuse more quickly.

Calculating the rate of diffusion

The rate at which a substance diffuses can be calculated from a simple formula which takes into account the factors that affect diffusion:

$$\text{Rate of diffusion} = \frac{\text{surface area} \times \text{concentration difference}}{\text{thickness of membrane}}$$

This relationship is known as **Fick's law**.

What limits diffusion?

Diffusion is not fast enough to be effective over distances greater than a few millimetres. To solve this problem, many organisms have developed a circulatory system which carries large amounts of material around their body in a stream of fluid. **Blood** is an obvious example of such a **mass flow** system, and plants operate on a similar principle with sap in their **vascular** (conducting) tissue.

The main function of any circulatory system is to bring materials within diffusing distance of living cells, and to take away cell products. In the human body, no cell is more than a fraction of a millimetre away from a capillary (see Chapter 18).

?

D Phosphorylated chemicals (those with a phosphate group added) cannot pass easily through cell surface membranes. Suggest why glucose molecules are phosphorylated as soon as they enter a cell.

?

E Name some organs whose main function is the exchange of materials. (Note: Think about plants, too.)

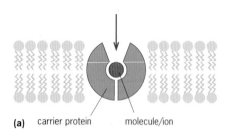

(a) carrier protein   molecule/ion

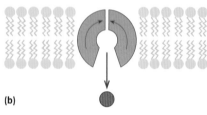

(b)

Fig 5.9 **Facilitated diffusion using a carrier protein. The diffusing molecule combines with the carrier protein, causing a shift in shape which 'squeezes' the molecule through the channel**

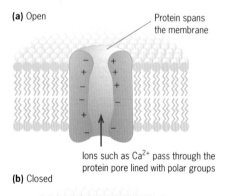

(a) Open   Protein spans the membrane

Ions such as Ca^{2+} pass through the protein pore lined with polar groups

(b) Closed

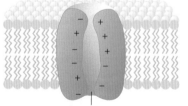

Protein changes shape, so pore becomes too narrow to allow ions through

4 FACILITATED DIFFUSION

Some substances enter and leave cells much faster than you would expect if only diffusion occurred. We now know that some membrane proteins assist, or **facilitate**, the diffusion of some substances across the cell membrane. Two types of protein are responsible for **facilitated diffusion**:

- Specific **carrier proteins** take particular substances from one side of the membrane to the other.
- **Ion channels**, or pores, open and close to control the passage of selected charged particles.

Carrier proteins

Until the 1970s, cell biologists thought carrier proteins worked by rotating within the membrane, like turnstiles. Newer research points to a different explanation (Fig 5.9). As soon as the diffusing molecule binds to the carrier protein, the protein undergoes a **conformational** (shape) change and the diffusing molecule ends up facing the other side of the membrane, where it is released.

Like diffusion, facilitated diffusion involves movement down a concentration gradient and requires no *input* of metabolic energy.

Ion channels

Ion channels are proteins with a central 'hole' lined with polar groups (Fig 5.10). Ion channels facilitate the diffusion of charged particles such as Ca^{2+}, Na^+, K^+ and Cl^- ions. Many are **gated**, so can open or close. Cells use ion channels to control the movement of substances between themselves and other cells, and to regulate the ionic composition of their cytoplasm.

A difference in concentration of ions may lead to a net positive or negative charge in a particular area. This type of concentration gradient is called an **electrochemical gradient**. Charged particles move towards areas of opposite charge. This is important in the process of transmission of a nerve impulse (see Chapter 21) and in the process by which red blood cells exchange materials with the tissues (see Chapter 18).

Fig 5.10 **An ion channel. Because they are charged particles, ions cannot easily pass through the non-polar lipid bilayer. Specific membrane proteins form polar pores through which ions can pass. These channels are usually specific for one type of ion and can open and close according to the needs of the cell. When fully open, over 1 million ions per second can flow through a single channel**

DIABETES AND CARRIER PROTEINS

IN MAMMALS, levels of blood sugar are kept relatively constant by controlling how much glucose passes from the blood into cells. After a meal, when blood glucose levels are high, the hormone insulin is released. This hormone activates a mechanism in the cell surface membrane which facilitates the diffusion of glucose into the cell, so lowering the blood sugar.

One form of diabetes in humans is caused by a gene mutation which alters the structure of glucose carrier proteins in the cell surface membranes. An affected person cannot get enough glucose into their cells. Unlike diabetes that results from the inability to make insulin (see Chapter 25), this form of the disease is difficult to treat, because injecting insulin has little effect.

5 OSMOSIS

Sea water, because of the salt it contains, is about three times more concentrated than human blood. If someone who is already dehydrating is desperate enough to drink sea water, the salt causes water to move out of their blood and into their stomach by **osmosis**, causing further dehydration. So victims of shipwrecks can die of dehydration, even though they are surrounded by water.

Osmosis is the diffusion of water only. It is the net movement of solvent (water) molecules from a region of their higher concentration to a region of their lower concentration, through a partially permeable membrane.

Look at Fig 5.11. The solute molecules cannot diffuse in either direction because they are too big to pass through the membrane. But water molecules *can* get through and water diffuses down its concentration gradient.

Water molecules move from the left, where there is a higher concentration of water molecules, to the right, where there is a lower concentration of water molecules. (The solution on the right is what we normally think of as the more concentrated solution, determined by the solute concentration.)

At this point, you might be thinking to yourself, 'Hang on, there are just as many water molecules on the right of the membrane as there are on the left, so how is there a concentration gradient?'

The key to understanding osmosis is to remember what happens when a substance dissolves in water. The solute molecules, because they carry a charge, become surrounded by a shell of water

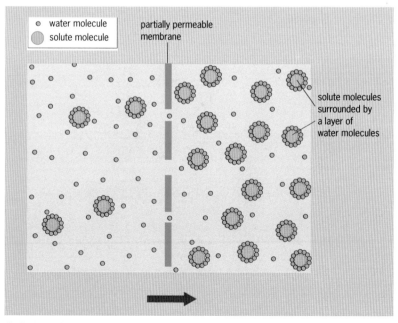

Fig 5.11(a) **A 'weak' solution is one with a low concentration of solute molecules but a high concentration of water molecules. It also has more 'free' water molecules than the 'concentrated' solution on the right and so has a** high water potential. **By high, we mean a** negative value close to zero

Fig 5.11(b) **A more 'concentrated' solution has a higher concentration of solute molecules but a lower concentration of water molecules. It therefore has a** low water potential. **By low, we mean a more negative value**

Overall, the water moves from an area of high water potential to an area of lower water potential

molecules (chemists say they are **hydrated**). When this happens, the water molecules that form the 'shell' are no longer free to move around as they were before: they have been 'tied up' by the solute molecules.

As we saw in the section on diffusion, movement of particles occurs down a concentration gradient, *from a region of high kinetic energy to a region of lower kinetic energy*. In Fig 5.11, the total kinetic energy of the water molecules in the solution on the left is high because most of them are free to move around. The total kinetic energy of water molecules in the solution on the right is lower because fewer of them are free to move around.

So, a second definition of osmosis is:

Osmosis is the movement of water molecules down a concentration gradient of water, from a region of high kinetic energy to a region of lower kinetic energy.

How water molecules cross cell membranes

We have already said that the cell surface membrane is impermeable to water. In fact, it is not 100 per cent impermeable. Some water (a tiny amount) should be able to pass through a lipid bilayer. But, recent studies have shown that water passes through purified phospholipid bilayers, with all the proteins removed, at rates approximately 100 to 1000 times faster than expected.

The exact reason for this is not known, but scientists think the constant sideways motion of the phospholipids, coupled with the flexing of the fatty acid tails, creates temporary holes (Fig 5.12) through which the water molecules can slip. These holes appear in only one half of the membrane at a time, so the water molecules 'wait' at the half-way point until a hole appears in the other side, rather like crossing a busy road, one lane at a time.

Membranes with a high proportion of cholesterol are less permeable to water. It may be that cholesterol reduces the number of temporary holes by stabilising the phospholipids (Fig 5.13).

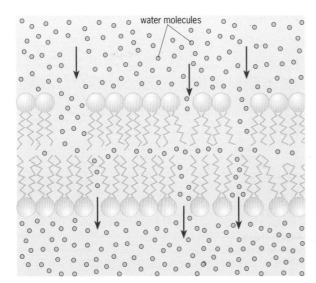

Fig 5.12 **A possible explanation for the fact that lipid bilayers are permeable to water. Temporary holes, created by the random sideways movement of the lipid molecules, act as channels through which water can pass**

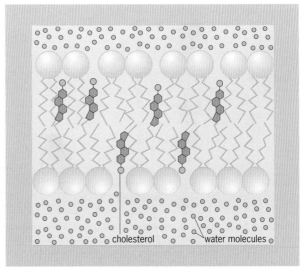

Fig 5.13 **Cholesterol is an important constituent of the membranes of many types of cell. The effect of cholesterol is to reduce leakage of water and ions through the lipid bilayer, probably by reducing the sideways motion of the lipid molecules**

Osmosis in living cells

All cells contain cytoplasm, a complex solution separated from its surroundings by a partially permeable membrane. So, all cells have the potential to gain or lose water by osmosis.

Solute potential, Ψ_s

If no other factors are involved, a cell that is placed in a solution of equal concentration will neither gain nor lose water. Water molecules pass equally in either direction and there is no *net* change in the cell volume: Fig 5.14(a).

A cell placed in a solution that contains less solute (more dilute) than the cytoplasm will gain water by osmosis: Fig 5.14(b). If the surrounding solution contains more solutes (is more concentrated) than the cytoplasm, the cell will lose water by osmosis: Fig 5.14(c).

> **The** solute potential, Ψ_s, **is the potential of any system to absorb water molecules by osmosis.**

This term replaces the older term **osmotic potential**, which you will, nevertheless, come across in some books. Solute potential is measured in units of pressure, **kilopascals, kPa**.

Solute potential is a negative scale. For example, a relatively weak solution of sucrose may have a solute potential of −200 kPa while a more concentrated solution may have solute potential of −500 kPa. Pure water has the highest possible solute potential: zero.

If the two solutions described above were separated by a partially permeable membrane, water would move from the more dilute solution into the more concentrated one. So, if there are no other factors involved (such as physical pressure), water will move from a region of high (less negative) solute potential to a region of lower (more negative) solute potential.

If you have a problem with negative scales (and many people do), compare it to temperature. Of two solutions, one at −40 °C and the other at −60 °C, it is obvious which is the colder. If the two solutions were placed in contact with each other, heat would pass from the warmer to the colder until they were both about equal.

Pressure potential, Ψ_p

In addition to solute potential, we must also consider the effects of physical pressure (**hydrostatic pressure**) on a cell that is subject to osmotic forces. If a cell surrounded by a less concentrated solution is put under pressure, water cannot enter because the osmotic forces causing water to flow in are opposed by the physical pressure. So the cell cannot enlarge (Fig 5.15).

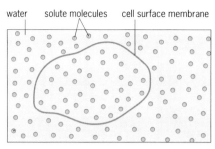

water   solute molecules   cell surface membrane

(a) **When the solution surrounding the cell contains the same amount of solutes as the cytoplasm, the cell neither gains nor loses water**

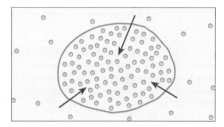

(b) **When the cell's cytoplasm contains more solutes than the surroundings, the cell gains water by osmosis**

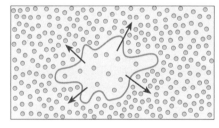

(c) **When the solution surrounding the cell has more solutes in it, the cell loses water by osmosis**

Fig 5.14 **The factors that determine whether water enters or leaves a cell by osmosis**

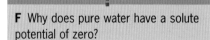

?

F Why does pure water have a solute potential of zero?

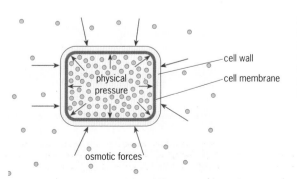

Fig 5.15 **This plant cell should gain water by osmosis but water cannot get in because the rigid cell wall resists any further increase in volume**

cell wall
cell membrane
physical pressure
osmotic forces

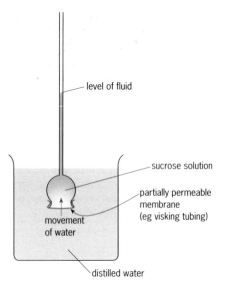

Fig 5.16 **The osmometer is a simple device which is used to demonstrate osmosis and to estimate the solute potential of a solution. The greater the solute potential, the higher the liquid rises up the tube**

Hydrostatic means 'force of fluid'. Water comes out of taps due to hydrostatic pressure.

?

G Preserves such as jam and marmalade have a high sugar content. Suggest why bacteria do not thrive when they land on these foods?

The **osmometer** can demonstrate the balance between osmotic and physical force (Fig 5.16). Water passes through the membrane into the sucrose solution until pressure from the weight of the column of water halts it. At this point an equilibrium is reached the osmotic and physical forces are equal and opposite.

Pressure potential, Ψ_p, **describes any hydrostatic (physical) forces that act on cells.**

In a plant cell, a pressure potential is created by the cellulose cell wall resisting expansion of the cell. Animal cells are also subject to pressures – those caused by the pumping of the heart, or by movement and posture that put pressure on particular areas of tissue, for example.

Since both plant and animal cells can be affected by hydrostatic pressure, the term *pressure potential* has replaced the terms **wall pressure** and **turgor pressure**, which meant exactly the same thing but referred solely to studies of cells with walls.

The concept of water potential, Ψ

When we look at how osmosis affects cells and organisms, there are two factors to consider:

- The solute concentrations (inside and outside the cell). These are the solute potentials.
- Any hydrostatic forces. These are the pressure potentials.

Solute concentrations and hydrostatic forces combine to give a measurement of the overall tendency of a cell or system to gain or lose water by osmosis. We call this overall tendency the water potential.

The relationship between water potential, solute potential and pressure potential is:

$$\text{Water potential} = \text{solute potential} + \text{pressure potential}$$
$$\Psi \quad = \quad \Psi_s \quad + \quad \Psi_p$$

This tells us that the overall tendency for water to enter or leave a cell usually depends on the relative sizes of the osmotic force drawing water into the cell and the hydrostatic pressure keeping water out.

Occasionally, pressure potential negative. In this case, water tends to enter the cell continuously. Xylem vessels in plants usually have a negative pressure potential. They contain a continuous column of water and dissolved minerals which is drawn up the plant because of the negative pressure potential generated by evaporation of water from the plant's leaves (see Chapter 18).

Now that we have seen the general rules, let's look at how osmosis affects specific cells: those of animals and those of plants.

Osmosis in plant cells

If you put a plant cell in a solution (eg a concentrated sucrose solution) that has a lower (more negative) water potential than the cell's cytoplasm and vacuole, water leaves the cell by osmosis. The vacuole shrinks and eventually the **protoplast** (the living part of the cell) becomes detached from the cell wall. The point at which the protoplast is just about to become detached is called **incipient plasmolysis**. When it has become detached, the cell is said to be **plasmolysed** (Fig 5.17).

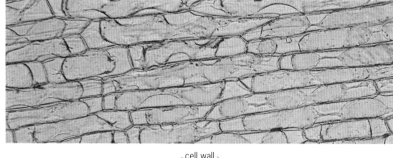

Fig 5.17 **When a plant cell becomes plasmolysed, sometimes the protoplast shrinks into a sphere in the centre of the cell. At other times a 'star' shape is produced because the cell surface membrane sticks to the inside of the wall at the points where there are connections with other cells**

cell wall

surrounding solution fills gap between wall and membrane

membrane

shrunken vacuole

points where membrane remains attached to walls

If the plasmolysed cell is placed in solution of higher water potential (eg distilled water), it absorbs water by osmosis and begins to swell. The vacuole and cytoplasm enlarge until they begin to press against the cell wall, which resists expansion beyond a certain limit.

An equilibrium is reached when the osmotic forces drawing water in are balanced by the wall pressure resisting further expansion. At this point the cell is said to be **turgid**, and the water potential is zero. In a normal healthy plant, most of the cells are turgid. In a **herbaceous** (non-woody) plant the straightness of the stem is due largely to turgor pressure (Fig 5.18).

Fig 5.18 **When a plant loses more water than it gains, the solution surrounding the cells becomes hypertonic, turgor pressure is lost and the plant wilts. Fortunately, the situation is reversible if more water soon becomes available**

Osmosis in animal cells

Animal cells have no cell wall, just a membrane. If you place an animal cell in distilled water, it absorbs water by osmosis and swells up. As membranes have virtually no mechanical strength, the animal cell often bursts.

The following terms are used to describe osmosis in animal cells, rather than plant cells:

Tonicity: the solute concentration of a solution
Hypertonic: a greater solute concentration
Hypotonic: a lower solute concentration
Isotonic: an equal solute concentration

These are relative terms and so can only be used to compare solutions. For example, sea water is *hypertonic* to human blood plasma, distilled water is *hypotonic* to human blood plasma.

Fig 5.19 All animal cells are prone to the same effects as these red blood cells

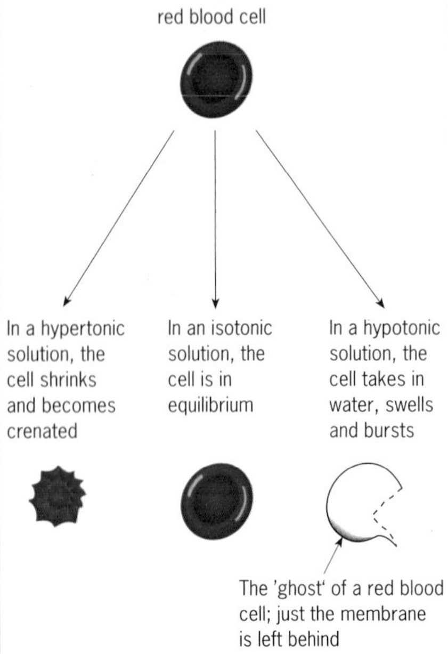

red blood cell

In a hypertonic solution, the cell shrinks and becomes crenated

In an isotonic solution, the cell is in equilibrium

In a hypotonic solution, the cell takes in water, swells and bursts

The 'ghost' of a red blood cell; just the membrane is left behind

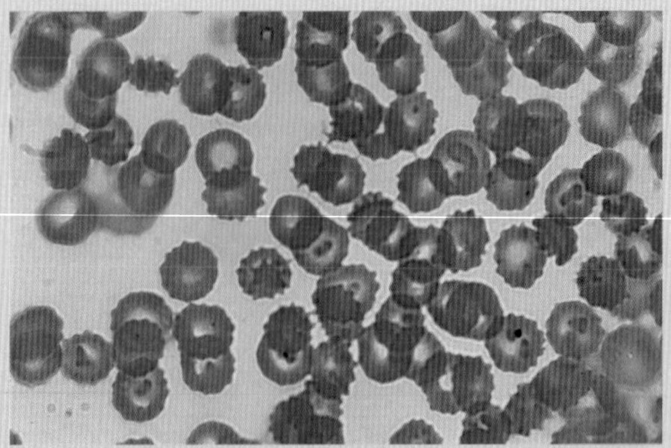

(a) When placed in a hypertonic solution, the cells gradually change from their normal shape and become crenated

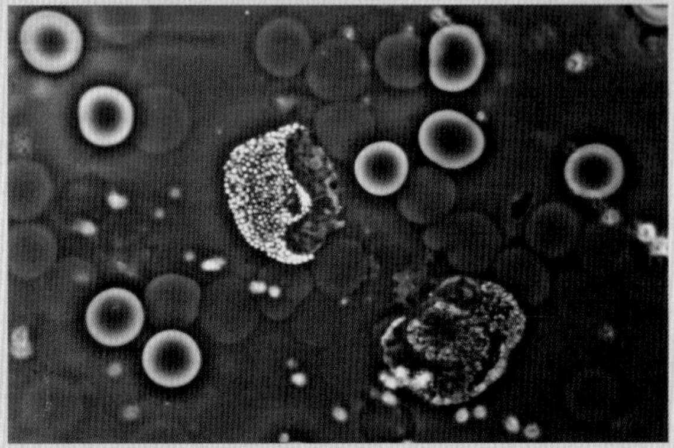

(b) In a hypotonic solution, they take in water and burst. This photo shows the faint ghost cells with some red cells that have not haemolysed. Two burst white cells are also present

Fig 5.19 shows what happens to red blood cells placed in hypertonic, isotonic and hypotonic solutions. When blood plasma becomes hypertonic to red blood cell cytoplasm, as it does during severe dehydration, water is lost from the red blood cells, which shrink and become crinkled, or **crenated**. At the other extreme, hypotonic plasma causes red blood cells to swell and burst, leaving 'ghosts' of plasma membrane. This process in known as **haemolysis**: literally 'blood splitting'.

In order to protect their cells from the effects of osmosis, animals must maintain their body fluids at the same solute potential as their cell cytoplasm. They do this by **osmoregulation**. In mammals, the kidneys are responsible for osmoregulation (see Chapter 27).

?

H Why is it incorrect to say that 'sea water is a hypertonic solution'?

THE AMOEBA AND ITS CONTRACTILE VACUOLE

THE AMOEBA (Fig 5.20) is a single-celled organism (a protoctist) found in freshwater and marine environments. Some species can infect the human intestine, causing amoebic dysentery.

The cytoplasm of the freshwater amoeba is hypertonic to the surrounding water, and so absorbs water by osmosis. To counteract this, the amoeba has a **contractile vacuole** – an organelle that fills up with water and periodically expels its contents, allowing the volume of the amoeba to remain relatively constant.

Fig 5.20 **The freshwater amoeba constantly gains water by osmosis. The organism actively pumps water into the contractile vacuole. When the vacuole is full, the water is emptied to the outside, rather like a mini bladder**

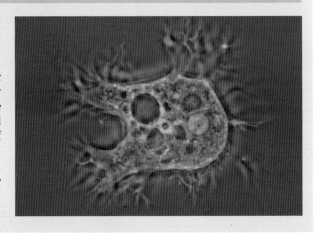

Water potential calculations

Look at Fig 5.21. In a plasmolysed cell, the protoplast is not pressing against the cell wall, so the pressure potential is zero. The tendency for water to enter or leave the cell depends entirely on the relative solute concentrations inside and outside the cell. In this situation, water potential is equal to solute potential.

A turgid cell is in equilibrium: its volume is constant. The solute potential (a negative value) is equal to the pressure potential (a positive value), and so they cancel each other out. Thus, the water potential of a turgid cell is zero.

?

I Suggest why the marine amoeba does not possess a contractile vacuole.

✔

$$\begin{array}{ccccc} \text{Water} & = & \text{solute} & + & \text{pressure} \\ \text{potential} & & \text{potential} & & \text{potential} \\ \Psi & = & \Psi_s & + & \Psi_p \end{array}$$

✔

Water potential, solute potential and pressure potential are all measured in kilopascals (kPa).

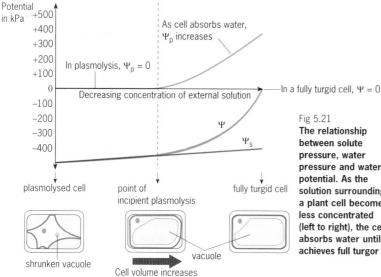

Fig 5.21
The relationship between solute pressure, water pressure and water potential. As the solution surrounding a plant cell becomes less concentrated (left to right), the cell absorbs water until it achieves full turgor

Now, let's look at some worked examples:

EXAMPLES

Q Use the water potential equation to predict movement of water between these two cells:

Cell A $\Psi_s = -500$ kPa
         $\Psi_p = 200$ kPa

Cell B $\Psi_s = -600$ kPa
         $\Psi_p = 100$ kPa

A This question is asking you to work out the water potential of the two cells. If we put values given into the equation:

$$\begin{array}{ccccc} \text{Water potential} & = & \text{solute potential} & + & \text{pressure potential} \\ \Psi & = & \Psi_s & + & \Psi_p \end{array}$$

For cell A: $-500 + 200 = -300$
For cell B: $-600 + 100 = -500$

Now, which one is closest to zero (pure water)? Water always passes from the highest value (closest to 0) to the lowest (more negative), so in this case water will pass from A to B.

Q A turgid plant cell was found to have a Ψ_s value of -350 kPa.
a) What was the Ψ?
b) What was the Ψ_p?

A You may think that you have been given only one value and that it is impossible to work out the other two, but the clue comes in the word *turgid*. In a turgid cell, the water potential is zero, and now you know two of the values it is easy to work out the third.
a) $\Psi = 0$
b) $\Psi_s = -350$ kPa, so $\Psi_p = +350$ kPa

Q A plasmolysed cell has a Ψ of -650 kPa.
a) What is the Ψ_p?
b) What is the Ψ_s?

A Here again, you are only given one value, but you know that in a plasmolysed cell the Ψ_p is zero, because the protoplast is not pushing against the wall. In this situation the solute potential is the only force acting, so the solute potential *is* the water potential:
a) $\Psi_p = 0$
b) $-650 = \Psi_s - 0$
    $\Psi_s = -650$ kPa

6 ACTIVE TRANSPORT

First we need to define active transport:

Active transport is movement *against* a concentration gradient.

Cell surface membranes have evolved specialised carrier proteins to transport a particular ion/molecule across the membrane, often from an area of low concentration to a region of much higher concentration. Active transport requires an *input of energy* from cell respiration, usually (but not always) provided by ATP.

These processes all involve active transport:

● Absorption of amino acids from the gut into the blood.

● Absorption of mineral ions (eg nitrate) by the roots of plants.

● Excretion of urea and hydrogen ions by the mammalian kidney.

● Pumping sodium ions out of cells.

● The exchange of sodium and potassium ions which allows conduction of nerve impulses.

● Filling of the contractile vacuole by amoeba.

The mechanism of active transport is similar to that of facilitated diffusion (Table 5.1). Specific carrier proteins combine with the substance to be transported, enabling it to pass rapidly across the membrane. The key difference is that, while facilitated diffusion is a passive process, the carrier protein complex in active transport is activated by an input of energy (Fig 5.22).

Fig 5.22 **An example of an active transport mechanism. The H⁺ ATPase pump removes hydrogen ions from a cell against a diffusion gradient. It is found in the membrane of the kidney tubule cells, where it contributes to the production of acidic urine**

Step 1 The carrier molecule binds to and hydrolyses ATP. The terminal phosphate group converts the molecule into a 'high energy' complex

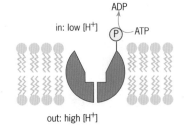

Step 2 The high energy complex combines readily with an H^+ ion

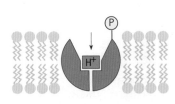

Step 3 The carrier protein undergoes a conformational (shape) change which exposes the H^+ to the other side of the membrane

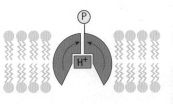

Step 4 The conformational change lowers the affinity of the carrier for H^+. The phosphate group is also lost and this induces a final conformational change, back to Step 1

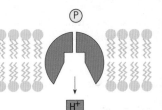

The rate of active transport is closely linked to the rate of respiration. We can assume that active transport is taking place if:

- movement of particles takes place against a diffusion gradient,
- the rate of transport increases as the metabolic rate increases,
- the process is inhibited by metabolic poisons such as cyanide.

The membrane proteins responsible for facilitated diffusion and active transport are similar to enzymes. Carrier proteins have particular binding sites for specific chemicals, they can only work at a particular rate, becoming saturated at high concentrations of transported substance, and they can be inhibited. They do, however, differ from enzymes in one important respect: *they do not alter the substances they transport.*

?

J Why do metabolic poisons such as cyanide prevent active transport?

Table 5.1 **Comparing diffusion, facilitated diffusion and active transport**

	Simple diffusion	**Facilitated diffusion**	**Active transport**
Type of membrane molecule involved	Lipids	Proteins	Proteins
Force driving the process	Concentration gradient	Concentration gradient	ATP hydrolysis
Direction of transport	With concentration gradient	With concentration gradient	Against concentration gradient
Specificity	Non-specific	Specific	Specific
Saturation at high concentration of transported molecules	No	Yes	Yes

7 ENDOCYTOSIS AND EXOCYTOSIS

The transport processes covered so far involve the movement of individual molecules or ions across the cell surface membrane. There are other processes, **endocytosis** and **exocytosis,** which transport larger volumes of material (solids or liquids) into or out of a cell. Fig 5.23 shows the main features of these two processes and highlights the difference between them.

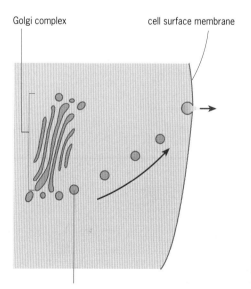

Golgi complex · cell surface membrane

Vesicle contains cell products

(a) Exocytosis

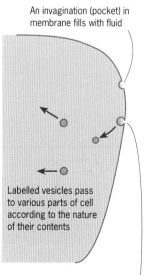

An invagination (pocket) in membrane fills with fluid

Labelled vesicles pass to various parts of cell according to the nature of their contents

Membrane closes round droplet, forming vesicle

(b) Endocytosis (pinocytosis)

Fig 5.23 **Exocytosis and endocytosis. These processes appear to be the reverse of each other, but endocytosis must be selective and the contents must be 'labelled' by the cell so that they can be processed in the correct way.**

In endocytosis, the cell surrounds material with cell surface membrane, bringing it into the cell inside a small vesicle. There are two main types of endocytosis:

- **Phagocytosis** (literally, 'cell eating'). Cells **phagocytose** (take in) solid particles such as bacteria and red blood cells.

- **Pinocytosis** (literally, 'cell drinking'). A process identical to phagocytosis but which takes in fluid.

Phagocytosis and pinocytosis

In the process of phagocytosis (Fig 5.24), a cell's cytoplasm flows around solid material. When this is completely surrounded, a segment of cell surface membrane is pinched off and a vesicle is formed within the cytoplasm. The membrane and the solid material inside the vesicle forms a **food vacuole**. Lysosomes fuse with the vacuole membrane and digestive enzymes enter the vacuole and begin to digest the contents. Soluble products, such as amino acids and sugars, are absorbed into the cytoplasm, while undigested remains are discharged from the cell by exocytosis.

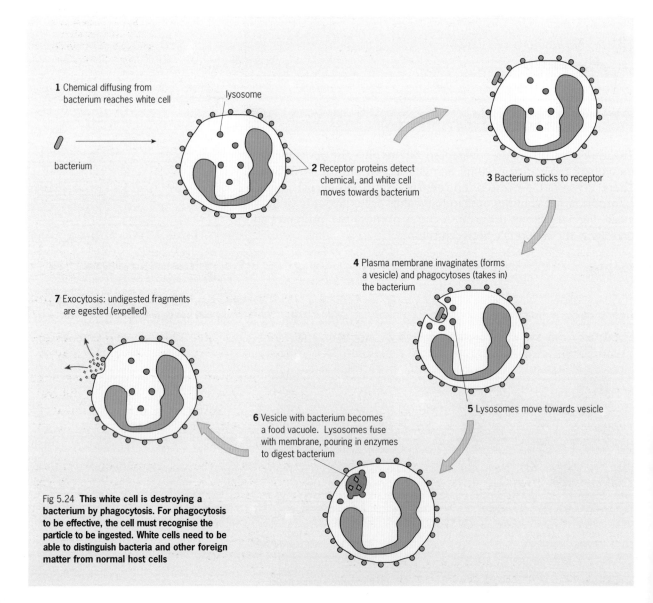

1 Chemical diffusing from bacterium reaches white cell

lysosome

bacterium

2 Receptor proteins detect chemical, and white cell moves towards bacterium

3 Bacterium sticks to receptor

4 Plasma membrane invaginates (forms a vesicle) and phagocytoses (takes in) the bacterium

5 Lysosomes move towards vesicle

6 Vesicle with bacterium becomes a food vacuole. Lysosomes fuse with membrane, pouring in enzymes to digest bacterium

7 Exocytosis: undigested fragments are egested (expelled)

Fig 5.24 **This white cell is destroying a bacterium by phagocytosis. For phagocytosis to be effective, the cell must recognise the particle to be ingested. White cells need to be able to distinguish bacteria and other foreign matter from normal host cells**

Examples of phagocytosis include the removal of foreign matter by white cells, the removal of old red blood cells from circulation by the Kupffer cells of the liver, and the feeding of single-celled organisms such as amoeba and related species.

Exocytosis – release of material from the cell

In exocytosis, materials encased in vesicles are expelled from the cell when the vesicle membrane fuses with the plasma membrane. When the process involves a liquid it is called **reverse pinocytosis** and this is how the cell secretes many of its synthesised products, such as digestive enzymes, mucus, hormones and the components of milk.

> **?**
>
> **K** Why can plant cells carry out pinocytosis but not phagocytosis?

PUS

When there is a heavy bacterial infection, such as a cut or infected sweat gland, white blood cells gather in great numbers and go into what can be described as a 'phagocytosis frenzy'. They engulf so many bacteria and dead cells that they become literally 'full'. Toxins secreted by the bacteria then kill the white cells which form pus. The lysosomes of dead white cells tend to liquefy the material surrounding them, forming a smelly, runny fluid.

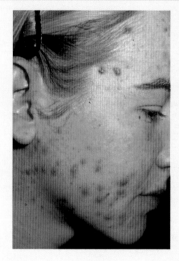

Fig 5.25 **This person's acne is caused by an over-secretion of sebum, a greasy substance that blocks sebaceous glands and hair follicles, trapping bacteria underneath. Pus is formed when white cells do their best to deal with the problem**

SUMMARY

By the end of this chapter you should know and understand the following:

■ **Diffusion** is the movement of particles down a concentration gradient until they are evenly distributed.

■ In order to maximise the process of diffusion, organisms have a large surface area, make membranes as thin as possible and maintain a diffusion gradient.

■ **Facilitated diffusion** is diffusion enhanced by specific proteins in cell surface membranes.

■ **Osmosis** is the diffusion of water from a region of high water potential to a region of lower water potential.

■ The overall tendency of a cell to absorb water (the **water potential**) depends on the relative sizes of the **solute potential, Ψ_s** and the **pressure potential, Ψ_p.**

■ Diffusion, facilitated diffusion and osmosis are passive processes: they do not require energy.

■ **Active transport** is the movement of particles across cell surface membranes against a diffusion gradient. This process requires energy, usually in the form of ATP. It is achieved by carrier proteins in the membranes.

■ **Endocytosis** is the passage of droplets of fluid (**pinocytosis**) or solid particles (**phagocytosis**) into the cell by becoming enveloped in cell surface membrane.

■ **Exocytosis** is the passage of material (usually a cellular secretion) out of the cell.

QUESTIONS

1 In a freshwater environment, the single-celled protozoan *Amoeba* takes up water by osmosis and would swell and burst if water was not expelled. An internal contractile vacuole fills with water and then empties to the outside at intervals. This vacuole is surrounded by mitochondria.

The rate at which the contractile vacuole emptied was determined for *Amoeba* placed in salt (sodium chloride) solutions of a range of concentrations. The results were as follows:

Salt concentration (%)	Rate of emptying (times hour^{-1})
0	20
0.5	14
1.0	8
1.5	3
2.0	0
2.5	0
3.0	0

a) Suggest why the rate of emptying of the contractile vacuole changes as the salt concentration rises.

b) Would you expect to see any difference between *Amoeba* in 2% compared with 3% salt solution? Explain your answer.

The diagram below represents an amoeba. The concentration of Na^+ and K^+ are given in mmol dm^{-3}.

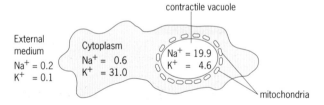

c) **(i)** What do the figures suggest about the role of the mitochondria?
(ii) How might you test your suggestion practically?

[NEAB 1994 Biology: Paper 1, q.7]

2 An artificial cell containing a solution of sucrose and glucose was suspended in a solution of four different sugars in a beaker. The diagram shows the concentration in mol dm^{-3} of each sugar inside and outside the cell.

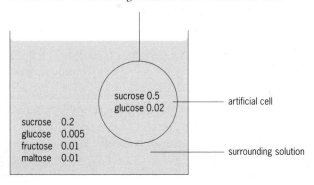

The membrane of this cell is permeable to water and to monosaccharides; it is not permeable to disaccharides.

a) At the start of the experiment, which solution would have the lower water potential? Give a reason for your answer.

b) How would you expect the volume of the cell to change? Explain your answer.

c) As the experiment proceeds, for which solute or solutes will there tend to be net diffusion: **(i)** out of the cell; **(ii)** into the cell?

[AEB 1997 Biology: Specimen Paper, q.2]

3 The graph below shows the relationship between the volume, water potential (Ψ_{cell}) and solute potential (Ψ_s) of a cell immersed in a series of sucrose solutions of increasing concentration. In each solution the cell was allowed to reach equilibrium with the bathing solution, so that water was being neither lost nor gained, before the measurements were made.

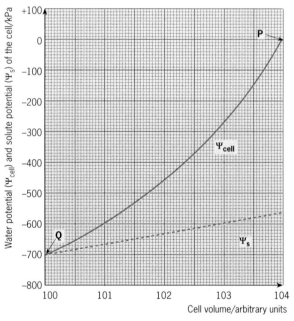

a) Explain what is meant by the following terms: **(i)** solute potential (Ψ_s); **(ii)** pressure potential (Ψ_p).

b) What terms are used to describe the condition of the cell at P and at Q?

c) What is the volume of the cell (in arbitrary units) when in equilibrium with a sucrose solution of solute potential −400 kPa?

d) Calculate the pressure potential (Ψ_p) of the cell when its volume is 103 units. Show your working.

e) Suggest *two* ways in which reversible changes in cell volume may be important in flowering plants.

[ULEAC 1996 Biology: Specimen Paper B2, q.9]

Assignment

THE SURFACE AREA AND VOLUME PROBLEM

Single cells exchange materials over the whole surface of their outer membrane. Multicellular organisms are larger and cannot exchange enough materials over their outside surface. In this assignment, we investigate why.

Surface area to volume ratio
The amount of material that an organism *needs* to exchange with its surroundings is proportional to its *volume*, but its *ability* to exchange material is proportional to its *surface area*. As organisms get larger, their surface area to volume ratio gets *lower*.

1 Imagine a cube-shaped multicellular organism that exchanges materials over its surface. (Obviously, no such organisms exist but this keeps the calculations simple.) Copy and complete the following table:

Size of each side of the organism (units)	1	2	5	10	?
Surface area (for a total of 6 sides (units^2)	6	24	?	600	60 000
Volume (units^3)	1	8	125	?	?
Surface area: volume ratio	6:1	?	1.2:1	?	?

So larger organisms can only survive if they evolve ways to optimise the exchange of materials. Organisms can:

- increase their surface area.
- make membranes/barriers as thin as possible.
- maintain a diffusion gradient. As soon as materials are absorbed they are moved away, often by a circulatory system to ensure that an equilibrium is never reached.

Functional design in organisms

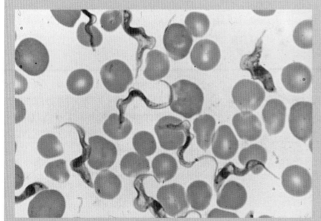

Fig 5.A1 **Trypanosomes in human blood**

2 Fig 5.A1 shows examples of a simple unicellular organism called a **trypanosome**, the parasite responsible for sleeping sickness. Does this organism need special adaptations to increase its surface area to volume ratio? State a reason for your answer.

3 Two relatively simple ways of increasing the surface of area are to become flat or to become long and thin (worm-like). The **tapeworm** is a parasite which lives in the intestines of animals such as fish, pigs, rats and even humans.
a) Suggest why the tapeworm has no mouth, intestine or excretory system.
b) Suggest why the tapeworm could not live on land.
c) Thin shapes, such as discs and ribbons, give the greatest possible surface area to volume ratio. What shape gives the least possible surface area to volume ratio?

4 Animals that have evolved beyond the worm stage have had to develop specialised organs to increase their surface area. The **nudibranc** (sea-slug) in Fig 5.A2 has feather-like gills on its back to increase the surface area for gas exchange.
  Suggest why some species of nudibranc constantly shake their gills.

Fig 5.A2 **A sea slug**   Fig 5.A3 **Gills of axolotl**

5 The **axolotl** shown in Fig 5.A3 is an unusual amphibian: it retains external gills into adult life and so remains confined to water. The gill filaments are red because of the thin membranes and good blood supply that runs through them.
  Suggest what would happen to the axolotl's gills on land.

6 Like most amphibians, the **frog** has three gas exchange surfaces. In addition to the moist, permeable skin, the frog can exchange gases across the lining of its large mouth. The frog also uses simple lungs when oxygen demand is high.
a) Suggest why the underneath of a frog's mouth is constantly moving up and down.
b) Why must an amphibian's skin remain moist?

Find out about gas exchange in insects, fish and mammals in Chapter 17.

6 The cell cycle

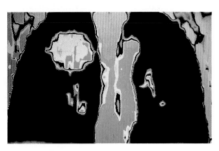

Highly sophisticated imaging techniques can be used to diagnose tumours. This CT (Computer Tomography) scan shows a large, spherical tumour in the person's right lung

Skin cancer is common, especially in countries such as Australia, where fair-skinned people are exposed to strong sunlight

IN A HEALTHY human adult, cells divide only when they should. Some cells, such as those that line the gut, are replaced at a remarkable rate. Other cells, such as muscle cells, live longer and need to be replaced far less often. Occasionally, the systems that control cell division break down, and a cell that should be stable divides uncontrollably. Soon, a mass of tissue called a tumour forms. Tumours can be either benign or malignant.

A benign tumour is not cancerous. Cells divide within a limited area and the growth is often surrounded by a membrane. New cells form in the centre of the mass. This type of tumour can be dangerous if it presses on important organs, or if it grows to a very large size. It is usually treated by surgery. Because the tumour cells are confined and do not invade other tissues, recovery and survival rates are good.

A malignant tumour is commonly known as cancer. The tumour grows at the edges, spreading out and invading the surrounding tissues (supposedly like a crab – hence the name cancer). This type of tumour is much more dangerous. Vital tissues and organs can be destroyed quickly and this can lead to death. Even if the tumour is removed from the body, actively dividing cells may have already broken off and set up secondary growths elsewhere in the body.

1 SOME BASIC PRINCIPLES

Multicellular organisms begin life as a single cell and go on to become more complex through a combination of **cell division** and **differentiation** (the process by which cells become specialised for a particular function). Even in adults, cell division is constantly providing new cells to replace those that die.

As we saw in Chapter 1, animals and plants are eukaryotic organisms: in eukaryotic cells, DNA is organised into chromosomes and is enclosed within a double nuclear membrane.

Before going on to study the mechanism of cell division in eukaryotic cells, you may find it useful to remind yourself of some of the basic terms and concepts shown in Table 6.1. Some of the terms will be new to you, but you will learn more about them as you read through the chapter.

An introduction to cell division

In many organisms, including humans, the nucleus of body cell contains two sets of chromosomes and therefore two sets of genes. These cells are therefore **diploid** (from Greek, meaning 'double number') and are called **somatic** or 'body' cells (Fig 6.1).

Table 6.1 **Some basic terms important in cell division**

Term or feature	What it is or what it means
Gene	A length of DNA that codes for the production of a particular polypeptide or protein
Genome	The name given to the full set of genes in a cell. The human genome consists of between 50 000 to 100 000 genes (the exact number is not yet known)
Diploid	Diploid cells contain two versions of every gene
Haploid	Haploid cells contain one version of every gene
Chromosome	A long single molecule of DNA, organised around proteins called histones. The largest human chromosomes contain about 4000 genes. Chromosomes exist in cells all the time but they can be seen only during cell division, when they condense and separate
Homologous chromosomes	Chromosomes exist in pairs – humans have 23 pairs. Homologous chromosomes have the same genes at the same positions, but not necessarily the same versions of each gene
Chromatid	During cell division, the DNA of a cell is replicated (copied). When the chromosomes condense, they therefore appear as double structures: each unseparated chromosome within such a pair is called a chromatid
Sister chromatid	When two chromatids are genetically identical (as in mitosis), they are called sister chromatids
Bivalent	A bivalent is a pair of homologous chromosomes that line up together, as they do during meiosis
Locus	The position of a gene on a chromosome
Transcription	The process in which a molecule of mRNA is assembled on an active gene. mRNA thus becomes a mobile copy of the gene
Translation	The process of converting the code on the mRNA into a protein. This is achieved by protein synthesis: amino acids are joined in a particular order to make a protein such as an enzyme

A 'DNA makes RNA makes proteins'. Explain this statement.

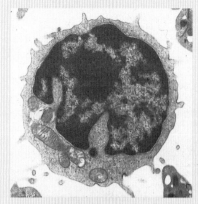

Fig 6.1(a) **A human white cell (a lymphocyte). The nucleus of this cell contains all the genes necessary to make a whole new organism, a clone of the person from whom this cell was taken. Although the technology exists to clone mammals, the cloning of humans would cause huge moral, ethical and legal problems**

Fig 6.1(b) **Dolly the sheep was the first ever mammal to be cloned using a body cell, in 1997**

Mitosis is the process of normal cell division. In mitosis, the chromosomes are copied and then divided equally between the two new daughter cells. So each mitotic division produces two cells, both diploid and each with exactly the same genes as the parent cell.

Some cells in the body of a multicellular eukaryotic organism are not diploid. Gametes (eg sperm and eggs) contain only one copy of each gene as they have only one set of chromosomes. These cells are haploid and are produced by a special type of cell division called **meiosis**.

A male and female gamete join together at fertilisation, to form a new diploid cell called a **zygote**. This one cell divides by mitosis to produce a complete new organism (Fig 6.2). Reproduction is covered in more detail in Chapters 28 and 29.

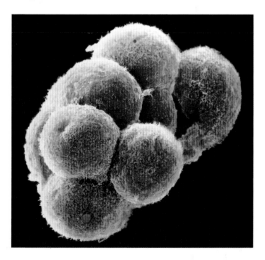

Fig 6.2 **This human embryo is 4 days old, but the fertilised egg has already divided mitotically 4 times, producing a cluster of 16 cells. The cells will soon specialise to form the tissues and organs of the body**

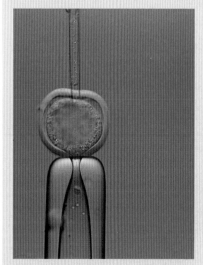

Fig 6.1(c) **A micropipette puts the nucleus from a body cell into a newly fertilised egg cell that has had its own nucleus removed**

CELL TURNOVER

AS CHILDREN, we grow because the cells in our bodies divide mitotically, and cell production outnumbers cell death. When we reach our adult size, our cell population stays constant. The rate at which cells die and are replaced is known as **cell turnover**.

Research has shown that different tissues and organs have very different turnover rates. Brain cells, for instance, are not usually replaced, and after our twenties there is a slow but steady decline in number. This is not normally something to worry about – you won't run out. Other cells have a low turnover. Liver cells, for example, may divide only once every one to two years although, unlike the brain, they can regenerate if some cells are damaged or destroyed. Most of the high turnover cells are epithelial cells (Fig 6.3 and Fig 6.4), but cells of the bone marrow (Chapter 26) and cells in the testes (Fig 6.5) also divide very frequently.

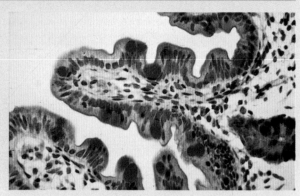

Fig 6.4 **Intestinal epithelial cells have a very rapid turnover: the entire gut lining is replaced every few days. Much of the content of these cells is digested and reabsorbed in an efficient recycling system, but even so, gut cells form a significant proportion of faecal matter**

Fig 6.3 **Scanning electron micrograph of human epidermis. Dead skin cells are constantly being lost and replaced by mitosis. A large proportion of household dust consists of human skin cells**

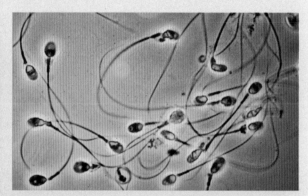

Fig 6.5 **The production of sperm is a remarkably intense process. The average healthy human male produces about 1000 new sperm every second. The process of sperm production,** spermatogenesis, **which involves meiosis, is described in Chapter 28**

2 THE CELL NUCLEUS

The DNA of a cell is contained within the nucleus. In a cell that is not actively dividing, DNA exists as **chromatin** (see Chapter 3: Fig 3.41). This has a granular appearance (Fig 6.6). Separate chromosomes are present, but the DNA is so spread out that we cannot tell one from another.

When a cell is about to divide, the chromosomes condense and separate. They become visible under the light microscope as short, dark, rod-like structures. Each pair of chromosomes is known as a **homologous pair**. This means that they both contain genes at the same positions, or **loci**. You received one chromosome of each pair from your mother and one from your father. So, although homologous pairs contain the same genes, they do not necessarily carry the same *versions* of each gene. This is the key to understanding why organisms vary. Alternative forms of genes are called **alleles** (see Chapter 32).

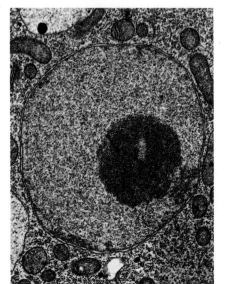

Fig 6.6 **In a non-dividing cell, the chromosomes are spread out as chromatin and appear as pale granules. (The darker granules are the nucleolus.) In this form, the DNA is used as a template to make mRNA in the process of transcription. mRNA passes to the cytoplasm and is itself used as a template for protein production in the process of translation**

The appearance, number and arrangement of chromosomes in the nucleus is called the **karyotype**. The human karyotype consists of 23 pairs of chromosomes (Fig 6.7). All diploid human cells contain 23 pairs of chromosomes and this is often given the notation $2n = 46$. This means that a diploid cell has 46 chromosomes (a haploid cell has 23). Other organisms have different numbers of chromosomes (Table 6.2).

?

B Which notation would you use to describe the number of chromosomes in human sperm?

Table 6.2 **The number of chromosomes in the cells of some plants and animals. Note that there is no connection between chromosome number and complexity of organism. Some ferns have hundreds of chromosomes**

Species	Number of chromosomes
Penicillin mould (*Penicillium notatum*)	5
Broad bean (*Vicia faba*)	12
Lettuce (*Lactuca sativa*)	18
Yeast (*Saccharomyces cerevisiae*)	34
Cat (*Felis cattus*)	38
Human (*Homo sapiens*)	46
Potato (*Solanum tuberosum*)	48
Chimpanzee (*Pan troglodytes*)	48
Horse (*Equus caballus*)	64
Chicken (*Gallus gallus*)	78
Dog (*Canis familiaris*)	78

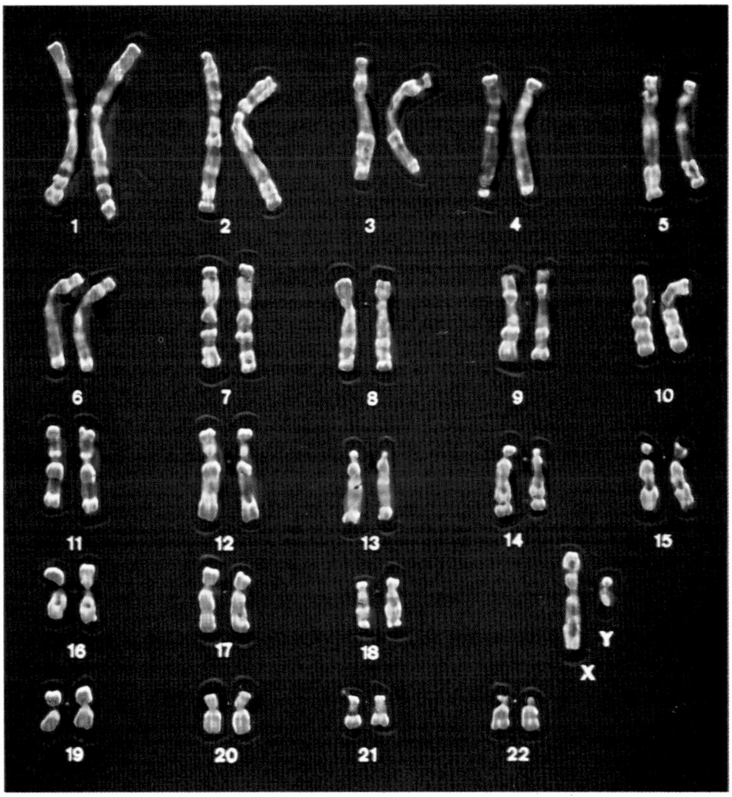

Fig 6.7 **The 23 pairs of chromosomes in the human karyotype. Karyotypes are used to detect chromosomal abnormalities and to perform sex texts on athletes. The presence of XX confirms the athlete is female, the presence of XY shows the athlete is male**

3 THE STAGES OF THE CELL CYCLE

The cell cycle is the complete sequence of events in the life of an individual diploid cell. The cycle starts and ends with cell division and consists of the stages of mitosis plus **interphase**, the interval between divisions in which the cell carries out its normal functions. Fig 6.8 shows the main stages of the life cycle of a cell.

Cells that do not divide, such as those in muscle and nerve tissue, are always in interphase: this is the normal state for a functioning cell. In interphase the DNA in the nucleus is spread out, and active genes are being read to produce proteins such as enzymes.

A new cell has three options:

- It may remain stable, in interphase for many months or years. Brain and other nerve cells rarely, if ever, divide.

- It may undergo mitosis within a short period of time. Skin and gut lining cells, and those at the growing points of plants (meristems), all have a very rapid turnover.

- It may undergo meiosis. Specialised germinal cells in the ovary and testes include meiosis in their cell cycle and produce egg cells or sperm.

Fig 6.8 **The four stages of the cell cycle: the three stages of interphase and the stage of active division in which mitosis occurs**

G1 – growth of cytoplasm and organelles
S – synthesis of DNA
G2 – second growth phase
M – mitotic phase (nucleus and cytoplasm divide to produce two diploid cells)

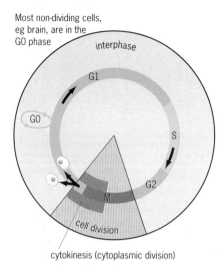

Most non-dividing cells, eg brain, are in the G0 phase

cytokinesis (cytoplasmic division)

?

C Name the three main stages of interphase.

D Without DNA replication, what would happen to the genetic material each time cell division took place?

✔

Exam hint – Remember the mnemonic:
IPMAT – **I**nterphase, **P**rophase, **M**etaphase, **A**naphase, **T**elophase.

Interphase

We shall start looking at the cell cycle at the point when a cell has just divided. The cell is now in interphase. Interphase has three distinct stages: G_1, **S** and G_2.

G_1 – the first growth phase

Just after it has been produced by division of its parent, a new cell is in the early part of the first **g**rowth phase, G_1, sometimes described as G_0. Cells that do not divide remain at this point in the cell cycle because they do not need to go further: they never replicate their DNA. 'New' cells are relatively small, with a full-sized nucleus but relatively little cytoplasm.

During G_1, protein synthesis starts and the volume of cytoplasm and the number of organelles increase rapidly. The process is actually quite complicated, not least because some organelles such as mitochondria (which have their own DNA) divide independently of the cell nucleus. In later G_1, the cell takes on more 'normal' proportions: the nucleus begins to look smaller as the surrounding cytoplasm increases in volume.

The S phase

The **S** phase – DNA **s**ynthesis phase – follows G_1. In this phase, the cell's DNA **replicates** (copies itself). Predictably, a cell only enters the S phase if it is going to divide. The point at which DNA replication starts is called the **restriction point**. After this, the cell becomes **restricted**, or locked in an automatic sequence that moves inevitably on to cell division.

G_2 – the second growth phase

Before the actual mitotic cell division (see below), the cell enters a second, shorter **g**rowth phase, G_2, in which the proteins necessary for cell division are synthesised.

Mitosis

Mitosis is a continuous sequence of events but, for clarity, they are divided into four distinct stages: **prophase**, **metaphase**, **anaphase** and **telophase**. The stages of mitosis are illustrated in Fig 6.9. You will need to refer to this constantly as you read the text.

The movement of chromosomes during cell division is controlled by microtubules (page 19), which form the **spindle**. In a non-dividing cell the microtubules are found as two bundles of fibres, the **centrioles**, in an area of cytoplasm known as the **centrosome**, or **microtubule organising centre (MTOC)**. During cell division the centrioles move to opposite sides of the nucleus, from where they form the spindle.

Prophase

Chromatin begins to condense into chromosomes: Fig 6.9(a). At this stage, each chromosome has replicated itself and now consists of two identical **chromatids**. These are known as **sister chromatids** and are joined by a **centromere**. Unless there has been a mutation (a fault in the DNA replication), these sister chromatids are genetically identical. The subsequent stages of mitosis organise and split the pairs of chromatids, so that one chromatid of each pair goes into each daughter cell.

Fig 6.9 **The stages of mitosis**

(a) **Prophase.** The DNA has already replicated during interphase and the chromosomes now condense. Each chromosome is a double structure made from two genetically identical chromatids

- chromatin threads
- nuclear membrane
- nucleolus
- cytoplasm
- cell surface membrane
- centrioles

- nuclear membrane
- nucleolus
- centriole
- centromere
- pair of chromatids

(b) **Metaphase.** The nuclear membrane has gone and the chromosomes arrange themselves on the equator (middle) of the spindle

- spindle fibres (microtubules)
- centromeres on 'equator' of spindle

(c) **Anaphase.** The chromatids are pulled apart and move to opposite poles

Daughter chromosomes move apart, led by their centromeres

(d) **Telophase.** Cytokinesis (cytoplasmic division) is beginning. The chromosomes becomes no longer individually visible

(e) **Cytokinesis.** Cytoplasmic division is achieved

- nuclear membrane
- nucleolus
- chromatin threads
- pair of centrioles

As the chromosomes condense, other changes occur in the cell:

- The nucleolus begins to break down.
- The centrioles move to opposite sides of the nucleus.
- The centrioles begin to assemble the spindle.
- The nucleolus disappears.
- The nuclear membrane begins to break up.

From prophase onwards, most 'normal' cell activity, such as protein synthesis and secretion, is halted until division is over.

Metaphase

The beginning of metaphase (meta = middle) is marked by the disappearance of the nuclear membrane, which breaks down into separate vesicles, moves into the surrounding cytoplasm and joins with the endoplasmic reticulum: Fig 6.9(b).

The spindle (Fig 6.10) becomes fully developed and fills the space that was occupied by the nucleus. Then the most obvious event of metaphase happens: the chromatid pairs attach themselves to individual spindle fibres and align themselves on the equator of the spindle.

Anaphase

At the start of anaphase (ana = apart), Fig 6.9(c), the chromatids are pulled apart by movements of the spindle fibres, so that sister chromatids are pulled to opposite poles. The newly separated chromatids are

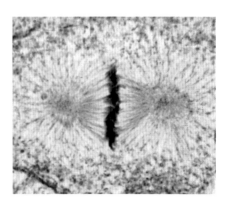

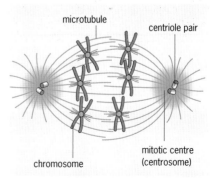

- microtubule
- centriole pair
- mitotic centre (centrosome)
- chromosome

Fig 6.10 **The spindle is a cradle of microtubule fibres which organise the chromosomes during cell division**

now called chromosomes, and are single structures. If you have ever seen a film of mitosis (highly recommended for learning purposes), you will know that anaphase is the most obvious event.

Telophase

In telophase (telo = final) the chromosomes reach the poles of the spindle: Fig 6.9(d). They then decondense and become indistinct, forming the familiar chromatin of interphase. The nuclear membrane re-forms and the nucleolus reappears. Soon afterwards, transcription resumes and the cell restarts protein synthesis, endocytosis and other normal cytoplasmic functions. This marks the end of mitosis.

Cytokinesis – division of the cytoplasm

In the events of mitosis just described, we have looked only at the splitting of the nucleus of the cell. Splitting of the cell itself is called **cytokinesis** and usually begins during anaphase. In most animal cells, microtubules form a furrow in a ring around the cell. These gradually constrict until the cells separate, as shown in Fig 6.11(a). In cells of higher plants, the Golgi apparatus produces a line of vesicles that fuse, causing the two daughter cells to separate, as shown in Fig 6.11(b).

Fig 6.11(a) **Cytokinesis in animal cells. As the events of mitosis come to an end, movement of microtubules causes a constriction around the centre of the cell. Eventually the cytoplasm divides, leaving two new daughter cells**

Fig 6.11(b) **Cytokinesis in higher plant cells. The cytoplasm is divided when vesicles arising from the Golgi apparatus fuse to form one continuous barrier. The vesicles contain cellulose and other chemicals that make up the cell wall**

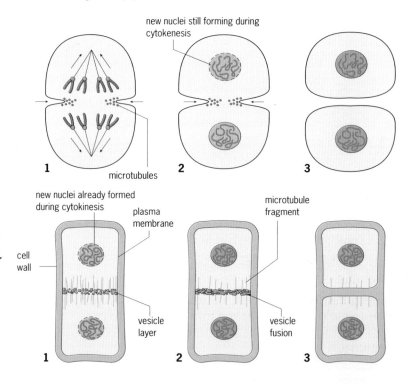

Meiosis: making the gametes

Meiosis is the type of cell division that halves the genetic material within cells. For this reason, it is known as a **reduction division**. In animals, meiosis makes haploid gametes (sex cells), but in many species of plants and fungi meiosis makes spores (see Chapter 29).

Meiosis is a remarkable process that produces haploid gametes *and* shuffles the genes, so that each gamete produced is genetically different. This is why children born to the same parents are usually not identical: they are produced by the fusion of a genetically unique sperm and egg. Identical twins are an exception because they originate from the same fertilised egg.

?

E (a) Where in the body does meiosis occur in **(i)** women and **(ii)** men?

(b) Why is meiosis sometimes referred to as reduction division?

The stages of meiosis

Meiosis has two separate divisions. Both divisions have stages that are given the same names as in mitosis, but they have a number to denote whether they refer to the first or second division, for example anaphase I, prophase II.

It is important to remember here that a normal diploid cell contains two copies of each chromosome, one from the organism's 'mother', and one from the organism's 'father'. Just before meiosis, the cell's DNA becomes organised into chromosomes and replicates to form pairs of chromatids in preparation for division, just as it does before mitosis. So in a human being, at the first stage of meiosis, the cell contains 92 (2 × 46) chromatids. The 23 chromosomes originally from the mother have been duplicated, making 23 identical pairs of chromatids, as have the 23 chromosomes originally from the father.

An overview of meiosis is shown in Fig 6.12. There is a basic difference between meiosis and mitosis: in meiosis, the homologous chromosomes pair up, while they stay apart in mitosis. During the first part of meiosis, each homologous pair can swap sections of DNA, so that each pair becomes 'genetically mixed'.

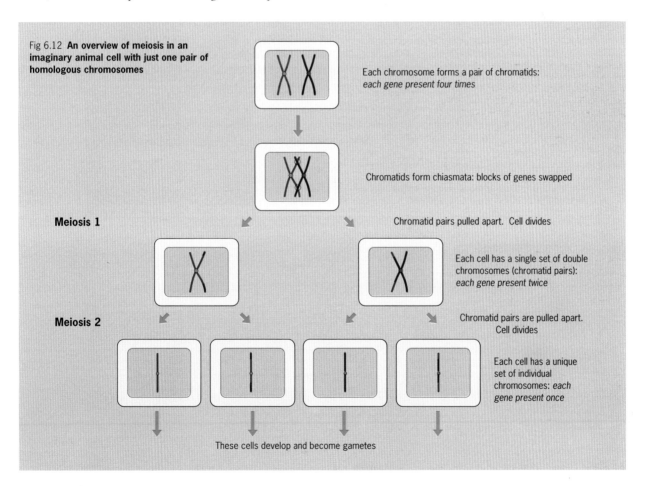

Fig 6.12 **An overview of meiosis in an imaginary animal cell with just one pair of homologous chromosomes**

Each chromosome forms a pair of chromatids: *each gene present four times*

Chromatids form chiasmata: blocks of genes swapped

Meiosis 1

Chromatid pairs pulled apart. Cell divides

Each cell has a single set of double chromosomes (chromatid pairs): *each gene present twice*

Meiosis 2

Chromatid pairs are pulled apart. Cell divides

Each cell has a unique set of individual chromosomes: *each gene present once*

These cells develop and become gametes

The individual stages of meiosis are shown in Fig 6.13. In the first meiotic division (prophase I to telophase I), one homologous pair of 'genetically mixed' chromosomes passes into each new cell. In the second meiotic division, the individual chromatids are pulled apart so that one chromatid (now a chromosome) goes into each daughter cell. The end result of meiosis is four haploid cells, each containing a single set of chromosomes. Each cell is genetically unique.

Fig 6.13 **The stages of meiosis in an imaginary animal cell showing just one of the pairs of chromosomes**

Stage of meiosis First division	What is happening	What it looks like
Interphase	Just before meiosis, DNA replicates so cells which contained two copies of each chromosome, now have four. Chromosomes not yet visible	
Early prophase I	Chromosomes become visible. Centromeres move to opposite sides of cell	
Mid prophase I	Each homologous pair of chromosomes comes together to form a bivalent	
Late prophase I	Each chromosome in a bivalent forms two chromatids. Genetic mixing occurs: chiasmata, the points of cross-over, are visible	
Metaphase I	The bivalents arrange themselves on the equator of the spindle	
Anaphase I	The chromatid pairs from each homologous chromosome split apart and move to opposite poles of the cell	
Telophase I	Cytokinesis begins, two new cells form, each has two copies of each chromosome. These chromosomes are genetically different from those in the original cell	
Interphase	A resting time (length varies between cell types)	
Second division		
Prophase II	A new spindle forms, at right angles to the first	
Metaphase II	Chromosomes, each of which is a pair of chromatids, align themselves on the equator of the spindle	
Anaphase II	Chromatids are pulled apart and move to opposite poles of the cell	
Telophase II	Cytokinesis begins. Four haploid cells, each with only a single chromosome, have been formed. Each chromosome is genetically different	

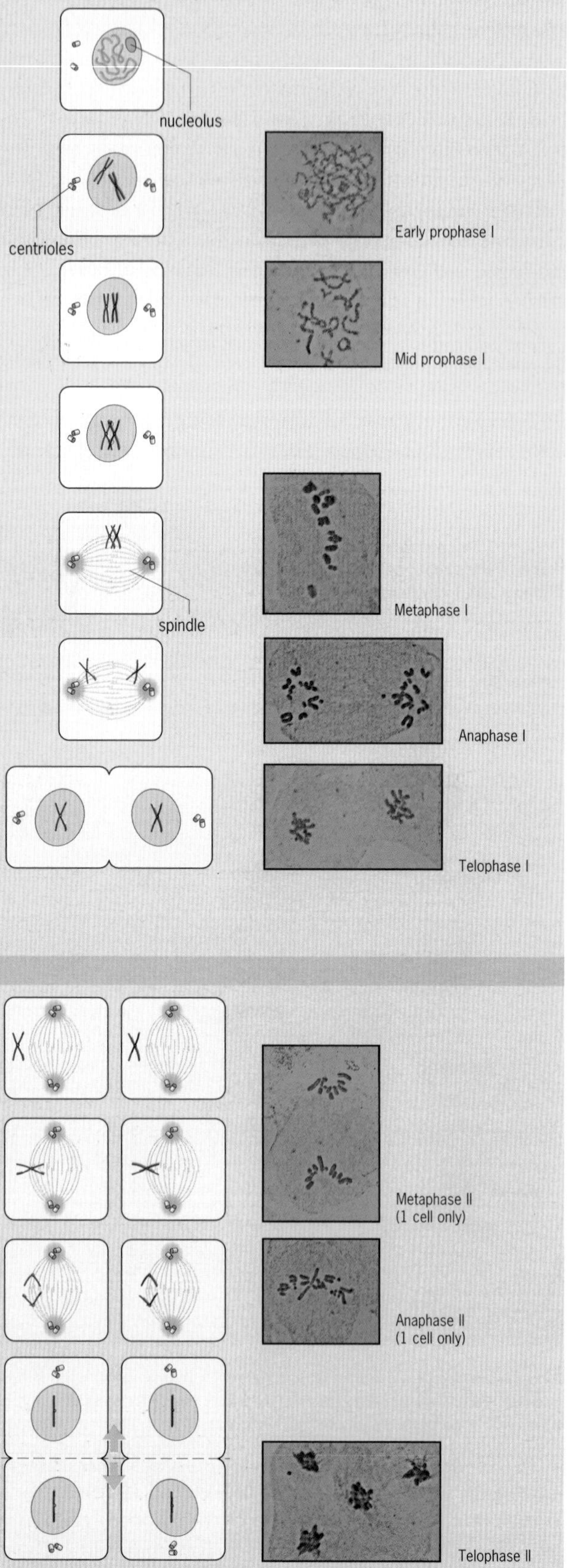

Prophase 1

Prophase 1 is more complex than the corresponding phase in mitosis, and can be subdivided into early, middle and late.

Early prophase I starts when the chromosomes condense and the nucleolus disappears.

In **mid prophase I**, the homologous chromosomes, one from each parent, pair up. Each pair forms a **bivalent**. This does not happen in mitosis. When a chromosome pair is exactly aligned, it is said to be at **synapsis**. Next, each chromosome in a pair divides into two chromatids, giving four chromatids per bivalent.

In **late prophase I**, **recombination** – or cross-over – takes place (Fig 6.14). One (or both) of the chromatids of the two homologous chromosomes breaks off at certain points and fuses with a chromatid of the other chromosome in the bivalent, forming joints called **chiasmata** (the singular, *chiasma*, means 'crosspiece'). This process ensures that blocks of genes are swapped between maternal and paternal chromosomes. The position of chiasma formation varies, even within the same species, and this produces a large variety of new gene combinations.

Prophase ends with the chromatids of each bivalent (pair of homologous chromosomes) entwined and joined by chiasmata.

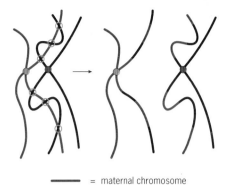

———— = maternal chromosome

Fig 6.14 **Chiasmata form between homologous chromosomes during late prophase I of meiosis. At these points, parts of the maternal chromosome separate and join with the paternal chromosome, and vice versa. This produces new chromosomes that are genetically different from each other, and from both the maternal and paternal chromosomes from which they are derived**

Metaphase 1

In metaphase 1, the nuclear membrane disappears and the spindle is fully developed. The bivalents move to the equator of the spindle in the same way as individual chromosomes do during the matching phase in mitosis.

Anaphase I

The chromatid pairs of a bivalent are pulled apart (Fig 6.15) because of the action of spindle fibres, a process that separates the entwined chromatids. At the end of anaphase 1, the chromatid pair from one of the original homologous chromosomes is positioned at one pole of the cell and the chromatid pair from the other homologous chromosome is at the other pole. Either the maternal or the paternal chromosome can pass into either cell. This is the process that allows **independent assortment of alleles** (see Chapter 32). It is another reason why meiosis increases variation.

Occasionally, this phase is not completed successfully and a pair chromosomes fails to separate. The result is that both homologous chromosomes pass into one daughter cell, the other receiving neither. This situation can lead to conditions such as Down's syndrome.

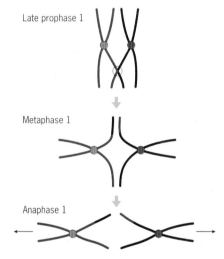

Late prophase 1

Metaphase 1

Anaphase 1

Fig 6.15 **Chiasmata form during late prophase I and cross-over takes place. The newly formed chromatid pairs separate during metaphase I and then the two pairs move to opposite poles of the cell during anaphase I**

Telophase I

The spindle disappears and the nuclear envelope re-forms around the two sets of chromosomes. At the same time cytokinesis separates the cytoplasm, forming two daughter cells, ready for the second meiotic division.

Interphase

The length of the resting interphase between the two meiotic divisions varies widely. It is sometimes short or even non-existent. If there is no interphase, the chromosomes remain condensed and the cell passes straight from telophase 1 into prophase II. In human females, however, ova may remain in interphase for decades. Basic egg cells are made in a girl before birth, but they do not complete meiosis until just before ovulation, anything up to fifty years later.

At the end of meiosis 1, there are two cells, each containing two copies of each chromosome on a chromatid pair. The second meiotic division separates the chromatids, so that each daughter cell formed is haploid (has one set of single chromosomes).

The second meiotic division

Prophase II
For each chromosome, the chromatid pair attaches itself to the new spindle, which forms at right angles to the first.

Metaphase II
Each chromatid pair lines up on the equator of the spindle.

Anaphase II
The chromatids are pulled apart and move to opposite poles.

Telophase II
The spindle disappears, the nuclear membrane re-forms, chromosomes expand and cytokinesis produces two separate cells.

The end product of meiosis
Meiosis produces four genetically different haploid cells, known as a **tetrad**. Genetic variation has been produced in three ways:

● The homologous chromosome pairs originate in different organisms, one maternal and one paternal, and so are genetically different.

● Blocks of genes are swapped between the chromatids of homologous chromosomes as the chiasmata form during prophase I.

● Each daughter cell can receive a copy of either chromosome from a pair, and each copy may have undergone cross-over and have different genes from the other three. This is called indpendent assortment and is covered in more detail in Chapter 32.

Learning meiosis
Meiosis is a complex process that many people find difficult to learn. Here are some hints that you may find useful.

First, divide your study of meiosis into three sections by looking at three questions:

What is the point of meiosis?
It makes sex cells by shuffling the genes to produce haploid cells that are all genetically different from either of the parents. Each fertilisation (combination of an egg and a sperm) produces an individual that is genetically unique.

What do each of the meiotic divisions achieve?
Start by learning an overview of each division: Fig 6.13 is a good place to start. In the first meiotic division, homologous chromosomes pair up, divide into chromatids, swap blocks of genes and then separate. In the second division the individual chromatids separate. The end result is four genetically different haploid cells from one diploid original.

What are the stages of the two divisions?
Once you are confident that you can answer the first two questions, you can put some flesh on the bones. Describe one stage at a time and then draw what you have just described. Try using different colours for paternal and maternal chromosomes. Alternatively, you could make some chromosomes out of modelling clay and work through the meiotic sequence yourself. Many students find this a valuable exercise.

?

F How many bivalents are formed in a human cell undergoing meiosis?

DOWN'S SYNDROME

IN THE UNITED KINGDOM, two children in every 1000 are born with Down's syndrome (Fig 6.16(a)), a genetic condition that arises from a fault in meiosis. Individuals with this condition have an extra chromosome 21 (Fig 6.16(b)) and usually have physical and mental disabilities. They often have a small mouth with a normal-sized tongue (making eating and speech difficult), reduced resistance to disease and heart abnormalities. Approximately one in three Down's syndrome children do not survive to their twelfth birthday.

The reason for the extra chromosome is usually a failure of the chromosomes to separate during anaphase 1 of meiosis (see Fig 6.13). Both chromosomes 21 pass into one daughter cell, and the other gets no copy at all. If a gamete with no chromosome 21 forms a zygote, the embryo fails to develop. But, if the gamete containing both chromosomes is fertilised, the resulting baby later develops Down's syndrome.

One reason for the non-separation may be the length of time the egg spends in the ovary. In females, egg cells do not complete their meiotic division until they are 'selected' for ovulation (see Chapter 28). The longer the eggs remain in the ovary, the 'stickier' the chromosomes become and the greater the probability that they will not separate during anaphase I. This explains why the likelihood of having a child with Down's syndrome increases as women get older. Women below the age of 25 have less than a 1 in 2000 chance, but this rises to 1 in 50 for women over 45.

The age of the father is also important: recent studies have shown that in a significant minority of cases the extra chromosome comes from the father, that is, a normal ovum is fertilised by a sperm containing two chromosomes 21s.

Fig 6.16(a) **A child with Down's syndrome**

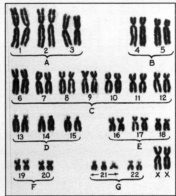

Fig 6.16(b) **The complement of chromosomes (the karyotype) of a person with Down's syndrome, showing the extra chromosome 21**

5 CANCER: CELL DIVISION OUT OF CONTROL

Cancer has been a major cause of death and disease throughout recorded history but it is now the second biggest cause of death in the western world. There are two probable reasons for this apparent increase. First, many people are surviving longer, free from other diseases, and so have an increased chance of developing cancer. Second, we are now more likely to be exposed to **carcinogens** (cancer causing agents) such as cigarette smoke and asbestos.

Cancer is not a single disease: over 200 different types of cancer have been identified. This has led to some bewildering jargon, with tumours named according to the tissue they are in, their growth pattern, their effect on the patient, their response to treatment or the person who discovered them. However, all cancers are basically similar: they all result from uncontrolled cell growth.

?

G Explain why older people are more likely to develop cancer.

What is cancer?

Cancer occurs when there is a breakdown in the cellular control mechanism which puts the brakes on cell division. So cells that should be stable begin to divide, forming a **tumour**. The rate of cell division varies greatly. Some tumours develop quickly, others may take ten years to reach a noticeable size. Cancer cells usually fail to **differentiate**: they cannot specialise for the particular function of the tissue they grow in.

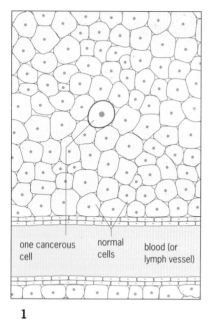

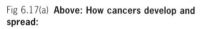

1

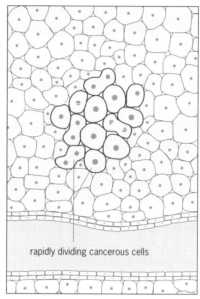

2

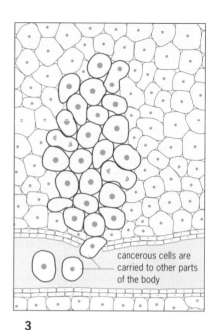

3

Fig 6.17(a) **Above: How cancers develop and spread:**

1 One cell starts to divide uncontrollable: it becomes cancerous

2 The cancerous cell divides rapidly forming a mass of cells which squash the neighbouring normal cells

3 Some of the cancerous cells split off from the main tumour and enter the blood. This carries them round the body and they can set up secondary tumours in other tissues. This process is called metastasis

Fig 6.17(b) **Below: a micrograph of cells taken from the cervix of a woman during a routine smear test. The tissue is then examined for signs of abnormal mitosis. The pathologist noticed some cancerous cells, because they stain red and have relatively large nuclei**

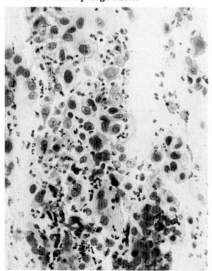

Fig 6.17 shows how cancerous cells spread and how they look different to normal cells. In 6.17(b), notice that the cancerous cells are smaller, with clear nuclei; the classic signs of active cell division.

Why is cancer so dangerous?

Tumours interfere with the activity of the cells and organs that surround them. They can compress tissues, preventing normal blood flow or nerve function. Even more seriously, they can invade surrounding tissues, killing normal cells in the process. Cancer can also spread to other parts of the body: if cancerous cells break free from their site of origin, they can travel in the blood or lymph and may set up secondary growths, or **metastases**, elsewhere in the body.

What causes cancer?

Smoking is now one of the major causes of cancer in the developed world. It leads to around 40 000 deaths from cancer each year in the UK (a large sports stadium holds about 40 000 people).

It is widely accepted that the underlying cause of cancer is a mutation of certain genes, **oncogenes**, which allow the cell to divide. The normal versions of these genes inhibit cell division, but the mutated versions fail to function. Mutations in these genes are biological accidents, and it is thought to take five or six of these accidents to override the brake on a cell's division.

This situation has been compared to the risk of having a car crash. The longer you drive a car, the more likely you are to have a crash, but the risk is increased by other factors such as tiredness, alcohol, a wet road and heavy traffic. So it is with the risk of getting cancer. The longer you live, the more chance there is that your cells will have these mutations, and the chances are increased if you are exposed to hazards such as smoking, asbestos or strong sunlight.

The following factors increase the chance of mutation:

- Exposure to carcinogens – chemicals such as asbestos, benzene.
- Infection by certain viruses.
- A genetic susceptibility.
- A long life.

SUMMARY

After reading this chapter, you should know and understand the following:

■ Almost all the cells in the human body are **diploid**: they contain a double set of chromosomes. The only exceptions are the gametes: sex cells – egg cells and sperm – are **haploid** and contain only one set of chromosomes.

■ The cell cycle consists of four stages: **G₁**, in which the cell grows; **S**, in which the DNA is doubled; **G₂**, a second growth phase and finally **mitosis** or **meiosis**. G₁, S and G₂ are collectively known as **interphase**.

■ Mitosis is normal cell division. One diploid cell divides to produce two genetically identical diploid cells.

■ Meiosis is a special type of cell division which produces sex cells or spores. It is also known as a 'reduction division'. One diploid cell will divide meiotically to produce four genetically different haploid cells.

■ **Cancer** occurs when the nuclear mechanisms that inhibit cell growth break down. Rapid growth of unspecialised cells results in a tumour.

QUESTIONS

1 The drawing of Fig 6.Q1 shows animal cells in different stages of mitosis

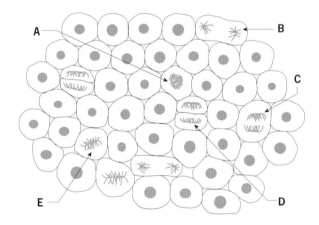

Fig 6.Q1

a) Using only the letters in the diagram, list the cells in the correct sequence, beginning with the cell in the earliest stage.

b) **(i)** Describe what is happening to the genetic material in cell **B**.
   (ii) How are the events you have described in **(i)** being brought about?

c) Explain why the cells produced by mitosis are genetically identical.

[AEB 1994 Human Biology: Paper 1, q.9]

2 The graph in Fig 6.Q2 shows the movement of chromosomes during mitosis. Curve **A** shows the mean distance between the centromeres of the chromosomes and the corresponding pole of the spindle.

a) What is represented by curve **B**?

b) **(i)** At what time did anaphase begin?
   (ii) Explain how **one** piece of evidence from the graph supports your answer.

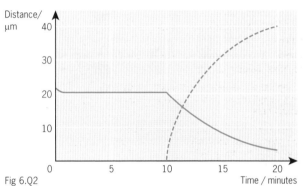

Fig 6.Q2

c) The number of cells in different stages of mitosis in a root tip were counted. The results are shown in the table.

Stage	Number of cells
Prophase	210
Metaphase	30
Anaphase	12
Telophase	48

What do these data tell you about the process of mitosis?

[AEB 1996 Biology: Specimen Paper, q.5]

3 The photographs in Fig 6.Q3 show cells from an aloe plant in different stages of mitosis.

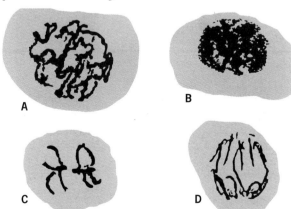

Fig 6.Q3 (photographs courtesy of the Royal Botanic Gardens, Kew)

a) Name the stage shown in each photograph.

b) Suggest **one** region of the plant in which these cells could be found.

c) In the photographs, the cell wall, cell membrane and spindle fibres are not visible.
    (i) Suggest why this is so.
    (ii) In the space below, make a drawing of the cell in photograph **C**. Include all of the structures visible in the photograph, and add to your drawing the cell wall, cell membrane and spindle fibres. Label your drawing fully.

[UCLES June 1994 Modular Biology: Foundation Module Paper, q.2]

4 Figures **A** and **B** in Fig 6.Q4 show two stages of meiosis in a diploid plant.

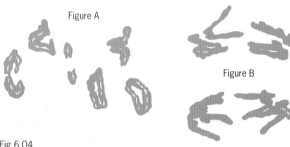

Figure A

Figure B

Fig 6.Q4

a) Give **one** piece of evidence from Figure **A** to support the fact that this cell is undergoing meiosis.

b) Giving **one** reason for your answer in each case, identify the stage of division shown in: **(i)** drawing **A**, **(ii)** drawing **B**.

c) What is the haploid chromosome number in this plant?

[AEB 1996 Biology: Specimen Paper, Section A, q.4]

5 Fig 6.Q5 shows chromosomes in a cell undergoing meiosis.

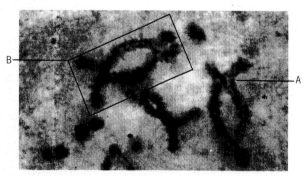

Fig 6.Q5

a) Name **(i)** the stage of meiosis shown in Fig 6.Q5; **(ii)** the process that has occurred at A.

b) Make a labelled drawing to interpret what has happened to the bivalent labelled **B**.

c) Explain how meiosis gives rise to variation.

[UCLES June 1994 Modular Biology: Applications Paper, q.1]

6 Fig 6.Q6 shows an animal cell in prophase of mitosis. The letters represent alleles of a gene that codes for eye colour, and a gene that codes for body colour. Allele **R** gives red eyes, and is dominant to allele **r**, which gives white eyes. Allele **B** gives black body, and is dominant to allele **b**, which gives grey body.

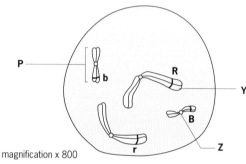

magnification x 800

Fig 6.Q6

a) Draw the diagram, and on it, **(i)** name structures **Y** and **Z**; **(ii)** shade with pencil **two** homologous chromosomes.

b) Calculate the actual length of chromosome **P**. Express your answer in micrometres (μm). Show your working.

c) What will be the genotypes of the cells produced as a result of mitosis of the cell in Fig 6.Q6?

d)   **(i)** If the cell in Fig 6.Q6 were to undergo **meiosis** to produce gametes, state the genotypes of the gametes formed.
    (ii) If gametes formed from several cells like the one in Fig 6.Q6 were to fuse randomly, state the phenotypic ratios expected in the resulting offspring.

[UCLES March 1995 Modular Biology: Foundation Module Paper, q.2]

ENERGETICS

T O SAY THAT energy is important is a bit of an under-statement: there can be no life without energy. Energy makes chemical reactions happen, it allows organisms to move, it lets them grow and reproduce themselves, it keeps them warm on a cold winter's night...

So, where does this energy come from? There are a few species of fairly weird bacteria that live deep in the oceans next to thermal vents and get their energy from chemical reactions powered by sulphur compounds and the heat of their surroundings. But virtually all other organisms on Earth get their energy from the Sun. This energy is harnessed by the process of photosynthesis which occurs only in green plants. All other forms of life get their energy second-hand, by eating plants or other animals that eat plants. So, ultimately, we all depend on photosynthesis for food. And, as oxygen is a by-product of photosynthesis, we are also in debt to plants for the oxygen we breathe.

In the early stages of the evolution of life on Earth, the atmosphere contained little or no oxygen. It was generated when photosynthesising prokaryotes appeared. This was a major event in evolution because it provided the conditions which selected for cells that could use oxygen in respiration. Aerobic respiration releases much more energy from a glucose molecule than anaerobic respiration. All multicellular organisms and many single-celled organisms alive today respire aerobically and, in fact, cannot live without oxygen.

In this section we first take an overview of energy and life and then go on to study the process of aerobic cell respiration in detail. Towards the end of Chapter 8, we look at anaerobic respiration. In the final chapter in the section we learn about the chemical reactions involved in photosynthesis and find out a bit more about those exceptional bacteria which have evolved to live without using energy from the Sun.

7 Energy and life

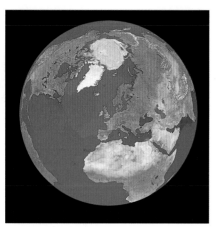

An understanding of energy is of fundamental importance in biology, from the study of individual molecules and reactions to global issues such as the use of fossil fuels, pollution, damage to the ozone layer and the 'greenhouse effect'

Introduction

All living organisms need energy. As you sit and read this book, you might not think that you are using up much energy. In fact, even at rest, your body needs a constant supply of energy. Think about some of the processes going on within your body:

- Your heart is beating.
- You are breathing.
- Food is being pushed along your intestine.
- Food is being absorbed from your intestine into your blood.
- Urine is being pushed from your kidneys to your bladder.
- Nerves are taking information from your sense organs to your brain, keeping you informed of changes in your environment.
- You are thinking; your brain is processing information coming in from the sense organs, sorting out what is important and what can be ignored.
- Unless you are sitting in a sauna, you are probably losing heat to your surroundings. As a warm-blooded organism, you need to replace any heat that you lose in order to maintain a constant internal body temperature.
- Some of the molecules from food you have digested, amino acids for example, are being used to make new cell components, new cells and new tissues.

All of these processes (and many more) require energy.

Fig 7.1 **Photosynthesis by blue-green bacteria was responsible for the appearance of oxygen in the Earth's atmosphere 2 billion years ago, and photosynthesis by plants maintains the oxygen level at 20 per cent of the atmosphere today**

1 SOLAR ENERGY, PHOTOSYNTHESIS AND RESPIRATION

This planet, and all the living things on it, must have energy and almost all of it is supplied by our nearest star, the Sun. A huge amount of sunlight reaches the surface of the Earth but only about 2 per cent of it is actually absorbed by plants. The rest of the energy is not lost; by heating up the land, air and water it prevents the planet from freezing.

In photosynthesis, plants exploit radiant energy from the Sun, using it to convert simple inorganic substances (mainly carbon dioxide and water) into larger organic molecules (glucose, starch, lipids and proteins). The plant tissues built from the products of photosynthesis form food for organisms which cannot make their own food (animals, fungi and most bacteria). Some of these become food for other organisms. Ultimately, therefore, the products of photosynthesis provide the energy which all living things need for growth, repair, movement and many other life processes. This energy is made available to living things through the process of **cell respiration**.

We will go on to look at cell respiration in detail in the next chapter (see pages 124 – 147) but, first, there are some key concepts that we need to understand before we can study the biochemistry of the process.

2 WHY DO LIVING THINGS NEED ENERGY?

Living things need a constant supply of energy to maintain their life processes. Let's look at the main processes.

Growth and repair of cells and tissues

Energy is required for the biochemical reactions which build large organic molecules from simpler ones. For example, energy is needed to build proteins from amino acids.

Active transport

Energy is required to move some substances in or out of cells. The transport of nitrate from the soil into the roots of a plant, and the transport of sugars and amino acids from an animal's small intestine into its blood are both examples of active transport.

Active transport often takes place against a diffusion gradient (see page 94) and it allows organisms to control their internal environment more efficiently.

Movement

Generally speaking, all organisms move, and movement requires energy. However, movement can be considered on several levels:
- within cells, eg chromosomes separating
- whole cells, eg white cells engulfing bacteria, sperm swimming (see Fig 7.2)
- tissues, eg muscles contracting
- whole organs, eg a heart beating
- parts or whole organisms, eg talking, walking

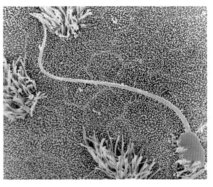

Fig 7.2 **Movement requires energy. The sperm cell shown here is on the ciliated epithelium of the uterus. To swim far enough up into the uterus to reach an egg, a sperm cell needs enough energy to enable it to swim over 7500 times its own length – the equivalent of a 10 kilometre swim for an adult human**

Temperature control

Warm-blooded animals, **endotherms**, use large amounts of heat energy to keep their body temperature constant. Humans use around 70 per cent of the energy from respiration to maintain the body at a constant 37 °C. Many smaller endotherms such as hummingbirds and shrews lose heat more rapidly because of their large ratio of surface area to volume (see also Chapter 17). Almost all of their energy of respiration is used to regulate body temperature.

Fig 7.3 **The shrew has a particularly high metabolic rate because it loses heat rapidly and must replace what is lost**

?

A Why do small animals have a problem with excessive heat loss? How do they compensate for it?

Making electricity and light

Some organisms are able to utilise the energy in food to produce light or electricity. Fig 7.4 shows a glow-worm producing light at night. Fig 7.5 shows an electric eel which uses rapid discharges of electricity to stun its prey.

Fig 7.4 **The glow-worm, really a type of beetle, is one of the few organisms that can use the energy in chemicals to produce light. The male, known as the firefly, uses light to signal to females. The wingless females – the glow-worms – reply in the same way**

Fig 7.5 **The electric eel can deliver enough electricity to kill other fish swimming nearby, or to stun anyone unlucky enough to step on it**

3 HOW DO LIVING THINGS GET THE ENERGY THEY NEED?

In order to respire and so release energy, organisms need a supply of food. Food can be obtained in two ways:

1 Some organisms can make their own food. Organisms that can do this are called **autotrophs** (meaning 'self-feeders'). They are able to use energy from the surroundings to make their food. There are two kinds of autotroph, the **photoautotroph** and the **chemoautotroph**. Photo-autotrophs, such as green plants, can photosynthesise, using sunlight to make their food. Chemo-autotrophs, all of which are bacteria, can synthesise their food using energy made available from chemical reactions other than those involved in photosynthesis. Bacteria responsible for some stages of nitrogen recycling are chemoautotrophs, see Chapter 37.

2 Some organisms need to obtain ready-made food because they cannot make their own. Such organisms are called **heterotrophs**. Heterotrophs must obtain their food, either directly or indirectly, from organisms which can synthesise food (usually plants). They either eat their food (ingesting it into a gut, as most animals do) or digest the food first and then absorb the nutrients, as decomposers, mainly fungi and bacteria, do. See Fig 7.6.

Nutrition is covered in more detail in Chapter 15.

Fig 7.7 **Sequoias are the largest trees on Earth. Apart from water, much of their weight is due to organic compounds such as cellulose, lignin and starch**

Fig 7.6 **All of these organisms have something in common – they are heterotrophs. They cannot make their own food and must take it, ready made, from an outside source. As you can see, there are many different ways of doing this**

?

D Look at Fig 7.7 and answer the following:

(a) Where did the plant get the carbon from to make the organic compounds which make up its enormous bulk?

(b) Where did it get the energy from to make the organic compounds?

(c) If the tree were burned, an enormous amount of energy would be released. Where would most of the energy go and which compounds would be produced?

4 ENERGY FLOW: WHERE DOES IT COME FROM, WHERE DOES IT GO?

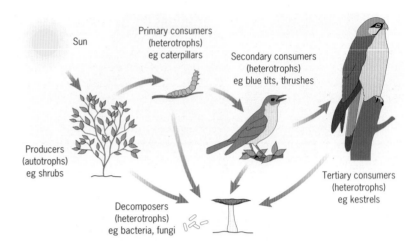

Fig 7.8 **The basic idea of energy flow. At the start of any food chain there must be an organism capable of making food – usually a green plant. From there, the energy is passed up the chain within food molecules. Some of this energy is eventually used to make the cells, tissues and organs of heterotrophic organisms**

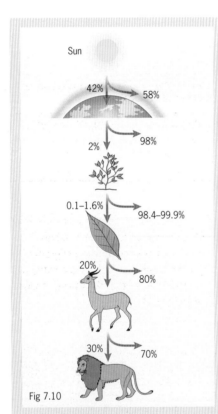

Fig 7.10

Of the light energy from the Sun that reaches the Earth, only about 42 per cent actually gets through to the surface of the planet. The rest is absorbed or reflected by the atmosphere, a process which warms the atmosphere.

Of the light energy reaching the Earth's surface, about 2 per cent is absorbed by plants. The rest goes to produce the heat which warms up the land and the oceans.

Of the energy absorbed by plants, only 0.1–1.6 per cent is ever incorporated into plant tissue – and even some of this is lost as heat because the plant itself must respire.

Of the plant tissue eaten by herbivores, only about 20 per cent (at most) will become incorporated into the tissues of the animal. Most of the rest is lost as heat as the animal respires.

Of the animal tissue eaten by a carnivore as meat, only a maximum of 30 per cent ever becomes incorporated into the body of the carnivore. Most of the rest is lost as heat.

Clearly, as so little energy is transferred at each point in a food chain, there is relatively little energy available for animals at the top, usually large carnivores. For this reason, such animals are quite rare.

To understand further how energy is lost, think of your own diet. You consume large amounts of food but very little of it becomes incorporated into your body. When you compare your own body weight to the weight of all the food you have ever eaten, you can see that a lot of energy must have produced heat as a result of respiration. This heat is used to keep your body temperature at a constant 37 °C. So much is needed because you are constantly losing heat to the air, your clothes and the objects you touch.

?

E When will you not lose heat to the surroundings?

5 ENERGY FROM FOOD – A HUMAN PERSPECTIVE

The recommended minimum daily intake of energy varies according to a person's age, sex, occupation and/or level of activity. Table 7.1 shows a variety of energy requirements for different people. These figures are the total requirement for all of an individual's metabolic processes for an average 24-hour period. The energy requirement must be provided by the food that they eat and is released for use by the process of cell respiration.

Table 7.1 **The approximate energy needs of different people**

Type of person and age	Energy requirements (kJ) per day
Newborn baby	2 000
Child 1 year	3 000
Child 2–3	6 000
Child 5–7	7 500
Girl 9–12	9 600
Boy 9–12	10 500
Girl 12–15	9 500
Boy 12–15	12 000
Girl 16–18	10 000
Boy 16–18	15 000
Adult male 18–35, moderately active	12 500
Adult female 18–35, moderately active	9 200
Adult male 18–35, very active	15 000
Adult female 18–35, very active	10 500
Pregnant woman	10 000
Breast-feeding woman	11 000
Tour de France cyclist (male)	30 000
Adult male 55–75	9 800
Adult female 55–75	8 600

If the total amount of food eaten every day contains more energy than the body needs, the excess is be stored as glycogen or lipid, as seen in Fig 7.10(a). If the food eaten contains less energy than is needed, the body makes up the difference by respiring these stored materials. When the lipid stores get low the body begins to respire an increasing amount of protein, and the actual fabric of the body starts to be used up, as illustrated by Fig 7.10(b).

F Look at Table 7.1.

(a) How do the figures given explain why people tend to have more of a weight problem as they get older?

(b) Compare the figures given for a moderately active adult woman of childbearing age and a pregnant woman. How much extra energy is needed in pregnancy?

(c) Note down the figure for the energy requirement of a breastfeeding woman and explain it in terms of two other figures from the table.

Table 7.2 **The approximate energy content of the three main food types**

Food type	Energy content
1 g of carbohydrate	17 kJ
1 g of lipid	38 kJ
1 g of protein	20 kJ

G Using the information in Tables 7.1 and 7.2, work out how many grams of **(a)** carbohydrate, **(b)** lipid is needed by an adult male to provide his daily energy requirement?

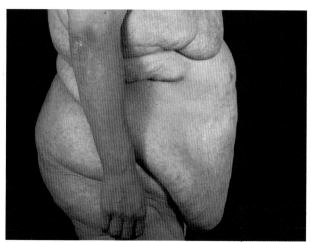

Fig 7.10 (a) **When the body takes in more energy from food than it needs, the excess is stored as fat under the skin and around the internal organs**

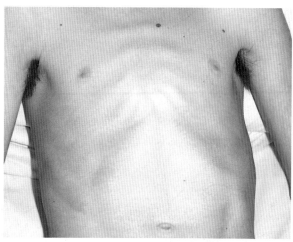

Fig 7.10 (b) **When the body does not get enough energy from food, it will be forced to respire stored fat and then protein. The very fabric of the body will start disappearing**

Of course, nutrition is not just about energy requirements. In theory, you could get all of your energy requirement by eating several chocolate bars every day. While you would avoid starvation, chocolate would not provide you with the correct balance of other substances your body needs to remain healthy – protein, vitamins, minerals and fibre.

Diet is covered in more detail in Chapter 15.

MEASURING THE ENERGY IN FOOD

THE SI UNIT of energy is the joule (J) but the amount of energy contained in food can be measured in either joules (J) or calories (cal).

1000 J = 1 kJ
1000 cal = 1 kcal

A calorie is an older unit of energy but is still in widespread use. A calorie is defined as the amount of energy needed to raise 1 cm³ of water by 1 °C.

1 calorie = 4.18 joules

Joules and calories are small amounts of energy so kilojoules (kJ) and kilocalories (kcal) are more frequently used when talking about human nutrition. In popular nutrition guides, the calories referred to are actually kilocalories and are often spelt Cal, with a capital C.

The energy value of food can be estimated by burning a known amount in oxygen and measuring the energy released as heat. This process is called calorimetry. The energy content information given on the packaging of many foods has been calculated in this way.

Fig 7.11 **Packaging for muesli showing energy given in kilojoules and kilocalories**

NUTRITION INFORMATION		
TYPICAL VALUES	Per 100g (3.5oz)	Per Serving 60g (2oz)
ENERGY	364 k.CALORIES	218 k.CALORIES
	1537 k.JOULES	921 k.JOULES
PROTEIN	11.8g	7.1g
CARBOHYDRATE of which	64.7g	38.8g
SUGARS	19.4g	11.6g
STARCH	45.3g	27.2g
FAT of which	6.4g	3.8g
SATURATES	0.7g	0.4g
MONO-UNSATURATES	4.3g	2.6g
POLYUNSATURATES	1.4g	0.8g
FIBRE	6.1g	3.7g
SODIUM	0.1g	less than 0.1g
Per 60g SERVING	218 CALORIES	3.8g FAT

INGREDIENTS
OATFLAKES, TOASTED WHEATFLAKES, VINE FRUITS, BROWN SUGAR, ROASTED HAZELNUTS, WHEY POWDER, MALT EXTRACT, SALT.

6 BASAL METABOLIC RATE (BMR)

The BMR is the rate of respiration/metabolism which the body needs to keep it 'ticking over' during periods of inactivity. For humans, BMR is defined as the rate of energy used by someone who has fasted for twelve hours and who is awake but resting, at a comfortable external temperature. The BMR is useful as it gives us a background rate against which we can compare the energy needs of activities such as swimming or walking. We can also measure and compare the BMRs of different organisms.

In warm-blooded animals the BMR is closely linked with the size of the organism and with the amount of heat that it loses. As a general rule, the smaller the animal, the higher the BMR. Chapter 8 gives more detail on the measurement of respiration and we cover thermoregulation in Chapter 25.

In practice, BMR can be measured by analysing the relative amounts of oxygen and carbon dioxide present in the exhaled air of a person under test. This allows calculation of the amount of energy that the person uses while at rest during a 24-hour period (see Fig 7.12).

Measurements of the BMR in many people of different ages has shown that BMR varies with age. In a one-year-old child, for example,

See questions 1 and 2 in Chapter 8.

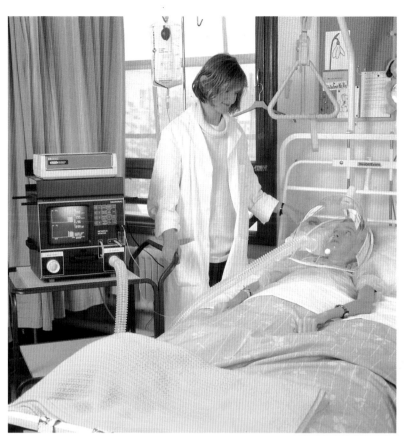

Fig 7.12 **Measuring metabolic rate in a young patient with bowel disease. The plastic canopy takes exhaled air to the monitor on the left. This measures the amount of oxygen and carbon dioxide present and calculates how much energy (in kilocalories per day) the patient is using**

Respiration or metabolism?

Metabolism is a general term which refers to the biochemical processes taking place in the body. Many of these processes require the energy which is provided by respiration, so metabolism and respiration are closely linked. In fact, the rate of metabolism is limited by the rate of respiration: the level of activity of an organism is limited by how fast its respiration processes can operate. Since most organisms require oxygen for respiration, we can get a good estimate the rate of metabolism by measuring oxygen uptake.

the BMR is higher at 220 kJ m^{-2} h^{-1}, than that of a young adult female at 150 kJ m^{-2} h^{-1} or a young adult male at 170 kJ m^{-2} h^{-1}. As we get older, our BMR tends to decrease but the values depend on individuals and their lifestyles. Taking plenty of exercise can significantly increase BMR.

See question 3 in Chapter 8.

BIOLOGY AND THERMODYNAMICS

THE FIRST LAW of thermodynamics states that:

Energy cannot be created or destroyed; it can only be transferred.

This law is relevant to biology because, as we have seen, all organisms obtain their energy by transferring it from the environment in different ways.

These are the most important energy transfers in biological systems:

- Energy from sunlight becomes energy in chemicals; this occurs in photosynthesis.
- Energy in food chemicals is used to: do mechanical work, eg to move muscles, produce heat, eg to maintain body temperature, build structures within cells and tissues.

THE SECOND LAW of thermodynamics states that:

When energy is transferred, some of the energy is always lost as heat.

This law means that no energy transfer will be 100 per cent efficient, and biological processes are no exception.

When we exercise, for example, the transfer of energy from chemicals to do mechanical work within the muscles is nowhere near 100 per cent efficient. Quite a lot of the energy is lost as heat, and this large amount of extra heat must be removed from the body. This explains why we go red and sweat during and after strenuous activity.

8 Cell respiration

IN HUMANS, as in all organisms, obtaining energy from food is a priority that cannot be ignored. Tragic scenes from parts of the world devastated by famine show the terrible consequences of malnutrition and starvation.

We obtain most of our energy by respiring the food that we have eaten recently. Carbohydrate is the primary source of energy, but we also respire small amounts of fat and protein. When the energy provided by recent meals runs out, as it does when we are asleep, for example, our bodies start to respire molecules stored in our tissues. These are replaced when we next eat, usually first thing in the morning when we *break* our *fast*.

During prolonged starvation, the body's stored carbohydrate runs out after about a day, and then the body must obtain its energy by respiring fat. An average 70 kg man will have about 10 to 15 kg of stored fat in his body. This is a large energy store, but one that is not readily available (fat can only be respired when carbohydrate is available). Increasingly the body is forced turn to protein, the very fabric of the cells, for the energy it desperately needs.

The body can withstand the loss of about half of its protein only. Complete starvation, where no food at all is available, leads to death within weeks. The exact cause of death varies. It can be an infection such as pneumonia (starvation weakens the immune system) or circulatory problems due to a much reduced blood volume.

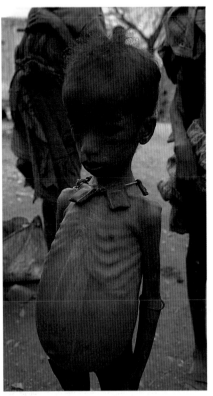

Children who are very short of food for long periods of time often have swollen abdomens. They are suffering from 'kwashiorkor'. This word means 'displaced child'. Their bodies swell because fluid builds up in the tissues instead of draining back into the blood, a problem caused by low levels of protein in the blood

1 CELL RESPIRATION: A VITAL LIFE PROCESS

Understanding cell respiration is a demanding task for any student. You will probably find it useful to read Chapter 7 first: this introductory chapter looks at key concepts that you will need to understand before you tackle the detailed biochemistry given in this chapter.

All living things obtain the energy they need from **respiration**, a chemical process which breaks down simple food molecules such as glucose. Animals, including humans, digest food to produce these molecules, which are absorbed into the blood and then transported round the body.

Multicellular organisms mainly respire **aerobically** (using oxygen). This is absorbed by lungs, gills or the body surface and is usually distributed around the body by the **circulatory system**. Only when food molecules and oxygen are together inside cells can the complex process of aerobic cell respiration begin.

Some important definitions

The terms 'respiration' and 'breathing' seem to be used interchangeably by many people, including scientists, and this can lead to confusion. As a basic guide, we use the following definitions.

> **Respiration is the complex series of reactions occurring in all living cells, which releases the energy in food and makes it available to the organism.**

To avoid confusion with breathing, it is often called tissue respiration or cell/cellular respiration. In this book we describe it as **cell respiration**.

Organisms that carry out aerobic cell respiration need to take in oxygen, and they produce carbon dioxide, a waste product that needs to be removed. Larger organisms have had to develop specialised organs with large surface areas to increase their capacity for **gas exchange** (Fig 8.1). (See also Chapter 17.)

> **Breathing is the mechanical process that supplies oxygen to the body to drive respiration and that removes the carbon dioxide produced.**

> **Aerobic respiration is respiration that requires oxygen.**

Most organisms respire aerobically: it releases a relatively large amount of energy and gives them a survival advantage over anaerobic organisms.

> **Anaerobic respiration is respiration without oxygen.**

Some organisms, mainly bacteria, can only respire anaerobically. A few more, yeast for example, can turn to anaerobic respiration when there is no oxygen. Some animal tissues (eg muscle during strenuous exercise) and plant tissues (eg plant roots in waterlogged soil) are able to respire anaerobically if circumstances demand it.

> ✔
>
> Although our muscles can work anaerobically for a short time, the brain must have a continuous supply of oxygen. A lack of oxygen for just 3 minutes is enough to cause brain damage.

> ?
>
> **A** Why is it important for organs involved in gas exchange to have a large surface area? What other features would you expect these organs to have?

Fig 8.1 **Living organisms have developed various organs to increase their ability to exchange gases. In mammals, this gas exchange occurs at the lung surface. In fish and the axolotl shown, it takes place at the gill membrane. In some animals, most worms, for example, it happens at the body surface. In plants it occurs mainly in the leaves**

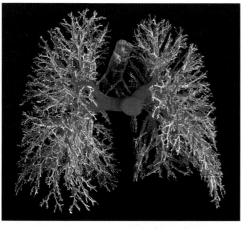

2 THE BIOCHEMISTRY OF AEROBIC GLUCOSE RESPIRATION

'Aerobic cell respiration' describes the cell process that require oxygen to release energy from all types of food molecules. In many organisms, cell respiration involves the breakdown of a variety of food molecules (see Fig 8.2). Humans respire mainly sugars with a small percentage of amino acids and fatty acids but, when the need arises, such as during starvation, the balance can change.

Fig 8.2 **Carnivores such as this brown bear obtain much of their energy from the respiration of amino acids derived from their high protein diet**

To make a very complex topic easier to study, we shall concentrate first on the aerobic respiration of only one type of food molecule: glucose. The aerobic respiration of other food molecules, and anaerobic respiration, are dealt with later (see pages 136 and 141).

For any chemical reaction to take place, energy is required to break existing molecular bonds. The process of forming new bonds may either require or release energy. For there to be a *net release of free energy* (the whole point of cell respiration), the products of cell respiration must be at a lower energy level than the reactants. Chemists say the products are more **thermodynamically stable** than the reactants.

The basic equation we use to sum up glucose respiration is:

$$C_6H_{12}O_6 + 6O_2 \rightarrow 6CO_2 + 6H_2O + ENERGY$$

The total energy contained in the reactants (the 6 molecules of glucose and the 6 molecules of oxygen on the left) is greater than the total energy contained in the products (the 6 molecules of carbon

?

B How is cell respiration similar to the working of a petrol engine?

dioxide and the 6 molecules of water on the right). The difference is the energy released by cell respiration.

In reality, glucose respiration is a sequence of many different reactions. Together, these can produce up to 36 molecules of a chemical called ATP per molecule of glucose. This large amount of energy can do useful work within the cell:

$$C_6H_{12}O_6 + 6O_2 \rightarrow 6CO_2 + 6H_2O + 36ATP$$

The steps involved in the production of ATP depend heavily on **redox reactions**. These are reactions that involve oxidation of one reactant and reduction of another. Before going on to look at the detailed biochemistry of glucose respiration, take a look at the box on redox reactions on page 129, to make sure that you are comfortable with this important concept.

Respiration or combustion?

It is often said that we 'burn up' food in respiration but this is not really true. The energy contained in glucose *could* be released by setting fire to it (Fig 8.3). The glucose would be oxidised to carbon dioxide and water, but the reaction would be rapid and uncontrolled.

During the process of cell respiration, glucose and other organic molecules are broken down in small stages, some of which release energy. This step-by-step breakdown (Fig 8.4) is more gradual than combustion. The whole process is controlled by enzymes which transfer energy in food molecules to ATP, a substance that is able to power cell processes by supplying on-the-spot, instant and usable energy in controlled amounts.

Fig 8.3 **The energy in food can be released by combustion, but almost all of it is lost as heat, raising the temperature of the surroundings to levels that could not be tolerated inside any living cell**

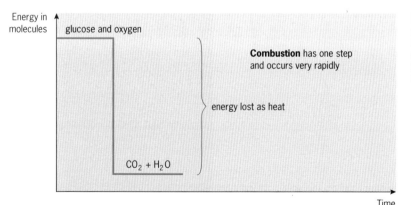

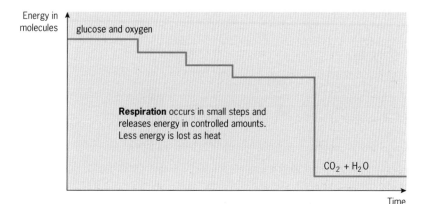

Fig 8.4 **The difference between combustion and respiration. Although the difference in energy between the reactants (glucose and oxygen) and the products (CO$_2$ and water) is the same in both cases, the way in which the energy is released is totally different**

All about ATP

ATP stands for **adenosine triphosphate**. ATP is a relatively small, soluble organic molecule which consists of a base (adenine), a sugar (ribose) and 3 inorganic phosphate (P_i) groups (Fig 8.5). ATP has a high **free energy of hydrolysis**, which means that when ATP is **hydrolysed** (Fig 8.6), a relatively large amount of energy is released. In hydrolysis, ATP reacts with water, producing ADP and inorganic phosphate (P_i). Because of its solubility and small size (relative molecular mass 507), ATP can be transported rapidly around cells and so can supply energy where it is needed. Metabolically active cells, such as those in muscle and in secretory tissue, transport and break down many ATP molecules.

Fig 8.5 **The structure of ATP. In the computer graphics image, adenosine is blue, pentose is white and the phosphate groups are red. This molecule is present in all cells and is used to provide energy. The purpose of cell respiration is to re-synthesise ATP at the same rate that it is used up**

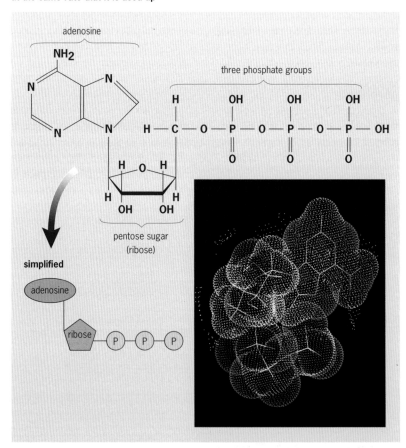

Fig 8.6 **The hydrolysis (splitting) of ATP. Under the control of the enzyme ATPase, the terminal phosphate group of ATP is removed and combined with water. This reaction releases free energy which can be used to drive energy-requiring reactions such as muscular contraction**

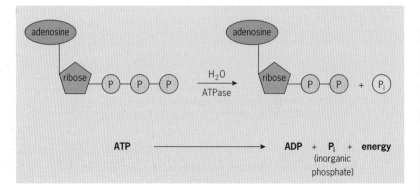

How does ATP release energy?

When ATP loses a phosphate group to become ADP, adenosine *di*phosphate, the reaction is **exergonic** (it *releases* energy). This can be used to do useful work within the cell. The same amount of energy is released when ADP loses another phosphate group to become AMP (adenosine *mono*phosphate). Less energy is released when the last phosphate group is lost.

Many of the reactions within cells are **endergonic**: they *require* energy. This energy is supplied by *coupling* them with reactions that involve ATP breakdown.

Endergonic reaction: amino acid A + amino acid B → dipeptide

Exergonic reaction: ATP → ADP + P$_i$ + energy

It is often said that ATP is a 'high energy molecule' or that is contains 'high energy phosphate bonds'. Neither statement is really true. Many molecules have more energy than ATP, and many others contain exactly the same phosphate-to-phosphate bonds. ATP is an important molecule in living systems because it can lose its terminal phosphate group *readily*, releasing just enough energy to power biological processes without producing excess heat.

ATP → ADP + P$_i$ + energy (30.6 kJ mol^{-1})

During strenuous exercise, when the muscles quickly use up all their ATP, energy is supplied by a back-up chemical called **creatine phosphate** (CP). Since CP has a higher free energy of hydrolysis than ATP, its breakdown releases enough energy to make more ATP instantly and so allows exercise to continue. See the Assignment, Energy and sport, at the end of this chapter.

Fig 8.7 **For explosive bursts of effort, such as a short sprint, most of the energy comes from ATP already present in the muscles**

?

C To move our muscles, we need energy to be available 'on demand'. Why do we use ATP as an immediate energy source, instead of glucose?

Redox reactions

Redox is short for **red**uction and **ox**idation, two chemical processes that often occur together in the same reaction.

Oxidation reactions may involve the addition of oxygen or the removal of hydrogen from molecules, but the important underlying rule is that a molecule that is oxidised *loses* electrons.

Conversely, reduction reactions involve a *gain* of electrons. As electrons carry energy, a molecule that has been reduced, and that therefore has gained one or more electrons, will usually carry more energy than the oxidised form of the same molecule.

Reduction and oxidation reactions usually occur together, because if one chemical loses electrons another must gain them (Fig 8.8).

The concept of redox reactions is a vital one in biology. The molecules that make up living things are produced, directly or indirectly, by photosynthesis. This process reduces carbon dioxide to form organic molecules (such as sugars) that contain energy. These can be later oxidised in the process of cell respiration to release energy. Lipids contain more hydrogen than carbohydrates: they have more C—H bonds than C—O bonds. For

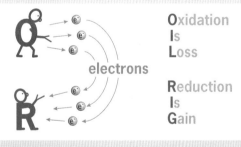

Fig 8.8 **Remember – OIL RIG: oxidation is loss, reduction is gain**

this reason, lipids are said to be more highly reduced chemicals than carbohydrates and more energy can be released by the oxidation of their many C—H bonds. Because they contain more energy per gram than other substances, lipids are ideal storage compounds (see Chapter 3).

Cell respiration is the release of energy from food by progressive oxidation.

In cell respiration, glucose is oxidised to form carbon dioxide and oxygen is reduced to form water. Most of the ATP created during this process comes from the final stage: a series of redox reactions that occur on the mitochondrial membrane.

The four main stages in glucose respiration

The complete process of aerobic respiration of glucose can be divided into four distinct processes:

- Glycolysis
- Pyruvate oxidation (the 'link' reaction)
- The Krebs cycle
- The electron transport chain of reactions

The way in which the four processes relate to each other is shown in Fig 8.9, and Table 8.1 gives the overall function of each stage.

Fig 8.9 **The overall process of cell respiration, showing the order of the four main stages**

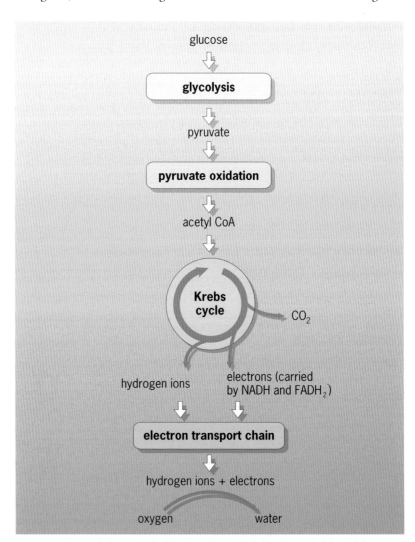

Table 8.1 The overall process of glucose breakdown. Stages 2, 3 and 4 can be thought of as the aerobic (oxygen requiring) reactions and take place in the mitochondria

Stage	Site within cell	Overall process	Number of ATP molecules produced*
1 Glycolysis	cytosol	glucose is split into 2 molecules of pyruvate	2 per glucose
2 Pyruvate oxidation	matrix (inner fluid) of mitochondria	pyruvate is converted into acetyl Co-A	none
3 Krebs cycle	matrix (inner fluid) of mitochondria	acetyl Co-A drives a cycle of reactions which produces hydrogen	2 per turn, so 4 per glucose
4 Electron transport chain	inner membrane of mitochondria	hydrogen drives a series of redox reactions which release enough energy to make ATP	up to 32 per glucose

* This table indicates that, for each molecule of glucose, 38 molecules of ATP are produced. The actual figure is nearer 36 molecules. See the Example on page 136.

Making ATP by phosphorylation

In the cell, ATP is made by adding a phosphate group to a molecule of ADP. This process is called phosphorylation and it is carried out in two different ways.

Substrate level phosphorylation is the process in which a phosphate group from a substrate molecule (a molecule other than ATP, ADP or AMP) is transferred to a molecule of ADP, giving a new molecule of ATP. This form of phosphorylation occurs in glycolysis and in the Krebs cycle.

In **oxidative phosphorylation**, ATP is synthesised using free phosphate groups. The energy required is obtained from a series of oxidation reactions. The process is complex and is intimately related to the internal structure of the mitochondrion.

Glycolysis

In this first stage a glucose molecule is converted, by a series of enzyme-controlled reactions, into 2 molecules of **pyruvate**, a 3-carbon compound. The process, shown in detail in Fig 8.10, yields relatively little energy (only 2 molecules of ATP) but this stage does not require oxygen and takes relatively little time to complete, so it can provide immediate energy.

?

D What is the difference between substrate level phosphorylation and oxidative phosphorylation?

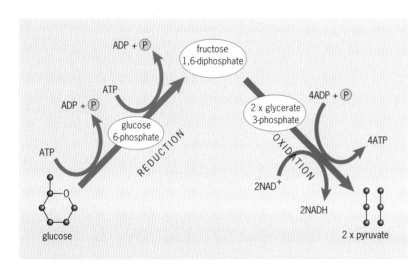

Fig 8.10 **Glycolysis is a two-stage process. The 'uphill' part involves raising glucose to a higher energy level by using ATP. In the 'downhill' part, the products are oxidised, yielding 2 molecules of pyruvate, 2 molecules of reduced coenzyme and a net gain in ATP**

Glycolysis takes place in the **cytosol**, the fluid part of the cytoplasm. The overall process consists of ten different reactions, each catalysed by a specific enzyme.

The first half of glycolysis actually uses ATP to raise the energy level of the original glucose molecules by adding phosphate groups to form **fructose 1,6-diphosphate**. This must be done if glucose is to have enough energy to complete the second half of the process.

Fructose 1,6-diphosphate is then split into 2 molecules of a 3-carbon compound, **glyceraldehyde 3-phosphate**. These are subsequently oxidised to pyruvate by a series of five reactions which also produce 4 molecules of ATP and 2 molecules of reduced coenzyme (NADH).

So, overall, per glucose molecule, glycolysis produces:

● 2 molecules of ATP (4 are created but 2 are used up in glycolysis)

● 2 molecules of NADH (reduced coenzyme which later feeds electrons into the electron transport chain)

● 2 molecules of pyruvate (which enters the 'link' reaction if oxygen is available)

Pyruvate oxidation: the 'link' reaction

This reaction links glycolysis with the Krebs cycle. It is often thought of as part of the Krebs cycle but we shall consider it to be a separate reaction.

In the presence of oxygen, pyruvate (the 3C molecule produced by glycolysis) moves from the cytosol to the mitochondrial matrix where it is oxidised into **acetate** (a 2C molecule), producing carbon dioxide as a by-product. This reaction also produces 2 molecules of NADH. The acetate is picked up by a carrier molecule, **coenzyme A**, and **acetyl coenzyme A** is formed.

Overall, the reaction is:

$$2 \text{ pyruvate} + 2\text{NAD}^+ + 2\text{H}_2\text{O} \rightarrow 2 \text{ acetate} + 2\text{NADH} + 2\text{H}^+ + 2\text{H}_2\text{O}$$

> **E** What does a reduced coenzyme have that an oxidised coenzyme does not?

Pyruvate or pyruvic acid?

These terms are used interchangeably but they refer to the same substance. Fig 8.11 shows the difference: pyruvic acid in solution becomes ionised and, like all acids in solution, it loses a hydrogen ion. In the ionised state it is known as pyruvate. The same applies to other organic acids such as lactic acid and acetic acid, which become lactate and acetate respectively.

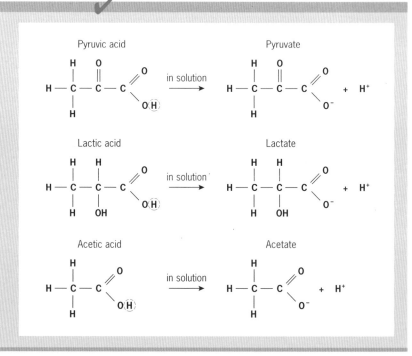

Fig 8.11 **Pyruvate, lactate and acetate and their corresponding acids**

The Krebs cycle

The **Krebs cycle** is a series of reactions named after Sir Hans Krebs, the biochemist who first worked out the details. It is also known as the **citric acid cycle**, the **tricarboxylic acid cycle** or simply the **TCA cycle**. The main purpose of the Krebs cycle is to provide a continuous supply of electrons to feed into the electron transport chain.

The details of the Krebs cycle, which takes place in the matrix of the mitochondria, can be seen in Fig 8.12. Overall, per turn, the Krebs cycle produces:

See questions 4 and 5.

> **F** In an animal, what happens to the CO_2 made by the reactions in the Krebs cycle?

- 3 molecules of NADH
- 1 molecule of FADH$_2$
- 1 molecule of ATP (by substrate level phosphorylation)
- 2 molecules of CO_2
- 1 molecule of oxaloacetate (to allow the cycle to continue)

The Krebs cycle will 'turn' twice for every glucose molecule that enters it, so, per glucose molecule, 6 molecules of NADH and 2 molecules of FADH$_2$ are produced. NADH and FADH$_2$ are particularly important because they carry the electrons which power the next stage of glucose respiration.

> **G** Why does the Krebs cycle turn twice for every glucose molecule?

Fig 8.12 **The process of cell respiration.** Most A-level students do not learn all the intermediate stages, but it is a good idea to look at the biochemical pathways involved to understand several important points about cell respiration:

● This is an example of a metabolic pathway. It looks complex, but, like a street map, it is easy to follow once you know where you are and where you would like to go. Each individual reaction is a relatively simple step.

● Each stage is catalysed by a specific enzyme.

● You can see how the respiration of different foods (glucose, fatty acids and amino acids) is closely interconnected. The central pathway, in black, shows the complete breakdown of a molecule of glucose. The path taken by fatty acids is shown in green and the path taken by amino acids is shown in red.

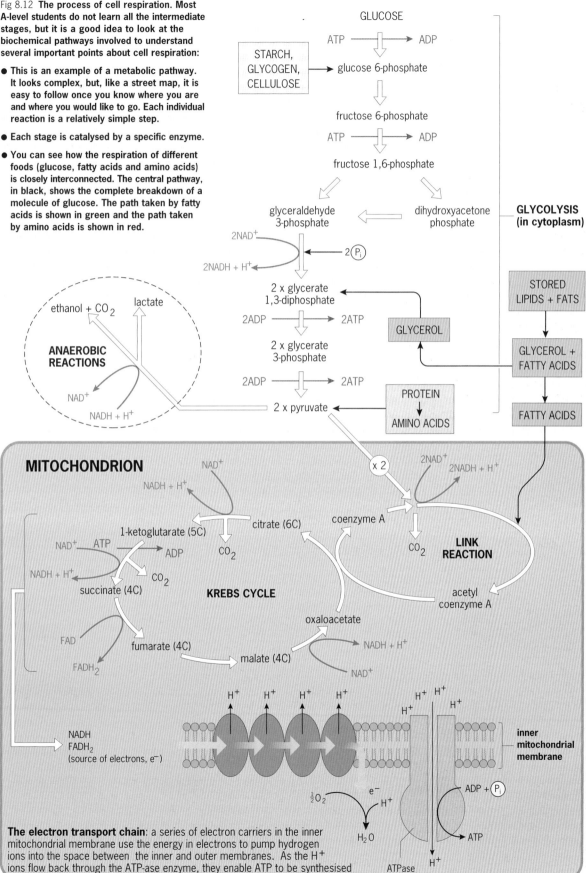

NAD and FAD

NAD (**nicotinamide adenine dinucleotide**) and FAD (**flavine adenine dinucleotide**) are **coenzymes**, organic compounds that are catalysts for reactions. As their name implies, coenzymes work closely with enzymes. Unlike enzymes, coenzymes are not proteins. NAD and FAD carry electrons from the **electron donors** in glycolysis, pyruvate oxidation and the Krebs cycle to **electron acceptors** in the electron transport chain in the mitochondrial membrane.

When NAD and FAD accept electrons, they are reduced, becoming NADH and $FADH_2$. In this form they are called **reduced coenzymes**.

Reduction of a coenzyme always requires association with a **dehydrogenase enzyme**, an enzyme that removes hydrogen from other molecules. The hydrogen removed by the dehydrogenase enzyme is split into an electron and a hydrogen ion. NAD (or FAD) accepts the electrons produced and the hydrogen ions play an important part in the electron transport chain.

In the human body, NAD is synthesised from vitamin B_3 (nicotinic acid) and FAD is made from vitamin B_2 (riboflavin): this explains why both vitamins are essential components of our diet.

The electron transport chain

This final phase of respiration produces the largest number of ATP molecules. The electron transport chain can be thought of as the 'pay day' for all the 'hard work' that has been done in the earlier parts of the process.

As Fig 8.13 shows, the electron transport chain consists of a series of carrier molecules, which will first accept an electron (thereby becoming reduced) and then lose it again (so becoming oxidised). At each of these transfers the electrons lose some energy, which can be used to power the active transport of hydrogen ions across the inner mitochondrial membrane (see Fig 8.14). This results in a high concentration of hydrogen ions in the outer mitochondrial space and a low concentration in the inner mitochondrial space.

Because of the difference in concentration, hydrogen ions leak back into the inner compartment. The only route that they have is through the middle of the stalked granules – ATPase enzymes. As the stream of hydrogen ions flows down the concentration gradient, enough energy is released to allow free inorganic phosphate molecules to be added to ADP, forming new molecules of ATP.

?

H Muscle cells and cells that manufacture and secrete large amounts of hormones have more mitochondria than, say, a skin cell. Can you say why?

I The function of a kidney tubule cell is to move substances by active transport. Suggest why these cells are packed with mitochondria?

Fig 8.13
The electrons and hydrogen ions made during the first stages of respiration are finally used to synthesise ATP

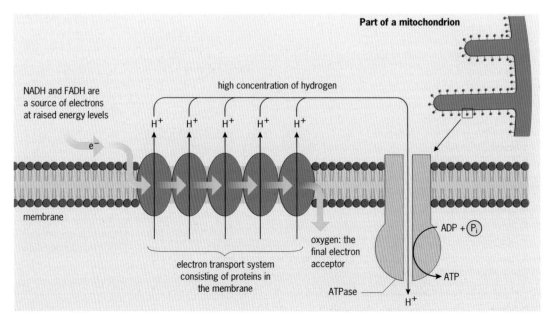

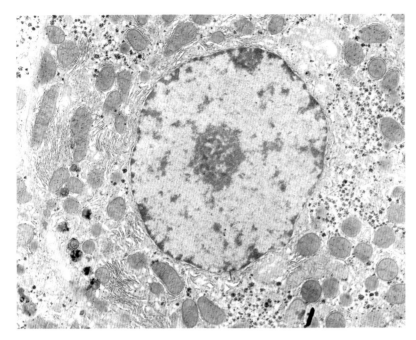

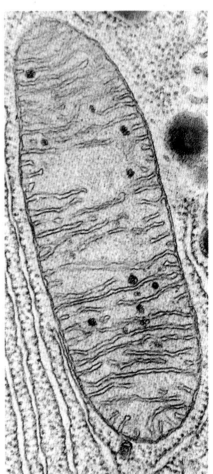

Fig 8.14 **Mitochondria are the organelles that house the enzymes, substrates and other chemicals associated with aerobic respiration. Pyruvate oxidation and the Krebs cycle take place in the matrix (fluid) centre of the organelle, while the electron transport chain takes place in the inner membrane itself**

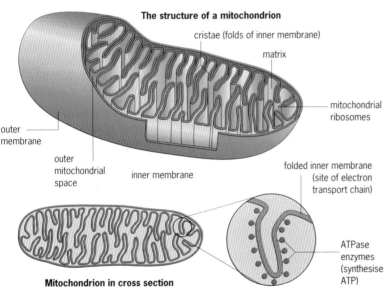

The structure of a mitochondrion

cristae (folds of inner membrane)

matrix

mitochondrial ribosomes

outer membrane

outer mitochondrial space

inner membrane

folded inner membrane (site of electron transport chain)

ATPase enzymes (synthesise ATP)

Mitochondrion in cross section

This model of ATP synthesis by **oxidative phosphorylation** is called the **chemiosmotic hypothesis** and was first proposed by Peter Mitchell in 1961. Mitchell received the Nobel Prize for Chemistry in 1978.

The end-product of the electron transport chain is spare electrons and hydrogen ions which combine with oxygen to form water. Although this reaction is at the end of the process, it is a key one: it is why most organisms need oxygen. If there is no oxygen to mop up the electrons and hydrogen ions from the electron transport chain, the pathway cannot be completed. This is a disaster for the cell as the intermediate compounds of glucose respiration build up and the cell loses its ability to make most of the ATP it needs.

Looking at the separate processes involved in the breakdown of one molecule of glucose makes it easy to lose track of the overall purpose of glucose respiration: ATP production. The following example gives you the chance to analyse the contribution to ATP production made by each of the different stages.

EXAMPLE

Q How many ATP molecules are produced, per molecule of glucose, in aerobic cell respiration?

A ATP can be made in one of two ways, by **substrate level phosphorylation** and by **oxidative phosphorylation** (see page 131). We shall take each process in turn.

Substrate level phosphorylation

Glycolysis and the Krebs cycle produce ATP in this way. We can complete the following table in this way:

Stage	ATP produced
Glycolysis	2 (2 used but 4 made)
Pyruvate oxidation	0
The Krebs cycle	2 (1 per turn)
Total	4

Oxidative phosphorylation

This is the production of ATP by the electron transport chain. It relies on a continuous supply of 'high energy' electrons from the preceding processes. Before we can calculate the final ATP total, we need to calculate the number of electrons being provided by NADH and FADH$_2$.

We can draw up a second table, as follows. Remember, pyruvate oxidation and the Krebs cycle occur twice for every glucose molecule put into the system.

Stage	NADH produced	FADH$_2$ produced
Glycolysis	2	0
Pyruvate oxidation	2	0
The Krebs cycle	6 (3 per turn)	2 (1 per turn)
Total per glucose	10	2

We know that;
> 3 ATP molecules are produced for every molecule of NADH fed into the chain,

and that;
> 2 ATP molecules are produced for every molecule of FADH$_2$ fed into the chain.

We can now estimate the final ATP total:

10 NAD molecules produce 3 ATPs, each giving a total of	30
2 FADH molecules produce 2 ATPs, each giving a total of	4

Therefore, the total number of ATP molecules produced by oxidative phosphorylation is **34**

The total number of ATP molecules produced by substrate level phosphorylation is **4**

The grand total per glucose molecule is therefore 38 molecules of ATP,
most of which are produced by the electron transport chain.

In practice, the figure is usually 36 molecules of ATP. Because the mitochondrial membrane is impermeable to NADH, the 2 molecules of NADH made during glycolysis cannot carry their electrons directly into the electron transport chain. To get over this, most cells have a shuttle system in which the electrons are released by the NADH and passed across the mitochondrial membrane where they are picked up by FADH$_2$. As we saw earlier, FADH$_2$ only produces 2 ATPs compared with the 3 made by NADH, thus reducing the final total by 2.

This is the ATP total under *ideal* conditions. Several other factors can also reduce ATP production even further, but these are beyond the scope of this book

3 OTHER FUELS IN RESPIRATION

So far, we have studied the respiration of glucose, but many organic molecules can be fed in the same central pathway to create ATP. Look back at Fig 8.12 and then read on to see how different substrates can be fed into the central pathway.

Lipid breakdown

The fat stores in the body can be used when carbohydrate is in short supply. Stored triglycerides are broken down into **glycerol** and **fatty acids**. The latter are further broken down to give several 2-carbon **acetyl fragments** by a process called **beta-oxidation**. These fragments are combined with coenzyme A and so enter the Krebs cycle and, eventually, the electron transport chain. Glycerol is converted into **dihydroxyacetone phosphate** so that it can be used as a fuel in glycolysis.

✔ Lipids can fuel aerobic exercise but not glycolysis. Aerobic exercise is therefore a good way to burn excess fat (see Assignment on page 146).

Protein breakdown

Adult humans need up to 60 g of protein per day. We cannot store protein so excess **amino acids**, the building blocks of proteins, are degraded in the liver by the process of **deamination**. Enzymes in the liver separate the amine group from the rest of the molecule, leaving an **organic acid**. The amine group is released as free **ammonia**, which enters the **ornithine cycle** (see Fig 25.15) to be incorporated into **urea** and excreted from the body in the urine. The organic acids are fed into the Krebs cycle and are respired (see Fig 8.12).

THE ORIGINS OF RESPIRATION: SOME SPECULATION

EVEN THE SIMPLEST life form needs to release the energy locked in organic molecules and it is thought that glycolysis was one of the first metabolic pathways to evolve.

Since there was very little free oxygen at the time, aerobic respiration was impossible. It seems reasonable to assume that the process of glycolysis, as practised by early, bacteria-like organisms (Fig 8.15), used up much of the available food, giving the organisms a major problem: that of energy supply. Solving this problem proved to be a massive evolutionary step: the development of photosynthesis. Not only did photosynthesis provide a new supply, but it also released free oxygen as a by-product. The oxygen in the atmosphere is due to photosynthesis.

The build-up of oxygen in the atmosphere kick-started another evolutionary leap: the appearance of organisms that could respire aerobically. This probably happened fairly quickly because of the massive survival advantage gained by organisms that respire aerobically to produce the much larger amounts of ATP.

It is at this stage that prokaryotic organisms, which could respire aerobically, may have developed a close association with large cells, so forming primitive eukaryotic cells. Find out more about the **endosymbiont theory**, on page xxx.

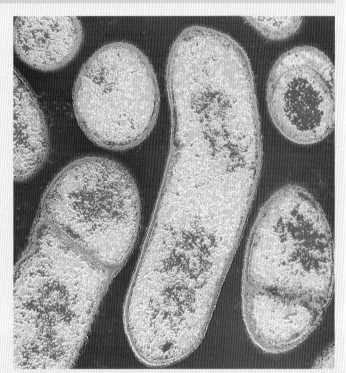

Fig 8.15 *Clostridium botulinum*, the anaerobic bacterium that causes botulism (serious food poisoning), is found in food such as meat and fish that have been badly canned. Anaerobic bacteria like these are believed to have been the only organisms that could have survived in the oxygen-poor environment 3500 million years ago

4 THE RESPIRATORY QUOTIENT (RQ)

The respiratory quotient, or RQ, is a ratio of gas exchange, worked out by comparing oxygen uptake with carbon dioxide production. The RQ can be used to determine the type of food being respired. The basic equation for the respiration of glucose:

$$C_6H_{12}O_6 + 6O_2 \rightarrow 6CO_2 + 6H_2O + energy$$

shows that, for every glucose molecule respired, 6 molecules of oxygen are required and 6 molecules of CO_2 are produced. Equal numbers of gas molecules occupy the same volume and so the volume of oxygen taken up by an organism is equal to the volume of CO_2 produced.

The RQ is worked out using the simple formula:

$$RQ = \frac{\text{volume of } CO_2 \text{ given out}}{\text{volume of } O_2 \text{ taken in}}$$

For the glucose equation, the RQ is: $6 \div 6 = 1$.

Using the respiratory quotient

If we measure the volumes of oxygen and carbon dioxide that are exchanged by an organism and find that the RQ is 1, we can infer that the organism is respiring mainly carbohydrate. Different substrates will produce different RQ values:

- Carbohydrate respiration gives an RQ of 1.
- Protein respiration gives an RQ of 0.8–0.9.
- Lipid respiration gives an RQ of 0.7.
- Fermentation gives an RQ of infinity.

The problem with trying to work out what substrate is being respired by looking at RQ values is that many organisms, including humans, respire more than one substrate. In humans, carbohydrate is the main fuel for respiration, but a small amount of protein is also respired. The RQ figure obtained is therefore variable.

Using a respirometer

The rate of aerobic respiration can be estimated by measuring how much oxygen is taken up or how much heat is produced. Gas exchange, including oxygen uptake, is measured using a sealed chamber called a **respirometer**.

A simple respirometer

A simple respirometer is shown in Fig 8.16. Organisms placed inside exchange gases and so alter the composition of the air around them.

To measure gas exchange due to a respiring organism, we could remove the air after a fixed time and analyse it. A simpler method is to place a carbon dioxide absorber such as sodium hydroxide in the respirometer chamber. Over a set time, a volume of oxygen is used up. Carbon dioxide is produced but is absorbed by the sodium hydroxide. So, overall, the volume of gas falls by the volume of oxygen used. The gas pressure inside the chamber falls and coloured liquid is drawn along the tube towards the chamber. The volume of oxygen used up equals the volume of the liquid moved in the fixed time. The following Example shows how the respiration rate is calculated from readings taken from the tube.

?

J (a) Why is there a range of values for the respiratory quotient of proteins?

(b) Why is the respiratory quotient for fermentation infinity?

Fig 8.16 **A simple respirometer**

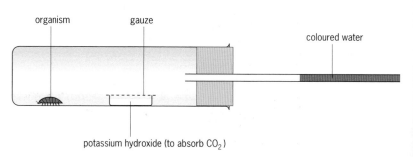

organism   gauze   coloured water

potassium hydroxide (to absorb CO_2)

EXAMPLE

Q

a) When measuring the gas exchange of an organism with a simple respirometer, how do you calculate the volume of oxygen used?

b) In what units do you express the rate of respiration?

A

a) The position of the liquid along the tube is recorded at the start and end of the set time and the distance the liquid moves is calculated. The volume of liquid (equal to the change in oxygen volume) is a cylinder inside the tube.

Therefore:

volume of oxygen used up (in cm³) = $\pi \times r^2 \times h$

where r = radius or diameter/2 (in cm) and h = distance moved by the liquid (in cm).

b) The rate of respiration is given as the volume of oxygen used per organism (or per unit weight, e.g. per gram) per unit time; for example, as cubic centimetres of oxygen per gram per hour. So:

rate of respiration = x cm³ O$_2$ g^{-1} h^{-1}

A more complex respirometer

The simple respirometer shows the principles of respirometry, but has these limitations:

- Changes in temperature or atmospheric pressure make gases expand or contract. These changes affect the distance the fluid moves, so the measured distance is not just due to the change in oxygen volume.

- The chemicals used might alter the composition of the gases in the chamber.

- It is difficult to restart the experiment without taking the apparatus apart.

- The volume change calculated using the diameter of the tube may be inaccurate.

To avoid these problems, a more sophisticated respirometer has been developed. This is shown in Fig 8.17.

This apparatus works on the same principle as the simple respirometer, but the syringe allows the volume change to be measured directly. After a set time, the syringe can be pulled up until the dye goes back to its starting point. This is a neat way to find the exact volume of oxygen used without having to work out the volume of a cylinder, and it allows the experimenter to move the experiment back to the start at the same time.

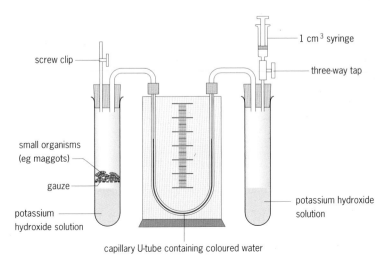

- 1 cm³ syringe
- three-way tap
- screw clip
- small organisms (eg maggots)
- gauze
- potassium hydroxide solution
- potassium hydroxide solution
- capillary U-tube containing coloured water

Fig 8.17 **A more complex respirometer. The left tube contains the organisms while the right tube acts as a control. In this way, any changes in fluid level which are not due to the organisms, such as changes in room temperature and pressure, are balanced out by the control tube. The experiment can be set up again using the syringe, which will also accurately measure the changes in volume due to the respiring organisms**

EXAMPLES
Using the respirometer

Q The respirometer in Fig 8.17 was set up with 50 g of blowfly (bluebottle) larvae and left for 1 hour. The fluid moved up the left side-arm of the U-tube by 5 cm. It was moved back to its original level by drawing 2 cm³ of air into the syringe.

a) What is the purpose of the syringe and why does it give this respirometer an advantage over the simple respirometer?

b) If there were 10 larvae, what is the respiratory rate per organism?

A
a) The syringe allows movement in the tube to be measured directly as a volume. This removes the need to calculate the volume of the cylinder to find out how much liquid is in the tube.

b) The volume of oxygen used up = 2 cm³. Per larva:

$$\text{respiratory rate} = 2/10 = 0.2 \text{ cm}^3 \text{ O}_2 \text{ h}^{-1}$$

Calculating the respiratory quotient

Q When the experiment was repeated using water instead of KOH, the fluid moved up 1 cm, not 5 cm.

a) What does this tell us about the volume of carbon dioxide produced?

b) What is the respiratory quotient for the larvae?

A
a) Carbon dioxide was not being absorbed, and its volume is one unit less than the oxygen volume. So:

$$CO_2 \text{ out} = 4 \text{ units}$$
$$O_2 \text{ in} = 5 \text{ units}$$

b)
$$\frac{CO_2 \text{ out}}{O_2 \text{ in}} = \frac{4}{5} = 0.8$$

Blowfly diet

Q
a) Assuming the larvae are only respiring one substrate, suggest which one.

b) From what you know about the diet of blow fly larvae, does this seem reasonable?

A
a) Proteins

b) Yes: they consume a large amount of protein in rotting meat.

Respirometer design

Q How does the addition of the tube on the right (known as a **thermobar**) help to overcome the problem of changes in atmospheric pressure and temperature?

A Any pressure changes in the tube on the left are mirrored and cancelled out by those in the right tube.

CYANIDE!

CYANIDE IS A FAMOUS poison, often featured in crime fiction and films (Fig 8.18). It is a very effective toxin because it inhibits one of the enzymes of the electron transport chain. This stops the flow of electrons and effectively prevents aerobic respiration, the source of most of the body's ATP. Without ATP, the muscles go into spasm and the body experiences convulsions. Spasm of the muscles that control breathing usually leads to death by asphyxiation (lack of oxygen to the brain).

Fig 8.18 'A faint smell of bitter almonds and the victim lay dead, the back arched, the eyes staring.'

5 RESPIRATION, ENZYMES AND TEMPERATURE

Like all metabolic reactions, respiration is controlled by enzymes. The Q_{10} law (page 70) states that the rate of an enzyme-catalysed reaction will approximately double with every 10 °C rise in temperature between 4 and 40 °C. So, with a temperature increase of 10 °C, much more ATP becomes available to power other metabolic processes.

A dramatic example is seen in **ectothermic** animals such as reptiles, which may have wide variations in body temperature during a 24-hour period (Fig 8.19). Their rate of respiration, and therefore their metabolic rate, is roughly twice as fast at 30 °C than it is at 20 °C. Such organisms are only active when their body temperature allows them to react and move quickly.

6 ANAEROBIC RESPIRATION

Anaerobic respiration does not require oxygen, so it can take place in oxygen-poor environments such as stagnant water or deep soil. It can even occur in parts of an aerobic organism that are starved of oxygen, such as muscles during strenuous exercise.

In anaerobic respiration, glucose is broken down into pyruvate but, because no oxygen is available, the pyruvate cannot be broken down any further. Instead of entering the link reaction and the Krebs cycle, the pyruvate is converted to a waste product. The nature of the waste product depends on the organism. Fig 8.20 shows the waste products of anaerobic respiration in different organisms.

Generally speaking, animals and some bacteria convert pyruvate into **lactate** by a simple reduction reaction. In contrast, plants and fungi such as yeast convert pyruvate to **ethanol** (alcohol) and carbon dioxide.

In animals, vigorous exercise leads to anaerobic respiration in the muscles and the resulting build-up of lactate causes fatigue: see the Assignment on page 146.

K Why do some plants produce cyanide in their leaves?

L If the rate of respiration of a lizard is 400 arbitrary units at 40 °C, what will the rate be at 10 °C?

Fig 8.19 **Reptiles are** ectotherms**: they cannot control their body temperature in the way in which mammals and birds do. This lizard must absorb sunlight in order to increase its body temperature to the level which allows its enzymes to function at their optimum efficiency**

See question 6.

Fig 8.20 **The main problem with anaerobic respiration is the build-up of NADH and the shortage of NAD+. To solve this, H+ is added to pyruvate to make lactate or ethanol and carbon dioxide.** Fermentation **is glycolysis plus the reduction of pyruvate to either lactate or ethanol and carbon dioxide**

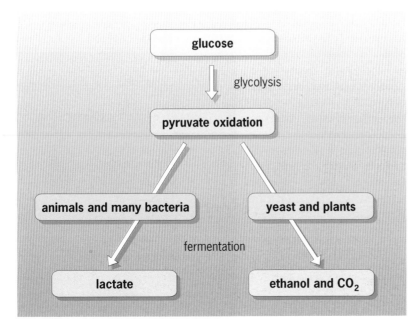

Fig 8.21 **Fermentation in different microorganisms can be used to make a variety of useful products**

Left: A vat of curds and whey in cheese making

Below: During fermentation, yeast produces CO_2 which makes bread dough rise

Above: Making red wine: yeast produces alcohol from the sugar in grapes

Right: Alcohol made from the fermentation of sugar cane is used as fuel in countries of South America

Anaerobic respiration in microorganisms is generally known as **fermentation**. As Fig 8.21 shows, this process can give rise to many useful products. Bacteria that produce lactic acid are used in the manufacture of dairy products such as yoghurt. The tangy taste is due to the high concentration of this organic acid. In addition to lactic acid and ethanol, other microorganisms can make solvents such as acetone and butanol.

Alcoholic fermentation

Anaerobic respiration in yeast (Fig 8.22) is also known as **alcoholic fermentation**. This process was discovered by accident and alcoholic drinks such as wine and beer were made for many centuries before the science behind the process was known. For fermentation, yeast simply needs a source of carbohydrate, anaerobic conditions and a suitable temperature. So wine can be made from many

Fig 8.22 **Yeast is a single-celled fungus that usually respires aerobically. When deprived of oxygen, it switches to anaerobic respiration: sugars are not completely broken down and the by-products are ethanol (alcohol) and carbon dioxide**

Fig 8.23 **Yeast occurs naturally on the surface of many fruits. For instance, yeast is partly responsible for the bloom on a grape. Fruit stored in anaerobic conditions will therefore ferment and produce liquids with 'interesting' effects!**

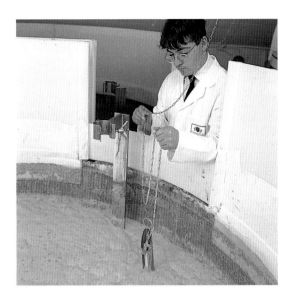

Fig 8.24 **Today, beer is brewed under controlled conditions using carefully selected strains of yeast. But the underlying process of fermentation, one of the oldest metabolic pathways known, is the same today as it was thousands of years ago**

organic materials, such as peaches, elderberries or even potato peelings, as well as grapes (Fig 8.23).

Fermentation stops when the yeast becomes poisoned by its own waste – the alcohol – and this is why most alcoholic drinks made by fermentation only are no stronger than about 14 per cent ABV (alcohol by volume). Spirits such as whisky and gin, which have a much higher alcohol content, are **distilled** from a fermented mixture.

Since alcohol is an intermediate product in the breakdown of sugar, it contains energy that can be released by respiration. Many drinks, particularly beer (Fig 8.24), are also rich in carbohydrates that have not been turned to alcohol. Heavy drinkers become overweight because they drink large amounts as well as eating meals that, by themselves, provide enough calories to satisfy the body's energy demands.

Anaerobes and disease

Many common bacteria can only thrive in anaerobic conditions. Two such bacteria are ***Clostridium tetani*** and ***Clostridium welchii***.

C. tetani usually thrives deep in soil, where there is little oxygen, but is present in small numbers almost everywhere. The bacteria can enter the human body through a deep cut or wound. Once inside, they multiply and produce a toxin that makes muscles go into spasm (Fig 8.25). This is **tetanus**, or lockjaw (named because of its effect on the jaw muscles), which can be fatal if it affects the muscles used for breathing.

C. welchii can cause **gas gangrene**. If it infects areas of dead tissue, such as frost-bitten toes or injured limbs that have a poor blood supply, it turns the tissues black (Fig 8.26). In such 'ideal' conditions, the bacteria multiply rapidly, producing a foul-smelling gas and several lethal toxins. Antibiotics help if the condition is treated early. If not, the affected limb may need to be amputated to prevent the toxins spreading further into the body.

As *C. tetani* and *C. welchii* can *only* respire anaerobically, they are known as **obligate anaerobes**. Organisms such as yeast can respire by either pathway and are known as **facultative anaerobes**. Organisms that thrive only when they are respiring aerobically are called **obligate aerobes**.

? M What is the difference between fermentation and glycolysis?

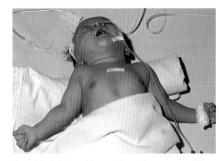

Fig 8.25 A **baby with muscle stiffness caused by tetanus. The toxins produced by the anaerobic bacterium, *Clostridium tetani*, are causing the baby's muscles to go into spasm**

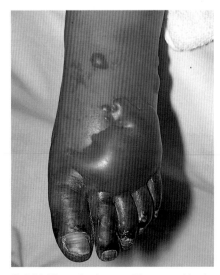

Fig 8.26 **Tissue death, caused by a poor blood supply, can lead to infection by the anaerobic bacterium *Clostridium welchii*. This organism causes gas gangrene and the toxins produced can be lethal if not treated quickly**

SUMMARY

By the end of this chapter, you should know and understand the following:

■ Cell respiration is the release of energy from food. The energy released is transferred to ATP, a chemical that can provide the cell with energy in small, instant, controllable amounts.

■ Cell respiration consists of many reactions. Each reaction is controlled by a different enzyme. Most of the reactions occur inside the mitochondrion.

■ Complete aerobic respiration of glucose has four stages: glycolysis, pyruvate oxidation, the reactions of the Krebs cycle and the reactions of the electron transport chain.

■ Glycolysis: Glucose is broken down into 2 molecules of pyruvate. Only 2 molecules of ATP are made. This happens in the cytosol of cells.

■ Pyruvate oxidation (the 'link' reaction): In the presence of oxygen, pyruvate enters the mitochondrion and is converted into acetyl coenzyme A.

■ The Krebs cycle: A series of reactions, fuelled by acetyl coenzyme A, which produces only 2 molecules of ATP but many electrons and hydrogen ions.

■ The electron transport chain: A series of redox reactions, fuelled by the electrons and protons from the preceding 3 processes, as supplied by the coenzymes NAD and FAD. Electron transport releases enough energy to synthesise most of the ATP produced by aerobic respiration.

■ Most food chemicals – carbohydrates, lipids and proteins – can be respired to provide energy.

■ The rate of respiration can be measured as a respiratory quotient, the ratio of CO_2 evolved to O_2 absorbed. It can be calculated from measurements made using a respirometer.

■ Most organisms respire aerobically (in the presence of oxygen). The substrate (eg glucose) is broken down completely to produce carbon dioxide, water and a lot of ATP.

■ Some organisms and some tissues respire anaerobically. In the absence of oxygen, substrates are only partially broken down, leading to waste products such as lactate (in animals and bacteria) or ethanol and carbon dioxide (in plants and yeast).

QUESTIONS

1 The diagram below shows a calorimeter used to determine the energy content of a sample of food.

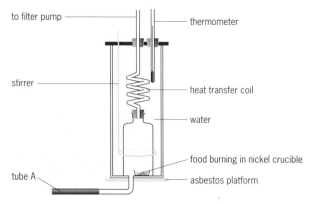

a) State the function of each of the following parts of the apparatus: **(i)** the filter pump, **(ii)** the heat transfer coil, **(iii)** the stirrer.

b) **(i)** Name the gas which enters the apparatus through tube A.
   (ii) What is the function of this gas?

c) Suggest *two* sources of error which may arise when using this apparatus to determine the energy content of food.

[ULEAC 1996 Biology/Human Biology: Specimen Paper 4, q.4]

2

a) Define the term *Basal Metabolic Rate*.

b) Assuming a low Physical Activity Level (PAL = 1.4) at work and leisure, the average requirements for energy for adults are given in the table below.

Age (years)	Average energy requirement/MJ day^{-1}	
	Men	Women
19–49	10.60	8.10
50–59	10.60	8.00
60–64	9.93	7.99
65–74	9.71	7.96
75+	8.77	7.61

   (i) What **two** general conclusions can you draw from this table?

   (ii) Average Energy Requirement (AER) can be calculated from the Basal Metabolic Rate (BMR) and the Physical Activity Level (PAL) using the formula:
   $$AER = BMR \times PAL$$
   Using data given in the table, calculate the BMR for a male adult under 60 years of age with a PAL of 1.4.

c) Sugars, complex carbohydrates, proteins and lipids are all eaten by adults in the UK. Although each type of food provides energy, they may be eaten in the wrong proportions. Suggest a proportion which each of these foods should occupy in a rational balanced diet for the average person. Give reasons for your answers.

[AEB 1996 Biology: Specimen Paper, q.12]

3 Write an essay on the conversion of energy from one form to another in living animals.

[AEB 1996 Biology: Specimen Paper, q.5]

4 The diagram below shows some of the stages in cell respiration.

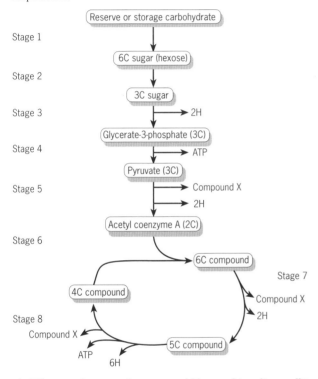

a) What respiratory substrate would be used in a liver cell?
b) State in which part of a cell Stage 6 occurs.
c) Identify compound X, removed at stages 5, 7 and 8.
d) Describe what happens to the hydrogen atoms removed at stages 3, 5, 7 and 8.

[ULEAC 1996 Biology/Human Biology: Specimen Paper, q.3]

5 The diagram summarises the process of cellular respiration.

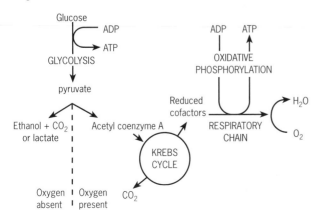

a) **(i)** Why is glycolysis sometimes referred to as 'the common pathway'?
    (ii) Where in the cell does glycolysis take place?
b) What is the significance of the Krebs cycle being a cyclical process?
c) **(i)** What is the importance of oxidative phosphorylation?
    (ii) Where exactly in the cell does it take place?

[AEB 1996 Biology: Specimen Paper, q.3]

6 The diagram below shows the apparatus used to compare the carbon dioxide production of different strains of yeast.

The yeast population to be investigated is suspended in sucrose solution in tube A. Nitrogen gas is bubbled through the apparatus during the experiments to ensure that respiration of yeast is anaerobic.

Tube C contains hydrogencarbonate indicator solution through which carbon dioxide has been bubbled. This allows the colour in the tube to develop and this tube is then used as a standard. Hydrogencarbonate indicator is red when neutral, purple in alkaline and yellow in acid conditions.

The time taken for the colour to develop in tube B to match the colour in tube C is recorded.

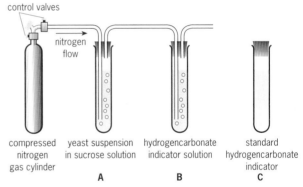

a) **(i)** State how yeast suspension in tube A can be maintained at a constant temperature during the experiment.
    (ii) State the colour change which would occur in the hydrogencarbonate indicator solution in tube B during the course of the experiment.
    (iii) Suggest why matching the colour by eye might not be a reliable method of determining the end-point.

b) **(i)** Name *one* product, other than carbon dioxide, of the reaction occurring in tube A.
    (ii) State *two* changes that occur in tube A as the reaction proceeds.

c) Describe, giving experimental details, how you would use the apparatus to compare the carbon dioxide production in two strains of yeast.

[ULEAC 1996 Biology: Specimen Paper q.6]

Assignment

EXERCISE PHYSIOLOGY, ENERGY AND SPORT

Muscle movement requires ATP. For prolonged exercise, muscles need a continuous supply of ATP. So, to a great extent, the stamina of athletes depends on their ability to supply their muscles with sufficient oxygen to make ATP by aerobic respiration.

In this chapter, you have studied aerobic and anaerobic respiration.

1 What is the essential difference between aerobic and anaerobic systems?

2 How many molecules of ATP are produced, per molecule of glucose:
a) from anaerobic respiration (glycolysis)?
b) from complete aerobic respiration (glycolysis, pyruvate oxidation, Krebs cycle and ET chain)?

For athletes, the aerobic pathway seems to be the most important, but we must also take into account the time taken to complete the process.

Complete aerobic respiration takes between 50 and 60 seconds, but glycolysis takes only about 10 seconds to produce ATP. Glycolysis is therefore very useful to an athlete because it is fast and requires no oxygen, and still allows the muscles to work when the body cannot supply them with enough oxygen for complete aerobic respiration.

If glycolysis takes at least 10 seconds to provide ATP, how is it that we can move instantly? The answer is that there is already some ATP stored in the muscles. When we begin to move, we use this store of ATP. During strenuous exertion, such as when we lift a heavy weight, we use this store of ATP up in about 3 seconds. That leaves 7 seconds, which we get through thanks to a back-up system, a chemical called **creatine phosphate** (CP). CP has an even higher free energy of hydrolysis than ATP and so can be broken down to allow instant resynthesis of ATP. This allows muscular contraction to continue for another 6 to 7 seconds.

3 Why do you think that the muscles have enough ATP/CP for about 10 seconds' worth of exercise?

4 Name the chemical produced as a consequence of anaerobic respiration. Explain its effect on the muscles.

Generally, the duration of a sport or event determines the main source of ATP used. The ATP/CP system is important for events that last less than 10 seconds, the anaerobic system (glycolysis) for events between 10 and 60 seconds and the aerobic system (complete aerobic respiration) for anything longer.

5 Copy and fill in the table below, identifying the energy system that provides most of the ATP for that particular activity.

Sport/event	ATP/CP	Anaerobic	Aerobic
javelin			
200 m sprint			
800 m run			
shot put			
marathon			
100 m sprint			
400 m sprint			
5000 m run			
soccer			

Fig 8.A1 **Sprinting is an anaerobic activity: athletes get the energy to sprint from the ATP/CP and glycolytic systems. Note that anaerobic specialists tend to be heavily muscled**

Fig 8.A2 **Prolonged exercise relies on the aerobic system to provide a continual supply of ATP. Generally, fatigue sets in only when fuels such as glycogen begin to run out (see page 44 in Chapter 3 for information about glycogen). Endurance athletes like this marathon runner tend to be of slight build, with small muscles**

Sports such as soccer, hockey and tennis use a mixture of all systems because the activity is prolonged but not constant. There are periods of intense activity such as a short sprint followed by periods of relative rest.

6

a) Why is the 400 m (world record under 40 seconds) thought by athletes to be a particularly painful event?

b) What use would the information in the completed table (on page 146) be to someone who wanted to improve his or her fitness for a particular event?

c) Why do you think sprinters usually have large muscles but marathon runners are generally quite thin?

d) A footballer wishes to get fit for the new season, so he jogs 5 miles each day. Explain why this would not be a complete preparation and suggest the additional training he would need to do.

e) Cats in the wild hunt their prey by pouncing or by short sprints. Hunting dogs rely on stamina in a long chase. Which energy system are cats and dogs using? Explain why dogs need more regular exercise than cats.

Aerobics and weight loss

Many people want to improve their overall fitness, but can exercising help them to lose weight as well? To understand more about aerobics and weight loss, we need to think about the type of fuel being burned in respiration. The way we use respiratory fuels has been likened to the way we use money.

We all need spending power, and the quickest and most universally accepted form of payment is cash. When we have no cash, we can go to the cashpoint or bank and get some more out instantly. When our bank account is low, we have to turn to our savings. This may take a while to get at, but it provides a useful cushion. When we are completely broke, we will have to sell our belongings and property just to survive. Eventually, with no injection of cash, we can be declared bankrupt.

7 Rewrite the last paragraph, applying the analogy to human energy sources, using the terms lipid, protein, glucose, glycogen, food, dead, energy. When you have done this, either write down or discuss in a group the limitations of this analogy.

It takes time to run down the glycogen store and start to use fats. This is a shame because otherwise you could simply go for a run and return slimmer, if a little baggy.

In practice, it takes several sessions – and often a reduction in calorific intake – to bring about an improvement. Generally, exercise is very beneficial because it strengthens the heart and circulatory system and raises the metabolic rate, so you use up more calories all the time, not just during exercise.

8 Imagine you are a GP faced with a rather fat couple in their forties. Write down or discuss how you would try to persuade them to take exercise. What exercises would you recommend and what precautions should they take?

Fig 8.A3 **Wild dogs chasing wildebeest**

9 Photosynthesis

Living stromatolites in Hamelin Pool, north-west Australia

IN HAMELIN POOL, a small bay on the north-west coast of Australia, the sea water is trapped every time the tide goes out. The weather is very hot and dry and the Sun evaporates the water in the bay. The salt concentration rises to a high level and this kills most living things: blue-green bacteria are the only organisms which manage to grow with any success. These photosynthesising prokaryotes bunch together to form mats which trap sand every time the tide comes in. Sediment builds up and a solid structure called a stromatolite grows, layer by layer.

When you slice through a stromatolite, you see a characteristic pattern which is found in fossilised stromatolites over 3.5 billion years old from Africa and Australia. Plant biologists now think that some of the very earliest photosynthesising organisms that lived on Earth were much the same as the blue-green bacteria found in Hamelin Pool today.

1 THE OVERALL PROCESS OF PHOTOSYNTHESIS

Most plants make their own food by photosynthesis. They use light from the Sun as an energy source to convert two colourless gases (carbon dioxide and hydrogen) and a colourless liquid (water) into sugars. The plant either respires these simple sugars to gain energy or uses them as raw materials to build more complex molecules such as starches and cellulose.

A useful summary of the overall process of photosynthesis is:

$$\text{carbon dioxide} + \text{water} \xrightarrow{\text{light}} \text{sugar} + \text{oxygen}$$

Since the sugar produced is usually glucose the equation is often written:

$$6CO_2 + 6H_2O \longrightarrow C_6H_{12}O_6 + 6O_2$$

In this chapter we look at the biochemistry of photosynthesis. Before going into the details, a good look at the summary in Fig 9.1 should help you see the process as a whole.

This chapter ends with a brief overview of **autotrophic nutrition**. As well as considering **photoautotrophs,** which are not green plants, we also investigate some of the ways in which **chemoautotrophs** make their own food using energy from chemical reactions.

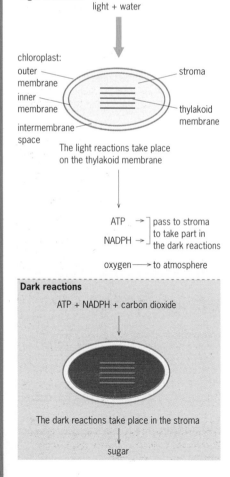

Light reactions

light + water

chloroplast:
outer membrane
inner membrane
intermembrane space
stroma
thylakoid membrane

The light reactions take place on the thylakoid membrane

ATP → pass to stroma to take part in the dark reactions
NADPH →

oxygen ⟶ to atmosphere

Dark reactions

ATP + NADPH + carbon dioxide

The dark reactions take place in the stroma

sugar

Fig 9.1 **Photosynthesis can be divided into two main stages. In the light reaction, ATP and NADPH are generated (for more details, see page 151). In the dark reaction, ATP and NADPH are used to make sugars (for more details, see page 155)**

Energy flow in photosynthesis

Let's start with an analogy. Water molecules in the sea remain together because of the electrostatic attraction between them. This is overcome when the molecules gain energy from the heat of the Sun and escape into the air (Fig 9.2(a)). They travel high into the atmosphere, come together again as water, and fall as rain. On the ground, the water drains down into streams and rivers and returns to the sea. If a dam is placed across a river, the potential energy of the water can be harnessed to generate electricity.

Fig 9.2 (a) **Energy flow in part of the water cycle. Water falling behind a dam has potential energy. As it falls down through the dam, its potential energy is harnessed to do useful work.** (b) **Energy transfer in photosynthesis. In a similar way, an electron becomes a 'high energy' electron when energy from light is transferred to it. As the electron returns to its ground state, it transfers this energy, which finally drives photosynthesis**

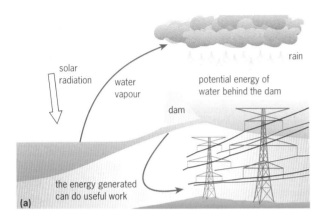

(a)

Fig 9.2(b) shows that, like water molecules, electrons can gain energy. Electrons circling in orbits around the atomic nucleus are held by the electrical attraction between their negative charge and the positive charge of the nucleus. If energy is supplied, an electron can gain enough energy to move further away from the nucleus. Later on, if it falls back towards the nucleus, the energy it gained may be given out again, sometimes as visible light.

But what if it were possible to add an imaginary 'dam' to trap the energy of the electron as it falls back to its original position. This is what plants do during photosynthesis. Light excites an electron out of its orbit and as the electron falls back, its energy is taken and used to make food. One of the key concepts in biology is:

Photosynthesis is the source of energy for virtually all living things.

The site of photosynthesis in the plant

Photosynthesis occurs in the green tissues of plants. These are green because they contain the green pigment, **chlorophyll**. The leaves, which have evolved to become efficient traps for light energy, are where most of the sugar is produced. Other parts of the plant above the ground can be green: some plants use their stems as their main site of photosynthesis as they have reduced leaves or no leaves at all (Fig 9.3). Roots are never green and never photosynthesise.

The site of photosynthesis in the cell

The chemical reactions of photosynthesis take place in **chloroplasts**. Chloroplasts are **plastids** (see page 22), plant cell organelles between 3 and 6 μm in diameter. The structure of chloroplasts was covered briefly in Chapter 1, but we will now look at it in more detail, to see exactly where the reactions of photosynthesis take place.

Fig 9.3 **In gorse, the main stems are covered with thorns. The leaves are reduced to spines. Photosynthesis occurs in the stems and the spines which are all green**

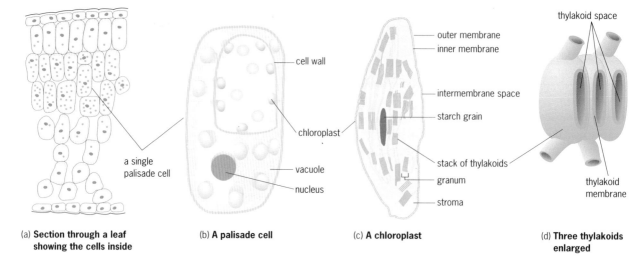

(a) **Section through a leaf showing the cells inside**

(b) **A palisade cell**

(c) **A chloroplast**

(d) **Three thylakoids enlarged**

Fig 9.4 **These diagrams show the site of photosynthesis within the leaf of a plant. (a) A palisade cell is shown in a leaf section. (b) The enlarged cell is cut away to show its chloroplasts. (c) The enlarged chloroplast shows the many thylakoids it contains. The thylakoid membrane carries the mechanism that provides energy for photosynthesis. (d) In a typical electron microscope section, the thylakoids appear to be separate but, as shown, they are actually connected. A cluster of stacked thylakoids is called a granum (plural grana)**

As Fig 9.4(c) shows, the chloroplast has two membranes which surround a space, the **stroma**. The stroma is rather like the mitochondrial space and contains various enzymes, ribosomes, RNA and DNA. In mitochondria there are folds on the inner membrane (cristae) where the reactions of the electron transport chain take place (see page 134). In chloroplasts the inner membrane connects with a much more elaborate arrangement, the **thylakoids**. These look like hollow discs and are stacked in groups or **grana**. The space inside a thylakoid is connected with every other thylakoid, forming a continuous internal compartment, the **thylakoid space**.

Most of the key reactions in photosynthesis occur on or across the thylakoid membrane, which is between the stroma and the thylakoid space and contains the pigments of photosynthesis including chlorophyll.

The pigments involved in photosynthesis

If you grind up a leaf in a solvent and then pass the mixture through a filter, you end up with a coloured solution. This contains a variety of pigments which can be separated by chromatography.

Sunlight (white light) contains all the wavelengths of the light spectrum that we see as colours in a rainbow. Most pigments absorb light of only certain wavelengths. Other wavelengths are reflected. Green plants are green because the chlorophyll they contain reflects green light. Leaves contain two very similar forms of chlorophyll, **chlorophyll A** and **chlorophyll B**. Some also have **accessory pigments** such as **carotene**. Carotene can trap energy from wavelengths of light which chlorophyll cannot, and so it helps the plant to make more efficient use of sunlight.

It is possible to measure how much of each wavelength is absorbed by a particular pigment. The graph this produces is called an **absorption spectrum** (Fig 9.5). Each pigment has a unique absorption spectrum which is a sort of 'fingerprint'.

Measuring the photosynthetic activity of a

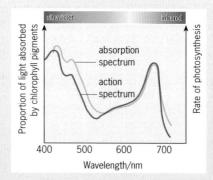

Fig 9.5 **Chlorophyll pigments in a green leaf are found to have an action spectrum for photosynthesis which is very similar to their absorption spectrum**

plant under light of different wavelengths gives a graph called an **action spectrum**. By taking a pigment and comparing its action spectrum of photosynthesis with its absorption spectrum, you can see whether or not the pigment takes part in the overall process.

The dark red pigment **anthocyanin**, sometimes found outside the chloroplasts in the cytosol, has an absorption spectrum which is completely different from the action spectrum of photosynthesis, showing that anthocyanin does not capture light for photosynthesis.

2 THE BIOCHEMISTRY OF PHOTOSYNTHESIS

See question 1.

Photosynthesis has two main stages:

- **Photosynthetic electron transfer**, also known as the 'light reactions'. Light energy becomes trapped in chlorophyll molecules. This energy is used to make ATP and NADPH, and oxygen is released.

- **Carbon fixation reactions**, also known as the 'dark reactions' or the 'light-independent reactions'. ATP and NADPH are used to convert CO_2 into carbohydrate.

Photosynthetic electron-transfer – the 'light reactions'

In the **light reactions** of photosynthesis, energy from sunlight excites electrons of chlorophyll molecules, as shown in Fig 9.2(b). The electrons lose their 'excitement' when their energy is passed on to enzymes. These enzymes make ATP and NADPH, used in the 'dark reaction' of photosynthesis to convert carbon dioxide to sugar.

The light reactions occur in the thylakoid membrane system of the chloroplast. Two chlorophyll molecules at the heart of the **antenna complex** shown in Fig 9.6 collect the energy from all of the light which falls on the antenna.

This energy is passed on to a **photochemical reaction centre**. Together, the antenna complex and the photochemical reaction centre form a system called a **photosystem.** Two types of photosystem, PSI and PSII, work together in photosynthesis.

Fig 9.6 **Each antenna contains an array of chlorophyll molecules. This increases the area over which light energy can be trapped. The antenna acts like a funnel. Energy is transferred from the outer chlorophyll molecules to a special pair of chlorophyll molecules near the centre of the antenna. Some antennae contain carotenoids which collect light of wavelengths outside the range for chlorophyll. Carotenoids hand their energy on to the chlorophyll**

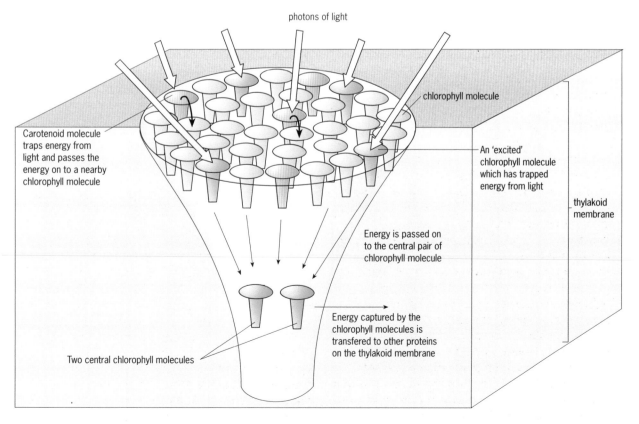

photons of light

chlorophyll molecule

Carotenoid molecule traps energy from light and passes the energy on to a nearby chlorophyll molecule

An 'excited' chlorophyll molecule which has trapped energy from light

thylakoid membrane

Energy is passed on to the central pair of chlorophyll molecule

Energy captured by the chlorophyll molecules is transfered to other proteins on the thylakoid membrane

Two central chlorophyll molecules

ANTENNA COMPLEX

The central pair of chlorophyll molecules in the antenna complex captures light, and an electron is excited. It moves immediately to the **photochemical reaction centre**, a series of proteins in the thylakoid membrane which act as electron carriers. Each protein accepts an electron and then hands it on to the next protein in the sequence.

When a molecule gains an electron it is **reduced**; when it loses an electron it is **oxidised**. So, since the electron moves from one electron carrier to the next, we describe the series of reactions as reduction and oxidation (**redox**) reactions (see page 129). In the light-dependent stage of photosynthesis, these reactions eventually lead to the production of ATP and NADPH – both needed later by the 'dark reactions'.

Let's follow the path of electrons which leave the antenna complex of the first photosystem in photosynthesis. For historical reasons, this is unhelpfully called photosystem II, or PSII.

Synthesis of ATP and NADPH by non-cyclic photophosphorylation

The arrangement of **electron carrier proteins** in the thylakoid membrane of the chloroplast is shown in Fig 9.7. You will need to refer to this diagram as you read on, so that you can follow the path of the electrons.

(a) Next to the antenna complex at PSII is a special enzyme that splits water. As each electron 'escapes' from chlorophyll to the photochemical reaction centre, another electron must take its place.

Fig 9.7 **The arrangement of the thylakoid membrane proteins which are involved in photosynthesis. From left to right, you can follow the sequence of events in the light-dependent phase of photosynthesis**

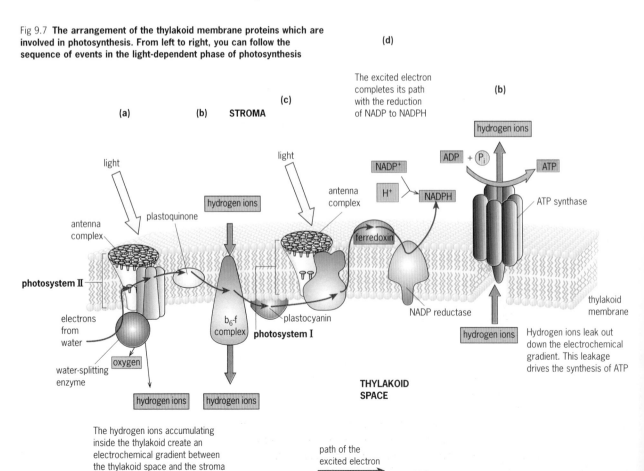

The replacement electrons come from water, removed by the water-splitting enzyme. When four electrons have been replaced in this way, hydrogen ions, H^+ (also called protons) are formed; also, a molecule of oxygen, O_2, is released:

$$2H_2O \xrightarrow{\text{energy of 4 photons}} 4H^+ + 4e^- + O_2$$

b) Meanwhile, the escaped electrons pass to the first protein in the chain, **plastoquinone**. They are then handed on to the **b_6-f complex** which is a proton pump, meaning it moves hydrogen ions from the stroma into the thylakoid space. The accumulating hydrogen ions make the fluid inside the thylakoid space more acidic than the stroma, creating an *electrochemical gradient* across the thylakoid membrane. This gradient represents a store of energy. The hydrogen ions can move down the gradient and leak out of the thylakoid space, but only by going through the enzyme **ATP synthase** (on the right in Fig 9.7). As they do so, they drive the phosphorylation reaction in which inorganic phosphate is added to ADP to give ATP:

$$ADP + P_i \rightarrow ATP$$

(c) While ATP is being produced, the electrons pass from the b6-f complex to the next electron carrier, a protein called **plastocyanin.** From here they pass to photosystem I (PSI). PSI operates in a similar way to PSII: the chlorophyll molecules at the heart of the antenna complex trap light, releasing excited electrons to a photochemical reaction centre.

The difference between PSI and PSII is the energy level of the electrons in their chlorophyll molecules. The electrons which pass from PSII to PSI still have some energy left over from their first boost in PSII. So, when light energy is trapped by the antenna complex in PSI, the electrons coming into the chlorophyll molecules are excited to a level higher than they reached in PSII.

(d) In fact, the electrons that leave PSI have been boosted to energy levels needed for them to pass to **ferredoxin** and then to the enzyme **NADP reductase**. This enzyme uses the energy from the electron to convert $NADP^+$ to NADPH.

$$NADP^+ + 2H^+ + 2e^- \rightarrow NADPH + H^+$$

Hydrogen ions pass out of the stroma, and so increase the electrochemical gradient across the thylakoid membrane.

This whole process – steps (a) to (d) – is called **non-cyclic photophosphorylation:**

- *phosphorylation* because ADP is phosphorylated to ATP,
- *photo*phosphorylation because the process is driven by light,
- *non-cyclic* photophosphorylation because the electrons lost by PSII do not return to PSII. Instead, they pass to PSI and so are not 'cycled'.

Overall, PSII produces ATP, and PSI produces NADPH. The electrons to produce NADPH come first from PSII, so NADPH and ATP production are linked. Just over one molecule of ATP is produced for every molecule of NADPH. Both chemicals are needed for the dark reactions of photosynthesis. But the dark reactions actually need 3 molecules of ATP for every 2 molecules of NADPH, so non-cyclic phosphorylation does not provide enough ATP. The rest comes from the process of **cyclic phosphorylation**.

Synthesis of ATP without NADPH by cyclic photophosphorylation

Electrons can leave the PSI chloroplast by another route. They are not used to make NADPH but instead pass back to the b_6-f complex via ferredoxin and plastoquinone (Fig 9.8). From here they go back to PSI and, in doing so, they provide energy to pump hydrogen ions from the stroma to the thylakoid space. This sequence produces ATP indirectly by increasing the electrochemical gradient which drives ATP synthase.

The end of the light reactions is the point when ATP and NADPH have been formed and oxygen has been released as a by-product. All reactions after this do not need light and belong to the second main stage, the 'light-independent' or 'dark' reactions.

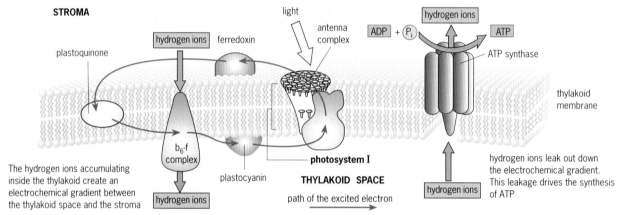

Fig 9.8 **The sequence of events in cyclic photophosphorylation**

CHLOROPHYLL, PLANKTON AND GLOBAL WARMING

PHYTOPLANKTON is the population of very small plants that live at the surface of seas and oceans. For years, scientists have wondered why phytoplankton is so unevenly distributed, even in areas which are rich in nutrients. It seemed that something vital was missing and one suggestion was iron. Some scientists suggested that if iron were added, phytoplankton might grow better and so soak up vast amounts of carbon dioxide to help prevent global warming. But how could this hypothesis be tested?

Tests were tried in the laboratory but, with so many variables to control, it could only be properly tested in the sea. In principle it sounds simple: just add iron and see if the plankton blooms. But how can you see where the iron that you put into the sea goes, and how do you measure the phytoplankton?

Scientists came up with two solutions. Vast amounts of iron sulphate were poured into the sea together with small amounts of radioactive sulphur hexafluoride. Oceanographers could test samples of sea water for radioactivity over a wide area and so track the flow of the currents carrying the iron.

Secondly, scientists surveyed the plankton from the air. If you illuminate chloroplasts strongly, the antennae absorb more photons than they can use and some of the energy is given back off as light (it **fluoresces**). By firing a laser at the sea and measuring chlorophyll fluorescence, scientists could estimate the amount of

chlorophyll (and hence phytoplankton) at a particular spot. Within two days they recorded a very rapid increase in the numbers of phytoplankton where the iron had been added. But this proved more short-lived than expected. Either the iron sulphate dropped quickly to depths too low for plankton to be able to use, or animal plankton also grew rapidly, taking advantage of the large feast of phytoplankton which suddenly became available.

So far, the scientists have not been able to find out which of these theories explains the results, and investigations are continuing.

Fig 9.9 **A bloom of marine phytoplankton**

Carbon fixation – the 'dark reactions'

In the dark reactions of photosynthesis, the ATP and NADPH made in the light reactions are the source of energy and reducing power to convert carbon dioxide to carbohydrate. The most important reaction in carbon fixing is when carbon dioxide is joined to the 5-carbon compound **ribulose bisphosphate**. This reaction is shown below in (Fig 9.10).

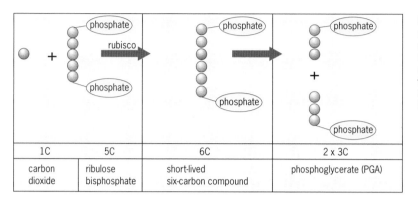

1C	5C	6C	2 x 3C
carbon dioxide	ribulose bisphosphate	short-lived six-carbon compound	phosphoglycerate (PGA)

Fig 9.10 **The key chemical reaction of the dark reactions: carbon dioxide joins with a 5-carbon compound. This is the *main carbon-fixing reaction*. Each circle represents a group of atoms containing carbon. Rubisco is the enzyme for this reaction. It is rather slow and so there is a lot of it in chlorophyll stroma: it makes up about 50 per cent of all the protein in a chloroplast**

Sugars are too unreactive to take part in many biochemical reactions: they need to be **phosphorylated** first. This simply means that a phosphate group is added on. Ribulose bisphosphate is a sugar with two phosphate groups. The enzyme **ribulose bisphosphate carboxylase** (often abbreviated to **rubisco**) catalyses the addition of carbon dioxide to form a very short-lived compound with six carbon atoms. It decays rapidly (within a tiny fraction of a second) to two 3-carbon compounds. A summary of the carbon fixation reaction is:

$$\text{ribulose bisphosphate} + CO_2 \rightarrow 2(\text{3-phosphoglycerate})$$

At this point, new carbon molecules enter a cyclical series of chemical reactions called the **Calvin-Benson cycle** (Fig 9.11). Follow it clockwise from the top.

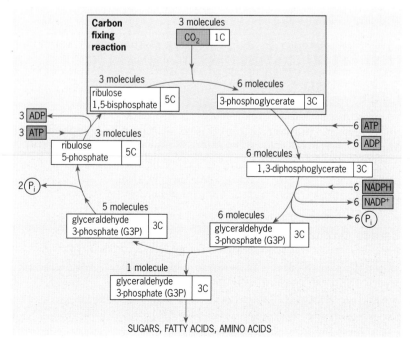

SUGARS, FATTY ACIDS, AMINO ACIDS

Fig 9.11 **The Calvin-Benson cycle: in each 'turn' of this cycle another carbon is added. It shows (from the top) that 3 atoms of carbon in carbon dioxide are fixed using 6 molecules of ATP and 6 hydrogens from NADPH, and that (at the bottom) one molecule of glyceraldehyde 3-phosphate is produced**

?

C Use Fig 9.11 to complete the following summary reaction for photosynthesis:

$3CO_2 + 9ATP + 6NADPH + \text{water} \rightarrow$ ____ + ____ + $8P_i$ + ____

?

D What is the first sugar phosphate produced as a result of photosynthesis?

The 3-phosphoglycerate formed in the first carbon-fixing reaction is not a sugar. It needs energy and hydrogen to convert it into a 3-carbon sugar phosphate, the triose phosphate **glyceraldehyde 3-phosphate** (G3P). Some of the G3P is converted into other sugars and some used to synthesise more ribulose bisphosphate. For every turn of the cycle, 3 molecules of ATP and 2 molecules of NADPH are needed to fix each CO_2 molecule into carbohydrate.

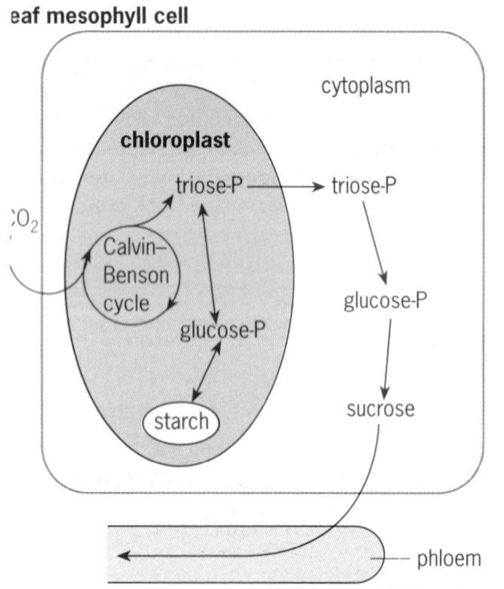

After the dark reactions: the export of sugar

Much of the G3P made in the stroma is exported from the chloroplast (Fig 9.12) to the cytoplasm where it is converted into glucose and then other sugars, fatty acids or amino acids. At peak periods of photosynthesis, G3P is produced so fast that it threatens to build up within the stroma. To avoid this, G3P can be converted into starch and stored for a short time in the chloroplast.

Once in the cytoplasm, sugars may be metabolised or exported to the phloem to be transported to other parts of the plant (see Chapter 18).

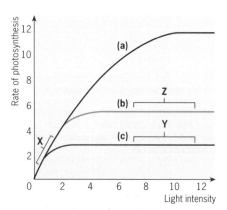

Fig 9.12 **A summary of the main routes taken by the triose sugars formed in the dark reaction of photosynthesis. Phosphorylated triose sugars (eg G3P) leave the chloroplast. In the cytoplasm they are converted into glucose and then sucrose. Sucrose is transported round the plant in the phloem**

See question 2. ■

Fig 9.13 **The effect of light intensity on the rate of photosynthesis:**
(a) **the effect of light intensity at 25 °C and 0.4 per cent carbon dioxide,**
(b) **the effect of light intensity at 15 °C and 0.4 per cent carbon dioxide,**
(c) **the effect of light intensity at 25 °C and 0.01 per cent carbon dioxide**

?

E In Fig 9.13, what is the main factor limiting respiration in each of the zones marked X, Y and Z?

3 WHAT LIMITS THE RATE OF PHOTOSYNTHESIS?

Photosynthesis is rather like a team of people making sandwiches. If the person cutting the bread into slices is very slow, he or she will control or 'limit' the number of sandwiches made per hour. If the bread cutter speeds up, or uses sliced bread, hard butter may limit production. With soft butter, cut bread and ready-made sandwich filling, production depends on the speed and stamina of the makers.

So what affects the rate of photosynthesis? Plant biologists agree that there are three main **limiting factors** (sometimes called **conditioning factors**). These are:

- light intensity
- concentration of carbon dioxide
- temperature

Water availability is not a limiting factor, but it can affect photosynthesis indirectly in some circumstances.

Light intensity

Photosynthesis cannot take place in the dark. As light intensity increases, the number of photons available to be caught by the antennae increases. The more light, the more ATP and NADPH is produced. But, as Fig 9.13 shows, increasing the light intensity does not increase the rate of photosynthesis indefinitely. At very high light intensities, other factors limit photosynthesis.

Carbon dioxide concentration

Photosynthesis needs only two raw materials: carbon dioxide and water. There is plenty of water in any living cell, but carbon dioxide has to diffuse in from outside the plant. The carbon dioxide level of normal air is low, at 0.03 per cent, and there are several barriers on the route to the chloroplast. The availability of carbon dioxide can easily limit the rate of photosynthesis.

Temperature

When neither light nor carbon dioxide is limiting, the rate of photosynthesis is limited only by the rate at which enzyme reactions can take place. The initial photochemical processes are not affected much by changes in temperature but the enzyme reactions of the Calvin-Benson cycle are temperature dependent.

You might expect to find a simple relationship between rate of photosynthesis and temperature. Increasing the temperature makes photosynthesis faster, but it also speeds up respiration even more. So the additional carbon fixed in photosynthesis is less than the extra carbon dioxide produced through faster respiration. The typical effect is to produce a broad, flattish curve as shown in Fig 9.14.

Water and the rate of photosynthesis

Water availability can indirectly affect the rate of photosynthesis. In dry conditions, plants close their **stomata**, the air holes in their leaves, preventing water getting out. But it also stops carbon dioxide getting in and this can slow photosynthesis down. If the plant is short of water for long periods, its new leaves will not grow to reach their full size, affecting the amount of photosynthesis that the whole plant can achieve.

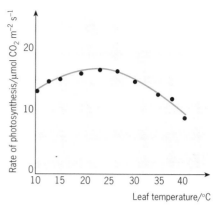

Fig 9.14 *Leucopoa* is a North American grass. Photosynthesis measurements were made on whole plants on cloudless days near noon

F (a) Look at Fig 9.14. What factor is likely to be limiting under the conditions described?

(b) What is the optimum temperature for growth of *Leucopoa*?

G Why is water is never a limiting factor in photosynthesis?

4 AUTOTROPHS: FEEDING THE WORLD

Organisms which appeared at an early stage of evolution had plenty of carbon-containing compounds to use as food. These had been produced as the Earth formed. But this food store was eventually used up, and organisms have had to manage without it for billions of years. So, how do organisms obtain their food?

All food comes from **autotrophs,** organisms which can make sugars from carbon dioxide using energy. Organisms which cannot make their own food are called **heterotrophs**. They need to feed on autotrophs or other heterotrophs. Fig 9.15 shows the difference between autotrophs and heterotrophs. Heterotrophic nutrition is covered in Chapter 15.

Fig 9.15 Strictly, an autotroph is an organism which adds hydrogen (from any source) to carbon dioxide to form sugars. The energy needed to drive this biosynthesis can be obtained from sunlight, or from chemical reactions. Heterotrophs are defined as organisms which cannot use carbon dioxide as a source of their carbon. Virtually all heterotrophs need to obtain carbon in ready-made food. (There are, however, a very few bacteria which use substances like ethanoate as a carbon source and so are, by definition, heterotrophs, even though they still make their own food!)

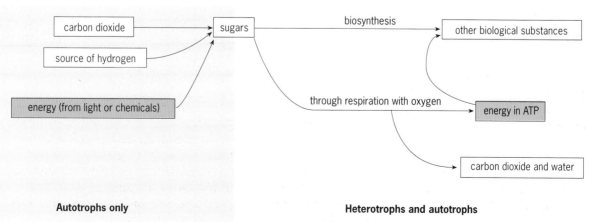

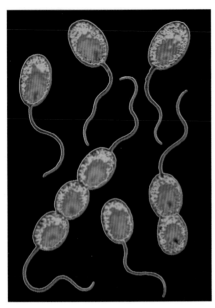

Fig 9.16 **A false colour electron micrograph of individual cells of purple bacteria. Two of the cells in the lower part of the photograph are dividing**

Some types of autotrophs

Earlier on in this chapter we saw that plants convert carbon dioxide to sugar in the process of photosynthesis. Light provides the energy to drive the process and hydrogen (from water) is added to carbon dioxide to form a sugar. However, not all autotrophs use light as a source of energy, water as a source of hydrogen or carbon dioxide as a source of carbon.

Photoautotrophs which are not plants

Autotrophs which use light as a source of energy are called **photoautotrophs**. Plants are photoautotrophs, but so are some bacteria, for example the **green bacteria** and the **purple bacteria** (Fig 9.16).

Like green plants, green bacteria need light and carbon dioxide for photosynthesis. However, unlike green plants, green bacteria use hydrogen or hydrogen sulphide as a source of hydrogen, rather than water. They release sulphur as a by-product. Because hydrogen and hydrogen sulphide are easily oxidised, these gases occur only in **anaerobic** conditions (when no oxygen is present). The green bacteria themselves therefore live only in anaerobic conditions.

There are two main types of **purple bacteria**: the **purple sulphur bacteria** and the **purple non-sulphur bacteria**. The purple sulphur bacteria are very similar to green bacteria. They too live in anaerobic conditions and use hydrogen or hydrogen sulphide for photosynthesis, but the sulphur they produce is retained inside their cells instead of being released into the environment.

The purple non-sulphur bacteria are very different. They use light as an energy source and they depend upon simple compounds of carbon (such as ethanoate) for their source of carbon, rather than carbon dioxide.

Chemoautotrophs

All **chemoautotrophs** are bacteria which have evolved enzyme systems to use chemicals such as ammonia or hydrogen sulphide as a source of energy. These specialised bacteria use a series of reduction and oxidation (**redox**) reactions (see page 129) to obtain the energy they need to add hydrogen to carbon dioxide and form sugar molecules.

Chemoautotrophs need a ready supply of a highly reduced substance to use as their energy source, but they must also grow close enough to aerobic conditions to allow them to carry out oxidation reactions. These needs limit their options, but they can live near the surface of a stagnant pool. Not far below the surface, the conditions are anaerobic and so substances like ammonia are available. At the surface, oxygen diffuses into the water and creates aerobic conditions. The interface is the best of both worlds.

Some bacteria obtain energy by oxidising sulphur or sulphur compounds. They are found in water discharged from mines that contain iron pyrites. Waterlogged soils are also a suitable environment for some chemoautotrophs.

These bacteria are often important in ecology. For instance, the bacteria *Nitrosomonas* and *Nitrobacter* are vital in the nitrogen cycle (see Chapter 37). They obtain energy from chemical reactions involving nitrogen compounds.

SUMMARY

By the end of this chapter you should know and understand the following:

■ **Photosynthesis** occurs in the green tissues of a plant. The usual site is the leaves, but plants with no leaves use their stems.

■ Inside the plant cell, the reactions of photosynthesis occur in the **chloroplast.** Most of the enzymes of the light reactions are embedded in the **thylakoid membrane.** The dark reactions take place in the **stroma.**

■ In the light reactions, energy from sunlight is trapped by chlorophyll molecules. It is used to boost electrons to a higher energy level. The extra energy in these electrons drives a series of **redox reactions** which produce ATP and NADPH.

■ The dark reactions do not need light. The enzyme **rubisco** catalyses the fixation of carbon in carbon dioxide by ribulose bisphosphate to produce a 3-carbon compound, which is converted into a 3-carbon sugar using ATP and NADPH.

■ The sugars produced by photosynthesis can be converted into other sugars.

■ Photosynthesis is affected by various factors, but particularly three **limiting factors**, light intensity, carbon dioxide concentration and temperature.

■ Water availability indirectly affects photosynthesis because dry conditions cause the stomata to close and so prevent the uptake of carbon dioxide.

■ Green plants are **autotrophs**, organisms which can make sugars from carbon dioxide using energy. Organisms which cannot make their own food are called **heterotrophs**.

■ Not all autotrophs use light as a source of energy, water as a source of hydrogen or carbon dioxide as a source of carbon.

QUESTIONS

1 Read through the following account of photosynthesis, then write it out and add at the dashed lines the most appropriate word or words to complete the account.

Photosynthesis is a type of ___ nutrition, involving the synthesis of organic molecules from inorganic materials. The process involves two types of reactions, light-dependent and light-independent.

  In the light-dependent reactions, light energy is absorbed by chlorophyll molecules located on the ___ of the chloroplasts; ___ and ___ are produced and oxygen gas is given off as a by-product.

  In the light-independent reactions, ___ accepts molecules of carbon dioxide, which together with the products of the light-dependent reactions, results in the formation of ___ . This compound can be converted to a hexose sugar or used to regenerate the carbon dioxide acceptor molecule.

[ULEAC 1996 Biology: Specimen Paper 2, q.1]

2 An investigation was carried out into the effect of carbon dioxide concentration and light intensity on the productivity of lettuces in a glasshouse. The productivity was determined by measuring the rate of carbon dioxide fixation in milligrams per dm^2 leaf area per hour

  Experiments were carried out at three different light intenities, 0.05, 0.25 and 0.45 (arbitrary units), the highest approximating to full sunlight. A constant temperature of 22°C was maintained throughout.

  The results are given in the table that follows.

Carbon dioxide concentration/ppm	Productivity at different light intensities/mg dm^{-2} h^{-1}		
	At 0.05 units light intensity	At 0.25 units light intensity	At 0.45 units light intensity
300	12	25	27
500	14	30	36
700	15	35	42
900	15	37	46
1100	15	37	47
1300	12	31	46

a) For the experiment at 0.25 units light intensity, described and comment on the effect on the productivity of the lettuces of increasing carbon dioxide concentration in the ranges **(i)** 300 to 900 ppm, and **(ii)** 900 to 1300 ppm.

b) Explain why the carbon dioxide concentration affects the productivity of plants.

c) State why the temperature should be kept constant during this experiment.

d) Suggest why, even with artificial lighting, glasshouse crops generally need to have more carbon dioxide added when temperatures are low, than when temperatures are high.

[ULEAC Jan 1994 Biology: Specimen Synoptic Paper, q.5]

Assignment

ARE PLANTS GOOD AT PHOTOSYNTHESIS?

One way to measure photosynthetic efficiency is to look at the proportion of light that plants can convert into stored energy in food.

Let us start with the energy in a glucose molecule. We can base our calculations on a mole of glucose molecules. A mole is the amount of a substance equal to its relative molecular mass in grams. A mole contains 6×10^{23} particles. When a mole of glucose is burned it releases about 2820 kJ.

1 How much does a mole of glucose weigh?

In the first half of this century, scientists believed that 24 photons of light were needed to synthesise one mole of glucose. Let us see if they were right. You have met the following summary reaction for photosynthesis:

$$6CO_2 + 6H_2O \rightarrow C_6H_{12}O_6 + 6O_2$$

Look back at Fig 9.7 and the text which describes it.

2
a) (i) How many electrons are needed to release one molecule of oxygen at PSII?
  (ii) How many electrons are needed to release 6 molecules of oxygen?
b) If one photon of light is needed for each electron excited, then how many photons would be needed to generate 6 molecules of oxygen at PSII?

But there is a complication to consider. The chloroplast has two photosystems which do not act independently of one another. Each electron released from PSII ends up at PSI, where another photon is absorbed.

3 Taking into account both photosystems, what is the minimum number of photons required to produce one molecule of glucose?

The latest research shows that the figure is closer to 54 photons per glucose molecule, but perhaps the difference you find may be caused by problems in making accurate measurements.

Scientists have compared the amount of energy that would be trapped in a glucose molecule with the energy in the photons needed to produce it. The actual efficiency, the **yield**, for most crop plants is below 1 per cent.

4 List as many factors as you can which could affect the efficiency of photosynthesis.

Crop researchers have estimated the energy in joules of light falling on a crop and compared it with the yield observed at harvest time. The UK receives sunlight at 100 joules per metre squared per second (100 J m^{-2} s^{-1}) in a 24-hour period.

5
a) Calculate the average energy input per hectare per year. (1 hectare = 10 000 m^2)
b) Assume the energy value of the crop of potatoes shown in Fig 9.A1 is 17.85 kilojoules per gram dry weight (17.85 kJ g^{-1}). If light were converted to chemical energy with an efficiency of 5 per cent, then what is the maximum crop (dry weight) you could expect in tonnes per hectare?
(1 tonne = 1000 kg)

Fig 9.A1 **Potato crop in Norfolk**

6 Copy and complete the following table:

Yield	% conversion of light to chemical energy	Total dry weight /tonnes hectare^{-1}
Possible maximum	5	
Record UK crop		22
Average		3

7 The values of the record and average conversion efficiencies are very low. Find out the strategies that scientists and farmers are using to improve on them.

THE VARIETY OF LIVING ORGANISMS

DO YOU KNOW how many different species there are on Earth?

No, neither does anyone else. But, while we don't know the exact number, we do know there are more than a million. And we have a good idea how they are grouped together, or classified. Understanding more about the relationships between organisms can help us to advance our knowledge of evolution, and it can help us work out how organisms are continuing to adapt and evolve.

In order to classify organisms, scientists use various systems: we take a broad overview of these in Chapter 10. The aspect of classification which will probably affect you most as you study biology is the naming system. All species have a double-barrelled Latin name: although this language has long-since lapsed from ordinary conversation, it is understood by biologists in all parts of the world.

And anyway, Latin names are fun. The gorilla, for instance, enjoys the name *Gorilla gorilla*; the orang utan is *Pongo pygmaeus*; the cat is *Felis cattus*; the grizzly bear is *Ursa horribilis*. There is pleasure in the sheer elegance of names such as *Enterobius vermicularis* (even if it is an obscure parasitic worm that infests the gut of some *Homo sapiens* - humans). There are a few bad ones of course: the famous *E. coli* bacterium's full name is *Escherichia coli* (pronounced esher-ish-eeyah). Difficult to say and difficult to spell - somebody should change it.

The main chapters in this section give a very brief glimpse of the vast array of living organisms that we share this planet with. As you read on, we hope that you will begin to appreciate this gallery of living organisms for the great art collection that it is. This immense work of art, the totally irreplaceable product of 350 000 000 000 years of evolution, is something to be cherished, looked after and marvelled at.

10 Biodiversity: the variety of life

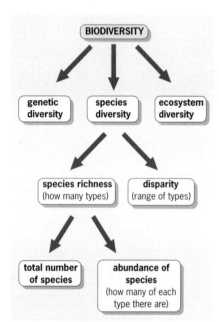

Fig 10.1 **What biodiversity means**

Fig 10.2 **The tropical rainforests of the world encompass a massive diversity of all forms of life**

1 INTRODUCING BIODIVERSITY

Biodiversity, a popular contraction of the term **biological diversity**, is used generally to refer to the variety of life on Earth. More specifically, we use it to describe the *numbers* of species present in a particular habitat or to show the *range* of types of organism found there (Fig 10.1).

Take the sea bed, for example. It has a large number of species but it is largely made up of a few basic types: annelid worms, molluscs and echinoderms. A rocky shore may have the same number of species but there is often a greater variety of animals and plants: arthropods, annelid worms, molluscs, echinoderms, cnidarians, fish, birds and seaweeds. We say that the shore has a greater **species diversity**, or range of different species types, when compared to the sea bed.

Diversity can be used to indicate the 'biological health' of a particular habitat. A slow increase in plant species diversity is normal for stable habitats such as hedgerows. If a habitat suddenly begins to lose its animal or plant types, ecologists become worried and search for causes (a pollution incident, for example).

You may also have heard of the term **genetic diversity**. Different species are obviously genetically different from each other, but there is also genetic variation within a species (no two people look identical, for example).

A third form of diversity is **ecosystem diversity**. The environment is an important factor that determines the number and type of species present in an ecosystem. At its simplest, ecosystem diversity is described as the number of different types of habitat in a given area.

Why biodiversity is important

When a species becomes extinct, many other species can be affected: no type of living organism exists independently. No one can really predict the long-term effects of repeated extinctions. Organisms have, of course, always become extinct, because of evolutionary pressures. What concerns us today is the rate at which species are disappearing because of the impact of human populations. Most people agree that it is very important to maintain biodiversity, to try to ensure the continued health and existence of other organisms.

A diverse world of animals and plants provides us with a massive resource of potential food and other necessities. It seems ridiculous that, although we have this great resource, we actually use around 30 species to provide 90 per cent of all the food we eat world-wide. Plants have already given us many useful medicines (Fig 10.3) and there are undoubtedly many more compounds still to be found. In addition, many different animals and plants stabilise soil and so minimise erosion, support fisheries, provide income through tourism and provide beauty and interest (Fig 10.4).

Fig 10.3 **Aspirin, the most commonly used drug in the world, was originally extracted from the bark of the willow**

Fig 10.4 **Ecotourism (travelling to see wildlife and areas of natural beauty) is now an important part of the world's economy. Tourism in general is now one of the world's largest industries**

Measuring biodiversity

Estimates for the total number of **extant species** (species that still exist) range from 1 million to 50 million. We base our estimates on the knowledge we already have of numbers and types of species found in specific habitats and on the rate at which our knowledge is increasing. Between 1978 and 1987, 367 new vertebrates, 173 new annelids and 7222 new species of insect were identified.

The actual figure for global biodiversity is likely to be enormous. Fig 10.5 gives a current species inventory of life on Earth. The problems of declining biodiversity and the steps that can be taken to maintain biodiversity are covered in more detail in Section 6 of this book (Chapters 35 to 39).

?

A The genetic diversity of vertebrates is much lower than that of invertebrates. Look at Fig 10.5. Is this also true for species diversity?

Organism	Species number		Organism	Species number
■ bacteria	4 000		☐ nematodes	15 000
▨ fungi	70 000		▩ molluscs	70 000
▦ protozoa	40 000		▧ crustaceans	40 000
■ algae	40 000		■ arachnids	75 000
☐ plants	250 000		▨ insects	950 000
■ vertebrates	45 000			

Fig 10.5 **The number of species present in the world today**

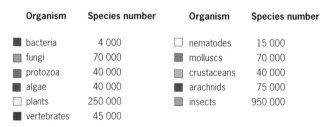

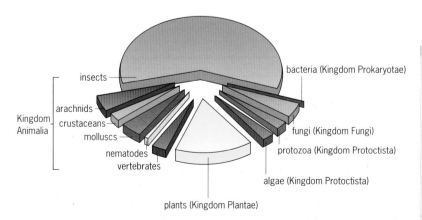

Kingdom Animalia: insects, arachnids, crustaceans, molluscs, nematodes, vertebrates

bacteria (Kingdom Prokaryotae)
fungi (Kingdom Fungi)
protozoa (Kingdom Protoctista)
algae (Kingdom Protoctista)
plants (Kingdom Plantae)

✔

Diversity consists of two components:

the variety of the species under study,

the relative abundance of the species.

Diversity is measured either by recording the number of species (the species richness), their relative abundance, or by using a measure that combines the two.

Data from *Global biodiversity*, World Conservation Monitoring Cenrtre (1992)

Why do we need to measure biodiversity?

There is more to measuring biodiversity than just being able to impress your friends by knowing how many species of living things there are on Earth. We need to measure the diversity of different environments for several reasons:

● To compare the diversity of the same habitat over time, to assess for example, whether pollution is damaging that habitat.

● To compare two different habitats at the same time, to find out which is the more diverse.

● To check whether a new habitat is being colonised by the number of species that it should. This would be important in the case of a polluted river or lake that had been 'cleaned up' and then repopulated with living organisms.

Putting a number on biodiversity – the Simpson diversity index

Being able to put a number on the diversity of a habitat makes the life of an ecologist much easier. A **diversity index** allows us to estimate the variety of living things in a particular area. However, we must make sure that the value we use accurately reflects all aspects of diversity. We could, for example, simply count the number of different species present:

Stream 1 (100 animals) 16 species

Stream 2 (100 animals) 10 species

The conclusion would be that stream 1 was more diverse. But there is a problem: this method gives an idea of *species richness*, but it does not reflect *how many* of each species are present. In stream 1, there may be 85 animals of one species and just one of each of the others. In stream 2, there may be 10 animals of each species. Ecologists would say that stream 2 was the more diverse habitat, and so our first conclusion would be wrong.

To avoid this problem, we can use the Simpson index. Its formula takes into account that diversity depends on the abundance of a particular species (the numbers of individuals of all the species present) and the species richness (the number of different species present). The higher the value of the index, the greater the variety of living organisms found in the area.

$$D = \frac{N(N-1)}{\Sigma n(n-1)}$$

where D is the diversity index, N is the total number of individuals of all species found, n is the total number of organisms of a particular species found, and Σ means 'the sum of'.

?

B The Simpson index was used to calculate the diversity of an oak wood and a conifer plantation. The values obtained were 19.6 for the oak wood and 12.04 for the conifer plantation. Which habitat is the more diverse?

EXAMPLE

Q Assume that you have studied a particular habitat and you have recorded the types of 20 animals you have found. The different types are represented by letters of the alphabet, and the record shows five species:

A A A A B B A C C B B A B B A D C C D E

What is the diversity of this habitat?

A Putting the figures into the equation for the Simpson index:

$$D = \frac{20 \times 19}{(7 \times 6) + (6 \times 5) + (4 \times 3) + (2 \times 1) + (1 \times 0)} = \frac{380}{86} = 4.42$$

2 CLASSIFICATION IN BIOLOGY

We can only hope to understand biodiversity if we have a thorough knowledge of **systematics**, the branch of biology that deals with classification of living things. We use classification to sort objects and place them in logical groups or sets. How would you set about classifying a collection of music CDs for a library? You might arrange them by date, or by type of music, or by the total length of time the CD played for. You might even want to arrange them using the colour of the paper in the casing. Some of these methods are obviously more useful than others.

Biological classification uses the same principles and has to overcome the same problems, but it can tell us a lot more. Any system of classification for CDs will always be artificial, no matter how clever it is. Living organisms have a natural order, which we discover rather than devise. By studying some organisms in detail and understanding more about the relationships between different organisms, we can learn to make predictions about organisms that we have not yet had the time to study in depth.

Why use systematics?

There are three reasons:

- We can find out more about the differences and the similarities between different organisms. This involves the basic science of classification, known as **taxonomy**. Taxonomists can work in museums, but some also work out in the field – pharmaceutical companies, for example, employ taxonomists to identify plant species that might be useful in drug development.

- Systematics allows us to suggest a **phylogeny** or pattern of evolutionary history of organisms that shows how different species are related, and so we can accurately place them into groups. This process is usually described as **classification**.

- We can give organisms names so that we all know we are talking about the same thing. The technical term for the name that we decide to use is the **nomenclature**.

Homologous and analogous structures

Organisms are grouped according to their similarities and differences. Some structures are similar because they have been inherited from a common ancestor. The leg of a horse and the human arm, for example, have both evolved from the limb of a common ancestral mammal. These are said to be **homologous structures**. However, a similar feature in another organism may look similar because it has developed for the same purpose. We say that this is an **analogous structure**. The wing of an insect and the wing of a bird are analogous. They *look* similar (they are thin, flat and broad), but they have *evolved independently* to perform the same function. Fig 10.6 shows some examples of homologous and analogous structures.

Distinguishing between the two types of structure is important when we think about evolution. Similar environmental pressures produce analogous adaptations in animals, while homologous structures are the result of evolution from a common ancestor.

**Homologous structures:
the front paws of the dog and the cat**

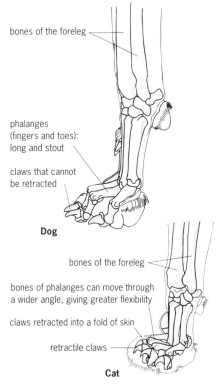

bones of the foreleg

phalanges
(fingers and toes):
long and stout

claws that cannot
be retracted

Dog

bones of the foreleg

bones of phalanges can move through
a wider angle, giving greater flexibility

claws retracted into a fold of skin

retractile claws

Cat

Adapted from Evolution Ed: P Skelton), Addison-Wesley (1993)

**Analogous structures:
the wing of a bird and a butterfly**

Fig 10.6 **Examples of homologous structures include the paws of dogs and cats and the forelimbs of other mammals. Analogous structures include the wings of birds and butterflies, and the tail fins of fish and porpoises**

Analogous features share a common function, homologous features share a common descent.

Natural and artificial classification systems compared

There are two broad types of classification. There is a **natural classification**, such as that devised by Linnaeus (see later in this chapter), which is based on *homology*. Natural groups are put together because they have common features that have an evolutionary significance. The advantage of a natural classification is that it gives information about the relationships between organisms. The more features they share, the more closely related they are.

An **artificial classification** uses only superficial features that have little or no evolutionary relevance, for example putting all the flying animals (bats, insects, birds) into one group, or all the yellow-coloured flowers. An artificial classification is one based on *analogy*. The advantage of an artificial system is that it simplifies classification and identification. Look at identification guides for flowering plants. Which would you find easiest to use: those describing plants in botanical detail or those grouping flowers together according to their colour, number of petals and so on?

The essential difference between these two schemes of classification is that a natural classification is discovered by studying organisms in depth and an artificial classification is invented to fit easily observable information.

Fig 10.7 **Carl Linnaeus developed a classification system based on structural similarities only. It did not take into account the evolutionary relationships that are used to classify organisms today**

?

C What features are shared by (a) birds, (b) fishes, (c) grasses? Try and give three characteristics for each group.

3 HOW WE CLASSIFY ORGANISMS TODAY

The system of classification we use today is based on one devised by Carl Linnaeus (1707–1778), a Swedish botanist (Fig 10.7).

Linnaeus placed animals and plants into **natural groups**. He defined a natural group as a group of organisms that showed very similar features. Birds, fish and ferns, for example, form natural groups. He subdivided to form smaller groups (the smallest group being the species) and combined to form larger ones (the largest being the kingdom).

He then used all the information available at the time to assign animals and plants to named groups (he could not know that fungi and microorganisms existed, because they had yet to be discovered). Today we use the same groupings, the same system of naming and the same language, Latin, but our knowledge has grown since the time of Linnaeus. Our ideas on classification are still changing, as you will see in later chapters in this section.

Putting living things into groups

Linnaeus was the first scientist to tackle the enormous job of systematically documenting and ordering all living organisms. Organisms were placed into large groups based on their shared features. These groups, or **taxa** (singular **taxon**), were then subdivided into smaller and smaller groups. In this way, Linnaeus established **taxonomic ranks** (different levels). We still use these ranks today:

- **Kingdom** – the largest natural group (eg the animal and plant kingdoms).
- **Phylum** – a major body plan (eg the arthropod plan or the seed-bearing plants).
- **Class** – a major division of a phylum (eg the Class Insecta, the Class Mammalia).

- **Order** – a group of related families (eg the orders of birds – ducks, falcons, perching birds; the orders of insects – beetles, grasshoppers, butterflies and moths).

- **Family** – a collection of related animal or plant types (eg the families of flowering plants – the violet, buttercup, geranium families; the families of beetles – ground beetles, ladybirds, weevils).

- **Genus** – a small group of very closely related forms, eg *Homo* (humans), *Ligustrum* (privet), *Patella* (limpet).

- **Species** – the basic unit of taxonomy. The concept of the species is examined in more detail later in this chapter.

We sometimes find that these divisions are too large and so we subdivide them again, introducing intermediate levels such as **sub-phyla** or **sub-classes**. Some species (including humans) are subdivided into **sub-species** (varieties or races).

Naming organisms

Animals and plants often have common names. For example, in the birds we find:

- Blackcap and spoonbill (names that describe physical features).

- Fieldfare and wood lark (names that relate to the birds' habitat).

- Chaffinch and herring gull (names that describe the food that the birds eat).

Common names are useful to identify individual animals and plants, but a problem arises if people call the same thing different names. Coltsfoot, for example, with its dandelion-like flower, may be called either coughwort or claywort, depending on where you live.

Biologists need a system of naming that is fixed and can be shared with scientists around the world who also speak different languages. The current **binomial system** of naming was developed by Linnaeus. (Latin, the scientific language of Linnaeus's day and a dead language, is just as acceptable to a Russian zoologist as to a biology student in Australia or a Portuguese botanist.) The word binomial means 'two names' and, in this system, each organism has two names: a **genus name** (a first name) and a **species name** (a second name). The names are unique to each type of organism. Today, international societies register the names of newly discovered animals and plants so that this information is freely available to people in all parts of the world. Table 10.1 gives some general rules that we use to name organisms.

> ?
>
> **D** To what order does the domestic cat belong?
>
> **E** Think of two specific examples of a species subdivision.

Table 10.1 **Naming some common organisms. When we name organisms, the following rules apply:**

1 The binomial system is used
2 The genus name comes first, the species name second
3 Within text, names should appear in italic type or be underlined
4 The genus name takes an initial capital letter, the species name does not (eg *Homo sapiens*)
5 The species name should always be written out in full when it is first encountered (eg *Escherischia coli*). Thereafter the name can be abbreviated (eg *E. coli*)
6 If the species is unspecified, the abbreviation 'sp.' may be used (eg *Amoeba* sp.)

The five Kingdom scheme

There are, currently, five Kingdoms (Figs 10.8):

Prokaryotae (Monera) consists of bacteria and blue-green bacteria. Members of this kingdom, the prokaryotes, lack a distinct nucleus and membrane bound organelles such as mitochondria and chloroplasts. Some have a flagellum but these lack the typical (9 + 2 microtubules) structure of the undulipodia found in eukaryotic cells (see Chapter 1).

Protoctista contains eukaryotic organisms only. It includes the algae (simple photosynthetic organisms) and the protozoans (single-celled organisms that are similar to animal cells). This kingdom has

It is important to remember that words such as 'invertebrate', 'protozoan' and 'algae' are useful and still used by many biologists, even though these are not correct taxonomic terms.

Fig 10.8 **Representatives of the five kingdoms illustrate the variety of body form and habits shown by life on Earth**

been described as a bit of a dustbin. If it is not an animal, a plant, a fungus or a prokaryote, then it ends up in the Protoctista.

Fungi also contains eukaryotic organisms that lack undulipodia. Fungi have a non-cellulose cell wall (generally made of chitin), and are **saprotrophs** (they absorb food from decaying organic matter). Fungi are made up of threads called **hyphae** which are not split into separate cells – a single hypha contains many nuclei.

Plantae. Plants are multicellular, eukaryotic, photosynthetic organisms. Most have stems, leaves and roots (some of the simpler forms have only a flat lump of cells called a thallus). Plant cells have cellulose walls and a vacuole.

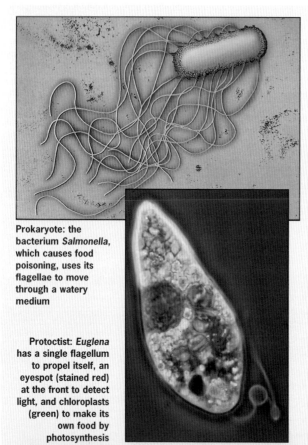

Prokaryote: the bacterium *Salmonella*, which causes food poisoning, uses its flagellae to move through a watery medium

Protoctist: *Euglena* has a single flagellum to propel itself, an eyespot (stained red) at the front to detect light, and chloroplasts (green) to make its own food by photosynthesis

Fungus: *Armillaria*, the honey fungus, feeds on dying and dead wood. Its fruiting bodies produce millions of spores

Plant: bright red bracts of this hot lips plant attract insects to pollinate its small flowers in the shaded under-storey of a tropical rain forest

Animal: the male peacock may not be very mobile, but its flamboyant tail feathers and piercing cry advertise its presence well

Animalia. Animals are multicellular, eukaryotic, heterotrophic organisms that show both chemical and nervous coordination. Animals actively **locomote** (move from place to place) at some stage during their life cycle, and their growth occurs all over the body surface.

What is a species?

Just as the metre is the basic unit of length and the cell is the basic unit of body structure, the species is the basic unit of classification. It a difficult term to define accurately, but there are three main concepts to remember:

- Organisms in the same species are similar in their structure and chemical make-up. In the past, organisms were classified mainly according to the appearance of dead specimens. Today it is more usual to use a combination of methods, including biochemical tests (protein and DNA analysis, for example) and studies to determine how living organisms behave in their natural habitat.

- Organisms within the same species can breed with one another to produce offspring that are **viable** (capable of normal growth and development) and **fertile** (able to reproduce). Occasionally, closely related species interbreed, but they produce offspring that are **infertile**. A cross between a female horse and a male donkey, which produces an infertile mule, is a good example of this.

- A species consists of all those individuals that are descended from a recent common ancestor. So, in a way, a species is a bit like an extended family.

4 BIOLOGICAL KEYS

When presented with a strange animal or plant, how do you identify it? Most biologists use a biological key to guide them (through a series of questions) to the name of the organism they are looking at. Traditionally, species are identified using structural features such as length or shape of body parts. But the modern taxonomist has at his or her disposal information on chromosome number and banding patterns, studies of behaviour (especially courtship behaviour) and special types of cell chemistry. Most of us, however, still use the standard key that relies on obvious body features.

Keys come in many forms and can use both pictures and words. All are based on the concept of placing the organism into a large group and then 'narrowing down' the search by progressively assigning them to smaller and smaller groups.

Fig 10.9 on the next page provides examples of two types of biological key. The flowers in Fig 10.9(a) are identified on the basis of colour and petal number (an artificial classification). The insects in Fig 10.9(b) are identified using structural features that place them into natural groups (a natural classification).

The **dichotomous key** is one of the most common types of biological key. It is so called because it uses paired statements (a **dichotomy** is a division into two parts). You have two features to choose from at each stage, and instructions lead you on to the next choice until you have correctly identified the organism.

The Extension box opposite to Fig 10.9 shows how to produce a simple key.

?

F Look at Fig 10.9. Describe how the authors have tried to simplify the task of identifying the different organisms.

Fig 10.9 A selection from two popular field guides demonstrates how biological keys can be used to name organisms

Fig 10.9(a) In this key, flowers are identified on the basis of colour

The Main Key

Individual flowers large or conspicuous   see below
Individual flowers small   see p. 19

(N.B. Composites, actually tight heads of small flowers, also have a separate key, pp. 240-1)

INDIVIDUAL FLOWERS LARGE OR CONSPICUOUS

Open, star-like or saucer-shaped flowers

Blue Woodruff **194**

Two petals

Enchanter's Nightshades **158**

Goldilocks Buttercup **74**

Mezereon **146**, Dwarf Cornel **164**, Field Madder **194**

Pearlworts **60**, Crucifers **86**, Allseed **140**, Water Chestnut **270**, May Lily **272**

Three petals

Mossy Stonecrop **106**, Water Plantains, Star-fruit, Arrowhead, Frogbit, Water Soldier **270**

Ranunculus hyperbereus **72**, Goldilocks Buttercup **74**, Rannoch Rush **232**

Water Plantains **270**

Water Plantains **270**

Poppies, Red-horned Poppy **84**, *Capsella rubella* **100**, Garden Cress **102**

Red-horned Poppy **84**, Wallflower **90**

Poppies, Greater Celandine **84**, Crucifers **86**, Tormentil **118**, Evening Primroses **158**, Yellow Centauries **190**

Mistletoe **38**, Pearlworts **60**, Narrow-leaved Pepperwort **102**, Golden Saxifrages **104**

Four petals

Bastard Toadflaxes **38**, Mossy Sandwort **52**, Water Tillaea **106**, Dwarf cornel **164**, Squinancywort, Woodruff **194**, Bedstraws, Cleavers **196**

Cranberry **182**, Squinancywort **194**, Slender Marsh Bedstraw **196**

Wall Bedstraw **196**

Crosswort, Lady's Bedstraw **196**

Spurge Laurel **146**, False Cleavers, Wall Bedstraw **196**, Herb Paris **280**

Opium Poppy **84**, Crucifers **86**

Meadow-rues, *Clematis recta*, Baneberry, **76**, Shoreweed **302**

Lesser Meadow-rue **76**, Roseroot **104**

Traveller's Joy **76**

Meadow-rues, Alpine Clematis **76**

Crucifers **86**, Willowherbs **160**

Willowherbs **160**

Willowherbs **160**

Speedwells **224**

Pink Water Speedwell **224**, *Veronica urticaefolia* **224**

Speedwells **224**

Heath Speedwell **224**

Five petals

Spring Beauty **42**, Chickweeds and Allies **50**, Catchflies **62** and **66**, White Campion **64**, Tunic Flower **66**, Roses **114**, Rock Cinquefoil **116**, Hollyhock **150**

Pink Purslane **42**, Spurreys **58**, Catchflies **62–6**, Red Campion **64**, Cow Basil, Proliferous Pink, Tunic Flower **66**, Roses **114**, Pink Barren Strawberry **116**, Cranesbills **142**, Mallows **150**, Birdseye Primrose **184**

Sticky catchfly **64**, Small-flowered Catchfly **66**

Cowslip **184**

Portulaca oleracea **42**, Cinquefoils **118**, Primrose, Oxlip, Cowslip **184**

Mallows **150**, Primroses **184**

Chickweeds and allies **50**, Bladder Campion **62**, Catchflies **62-6**

Catchflies **62-6**, Ragged Robin **64**

Spanish Catchfly **62**

Berry Catchfly **66**

Northern Catchfly **62**

Wild Pink **68**, White Campion **64**

Red Campion **64**, Pinks **68**, Cranesbills, Herb Robert **142**, Mallows **150**

Plnks **68**

Bastard Toadflax **38**, Orpine, Stonecrops **106**, Purging Flax **140**, Burning Bush **146**, Chickweed Wintergreen **186**, Bogbean, Vincetoxicum **188**, Nightshades **216**

Orpine, Pink Stonecrop **106**, Burning Bush **146**, Bogbean **188**

Bastard Toadflax **38**, Mousetail **78**, Orpine, Stonecrops **106**, Vincetoxicum **188**, Wild Madder **196**, Tomato **216**

Hop **38**, Saltwort **48**, Lesser Chickweed **54**, Lady's Mantles **112**, Gooseberry **148**, Ivy **164**, False Cleavers, Wall Bedstraw **196**

Lomatogonium **192**

Fastigiate Gypsophila **66**, Marsh Cinquefoil **116**, Marsh Felwort **192**, Bittersweet **216**

Knotgrass **40**, Gypsophila **66**, Buttercups, Water Crowfoots **74**, Sundews, Grass of Parnassus **104**, Saxifrages **110**, Meadowsweet Goatsbeard Spiraea **112**, Strawberries **116**, Brambles **114**, *Linum tenuifolium*, White Flax **140**, White Rockrose **158**, Common Wintergreen **178**, Labrador Tea **182**, Brookweed **186**, Heliotrope **198**, White Mullein **218**

Knotgrass **40**, Soapwort **64**, Annual Gypsophila **66**, Bramble **114**, Cotoneasters **122**, Pink Oxalis, *Linum viscosum*, *L. tenuifolium* **140**, Cranesbills **142**, Sea-heath **158**, Umbellate Wintergreen **178**, Sea Milkwort **186**, Nonea **198**, Lungworts **200**, Viper's Bugloss, Oyster Plant **202**

Fig 10.9(b) In this key insects are identified using structural features

INSECTS OF BRITAIN & NORTHERN EUROPE

Key to the Orders of European Insects
* Denotes orders not found in the British Isles

1. Insects winged — 2
   Insects wingless or with vestigial wings — 29

2. One pair of wings — 3
   Two pairs of wings — 7

3. Body grasshopper-like, with enlarged hind legs and pronotum extending back over abdomen — Orthoptera p. 68
   Insects not like this — 4

4. Abdomen with 'tails' — 5
   Abdomen without 'tails' — 6

5. Insects <5mm long, with relatively long antennae: wing with only one forked vein — Hemiptera p. 97
   Larger insects with short antennae and many wing veins: tails long — Ephemeroptera p. 52

6. Forewings forming club-shaped halteres — Strepsiptera p.161
   Hind wings forming halteres (may be hidden) — Diptera p. 170

7. Forewings hard or leathery — 8
   All wings membranous — 13

8. Forewings horny apart from membranous tip — Hemiptera p. 97
   Forewings of uniform texture throughout — 9

9. Forewings (elytra) hard and veinless, meeting in centre line — 10
   Forewings with many veins, overlapping at least a little and often held roofwise over the body — 11

INSECTS OF BRITAIN & NORTHERN EUROPE

10. Abdomen ending in a pair of forceps: elytra always short — Dermaptera p. 78
    Abdomen without forceps: elytra commonly cover whole abdomen — Coleoptera p. 134

11. Insects with piercing and sucking beaks — Hemiptera p. 97   ← beak
    Insects with chewing mouths: cerci usually present — 12

12. Hind legs modified for jumping — Orthoptera p. 68
    Hind legs not modified for jumping — Dictyoptera p. 82

13. Tiny insects covered with white powder — 14
    Insects not like this — 15

14. Wings held flat at rest: mouth-parts adapted for piercing and sucking — Hemiptera p. 97
    Wings held roofwise over body at rest: biting mouthparts — Neuroptera p. 127

15. Small, slender insects with narrow, hair-fringed wings: often found in flowers — Thysanoptera p. 125
    Insects not like this — 16

16. Head extending downwards into a beak — Mecoptera p. 163   ← beak
    No such beak — 17

Constructing a biological key

A step-by-step guide:

1 Select only clearly visible features. Avoid using information that is not visible in the specimen or the diagram.

2 Do not use features that vary. For example, colours might vary between males and females or between adults and their young. Size can also vary dramatically between individuals of the same species.

3 Prepare a table of similarities and differences.

4 Aim to divide your collection into two equal-sized groups. Then split those groups into halves (approximately).

5 Select the most appropriate features and write the key as a series of paired statements. If three specimens are to be separated, you need only two pairs of statements. If four specimens are to be separated, you need three pairs of statements).

For example, imagine you need to construct a simple key to separate three insects. First, you would draw up a table like Table 10.2 to compare them:

Table 10.2 **Comparing insects**

Specimen	Mouthparts	Wings		Body
		Number	Type	
A	fleshy lobe	2	clear	no waist
B	long tongue	4	clear	distinct waist
C	coiled tube	4	scaly	no waist

Then you would draw up a series of paired statements: look at the three insects below. In the example that follows, only statements about the wings have been used.

1   Insects with two wings   Housefly
   Insects with four wings   Go to 2

2   Insects with clear wings   Honey bee
   Insects with scaly wings   Moth

Fig 10.10

11 Simple organisms

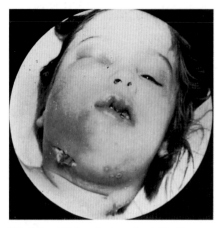

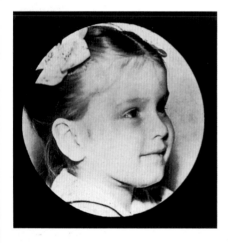

This little girl (above) was seriously ill with a massive infection and would have died without antibiotics. She was one of the first people to receive penicillin. The second photograph was (below) taken six weeks later

MANY BACTERIA are harmless to humans, but some can cause disease. Until the 1930s there were no effective treatments for bacterial infections. Something as simple as a boil on the skin could lead on to a more extensive infection (septicaemia, or blood poisoning) which was often fatal. In the 1930s and 1940s, scientists discovered two important classes of modern medicine. The sulphonamide Prontosil was first used to treat bacterial infection in 1936 and penicillin, the first antibiotic, was developed in 1940.

An antibiotic can stop an infection either by killing bacteria or by preventing them from growing (this gives the immune system of the host time to get to work). Individual antibiotics of either type work by many different biochemical mechanisms. Penicillin, for example, interferes with cell wall production. Affected bacteria have weak cell walls and tend to burst. Streptomycin binds to bacterial ribosomes and stops normal protein synthesis. Without new proteins, the bacteria cannot carry out other chemical reactions and begin to die. Streptomycin does not bind to mammalian ribosomes and so does not affect protein synthesis in the patient's cells.

1 CLASSIFYING SIMPLE ORGANISMS

As we saw in Chapter 10, living organisms are classified into five kingdoms: the **Prokaryotae** (also called the **Monera**), the **Protoctista**, the **Fungi**, the **Plantae** and the **Animalia**. In this chapter we take a detailed look at viruses and at organisms in the first three kingdoms. The diagram below summarises the groups of organisms covered.

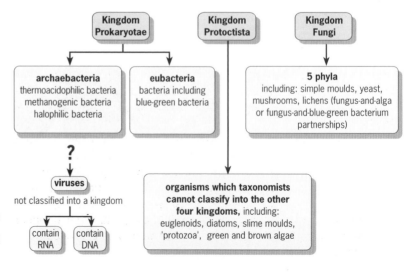

Fig 11.1 **A summary of the organisms in the Kingdoms Prokaryotae, Protoctista and Fungi, which are described in this chapter. Viruses (page 173) are not classified into any of the five kingdoms. The other two kingdoms are the Kingdom Animalia, covered in Chapter 12, and the Kingdom Plantae, Chapter 13**

2 VIRUSES

Viruses do not fit into any of the five kingdoms. In fact, it is difficult to decide whether viruses are actually alive. They do not show many of the characteristics that define a living organism (see Chapter 14) but they are more than a collection of very complex chemicals. Scientists are still unsure about the origin of viruses and don't yet understand fully how viruses are related to each other, or to other organisms (see page 175).

All viruses consist of a piece of genetic material (DNA or RNA), usually surrounded by a protein coat and sometimes also by a membrane. In contrast to bacteria, viruses have distinctive structures and show a wide variety of shapes and sizes, as Fig 11.2 illustrates.

Fig 11.2 **In the box on the lower left of the diagram, six different viruses have been drawn to the same scale as two bacterial cells (*E. coli* and *Chlamydia*) and an animal cell. In general, viruses are much smaller than bacteria but there is some overlap:** *Herpes simplex*, **one of the larger viruses, is similar in size to** *Chlamydia*, **one of the smaller bacteria. The magnified views of four of the viruses show the complex molecular structure of individual virus particles**

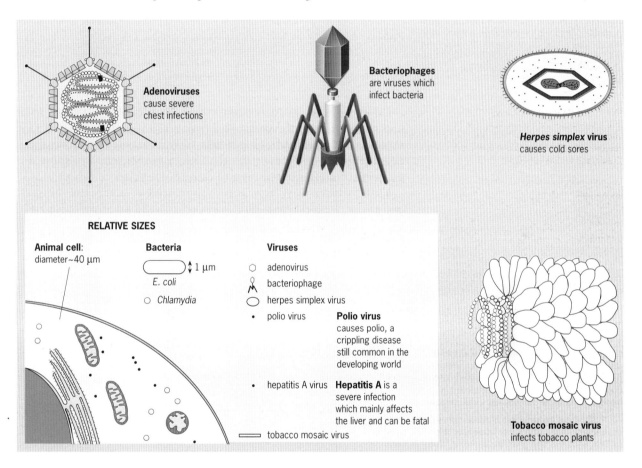

Adenoviruses cause severe chest infections

Bacteriophages are viruses which infect bacteria

Herpes simplex virus causes cold sores

RELATIVE SIZES

Animal cell: diameter~40 μm

Bacteria
⬭ ↕ 1 μm
E. coli
○ *Chlamydia*

Viruses
○ adenovirus
⚲ bacteriophage
⬭ herpes simplex virus
• polio virus
  Polio virus causes polio, a crippling disease still common in the developing world

• hepatitis A virus **Hepatitis A** is a severe infection which mainly affects the liver and can be fatal

▭ tobacco mosaic virus

Tobacco mosaic virus infects tobacco plants

Parasites of the cell

Viruses are parasites. To reproduce, they must take over the machinery of a living cell. Otherwise, they cannot survive. Unlike other parasites, viruses infect cells in organisms in each of the five kingdoms.

Once a virus gets inside its host cell, it makes use of the cell's enzymes to reproduce. The details of viral reproduction vary from virus to virus. Some viruses start to take over the machinery of a host cell immediately. The cell makes large numbers of new virus particles, which are then released, ready to infect other host cells. Fig 11.3 shows, in a much simplified diagram, what happens when herpes simplex virus, the virus that causes cold sores, infects an animal cell.

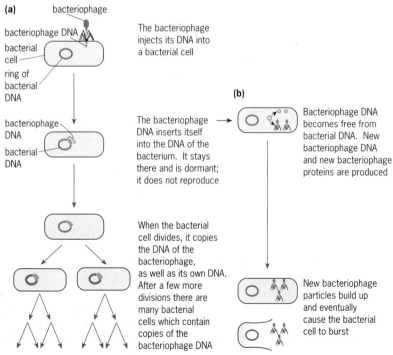

Fig 11.3 **The DNA of the herpes simplex virus goes into the nucleus of the cell. The cell cannot tell the difference between viral DNA and its own DNA and so its enzymes produce viral messenger RNA. This passes out of the nucleus and is used as a template for protein synthesis by the cell's ribosomes. Some of the proteins produced form the coats of new viruses; others go back into the nucleus to force the cell's enzymes to produce more copies of the original viral DNA. These leave the nucleus, are packaged into the coats, and leave the cell. One virus can produce many thousands of copies of itself and, as these build up in the cytoplasm, the cell can be damaged or even killed**

1 virus enters cell by joining its membrane with the cell surface membrane. The DNA of the virus enters the cytoplasm

2 DNA of the virus goes into the nucleus of the cell

3 The viral DNA is 'read' by the cell's enzymes to produce viral mRNA

4 The cell uses the mRNA to make viral proteins. Some of these make up the coats for the new viruses

5 Some of the viral proteins direct the enzymes of the cell to make more copies of viral DNA

6 The viral DNA is packaged into the proteins made to surround it

7 The newly made viruses are released as the cell bursts open

cell

cell surface membrane

Herpes simplex virus

Other viruses can incorporate their DNA into the DNA of their host cell, remaining **dormant** (inactive) until they are reactivated. One example is **bacteriophage lambda**, a viral parasite of the bacterium *E. coli* (Fig 11.4). Some viruses that infect animal cells also show this property and are now known to cause some types of cancer.

Viruses have very few genes compared to even the simplest bacterium. Nevertheless, they may have played an important part in the evolution of other organisms. When viruses insert their DNA into host DNA, small pieces are sometimes left there, when the virus is reactivated. In this way, the host cell gains viral genes and its characteristics can be changed. Bacteria of the species *Corynebacterium diphtheriae*, for example, cannot cause diphtheria unless they are infected by a bacteriophage that carries the gene for a potent toxin.

More information about viruses and disease is given in Chapter 26.

?

A What advantage is it to a virus to integrate its genetic material into the host's DNA?

Fig 11.4(a) **Bacteriophage lambda inserts its DNA into the DNA of *E. coli*. Every time an infected bacterial cell divides, the bacteriophage DNA is copied. It therefore reproduces without the trouble of having to make whole virus particles.**

Fig 11.4(b) **If the bacterial cell becomes 'stressed', the bacteriophage effectively 'abandons ship'. Its DNA becomes free from the bacterial DNA and duplicates itself and makes new proteins. Large numbers of new viral particles build up and eventually burst the bacterial cell. Each one is then able to infect a new host**

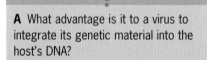

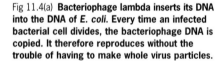

(a)

bacteriophage

bacteriophage DNA

bacterial cell

ring of bacterial DNA

The bacteriophage injects its DNA into a bacterial cell

bacteriophage DNA

bacterial DNA

The bacteriophage DNA inserts itself into the DNA of the bacterium. It stays there and is dormant; it does not reproduce

When the bacterial cell divides, it copies the DNA of the bacteriophage, as well as its own DNA. After a few more divisions there are many bacterial cells which contain copies of the bacteriophage DNA

(b)

Bacteriophage DNA becomes free from bacterial DNA. New bacteriophage DNA and new bacteriophage proteins are produced

New bacteriophage particles build up and eventually cause the bacterial cell to burst

Where did viruses come from?

Viruses definitely did not exist before either prokaryotic or eukaryotic cells: they have appeared much more recently. In speculating about the origin of viruses, scientists recognise that many non-viral cells contain fragments of DNA or RNA that can reproduce independently of the cells themselves. Fragments of naked DNA (**plasmids**) or naked RNA (**viroids**), may, at some time during evolution, have become surrounded by a protein coat and a membrane, giving rise to viruses.

If viruses are pieces of DNA or RNA that 'escape' from host cells, it would make sense for their sequences to be similar to sequences found in the host genome. It is therefore possible that different viruses are actually more closely related to their hosts than to each other.

Today, both viruses and plasmids are essential tools in the development of DNA technology, including the techniques used in genetic engineering (see Chapter 34).

3 KINGDOM PROKARYOTAE (MONERA)

The Kingdom Prokaryotae contains all species of bacteria (including the blue-green bacteria, once considered to be algae, see page 6). Bacteria are all **prokaryotes**: they have no defined nucleus and show little organisation and do not have the complex organelle systems of eukaryotic cells. Eukaryotes make up the four other kingdoms.

Prokaryotes and eukaryotes show several important differences:

- Cells of prokaryotes do not have a clearly defined nucleus with a nuclear membrane; cells of eukaryotes do.

- Cells of prokaryotes have a single length of genetic material that consists only of nucleic acid. Eukaryotic cells have several lengths of genetic material called **chromosomes.** Their DNA is wound around proteins.

- Prokaryotes have small ribosomes but no other cell organelles. Eukaryotic cells have larger ribosomes and several membrane-bound organelles, such as mitochondria, chloroplasts and endoplasmic reticulum.

- Prokaryotes that photosynthesise have chlorophyll, but it is not contained within chloroplasts. Eukaryotic cells that photosynthesise have their chlorophyll inside chloroplasts.

The structure of bacterial cells and eukaryotic cells, and the differences between prokaryotes and eukaryotes, are covered in more detail in Chapter 1.

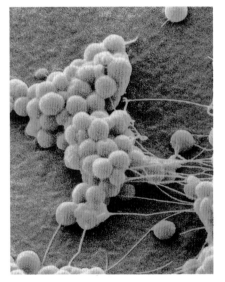

Fig 11.5 **The surface of the human skin is covered with many different bacteria, including** *Staphylococcus epidermidis* **shown here. Many bacteria are useful since they create acidic conditions on the surface of the body which prevent other, harmful bacteria from gaining a foothold and causing an infection**

■ See question 1.

Classifying bacteria

We know of over 10 000 living species of bacteria, but there are probably many times this number that we still have not come across. For years, bacteriologists were the only scientists who studied bacteria and they had their own ways of describing and grouping individual types. These were not always in line with the views of taxonomy specialists and, even today, scientists have still not come up with a classification system that everyone agrees with.

The **phylogeny**, or **phylogenic inheritance** of an organism refers to its evolutionary 'family tree'. Phylogeny is important because it helps taxonomists to understand how organisms are related to each other.

When they classify living organisms, taxonomists try not to group together organisms just because they look similar. (This would be as ridiculous as saying all people with ginger hair belong to the same family.) They also take into account what is known about the evolutionary relationships, or **phylogenies,** of the groups. It is impossible to make this sort of family tree for bacteria on the basis of physical characteristics only, because they all have a very similar internal structure. Instead, taxonomists use chemical tests to identify **metabolic differences** – differences in the enzymes of different types of bacteria.

On this basis, many different bacteria have been identified. Today, the Kingdom Prokaryotae is split into two unevenly sized groups, the **archaebacteria** and the **eubacteria**. It is then divided further into seventeen phyla, but this may change as the debate about classifying bacteria continues.

The archaebacteria

Bacteria were the earliest organisms to evolve. They appeared almost 4000 million years ago. As the Earth's atmosphere contained no oxygen, the early bacteria survived without it. Later, as new bacteria evolved, some of them developed the ability to photosynthesise. One particular group, that of the **blue-green bacteria** (which belong to the eubacteria) was very successful. The oxygen they released as a by-product of photosynthesis changed the Earth's atmosphere. As oxygen levels rose, many of the archaebacteria died out: they were 'poisoned' by the oxygen.

Only a small number direct descendants of the earliest non-photosynthesising bacteria still exist today. They are the archaebacteria and they grow in unusual environments, particularly where there are low levels of oxygen. There are three main types:

- The **methanogenic** bacteria are found world-wide in sewage and in the intestinal tracts of animals. Methanogenic bacteria produce about 2 billion tonnes of methane per year.

- The **thermoacidiphilic** bacteria (Fig 11.6) live in very hot springs and can survive in strong acids, even in concentrated sulphuric acid!

- The **halophilic** bacteria require very salty conditions, such as salt flats where sea water becomes concentrated by the hot sun, and in the Dead Sea.

?

B Explain why some bacteria could not survive when the blue-green bacteria became plentiful.

Fig 11.6 **The bacterium *Thermoplasma acidophilum* that lives in this hot, acidic geyser pool would die at room temperature. Its DNA is protected from the acid by a protein coat**

The eubacteria

All other living bacteria (including blue-green bacteria) belong to the eubacteria and are usually referred to in this book simply as bacteria.

The basic structure of bacterial cells is covered in Chapter 1. Fig 11.7 shows some examples. Bacteria form a very diverse group and their different methods of obtaining energy is covered in Chapter 9.

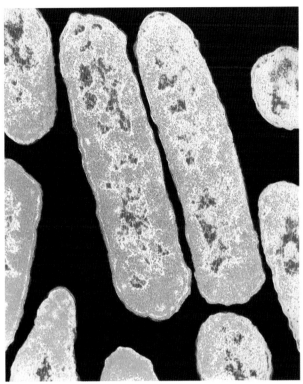

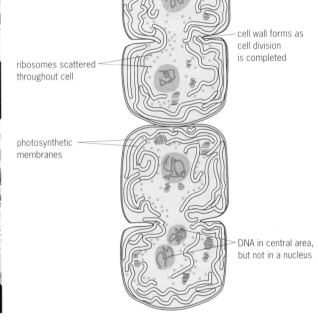

ribosomes scattered throughout cell

cell wall forms as cell division is completed

photosynthetic membranes

DNA in central area, but not in a nucleus

Fig 11.7(a) *Haemophilus influenzae* (above) is a small bacterium that used to be the most common cause of meningitis in children under 4 years old. A genetically engineered vaccine to *H. influenzae* was developed in the early 1990s and this is now given routinely to babies in the UK and other parts of the developed world, saving many lives

Fig 11.7(b) *Anabaena* (above right) is a common filamentous blue-green bacterium which photosynthesises in the same way as plants and algae

Fig 11.7(c) *Staphylococcus aureus* (right), the bacterium often responsible for boils and abscesses on human skin

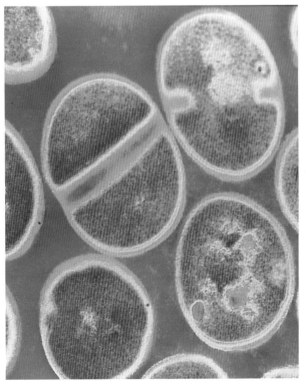

The role that bacteria play in health and disease is well known, but bacteria also help to regulate the environment. They can use and produce all the major environmental gases including nitrogen, dinitrogen oxide, oxygen, carbon dioxide, carbon monoxide, several sulphur-containing gases, hydrogen, methane and ammonia.

Many scientists now think that eukaryotic cells evolved from **symbiotic** associations between different species of bacteria. They describe this idea in the endosymbiont theory (endo = inside), which is covered in more detail in Chapter 1. The earliest associations could have produced the protoctists that, over millions of years, evolved into plants, animals and fungi. Fig 11.8 summarises how leading taxonomists think this might have happened.

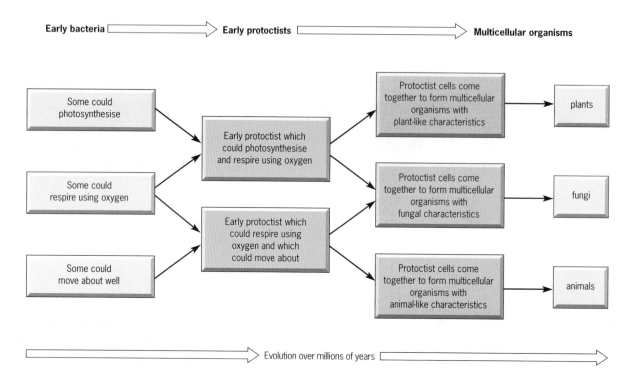

Fig 11.8 **A scheme suggested by taxonomists to explain the evolution of organisms in the present-day kingdoms (including the Monera) from early forms of bacteria**

Using RNA as a tool in classification

Diagrams which represent the phylogeny (relationships) between different groups of organisms can be found in many books about classification. In one direction (usually horizontally), the diagrams show relationships between groups. In the other direction (vertically) they show how organisms may have evolved over time. So, what information has been used to devise these diagrams?

Sometimes fossil evidence is used. Some organisms, mainly those with hard structures such as bones, have left fossils that tell us a lot. But for others, such as bacteria, we have little or no fossil clues to go on.

Despite this, taxonomists have been able to piece together bacterial family trees by studying living bacteria. Parts of the cell machinery are so vital that they change very little over long periods of time. Ribosomal RNA is one of these vital components and this has been used as a tool to trace the evolutionary relationships between bacteria and other organisms. There are obvious differences between bacteria, archaebacteria and eukaryotes (Fig 11.9). This tells us that these three groups **diverged** (split away from each other) a very long time ago.

When we look at different bacterial species, we find that they have small differences in the sequence of bases in their RNA. In general, the more differences there are between groups, the less well related they must be, and the longer ago they must have diverged. This technique is quite recent, and so many phyla have not yet been analysed. Taxonomists may be in for some surprises.

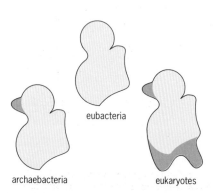

Fig 11.9 **Taxonomists can now extract ribosomes from cells and break them up into their constituent subunits. One, the 30S subunit, is different enough in bacteria, archaebacteria and eukaryotes to be used as a classification tool**

4 KINGDOM PROTOCTISTA

What is a protoctist? It is easier to say what a protoctist is not than to say what it is:

- Protoctists are not bacteria.

- Protoctists are not animals (see Chapter 12 for the definition of an animal).

- Protoctists are not plants (see Chapter 13 for the definition of a plant).

- Protoctists are not fungi (see the next section in this chapter for the definition of a fungus).

Protoctists include everything else! All protoctists are eukaryotes. Some examples are shown in Figs 11.11 to 11.14. As Fig 11.8 illustrates, their ancestors probably arose by symbiosis between different species of bacteria. This kingdom includes all the algae, all the motile phytoplankton, all the protozoans and some weird organisms, the slime moulds and slime nets. Some members of this kingdom photosynthesise, some capture food, and others are decomposers. It is tempting to think that the protoctists can be divided into 'plant-like' members, 'animal-like' members and those that are like fungi, but this would be too simple. Some of its members show characteristics of both animal and plants cells. **Euglena**, for example, can photosynthesise *and* engulf prey.

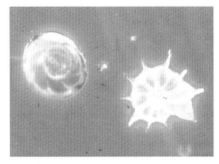

Fig 11.11 *Foraminifera* are small protoctists with a shell that live in the sea. They feed on ciliates, algae and other small organisms (an animal-like feature) but their life cycle shows alternation of generations (a plant-like feature, see page 204). This combination of features is often found in the Kingdom Protoctista

Fig 11.12 *Euglena* also shows plant-like and animal-like features. Most species of *Euglena* can photosynthesise, and they have flagella and so can move. They can also change shape. Some types have no chloroplasts and live solely by eating other organisms

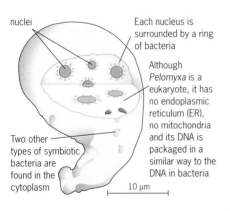

nuclei

Each nucleus is surrounded by a ring of bacteria

Although *Pelomyxa* is a eukaryote, it has no endoplasmic reticulum (ER), no mitochondria and its DNA is packaged in a similar way to the DNA in bacteria

Two other types of symbiotic bacteria are found in the cytoplasm

10 µm

Fig 11.10 *Pelomyxa* is one of the most primitive living eukaryotes. It contains three types of endosymbiotic bacteria – bacteria thought to have existed independently at one time but which have come to live inside *Pelomyxa*

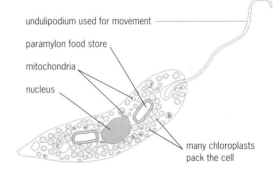

undulipodium used for movement

paramylon food store

mitochondria

nucleus

many chloroplasts pack the cell

Fig 11.13(a) *Fucus vesiculosus* is a bladder wrack, a large seaweed. You might think it odd that a multicellular alga should be classified with single-celled organisms such as *Pelomyxa*, rather than being included in the Kingdom Plantae. The reason is to do with the life cycle; see Fig 11.13(b)

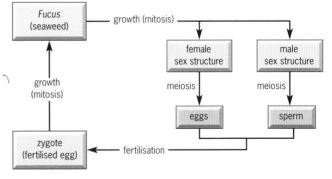

Fig 11.13(b) **The life cycle of *Fucus*. *Fucus* is unusual among protoctists: it has a well developed cycle of sexual reproduction. The female sex structures release eggs into the water and these are fertilised by sperm released from the male parts of the seaweed. The fertilised egg grows into the mature seaweed, which is called a thallus, not a plant. A thallus is a multicellular structure that shows little or no differentiation: it has no definite leaves, shoots or roots, even though it looks plant-like**

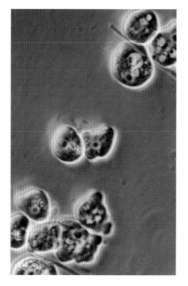

Fig 11.14 **Slime moulds would be very difficult to classify if the Kingdom Protoctista did not exist. Left: The cellular slime mould takes the form of individual amoebae at one stage in its life cycle. Above: These later come together to form a single multicellular 'slug', which can move about. Cells within the slug then differentiate to form a spore body, with a base, stem and cap. The spores that are released develop into amoeboid cells and the cycle begins again**

?

C What would you expect to happen if *Pelomyxa* were treated with antibiotics?

D Give a reason why *Pelomyxa* is put into the Kingdom Protoctista, rather than into the Kingdom Prokaryotae.

Difficulties with classification

When is a plant not a plant? When it's a protoctist!

In everyday language we use the words 'plant' and 'animal' quite loosely. Plants are large green things that photosynthesise, and animals eat plants or other animals. Most people don't lose sleep over whether a fungus is a plant or not. But in science, and particularly in taxonomy, we need to be much more precise.

When the microscope was developed, scientists could study very small organisms for the first time. Some of them were green and obviously 'plants', some engulfed prey and were obviously 'animals' and some fed on dead material and were obviously 'fungi'. But many had a mixture of plant, animal and fungal characteristics. How should they be classified?

A key principle of taxonomy is that any classification system should take account of the evolutionary relationships between groups (their phylogeny). As we discover more about the ancestry of organisms, we change our ideas about how we should classify them. Changing minor details of a classification system (families, species, etc) is not too much of a problem but, occasionally, new information forces us to change our ideas about how the big groups (kingdoms and phyla) are organised. This is exactly what has happened with the protoctists.

The old Kingdom Protista and the new Kingdom Protoctista

Clearly, evolution did not give us two big groups, plants and animals. The earliest eukaryotes arose through associations between different types of prokaryotic cell to give a great diversity of single-celled life. Taxonomists originally placed all of these single-celled organisms in the Kingdom Protista.

But this created problems. For example, some single-celled plants (algae) were placed in the Kingdom Protista, yet multicellular algae were placed in a different kingdom, together with the land plants. Under this system many very similar organisms were placed in different kingdoms. For example, the single-celled photosynthesising green alga *Chlamydomonas* and the eukaryote *Pandorina* were placed in separate groups (see Fig 11.15). This did not make evolutionary sense and so all the unicellular protists, together with all their immediate multicellular 'descendants', were placed in a new kingdom, the

Kingdom Protoctista. This is the classification system that biologists use now. It is more helpful taxonomically, even though it still doesn't solve all the difficulties!

Choose your terms carefully!

Some older terms no longer apply to modern taxonomic groups, but many biologists still find them useful. 'Protozoa', for example, can be thought of as a collective noun for four Protoctist phyla, the rhizopods, flagellates, apicomplexans and ciliates – all single-celled 'animals'. The term 'algae' covers a range of phyla that are all now within the Kingdom Protoctista rather than the Kingdom Plantae.

So terms such as 'protozoa' and 'algae' are still useful when talking about ecology, for instance, but you should not use them when talking about how such organisms are classified. It is also worth mentioning that some books still use the general terms 'animal', 'plant' and 'fungus' as they always have. You need to remember that some of the organisms they are referring to may now be members of the Kingdom Protoctista.

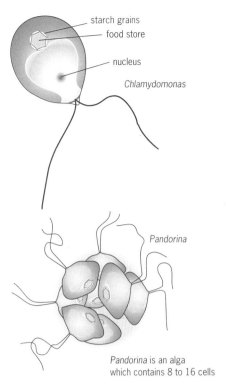

Fig 11.15 **Although *Chlamydomonas* is a single-celled organism and *Pandorina* is multicellular, the individual cells of the two organisms are very similar. It does not make sense to put them in different kingdoms**

5 KINGDOM FUNGI

The Kingdom Fungi includes moulds, mushrooms, yeasts and other spore-forming organisms. (Biologists do not distinguish between mushrooms and toadstools.) It does not include organisms that show fungus-like characteristics but that have cells that are capable of movement. These organisms are now classified as protoctists.

Characteristics of fungi

Fungi cannot photosynthesise and they are **non-motile**, meaning they cannot move independently. For these reasons they have evolved two particular characteristics:

- They produce very large numbers of tiny **spores** to maximise their chances of colonising new food sources. For example, a single giant puffball can contain over 7 million spores.

- They are extremely resistant to adverse conditions such as drying out. Some have even been found, alive and thriving, in concentrated sulphuric acid.

Fungi secrete powerful enzymes to break down the material of living or dead organisms. This becomes their food. They absorb it through their cell walls rather than ingesting it as animals do (see Chapter 15 for more details on how different heterotrophic organisms obtain their food). Fungi therefore have to grow in or on their potential food. Since they surround themselves with food-digesting enzymes, it is hardly surprising to find that their cell walls include a very tough protective material, **chitin**. (This material is also found in animal exoskeletons, see page 45.)

How fungi reproduce

Most of the time, fungi reproduce **asexually** by producing spores. Many of the structures that most people think of as fungi are just the temporary spore-forming **fruiting bodies** of the fungus. The rest of the fungus consists of fine threads called **hyphae** (Fig 11.16), which grow out of sight, underground or inside a tree, for example.

Fig 11.16 **Fungal hyphae spread through dead material, absorbing nutrients as they go. The network formed is called a** mycelium

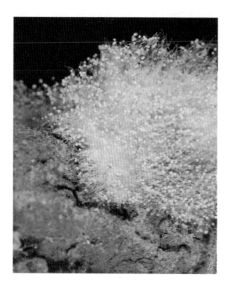

Fig 11.17(a) **The mould growing on the bread in the photograph above is *Mucor mucedo*. Spores travel through the air and land on the bread. They grow into hyphae and soon a mycelium develops. The fungus reproduces asexually by forming spores by mitosis. The photograph on the left shows the sporangia that contain the spores**

Fig 11.17(b) ***Mucor* can also reproduce sexually. The series of diagrams below shows how two hyphae from different strains fuse and produce spores by meiosis**

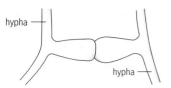

hypha

hypha

In sexual reproduction, swellings from hyphae of two different strains of *Mucor* grow towards each other

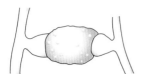

The two cells fuse. Nuclei inside the cells fuse. Later, meiosis occurs

zygospore

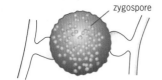

All the nuclei produced by meiosis degenerate except one. The zygospore which contains this nucleus becomes surrounded by crystals

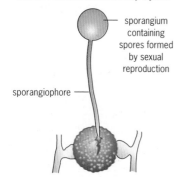

sporangium containing spores formed by sexual reproduction

sporangiophore

The zygospore germinates and produces a sporagium. This releases spores which, if they land on a food source, produce a new mycelium

Fig 11.18 **This bracket fungus could have been living inside this birch tree for years, but the appearance of the fruiting body of the fungus is a sign that the tree is now dying. By the time the fungus produces spores, its hyphae will have already spread throughout the tree, using up all the available nutrients. Producing spores gives the fungus a chance to establish itself in another tree, before it runs out of food**

In *Mucor*, large numbers of spores are produced and released from fruiting bodies called **sporangia** (Fig 11.17). Each **sporangium** forms at the tip of a special vertical hypha known as a **sporangiophore**, which keeps spores clear of the material on which the fungus is growing. When released, the spores are light enough to be dispersed by the wind. In some fungi, the fruiting body is quite large and complex. It may be short lived (mushrooms, for example, soon shrivel) or it may persist for a long time (bracket fungi on trees, for example, see Fig 11.18).

Fungi can also reproduce sexually (see Fig 11.17(b)). Hyphae from different strains of a fungus fuse together in a process known as **conjugation**. In *Mucor*, for example, a **zygospore** is formed, which can resist unfavourable conditions such as drought. It develops into a fungus when conditions improve.

Fungi and disease

Some fungi can infect humans and cause disease. Athlete's foot and thrush are common fungal infections that tend to be minor and easily treatable. Fungal infections cause more serious problems in AIDS patients. A species of *Mucor*, which is the same genus as bread mould, infects the lungs of AIDS patients (Fig 11.19). Because their immune systems are weakened by infection by the human immunodeficiency virus (HIV), they cannot resist the growth of the hyphae. Such infections cannot be treated easily and can lead to the death of the AIDS sufferer.

Classification of fungi

Fig 11.20 shows the phyla into which fungi are classified. Members of each phylum are decided according to their physical characteristics:

- The **Zygomycota**, which includes *Mucor*, have hyphae that are not divided into individual cells by cross-walls. The nuclei are spread out at intervals along each hypha.

- The **Ascomycota,** which includes yeast, bread mould, and truffles, have hyphae divided by cross-walls and form spores in flask-shaped structures called **asci** (Fig 11.21)

- The **Basidiomycota** includes mushrooms, puffballs, stinkhorns, jelly fungi and also the smuts and rusts, which cause some damaging diseases of crops. Basidiomycetes form spores on a club-shaped structure called a **basidium** (Fig 11.22).

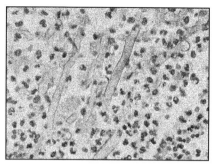

Fig 11.19 **Lung tissue from an AIDS patient suffering from an infection by *Mucor*. The hyphae of the fungus grow between the cells in the lungs. Although the patients cannot kill the fungus, their bodies do put up some response: white blood cells flood into the area, causing an intense inflammation**

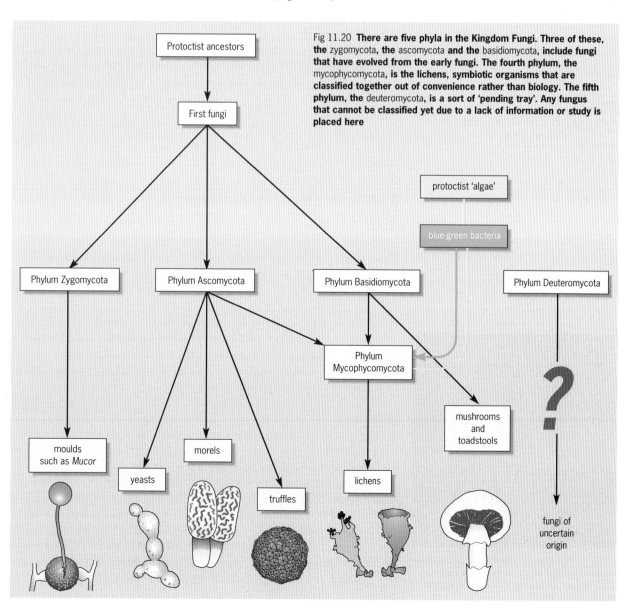

Fig 11.20 **There are five phyla in the Kingdom Fungi. Three of these, the** zygomycota, **the** ascomycota **and the** basidiomycota, **include fungi that have evolved from the early fungi. The fourth phylum, the** mycophycomycota, **is the lichens, symbiotic organisms that are classified together out of convenience rather than biology. The fifth phylum, the** deuteromycota, **is a sort of 'pending tray'. Any fungus that cannot be classified yet due to a lack of information or study is placed here**

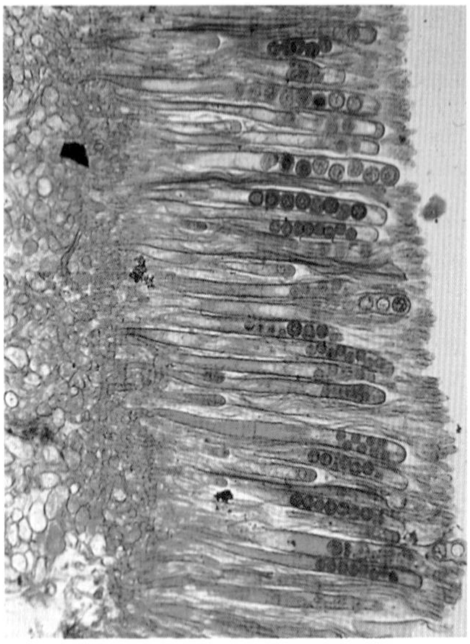

Fig 11.21 **The Ascomycota are named after their reproductive structure, the ascus, meaning flask. The asci, long tube shapes, contain the developing spores**

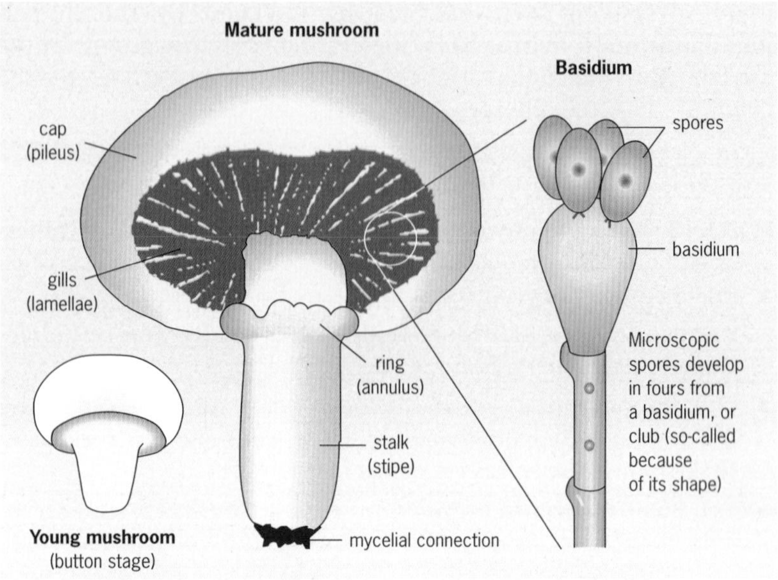

Mature mushroom

Basidium

cap
(pileus)

spores

basidium

gills
(lamellae)

ring
(annulus)

Microscopic
spores develop
in fours from
a basidium, or
club (so-called
because
of its shape)

stalk
(stipe)

Young mushroom
(button stage)

mycelial connection

Fig 11.22 **The surface of the gills of mushrooms is covered with basidia. Each club-shaped basidium produces four** basidiospores. **A spore contains only one set of chromosomes, so is** haploid

See question 2. ■

In addition, taxonomists have devised two other groups, the **lichens** (**Mycophycomycota**) and the **Deuteromycota** (see Fig 11.20). The phylum Deuteromycota reflects one of the difficulties of classifying rather than just 'cataloguing'. The Deuteromycota is a mixture of fungi, probably ascomycetes and basidiomycetes, that have lost the ability to form either asci or basidia, and no-one is sure where they should be placed. *Penicillium,* the mould from which penicillin was originally extracted, is in this group.

Symbiotic fungus relationships

Fungi can't make their own food and they can't move to find new food sources. In order to survive when their food supply runs out, some have evolved to become partners in **symbiotic** relationships. Two such associations are the lichens and the mycorrhizae. As you will see, the biological importance of mycorrhizae is immense, but such a relationship cannot, by itself, be used as a basis for classification (this is why mycorrhizae do not appear in Fig 11.20).

Lichens, on the other hand, are so different from their constituent partners that they have always been treated as a separate group. This is true even though some of the fungi and partners can, at times, grow independently of each other.

Lichens

A lichen is an association between a fungus and an alga (sometimes a blue-green 'alga', which is strictly a blue-green bacterium). The lichens *Cladonia* and *Parmelia* are shown in Fig 11.23 and 11.24.

Lichens are slow growing and highly resistant to drying out. Gravestone surveys show that some lichens grow as little as a few millimetres in a century. The alga photosynthesises and produces sugar or alcohols that the lichen can use as a food source. In turn, the lichen surrounds and protects the alga. Most lichens have evolved special (asexual) reproductive structures that ensure that fungus and alga are dispersed together. Each partner can also reproduce sexually.

Fig 11.23 **The lichen *Cladonia* is an association between an alga and an ascomycete fungus. The red patches are the tissue where spores are formed in asci**

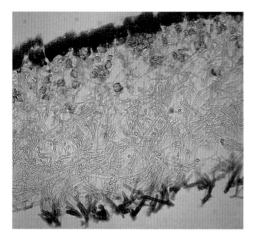 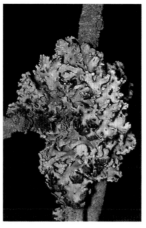

Fig 11.24 **The lichen *Parmelia*. Left: The micrograph shows the network of fungal hyphae which supports the unicellular algae (yellow) and gives shape to the lichen, seen right**

E Which partner is reproducing in the red patches of *Cladonia* in Fig 11.23?

Different algae can be symbiotic with the same fungus, and vice versa. Sometimes, more than one alga is present in a lichen and this has made classification difficult.

Mycorrhizae

A mycorrhiza (literally 'fungus-root') is a partnership between the roots of a plant and a fungus (Fig 11.25). Over three-quarters of all living plants have an association with a fungus partner.

The oldest fossil fungi, which date back to the Devonian Period, have been found associated with fossilised plant material. Some researchers have suggested that fungal partners might have enabled water-living plants to evolve to become land-living plants.

In mycorrhizae, the fungal threads penetrate the root of the plant and spread into the soil, to form a network that can gather mineral nutrients. The fungus delivers minerals to the plant and the plant provides food for the fungus. Without the fungus, plants can only extract small amounts of nutrient minerals from the soil and so grow much more slowly. Some species cannot grow at all without their mycorrhizal partner.

F Experiments on mycorrhizal partnerships have shown that when a radioactive tracer is injected into one tree, it appears shortly after in another nearby tree. Suggest how this is possible.

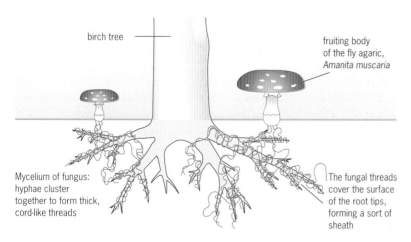

birch tree

fruiting body of the fly agaric, *Amanita muscaria*

Mycelium of fungus: hyphae cluster together to form thick, cord-like threads

The fungal threads cover the surface of the root tips, forming a sort of sheath

Fig 11.25 **The fly agaric, a poisonous fungus, is usually found near to a birch or pine tree with which it forms a mycorrhizal association**

In some North American pine forests, the forest floor is too dark for young seedlings to photosynthesise, and their early development depends on the fungus that their parent tree associates with. The seeds that germinate are nourished by the fungal network until the young trees grow to reach the light and photosynthesise.

G Young pine trees may be nourished by food from the mycorrhiza, but what is the source of food for the fungus?

SUMMARY

When you have finished this chapter you should know and understand the following:

■ Viruses are an exception to cell theory; they can reproduce themselves by taking over the machinery of host cell, but they cannot exist as independent living organisms.

■ Simpler organisms (those that are not animals or plants) are classified into one of three kingdoms: the **Kingdom Prokaryotae**, the **Kingdom Fungi** and the **Kingdom Protoctista**.

■ Simpler organisms are classified on the basis of their physical characteristics and also on the basis of their evolutionary relationships with other organisms.

■ The Kingdom Prokaryotae includes all bacteria, including the blue-green bacteria and three types of archaebacteria (the halophilic bacteria, the thermophilic bacteria and the methanogenic bacteria).

■ The Kingdom Protoctista includes organisms such as protozoa or algae that cannot be placed in any other kingdom. This group includes organisms which have both plant and animal characteristics.

■ The Kingdom Fungi includes moulds, mushrooms, yeasts and other spore-forming organisms. These are all eukaryotic organisms that feed on dead material.

■ Lichens and mycorrhizae are associations between organisms from different kingdoms.

■ It is often difficult to classify living organisms: ideas change as new evidence is discovered.

QUESTIONS

1

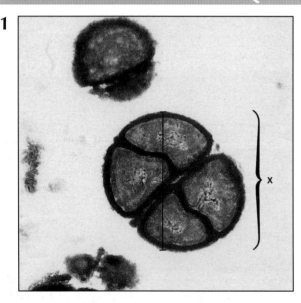

a) Make an enlarged drawing of the group of dividing cells labelled **X** in the photograph. There is no need to label your drawing or to show internal features of the cells.

b) Give **one** piece of evidence, visible in the photograph, which suggests that these cells are members of the Prokaryotae.

c) The cells in the photograph are magnified 32 000 times. Calculate the diameter in micrometres of the group of cells labelled **X** along the line shown in the photograph.

[AEB winter 1994 Biology: Paper 3, q.2]

2 The diagram below shows stages in the life cycle of a basidiomycete fungus (Basidiomycota), *Agaricus*.

a) Identify the structures labelled A, B and C.

b) Explain the function of structure D in spore dispersal.

c) State *two* features which distinguish the basidiomycete fungi from the zygomycetes and the ascomycetes.

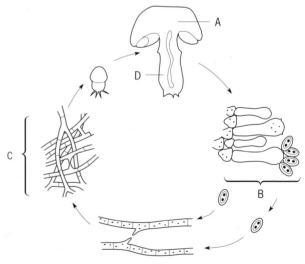

[ULEAC 1992 Biology: Paper 4A, q.5]

Assignment

MICROORGANISMS IN SEWAGE TREATMENT

Practically anything can end up in our sewerage system and the process of sewage treatment must safely dispose of all potentially harmful waste. A summary of one type of process is shown in Fig 11.A1.

Filters remove unwanted inorganic matter (grit, cans, bits of motorbike, etc) leaving a mixture of organic solids, a thick **sludge,** swimming around in foul-smelling liquid. The solid and liquid parts of the sewage are separated by allowing the solid matter to settle to the bottom

of large tanks (the **primary settlement tanks**). The sludge is pumped into **sludge digester tanks** and kept here for up to a month while it is digested by microbes that thrive in anaerobic conditions. The large amounts of methane gas that are produced are collected and burned to generate electricity for pumping.

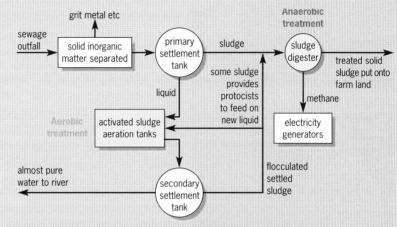

Fig 11.A1 **A summary of the sewage treatment process**

(a) *Squatinella* has clusters of cilia that waft food towards a primitive mouth containing two grinding plates. The food then passes into its gut

(b) *Paramecium* is covered with cilia

nuclei
contractile vacuole

contractile vacuole
tubes to grasp prey
nucleus

(c) *Tokophrya* is a parasite that sucks the cytoplasm out of other protoctists

cilia
nucleus

(d) *Saprodinium* is an obligate anaerobe

two nuclei

(e) *Oxytricha* has cilia that are grouped together to form structures called cirri

cirri
100 μm

Fig 11.A2 **Four of these organisms are found in aeration tanks**

1

a) What type of microorganisms might be used in the sludge digester?
b) 'Waste' heat is used to keep the sludge tanks at about 35 °C. Why?

Liquid sewage is treated by protoctists ('protozoa') that feed on the potentially harmful bacteria it contains. This takes place in **aeration tanks** through which compressed air is pumped continuously. (Without aeration, the bacteria would use up all the oxygen and the protoctists would die.) Many of the protoctists produce thick mucus, which causes the suspended organic matter to clump together, or **flocculate**. After a few hours, the mixture is allowed to settle and the flocculated sediment is pumped to the sludge digester, leaving behind clean water.

2

a) Which organism shown in Fig 11.A2 is not found in aeration tanks, and why?
b) Which of the four organisms that *are* found in the aeration tanks is not a protoctist, and why?

3 As a result of an accident at an electroplating works, some cyanide was released into the sewerage system. Soon afterwards, many fish were found dead in the local river downstream of the sewage treatment works. The fish had not been poisoned – they had suffocated from lack of oxygen. Explain how.

4 Not all sewage treatment works use the activated sludge process to treat liquid sewage. Find out other methods that are used to aerate the tanks during treatment. Describe the advantages of each method.

IN DECEMBER 1995, a tabloid newspaper introduced a completely new animal to its unsuspecting readers. *Symbion pandora*, a small jelly-like blob less than a millimetre in length, lives on the 'lips' of the Norwegian lobster. You may know this lobster by its more common name: scampi.

So why is this tiny creature of so much interest – to zoologists in particular? Strangely, *Symbion pandora* does not belong to any of the recognised animal phyla – it is definitely not a vertebrate, an arthropod or mollusc, or any other type of animal that we know about. And its structure, behaviour and life-style make other stories that you read in tabloid newspapers seem distinctly tame.

It has no head, but can have two mouths at some stages of its life cycle. These have circular mouthparts and are sited next to the creature's anus – an unusual design feature by anyone's standards. Its sex life is peculiar, too. The female reproduces asexually most of the time, budding off male 'babies' from the surface of her body. Males have two penises and cling on to the first female they come across. Female 'babies' are also produced: these grow inside the female's body. Some of these babies have their own babies while still there, and then two generations of females burst out of the mother's/grandmother's body.

When the scampi sheds its skin, *Symbion* is shed with it. The female *Symbion* then needs to find a mate to reproduce sexually. She literally dissolves and reforms inside her 'skin', and becomes a new form that is more independent and can move about in search of a mate.

Symbion pandora – it may not be pretty, but it can turn heads in the world of zoology

1 WHAT IS AN ANIMAL?

You might wonder why we need a definition of an animal. After all, anyone can recognise one. But not all animals look like our idea of what an animal should look like, so we need a set of features that defines exactly what an animal is:

- Animals have cells that are eukaryotic (see Chapter 1). This distinguishes them from bacteria and blue-green bacteria. These organisms are prokaryotes, members of the Kingdom Prokaryotae.

- Animal cells do not have cell walls. This distinguishes them from organisms in the Kingdoms Plantae and Fungi.

- Animals are multicellular. We now place single-celled organisms with animal-like features in the Kingdom Protoctista (see Chapter 11).

- Animals have nerves and muscles. These structures enable them to respond to their environment and to **locomote** (move from place to place) at some stage during their life cycle.

- When animals reproduce sexually, a microscopic sperm cell fertilises a larger female egg cell, forming a **diploid zygote**. During embryological development, the zygote divides and forms a hollow ball of cells called a **blastula** (see page 33).

- Animals do not have growing points like plant meristems. Growth takes place throughout the body. Animal growth usually results in a compact body that is easy to move about. Most adult animals are mature, reproductive organisms whose overall growth has slowed down or has stopped altogether.

A Name three differences between animals and fungi.

How animal bodies are organised

Animal bodies show a range of different forms. These patterns of body organisation are **progressive** (they have developed through evolution) and **adaptive** (they fit the life-style of the organism). Zoologists recognise three basic patterns of body organisation: **symmetry**, **tissue organisation** and possession of a **coelom**. We now look at all three in detail.

Symmetry

Some animals, such as sponges, are **asymmetrical**: the shape of their bodies is not regular. In general, asymmetry does not allow complex body functions to develop. A **symmetrical** animal body is arranged around a **point** or an **axis** (Fig 12.1). There are two main types of symmetry:

- Slow moving or stationary animals such as jellyfish and fan worms are **radially symmetrical**. Their sensory, feeding and defensive structures are evenly spaced over the body surface. This allows them to sense what is going on in their environment, enables them to capture prey more easily and helps them to avoid predators (they don't have a 'blind side').

- Most animals move around much more actively because they need to search for their food – mobile animals tend to be **bilaterally symmetrical**. Their body has a definite front, or **anterior end**, and a back, or **posterior end**. The mouthparts and the main sense organs are usually at the front because this end usually reaches new situations first, and most animals have a distinct **head** (see Chapter 21).

The **dorsal** surface is the 'back' of an animal, or, more generally, the surface that faces away from the ground. So, a camel has humps on its dorsal surface.

The **ventral** surface is the 'front', or more generally, the surface that faces the ground. So, the teats of a female pig are on her ventral surface.

The **anterior** end of an animal is the front end, usually where the head is. The **posterior** end is the rear end of an animal, usually where the anus is.

An **internal** structure lies inside the body. An **external** structure lies outside.

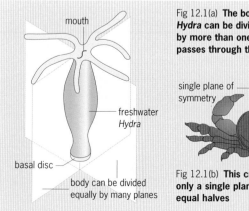

mouth

freshwater
Hydra

basal disc

body can be divided
equally by many planes

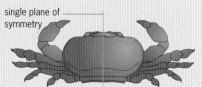

single plane of
symmetry

Fig 12.1(a) **The body of this radially symmetrical *Hydra* can be divided into roughly equal halves by more than one straight line, or plane, which passes through the central point of the body**

Fig 12.1(b) **This crab is bilaterally symmetrical; only a single plane can divide its body into equal halves**

Fig 12.1 **Radial and bilateral symmetry in animals**

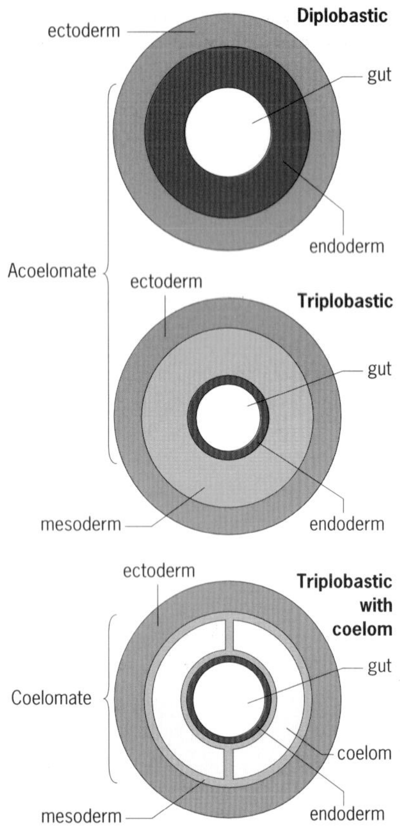

Fig 12.2 **Diploblastic and triploblastic bodies have two and three cell layers respectively. Only triploblasts can develop a true coelom, a body cavity that forms within the mesoderm**

Tissue organisation

Sponges are primitive animals which have specialised cells, but no tissues – if a sponge is put into a blender and pulverised, it will later recover as its cells rearrange themselves. In most animals, however, cells are organised into tissues (see Chapter 2). We divide animals into two types, depending on the number of layers of tissue present in their embryo:

● The **diploblastic** (two-layered) body plan is the simplest tissue level of organisation (Fig 12.2). Such animals develop from a two-layered embryo. The outer layer, or **ectoderm**, forms the epidermis of the animal and inner layer, or **endoderm**, forms the internal structures. Hydra, jellyfish and sea anemones (all from the Phylum Cnidaria) are diploblastic.

● All animals from flatworms to humans are **triploblastic**: they develop from a three-layered embryo (Fig 12.2). In the extra middle layer, the **mesoderm**, the animal can develop complex body organs and organ systems.

Possession of a coelom

The three body layers of a triploblastic animal develop during **gastrulation**, a process of infolding in the embryo (see Fig 2.8 on page 33). In some animals the internal organs are simply crammed into the body and the overall effect is of a fairly dense and solid body. In most animals, however, the organs are suspended in a fluid-filled space called the **coelom** (see Fig 12.2). A coelom allows an animal to move its internal organs independently. The gut can therefore perform peristalsis without causing the outer part of the body to contract.

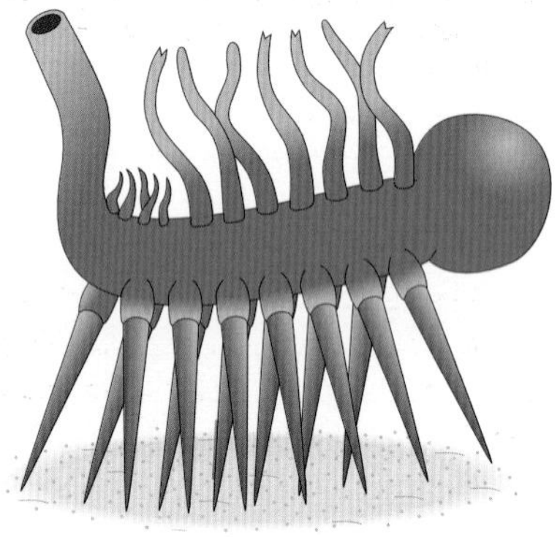

Fig 12.3 *Hallucigenia*. **This truly bizarre animal has been drawn from finds made in the Burgess shale**

2 HOW WE CLASSIFY ANIMALS

In 1909, the American palaeontologist Charles Walcott discovered some invertebrate fossils high in the Canadian Rockies. Studying rock sediments called the Burgess shales, he found the remains of ancient animals that had lived on a muddy sea floor, 500 million years ago. They had been perfectly preserved: he could see the soft parts as well as the hard parts.

Such well-preserved fossils are unusual enough, but even more significant was the amazing diversity of these animals' body plans. Most zoologists recognise about 30 animal phyla. This one small quarry in British Columbia revealed 15 to 20 organisms so bizarre and different from anything alive today that they have been placed in several new phyla (Fig 12.3).

The rest of this chapter looks at the variety of the more usual animal types. Complete coverage of all types is outside the scope of this book: we cover only the main seven.

The Phylum Cnidaria

The **Cnidaria** (also called the **Coelenterata**) contains soft-bodied animals including jellyfish, hydras, corals and sea anemones. They can exist in two forms: a fixed **polyp** such as a sea anemone, and a floating **medusoid** form such as a jellyfish (Fig 12.4). All have a

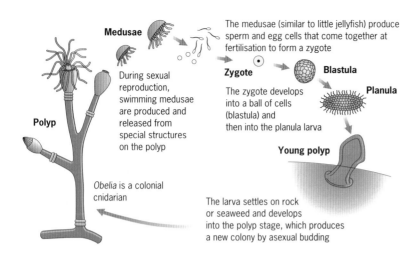

Fig 12.4 **Life cycle of** *Obelia* **showing the polyp (a colony of many individuals) and the medusoid stages**

Medusae

The medusae (similar to little jellyfish) produce sperm and egg cells that come together at fertilisation to form a zygote

During sexual reproduction, swimming medusae are produced and released from special structures on the polyp

Polyp

Zygote

The zygote develops into a ball of cells (blastula) and then into the planula larva

Blastula

Planula

Young polyp

Obelia is a colonial cnidarian

The larva settles on rock or seaweed and develops into the polyp stage, which produces a new colony by asexual budding

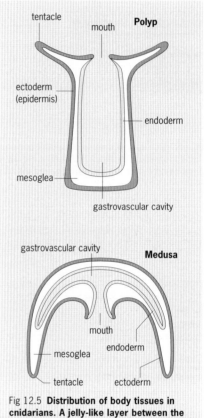

Fig 12.5 **Distribution of body tissues in cnidarians. A jelly-like layer between the outer and inner layer can form a large part of the jellyfish cnidarians. This jelly is called the** mesoglea

relatively simple body construction: two cell layers enclosing a simple gut (Fig 12.5). Cnidarians show simple radial symmetry.

All cnidarians are carnivorous: they kill prey, such as small crustaceans, using tentacles armed with **stinging cells** (Fig 12.6). A cnidarian paralyses prey, pushes it into its mouth and then digests it in its gut. Its behaviour is limited by its simple nervous system. A **nerve net** coordinates movements such as the feeding responses of *Hydra* and the escape responses of jellyfish (see Chapter 21).

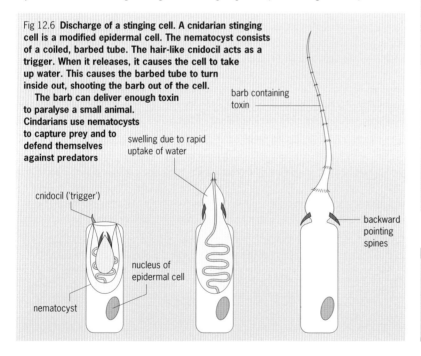

Fig 12.6 **Discharge of a stinging cell. A cnidarian stinging cell is a modified epidermal cell. The nematocyst consists of a coiled, barbed tube. The hair-like cnidocil acts as a trigger. When it releases, it causes the cell to take up water. This causes the barbed tube to turn inside out, shooting the barb out of the cell.**
 The barb can deliver enough toxin to paralyse a small animal. Cindarians use nematocysts to capture prey and to defend themselves against predators

B A sea anemone lives in one place all its life, fixed firmly to a rock in a pool. How does it colonise other parts of the rock pool?

The Phylum Platyhelminthes

This phylum contains the flatworms, which include the free-living planarian worms (**Class Turbellaria**) and two types of internal parasite, the tapeworms (**Class Cestoda**) and the flukes (**Class Trematoda**). All three groups have flattened bodies and, as a result, a high ratio of surface area to volume.

The platyhelminths have more complex organ systems than the cnidarians because they are triploblastic (they have three body layers, rather than two). They are bilaterally symmetrical and **acoelomate** (they have no coelom).

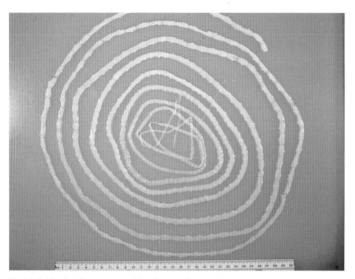

Fig 12.7 **The beef tapeworm, *Taeniarhynchus saginatus*, usually infects humans when they eat undercooked meat, although strict control measures have made it very rare in more developed countries like Britain**

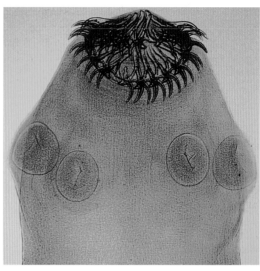

Fig 12.8 **The scolex of the pork tapeworm, *Taenia solium*, showing the hooks and the four suckers which attach it to the intestine of its host**

Platyhelminths have complex life cycles. Tapeworms are highly adapted parasites that live in the intestines of many vertebrates, including fish, rats, pigs, sheep and humans. They can reach several metres in length (Fig 12.7). An adult tapeworm is impressively simple. It anchors itself to the intestine wall using a **scolex** (Fig 12.8) and then simply absorbs the digested food generously provided by the host. It does not need a mouth, gut or excretory system: exchange of all materials occurs over the whole body surface.

As its host provides food, warmth and protection from predators, the tapeworm is free to concentrate on reproduction. The reproductive capacity of these organisms is incredible: each day several **proglottid segments** break off and are pushed out of the host's body in faeces. Each proglottid contains thousands of eggs. The life cycle varies from species to species, but usually involves an intermediate host, such as a pig, cow or sheep, as well as a main host. Humans can become infected by eating under-cooked meat that contains tapeworm larvae.

Flukes, such as the liver fluke ***Hepatica fasciola***, are parasites of sheep, cattle, pigs and occasionally people. Like tapeworms, the leaf-shaped flukes are highly adapted for a parasitic lifestyle, having suckers at both ends of the body (Fig 12.9). When ingested, liver flukes swim up the bile duct and attach to the wall of the duct. A few individual flukes produce no symptoms but a heavy infestation causes an illness known as **fascioliasis** or liver-rot.

?

C Why do tapeworms need to anchor themselves to the intestine wall?

?

D Some basic body features enable animals to develop complex body systems. Name two of these features.

Fig 12.9 **The liver fluke *Hepatica fasciola* has suckers at both ends of its body. It uses these to attach to the walls of the bile duct of its host. The ventral sucker can be seen clearly in this photograph**

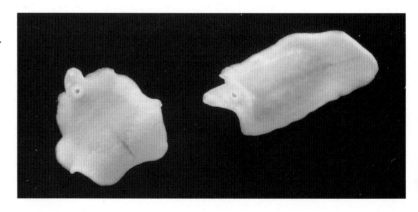

BILHARZIA

THE DISEASE OF BILHARZIA, or schistosomiasis, is less well known than malaria, but it is a major cause of suffering and death. It affects an estimated 200 million people world-wide, most of them in the tropics.

The organism responsible for bilharzia is the blood fluke or **schistosome** (Fig 12.10). This remarkable parasite can bore its way straight through skin and so anyone bathing in infested water is likely to be infected. Once it has entered the blood, the schistosome migrates to the liver, where it breeds. The eggs eventually reach the rectum or bladder (depending on host species) and leave the body. Once in the water, the schistosomes infect various species of water snail and start to multiply. The snails release large numbers of parasites, which are able to reinfect a human host.

Symptoms of bilharzia include blood passing in the urine and faeces, after the bladder or intestinal wall ruptures. Heavy blood loss can lead to anaemia and death.

We fight the disease by destroying the intermediate host (the snail) and by warning people to avoid infected water. Once people are infected they can be treated effectively with drugs, but only if taken in time.

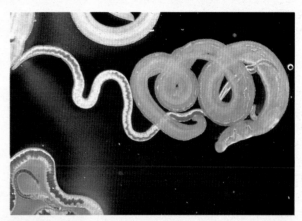

Fig 12.10 **The schistosome is a platyhelminth with the characteristic parasitic ability to reproduce in huge numbers. Here, the male (thick and blue) lies permanently attached to the female, fertilising eggs that pass out of her body**

The Phylum Nematoda

Nematodes are free-living or parasitic roundworms. They may be the most abundant multicellular animals on the planet: one cubic metre of soil contains, on average, about 1 billion individual nematodes – but how many nematodes can you name? Pet-owners might have heard of *Toxocara canis* (which infects dogs) and those familiar with farming may know of *Ascaris lumbricoides*, a parasite of pigs (Fig 12.11), but the rest of this enormous group remains quietly anonymous in the developed world. In other parts of the world, nematode infections can cause severe disease in humans (Fig 12.12).

Unlike the platyhelminths, the nematodes have a **through gut** that starts at a mouth and ends with an anus. This is a major step forward in food processing, enabling food to be eaten and digested continually. Another important feature of body organisation is the presence of a body cavity, rather like the coelom in other groups.

Fig 12.11 **The pig roundworm *Ascaris lumbricoides* can grow to be 30 cm long or more. It is widespread and relatively harmless, but a large infestation can block the gut of an affected animal**

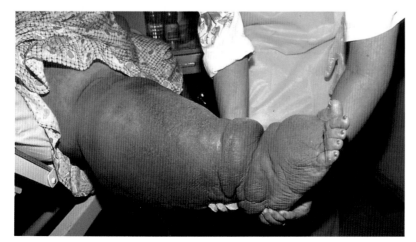

Fig 12.12 **The swelling disease elephantiasis is caused by an accumulation of fluid due to an infection of the nematode worm *Wuchereria bancrofti*. The worm blocks lymph vessels and so stops lymph draining away**

The Phylum Annelida

Annelid worms include the earthworms, leeches and marine rag-worms. Some swim in pursuit of food but most burrow in the sand, mud or garden soil. They move by contracting alternate body segments (see page 308). Annelids have a true coelom (a fluid-filled body cavity) and a through gut. The coelom is important for the transport of body fluids and discharge of gametes, and it also acts as a **hydrostatic skeleton** (see page 308).

Annelids show true **body segmentation**: their bodies are divided into definite sections or **segments** (Fig 12.13). Annelids can control each segment separately and continue to add segments throughout life.

?

E Why is it better for an animal to eliminate undigested food through a second opening, an anus, rather than through the mouth?

F What is the main difference between an earthworm and a ragworm? Hint: the name of the class in which each organism is placed is a big clue.

(a)

Fig 12.13(a) **Leech (Class Hirudinea)**

Fig 12.13(b) **Earthworm (Class Oligochaetae)**

Fig 12.13(c) **Ragworm (Class Polychaetae)**

(b)

(c)

Fig 12.13 **The three main classes of annelid worm. All annelids have a segmented body and most have bristles or chaetae (also called setae). The parasitic leech has suckers to attach to its host**

The Phylum Mollusca

Molluscs are an extremely diverse and successful group of animals. They form the second largest animal phylum (arthropods are the largest) with almost 100 000 species. The common snail is a typical mollusc: it has a soft body, covered by protective mucus, or slime, and it has a hard shell into which it can retreat when in danger. Slugs and octopuses, which do not have a shell, are also molluscs.

The snail belongs to a group of molluscs that make up the **Class Gastropoda**. This term literally means 'stomach-foot' and refers to the snail's body plan (Fig 12.14(a)): the guts and other internal organs are twisted and piled on top of the muscular foot that enables the snail to move around. The internal organs are covered by a 'coat' or **mantle**, which secretes the shell and forms a cavity that contains the organs that are used for gas exchange.

The most advanced molluscs belong to the **Class Cephalopoda**, which contains the squids and octopuses, see Figs 12.14(b) and 12.14(c). These animals move through the water using a simple form of jet propulsion. By forcing water out from their **mantle cavity**, these animals can move fast, sometimes leaping out of the water. Molluscs that live on the sea bottom have exceptionally well-developed camouflage techniques: some can change colour completely in less than two-thirds of a second.

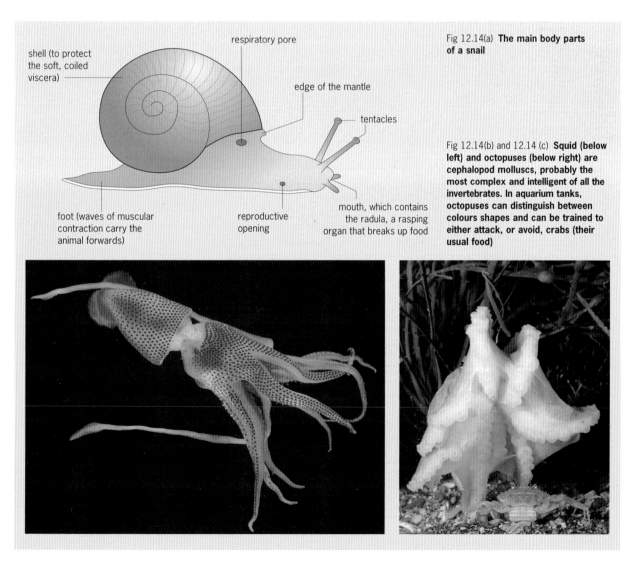

shell (to protect the soft, coiled viscera)

respiratory pore

edge of the mantle

tentacles

foot (waves of muscular contraction carry the animal forwards)

reproductive opening

mouth, which contains the radula, a rasping organ that breaks up food

Fig 12.14(a) **The main body parts of a snail**

Fig 12.14(b) and 12.14 (c) **Squid (below left) and octopuses (below right) are cephalopod molluscs, probably the most complex and intelligent of all the invertebrates. In aquarium tanks, octopuses can distinguish between colours shapes and can be trained to either attack, or avoid, crabs (their usual food)**

The Phylum Arthropoda

The world arthropod population, which includes spiders, insects, centipedes, millipedes, shrimps and other crustaceans, contains about 1 billion billion (10^{18}) individuals.

All arthropods share a common body plan. They have a segmented body, a hard **exoskeleton**, a skeleton on the outside that is a bit like armour plating, **haemocoeles** (body cavities filled with blood) and pairs of jointed limbs. This body design is very adaptable and a wide variety of running, jumping, burrowing, swimming and flying arthropods have developed.

The arthropod exoskeleton (Fig 12.15(a)), also called the **cuticle** is secreted by the outer layer of body cells. It covers the whole of the outside of the body and lines the air passageways and part of the digestive tract. It is made of **chitin**, a complex carbohydrate (see page 45 in Chapter 3). Some arthropods, such as crabs and lobsters, have a cuticle that also contains calcium salts. The cuticle provides support for the body and allows the animal to move – it is thin at joints – and provides points to which muscles attach. One disadvantage of a cuticle is that it cannot grow with its occupant – it needs to be shed and reformed at regular intervals in a process called **ecdysis**, or **moulting** (Fig 12.15(b)).

Fig 12.15(a) **A section of arthropod cuticle. It is suggested that arthropods are such a successful group because of their tough outer cuticle that forms the exoskeleton**

Fig 12.15(b) **The release of moulting hormones causes moulting of the cuticle, as shown here for a dragonfly. As the old cuticle is shed, a new one forms underneath**

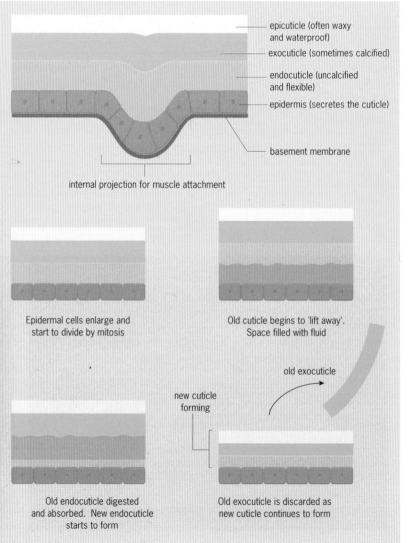

epicuticle (often waxy and waterproof)

exocuticle (sometimes calcified)

endocuticle (uncalcified and flexible)

epidermis (secretes the cuticle)

basement membrane

internal projection for muscle attachment

Epidermal cells enlarge and start to divide by mitosis

Old cuticle begins to 'lift away'. Space filled with fluid

old exocuticle

new cuticle forming

Old endocuticle digested and absorbed. New endocuticle starts to form

Old exocuticle is discarded as new cuticle continues to form

Table 12.1 shows how arthropods are classified into subgroups.

Table 12.1 **How we classify arthropods**

Class	Number of species	Features
Insecta	over 1 million	Insect species occupy all habitats except open oceans. The body is divided into head, thorax and abdomen. They have three pairs of legs, which arise in the thorax, and a single pair of antennae. Most adults have wings
Crustacea	around 32 000 divided into many different sub-classes	Crustaceans lack a waterproof cuticle and so all live in water or damp places. The body has two parts: the cephalothorax and the abdomen. They have five or more pairs of legs and two pairs of antennae
Arachnida	around 60 000	Includes the spiders, scorpions and mites. Nearly all live on land and all are carnivorous. The body has two parts: the prosoma and the opisthosoma. They have four pairs of legs but no antennae and do not have compound eyes
Chilopoda	3 000 centipede species	Centipedes are active carnivores and have a distinct head that bears poison claws. They have one pair of antennae on the head and one pair of legs per body segment
Diplopoda	10 000 millipede species	Millipedes are herbivorous and have no distinct head or poison claws. They have one pair of antennae and have two pairs of legs per segment

?

G Which class of animal is the only invertebrate that can fly?

See questions 1 and 2. ▥

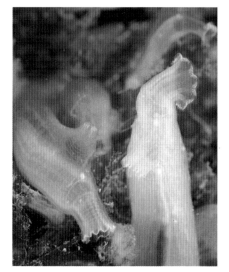

Fig 12.16 **This sea squirt, or tunicate, looks like a sea anemone, but it is in fact a chordate. Although the adult stage is** sessile **(fixed to a rock), the juvenile stage is a free-swimming tadpole-like creature that has a definite head,** notochord **and** dorsal nerve cord**. It swims by moving its muscular body like a fish. It is possible that an organism like this gave rise to the first primitive fish**

The Phylum Chordata

The chordates include the vertebrates and some less well-known invertebrates, such as sea squirts (Fig 12.16). Members of the Phylum Chordata have a **notochord**, or stiffening rod. This is a skeletal rod is made of tightly packed cells and forms a rigid internal support. The vertebrates are a subgroup of the chordates in which the notochord has been replaced by a bony backbone. The backbone consists of separate structures called **vertebrae**.

Zoologists classify vertebrates into five main groups, which are covered in detail below.

Fish

The 30 000 species of fish are divided into three recognised classes: the bony fish (**Class Osteichthyes**), the cartilaginous fish (**Class Chondrichthyes**) and the jawless fish (**Class Agnatha**).

The Class Chondrichthyes includes the largest of all fish – the whale shark. Other members are ordinary sharks, dogfish and rays. All have fleshy fins and a skeleton made from cartilage (Fig 12.17)

Fig 12.17 **The great white shark shows the characteristic features of the chondrichthyes: fleshy fins and a ventral mouth**

Fig 12.18 **This brown discus fish from the Amazon shows the characteristic features of the osteichthyes: bony fins, an operculum and a terminal mouth. Discus are part of the cichlid family. Cichlid parents spend a lot of time and effort caring for their young. Here the newly hatched fry are feeding on the nutritious slime that the parent fish produces**

See question 4. ▮

Fig 12.19 **All frogs and toads secrete poisons through their skin, but most poisons are not dangerous to humans. These Kokoi poison arrow frogs, however, secrete batrachotoxin, one of the most powerful poisons known. Just one nanogram (a thousand millionth of a gram) is enough to kill an average adult human, and the Choco Indians of Columbia use it to poison the tips of their arrows**

The Class Osteichthyes includes most of the familiar fish, such as salmon, tuna, stickleback, piranha and cichlids (Fig 12.18). They have a bony skeleton, **fins**, an **operculum** (gill cover) and a **swim bladder**.

The Class Agnatha includes what is left of a very primitive group of jawless fish, including the lampreys and hagfish, which are eel-like parasites of other fish.

Class Amphibia

There are 4 000 species of amphibians, which include frogs, toads, newts and salamanders (Fig 12.19). All members of this class have soft, permeable skin that can be used for gas exchange but that leaves the animal vulnerable to water loss, so most amphibians live in moist habitats. They depend on water for reproduction (Fig 12.20). Eggs hatch into aquatic larvae, **tadpoles**, which have external gills. Most species undergo **metamorphosis** (change of form) into land-living adults that have lungs. Adults use their entire skin,

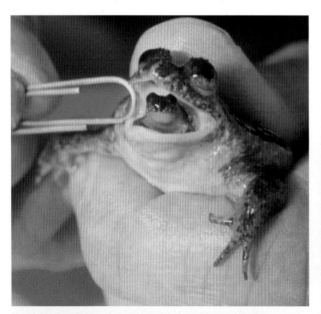

Fig 12.20 **Amphibian eggs must be incubated in an aquatic environment, and several species go to extraordinary lengths to achieve this. This Australian gastric-brooding frog incubates the eggs in her stomach, and vomits up the tadpoles or froglets. The tadpoles secrete a mucus containing a hormone which inhibits gut movements and acid production. Unfortunately, this frog is very rare and no wild specimens have been found since 1981**

Fig 12.21 **Like many cold-blooded animals, crocodiles grow throughout their life. This species can grow to a length of 7 metres and can kill animals as large as these wildebeest.**

Their metabolic rate is so low that one large meal can last a crocodile many months and possibly for a whole year, until the wildebeest migrate the same way again. This low food demand allows reptiles to become the dominant species in habitats where food is scarce

mouth and lung surface for gas exchange.

Class Reptilia: the reptiles

The 6 500 species of reptile consist of four main groups: the snakes, lizards, crocodilians and turtles/tortoises (Fig 12.21). Reptiles have developed ways of minimising water loss and have become truly **terrestrial** (land living). They have waterproof scales, made largely of keratin, and they lay eggs that can be incubated on land. These eggs have protective leathery shells and internal membranes that are permeable to oxygen, carbon dioxide and water vapour.

Class Aves: the birds

All 9 000 species of bird have beaks, feathers, lay eggs and are **endotherms** (warm blooded – see Fig 12.22). Many can fly and most have scaly legs and feet. The skeletons of birds and reptiles are very similar, and the chemical composition of feathers is virtually identical to reptilian scales, all of which indicates that these two classes descended from a common ancestor.

Class – Mammalia

The name mammal is derived from mamma which means 'milk gland'. All 4 000 species of mammal feed their young on milk (Fig 12.23). They are also endothermic (warm blooded) and have fur. Mammals protect their developing embryos in three different ways: this difference is used to divide them into three sub-classes:

- The **protheria**, or **monotremes**, are curious creatures. They include the duck-billed platypus (Fig 12.24) and the spiny anteater. Unlike most mammals, monotremes lay eggs. After a brief incubation, the eggs hatch and the young complete their development in the mother's pouch where they feed on milk. These animals are found only in Australasia.

- The **metatheria**, or **marsupials**, do not lay eggs, but they give birth to a very small and helpless fetus which also completes its development in a pouch, nourished by milk. Familiar marsupials include kangaroos, koalas, wombats and opossums (Fig 12.25).

Fig 12.22 **Birds have the highest core body temperature of all endotherms: 41–43 °C. To maintain this level, they have a high metabolic rate fuelled by an energy-rich diet and rapid digestion. In this Anna'a hummingbird, the relatively large heart pumps sugar-rich blood at the rate of over 10 beats per second**

Fig 12.23 (above) **Milk is a complete food that contains all the nutrients a young mammal needs. Milk ducts are thought to be modified sweat glands that have become adapted to produce nutritious sweat**

Fig 12.25 (right) **The kangaroo is a marsupial. Just before giving birth, the mother licks her fur, creating a trail that the grub-like baby can follow to the pouch. It stays here for a further 8 months**

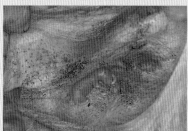

Fig 12.24 (left) **The duck-billed platypus. This curious mammal lays eggs that it incubates in a burrow. The young are born at a very early stage in their development, and are nourished by milk that oozes from the mother's skin and is lapped up from the fur**

Fig 12.26 (above) **The birth of a mammal. This lamb has developed in the uterus of the mother, protected by membranes and nourished via a placenta**

Marsupials are also only found in Australasia, except opossums, which also live in North and South America.

● **Eutherian**, or **placental mammals**, nourish their young through a placenta. This allows the developing fetus to be protected by the mother's body and allows birth at a relatively advanced stage (Fig 12.26). Most mammals belong to this sub-class, some major groups being the primates (including humans), rodents, cetaceans (whales and dolphins), carnivores (cats, dogs, bears, weasels/otters), ·chiroptera (bats), ungulates (hoofed animals) and the insectivores (moles, hedgehogs, shrews).

SUMMARY

The following summary highlights the characteristic features of the animal types covered in this chapter:

▨ Cnidaria
Diploblastic body organisation.
Radially symmetrical.
Simple body cavity leading from the mouth. No anus.
Two different body forms: the stationary polyp and the free-swimming medusoid forms.
Have stinging cells.
All are aquatic, most are marine.
Simple behaviours, no brain, just a nerve net.

▨ Platyhelminthes
Triploblastic body organisation.
Bilaterally symmetrical.
Flattened in shape – large surface area to volume ratio.
No coelom (body cavity).
A mouth, but no anus.
Many show hermaphrodite reproduction (both sexes in one animal).

▨ Nematoda
Triploblastic body organisation.
Bilaterally symmetrical.
Do not have a true coelom, but have a fluid-filled cavity.
Have a muscular gut, with a mouth and anus
Separate sexes.

▨ Annelida
Triploblastic body organisation with tissues and organs.
Bilaterally symmetrical.
Have a coelom.
Have a muscular gut with mouth and anus.
Body divided into segments.
Outer, soft cuticle.
A closed blood system (blood confined to vessels).
Ventral nerve cord and a very simple brain.

▨ Mollusca
Triploblastic body organisation with tissues and organs.
Bilaterally symmetrical.
Have a muscular gut with mouth, digestive tract and anus.
Body divided into a muscular head-foot together with internal organs above.
Have a fleshy mantle, gills lungs.
Many have a shell.

▨ Arthropoda
Triploblastic body organisation with tissues and organs.
Bilaterally symmetrical.
Segmented.
Have a coelom.
Most have a well developed head.
Have a hard exoskeleton for protection and support.
Have a heart and closed circulatory system.
Have jointed limbs.
Growth occurs in stages following moulting of the old cuticle.

▨ Chordata
Triploblastic body organisation with tissues and organs.
Bilaterally symmetrical.
Have a coelom.
Have a notochord at some stage of their life cycle.

▨ Vertebrata
These have the same characteristics as the Chordata, plus:
An endoskeleton (see Chapter 19).
A well-developed nervous system and brain (see Chapter 21).
A muscular heart and closed blood system (see Chapter 18).
Kidneys for excretion and osmoregulation (see Chapter 27).
Two pairs of limbs.

QUESTIONS

1 Complete the table below, naming *one* organism from each group in a single terrestrial ecosystem. For each organism, state *one* external feature characteristic of its group.

Group	Name of organism	Characteristic external feature
Arthropods		
Ferns (Filicinophyta) [see Chapter 13]		
Molluscs		

[ULEAC 1996 Biology: Module Test B2, q. 4]

2 Name **two** features which would confirm that an animal belongs to the phylum Arthropoda. The drawings show the structure of a barnacle, a sessile arthropod found commonly on rocks on the seashore.

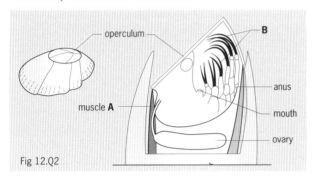

Fig 12.Q2

a) Explain **one** reason why this organism is likely to require less food than a motile arthropod of similar size.

b) Suggest the function of **(i)** muscle **A**; **(ii)** the structures labelled **B**.

[AEB Nov 1993 Biology: Paper 1, q. 2]

3 Some arrangements of animal body layers are shown in the diagrams.

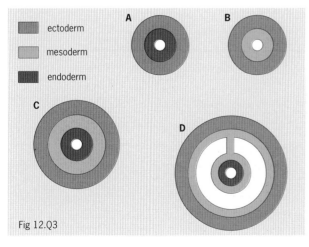

Fig 12.Q3

Using each letter once, more than once or not at all, identify the section which represents:

a) a triploblastic acoelomate;

b) an animal with a chitinous exoskeleton

c) a member of the phylum Chordata

d) a member of the phylum Cnidaria

[AEB June 1993 Biology: Common Paper 1, q. 1]

4 Fig 12.Q4 shows the external features of a cartilaginous fish and a bony fish.

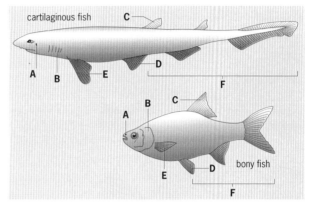

a) Write down the names of the structures labelled **A**, **B**, **C**, **D** and **E** for **either** the cartilaginous fish **or** the bony fish, stating the fish you have chosen.

b) What is the name of the part of the fish indicated by label **F** in Fig 12.Q4?

[UCLES 1996 Biology: Specimen Paper 1, Option 1: Biodiversity, q. 1]

5 Which diagram **A** to **E** in Fig 12.Q5 represents the relationship between the taxa shown?

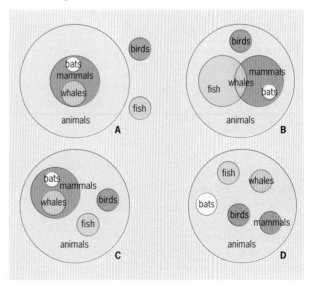

[NEAB June 1995 Biology: Paper 1, Section A, q. 1]

13 The plant kingdom

Scientists who travel to remote areas of rainforest to search for and identify plants which may be useful in medicine are called ethnobotanists. They work with local traditional healers to narrow down the choice of plants to study. Samples are sent back to laboratories for testing. Research programmes are now being set up to share the profits from successful compounds, so that indigenous people can benefit from the discovery, instead of suffering exploitation

THE WEATHER CONDITIONS in the tropical rainforests around the equator have encouraged the evolution of areas of high plant and animal diversity. The indigenous populations who live and work in the rainforests need to make a living and working for developers who sell timber and other products to the developed world is an obvious means of doing this. The consequences for the environment are serious. Some scientists have estimated that three species of living thing are being made extinct every *hour*, because the rainforest is being destroyed.

But what difference can a few plants make? Possibly the difference between life and death for many people. Screening of plants from the rainforest has shown they contain many compounds that protect them against predators and parasites, and which may also be useful in human medicine. Since 1990, at least four compounds have been identified which may provide drugs for use against the virus which causes AIDS (Acquired Immuno-Deficiency Syndrome).

The work of developing medicines needs sophisticated chemical and pharmaceutical techniques, but it cannot be done unless plants that are collected for screening can be identified accurately and consistently. Speed is also important. If destruction continues at its present rate, the rainforests will have been completely wiped out by the end of the next century.

1 THE KINGDOM PLANTAE

The earliest ancestors of plants lived in water and, compared with the land plants of today, had an easy time. Water supported them, allowed light for photosynthesis to reach them and provided them with carbon dioxide and essential minerals.

When plants started to live on land, they had to develop strategies to avoid drying out, to support themselves in air and to reproduce without plenty of water to carry sperm from one plant to the eggs of another.

Plants that have evolved to colonise the land are classified as members of the **Kingdom Plantae**. Not everything that appears plant-like is a member of the Kingdom Plantae. In some habitats, for example, algae are common, but algae are protoctists, not plants (see Chapter 11).

In this chapter, we look at what distinguishes land plants from organisms in the four other kingdoms.

Characteristics of land plants

Land plants have the following characteristics:

- They are multicellular.
- They can photosynthesise (a few exceptional parasitic plants have lost the capacity to photosynthesise, but they are so obviously related to similar photosynthesising plants that they are placed in this kingdom).
- Many reproduce asexually.
- All land plants reproduce sexually and have life cycles which involve **alternation of generations** (see later in this chapter).
- They are eukaryotic – this distinguishes them from *bacteria*.
- Their cells have cellulose walls – this distinguishes them from *animals* (which lack cellulose) and from *fungi* (which have cell walls made from a similar but chemically distinct substance).

It is more difficult to distinguish land plants (Kingdom Plantae) from members of the Kingdom Protoctista because many protoctists share some of the characteristics mentioned above. Some 'algae' have all the features. However, there is one further feature which truly defines membership of the Kingdom Plantae.

The characteristic that defines a land plant

Many aquatic organisms reproduce by shedding gametes into water. Plants, like animals, have swimming male and female gametes which join to form a zygote. This develops into a tiny **embryo** (Fig 13.1) which cannot photosynthesise at first. So, how does it obtain food? Unlike an animal embryo, it does not have a yolk. Instead, it is fed by the **maternal plant** (the plant which formed the egg). The plant may provide food directly (if the embryo remains attached to the 'mother') or indirectly by storing food in **seeds**.

So the definition which distinguishes land plants from protoctists and organisms from all the other kingdoms is:

> **All land plants develop from embryos which are supported by maternal tissue (such as in a seed).**

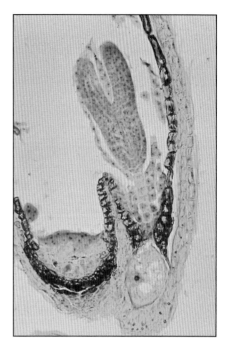

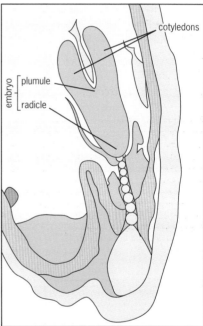

Fig 13.1(a) **The embryo of shepherd's purse, a dicotyledonous plant**

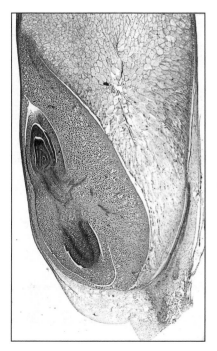

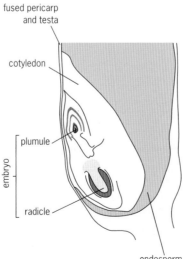

Fig 13.1(b) **The longitudinal section of a maize seed showing the embryo**

Animals (Kingdom Animalia) all develop from a blastula (a hollow ball of cells), while plants do not. Algae, other protoctists and fungi do not develop from embryos.

Alternation of generations

Land plants show **alternation of generations** – they exist in two forms, a haploid form and a diploid form. In most plants, one of these forms often goes unnoticed because it is small or concealed.

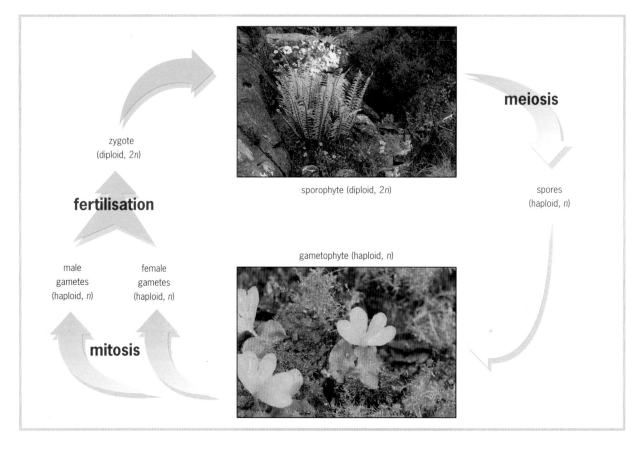

zygote
(diploid, 2*n*)

meiosis

sporophyte (diploid, 2*n*)

spores
(haploid, *n*)

fertilisation

gametophyte (haploid, *n*)

male
gametes
(haploid, *n*)

female
gametes
(haploid, *n*)

mitosis

Fig 13.2 **The two photographs, of the male fern *Dryopteris*, show how different the gametophyte and sporophyte stages of the same plant can be. The gametophyte is heart-shaped, flat on the ground**

See question 3. ■

The haploid stage produces haploid sex cells, the **gametes**, by mitosis and is called a **gametophyte** ('gamete plant'). When the female gamete, the egg, is fertilised by a male gamete, the sperm, a diploid **zygote** is formed. This grows up into a plant which produces haploid **spores** by **meiosis.** This plant is called a **sporophyte** ('spore plant'). The haploid spores germinate to produce gametophyte plants, and so the two stages, gametophyte and sporophyte, alternate with each other.

As Fig 13.2 shows, the gametophyte and sporophyte stages, which are different genetically, usually look completely different as well.

Alternation of generations is not unique to the plant kingdom: some protoctists show alternation of generations. (In protoctists the gametophyte and sporophyte stages are very similar.) In true plants (members of the Kingdom Plantae) the stages are always very different and one stage is always larger and more conspicuous than the other. In mosses and liverworts (**bryophytes**), the gametophyte is the main stage whilst in all the other groups the main stage is the sporophyte.

Alternation of generations occurs in all plants, including the more recently evolved seed plants. In the **gymnosperms** (conifers) and the **angiosperms** (flowering plants), alternation of generations is not obvious but still happens. The gametophyte is reduced to just a few cells that are entirely dependent on the sporophyte.

✔

The term alternation of generations was once applied to animals such as the coelenterates (eg jellyfish) in which two very different forms exist (see page 191). The polyp and medusa stage of coelenterates are both diploid, however, and so their case is not the same. It is called **metagenesis**, rather than alternation of generations.

2 SUCCESS ON LAND

Plants which have successfully overcome the problems of adapting to life on land show the following features:

- All have a branching **shoot system**. Simple plants such as the liverwort *Marchantia* (see Fig 13.6) have a primitive shoot. This is not differentiated into what we would recognise as typical plant structures, but grows as a flat branching **thallus**. More advanced plants have branching **stems** and most bear **leaves**.
- Most have a **root system**. (Bryophytes – liverworts and mosses – are an exception: they have only simple **rhizoids**.)
- All have a waxy surface layer (**cuticle**) to cut down water loss from parts exposed to the air.
- Most (all but a few liverworts) have tiny pores, **stomata**, which can open or close, depending on the conditions. Open stomata allow for gas exchange and they close to help reduce water loss.
- Most have developed a system of tubes – the **vascular system**. This transports water and dissolved substances such as minerals and food. The evolution of a vascular system, particularly the woody **xylem**, has enabled land plants to grow to great heights. Where growth is dense, taller plants can compete more successfully for light.
- They can reproduce without being submerged in water: most algae cannot. Simpler land plants (bryophytes and ferns) still have swimming sperm, as their aquatic ancestors did. The egg is held in a flask-like structure called an **archegonium**. This is filled with watery fluid that the sperm swim in to reach the egg (Fig 13.3). The more advanced plants do not have swimming sperm and avoid the need for archegonia. Their sophisticated reproductive techniques have helped the conifers and the flowering plants to become the most successful plants on Earth.

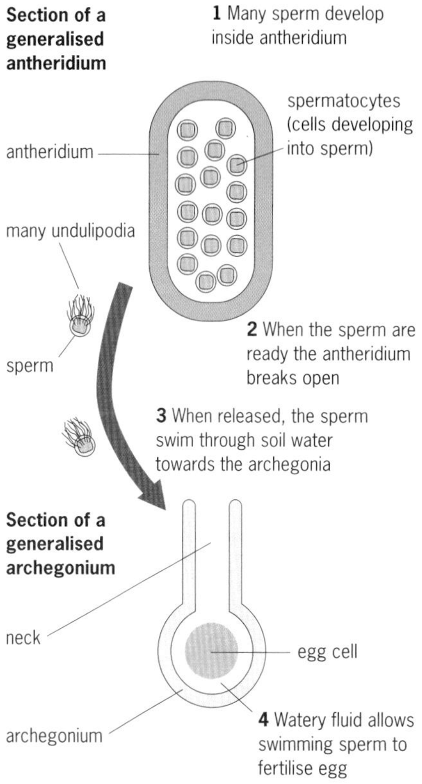

Section of a generalised antheridium

1 Many sperm develop inside antheridium

antheridium

spermatocytes (cells developing into sperm)

many undulipodia

sperm

2 When the sperm are ready the antheridium breaks open

3 When released, the sperm swim through soil water towards the archegonia

Section of a generalised archegonium

neck

egg cell

archegonium

4 Watery fluid allows swimming sperm to fertilise egg

Fig 13.3 **Archegonia and antheridia are the sex organs of all the land plants except conifers and flowering plants. Sperm are produced in the antheridium, the male sex organ. These swim through soil water until they reach the 'swimming pool' provided by the archegonium, the female sex organ. So, fertilisation takes place in an aquatic environment that mimics the conditions in which the aquatic ancestors of modern plants lived**

Where did the land plants come from?

The earliest land plants evolved from green algae and would have been small, without roots and unable to survive dry conditions.

The earliest land plants died out long ago and few living plants bear much resemblance to them. Members of the present-day plant phyla vary widely in their adaptations to living on land. Some of the important developments in the evolution of the land plants are summarised in Fig 13.4.

A In what sort of conditions would you expect the earliest land plants to have evolved?

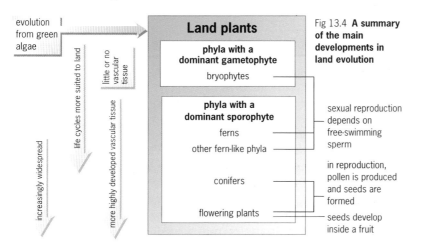

Fig 13.4 **A summary of the main developments in land evolution**

evolution from green algae

Land plants

phyla with a dominant gametophyte

bryophytes

phyla with a dominant sporophyte

ferns

other fern-like phyla

conifers

flowering plants

little or no vascular tissue

life cycles more suited to land

more highly developed vascular tissue

increasingly widespread

sexual reproduction depends on free-swimming sperm

in reproduction, pollen is produced and seeds are formed

seeds develop inside a fruit

The phylum Bryophyta (the liverworts, mosses and hornworts) could be described as the least well adapted group. Like the earliest land plants they thrive only in very damp conditions. The group which is most successful and which has evolved most recently is the phylum Angiospermophyta, the flowering plants. Flowering plants outdo all the other plant phyla in number, variety, range of habitats and the area of the world that they cover.

We are going to look at how living plants are grouped into phyla and at the way these phyla have evolved. It is more important to appreciate the main characteristics of each phylum and to understand the significance of these characteristics than it is to remember the details of individual species.

3 CLASSIFYING LAND PLANTS

Fig 13.5 **The main plant phyla have a long history. Each shape represents a phylum, and its width relates to the number of species of that phylum over the ages. Some phyla were more abundant in the past when conditions on Earth were different. The most recent group, the flowering plants, is the most successful**

Today, there are about 300 000 living plant species in ten plant phyla. Four phyla are small and represent few plants and so are not covered here. The six main plant phyla are named in Fig 13.5. You may find it useful to refer to this diagram, and to Fig 13.4 as you read through the rest of the chapter.

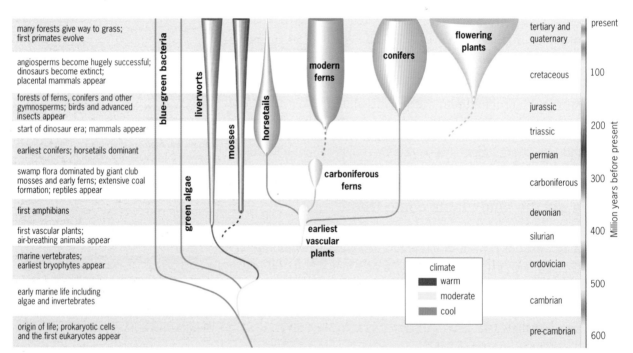

The least well adapted land plants

It is sometimes suggested that modern **bryophytes**, the liverworts and mosses, are poorly adapted to living on land, but this is an overstatement. Bryophytes are an obviously successful group of plants with many species. They have been around for millions of years and occupy many different habitats. While many bryophytes live only where it is permanently damp, others survive in habitats which dry out completely from time to time: they have become *physiologically* adapted to their environment. Yet bryophytes lack many of the *anatomical* adaptations that we associate with the more successful land plants (the conifers and the flowering plants). For this reason we say the bryophytes are *less* well adapted to living on land.

Phylum Bryophyta: the liverworts, mosses and hornworts

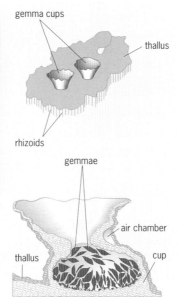

Fig 13.6 (a) **Many liverworts are not differentiated into leaves and stems but grow as a flat structure called a thallus. This is a haploid gametophyte. Little balls of cells** (gemmae) **grow in cups** (gemma cups) **on the plant surface. Gemmae are scattered by splashes of rain and each one grows into a new gametophyte. This is** asexual reproduction **and the offspring are genetically** *identical* **to the parent**

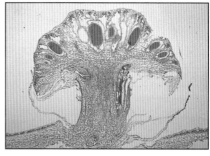

Fig 13.6(b) **Liverworts can also reproduce sexually.** *Marchantia* **shown in the four pictures below has separate sexes**

Male plants produce sperm from antheridia **(above) which develop on umbrella-like, star-shaped** antheridial heads **(below).**

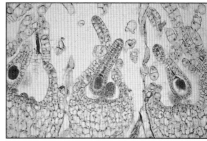

Eggs develop in archegonia **(above) on spider-shaped** archegonial heads **(below) on female plants.**
Sperm swim through a film of water and fertilise the egg. Each embryo develops into a diploid sporophyte. Sporophytes grow as tiny capsules on the under-surface of the archegonial head

The three classes in the **Bryophyta** are the liverworts of class **Hepaticae** (see Fig 13.6), the hornworts of class **Anthocerotae** and the mosses of class **Musci**.

The life cycle of bryophytes shows the pattern of alternating generations clearly. Taking the moss *Polytrichum* as an example, the main moss plant is the **haploid gametophyte** which produces eggs or sperm (Fig 13.7). The sperm swim through the soil water and fertilise an egg.

The fertilised egg develops through an embryo stage into a **diploid sporophyte**. The sporophyte is little more than a stalk bearing a **capsule**, and, although it is green, most of the nutrients which allow it to grow come from the gametophyte below. The capsule looks like part of the gametophyte but, because it is diploid, it is genetically different from its leafy parent.

Inside the capsule, meiosis (see Chapter 6) produces spores which are shed and dispersed by wind. In many mosses, the capsules open to release their spores only when the air is dry. Spores that fall on moist soil germinate to form threads which later develop into the leafy haploid shoots of the next gametophyte.

Although the bryophytes depend on water for fertilisation, they have managed to survive for over 400 million years. This is partly because their spores can withstand periods of drying out and partly because they can thrive in environments which other plants find difficult. These include surfaces which have no soil, such as rocks and tree bark.

Plants that prefer to live on land

The three phyla, the **Sphenophyta** (the horsetails), the **Lycopodophyta** (the club mosses and their relatives) and the **Filicinophyta** (the ferns) are also ancient groups of plants. But they show some adaptations to land living which represent evolutionary steps on the road to the most advanced plants.

In all three phyla, vascular tissue and leaves have developed but, like the liverworts and mosses, all still need a wet environment to transport their motile sperm.

?

B What advantages does a tall plant have over a small one? Suggest conditions that are necessary for a plant to grow tall.

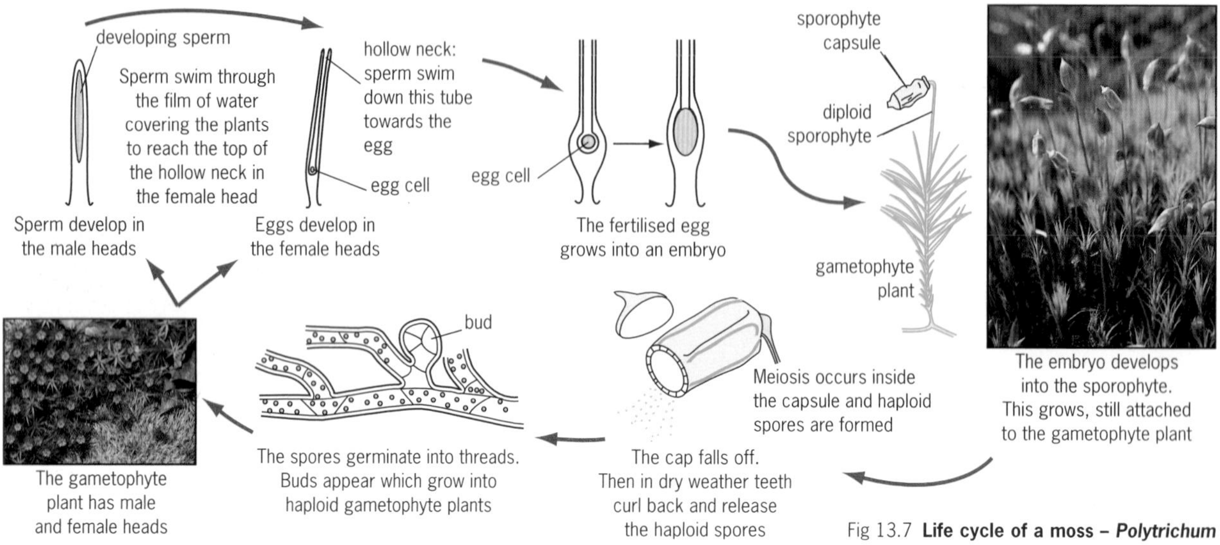

developing sperm

Sperm swim through the film of water covering the plants to reach the top of the hollow neck in the female head

Sperm develop in the male heads

hollow neck: sperm swim down this tube towards the egg

egg cell

Eggs develop in the female heads

egg cell

The fertilised egg grows into an embryo

sporophyte capsule

diploid sporophyte

gametophyte plant

The embryo develops into the sporophyte. This grows, still attached to the gametophyte plant

bud

The spores germinate into threads. Buds appear which grow into haploid gametophyte plants

Meiosis occurs inside the capsule and haploid spores are formed

The cap falls off. Then in dry weather teeth curl back and release the haploid spores

The gametophyte plant has male and female heads

Fig 13.7 **Life cycle of a moss – *Polytrichum***

Fig 13.8 **A portion of the fossilised trunk of a giant club moss. During the Carboniferous period, most of the swamps that became coal deposits were formed from giant club moss trees such as *Lepidodendron*, which probably grew to 30 metres or more**

Fig 13.9 **The modern club mosses are much smaller plants, such as this *Lycopodium* which grows widely on mountain grassland in the United Kingdom**

✔

The vascular tissues of a plant are the **xylem** and **phloem** (see page 292). Both are specialised cells. The phloem allows the transport of the dissolved products of photosynthesis around the plant, and the xylem carries water and dissolved minerals up from the soil to the photosynthetic parts of the plant.

Sphenophyta: the horsetails

Horsetails were a prominent feature of the vegetation between 150 and 250 million years ago, when the dinosaurs lived on Earth (see Fig 13.5). They had woody vascular tissue and so could grow many times taller than their bryophyte neighbours. Their height allowed them to compete more successfully for sunlight. Today, only a single genus, *Equisetum*, survives. It is easily recognised by its jointed hollow stems and ribbed texture.

Lycopodophyta: the club mosses and their relatives

Like horsetails, the club mosses (Fig 13.8) were common during the Carboniferous period, but only five genera exist today. They all have vascular tissue and also produce small leaves (Fig 13.9).

The club mosses all show alternation of generations but their individual life cycles are beyond the scope of this book. It is worth mentioning that one genus, *Selaginella*, produces two types of spore; this is called **heterospory**. The larger **megaspore** develops into a female gametophyte which produces eggs, and the smaller **microspore** becomes the male gametophyte which releases sperm. As you will see later (page 210), heterospory is still a feature of conifers.

Vascular tissue and its role in support

Vascular tissue acts as a transport system; it also provides land plants with physical support. Xylem, in particular, becomes **lignified** (thickened by the laying down of woody material called **lignin**), making the tissue strong and supporting.

In a crowded community of plants, those that can grow taller than the others can compete more successfully for light and so have a great advantage over their smaller neighbours. Plants from different phyla have independently developed into trees or tree-like structures. As long ago as the carboniferous period, there were land plants over 30 metres high.

Woody flowering plants and conifers are tall because of the strength provided by the vascular tissue. In the past, other groups of plants have achieved the same result by other methods. When different organisms evolve similar solutions to the same problem, this is known as **convergent evolution** (see Chapter 33).

Phylum Filicinophyta: the ferns

Plants in the phylum Filicinophyta, called simply ferns from now on, have well developed vascular tissue, an extensive root system and produce leaves called **fronds** (see Fig 13.10). Ferns greatly outnumber both the horsetails and the club mosses; there are about 10 000 known species.

The life cycle of ferns also shows alternation of generations. The dominant phase is the **diploid sporophyte** plant. The sporophyte lives for many years. Each year it produces a new crop of haploid spores which are usually formed on the underside of the fronds (see Fig 13.11).

In the life cycle of a typical fern such as *Dryopteris* (Fig 13.11), each haploid spore develops into a tiny **haploid gametophyte** plant, a **prothallus**, which has no vascular tissue or roots. The prothallus is attached to the ground by tiny anchoring threads called **rhizoids** and it usually forms a symbiotic association with a fungus (mycorrhizae: see below and Chapter 11). Each prothallus produces eggs and sperm. When a sperm fertilises an egg, the zygote develops into an embryo which is nourished by the parent prothallus. The embryo develops into a **diploid sporophyte** plant which is soon able to photosynthesise. The gametophyte withers and dies.

Fig 13.10 **Like all ferns, tree ferns often grow in the shady under-storey of forests. They can photosynthesise well enough at low light levels, and the damp conditions suit their reproductive needs**

?

C What sort of division process gives rise to the spores seen in Fig 13.11?

✔

Mycorrhizae (literally 'fungus-root') are mutually beneficial partnerships between plants and fungi. Fungal threads penetrate the root of the plant and spread into the soil to form a network which can gather mineral nutrients. The fungus delivers minerals to the plant and the plant provides food for the fungus.

Fig 13.11 **Because ferns have motile sperm in their life cycle, they continue to be restricted to sites which are damp. However, because the sporophyte has a good root system and a well developed vascular system, ferns are not as dependent on surface water as the bryophytes and can survive long periods of dry weather**

Life cycle of the fern *Dryopteris*

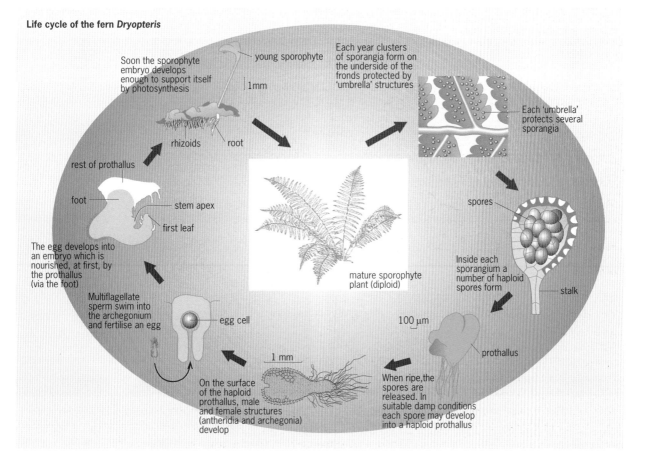

Soon the sporophyte embryo develops enough to support itself by photosynthesis

young sporophyte

|1mm

Each year clusters of sporangia form on the underside of the fronds protected by 'umbrella' structures

Each 'umbrella' protects several sporangia

rhizoids   root

rest of prothallus

foot

stem apex

first leaf

spores

The egg develops into an embryo which is nourished, at first, by the prothallus (via the foot)

mature sporophyte plant (diploid)

Inside each sporangium a number of haploid spores form

stalk

Multiflagellate sperm swim into the archegonium and fertilise an egg

egg cell

100 μm

prothallus

On the surface of the haploid prothallus, male and female structures (antheridia and archegonia) develop

1 mm

When ripe, the spores are released. In suitable damp conditions each spore may develop into a haploid prothallus

In any haploid organism, a mutation is likely to be apparent in the organism and may reduce the organism's chance of survival. In a diploid organism, a second 'normal' allele can mask the effect of a mutation.

The **gymnosperms** include the phylum **Coniferophyta** - the conifers. Other gymnosperms include the cycads and other plants beyond the scope of this book. The **angiosperms** are the phylum **Angiospermophyta** – the flowering plants.

D a) What is the function of the cuticle in a conifer?

b) Suggest reasons why conifers can grow where flowering plants cannot.

Seed-bearing plants – the most successful land plants

The seed-bearing plants dominate the Earth's land-based vegetation. These are some of the evolutionary advances which give them survival advantages over plants from other phyla:

- They do not depend on free-swimming sperm, and so are not limited to wet environments. They can therefore live and reproduce in dry conditions.

- Although they show alternation of generations, the diploid sporophyte is dominant and the haploid stage is reduced and dependent on the sporophyte. Haploid organisms are a weak link because any mutation that arises in them will be fully expressed.

The two main phyla of seed-bearing plants found today are the Coniferophyta and the Angiospermophyta. They make up the majority of plants we see in our countryside, gardens and homes.

Phylum Coniferophyta – the conifers

Conifers have an advanced vascular system and most have narrow needle like leaves. The leaves of the plant are protected by a cuticle (see page 265). They dominate the landscape in colder climates where flowering plants seem less able to compete (Fig 13.12).

The conifer tree is the diploid sporophyte. It produces two types of diploid spore (this is called **heterospory**): both form in cones. In the female cone, each **ovule** produces a single diploid **megaspore** (large spore). The ovule remains attached to the cone and a female gametophyte develops in the megaspore. The gametophyte has no independent existence and is tiny. The male cone produces huge numbers of haploid **microspores** (small spores) which are shed as pollen grains (Fig 13.13).

Fig 13.12 **Conifers can survive freezing temperatures better than flowering plants. (Flowering plants probably first evolved when the world was warmer since their large open vessels are less able to recover from frost damage)**

Fig 13.13 **Most of the pollen from a male pine cone will never reach a female cone. Huge amounts of pine pollen must be released to make pollination likely**

A seed consists of an embryo plant and a food reserve enclosed in a seed coat (the **testa**).

Fig 13.14 **The development of a pollen grain from a pine tree. In the conifers, the gametophyte stage does not grow independently. When pollination has been successful, a pollen tube grows towards the female gametophyte. The male gametophyte is very much reduced: all that remains is the few cells which divide inside the pollen tube, just before fertilisation occurs**

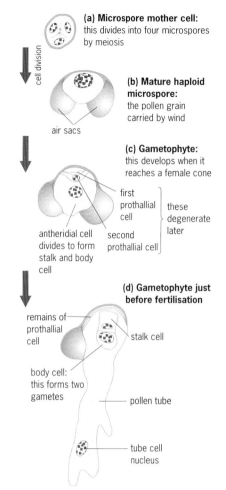

cell division

(a) Microspore mother cell: this divides into four microspores by meiosis

(b) Mature haploid microspore: the pollen grain carried by wind

air sacs

(c) Gametophyte: this develops when it reaches a female cone

first prothallial cell — these degenerate later

second prothallial cell

antheridial cell divides to form stalk and body cell

(d) Gametophyte just before fertilisation

remains of prothallial cell

stalk cell

body cell: this forms two gametes

pollen tube

tube cell nucleus

A microspore (Fig 13.14) is carried by the wind until it reaches a female cone. Here, and only here, it develops into a male *gametophyte*. A pollen tube is formed as an outgrowth, while inside, the two male gametes are formed by meiosis. These then pass into the pollen tube and are transported to the female gametophyte inside the ovule. After fertilisation by one of the male gametes, a new diploid sporophyte embryo develops, still within the ovule. This embryo, together with a small food supply, is released as a winged seed.

Phylum Angiospermophyta – the flowering plants

Conifers have an enormous range of size: some are amongst the tallest trees in the world (the redwoods), while mosses are tiny and ferns lie in between. So, you might think that evolutionary advance means getting bigger. However, in flowering plants, the evolution of height has almost gone into reverse. Many of the more primitive flowering plants are trees, and some of the most highly evolved species such as the grasses are low growing.

The main reason for this is the success of the flowering plants in producing a huge variety of species to colonise all sorts of different environments: four out of five plants living on Earth today are flowering plants. The most advanced species have had the advantage of growing in a world populated by trees from their own phyla and from the Coniferophyta. The presence of trees themselves makes the landscape more complex, creating more opportunities for obtaining a living and providing more niches. Flowering plants have evolved strategies to take advantage of these 'new' niches:

- Weed species exploit disturbed ground.
- Climbers, such as ivy, reach the light without using up precious resources on support tissues.
- Parasites such as dodder give up photosynthesis and take to robbing other plants of nutrients.
- Epiphytes, like orchids, get closer to the light without wasting time and resources on growing tall; they live their whole life on the surface of other tall plants.
- Some plants have returned to the water, with some aquatics living completely under water.

For many of these 'lifestyles', it pays to be smaller.

The flowering plants show other clear developments when compared to the conifers:

- Flowering plants first evolved in warm climates in which it was an advantage to have vascular tissue with wide, open ended xylem vessels, and well connected phloem sieve tube cells. These characteristics may have given the early species a competitive edge in all but the coldest parts of the Earth.
- Flowering plants have improved their reproduction strategies: they produce flowers. While conifers depend entirely on the

Fig 13.15 **The first leaves, the cotyledons, of this beech seedling are quite different from all the subsequent leaves produced by the plant**

wind to disperse their pollen and seeds, many flowering plants have enlisted animals to help in these processes. It is no coincidence that flowering plant populations expanded rapidly as insects and mammals flourished, since both these groups play an important role in dispersing pollen and seeds.

Classification within the Angiospermophyta

The enormous diversity of flowering plants is also seen in the great variety of their flowers and fruits and so the classification we use is complex. We shall not go into it in detail, but we will look at the two main groups which diverged very early in flowering plant evolution.

These groups are often referred to by their old taxonomic names, the **monocotyledons (monocots)** and the **dicotyledons (dicots)**. In fact, the number of cotyledons does not always reliably indicate which plants belong to which group. It is better to use the proper taxonomic names – **Liliidae** and **Magnoliidae** – and to also look at other characteristics. These are shown in Fig 13.16.

?
F Here is a list of common flowering plants: daffodil, grass, fuchsia, an oak tree, a yucca, a beech tree. Which are Liliidae and which are Magnoliidae?

?
E Give three examples of animals which help to disperse seeds.

Magnoliidae (dicots)

Cotyledons

two cotyledons

eg cress

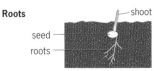

Two but rarely one, three, four cotyledons or undifferentiated

Roots

shoot

seed
roots

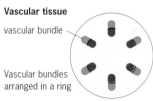

Primary root develops from the **radicle** in the embryo

Vascular tissue

vascular bundle

Vascular bundles arranged in a ring

Leaves

veins spread out in a network

broad lamina

petiole

Flowers

eg primula

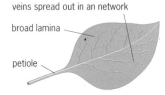

Flower has 5 petals, or multiples of 5 petals. Less commonly 4 or multiples of 4, rarely 3 or multiples of 3 petals

Liliidae (monocots)

one modified cotyledon remains in the seed

developing true leaf

eg grass

coleoptile

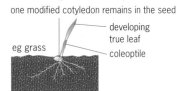

One cotyledon or undifferentiated

Adventitious roots develop from tissue in the embryo that is not part of the radicle

vascular bundle

Vascular bundles are scattered throughout stem

Leaf is formed from petiole: lamina is vestigial

parallel veined leaf typical of grasses

eg daffodil

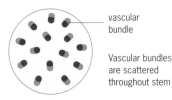

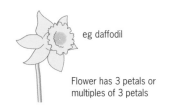

Flower has 3 petals or multiples of 3 petals

Fig 13.16 **Differences between the Magnoliidae and the Liliidae (dicots and monocots)**

SUMMARY

When you have finished this chapter, you should know and understand the following:

▦ The **Kingdom Plantae** contains plants which have become adapted to live on land.

▦ All land plants develop from **embryos** (ie diploid multicellular young organisms) which never go through a blastula stage. This feature distinguishes them from organisms in the four other kingdoms.

▦ Different phyla within the Kingdom Plantae show different degrees of adaptation to the land: liverworts, hornworts and mosses are the least well adapted, conifers and flowering plants are the most highly adapted.

▦ The features of adaptation to living on land include: **supportive stems**, a **vascular system**, a **root system, leaves, stomata**, a **cuticle,** and the development of reproductive mechanisms which do not need significant amounts of surface water.

▦ There are six main plant phyla: the **Bryophyta** (liverworts, hornworts and mosses), the **Sphenophyta** (horsetails), the **Lycopodophyta** (club mosses), the **Filicinophyta** (ferns), the **Coniferophyta** (conifers) and the **Angiospermophyta** (flowering plants).

▦ Flowering plants are the most advanced land plants. They are classified into two main groups: the **Liliidae (monocots)** and the **Magnoliidae (dicots)**.

QUESTIONS

1 Write an essay on the occurrence and importance of haploid and diploid phases in the life cycles of plants and animals.

[ULEAC January 1995 Biology: Paper 3, q.9]

2 Write an essay on the alternation of generations in mosses, ferns and flowering plants.

[ULEAC June 1993 Biology: Paper 3, q.9]

3 The diagram below summarises the life cycle of a fern, which has alternation of generations.

a) On [a copy of] the diagram, write the letter **M** to show where meiosis occurs.
  (i) which stage requires dry conditions? Give a reason for your answer.
  (ii) Give *two* stages which require the presence of water and, in each case, give a reason for your answer.

b) State *three* ways in which the sporophyte of a fern differs from the gametophyte.

[ULEAC January 1994 Biology: Paper 3, q.2]

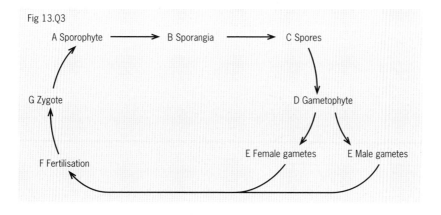

Fig 13.Q3

Assignment

BRACKEN

Bracken, a fern, is more widespread than many flowering plants (the most successful group of land plants).

In this assignment you are going to identify how bracken completes so successfully against other plants and how it manages to avoid being eaten by insects.

Outwitting other plants

Bracken grows by sending out horizontal shoots called **rhizomes** just beneath the surface of the ground. These branch and spread, and one plant can eventually colonise hundreds of square metres of land. In a way, bracken grows like a tree on its side, sending up leaves into the light at many points along its rhizomes.

1 Why does this sort of growth habit give bracken an advantage over plants which grow in a mainly vertical direction?

Bracken is not an evergreen plant. In autumn its fronds die and their brown remains fall to the ground, creating a thick carpet of rotting material. In spring, the rhizomes produce new curled fronds which soon open out to form a thick canopy.

2 How might these two features of the bracken's life cycle help it to compete with other plants in the same area?

Outwitting insect predators

Since bracken is so common, you might expect it to be food for many insect larvae. In fact, caterpillars from only a very few insect species are found on bracken. To find out why this is, look at Fig 13.A1.

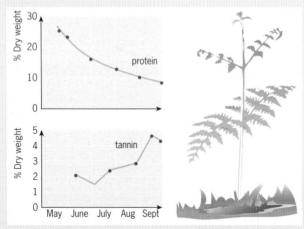

Fig 13.A1 **Seasonal variation in protein and tannin levels in bracken fronds during the summer**

3 How does the proportion of **(i)** protein and **(ii)** tannin in bracken change over the course of the summer?

High levels of tannin make the bracken fronds unpalatable to most caterpillars. They choose to eat other plants instead. This obviously reduces the number of insects which the bracken supplies with precious food resources, but some insect larvae do not seem to mind the tannins.

4 Look at Fig 13.A1. How might the change in protein deter these feeders from eating bracken?

Any protection which the bracken gains from the changes in its tannin and protein levels is available only towards the end of the summer. Earlier, when the fronds are just emerging, they are vulnerable to attack. To help protect itself, the bracken has formed a beneficial relationship with ants. Its young fronds are covered in external **nectaries** which produce a sweet fluid which is high in nutrients. Ants find it irresistible and use it as a valuable source of sugars. What bracken gains from the relationship only became clear a few years ago when researchers set up an experiment using two groups of bracken fronds.

In the first group, ants were present on the young fronds. In the control group, the ants had been removed from the plants (and were prevented from re-colonising). The researchers placed the same species of caterpillar (a species which does not normally live on bracken) on both groups of plants, and measured how many of the caterpillars survived over a period of 5 hours. The results of one experiment are shown in Fig 13.A2.

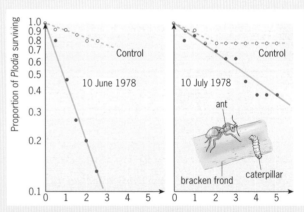

Fig 13.A3 **The rate of removal of caterpillars from bracken fronds in the presence and absence of ants**

5 Look at the results of the ant and caterpillar experiment shown in Fig 13.A3.

a) How many caterpillars in each group are left on the bracken fronds after 2 hours?

b) What do you think the ants are doing?

c) What precaution do you think the researchers took when they first placed the caterpillars on the two groups of plants? Hint: Think about making sure the experiment is properly controlled.

LIFE PROCESSES

'ANIMALS ARE INTERESTING, plants are boring.' This sums up the view of many biology students. It is certainly true that plants and animals are fundamentally different: plants make their own food and so do not have to go off to search for it – often they are literally 'rooted to the spot'. They do not need muscles, a movable skeleton or elaborate sense organs. With no need to live life on our time scale, it is not surprising that we humans often fail to appreciate many of the subtle wonders of the plant world.

Despite their differences, the ultimate aim of plants and animals and all organisms is survival. They must obtain food and avoid being eaten themselves, long enough to pass their genes on to the next generation.

In order to survive, organisms must respond to changes in their external environment, while at the same time keeping their internal environment as constant as possible. It is the study of these abilities which forms the bulk of this section. How can water reach the top of a tree that is 100 metres tall? How can the kangaroo rat live on dry food for months on end? How do humans keep the pH of their blood constant at pH 7.35? How can the Arctic fox survive temperatures of −500 °C? Answering these questions is part of the branch of biology known as physiology.

In this section, we study the physiology of plants and animals, focusing on mammals and flowering plants. The first few chapters concentrate on nutrition and on the exchange of materials that occurs in digestion and gas exchange. In Chapter 18, we go on to study how these materials are moved around by transport systems. Later in the section we study coordination – the way in which organisms detect and respond to changes in their internal and external environment.

Finally, as a prelude to the genetics section, we look at the ultimate obsession of most plants, animals and students – reproduction.

14 Characteristics of living organisms

Defining life

In this and the next 15 chapters, we look at the processes that are characteristics of all living organisms and see how these characteristics define what is living and non-living. This chapter pinpoints the characteristics and the chapters in which they are covered, both in this section and in other sections.

Living organisms are different from dead organisms and from objects that have never been alive. This seems obvious, but actually defining life is more complicated than you might think at first. Living organisms are defined at a basic level by their need to:

- obtain and maintain a supply of energy,
- copy genetic material and pass this on to the next generation.

They also carry out several life processes. To remember these, you might be familiar with the mnemonic (memory aid) – 'MR GREEN':

M – movement

R – reproduction

G – growth

R – respiration

E – excretion

E – excitability

N – nutrition

This is a convenient way of learning the main characteristics of living things (quite useful in exams!). But living things are much more complex than this list suggests.

Living organisms are objects which can be described as **animate** (having life). Non-living objects and dead organisms are said to be **inanimate** (without life).

A Is a wooden desk alive? Give a reason for you answer.

What living things have in common

All living organisms are made up from the same basic chemical building blocks. They all carry out the same life processes using remarkably similar biochemistry (the same genetic code, the same enzyme systems and very similar biochemical pathways).

Living things grow and develop

All organisms must grow and develop so that they reach the size, complexity and maturity needed to complete their life cycle. Growth in plants is covered in Chapter 20. The main stages in the growth of a human baby in the uterus is shown on page 460 in Chapter 28.

Living organisms can reproduce

Each cell has **genetic information** in the form of **DNA** (look back to Chapter 3). DNA is the 'plan' which the cell uses to make more copies of itself when it **reproduces**. Chapter 6 describes the process of cell division in detail. Chapters 28 and 29 in this section look at how animals and plants reproduce.

Living things use energy

Useful energy comes from the process of cell respiration (Chapters 7 and 8). Energy is used to do work within the cell and to maintain the structure of individual cells and of the organism as a whole. Reactions which require energy are usually coupled with the breakdown of ATP (adenosine triphosphate – see page 128), a reaction that releases energy. ATP is made in the cell by the respiration of simple molecules – sugars, amino acids, fatty acids and glycerol – which come from food.

Living things excrete their waste

Every living cell, whether it exists independently or is part of a multicellular organism, must eliminate waste products that might poison it if they were to build up. The process of excretion in animals is covered in Chapter 27.

Living things need food

Some organisms need to take in food in a ready-made form (see Chapters 15 and 16). Others make their own food by trapping energy from sunlight, or from chemical reactions. Most organisms get their energy, ultimately, from the process of photosynthesis (see Chapter 9).

Living things respond to their surroundings

Organisms from all of the five kingdoms react to their surroundings. Whether this reaction is simple or complex depends on the type of organism: a bacterium can detect a food source and move towards it; people can perform exceptionally complex tasks and can modify their environment as well as respond to it. Coordination in plants is covered in Chapter 20, nervous and chemical control in animals can be found in Chapters 21 and 22 respectively.

Living organisms control internal conditions

All organisms try to maintain a constant environment for their cells. Cells that experience a constant pH, temperature, pressure, etc are more likely to survive than cells subjected to external condition changes too great to compensate for. When an organism maintains a balanced internal environment for each of its cells, we say that it is practising **homeostasis**. This is covered in more detail in Chapter 25.

Living organisms are highly organised

The cell (see Chapter 1) is the basic unit of life: it is the smallest and simplest unit which can carry out all the life processes already listed. Some live independently of other cells, as single-celled organisms, others coexist as part of a multicellular organism. Even the simplest cells show an incredibly high degree of order and organisation compared to their non-living surroundings.

Viruses cannot live independently from living cells, do not show such high levels of organisation and do not carry out many of the life processes (except reproduction). Because of this, many scientists consider them to be non-living (see Chapter 11).

As you study biology, you will see that multicellular organisms show different levels of organised complexity. Collections of cells form tissues, which form organs, which form body systems, and several of these make up a whole organism (see Chapter 2). Organisms themselves are also units in a **population**.

Living things can move

We can identify movement in living organisms at different levels. Whole organisms can move – a mammal can run; organs inside an organism can move – a heart beats and intestines move food along them using squeezing movements; individual cells can move – an amoeba can move from place to place using pseudopodia; organelles within cells can move – the chromosomes of a dividing cell move to opposite ends of the cell before it splits into two; and individual molecules can move – potassium ions and sodium ions moving into and out of a nerve cell cause an nerve impulse to be tranmitted along that nerve cell.

Animal locomotion is covered in Chapter 19. Movement of molecules in and out of cells is covered generally in Chapter 5 and nervous transmission is covered in detail in Chapter 21. Chapter 20 looks at movement in plants.

Populations of living organisms can change

Evolution is a central theme of biology. The theory of evolution says that life on Earth shows such variety because of the process of natural selection. This process attempts to explain how organisms appear to be so well suited their environments and habitats. Evolution and natural selection are covered in Chapter 33.

The characteristics of living organisms

Our list of characteristics has itself evolved. The new version does not make such a neat mnemonic, but it is more accurate. We can now say that all living organisms:

G – **g**row and develop,

R – **r**eproduce,

E – need **e**nergy

E – **e**xcrete waste

N – **n**ourish themselves

E – show **e**xcitability

H – show **h**omeostasis

O – show **o**rder and organisation

M – **m**ove

E – **e**volve

Biology: its breadth and popularity

Biologists investigate organisms at all levels, from the molecular level at one extreme, to the population level at the other. Early biologists recognised that trying to understand something as complex as a living organism is difficult without taking it apart and seeing how the bits work. But a dissected animal no longer functions, and a cell broken down into its different molecules no longer works either, and so biologists also study whole organisms to see how they function and interact with each other.

The scope of biology is immense. Most of the biologists who have ever lived are alive today, and they write over 450 000 new research articles each year. Biology is divided into many areas: ecology, genetics, microbiology, palaeontology, horticulture, cytology, parasitology, immunology, embryology, physiology, pharmacology and many others. Fig 14.1 shows the further sub-divisions of physiology, itself a branch of biology.

B To an alien arriving from space, a Ferrari might appear to move, respire, excrete, respond and require food. How would you argue this car is a non-living object?

Fig 14.1 **Some of the many sub-divisions of physiology, itself just one branch of biology. New disciplines are still developing: in fifty years this tree might need several new branches**

THE DIFFERENCE BETWEEN LIFE AND DEATH

IN A HOSPITAL, deciding whether someone is alive or dead is crucially important. Paramedics have to look for vital signs to check if someone who has been seriously hurt is still alive, and a doctor needs to certify when someone has died.

The first thing to do is to look at the patient. Are they very still? Is their skin pale and cold? The next thing is to check for any signs breathing or a pulse, and to test for reflexes. If someone is dead, the pupils of their eyes do not react to light and their muscles no longer react when touched at specific reflex points. Their body feels cold as it loses heat to the environment (at the rate of about 10 °C per hour).

With advanced medical technology, doctors can use electronic equipment to investigate brain activity and heart movement. Sometimes, if someone is badly injured, their brain can be dead, even though their breathing and circulation is being maintained artificially by a life-support machine. In these cases, the doctors and relatives may have to take the difficult decision about whether to turn the machine off.

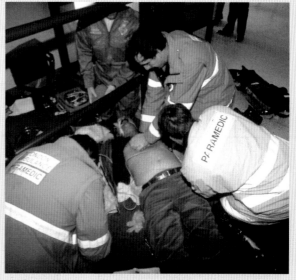

Fig 14.2 **Paramedics need to make fast decisions – these are matters literally of life and death**

15 Heterotrophic nutrition and feeding designs

The adult crown-of-thorns starfish are large, up to 30 cm in diameter. Their feeding habits cause serious damage to coral reefs

THE CROWN-OF-THORNS STARFISH is a natural marine predator. Adult starfish live on coral reefs in the Indian and Pacific Oceans, where they feed predominantly on the soft body tissues of corals. They obtain food by turning their stomach inside out to release enzymes directly onto the living coral. Digestion is complete in three to five hours. The starfish then absorb the soluble products, leaving behind a white scar of coral skeleton. The damage is particularly serious because some starfish are choosy feeders: one species feeds only on one very rare type of coral.

Scientists have watched this starfish destroy corals since the early 1960s. In large 'outbreaks' of starfish activity, up to 90 per cent of the coral is killed. The habits of the starfish have important consequences for humans, as they affect the economy of the area (they affect tourism, for example) *and* its ecology. However, no one has yet found out exactly what is going on. Are population outbreaks natural phenomena, or are they caused by human disturbance of tropical systems? The controversy has raged for many years but we still do not know enough about the feeding habits of this predator to reach any definite conclusions.

1 THE IMPORTANCE OF FOOD

All organisms need food:

- as a source of energy,
- to supply raw materials for repair, growth and development of body tissues,
- to supply vital vitamins and minerals.

As we saw in Chapter 9, **autotrophs** manufacture their own food using an external energy source (usually sunlight) and a simple inorganic supply of carbon (usually carbon dioxide).

Animals are **heterotrophs**, as are fungi, some protoctists and many bacteria. Heterotrophs cannot make their own food: they must take in food molecules from their surroundings (Fig 15.1). They use an **organic carbon source** (carbon that was once in living things), and so they are totally dependent upon organisms like green plants that are capable of making their own food. Heterotrophs use some of their food energy to obtain more food.

In this chapter we concentrate mainly on human nutrition, and then look at some of the different strategies that heterotrophs use to obtain food. Digestion is covered in Chapter 16.

Fig 15.1 **Amoeba, a single-celled organism, is a heterotroph. It surrounds food particles, such as bacteria, and takes them into its cytoplasm. It then pours digestive enzymes into the food vacuole and absorbs the simple molecules that are produced**

2 HUMAN NUTRITION: FOOD AND DIET

Humans and other animals take materials into the body by feeding, drinking and breathing. Waste materials leave by **excretion** (the removal of metabolic or cell waste, mainly in the urine), breathing out and **defecation** (the elimination of undigested food).

A healthy diet provides us with a balanced selection of nutrients that our bodies need in order to carry out vital life processes. People who are **malnourished** are short of some of the types of nutrients that they need to stay healthy. Someone who is **starving** cannot obtain enough energy to meet the demands of the body (see Chapter 7).

The foods we eat

Humans are **omnivores**: we eat food derived from both plants and other animals. Three-quarters of our food comes directly from plants.

The amount of food a person needs depends on:

- their age,
- the type of activity he or she carries out,
- their body size,
- their genetic background (some people inherit a very high or very low metabolic rate, for example).

We need to eat *enough* food to maintain an energy balance: our **energy intake**, the amount of energy we get from food, must balance our **energy expenditure**, the amount of energy we use. Energy balance in humans is covered in more detail in Chapter 7.

We also need to eat a *range* of foods to keep the chemical composition of the body more or less constant. Table 15.1 shows the chemical composition of the human body.

A What happens when a person consistently takes in more energy from their food than they use up in their everyday activities?

Table 15.1 **Humans are composed mainly of water with roughly equal amounts of protein and fat. We are little more than a mixture of carbon, oxygen, hydrogen and nitrogen together with a pinch of calcium, sulphur and other elements**

(a) **Compounds: the percentage of chemical compounds by weight in the human body**

Compound	Men (%)	Women (%)
Water	62	54
Protein	17	15
Lipids	14	25
Carbohydrates	1	1
Minerals	6	5

(b) **Elements: the percentage of chemical elements by weight in the human body**

Chemical element	% in an adult human
Oxygen	65.0
Carbon	18.0
Hydrogen	10.0
Nitrogen	3.0
Calcium	1.4
Phosphorus	0.7
Sulphur	0.2
Potassium	0.16
Sodium	0.11
Chlorine	0.11
Magnesium	0.045
Trace elements (iron, copper, etc)	0.015

Carbohydrates

Carbohydrates provide the bulk of the energy in most human diets and are also necessary for the metabolism of proteins and fats. Plant foods have a relatively high carbohydrate content compared to meat. Starchy vegetables like potato and foods made from wheat flour, such as pasta and bread, are particularly rich in carbohydrate, as are foods that contain a lot of sugar.

B What sort of animal has a relatively low carbohydrate intake?

Fig 15.2 **This bar chart shows the different proportions of carbohydrate, protein and fat eaten by an average person living in Greenland, Japan, the United Kingdom and Russia**

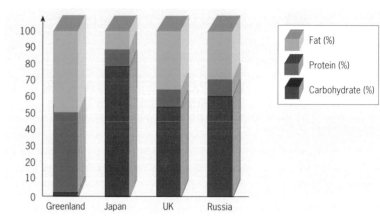

?

C Look at Fig 15.2.

(a) Suggest a possible food source for the high carbohydrate content of the Japanese diet.

(b) Why do you think there is so little carbohydrate in the Greenland diet?

Proteins

Proteins make up more than 50 per cent of the dry weight of animals and bacteria (cellulose is the major constituent of plants) and they carry out a wide variety of functions in living cells. Proteins are polymers that contain many amino acids (see Chapter 3).

There are 20 amino acids and they can be divided into two dietary groups: the **non-essential amino acids**, which the body can make and the **essential amino acids**, which can only be obtained from food. As a general rule, proteins of animal origin contain all the essential amino acids, most plants contain only some of them. If you are a vegetarian, you need to combine proteins from different plants in the same meal to take in the full range of amino acids. This is not as difficult as it sounds: a simple meal of beans on wholemeal toast contains all 20 amino acids.

A food contains the same amount of protein, whether it is raw or cooked. When food is heated above 100 °C, many proteins are **denatured** (their normal three-dimensional structure is destroyed) and they may also **coagulate** (stick together in lumps).

NEW FOODS

BIOTECHNOLOGY (the use of living organisms, or the application of a biological principle, in a technological process) is important in the manufacture of new foods. Bacteria and fungi are useful organisms in food production. They can be grown on a wide range of inexpensive or waste materials, harvested and then processed to provide food for humans or for livestock.

'Pruteen' is an animal feed supplement made from methanol by the bacterium *Methylophilus methylotrophus*. The bacteria are grown in a huge fermenter that can hold 600 tonnes of fluid. Methanol and ammonia enter continuously at the top of the fermenter, while bacteria leave at the bottom. The **single cell protein** produced by this process is fine and powdery. It can be made into blocks for feeding to cattle but it does not have enough texture to make it acceptable food for people.

Quorn, another microbial food, is a **mycoprotein**, or fungal protein, prepared from the fungus *Fusarium graminearum*. Quorn has a protein to fat ratio similar to that of meat but has the advantage that it is cholesterol-free. Its meaty texture depends on the type of fibres that

the fungus produces. Different lengths of fibres can be produced under different conditions. Short fibres produce a chicken-like meat substitute, longer fibres have the texture of lamb or beef. Quorn can be flavoured and shaped to form meat pieces, burgers and sausages.

Fig 15.3 **Microbial foods, such as Quorn, are used in a range of meat substitutes. They are suitable for vegetarians and people who need to eat foods low in cholesterol**

Lipids

Dietary lipids consist of the edible oils, fats, waxes and related compounds. Approximately 95 per cent of the lipids in the human diet are **triglycerides** in which three fatty acids are attached to a glycerol molecule. The remaining 5 per cent comprise mainly **cholesterol** and the **phospholipids**. The body can synthesise many of its fatty acids from carbohydrates and proteins: only a few, the **essential fatty acids**, must come from food.

The lipid content of most plant foods is low, but there are exceptions, such as nuts, seeds and olives. In general, animal products have a higher lipid content as they store fat in their bodies. Animal sources include fatty meats, fish such as herring and pilchard, milk and milk products.

Vitamins

Vitamins are essential organic compounds needed by the body in small amounts (Table 15.2). Vitamins help to regulate cell chemical processes, often by interacting with important enzyme systems. Lack of a particular vitamin in the diet leads to a **deficiency disease**.

(a) **Fat-soluble vitamins**

Table 15.2 **Vitamins important in the human diet**

Vitamin	Name	Minimum daily requirement	Rich food source	Function	Deficiency disease
A	Retinol	1000 µg	Fish liver oils, dairy products, liver. Most leafy vegetables and carrots contain carotene that can be converted into retinol	Needed for healthy epithelial cells. Needed for regeneration of rhodopsin in rod cells of the eye	Dry skin. Night blindness
D	Calciferol	10 µg	Fish oils, egg yolk, butter. It can be made by the action of sunlight on skin	Promotes absorption of calcium from intestines. Necessary for formation of normal bone and reabsorption of phosphate from urine	Rickets in children ('soft' bones that bend easily). Painful bones in adults
E	Tocopherol	10 µg	Vegetable oils, cereal products and many other foods	In rats, formation of red blood cells, affects muscles and reproductive system. Unclear in humans	Mild anaemia and sterility in rats. Deficiency is rare in humans
K	Phylloquinone	70 µg	Fresh, dark green vegetables. Also made by gut bacteria	Formation of prothrombin (involved in blood clotting)	Delayed clotting time. May occur in new-born babies before their gut bacteria become established

(b) **Water-soluble vitamins**

The B vitamins form a vitamin complex. They are chemically unrelated, but tend to occur together and have similar functions.

B1	Thiamine	1.5 mg	Yeast, cereals, nuts, seeds, pork	Coenzyme in cell respiration, necessary for complete release of energy from carbohydrates	Beriberi (severe muscle wastage, stunted growth, nerve degeneration)
B2	Riboflavin	1.8 mg	Liver, milk, eggs, green vegetables	Coenzyme in cell respiration. Precursor of FAD	Cracked skin, blurred vision
B3	Niacin	20 mg	Liver, yeast, whole cereals, beans	Coenzyme in cell respiration. Precursor of NAD/NADP	Pellagra (severe skin problems, diarrhoea, dementia)
B6	Pyridoxine	2 mg	Meat, fish, eggs, cereal bran, some vegetables	Interconversion of amino acids	Skin problems, nerve disorders
B12	Cyanocobalamin	2 µg	Liver, milk, fish, yeast. None in plant foods	Maturation of red blood cells in bone marrow. Maintenance of myelin sheath of nerves	Pernicious anaemia. Nerve disorders
	Folic acid	200 µg	Liver, raw green vegetables, yeast, gut bacteria	Formation of nucleic acids. Formation of red blood cells	Anaemia, especially during pregnancy
C	Ascorbic acid	60 mg	Blackcurrants, peppers, sprouts, citrus fruits	Formation of collagen and intercellular cement	Scurvy. Poor wound healing

Food is the main source of vitamins for most animals, but it is not the only one. Bacteria in the rumen of cattle supply the animals with B vitamins. Carnivores such as dogs and cats do not eat citrus fruits and other sources of vitamin C but they do not suffer from scurvy: their livers can actively manufacture vitamin C. Humans can make vitamin A from carotene, a substance found in most plant foods, particularly carrots. Also, ultraviolet radiation from sunlight acts on a compound in our skin called **ergosterol**, converting it to vitamin D.

We divide vitamins into two groups: **fat-soluble vitamins** (A, D, E and K), which contain only carbon, hydrogen and oxygen, and **water-soluble vitamins** (vitamin C and the B complex of vitamins), most of which also contain nitrogen. Fat-soluble vitamins can be stored in the body but water-soluble ones cannot, and are excreted in the urine. Water-soluble vitamins need to be taken into the body every day, and it is virtually impossible to 'overdose' on them. Taking too much fat-soluble vitamins can cause problems: they can build up in fatty tissues until they reach toxic levels.

?

D Early Arctic explorers learned from the local Inuit people not to eat the livers of polar bears. It causes headache, drowsiness, vomiting and extensive peeling of the skin. Since polar bear liver contains up to 0.6 g of vitamin A per 100 g, what is the likely explanation for this observation?

FINDING OUT ABOUT VITAMINS

Vitamin A

FREDERICK GOWLAND HOPKINS worked out the importance of vitamins in the diet in the early 1900s. His classic experiment, described in Fig 15.4, helped people understand the need for vitamins as additional food factors in a healthy diet.

He showed that rats deprived of milk did not grow properly. He suggested that the milk contained an 'accessory food factor', or 'vital amine' – hence the name *vitamin* – which the rats need to stay healthy. American researchers later showed this compound to be fat soluble. We now know it as vitamin A.

Vitamin C

The need for fruits like oranges, lemons and limes was first discovered in the 17th century when sailors on long sea voyages developed scurvy after many months living on dried and stale meat.

But it was not until the 1940s that anyone showed that the substance in fruit that prevented these symptoms was vitamin C. In 1940, a young American surgeon, John Crandon, experimented on himself to investigate the effects of a diet deficient in ascorbic acid. After 3 months of deficiency, a skin cut healed normally, but after 6 months a cut would not heal at all. We now know that long-term vitamin C deficiency leads eventually to a severe reduction in the body's ability to produce collagen and intercellular cement, making skin damage difficult to repair.

Today, some scientists think that some vitamins, including vitamin C, may help protect the body against cancer, but there is no hard evidence for this.

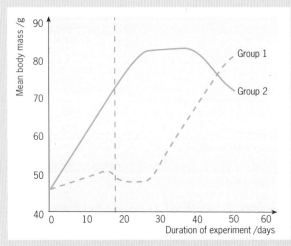

Fig 15.4 **Frederick Gowland Hopkins studied accessory food factors between 1906 and 1912. He used two groups of rats in the experiment shown here. For the first 18 days, he fed the rats in Group 1 on a diet of purified milk protein (casein), starch, cane sugar, lard and salts only. He fed the rats in Group 2 the same diet with an added milk supplement.**

As the graph shows, rats in Group 1 did not gain weight at the same rate as rats in Group 2. On day 18, the diets of the two groups were swapped. The rats in Group 1 began to catch up rapidly, but the rats in Group 2 stopped growing and then began to lose weight

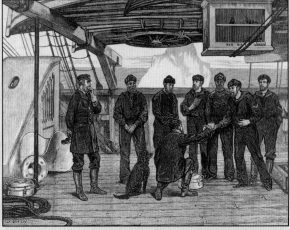

Fig 15.5 **Sailors on an Arctic expedition in 1875 were given lime juice to prevent scurvy**

Minerals

The minerals essential to human health are listed in Table 15.3. Minerals have four main functions in the body. They:

- are raw materials needed for the construction of body tissues,
- are metabolites for various cell processes,
- have a homeostatic role, providing the correct chemical environment for cells,
- are essential partners for enzymes and vitamins, and so help to control cell activity.

Mineral	Major food source	Needed for	Total body content (g)
Major minerals: macronutrients			
Calcium	Milk, cheese, bread, watercress	Muscle contraction, nerve action, blood clotting and the formation of bone	1000
Phosphorus	Cheese, eggs, peanuts, most foods	Bone and tooth formation, energy transfer from foods, DNA, RNA and ATP formation	780
Sulphur	Dairy produce, meat, eggs, broccoli	Formation of thiamin, keratin and coenzymes	140
Potassium	Potatoes, meat, chocolate	Muscle contraction, nerve action, active transport	140
Sodium	Any salted food, meat, eggs, milk	Muscle contraction, nerve action, active transport	100
Chlorine	Salted foods, seafood	Anion/cation balance, gastric acid formation	95
Magnesium	Meat, green vegetables	Formation of bone, formation of co-enzymes in cell respiration	19
Trace elements (micronutrients)			
Iron	Liver, kidney, red meat, cocoa powder, watercress	Formation of haemoglobin, myoglobin and cytochromes	4.2
Fluorine	Water supplies, tea, sea food	Resistance to tooth decay	2.6
Zinc	Meat, liver, beans	Enzyme activation, carbon dioxide transport	2.3
Copper	Liver, meat, fish	Enzyme, melanin and haemoglobin formation	0.072
Iodine	Seafood, iodised salt, fish	Thyroxine production	0.013
Manganese	Tea, nuts, spices, cereals	Bone development, enzyme activation	0.012
Chromium	Meats, cereals	Uptake of glucose	<0.002
Cobalt	Meat, yeast	Synthesis of vitamin B12, formation of red blood cells	0.0015

Table 15.3 **Minerals essential in the human diet**

The actual amount of a mineral that the body takes up can be affected by the presence of other vitamins and minerals. For example, calcium, the most abundant mineral in the body, is taken up more quickly in the presence of vitamins C and D, and iron is better absorbed in the presence of vitamin C.

Water

Just under 70 per cent of the human body is water. Two-thirds of this is inside cells, the other third is outside the cells in tissue fluid and blood plasma. It is essential in the diet and can be obtained from three main sources:

- as a drink,
- as food (lettuce, for example, is 94 per cent water),
- as metabolic water (water that is released from cellular chemical reactions, especially cell respiration).

Animals must take in enough water to balance what they lose in urine, sweating, breathing out and in faeces. The role of the kidneys in water balance is covered in Chapter 27.

Dietary fibre

Fibre, the indigestible component of food, provides bulk to the digestive tract, stretching the gut wall and stimulating a faster throughput of food. If we eat a diet based on refined, sweet and high-fat foods with little fresh fruit, vegetables or whole grains, we do not get enough fibre. In the short term this means chronic constipation, and it increases our risk of developing gallstones and bowel cancer later in life.

See questions 1 and 2.

3 HOW HETEROTROPHS OBTAIN FOOD: FEEDING DESIGNS

Heterotrophs use different strategies for obtaining food. Heterotrophic nutrition is either **saprotrophic**, **parasitic** or **holozoic** (Fig 15.6). **Saprotrophs** (such as fungi) feed exclusively on dead and decaying material, **parasites** (such as tapeworms) obtain their food from the living body of another organism. **Holozoic feeders** (most animals fall into this class) take in solid organic material, break the food down by **digestion** and absorb the soluble products.

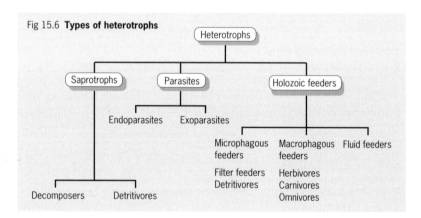

Fig 15.6 **Types of heterotrophs**

Saprotrophic nutrition

Why aren't we knee deep in dead leaves, insects, birds and mice? The answer lies with a specialised group of heterotrophic organisms known as **saprotrophs**. The term saprotrophic or **saprobiotic** describes organisms that feed on dead and decaying material (*sapro* = rotten). The term saprotroph replaces the older label **saprophyte**.

There are two types of saprotroph. The first is a vital group of organisms, the **decomposers**, which rot the dead bodies of animals and plants. Most decomposers are bacteria or fungi that live on or near decaying organic matter. They absorb soluble molecules, such as amino acids, organic acids and mineral salts, unchanged through their cell walls. Insoluble materials, such as starch, cellulose, lignin, fats, waxes and resins, are broken down first. Saprotrophs digest complex molecules by secreting digestive enzymes onto their food source, digesting it externally and then absorbing the breakdown products.

Decomposers therefore have several defining features. They:

- produce powerful protease, amylase, lipase, cellulase and lignase enzymes.
- show external digestion.
- absorb simple food molecules through a permeable cell wall.
- are small or have thin structures (often microscopic) to aid diffusion.
- grow rapidly, multiplying or spreading to colonise an area quickly.

Detritivores, the other important saprotrophic group, live in water and include the marine worms (Fig 15.7). They feed on the **detritus** (dead organic matter) at the bottom of seas or lakes.

Fig 15.7 **Detritivores, such as many marine worms, feed on organic sediments on the sea bed or the lake bottom**

Saprotrophic bacteria and fungi

Some saprotrophic bacteria and fungi are economically important to humans (Fig 15.8).

Fig 15.8
Examples of saprotrophs that are important economically

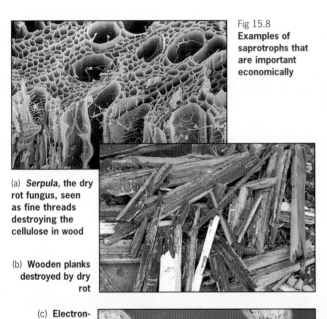

(a) **Serpula**, the dry rot fungus, seen as fine threads destroying the cellulose in wood

(b) **Wooden planks destroyed by dry rot**

(c) **Electron-micrograph of Clostridium, a bacterium important in sewage breakdown and food poisoning**

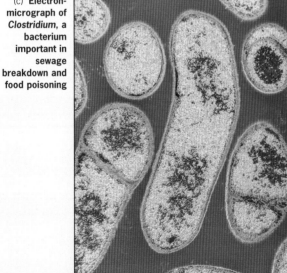

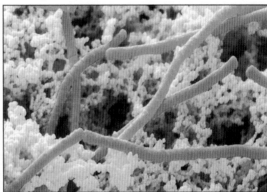

(d) **Scanning electronmicrograph of Lactobacillus, a bacterium used in the manufacture of yoghurt**

(e) **Familiar Lactobacillus products**

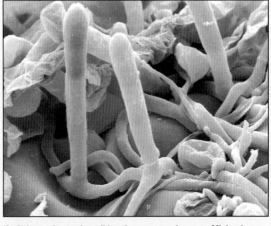

(f) **Sphaerotheca, the mildew fungus, growing on a Michaelmas daisy leaf (mildews attack chrysanthemums and fruit such as gooseberries and peaches)**

When saprotrophic fungi colonise a piece of dead or waste material, the process is not as random as you might think (Table 15.4). A freshly produced mound of herbivore dung is first colonised by sugar fungi, which use up any available soluble sugars. Then the lignin and cellulose decomposers move in, often followed by a second flush of the sugar fungi, which sneak back to feed on sugars produced by the breakdown of cellulose and lignin. This process is called **nutritional succession**.

Table 15.4 **Nutritional succession on herbivore dung**

Fungus	When it appears	How long it lasts	Feeding activity
Mucor	1–3 days	7 days	primary sugar fungus
Coprobia	5–6 days	3–4 weeks	cellulose-digesting fungus
Coprinus	9–10 days	3–4 weeks	lignin-digesting fungus

Parasitic nutrition

A parasite is an organism that feeds from another living organism for most of its life cycle. The parasite benefits as it obtains most of its metabolic requirements, but the host is usually harmed. A successful parasite minimises this harm so that it can continue to use its host for longer periods of time and so that it has more chance of spreading to other hosts.

Endoparasites live inside the body of their host: **ectoparasites** live outside. Both types show adaptations which allow them to occupy a wide variety of habitats on or in another living body. Endoparasites usually show extreme specialisation with organ systems, such as digestive and sensory systems, that are very much reduced. They also tend to have complex life cycles to enable the organism to spread from host to host. Ectoparasites generally have less specialised features.

See question 3.

Feeding adaptations in endoparasites

Endoparasites, such as tapeworms (Fig 15.9), live in the gut of their host. Once in position, they are continually bathed in predigested food and so do not need either a mouth or a digestive tract – they absorb the simple food molecules across the membrane that makes up their body surface.

Tapeworms make the most of their situation because they:

- have a body wall with a high degree of folding (microscopic folds called **microtriches**.
- have many mitochondria, which provide the energy required to actively transport some food products.
- produce some enzymes to aid external digestion close the body wall. This enables the tapeworm to compete with its host for the food available.

Fig 15.9 **Tapeworms are highly adapted to living in the gut of their host**

Holozoic nutrition: feeding strategies in animals

Animals are described as **microphagous feeders**, **macrophagous feeders** or **fluid feeders**, according to the type of food they eat or the feeding methods they use.

Microphagous feeders

Microphagous feeders, or **microphages** (Fig 15.10), feed on tiny food particles (tiny even when compared to their own microscopic body size). Most microphages feed continuously.

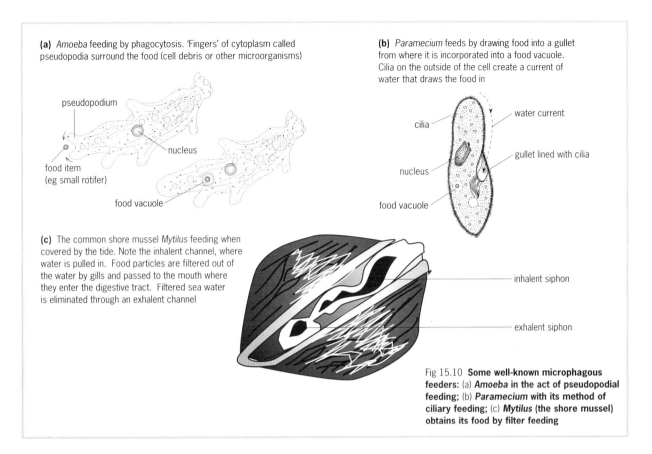

(a) *Amoeba* feeding by phagocytosis. 'Fingers' of cytoplasm called pseudopodia surround the food (cell debris or other microorganisms)

pseudopodium

nucleus

food item (eg small rotifer)

food vacuole

(b) *Paramecium* feeds by drawing food into a gullet from where it is incorporated into a food vacuole. Cilia on the outside of the cell create a current of water that draws the food in

water current

cilia

gullet lined with cilia

nucleus

food vacuole

(c) The common shore mussel *Mytilus* feeding when covered by the tide. Note the inhalent channel, where water is pulled in. Food particles are filtered out of the water by gills and passed to the mouth where they enter the digestive tract. Filtered sea water is eliminated through an exhalent channel

inhalent siphon

exhalent siphon

Fig 15.10 **Some well-known microphagous feeders:** (a) *Amoeba* **in the act of pseudopodial feeding;** (b) *Paramecium* **with its method of ciliary feeding;** (c) *Mytilus* **(the shore mussel) obtains its food by filter feeding**

There are two major types of microphagous feeder: **filter feeders** and **deposit feeders**.

Filter feeders consume organic material, often plankton, in open water. This type of feeding is characteristic of **sedentary animals** (animals that move very little), such as those that live on a rocky shore, on the bottom of the ocean, or on a river bed. Different filter feeders filter their food from the water using different techniques. The fan worm, for example, filters water through tentacles, while the mussel filters water through modified gills. All filter feeders transfer particles of food to their digestive tract using cilia.

Deposit feeders feed on rich organic material on the sea bottom. As organisms die, their bodies accumulate as sediment, often called detritus, on the surface of the sea bed. This provides a rich source of food for detritivores such as worms and molluscs. Most deposit feeders do not have elaborate feeding structures: many worms that feed on the sea bottom do not have a specialised mouth. However, some molluscs and echinoderms do have tentacles and specialised feeding structures.

Macrophagous feeders

Macrophagous feeders feed on larger particles of food, selecting food on the basis of nutritional quality and previous experience. There are three groups of macrophagous feeders, defined by the origin of the food that the animals eat. Those that feed exclusively on plant food are **herbivores**, those that consume other animals are **carnivores** and those that eat a variety of food types are **omnivores**. A **scavenger** is a carnivore that feeds on **carrion** (dead and rotting flesh). Technically these terms can be used to describe any animal, but in practice, they are most often used to describe mammals.

?

E Bivalves (two-shelled molluscs like *Mytilus*) that burrow in the sand have long, extended siphons for moving water in and out. In the shore mussel, this tissue has degenerated and siphons are absent. Can you think of a reason for this?

F Why is deposit feeding likely to provide less raw material and energy per day than filter feeding?

?

G Both decomposers and scavengers feed on dead and decaying material. What are the main differences between the two?

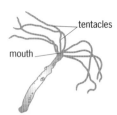

(**a**) *Hydra* feeds by capturing prey with its tentacles which are armed with stinging cells. It then pushes the food into its mouth

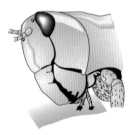

(**b**) A locust feeding on grass. The mandibles (mouthparts) bite the leaf and put the strip of leaf into the mouth

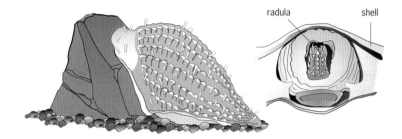

(**c**) A dog whelk. The diagram above left shows the shell of this gastropod. The diagram above right shows the file-like radula which allows it to scrape and shred the tissues of the animals on which it feeds. This ensures that the whelk swallows only small and easily digestible pieces of food

Fig 15.11 **Macrophagous feeders show a wide range of body designs that fit their methods of feeding**

As Fig 15.11 shows, macrophagous feeders use a wide variety of techniques to capture, immobilise and eat food:

- Some, such as *Hydra* and jellyfish, use tentacles to capture food and then transfer it to a simple mouth opening. This technique is called **tentacular feeding**.

- Many others, such as humans and locusts, use jaws, often armed with teeth, to **bite** their food and modify it mechanically by **chewing**.

- Those such as the garden snail and the dog whelk use a file-like scraping organ, the **radula**, to scrape away at rocks, leaf surfaces or animal tissue in an action known as **rasping**.

Fluid feeders

Fluid feeders feed on a liquid diet, which can consist of nectar (butterflies), plant phloem sap (greenfly) or blood (mosquitoes). They all have specialised mouthparts (for more detail, see page 233). Sap-feeding insects such as greenfly avoid the need to break down cellulose in plant cell walls by using the plant's fluid transport systems to supply food 'on tap'. This technique can lead the insect to develop nitrogen deficiency but, to compensate, many sap-feeding insects have nitrogen-fixing bacteria in their gut.

Feeding adaptations in mammals

Mammals have mouthparts and digestive systems that are adapted to the types of food they eat. We shall look at just two groups: the herbivores and the carnivores (Table 15.5).

Table 15.5 **A summary of the physical adaptations in herbivores and carnivores**

H Suggest the feeding methods (microphagous, macrophagous or fluid-feeding) used by the following animals: (**a**) spider, (**b**) baleen whale, (**c**) garden snail, (**d**) hummingbird, (**e**) magpie, (**f**) cat flea, (**g**) barnacle, (**h**) earthworm, (**i**) leech, (**j**) locust.

Carnivores	Herbivores
Incisors are sharp and thin for nipping and biting	Incisors are sharp and broad for cropping
Upper incisors never absent	Upper incisors may be absent (sheep, cattle)
Canines are pointed (for grasping)	Canines are small
Molars are pointed, adapted for cutting	Molars are heavily ridged for crushing and grinding
Teeth stop growing at adulthood	Teeth continue to grow to replace worn material
No diastema	Diastema present to allow efficient action of the tongue
Digestive tract short	Digestive tract long
No specialisation of stomach and caecum	Modification of stomach in ruminants with enlarged caecum and appendix

Herbivores

Although plant tissues are mainly water, with relatively indigestible materials such as cellulose and lignin, their roots, seeds and fruits are often rich in nutrients. Herbivores need to overcome the problems of eating these tough, fibrous foods:

- All herbivores have strong jaws, broad incisors and continuously growing ridged molars to deal with the mechanical properties of leaves (Fig 15.12). Grass, for example, has a high silica content and tough cellulose cell walls, which makes it difficult to break up into small pieces ready for digestion.

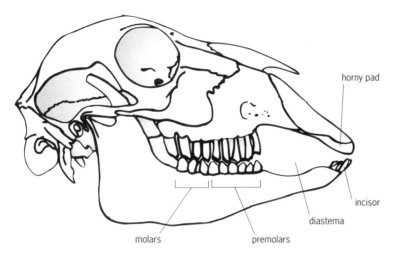

Fig 15.12 **The skull of a sheep, showing the arrangement of its mouthparts and teeth**

- The tongue in a herbivorous animal is usually large to help with food collection and chewing. Herbivores that eat plants with tough defensive structures such as thorns (acacias), spines (roses) and defensive hairs (stinging nettles) have thickened mouthparts. A camel, for example, chews at acacia bushes, apparently oblivious of the thorns.

- Herbivores have specialised digestive tracts that tend to be much longer than those found in carnivores. The extra length, combined with other modifications such as the **ruminant** stomach system, ensures that the tough cellulose- and lignin-containing plant food stays inside the animal long enough to be digested.

- The bacteria usually present in the intestines of a herbivore are different from those commonly found in carnivores. Many of the bacteria produce **cellulases** enzymes that help to break down cellulose. These are found in the **rumen** and **reticulum** of a ruminant (see the feature box on the opposite page) and in the **caecum** of a non-ruminant such as a rabbit (Fig 15.13).

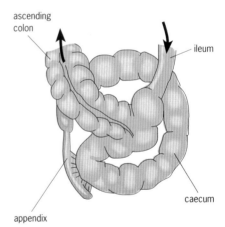

Fig 15.13 **The caecum of a rabbit is much larger than the caecum found in a carnivorous mammal of similar size. It contains microorganisms that produce cellulase**

Carnivores

Although food of animal origin is highly nutritious, it is not as plentiful as plant food and it has to be caught. Consequently, carnivores have well-developed senses and often show complex social behaviour associated with hunting. Like herbivores, carnivores have adaptations that enable them to deal with their food effectively:

- Carnivores have a shorter digestive tract because their food is more readily digested and does not need to stay inside the animal for an extended time.

- They also have modified teeth (Figs 15.15 and 15.16). The upper canines are well developed and are used to pierce and cut skin.

?

I Herbivores eat more or less continuously. Some carnivores eat only once every two or three days. Give two possible reasons for this difference.

RUMINANTS

RUMINANT MAMMALS such as cows, sheep and deer have a four-chambered stomach (Fig. 15.14). When they eat, food enters the first two of these chambers, the rumen and the reticulum, and is reduced to a workable pulp by the addition of liquid. Bacteria and other single-celled organisms (protoctists) that live in these two chambers produce digestive enzymes such as cellulase, which help to break down plant cell walls. These microbes also contribute to the animal's metabolism by manufacturing useful organic compounds such as amino acids and fatty acids, which are absorbed directly.

A ruminant animal 'eats' its food twice. It is first broken up and partially digested in the rumen and reticulum. Later, the animal regurgitates the 'cud' into the mouth for further chewing. On its second passage down the oesophagus, food enters the **omasum** for further physical reworking, and finally reaches the **abomasum**, which is the stomach proper (it has **cardiac**, **fundic** and **pyloric** regions similar to the human stomach – see Chapter 16).

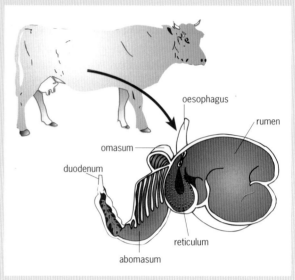

Fig 15.14 **The four-chambered stomach of the cow**

See question 4.

Carnassial teeth have a scissor-like action: they can cut through flesh and will also crack bones. The incisors are sharp and can tear meat away from bone and connective tissue. Grinding teeth are small: much of the food is swallowed whole.

- Carnivores have only one stomach where food is partially digested before passing into the small intestine to be completely broken down. Soluble food products are absorbed, and tough, indigestible fragments of bone, feathers and fur pass through to the large intestine for **egestion**.

- Living exclusively on meat means than carnivores do not take in some of the water-soluble vitamins found predominantly in plant material: for example, vitamin C. Instead, carnivores synthesise their own vitamin C and other water-soluble vitamins.

Fig 15.15 **Carnivores, such as this wolf, have specialised teeth, which are used to deal with a kill as quickly as possible, before rival predators and scavengers close in for their share**

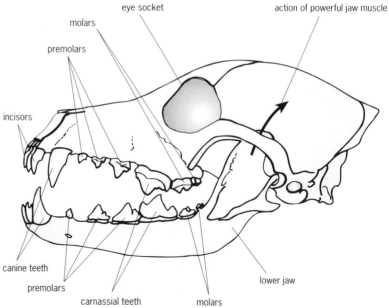

Fig 15.16 **The skull of a dog showing the arrangement of teeth**

Feeding adaptations in arthropods

Arthropods feed on a wide variety of animal, plant and dead organic materials. Although different arthropods look superficially very different, they all have one of two basic mouthpart designs: the biting mouthparts typical of locust and cockroach (Fig 15.17) or the sucking mouthparts typical of the aphid, mosquito and butterfly (Fig 15.18).

Arthropods can also have:

- a muscular **pharynx** and a large **crop**. These are present in sucking insects, such as mosquitoes.
- a grinding region called a **proventriculus**. This occurs in plant leaf feeders, such as locusts.
- microbes in their gut. These are common in wood-eaters.

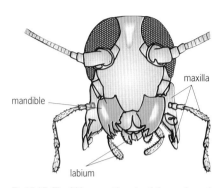

Fig 15.17 **The biting mouthparts of the cockroach**

Feeding designs in birds

Birds use a large proportion of the energy they obtain from food to maintain their body temperature and so they must feed frequently. A small bird, which loses heat very quickly because it has a large surface area to volume ratio, needs to consume around a third of its body weight in food per day. A larger bird needs to eat only one-seventh of its body weight.

The feeding habits of birds are reflected in the size, shape and strength of their beaks (Fig. 15.19). Because they have no teeth, birds must rely on modifications of the gut to break up food. They digest food mechanically in the stomach or **crop**. This has a muscular **gizzard** with a ridged lining of tough keratin to grind up the food in much the same way that mammals grind food in their mouths. Bird swallow small stones and grit with their food to help the grinding process.

Bird intestines are similar to those of mammals and, as with mammals, the length varies with diet (intestines are longer in seed- and plant-feeding species). Very active birds need to eat a large volume of food in a short time and so food passes through the gut very quickly. A shrike (*Lanius*) can digest a mouse in 3 hours.

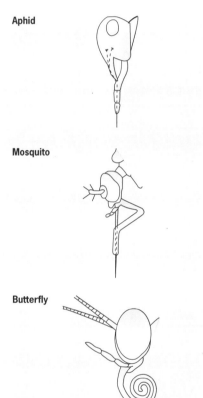

Fig 15.18 **Mouthparts of some common liquid-feeding insects. Different insects feed on different fluids: the aphid takes in phloem sap, the mosquito feeds on blood, the butterfly drinks nectar**

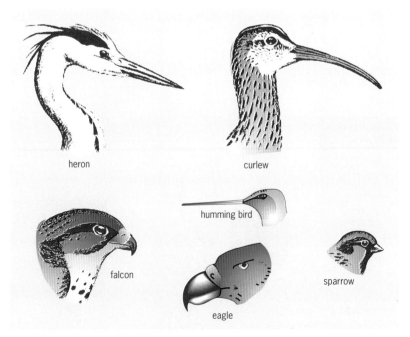

Fig 15.19 **The beak (or bill) of a bird is adapted to the type of food it normally eats. You can probably guess the type of food eaten. The long, thin beaks are for probing (curlew, hummingbird) or spearing fish (heron), the vicious, hooked beaks rip and tear flesh (eagle, falcon), whereas short, squat beaks are ideal for crushing (the seed-eating sparrow)**

SUMMARY

When you have read and studied this chapter you should know and understand the following:

■ Organisms need food to obtain the energy and raw materials they need for repair, growth and development of body tissues.

■ Humans need a balanced diet that contains **carbohydrate**, **protein**, **fat**, water, **fibre**, **vitamins** and **minerals**. Vitamins are either water soluble or fat soluble. Lack of particular vitamins or minerals in the diet can lead to **deficiency disease**.

■ Heterotrophic nutrition can be **saprotrophic** (saprotrophs feed on rotting organic matter), **parasitic** (parasites use their host to supply their food needs and give nothing in return) or **holozoic** (holozoic feeders feed on particles of solid food, which they break down by digestion).

■ Feeders in each group show adaptations to their lifestyles.

QUESTIONS

1 The composition of three items of food is shown in the following table.

	Composition per 100 g of edible portion		
	Brown bread	**Butter**	**Cheese**
Energy (kJ)	980	3000	1700
Protein (g)	9	0.5	25
Fat (g)	1	80	35
Carbohydrate (g)	48	0	0
Calcium (mg)	88	15	800
Iron (mg)	3	0.2	0.6
Vitamin A (µg)	0	1000	420
Thiamin (mg)	0.3	0	0.04
Riboflavin (mg)	0.1	0	0.05
Nicotinic acid (mg)	2.7	0.1	5.2
Vitamin C (mg)	0	0	0
Vitamin D (µg)	0	1	0.4

To what extent does a meal of brown bread, butter and cheese satisfy dietary requirements in humans.

[NEAB June 1994 Biology: Paper 1, Section B, q.5]

2 The table shows the average energy stores in an adult male and the time, when walking or running in a marathon (approximately 42 km), each store would last.

Energy store	Mass/ grams	Energy value/ kilojoules	Time reserves would last/minutes	
			Walking	**Marathon running**
Adipose triglyceride	9000	337 500	15 500	4018
Liver glycogen	100	1 660	86	20
Muscle glycogen	350	5 800	288	71
Blood glucose	3	48	2	<1

(From *The Runner*, Eric Newsholme and Tony Leach)

a) Give **one** site for the adipose triglyceride store.

b) **(i)** What is the ratio of energy available in a gram of triglyceride compared to that available in a gram of liver glycogen? (Show your working.)

(ii) What is the advantage of an energy store in the form of triglyceride rather than glycogen?

c) The table shows that blood glucose is an insignificant energy store. What is the importance of blood glucose and how would its level be maintained, apart from drinks of glucose, during a marathon run?

[AEB June 1994 Human Biology: Paper 1, q.5]

3 The drawings below show three parasitic worms that may be found in the alimentary canal of a frog.

a) **(i)** What are the main problems that a parasite must overcome in order to live in the gut of a frog?

(ii) Briefly describe how **three** features shown in the drawings enable the parasites to overcome these problems?

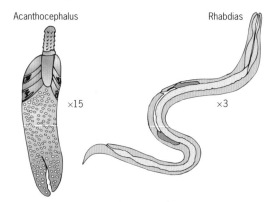
Acanthocephalus ×15   Rhabdias ×3

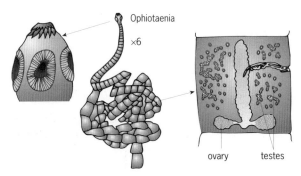
Ophiotaenia ×6   ovary   testes

b) **(i)** What is the main difference between the ways in which *Rhabdias* and *Ophiotaenia* obtain their food? Give the evidence from the drawings that supports your answer.

(ii) Suggest an explanation for the fact that *Ophiotaenia* is confined to the intestinal region of the host's gut, while *Rhabdias* may be found throughout its length.

c) A particular species of parasite may be found in the lens of its fish host. The table shows the percentage of parasites in various parts of the fish's body at different times after infection.

Time after infection	Percentage of parasites in:			
	Epithelium of pharynx	Connective tissue and muscle	Blood	Lens
15 minutes	84.2	0	0	0
1 hour	19.5	59.0	3.1	0
5 hours	12.8	20.7	2.8	5.1
19 hours	3.4	3.4	3.4	72.6
139 hours	0	0	0	92.7

Use the information in the table to answer the following questions.

(i) Where does the parasite penetrate the body?

(ii) Giving evidence for your answer, describe the route taken by most of the parasites once they have penetrated the body.

[AEB June 1991 Biology: Paper 2, q. 1]

4 The drawing shows the jaws, dentition and jaw muscles of a sheep, a typical herbivorous mammal.

a) **(i)** Describe how **two** features visible in the drawing (below) adapt the dentition to a herbivorous diet.

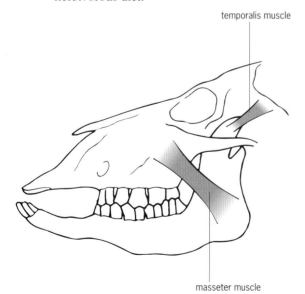

temporalis muscle

masseter muscle

(ii) Explain how the jaw muscles are involved in the particular jaw action found in a herbivorous mammal like a sheep.

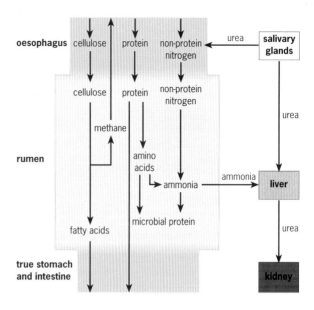

A cow obtains most of its nutritional requirements from fermentation by mutualistic (symbiotic) microorganisms in its rumen. The diagram (above) summarises the biochemical processes carried out by these microorganisms.

b) With the aid of information in the diagram, explain why:

(i) the relation between the cow and the organisms living in its rumen may be described as mutualistic;

(ii) it is possible for cattle to survive on a diet that contains no protein for a considerable period of time;

(iii) ruminants such as the cow are less efficient than non-ruminant animals at converting energy in their food into energy in their tissues.

c) **(i)** What is likely to be the main respiratory substrate of the cow?

(ii) How does the cow obtain ATP from this respiratory substrate? Details of biochemical pathways are **not** required.

Fermentation in the rumen is sometimes likened to the process in an industrial fermenter.

An industrial fermenter may be used for the continuous production of substances such as penicillin by the fungus *Penicillium*. Sterile medium is continuously added and the penicillin harvested. Temperature and pH are carefully controlled.

d) **(i)** Suggest **two** important differences between the fermentation process in an industrial fermenter and fermentation in the rumen.

(ii) How are constant conditions of temperature achieved in the rumen?

[AEB June 1994 Biology: Paper 2, q. 1]

16 Digestion and the digestive system in mammals

William Beaumont, army surgeon, and his drawing of Alexis St Martin's stomach, as seen through the hole in Alexis' abdomen

IN JUNE 1822, Alexis St Martin, an American army porter, was shot accidentally by a musket fired at close range. He sustained several wounds, including a large abdominal hole out of which poured his recently eaten breakfast. This horrific injury was a turning point in our understanding of digestion.

Alexis was strong and, apart from the large hole in his abdomen, the wounds healed. William Beaumont, an army surgeon, tried to close this persistent hole for nearly 10 months. He used a tightly rolled bandage to plug the wound and eventually a skin flap grew over the hole, forming a sort of valve which kept the stomach contents in. This, however, could be pressed to reopen the hole at any time.

Beaumont had a walking, talking experiment! He used the opportunity to the full, conducting many investigations and keeping meticulous records. He tells of how he could press on the abdomen above the liver to obtain bright yellow bile that had been released into the duodenum. And how he put different kinds of food in the stomach and checked them at various times afterwards to follow the digestive processes. This strange partnership between doctor and patient lasted for nine years.

1 THE ALIMENTARY CANAL

Heterotrophs take nutrients into their bodies from their surroundings. They **digest** (break down) food, **absorb** the simple molecules that are produced and then incorporate them into body tissue. To digest food, an animal needs a **gut**. This can be a simple **gut cavity**, as in *Hydra*, or a much more complex **alimentary canal**, as in mammals.

An alimentary canal, a muscular tube that leads from the mouth to the anus, enables a mammal to:

- **ingest**, or take in food.
- pass food through its body, propelling it along the alimentary canal by **peristalsis:** rhythmic contractions of the gut wall.
- break down food material into smaller pieces. This is **mechanical breakdown**.
- digest complex food molecules such as carbohydrates, proteins and fats into simpler ones. This is **chemical breakdown**.
- absorb simple food molecules such as amino acids, sugars and fatty acids.
- absorb water and some mineral salts (before they are lost from the body as waste).
- **egest**, or eliminate, undigested food material from the body.

Heterotrophs cannot make their own food: they must take in food molecules from their surroundings. They use an **organic carbon source** (carbon that was once in living things) so they are totally dependent on organisms like green plants, which can make their own food.

2 THE HUMAN DIGESTIVE SYSTEM

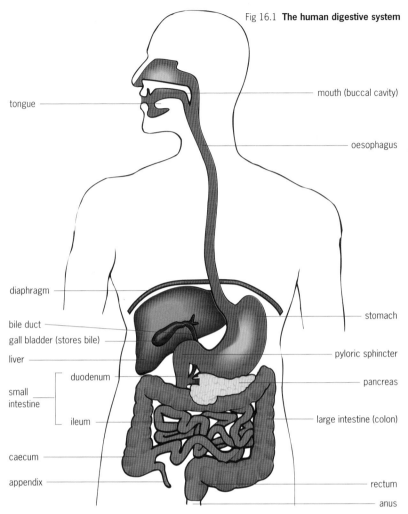

Fig 16.1 **The human digestive system**

Labels: tongue, mouth (buccal cavity), oesophagus, diaphragm, bile duct, gall bladder (stores bile), liver, duodenum, small intestine, ileum, caecum, appendix, stomach, pyloric sphincter, pancreas, large intestine (colon), rectum, anus

The human **digestive system** (Fig 16.1) consists of the alimentary canal and its associated glands, the **salivary glands**, the **liver** and the **pancreas**.

The alimentary canal begins at the **mouth** and ends at the **anus**. Between the two openings is a long, convoluted tube, organised into several distinct regions. The **oesophagus** carries food from the mouth to the **stomach**. Beyond the stomach is the **small intestine** which has three parts: the **duodenum**, the **jejunum** and the **ileum**. The **large intestine** includes the **caecum,** the **colon** and the **rectum** and ends at the anus.

In this chapter we concentrate on digestion in the human alimentary canal. Table 16.1 provides an overview of the digestive enzymes produced: you will find it useful to refer to this table as you read on.

Table 16.1 **A summary of the main human digestive enzymes**

Secretion	Enzymes produced	Site of production	Site of activity	pH	Substrate	Products
Saliva	salivary amylase	salivary glands	mouth	6.5–7.5	starch	maltose
Gastric juice	pepsin rennin	stomach	stomach	2.0	proteins milk protein	polypeptides
Pancreatic juice	trypsin chymotrypsin carboxypeptidase pancreatic amylase pancreatic lipase nuclease	acini (secretory cells of the pancreas)	duodenum	7.0	proteins proteins polypeptides starch fats and oils nucleic acids	polypeptides polypeptides amino acids maltose fatty acids, glycerol nucleotides
Intestinal juice*	pancreatic amylase maltase sucrase lactase peptidase mixture lipase nucleotidase	cells of villi lining the small intestine	ileum	8.5	starch maltose sucrose lactose polypeptides fats and oils nucleotides	maltose glucose glucose, fructose glucose, galactose amino acids fatty acids, glycerol sugar, base, phosphoric acid

* The enzymes produced by cells of the small intestine act within cells or close to the wall of the intestine. Consequently, intestinal juice itself is mainly water: it creates the conditions which the cells require to carry out their digestive functions.

3 FROM THE MOUTH TO THE STOMACH

The mouth or **buccal cavity** is essentially a box which acts as the reception area for food. The **palate** is the top of the box, the **tongue** and **lower jaw bone**, or **mandible**, form the bottom and the muscles of the **cheeks** are the side walls. The mouth is lined with **stratified epithelium** (layered epithelial cells), which protects the mouth from friction damage as food is chewed.

The mouth receives ingested food and is the place where food is chewed or **masticated**. It also contains receptors that taste food (smell receptors are close by). The tongue has a rough surface caused by raised projections, **papillae,** that help it to grip food pieces.

By chewing, we break down food mechanically, making it easier to swallow. The powerful muscles in the jaw generate enormous force. Large carnivores, such as lions, use this force to break large bones. Saliva lubricates and softens the food. We all produce between 1 and 1.5 litres of saliva each day. We secrete saliva continuously, but we produce more if we see, smell, taste, or even just think about food. (A summary of the mechanisms that control the release of all digestive secretions is shown later in Fig 16.14.)

Saliva is mainly water (99.5 per cent), with some dissolved substances (0.5 per cent) including:

- **mineral salts** (phosphates and hydrogen carbonates),
- **salivary amylase** (a starch-splitting enzyme), which breaks molecules of starch into maltose sugars,
- **mucin** (a slimy, glycoprotein lubricant),
- **lysozyme** (an enzyme that kills bacteria).

Salivary amylase, the starch-splitting enzyme in saliva, acts on food pieces that are produced by chewing, beginning the process of chemical breakdown. Food leaves the mouth when we swallow.

The senses of taste and smell

Gustation (taste) and **olfaction** (smell) are the body's chemical senses. There are four basic sensations of taste: *sweet, salt, sour* and *bitter*. Try this experiment: pinch a willing friend's nostrils and place a small slice of apple and then a small piece of raw onion on his or her tongue. You should find that he or she will taste both as sweet, and cannot identify the food exactly. All the other 'flavours' are odours produced as food is crushed by the teeth.

Receptor cells for taste are mainly in taste buds on the upper surface of the tongue. Certain regions of the tongue appear to react more strongly than others to particular taste sensations (Fig 16.2).

We can only smell a substance that is soluble in water and is **volatile** (evaporates easily). **Olfactory receptor cells** (smell detectors) are located in the nose. Molecules of the odour stimulate these olfactory cells, causing them to send impulses to the brain for processing.

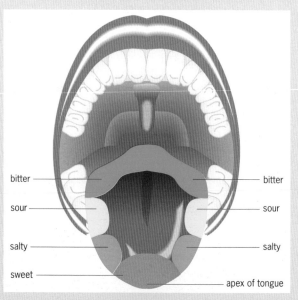

Fig 16.2 **The taste receptors for sweet, salty, bitter and sour are located in well-defined regions round the rim of the tongue**

Swallowing

Swallowing is not a voluntary action. It is a reflex response to food or liquid touching the back of your throat (the act of moving food to the back of your throat is the voluntary action). The act of swallowing is illustrated in Fig 16.3.

Theoretically, when the tongue forces food or liquid to the back of the throat, the food is able to travel in one of four directions:

- back out of the mouth,
- into the nasal cavity,
- into the windpipe,
- into the oesophagus (where it should go).

When we swallow normally, the pathways that would allow the first three options are closed off, and food or liquid is forced into the oesophagus. But sometimes it doesn't work out like that. If you have ever been laughing and drinking fizzy liquids or eating crisps simultaneously you may have experienced fizzy liquid going 'up your nose' or crisp particles going 'down the wrong way' (into your windpipe).

Looking at teeth

The arrangement and number of teeth in an animal is called its **dentition**. Humans have two sets of teeth during their life. **Milk dentition** appears from the age of about 6 months and lasts until the age of 5 or 6 (Fig 16.4). Humans have 20 milk teeth, which are gradually replaced (they are actually forced out) by a second set, the **permanent dentition**. Our 32 adult teeth should last for the rest of our lives, but rarely do because of **tooth decay**.

Tooth structure

A tooth typically consists of three parts: the **crown**, the part above the **gum**; the **root**, the part hidden below the gum (embedded in a **socket**); and the **neck** of the tooth in between (Fig 16.5).

Teeth are made of a bone-like material, **dentine**. This surrounds a **pulp cavity** with blood vessels and nerves running through it. The blood supply nourishes the living bone with food molecules and oxygen, and also removes waste products. The nerves allow us to sense pressure and touch, and, if the nerve is exposed by damage, we experience intense pain.

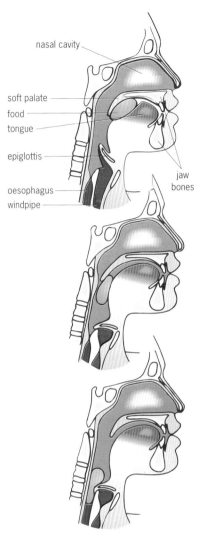

Fig 16.3 **Swallowing food or liquids is quite a tricky procedure. Since you can drink a glass of water standing on your head, it's not just a question of allowing material to drop into the stomach. The food or liquid is forced out of the mouth by the action of muscles and into a thin flexible tube, the oesophagus**

A Look at Fig 16.3. Describe how food is prevented

(a) from entering the nasal cavity,

(b) from entering the windpipe,

(c) from being forced back out of the mouth.

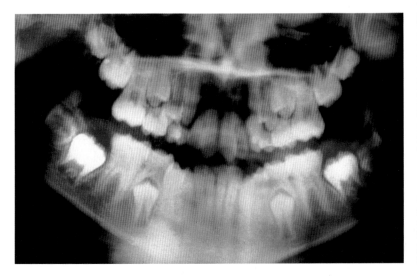

Fig 16.4 **An X-ray picture of a young child showing the current (milk) teeth and the new (adult) teeth waiting to push through. The two lower-jaw teeth highlighted in yellow have just pushed through the gum, but their crowns are not up to the level of the other (milk) teeth**

Fig 16.5 (a) **A human adult incisor tooth and** (b) **a molar tooth are shown in vertical section**

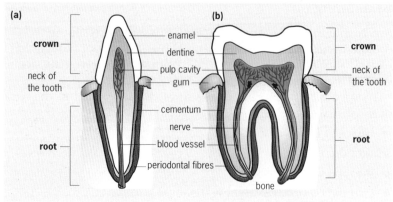

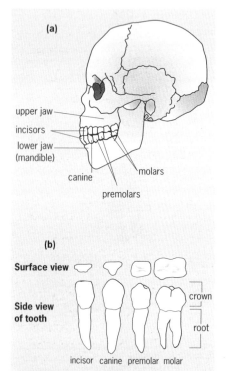

Fig 16.6 **Humans have** heterodont dentition (different types of teeth): (a) shows the position of the four types of human teeth in the skull; (b) shows what each type looks like (surface and side views)

Tooth enamel, the hardest material in the body, is the surface layer of the crown. **Periodontal fibres** help secure the tooth, and a layer of **cement** around the root fixes the tooth in its bony socket.

Types of teeth

The number, size and shape of teeth that a mammal has depends on the species. As Fig 16.6 shows, humans have four types of teeth:

Incisors: chisel-shaped front cutting teeth.

Canines: pointed 'corner' teeth used for grasping and tearing.

Premolars: small chewing teeth with two **cusps** (projections) on the crown.

Molars: large back teeth, with four cusps for chewing and grinding.

The oesophagus, route to the stomach

The human oesophagus is about 25 cm long and 2 cm in diameter. Muscles in the oesophageal wall contract rhythmically to propel food from the **pharynx** (throat) to the stomach. These rhythmic movements are called **peristalsis**. Elastic tissue in the walls enables the oesophagus to expand as food passes along it. As in the mouth, stratified epithelium helps prevent friction damage.

TEETH AND DISEASE

THERE ARE TWO main types of dental disease, **dental caries** and **periodontal disease**.

Dental caries, or tooth decay, occur when the dentine and enamel on a tooth are gradually broken down. The process begins when bacteria such as *Streptococcus mutans* release acids onto teeth as they feed on the sugars that remain there after we have eaten. The bacteria become fixed to the teeth in a capsule formed by the sticky polysaccharide **dextran**. The bacteria, dextran and other debris are collectively known as **dental plaque**.

The acids produced by bacterial digestion start to break down tooth enamel, and plaque may react with chemicals in saliva, hardening to form **calculus**. If this condition remains untreated, microbes can reach the pulp cavity of the tooth causing inflammation, infection and pain.

Periodontal disease is a gum disease that affects the **cement** and **periodontal fibres**. Teeth eventually become loose as the gums recede.

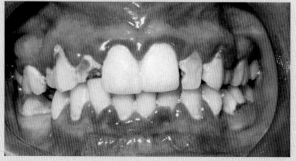

Fig 16.7 **Lack of proper care and attention to dental hygiene can lead to** dental caries **(cavities that need filling) and** periodontal disease **(loosening of the teeth)**

LOOKING INSIDE THE ALIMENTARY CANAL

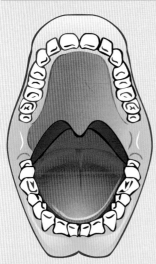

WHEN YOU LOOK INSIDE the mouth (left), you can see the apparatus which breaks up food and then pushes it into the oesophagus. It is possible to look deeper inside the alimentary canal of a living person using an **endoscope** (a flexible tube with a light and miniature camera at the end). The photograph above right shows the inside of the oesophagus: the diagram on the right shows the structure of the oesophagus wall. The rest of this Feature box continues this remarkable journey through the alimentary canal.

stratified epithelial cells

mucosa

submucosa

muscle layer

serosa

blood vessels gland secreting lubricating fluid

After the oesophagus is the stomach. As you can see from the endoscope picture on the right, the stomach has deep ridges called **rugae**. The gastric pits produce a strong acid. The stomach lining also secretes large quantities of mucus, to prevent its own tissues from being damaged. The structure of the stomach wall is shown in the diagram, far right.

blood vessels rugae

mucosa

submucosa

muscle layer

serosa

Next is the small intestine. As you can see from the endoscope picture on the left, the walls of the intestine are more gently folded. They are also awash with digestive secretions. The velvety appearance of the lining is due to the 4 or 5 million tiny projections, or villi, which line the wall. Villi are only 0.5 to 1 mm high, but they can move up and down to make intimate contact with food in the intestine. They increase the surface area of the lining and so ensure efficient absorption of nutrients. The diagram below left shows that the lining of the intestine, like the lining of the entire canal, has four basic layers:

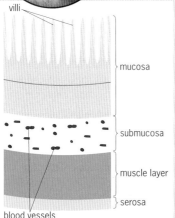

villi

mucosa

submucosa

muscle layer

serosa

blood vessels

An **inner mucosa**. This secretes digestive juices or absorbs food products. In areas prone to damage, the cells of the lining are stratified, or layered, for protection.

A **submucosa**. This contains lots of blood vessels to take away absorbed food. It is rich in nerves which coordinate the actions of muscles that perform the squeezing movements of peristalsis.

A thick muscle layer, the **muscularis mucosa**. Muscle fibres are positioned in rings around the intestine and along the length of the wall. This muscle is smooth muscle (see page 32).

An **outer serosa**. This is a tough layer of connective tissue which binds the canal to the wall of the abdomen. The layers which support the intestine are called the **mesenteries**, and the layers which line the abdomen are called the **peritoneum**.

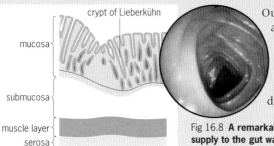

crypt of Lieberkühn

mucosa

submucosa

muscle layer

serosa

Our journey ends at the colon (endoscope picture, left). There are fewer secretions, but lots of mucus here. The colon does not have villi or permanent folds like the small intestine. Instead, it has a 'pinched' appearance due to thin bands of muscle. These produce a strong churning movement. Bacteria on the walls of the colon cause some of the non-digested food to ferment.

Fig 16.8 **A remarkable journey through the alimentary canal. Note the rich blood supply to the gut wall and the glistening digestive secretions. The diagrams adjacent to each photograph show the structure of the canal wall at each of the locations**

4 THE STOMACH

Food remains in the stomach (Fig 16.9) for between 30 minutes and 4 hours depending on the type of meal (fatty meals stay there the longest). It is digested mechanically and chemically. The remaining semi-liquid material, **chyme**, passes into the duodenum, the first part of the small intestine.

Fig 16.9 **Structure of the human stomach**

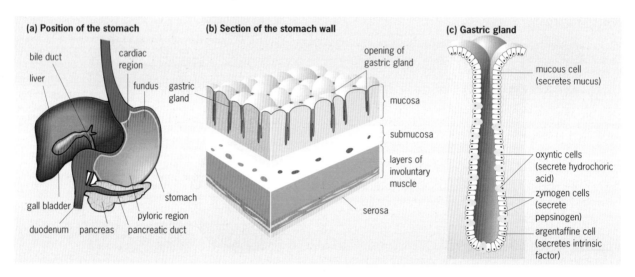

The stomach is a muscular sac under the **diaphragm**. When empty, it is the same size as a large sausage, but it can stretch to the size of a melon. The stomach:

- mixes food with gastric juice by muscular action.
- retains food, giving enzymes time to act.
- digests proteins through the action of the enzyme **pepsin**.
- performs some absorption of salts and alcohol.
- produces **gastric intrinsic factor** (GIF), which binds to and helps to absorb vitamin B12.

Gastric juice is secreted by specialised groups of cells inside **gastric pits** in the mucosa (see Fig 16.9). Each type of cell produces a specific secretion:

- **Oxyntic cells** secrete a solution of hydrochloric acid, which brings the pH of gastric fluid down to between pH 2.0 and 3.0.
- **Zymogen cells** (peptic cells) secrete the enzyme **pepsinogen**, which is later converted into the protein-splitting enzyme **pepsin**.
- **Argentaffine cells** produce **gastric intrinsic factor (GIF)**, which aids absorption of vitamin B12.
- **Mucous cells** secrete the **mucus** that protects the stomach lining from the digestive action of its own secretions.

Pepsin and pepsinogen

One of the main functions of the stomach is to begin to digest proteins. In doing so, it must avoid digesting its own tissue. It does this by secreting an *inactive* form of the protein-splitting enzyme pepsin. The conversion of pepsinogen to pepsin occurs only in the acidic conditions that are created by the hydrochloric acid in the stomach lumen. The hydrogen ions in the acid cause the pepsinogen to unfold to form the active form of the enzyme. The stomach protects itself from the strong acid by secreting large quantities of mucus.

Pepsin is a powerful **endopeptidase enzyme**: it breaks specific peptide bonds in the middle of the protein chain, turning protein molecules into polypeptides (see Chapter 3 for more detail on the structure of proteins).

See question 3.

Milk digestion

Caseinogens, the proteins in milk, are *soluble*. They are valuable nutrients (milk is the sole source of food for young mammals) but, if they remained in their soluble form, they would leave the stomach before protein digestion had finished. To avoid this, the stomach produces **rennin**. This **curdles** milk, converting soluble caseinogen into insoluble **casein**. Like pepsin, rennin is secreted in an inactive form. **Prorennin** is converted to its active form by contact with stomach acid.

B (a) Would you expect rennin production to increase or decrease with age in mammals? Give a reason for your answer.

(b) Would you expect rennin to be found in animals other than mammals?

The acid conditions of the stomach

The hydrochloric acid in gastric fluid:

● provides the optimum pH for pepsin and rennin.

● begins the digestion of some carbohydrates and lipids through the chemical reaction of hydrolysis.

● denatures proteins and helps to soften tough connective tissue in meat.

● is a strong **bacteriocide** (it kills bacteria) and so protects the body from some of the harmful microbes which might enter the body in food.

Control of gastric secretion

Since we do not eat food continuously, it is important that we produce digestive enzymes only when food is in the gut. If large quantities of the acids and protein-attacking enzymes were released into an empty stomach, there would be a danger of **autolysis** (self-digestion). Nerves and hormones control the secretion of gastric juice in a process that has three distinct phases (Fig. 16.10):

● **Nervous phase.** The sight, smell or taste of food initiates a **nerve reflex**: impulses from the brain trigger gastric glands to release their secretions.

● **Gastric phase** (hormonal). Food in the stomach stimulates the lining to secrete the hormone **gastrin**. Gastrin increases gastric juice secretion through direct action on the gastric glands.

● **Intestinal phase** (hormonal). Stimulated by partially digested food, the duodenal lining produces a second hormone, **enteric gastrin**. This hormone also acts on the gastric glands, producing further small amounts of gastric juice.

A summary of the mechanisms that control the release of all digestive secretions is shown later in Fig 16.14.

Pressure sensors and chemical sensors in the stomach detect stomach stretching and the presence of chyme. The resulting nerve impulses, together with the action of gastrin, direct the stomach to start emptying its contents into the duodenum.

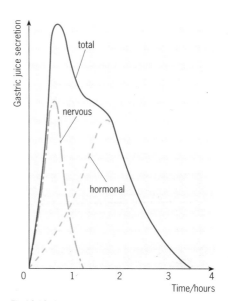

Fig 16.10 **Secretion of gastric juice is partly controlled by the** vagus nerve **and partly by the hormone** gastrin. **Gastrin stimulates gastric glands causing them to release their secretions. It also increases movement in the digestive tract and changes the rate at which food moves into and out of the stomach**

5 THE SMALL INTESTINE

The human small intestine is about 6 metres long and is made up of three main parts, the **duodenum**, the **jejunum** and the **ileum**.

The duodenum is about 25 cm long and receives secretions both from the liver (through the bile duct) and the pancreas (through the pancreatic duct). The bulk of digestion takes place here. Most absorption occurs in the jejunum and ileum, which make up the remainder of the small intestine. The ileum joins the large intestine at the caecum.

The small intestine:

- moves food from the stomach to the large intestine by peristalsis at the rate of about 1 cm per minute (Fig 16.11).

- secretes intestinal juice to provide the water necessary for most of the chemical reactions involved in digestion.

- completes the digestion of carbohydrates, proteins and fats.

- absorbs the vast majority of small soluble food molecules produced by digestion.

- secretes the hormones **cholecystokinin** and **secretin**.

- protects the body against infection: it is closely associated with intestinal lymph nodes.

The walls of the small intestine are structured in much the same way as those of other parts of the alimentary canal (see box on page 241), but with some obvious differences that enable it to perform its function as the main site of digestion and absorption. The lining of the small intestine contains many glandular pits. Within these pits, intestinal glands produce intestinal juice, a cocktail of digestive enzymes. **Brunner's glands** (found only in the duodenum) release an alkaline, mucous secretion. Mucus helps to protect the walls of the small intestine from the action of digestive enzymes. The alkaline secretion neutralises the acidic chyme from the stomach.

?

C The average length of a human adult's small intestine is about 6.35 metres. How long does it take for food to travel from one end to the other?

Fig 16.11 **Food is propelled through the gut mainly by peristalsis. The process of** segmentation **helps to mix the food**

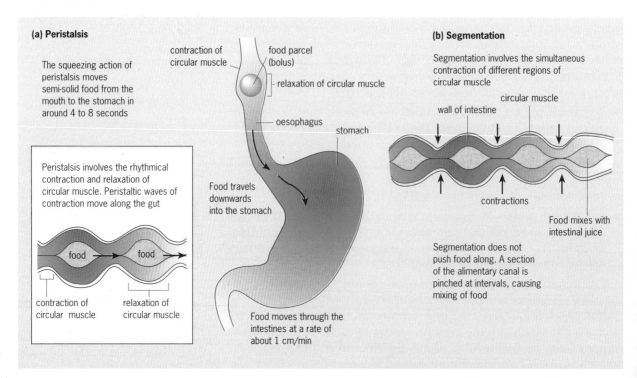

(a) Peristalsis

The squeezing action of peristalsis moves semi-solid food from the mouth to the stomach in around 4 to 8 seconds

Peristalsis involves the rhythmical contraction and relaxation of circular muscle. Peristaltic waves of contraction move along the gut

contraction of circular muscle

relaxation of circular muscle

food   food

contraction of circular muscle   food parcel (bolus)

relaxation of circular muscle

oesophagus

stomach

Food travels downwards into the stomach

Food moves through the intestines at a rate of about 1 cm/min

(b) Segmentation

Segmentation involves the simultaneous contraction of different regions of circular muscle

circular muscle

wall of intestine

contractions

Food mixes with intestinal juice

Segmentation does not push food along. A section of the alimentary canal is pinched at intervals, causing mixing of food

Localised digestion and the Paneth cells

Paneth cells at the base of the crypts of Lieberkühn produce the digestive enzymes of the small intestine. Paneth cells are formed rapidly in the crypts, and they gradually migrate upwards to the tips of the villi. From here, they are shed into the lumen. Digestion by the Paneth cell enzymes occurs inside these cells, and in the epithelial cells lining the villi. These enzymes also digest food in the lumen, in the fluid that immediately surrounds them, but *only* when Paneth cells are dislodged, break up and release the enzyme.

The rate of loss of these and other cells that line the small intestine is phenomenal. The cells lining the intestinal glands and villi multiply faster than any other type of cell in the body: 1 per cent of these cells are replaced every hour of your life and the whole lining is completely renewed every 2 to 4 days.

Digestion in the small intestine

Three main secretions digest food chemically in the small intestine:

- **Intestinal juice** is a clear yellow fluid secreted by glands in the walls lining the duodenum. Fig 16.12 shows the position of these glands, the **crypts of Lieberkühn**. Intestinal juice contains water, a protective mucus, and the enzyme **enterokinase**. This activates the enzyme **trypsinogen**, which is secreted into the small intestine by the pancreas.

- **Pancreatic juice** is a secretion of the pancreas that enters the duodenum via the **pancreatic duct**. The human pancreas secretes about 1.5 litres of alkaline pancreatic juice every day. The fluid contains about 15 enzymes.

- **Bile**, a secretion of the liver, is stored by the **gall bladder** and released into the small intestine via the **bile duct**.

Intestinal juice contains only small quantities digestive enzymes (see Table 16.1): it mainly provides the water in which the chemical reactions of digestion take place. The range of digestive enzymes that the small intestine produces act within cells or close to the wall of the intestine, at the borders of the epithelial cells (see the Extension box above). They are not secreted into the lumen.

See question 4.

Fig. 16.12 **The small intestine is adapted both for digestion and absorption of the digested food products**

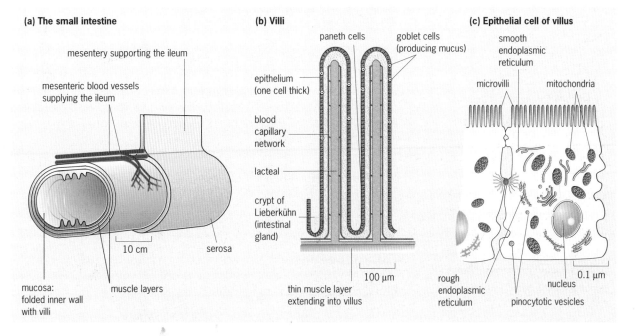

(a) The small intestine

mesentery supporting the ileum

mesenteric blood vessels supplying the ileum

10 cm

serosa

mucosa: folded inner wall with villi

muscle layers

(b) Villi

paneth cells

goblet cells (producing mucus)

epithelium (one cell thick)

blood capillary network

lacteal

crypt of Lieberkühn (intestinal gland)

100 μm

thin muscle layer extending into villus

(c) Epithelial cell of villus

smooth endoplasmic reticulum

microvilli

mitochondria

nucleus

0.1 μm

rough endoplasmic reticulum

pinocytotic vesicles

The liver and pancreas both have other functions that are not connected directly with digestion. The pancreas produces a number of important hormones, such as insulin (see Chapter 25), and the liver has many regulatory, or **homeostatic** functions (see also Chapter 25).

The role of the pancreas and liver in digestion

The pancreas and liver (and to a lesser extent the salivary glands – see page xxx) all produce an **exocrine secretion** (one released through a duct) that acts on food in the alimentary canal.

The pancreas

The human pancreas (Fig 16.13) produces over a litre of a clear colourless **pancreatic juice** every day. This fluid contains water, some salts, sodium hydrogencarbonate and a mixture of about 15 different digestive enzymes. The main enzymes in pancreatic juice are listed in Table 16.1 and are discussed in detail below.

Secretory cells, **acini**, make up 99 per cent of the glandular tissue of the pancreas. They secrete digestive juice, discharging their contents into the further pancreatic duct. Hydrogencarbonate ions make pancreatic juice slightly alkaline (pH = 7.1 to 8.2), allowing it to neutralise stomach acid and so create optimum conditions for the digestive enzymes.

Fig 16.13 **The pancreas is a compound gland. This means it is both an endocrine gland and an exocrine gland: it produces a hormone that is released into the blood to act away from the site of production and it also produces digestive secretions that pass into the gut though a duct.**

This photograph shows a section through a human pancreas: the main features are labelled

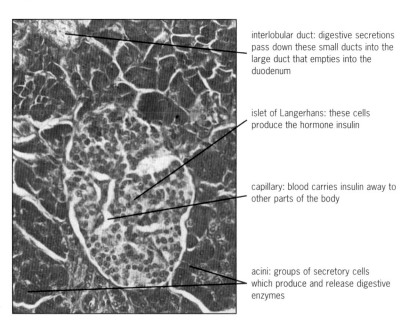

interlobular duct: digestive secretions pass down these small ducts into the large duct that empties into the duodenum

islet of Langerhans: these cells produce the hormone insulin

capillary: blood carries insulin away to other parts of the body

acini: groups of secretory cells which produce and release digestive enzymes

Enzymes produced in the pancreas

Trypsin is a powerful **endopeptidase enzyme** secreted in an inactive form called **trypsinogen**. **Enterokinase,** a substance released by the intestinal wall, converts the inactive trypsinogen into its active form. Trypsin breaks down proteins into polypeptides.

$$\text{trypsinogen} \xrightarrow{\text{enterokinase}} \text{trypsin}$$

Chymotrypsin continues the digestion of proteins into polypeptides. This enzyme is secreted in an inactive form and converted to its active form by trypsin.

$$\text{chymotrypsinogen} \xrightarrow{\text{trypsin}} \text{chymotrypsin}$$

Carboxypeptidase is also activated by trypsin in the lumen of the small intestine. An exopeptidase enzyme, it converts polypeptides into smaller peptides and amino acids.

$$\text{procarboxypeptidase} \xrightarrow{\text{trypsin}} \text{carboxypeptidase}$$

D Why are the protein-splitting enzymes trypsin and chymotrypsin produced in an inactive form?

E What is the difference between an endopeptidase and an exopeptidase enzyme?

Pancreatic amylase completes the breakdown of starch to maltose, started by salivary amylase in the mouth. A second amylase enzyme is needed because salivary amylase is inactivated by stomach acid.

Pancreatic lipase continues the breakdown of fats into fatty acids and glycerol.

Nuclease digests nucleic acids into their constituent nucleotides.

Control of pancreatic secretions

Pancreatic secretion, like gastric secretion, is controlled by both nervous and hormonal mechanisms. A summary of the mechanisms that control the release of all digestive secretions is shown below in Fig 16.14. Impulses travel from the brain along the vagus nerve to trigger the secretion of pancreatic juice.

The lining of the duodenum secretes two hormones in response to the presence of acidic chyme in the lumen of the duodenum. **Secretin** stimulates the pancreas and liver to secrete pancreatic juice rich in hydrogencarbonate ions. **Cholecystokinin** stimulates the gall bladder to release bile and the pancreas to release digestive enzymes.

Fig 16.14 **The digestive tract is supplied by nerves from the autonomic nervous system (see Chapter 21). Particularly important is the vagus nerve, which stimulates the secretion of digestive juices.**

Information coming into the body from food (its appearance and smell) travel to the hypothalamus and cerebral cortex in the brain. These send signals back to the body to stimulate the salivary glands, stomach, liver and pancreas to start releasing their secretions. As food enters the stomach, the hormone gastrin is released and this encourages more gastric juice secretion. Similarly, the presence of food in the duodenum causes the release of gastrin and two other hormones, secretin and cholecystokinin.

We don't carry on eating until we burst because in humans, as in other animals, food intake is controlled by centres for hunger and satiety (fullness) in the hypothalamus. A full stomach and high blood glucose level stimulate the satiety centre so that we do not feel hungry any more

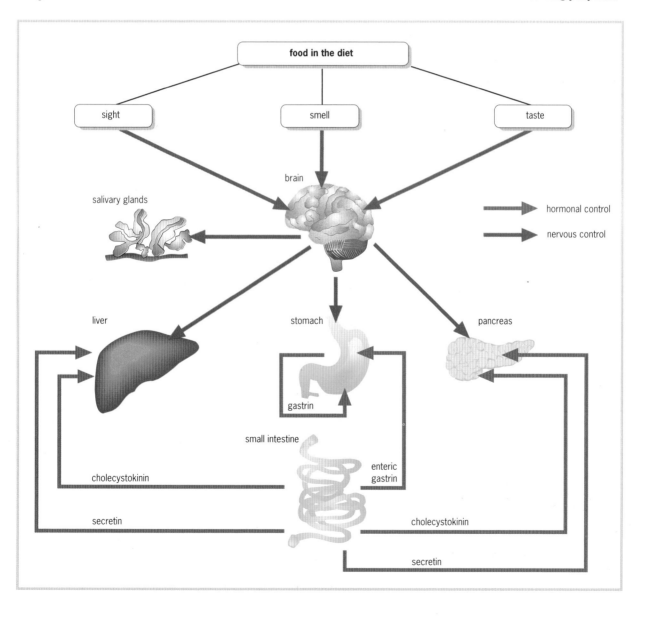

The liver

Liver cells, **hepatocytes**, produce thick yellow-brown or olive-green bile containing:

- water,
- **bile salts** (sodium glycocholate and sodium taurocholate),
- **bile pigments** (the breakdown products of red blood cells, eg **bilirubin**),
- some mucus,
- cholesterol.

Liver cells produce around 0.8 to 1.0 litre of bile daily. Secretions from individual cells pass into tiny canals called **bile canaliculi**. These lead to the gall bladder, a small sac-like organ, which stores the bile until it is needed. Bile is released into the duodenum when cholecystokinin stimulates the muscle wall of the gall bladder to contract. Bile reaches the duodenum through the **common bile duct**. Bile:

- **emulsifies** fats (breaks large fat or oil droplets into an emulsion of microscopic droplets), a process that massively increases the surface area available for fat-digesting enzymes to attack.
- neutralises the acidic chyme from the stomach and creates the ideal pH for intestinal enzymes.
- stimulates peristalsis in the duodenum and ileum.
- allows the excretion of cholesterol, fats and bile pigments.

Control of bile secretion

Bile secretion is *controlled* in a number of ways. The hormone **secretin** acts together with nervous stimulation by the vagus nerve to increase the rate of bile secretion. The acidity of chyme in the duodenum and the hormone cholecystokinin stimulate the gall bladder to contract (see Fig 16.14).

Food absorption in the small intestine

Digestion breaks up large insoluble food molecules into small soluble molecules that can be easily absorbed through the gut lining. About 90 per cent of all absorption takes place within the small intestine. (A small amount occurs in the stomach and the rest in the large intestine.) Undigested material travels through the large intestine and is expelled from the body.

The internal lining of the small intestine is well adapted to its function. The epithelial lining has a much greater surface area than that of an equivalent smooth-sided tube because of the three structural features shown in Fig 16.15. The cells absorb amino acids, monosaccharides, fatty acids, glycerol, water and other substances including vitamins, nucleic acids, ions and trace elements (Fig 16.16). The type of transport process depends on the substance, but diffusion, facilitated diffusion and active transport are all involved (see Chapter 5).

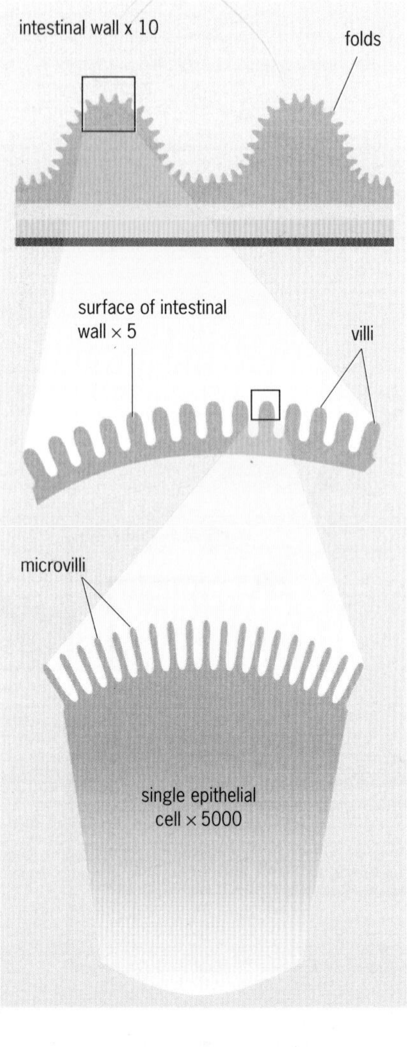

intestinal wall × 10

folds

surface of intestinal wall × 5

villi

microvilli

single epithelial cell × 5000

Fig 16.15 **An average 70 kg man has approximately 100 m² of absorbing surface in his small intestine. This increased surface area of the intestinal lining is due to:**
1 **folds in the inner surface of the intestinal wall.**
2 **movable projections (0.5–1.0 mm in length) called villi, which are present on the folded surface of the wall.**
3 **microscopic projections called microvilli in the cell membranes of epithelial cells that line the villi**

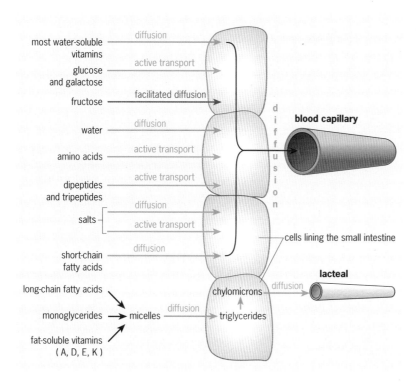

Fig 16.16 **Absorption of digested nutrients through the lining of the intestine**

Absorption of carbohydrates

Glucose and galactose are actively transported into the cells of the small intestine and across the walls of blood capillaries. Active transport is necessary because, for these small molecules, diffusion alone would be too slow to supply the body's needs. Also, diffusion would occur *out of* the epithelial cells if their concentration in the intestines was very low. Energy for the transport is obtained from the hydrolysis of ATP (see Chapter 8). Fructose, a larger molecule, enters intestinal cells more slowly by facilitated diffusion.

Absorption of proteins

Amino acids and small peptide molecules (di- and tripeptides) are actively transported into epithelial cells. The small peptides then undergo **intracellular digestion** (digestion inside the cells) so that, eventually, every protein is broken down into amino acids.

Absorption of fats

Bile salts and lipase enzymes break up complex lipid molecules into monoglycerides and free fatty acids. The monoglycerides and some of the fatty acids combine with bile salts to form microscopic particles called **micelles**. Micelles diffuse easily into epithelial cells, along with the remaining fatty acids. Inside the cells, the fatty acids and glycerol recombine to form triglycerides, which acquire a protein coat to stop them sticking together. These particles are called **chylomicrons**. It is in this state that lipids leave epithelial cells and enter, not a blood capillary this time, but a branch of the lymphatic system called the **lacteal** (see Fig 16.16).

Transport of absorbed food products

Amino acids, dipeptides and simple sugars diffuse out of epithelial cells in the small intestine, and directly into blood capillaries within

BACTERIA IN THE GUT

AROUND A HUNDRED million million (10^{14}) bacteria live in a healthy human body, mostly in the digestive tract. The bulk of these, at least 400 species, inhabit the large intestine. These microorganisms are opportunistic 'partners' that share our food and in return may supply useful products such as vitamins, or help the digestive process, as in ruminants such as cattle. They can also contribute to an animal's resistance to disease by competing with potentially harmful bacteria for sites in the gut.

Successful digestive tract bacteria tolerate a wide range of pH conditions, resist the effects of antibodies secreted by intestinal tissue and survive the constant motion of the intestinal contents.

Although gut bacteria are **non-pathogenic** (they do not cause disease in a healthy body), infection can arise, for example, after damage to the large intestines. Bacterial contamination of the **peritoneum** (the lining of the abdominal cavity) can lead to **peritonitis**, a potentially fatal infection.

?

F Why are the nutrients produced by colonic bacteria in the large intestine of less use to mammals than nutrients provided by bacteria in the small intestine?

the villi. From here they pass into the **mesenteric veins**, which take blood away from the intestines. Nutrient-rich blood enters the liver via the **hepatic portal vein**. The liver processes absorbed food, converting some into storage products. Glycogen, copper, iron, and vitamins A, D, E and K can all be stored. The liver breaks down other food products, including excess amino acids which are **deaminated** – their **amine groups** are removed and passed to the kidneys for excretion. See Chapter 25 for more about liver function and Chapter 27 for details on excretion.

The lymphatic system plays a major role in the transport of absorbed lipids. Chylomicrons that have entered the lacteal remain in suspension, giving this **lymph fluid** a milky white appearance. From here, the chylomicrons move into larger lymphatic vessels, which eventually drain into a large duct that empties into the blood. In the blood, chylomicrons are broken down into fatty acids and glycerol and enter cells to be used in the synthesis of complex lipids.

6 THE LARGE INTESTINE: DEALING WITH UNDIGESTED FOOD

The human large intestine is about 1.5 metres long and extends from the ileum to the anus. It has four sections, the **caecum**, the **colon**, the **rectum** and the **anal canal**. The large intestine:
- absorbs water.
- is the site of manufacture of certain vitamins (microbes in the colon of some animals produce vitamin K and folic acid).
- forms and expels undigested food residue in the process of **egestion**.

The caecum receives material from the ileum. In humans, the bottom of the caecum is attached to a small tube, the **appendix**. This twisted and coiled tube is about 8 cm long and closed at the end. The human appendix plays no part in digestion but can become inflamed, causing **appendicitis**.

Much of the 6 litres of fluid that is poured onto food during its passage through the digestive tract each day (as saliva, gastric juice, intestinal juice) is reabsorbed. This is a vital process: if we lost all this fluid we could dehydrate. Most of the water is reabsorbed in the small intestine. Of the litre or so that enters the large intestine, all but 100 cm^3 is reabsorbed by the colon. Minerals also diffuse or are actively transported into the bloodstream from the colon.

The final sections of the large intestine, the rectum and anus, compact and store **faeces** prior to **defecation**.

✔

Faeces contain some water, mineral salts, undigested food, bile pigments, products of bacterial decomposition and epithelial cells that have become detached from the digestive tract. Defecation is the expulsion of faeces that occurs in mammals; it is a form of egestion, a more general term which describes the expulsion of undigested food that occurs in any animal.

SUMMARY

By the end of this chapter, you should know and understand the following:

▓ The human **alimentary canal** is a tube that leads from the **mouth** to the **anus**. It is associated with glands such as the **salivary glands**, **pancreas** and **liver**.

▓ The alimentary canal or digestive system is responsible for the **ingestion**, **digestion**, **absorption** and **egestion** of food.

▓ Food is broken down *mechanically* by the chewing action of teeth and by the muscular churning of the alimentary canal.

▓ Digestive juices are poured onto food as it travels through the alimentary canal. These soften and lubricate the food and also contain digestive enzymes that break the food down *chemically*.

▓ The muscles of the alimentary canal 'squeeze' the food – a process called **peristalsis** – and push it along from one end to the other.

▓ Large food molecules are broken down to their building blocks: **carbohydrates** are digested to form simple **sugars**, **proteins** are broken into **amino acids**, and **fats** are converted to **fatty acids** and **glycerol**.

▓ Digested food products are absorbed into the body across the wall of the alimentary canal. Most enter the bloodstream, but the products of fat breakdown enter the **lymphatic system**.

▓ A sophisticated control system ensures that powerful digestive enzymes are released in the right place at the right time.

QUESTIONS

1 Copy and complete the following table.

Enzyme or enzyme precursor	Activator	Details of substrate	Main product
amylase	none	starches	
trypsinogen			peptides
chymotrypsinogen	trypsin	protein	
lactase	none		
ribonucleotidase (ribonuclease)	none		

[AEB 1987 Biology: Paper 1, q. 8]

2 Copy and complete the table which refers to enzyme in the mammalian gut.

Name of enzyme	Site of production	Substrate	Product(s)
	duodenal mucosa		trypsin
	pancreas	fat	fatty acids, glycerol
lactase		lactose	

[AEB June 1991 Biology: Paper 1, q. 7]

3 Which of the following statements is/are correct? Proteins are:

a) digested in the stomach and in the small intestine to amino acids.

b) digested in the small intestine by the action of pepsin.

c) not hydrolysed in the stomach.

d) enzymically degraded during germination of bean seeds.

[NEAB June 1993 Biology: Paper 1, Section A, q. 3]

4 The diagram below shows a longitudinal section of part of the ileum wall.

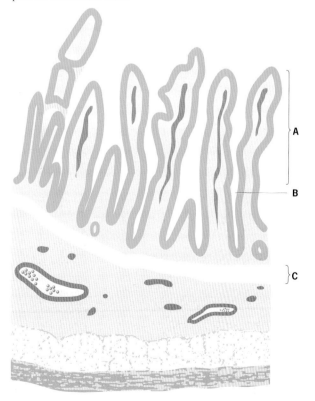

a) Name the structures labelled **A**, **B** and **C**.

b) Describe **one** way in which the structure of the ileum is adapted to the function it performs.

[ULEAC 1996 Human Biology: Specimen Paper, Module Test HB2, q. 5]

17 Gas exchange

This sea lion is well adapted to life under water. Before a dive, the nostrils close, preventing entry of water

When they return to the surface after a dive, whales clear their blow-hole before inhaling. This produces the familiar plume of spray

AS A SPECIES we are not very good at diving without special equipment: most people cannot manage to hold their breath for more than a minute or so. But diving mammals such as dolphins, whales and seals can stay under water for much longer periods. Sperm whales, for instance, can dive for over an hour. How do they do it?

One reason is that the muscles of diving mammals contain more myoglobin (the oxygen-storage pigment which makes muscle red) and so can store more oxygen than those of non-diving mammals. Another reason is that diving mammals breathe out just before a dive. This makes them less buoyant and means there is less chance of gases in the lungs being forced into the blood at high pressure. Dissolved gases in the blood could form bubbles as the animal surfaced, causing the 'bends': a well known problem for human divers.

During prolonged dives, diving mammals also move oxygenated blood away from organs that can survive without it for a short time (such as the skin) to those organs that need a constant supply (such as the brain and heart). The decreased flow to the skin is so effective that cuts sustained during a dive do not bleed until the animal surfaces.

1 WHY EXCHANGE GAS?

All living things need energy to carry out the processes of life. Most organisms obtain their energy from the oxidation of food in a process known as cell respiration (see Chapters 7 and 8). Cell respiration uses oxygen and produces carbon dioxide. To keep the process going, a living organism needs to obtain oxygen and expel waste carbon dioxide. This process is known as **gas exchange**.

Usually, small, simple organisms, such as those made up of only one cell, have a relatively small oxygen demand. All the oxygen they need diffuses into their body through its surface. (See Chapter 5 for an explanation of diffusion.) They expel carbon dioxide in the same way. But this strategy would not work in larger animals.

A summary of the overall process of cell respiration:

glucose + oxygen → carbon dioxide + water + energy held in the ATP molecule

Photosynthetic organisms carry out the process of photosynthesis and respiration at the same time. If the rate of photosynthesis is greater than the rate of respiration, there is a net uptake of CO_2 and a net output of O_2.

The surface area and volume problem

Although the amount of material an organism *needs* to exchange is proportional to its *volume*, the amount of material it *can* exchange is proportional to its *surface area*. Fig 17.1 shows that as organisms get larger, their volume increases at a faster rate than their surface area and so their need to exchange materials could quickly outstrip their ability to do so. (The Assignment in Chapter 5 should help you to understand this vital concept.)

Organisms with a diameter larger than a few millimetres can survive only if they evolve strategies to increase their surface area: volume ratio. Many species of worm have evolved a flat or cylindrical body. However, this sort of adaptation allows only a limited size increase. Much larger organisms have developed specialised organs to increase the surface area that is available for exchanging materials. Lungs, gills, intestines and kidneys are all examples of such organs.

The evolution of specialist exchange organs is linked to the development of a **transport system** (see Chapter 18). Without it, exchanged materials cannot reach other parts of the body quickly enough.

The first part of this chapter looks at gas exchange in invertebrates, insects, fish and mammals. Part II covers gas exchange in plants.

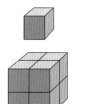

(a) A 1 cm³ cuboid organism. Surface area: volume ratio = 6:1

(b) A cuboid organism with a volume of 8 cm³. Surface area: volume ratio = 3:1

(c) A cuboid organism with a volume of 27 cm³. Surface area: volume ratio = 2:1

Note that not all the cells of this organism are on the surface

Fig 17.1 **These simple shapes illustrate the problems faced by an organism as it gets larger. Think of each block as a cell and each complete cube as an organism. You can see that increasing the size of the organism lowers the surface area available to each cell. If the third 'organism' could not evolve a system to transport materials to the cell in the middle, it would not have much chance of survival**

2 RESPIRATORY ORGANS

Oxygen passes into the body of an organism by diffusion, and carbon dioxide leaves in the same way. Respiratory organs increase the rate at which gases can be exchanged. They maximise the efficiency of diffusion by having the following features in common:

- **A large surface area.** The greater the area of tissue in contact with the environment, the greater the rate of diffusion. Many aquatic animals increase the surface area for gas exchange by having projections, or **gills**. Most land animals increase surface area by having little pockets called **alveoli** in the lungs.

- **Thin, permeable walls.** These ensure that the diffusion distance is as short as possible.

- They must be **moist**. Only gases in solution can pass into cells.

- **A good blood supply.** Organisms which have a circulatory system supply their respiratory organs with plenty of blood. Blood quickly takes incoming oxygen away to a different part of the body. This ensures that a diffusion gradient is always present: there is always less oxygen in the blood which flows through the gas exchange organ than there is in the air or water.

- **Good ventilation.** Many animals increase the efficiency of gas exchange by ventilation. This is a physical pumping mechanism which continually brings fresh water or air to the gas exchange surfaces to keep the diffusion gradient as large as possible.

The problems of gas exchange in air and water

Animals obtain oxygen from the air or water that surrounds them. Air and water are completely different gas exchange mediums and so pose a different set of problems (Table 17.1).

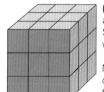

The oxygen content of atmospheric air is remarkably constant, but the amount of oxygen dissolved in water varies considerably. Cold, turbulent water is highly oxygenated, but warm, still water contains much less oxygen. Polluted water often has very little oxygen (see Chapter 39).

	Water	Air	Water:air ratio	Practical consequence
Oxygen content (%)	0.7	20	1:30	far more oxygen is available from air than from water
Density (kg per litre)	1.0	0.0013	800:1	more energy is required to move water than to move air
Viscosity	1.0	0.02	50:1	air flows much faster than water
Diffusion constant for oxygen	0.000034	11	1:300 000	oxygen can be absorbed from air far more quickly than it can from water
Diffusion constant for carbon dioxide	0.00085	9.4	1:11000	carbon dioxide is taken away more quickly by air than by water

Table 17.1 **Some of the different physical and chemical properties of water and air. The diferences between the two mediums affect the way in which organisms exchange gases. For example, water flows over gills as the animal moves, instead of being drawn in and out. It would take far too much energy for water dwellers to suck water in, stop it and then reverse the flow, as air-breathers do**

Fig 17.2 **There is usually about thirty times more oxygen in air than in water. Warm-blooded animals like these harvest mice can meet their large oxygen demand only by breathing air. The oxygen consumption of this species has been measured at 2.5 cm² of O₂ per gram of body weight per hour**

Fig 17.3 **The air we exhale is saturated with water vapour. When you wake up on a cold day to see condensation on the windows, it is likely that much of that water came from your lungs.**

Loss of water in exhaled air represents a significant problem for land-using organisms – they must replace it continually to avoid dehydration

Fig 17.5 **The coconut crab has become adapted for life on land: it can breathe air with its gills. It gets its name from its habit of breaking open coconuts with its fearsome claws**

Organisms that breathe air have an easier time than those that live in water. It is no coincidence that mammals and birds, the animals with the highest metabolic rate, are air breathers (Fig 17.2). They could not obtain the amount of oxygen they use simply by moving water over a system of gills. However, a major drawback associated with breathing air is water loss. The combination of a large surface area and moist membranes means that exhaled air is saturated with water vapour (Fig 17.3).

Lungs and gills compared

Gills usually project into the organism's external environment, see Fig 17.4(a), but, as Fig 17.4(c) shows, some organisms do have internal gills. Generally, gills are used for gas exchange only in water because the **gill filaments**, which need to be thin-walled, tend to collapse out of water. Even so, the coconut crab *Birgus latro*, in Fig 17.5, has rigid gills that do not collapse in air, and this crustacean drowns if kept underwater.

Lungs are cavities that allow the environment to enter the organism – see Fig 17.4(b) and (d). In general, most animals with lungs live on land and breathe air, but, as ever, there are exceptions. Some molluscs breathe under water using lungs, but these tend to be simple pockets that supply small amounts of oxygen to meet the animal's low demand.

Air-breathing animals drown in water because, although water contains some oxygen, it does not contain enough. The effort required to move the amount of water in and out of lungs to gain sufficient oxygen would be impossibly large: they would use more oxygen than they would gain. Also, the greater density of water would burst the delicate air-sacs.

So, why do animals with gills die when removed from water? After all, there is a lot more oxygen in air. Unfortunately, the gill filaments tend to collapse out of water and the moist filaments stick together, rather like the bristles on a paint brush, drastically reducing their surface area.

Fig 17.4 **Gas exchange organs**

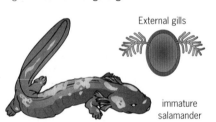

External gills

immature salamander

(a) Gills are projections over which water flows; some are completely external; some are protected by covers

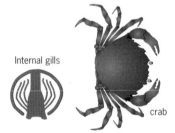

Internal gills

crab

(c) Internal gills are inside the body of the organism

Lungs

tiger

(b) Lungs are internal cavities which bring air deep inside the body of the organism

Tracheae

wasp

(d) Insects have a series of tubes called tracheae which take air from the outside to the internal organs

3 GAS EXCHANGE STRATEGIES IN ANIMALS

Later in this section, we concentrate on the highly developed gas exchange organs of mammals, insects and fish. But let's look first at how some simpler organisms overcome the problem of gas exchange. Fig 17.6 shows the gas exchange strategies of several different invertebrates.

A What would happen to the external gills of a sea slug in air?

Fig 17.6(a) **This sea anemone is a cnidarian, a simple animal. It has no special respiratory organs but all of its living cells are within a millimetre or so of its surface and so each cell exchanges gases with its surroundings by diffusion**

Fig 17.6(b) **This flatworm is a platyhelminth: it belongs to a phylum of animals that also includes the tapeworms. Its flat, ribbon-like shape provides a large surface area:volume ratio, so it is able to exchange sufficient gases with its environment by diffusion**

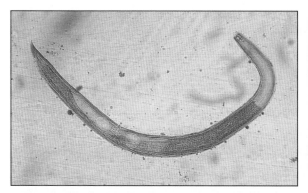
Fig 17.6(c) **The long, cylindrical shape of nematode worms gives them a large surface area, further increased by the gut that passes through the middle. All active cells are very close to the external environment**

Fig 17.6(d) **Larger and more complex organisms such as ragworms are also elongated, but this does not provide a large enough surface area to meet their gas exchange needs. Extensions known as** parapodia **further increase their surface area. These gill-like structures absorb oxygen, which the circulatory system then distributes around the body**

Fig 17.6(e) **This sea slug is a mollusc. Its external gills increase its surface area**

Fig 17.6(f) **Tadpoles, the larvae of amphibians, have gills. Adult frogs and toads exchange gases using a combination of simple lungs and skin diffusion**

Most fish are bony fish, the **osteichthyes**, which have swim bladders and spiny fins. The other main group of fish are the cartilaginous fish, the **chondrichthyes,** which include sharks and rays (see Chapter 12).

Gas exchange in fish

The gills of bony fish such as trout and tuna are particularly efficient structures for gas exchange in water. Fig 17.7 shows the basic structure of the gills of a typical bony fish. Each gill consists of several large **gill arches**, and each arch has two rows of gill filaments. The ends of the filaments interlock, forming a continuous mesh through which water must pass. This arrangement ensures that water passes within a fraction of a millimetre of every gill filament, allowing the maximum amount of oxygen to pass into the fish.

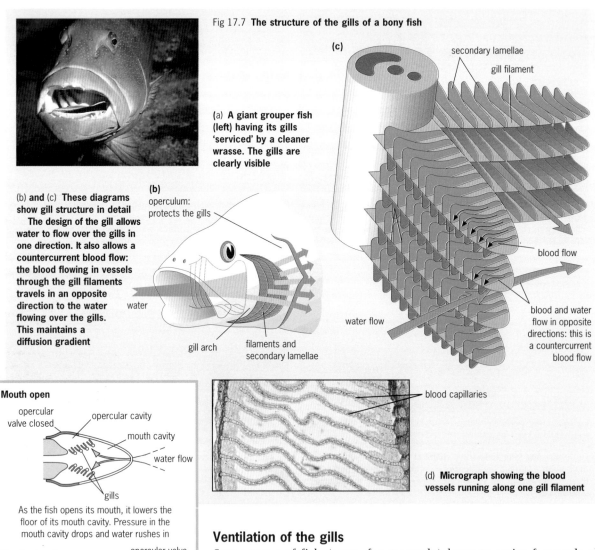

Fig 17.7 **The structure of the gills of a bony fish**

(a) **A giant grouper fish (left) having its gills 'serviced' by a cleaner wrasse. The gills are clearly visible**

(b) **and** (c) **These diagrams show gill structure in detail**
   The design of the gill allows water to flow over the gills in one direction. It also allows a countercurrent blood flow: the blood flowing in vessels through the gill filaments travels in an opposite direction to the water flowing over the gills. This maintains a diffusion gradient

(b) operculum: protects the gills

water

gill arch

filaments and secondary lamellae

(c) secondary lamellae

gill filament

blood flow

water flow

blood and water flow in opposite directions: this is a countercurrent blood flow

blood capillaries

(d) **Micrograph showing the blood vessels running along one gill filament**

Mouth open

opercular valve closed

opercular cavity

mouth cavity

water flow

gills

As the fish opens its mouth, it lowers the floor of its mouth cavity. Pressure in the mouth cavity drops and water rushes in

Mouth closed

opercular valve open

As the fish closes its mouth, it raises the floor of its mouth cavity. This forces water over the gills

Fig 17.8 **In the double pumping system, water flows into the fish's open mouth. When the mouth closes, a series of valves ensures that water is forced out over the gills. Once water has passed the gills, the opercular pump expels the 'old' water and draws 'new' water in**

Ventilation of the gills

Some types of fish (tuna, for example) have to swim forwards all the time to make sure that water flows continuously over their gills. Stationary bony fish achieve a continuous flow of water over their gills using a double pumping system involving the **buccal** (mouth) cavity and the **opercular** (gill) cavity (Fig 17.8).

The countercurrent system of the fish gill

A countercurrent system occurs when two substances flow through the same body part in opposite directions. Fig 17.9 shows the countercurrent system in the gill filaments of a fish. To understand the advantage of blood and water flowing in opposite directions, consider what would happen if they both flowed in the same direction – see Fig 17.9(a). The two fluids would quickly reach equilibrium

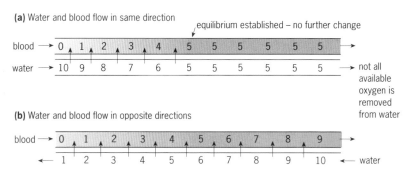

(a) Water and blood flow in same direction

equilibrium established – no further change

blood → 0 1 2 3 4 5 5 5 5 5 5 →

water → 10 9 8 7 6 5 5 5 5 5 5 → not all available oxygen is removed from water

(b) Water and blood flow in opposite directions

blood → 0 1 2 3 4 5 6 7 8 9 →

← 1 2 3 4 5 6 7 8 9 10 ← water

Fig 17.9 **A countercurrent system is effective because it *maintains a diffusion gradient* along the whole length of the gill filament**

See question 1.

and the blood would extract much less of the oxygen available in the water. A countercurrent system ensures that there is always a diffusion gradient – see Fig 17.9(b). Blood gathers oxygen as it flows along the lamellae, but it keeps on encountering fresh water that contains even more oxygen than the blood itself.

Gas exchange in insects

Insects tackle the problem of gas exchange in a unique way. An insect has a **tracheal system** consisting of tubes leading from the inside to the outside of the body. These branch and supply air directly to the internal organs.

Fig 17.10 shows the tracheal system of an insect. The individual tubes, the **tracheae**, lead to openings called **spiracles**, which can open or close to control the level of ventilation. Some of the smallest branches, the **tracheoles**, are less than 1 μm in diameter and supply oxygen to individual cells. The tiniest tracheoles, the **air capillaries**, pass within a few micrometres of every actively respiring cell.

Small insects, and large but inactive insects, can meet their oxygen needs by diffusion alone. Insects that need more oxygen have developed physical mechanisms of ventilation that pump air in and out of their bodies. Next time you see an angry wasp, look more closely and watch for the pumping movements of the abdomen. These muscular contractions expel air from the tracheae and the air sacs, and passive inspiration automatically follows.

Fig 17.10 **The tracheal system in insects**

(a) **A generalised tracheal system of an insect. The spiracles open into the tracheoles, which form a network of tubes extending right through the body of the insect**

(b) **The spiracles are clearly visible on this caterpillar. Generally, spiracles open in response to a lack of oxygen or an increase in carbon dioxide. Experiments in which a tiny jet of carbon dioxide is aimed at one spiracle have shown that only that spiracle will open: each is controlled independently of all the others**

(c) **This micrograph of insect tracheae shows how they branch out into tiny individual tracheoles. The supportive rings of chitin are clearly visible. These spiral rings can squeeze together as in a concertina, to expel air**

(d) **Individual tracheoles reach right in between cells, delivering oxygen to the mitochondria and removing carbon dioxide. Many species have air sacs which can be inflated and deflated, like bellows, drawing air into the insect's body**

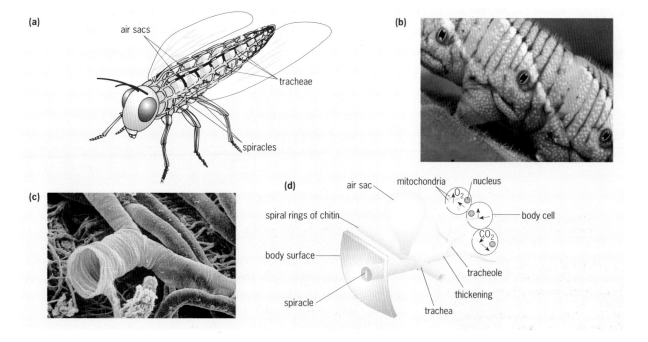

4 GAS EXCHANGE AND THE RESPIRATORY SYSTEM IN HUMANS

Fig 17.11 **The structure of the human respiratory system**

(c), (d) **and** (e) **The fine structure of the alveolus and the mechanism by which gas is exchanged**

(c)

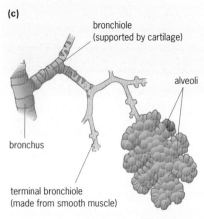

(d)

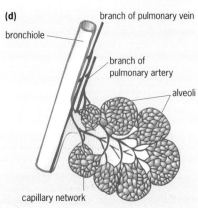

(e)

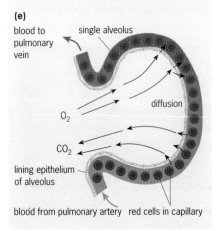

(b) **The overall structure**

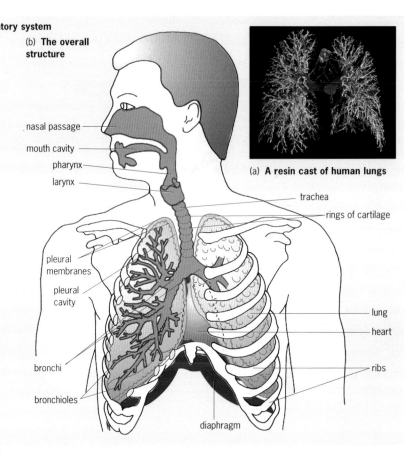

(a) **A resin cast of human lungs**

Alveoli don't collapse when we breathe out because their surface is covered by an anti-sticking chemical called **surfactant** (see Feature box on page 261).

Mammalian lungs have developed to allow efficient gas exchange in air. In this section we are going to look in detail at the human respiratory system. Fig 17.11 shows its overall structure. The intricate structure of the lungs greatly increases the surface area available for gas exchange.

The mechanism of breathing

Lungs contain no muscle, so how do we breathe in and out? Lungs **inflate** and **deflate** because they are *expanded* and *compressed* by movements of the ribcage and diaphragm. This is possible because two **pleural membranes** attach the lungs to the ribcage and diaphragm. The outer membrane lines the **thoracic cavity**; the inner membrane encloses the lungs. Between the two is a narrow space, the **pleural cavity**, filled with **pleural fluid**. This allows the pleural membranes to slide smoothly over each other during breathing and at the same time prevents them from separating. Fig 17.12 shows how we breathe in and out.

To understand why two pleural membranes continue to stick together during breathing, imagine two wet pieces of glass pressed together. They can easily slide over each other, but it is virtually impossible to pull them apart without introducing air into the middle. If air is introduced into the pleural cavity, after a stab wound for example, the lung collapses, a situation known as a **pneumothorax**.

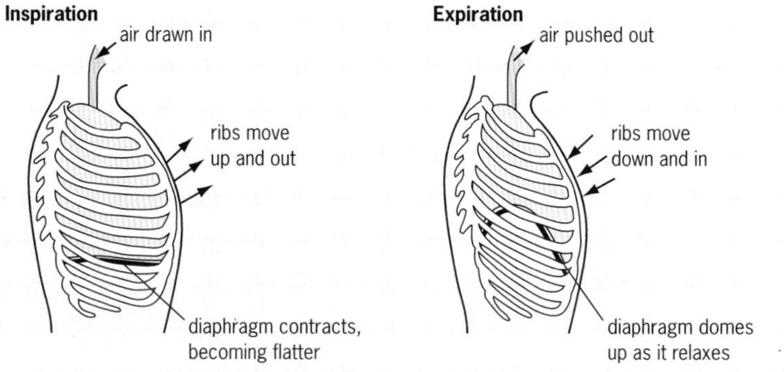

Inspiration

air drawn in

ribs move up and out

diaphragm contracts, becoming flatter

Expiration

air pushed out

ribs move down and in

diaphragm domes up as it relaxes

Fig 17 12 **The mechanism of breathing:**

Fig 17 12 **The mechanism of breathing:**

(a) **Inspiration (breathing in) happens when the external intercostal muscles contract and pull the ribcage upwards and outwards, away from the spinal column. At the same time, the diaphragm contracts and flattens, pushing down on the abdominal organs. These movements increase the volume and therefore lower the pressure in the thorax. As the pressure in the thorax falls below that of the atmosphere, air is forced into the lungs to equalise the pressure**

(b) **Expiration (breathing out) is normally a passive process: it uses no energy. When inspiration is over, the muscles in the thorax relax and breathing out follows due to a combination of gravity, the elastic recoil of the connective tissues of the lung and the pressure exerted by the abdominal organs such as the liver. Of course, we can consciously speed up expiration by forcing air out of our lungs using our** internal intercostal muscles. **This happens when we do something like playing a wind instrument**

Using a spirometer to determine lung volume

Fig 17.13 shows a **spirometer**. The trace that this apparatus produces tells us a lot about the lungs.

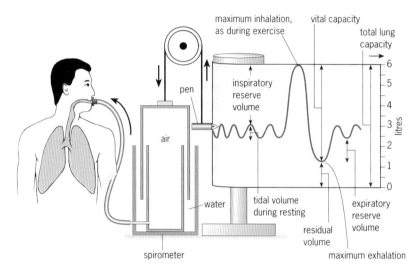

Fig 17.13 **A spirometer and the trace it produces**

First of all, it shows that the lungs have a **total lung capacity**, a maximum amount of air that the lungs can hold during the deepest possible breath. This is not the total volume of the lungs, however, because the lungs are never totally emptied. Even when you have exhaled as much as possible, there is still some air in the alveoli and in the bronchi and tracheae (these are held open permanently by rings of cartilage). The volume of air that remains in the lungs after breathing out is called the **residual volume**. The maximum usable volume (the total lung capacity minus the residual volume) is called the **vital capacity.** The average vital capacity for men is 4.5 to 5 litres, for women, 3.5 to 4 litres.

Secondly, we can see that, during normal breathing, the volume of air that moves in and out of the lungs in each breath is called the **tidal volume**. In a normal adult at rest, this is about 0.5 litres.

Thirdly, the trace shows that, after breathing in at rest, the subject could inhale an extra 1.5 litres, the **inspiratory reserve volume (IRV)**. He or she could also breathe out another litre: the **expiratory reserve volume (ERV).** These values represent the extra volumes of air that we can breathe in and out during exercise.

Finally, we can work out the **ventilation rate**: the volume of air taken into the lungs in one minute. We can do this by multiplying the number of breaths taken in a minute by the tidal volume.

$$\text{Vital capacity} = \text{total lung capacity} - \text{residual volume}$$

$$\text{Ventilation rate} = \text{number of breaths taken per minute} \times \text{tidal volume}$$

B If the average breathing rate is 15 breaths per minute, and the tidal volume is 0.5 litre, calculate the ventilation rate.

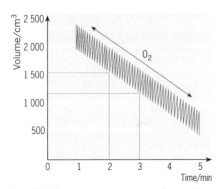

Fig 17.14 **The spirometer measures lung volume. The trace is produced on a revolving drum called a kymograph. We can use the trace to see the effects of exercise on breathing rate**

See question 2.

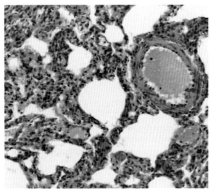

Fig 17.15(a) **In this micrograph of mammalian lung tissue, the alveoli and two of the larger blood vessels can be seen clearly**

Fig 17.15(b) **Gas exchange at the alveolar surface. In the human lung, the** alveolar endothelium, **the membrane which lines the air sacs, is only 0.2 μm thick. This means that oxygen has to diffuse only a very short distance to enter the blood. Breathing maintains a fresh supply of air and the circulatory system takes away oxygenated blood so that constant gas exchange is possible**

?

C The alveolar membrane is 0.2 μm thick. What fraction of a millimetre is that?

Table 17.2 **The composition of inhaled, alveolar and exhaled air**

Using a spirometer to determine oxygen consumption

We can use a spirometer to estimate a person's rate of oxygen consumption. If the person rebreathes the air in the closed system of the spirometer, the composition of that air will change: oxygen levels decrease and carbon dioxide levels increase. If we include a cylinder of soda lime in the apparatus, this absorbs the carbon dioxide and so the volume of air in the spirometer decreases in volume as the oxygen is used up. Fig 17.14 a shows a trace produced in this way.

To calculate the amount of oxygen used, we simply measure the volume decrease in a given time. In our example, the trace shows that in 1 minute the air volume fell from the 1600 cm³ mark to the 1300 cm³ mark. So, this person used up 300 cm³ of oxygen in 1 minute.

WARNING! Re-breathing your own air can be dangerous. You should never do investigations like this without close supervision.

Gas exchange at the alveoli

Gas exchange between air and blood occurs at the alveoli (Fig 17.15(a)). These tiny air sacs create a huge surface area: 1 cm³ of frog lung tissue has a surface area of about 20 cm², but the corresponding figure for a mouse lung is over 800 cm². Other mammals have a similar value. The total surface area of one human lung is about 100 m².

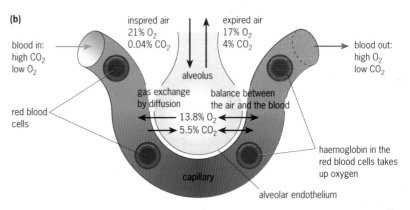

As we breathe in, fresh air enters the lungs and passes into individual alveoli (Fig 17.15(b)). Oxygen diffuses rapidly from the inhaled air through the walls of the alveoli and into the blood. Here, most of it combines with haemoglobin in the red blood cells. At the same time, carbon dioxide diffuses out of the blood and into the alveoli. It is breathed out during the next few expirations.

	Percentage of total volume				
	O_2	CO_2	N_2 + inert gases	H_2O vapour	Temperature (°C)
Atmospheric	21	0.04	79	variable	variable
Alveolar	13.8	5.5	80.7	saturated	37
Exhaled	17	4	79.6	saturated	37

Table 17.2 shows the composition of atmospheric, alveolar and exhaled air. As you can see, each of the values for exhaled air is an average of the values for inhaled and alveolar air. This is because exhaled air is a mixture of the two.

THE LUNGS OF PREMATURE BABIES

ALVEOLI ARE MINUTE bubble-like air sacs lined with moisture, and are liable to collapse because of surface tension: if their sides touch, they could stick together. To prevent this, the alveolar epithelium secretes a **surfactant**, a mixture of phospholipids, which greatly reduces the surface tension and keeps the alveoli open.

Without surfactant the lungs cannot function effectively and severe breathing problems can develop. An unborn baby does not start to secrete surfactant until about the 22nd week of pregnancy and the lungs have not accumulated enough surfactant to cope with breathing until about the 34th week.

Any babies born before this have immature lungs and suffer from a condition called **respiratory distress syndrome**. The effort needed to inhale and inflate the collapsed alveoli becomes too great and, without medical help, the baby can die from exhaustion and suffocation. Surfactant can now be made artificially. It is introduced into the lungs of premature babies to help them to continue breathing. This is a major break-through which means that babies as young as 23 weeks (17 weeks premature) now have more of a chance of survival.

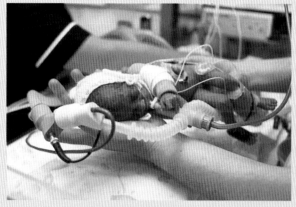

Fig 17.16 **This baby was born at 24 weeks, but survived due to the treatment of her lungs with surfactant. Only a tiny amount (0.5 cm³) of surfactant is needed, enough to line all the alveoli with a layer one or two molecules thick**

The composition of exhaled air varies during the course of a single expiration. The first air to emerge has a very similar composition to atmospheric air because it has been nowhere near the alveoli – it has simply filled the **dead space** in the trachea and bronchi. As the exhalation continues, air that has been deep inside the alveoli is breathed out. The composition of this air has been altered by gas exchange and it contains more CO_2 and less O_2 than atmospheric air.

The control of breathing

Control of breathing is **involuntary**: we don't have to think continually about breathing in and out and our breathing rate is automatically matched to our needs. The rate changes as the brain detects the physical and chemical variations that occur in the body as we carry out different activities.

Setting a regular pattern

Breathing is controlled by a bundle of nerves called the **respiratory centre**. This is located in the brain in an area called the **medulla oblongata** (Fig 17.17).

Regular nerve impulses travel down **efferent** nerves that pass from the respiratory centre out to both the external intercostal muscles and the diaphragm. These muscles then contract, starting off inhalation. As air enters the lungs, stretch receptors

?

D Why would it be impractical to breathe under water using a 6 metre snorkel?

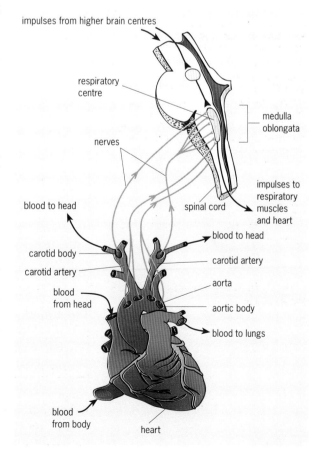

Fig 17.17 **Chemoreceptors in the blood vessels above the heart are sensitive to the CO_2 content and pH of the blood. They are less sensitive to oxygen content, but will respond to very low levels**

in the airways start firing and feed information to the brain about how the inflation of the lungs is progressing. The more the lungs inflate, the faster the stretch receptors feed back impulses. When the lungs are inflated sufficiently, signals from the respiratory centre stop for a short time and exhalation follows automatically.

Changing the breathing rate to meet demand

Ensuring the body has a constant supply of oxygen is obviously an important aspect of homeostasis (see Chapter 25) but, surprisingly, the body is relatively insensitive to falling oxygen levels. It is much more sensitive to an increase in CO_2 and so this is the indicator of the need for oxygen. The levels of oxygen in the arterial blood vary very little, even during exercise, but the CO_2 levels vary in direct proportion to the level of exertion. The heavier the exercise, the greater the CO_2 concentration. Lactic acid levels also increase during exercise. Any increase in CO_2 or lactic acid concentration in the blood lowers its pH. **Chemoreceptors** (see Fig 17.17), which are extremely sensitive to the composition of the blood that flows past them, can detect very small changes in pH.

There are three types of chemoreceptor:

- **Central receptors** in the medulla oblongata. These are sensitive to the carbon dioxide concentration in the blood that flows through this region of the brain.
- The **carotid bodies** in the wall of the **carotid arteries.**
- The **aortic bodies** on the **aortic arch**, just above the heart.

Together, the carotid and aortic bodies are described as the **peripheral chemoreceptors**. These cells sense changes in CO_2 and pH levels and, to a lesser extent, they are also sensitive to changes in oxygen levels.

When chemoreceptors register and change in CO_2 levels or pH, they send nerve impulses to the respiratory centre in the brain. This responds by sending more frequent impulses to the external inter-costal muscles and diaphragm. When this happens, our ventilation rate increases: we breathe harder and faster. Heart rate also increases (see page 287) and so the body automatically increases oxygen delivery at the same time as removing the extra CO_2.

The control of breathing rate is very similar to the control of heart rate, with one important difference: we can control our breathing rate by thinking about it. This suggests that the higher, 'conscious' centres of the brain are more closely linked to the respiratory centre than they are to the cardiovascular centre. Also, research shows that pulse and ventilation rate change dramatically during exercise, even *before* the concentration of blood gases has a chance to change. It is as if the body predicts what is about to happen. How this works is not understood and is an active area of research.

The effects of oxygen deprivation

In some situations, at high altitudes for example, oxygen levels can fall without CO_2 levels increasing. In a rarefied or artificial atmosphere, normal breathing flushes CO_2 out of the blood via the lungs but there may not be enough oxygen to replace it. When this happens, the chemoreceptors often fail to register that anything is wrong, and the brain can become starved of oxygen.

The first symptoms of oxygen starvation are feeling ridiculously happy, having impaired senses and lacking judgement. When mountaineers, fighter pilots or deep sea divers start giggling and making stupid mistakes, it is a sure sign that they are not getting enough oxygen. It is also a signal for their colleagues to act fast, if they can, and provide them with emergency oxygen. If they don't get help quickly, they soon lapse into unconsciousness and brain damage and death follow (Fig 17.18).

Athletes and VO_2(max)

Many physical activities – such as jogging, swimming, team sports – rely on energy released by the aerobic pathway of cell respiration (see Chapter 8). The level of performance an athlete can achieve is largely governed by how fast oxygen gets to the muscles.

The rate at which a person uses oxygen is called the VO_2 and is measured in terms of the volume of oxygen consumed (cm^3), per kg of body weight, per minute. The VO_2(max) is the maximum rate at which oxygen is consumed and is the amount of oxygen that can be delivered to the tissues when the lungs and heart are working as hard as possible. Athletes use a knowledge of VO_2(max) in their training, as a measure of how hard they are working. A training schedule, for example, may require the athlete to work at 55 per cent of their VO_2(max) for a set length of time.

The worked example shows how to calculate VO_2 and VO_2(max) for an average person. (You can work out your own values if you are able to measure your tidal volume, the number of breaths you take per minute and your weight in kilograms.) You can see that the amount of oxygen consumed during exercise increases from 4.28 to 51.42 cm^3 O_2 kg^{-1} min^{-1}, an increase of over 1000 per cent.

Fig 17.18 **In 1875, three French physiologists decided to investigate the effects of low oxygen levels on the human body. The easiest way to get oxygen-poor air was to go up in a hot-air balloon to observe the effects the rarefied atmosphere on each other. At first there were no obvious effects and they happily continued to throw out ballast, going up to 8000 metres. At this point they all fainted. The balloon eventually came down on its own, and one of the scientists woke up to find the other two dead**

EXAMPLES

Q What is the rate of oxygen consumption of a normal 70 kg adult at rest? Assume that the tidal volume of a normal adult at rest is 0.5 litres and that they take 15 breaths per minute.

A We can work out their ventilation rate from the formula:

$$\text{Ventilation rate} = \frac{\text{number of breaths taken per minute}}{} \times \text{tidal volume}$$

$$\text{Ventilation rate} = 15 \times 0.5 \text{ litres min}^{-1} = 7.5 \text{ litres min}^{-1}$$

Atmospheric air is one-fifth oxygen and from Table 17.2 we know that only one-fifth of the available oxygen is absorbed. So:

amount of oxygen used per minute

$$= 7.5 \times 0.2 \times 0.2 \text{ litres min}^{-1}$$

$$= 0.3 \text{ litres min}^{-1}$$

And the VO_2 for this person is:

$$\frac{0.3}{70} = 0.00428 \text{ litres } O_2 \text{ kg}^{-1} \text{ min}^{-1},$$

or: $\qquad$ 4.28 cm^3 O_2 kg^{-1} min^{-1}

Q What is the VO_2(max) for the same adult? Assume that the volume of air taken in during each breath during strenuous exercise is 3 litres and that the number of breaths per minute increases to 30.

A Knowing that the tidal volume of the same adult during exercise is 3 litres and that they take 30 breaths per minute, we can work out their ventilation rate:

$$\text{Ventilation rate} = \frac{\text{number of breaths taken per minute}}{} \times \text{tidal volume}$$

$$\text{Ventilation rate} = 30 \times 3 \text{ litres min}^{-1} = 90 \text{ litres min}^{-1}$$

Atmospheric air is one-fifth oxygen and from Table 17.1 on page 253 we know that only one-fifth of the available oxygen is absorbed. So:

amount of oxygen used per minute

$$= 90 \times 0.2 \times 0.2 \text{ litres min}^{-1}$$

$$= 3.6 \text{ litres min}^{-1}$$

And the VO_2(max) for this person is:

$$\frac{3.6}{70} = 0.05142 \text{ litres } O_2 \text{ kg}^{-1} \text{ min}^{-1},$$

or: $\qquad$ 51.42 cm^3 O_2 kg^{-1} min^{-1}

PART II: GAS EXCHANGE IN PLANTS

5 WHAT GASES DO PLANTS EXCHANGE?

Like animals, plants exchange gases with the atmosphere. It is easy to get the idea that gas exchange in plants is the opposite of that in animals, but this is an oversimplification.

Like all organisms, plants respire so that they can obtain energy from food. This process uses oxygen and produces carbon dioxide as a waste product. However, during daylight, they also photosynthesise to make food. This process also uses carbon dioxide and makes oxygen. Usually, in daylight, the rate of photosynthesis is faster than that of respiration and so the green tissues of a plant need to take in carbon dioxide and allow excess oxygen to escape. Most photosynthesis happens in the leaves and so most gas exchange takes place here, too.

6 GAS EXCHANGE AND THE LEAF

Flowering plants are divided into two main groups: the **monocots** and **dicots**. Grasses are monocots, oak and walnut trees and dandelions are dicots. (See Fig 17.19 and Chapter 13.)

Plant leaves are rather like solar panels: they are broad and flat to trap the light from the Sun. Leaves have large numbers of tiny holes called **stomata**. Carbon dioxide enters the leaf through these pores and diffuses directly into the cells that need it. Oxygen escapes by the same route. Before we go on to look at the process of gas exchange in detail, let's look at the structure of a typical leaf from a **dicot**, a type of flowering plant.

Structure of a leaf from a dicot plant

A typical dicot leaf is broad and flat with a network of branching veins (Fig 17.19).

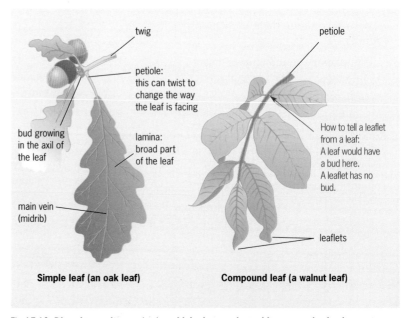

twig

petiole

petiole:
this can twist to
change the way
the leaf is facing

bud growing
in the axil of
the leaf

lamina:
broad part
of the leaf

How to tell a leaflet
from a leaf:
A leaf would have
a bud here.
A leaflet has no
bud.

main vein
(midrib)

leaflets

Simple leaf (an oak leaf)

Compound leaf (a walnut leaf)

Fig 17.19 **Dicot leaves have a** lamina **which shows a branching network of veins. In compound leaves, the lamina is divided into several smaller** leaflets

When you look at the leaf in cross-section, you can see four main layers (Fig 17.20).

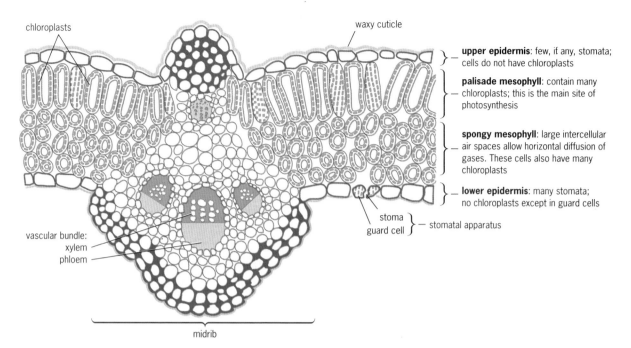

chloroplasts

waxy cuticle

upper epidermis: few, if any, stomata; cells do not have chloroplasts

palisade mesophyll: contain many chloroplasts; this is the main site of photosynthesis

spongy mesophyll: large intercellular air spaces allow horizontal diffusion of gases. These cells also have many chloroplasts

lower epidermis: many stomata; no chloroplasts except in guard cells

stoma
guard cell } — stomatal apparatus

vascular bundle:
xylem
phloem

midrib

Fig 17.20 **A section through a typical dicot leaf, showing the four main layers**

The upper epidermis

Most cells in this single layer do not have chloroplasts and so do not photosynthesise. Light passes through the upper epidermis to the layers below. Epidermal cells secrete a protective waxy layer, the **cuticle**, which helps reduce water loss. Often there are stomata, but usually fewer than on the lower epidermis. Stomatal pores are flanked by a pair of **guard cells**, the only epidermal cells that contain chloroplasts.

Like the lower epidermis, the upper epidermis can have hairs called **trichomes** (see page 269). These help reduce water loss and protect the plant against herbivores by preventing access to the leaf.

The palisade mesophyll

Most photosynthesis occurs in this layer. It contains long cylindrical cells, sometimes several layers of them, packed with chloroplasts. The large spaces between the cells allow effective gas exchange between the cells and the air spaces inside the leaf.

The spongy mesophyll

This layer has even larger spaces between cells. These allow sideways movement of gases within the leaf so that, for example, carbon dioxide entering through the lower epidermis can spread out and diffuse more easily to all the cells of the palisade layer. The cells of the spongy layer are less important for photosynthesis.

The lower epidermis

This layer resembles the upper epidermis, but with two important differences. It lacks a waxy cuticle and usually has many more stomata. Floating leaves are an exception: they have functional stomata in the upper epidermal surface only (the water in contact with the lower epidermis prevents gas exchange).

There are more stomata on the shaded underside of a leaf so that gas exchange can occur without excessive water loss.

E Why do you think the spongy cells in a dicot leaf have fewer chloroplasts than the palisade cells?

The transport system in a leaf

The leaf is supported by a stalk or **petiole** which can grow or twist to position the leaf, usually to get the most light. The **main vein** runs through the petiole. This vein divides to form finer and finer branches (Fig 17.21).

Substances move through and to the leaf in the **vascular tissue**. There are two main types of vascular tissue in the veins of a leaf:

● **Xylem** conducts water and dissolved minerals (see page 292).

● **Phloem** transports food substances (see page 292). Phloem brings food to the leaf when it is first growing and takes food away from the leaf when it is mature enough to be a net food producer.

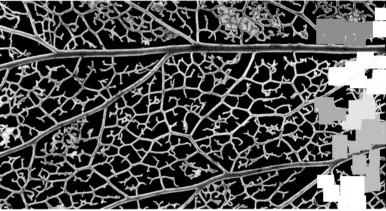

Fig 17.21 **This dicot leaf has been chemically treated to make it transparent. The veins that branch and form a network are clearly visible. The islands of mesophyll created by the network of veins are called** areoles. **The smallest veins branch still further to form vein endings within the areoles. In this way, almost every cell is in contact with a vein and so water can be delivered by the xylem and food taken away by the phloem**

How gases are exchanged

As we saw in the first part of this chapter, a large surface area is more effective for gas exchange than a smaller surface area. Altogether, the cell surfaces of the palisade and spongy mesophyll are between 10 and 40 times greater than the external surface area of the leaf.

Fig 17.22 summarises carbon dioxide movement in the leaf. Carbon dioxide in air enters the leaf via the stomata. Because leaves are thin, the distances involved are small and so the CO_2 diffuses quickly through the spongy mesophyll, and up between the palisade cells. For each cell it passes, some CO_2 dissolves in water held between the microfibrils of the cell wall and then goes through the cell membrane as CO_2, H_2CO_3 or as the bicarbonate ion, HCO_3^-. In bright light, the carbon dioxide is used quickly in photosynthesis. This maintains the concentration gradient between the inside of the leaf and the atmosphere outside.

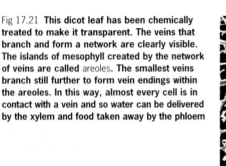

F In what three main ways is the leaf adapted for gas exchange?

Fig 17.22 **Diffusion of carbon dioxide into a leaf cell. The proportion of carbon dioxide in the air is about 0.04 per cent. In an actively photosynthesising leaf the carbon dioxide metabolised must be replaced or photosynthesis will stop. Oxygen, formed as a by-product of photosynthesis, diffuses in the opposite direction and escapes from the leaf via the stomata**

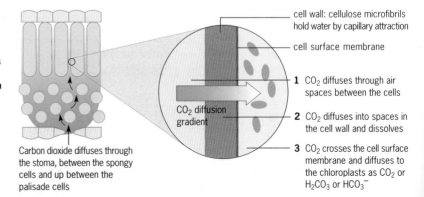

Carbon dioxide diffuses through the stoma, between the spongy cells and up between the palisade cells

CO_2 diffusion gradient

cell wall: cellulose microfibrils hold water by capillary attraction

cell surface membrane

1 CO_2 diffuses through air spaces between the cells

2 CO_2 diffuses into spaces in the cell wall and dissolves

3 CO_2 crosses the cell surface membrane and diffuses to the chloroplasts as CO_2 or H_2CO_3 or HCO_3^-

Stomata

As we have seen, stomata are pores surrounded by two specialised cells called guard cells (Fig 17.23). These are roughly banana shaped and, unlike other epidermal cells, they contain chloroplasts. Most of the cellulose microfibrils of the cell wall run across the width of the cell: this allows the two cells to 'bend' and so open and close the pore (see page 268).

Successful exchange of gases by the leaf depends on stomata being open. There is a problem: these pores also allow water to escape by diffusion into the air. Water in the gaps between the cellulose microfibrils of the cell walls readily evaporates into the spaces between the cells. It is then free to diffuse out of the stomata. In hot environments the evaporation of water in this way helps to cool the plant. However, water lost has to be replaced and if water is in short supply, the plant can suffer water stress. Water loss can be reduced by closing the stomata but this also, inevitably, reduces gas exchange. (Transpiration and water loss is covered in more detail in Chapter 18).

So, leaf design turns out to be a compromise between the requirements of photosynthesis and the need to conserve water. Stomata are involved in a continual balancing act: they open and close, according to the external conditions and the needs of the plant.

When stomata open and close

In general, stomata open in the light and close in the dark.

Like all living organisms, plants respire 24 hours a day. During daylight the rate of photosynthesis usually exceeds the rate of respiration, and plants are net producers of oxygen. At night when there is no photosynthesis, plants must take in oxygen from the air. You might think that if the stomata are closed, the plant might 'suffocate'. However, plant metabolism is fairly slow and the proportion of oxygen in the air is quite high. Enough oxygen can leak in through the leaf surface, even when the stomata are closed to conserve water.

As the stomata open, water is lost. Water uptake replaces this water as fast as possible, but it can lag behind (Fig 17.24). If a plant loses too much water it **wilts**. If it continues to lose water it can be permanently damaged because its cells suffer **plasmolysis** (see Chapter 5). On hot bright days it is much safer for plants to conserve water, even though this means that the rate of photosynthesis slows. On such days, stomata often close up around noon and open again later on, when the Sun is not so strong.

Fig 17.23 **Scanning electron micrograph showing stomata on the lower epidermis of the elder (Sambucus nigra). The red lines show the direction in which the cellulose microfibrils run in the cell wall of the guard cells**

See question 4.

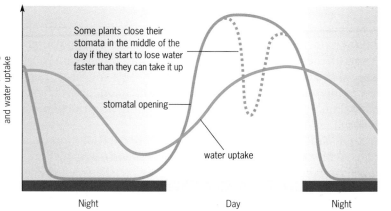

Fig 17.24 **Stomatal opening and water uptake are closely linked. Soon after the stomata open the plant starts to take up water more quickly. In some species, if the plant loses water too quickly the stomata can close at around midday**

How stomata open and close

We know that stomata do open and close, and that small movements of the guard cells are responsible. If water enters the guard cells, causing them to become fully **turgid** (swollen), the size of the cells increases. As their cellulose fibres run across the cell (see Fig 17.23) they cannot expand widthways. Instead, they are forced to elongate slightly. The guards cells are fixed at each end and so, as their volume increases, they bow outwards. This forms a gap, the **stomatal pore**, between the two cells. We know the mechanism, but the exact stimulus that causes stomata to open and close is not fully understood and is still being studied.

An early theory proposed that, when the guard cells photosynthesise, they make enough sugar to cause water to enter them by osmosis from the surrounding cells. In the 1960s, researchers realised that this theory could not explain how some plants manage to close their stomata at midday in hot sunlight or why stomata that have guard cells with no chloroplasts can open and close.

Further studies showed that the stomatal opening and closing depends on the generation of a potassium ion gradient (Fig 17.25). During the day, chloroplasts in the guard cells photosynthesise, which enables them to produce ATP. The ATP is used to actively pump potassium ions (K^+) into the guard cells from adjacent cells. The accumulation of potassium ions causes water to be drawn in by osmosis and the guard cells become more turgid. At night the pumping stops, the ions leak out and, as the turgor of the guard cells reduces, the stomata close.

Fig 17.25 **The potassium pump mechanism that explains how stomata open and close**

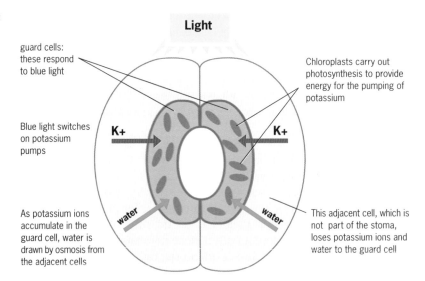

Light

guard cells: these respond to blue light

Chloroplasts carry out photosynthesis to provide energy for the pumping of potassium

Blue light switches on potassium pumps

K+   K+

As potassium ions accumulate in the guard cell, water is drawn by osmosis from the adjacent cells

water   water

This adjacent cell, which is not part of the stoma, loses potassium ions and water to the guard cell

So, how do some plants manage to close their stomata at midday? An 'override' mechanism must be involved. It turns out that when plants are deprived of water, the concentration of a plant 'hormone' **abscissic acid** (ABA) rises dramatically. This triggers a potassium ion pump which works in the opposite direction, actively excreting potassium ions from the guard cells. The turgor in the guard cells is reduced, and so the stomata close.

Noon-time closure of stomata is not a perfect solution because it reduces the capacity of the plant to cool itself. But, evolution has taken care of this – plants that live in the hottest environments have evolved enzymes which can withstand temperatures up to 60 °C before they are denatured.

?

G Why should noon time closure of stomata affect the plant's temperature?

See question 5. ■

Leaves that can survive dry conditions

To reduce water loss, dry adapted, or **xeromorphic**, leaves have many of the following features:

- **A reduced surface area: volume ratio**. This reduces the area from which water can be lost.

- **Sunken stomata** (stomata that are located in grooves or depressions). Water vapour that accumulates in the groove forms a barrier to diffusion and so reduces water loss.

- **Leaf hairs**. These trap moist air round the leaf and prevent drying out.

- **A thickened cuticle**. Since water can be lost directly from the epidermis, a thick waterproof layer can reduce evaporation from the leaf surface.

- **A thickened epidermis.** This can consist of several layers of cells, which may be lignified. The extra thickness helps prevent water escaping from the mesophyll.

- **The development of sclerenchyma** (a lignified but non-conducting supporting tissue) in the leaf (Fig 17.26). This helps support the leaf in conditions where it might wilt. If it makes a continuous layer inside the leaf, it can form a barrier to reduce water loss.

- **The ability to roll up in dry conditions**. Rolled leaves have a smaller surface area from which water can evaporate (Fig 17.27).

Fig 17.28 shows a section through a dry adapted leaf showing many of these features.

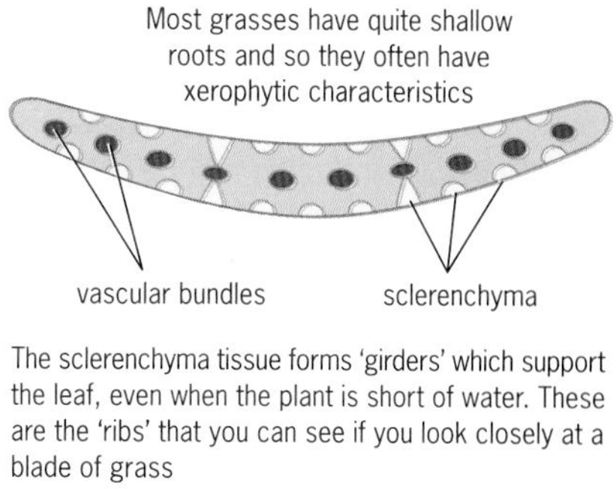

Most grasses have quite shallow roots and so they often have xerophytic characteristics

vascular bundles   sclerenchyma

The sclerenchyma tissue forms 'girders' which support the leaf, even when the plant is short of water. These are the 'ribs' that you can see if you look closely at a blade of grass

Fig 17.26 Sclerenchyma (a lignified tissue) is useful because water-stressed leaves cannot depend entirely upon cell turgor pressure for support. Many monocot leaves have 'girders' of sclerenchyma above and below the leaf veins

?

H Look at the section of the oleander leaf shown in Fig 17.28. Write down the features of adaptation to dry conditions that you can see in *Oleander*.

Fig 17.27 **In some grass plants such as *Marram*, shown here, there are groups of special 'hinge' cells. When the plant loses water, these hinge cells shrink and the leaf rolls up. The stomata, which are only found on the upper surface, become hidden, slowing down water loss**

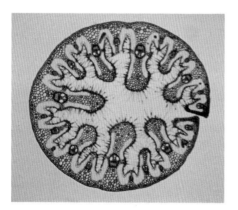

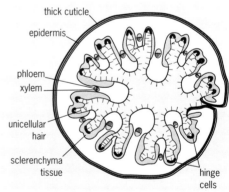

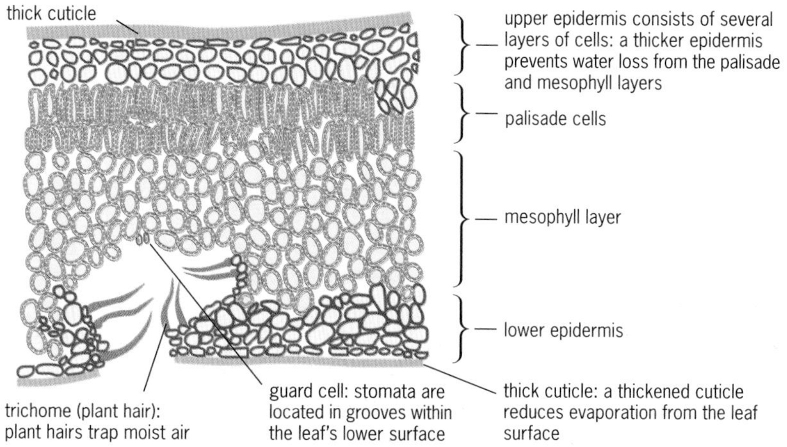

thick cuticle
epidermis
phloem
xylem
unicellular hair
sclerenchyma tissue
hinge cells

Fig 17.28 ***Oleander* is an evergreen dicot shrub. It is quite common in Mediterranean countries where it thrives in dry summers. The diagram shows how its leaf is adapted to conserve water**

thick cuticle

upper epidermis consists of several layers of cells: a thicker epidermis prevents water loss from the palisade and mesophyll layers

palisade cells

mesophyll layer

lower epidermis

trichome (plant hair): plant hairs trap moist air

guard cell: stomata are located in grooves within the leaf's lower surface

thick cuticle: a thickened cuticle reduces evaporation from the leaf surface

7 GAS EXCHANGE IN OTHER ORGANS

Although the leaves are the major site of photosynthesis, other structures such as stems can be green and need carbon dioxide for photosynthesis. All plant tissues need to exchange gases for cell respiration and the usually do so by simple diffusion.

Lenticels allow bark to 'breathe'

Woody plants produce a protective layer of bark. But the cells underneath are still alive and need to respire. In most species, young bark has patches of loose cells called **lenticels** (Fig 17.29), which allow gases to diffuse in and out. Lenticels often form on the sites of old stomata but do not themselves open or close.

> Photosynthesis occurs in the green tissues of plants. The leaves, which have evolved to become efficient traps for light energy, are where most of the food is produced. Other parts of the plant above the ground can be green: some plants use their stems as their main site of photosynthesis. Roots are never green and never photosynthesise.

Fig 17.29 **A lenticel: a break in the bark which allows gas exchange to occur through the loosely packed cells**

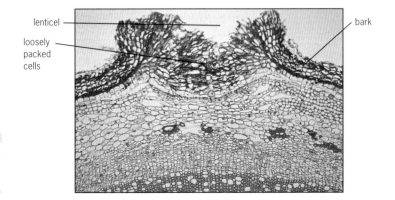

lenticel

loosely packed cells

bark

> **?**
>
> **I** What is the direction of gas diffusion in non-green plant parts, and why?

Water plants

Most plant parts are relatively thin and diffusion distances are not great, so the gas exchange necessary for cell respiration is usually rapid. There are two important exceptions to this:

- **Woody stems and roots**. These contain mainly dead cells that do not respire.

- **Water plants.** The rhizomes and roots buried in mud may depend on oxygen that gets in through stomata above the water level (Fig 17.30). Such plants often have huge spaces between their cells to allow the gases to diffuse over very long distances (Fig 17.31).

Fig 17.30 **The mangrove tree (right) lives in muddy swamps. Its roots are submerged and cannot get oxygen from the mud. So it has 'breathing roots' called** pneumatophores **which grow upwards into the air**

Fig 17.31 **Cross section through a stem that carries the leaf of** Nuphar luteum, **a water plant (far right). Inside this plant, the** aerenchyma **tissue has large air spaces which give buoyancy and allow gases to diffuse easily over long distances**

After reading this chapter, you should know and understand the following:

■ Animals must **exchange gases** to provide oxygen for cell respiration, and to take away the waste carbon dioxide that this process produces.

■ Small animals have a high **surface area: volume ratio** and gas exchange over the body surface is enough to satisfy their needs. Adopting a flat or cylindrical body is a simple strategy that increases surface area.

■ Larger organisms have **respiratory organs** that increase the surface area for gas exchange. These organs have a large surface area and thin, moist membranes. Ventilation and a good blood supply help to maintain a large diffusion gradient.

■ **Gills** are organs that are adapted to remove dissolved oxygen from water. Water flows in one direction over the gills, and the blood in the gill filaments flows in the opposite direction.

■ Insects have a **tracheal system** in which air-tubes branch out from **spiracles** on the surface of the insect and lead to individual cells.

■ **Lungs** are organs that have evolved to allow very efficient gas exchange in air. In mammals, breathing movements ventilate air sacs called **alveoli**, bringing in a continuous supply of fresh air and flushing out waste carbon dioxide.

■ The rate of breathing is controlled by the **respiratory centre** in the **medulla oblongata** of the brain. This sets a regular breathing pattern, which is modified according to information received from **chemoreceptors**, which detect changes in the blood that indicate the body's need for oxygen.

■ The leaf is the plant organ best adapted for photosynthesis. It is the major site of gas exchange where carbon dioxide enters and excess oxygen leaves.

■ A **dicot** leaf has four main layers: the **upper epidermis**, the **palisade layer**, the **spongy mesophyll** and the **lower epidermis**. Each layer has its own function.

■ A network of veins in the leaf contains **xylem** and **phloem** and ensures effective water delivery and food removal.

■ Gases diffuse in and out of leaves along a concentration gradient.

■ The direction of diffusion of oxygen and carbon dioxide depends on whether the rate of photosynthesis exceeds respiration or not.

■ **Stomata** open during the day to let carbon dioxide in, but this leads to increased water loss.

■ Under conditions of water stress, **abscissic acid** can cause stomata to close in the light.

1 The drawing in Fig 17.Q1 shows part of a fish's gill

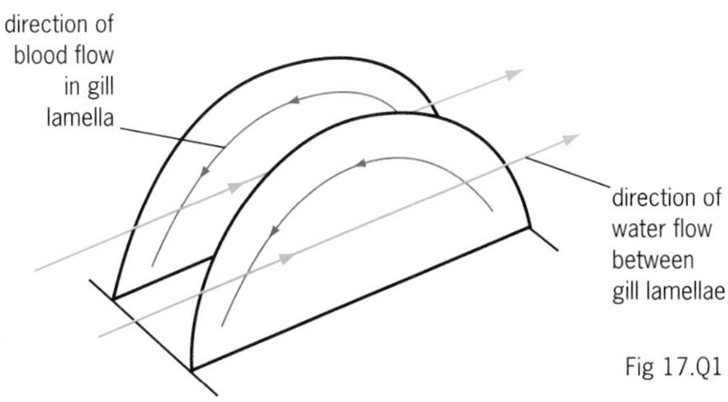

direction of blood flow in gill lamella

direction of water flow between gill lamellae

Fig 17.Q1

a) Use information from the drawing to explain how the direction of flow of blood and of water increase the efficiency of the gill filament as an exchange surface.

The table shows some features of gills in several species of fish.

Species	Thickness of lamellae /μm	Number of lamellae per mm	Distance between lamellae /μm	Distance between blood and water/μm	Activity
Icefish	35	8	75	6	slow-moving
Bullhead	25	14	45	10	slow-moving
Sea scorpion	15	14	55	3	active
Trout	15	20	40	3	active
Roach	12	27	25	2	active
Coalfish	7	21	40	<1	active
Herring	7	32	20	<1	very active
Mackerel	5	32	20	<1	very active

b) Use information from the table to describe **two** ways in which the structure of the gill is related to activity of the fish.

c) Explain how the structure of the gills in the mackerel enables it to be far more active than the bullhead.

[NEAB February 1995 Modular Biology: Physiology Test, q. 3]

2 Fig 17.Q2(a) shows a spirometer which is used to measure various aspects of breathing. An oxygen chamber floats in a tank of water. A student breathes through a length of tubing connected to the oxygen chamber. Soda lime is used to absorb all the carbon dioxide in the expired air. As the student breathes in and out, the chamber goes down and up. These movements are recorded on a revolving drum.

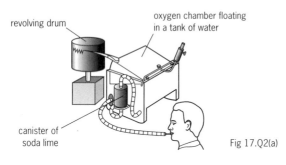

Fig 17.Q2(a)

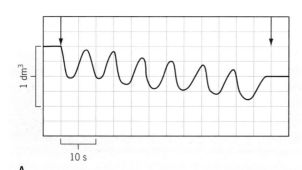

A

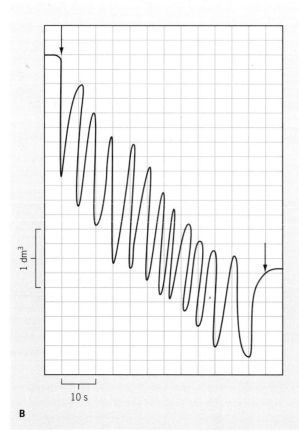

B

Fig 17.Q2(b)

Fig 17.Q2(b) shows two spirometer traces made by the student. Trace **A** is a record of breathing at rest. (The arrows indicate when the student was breathing from the spirometer.) Trace **B** is a record of breathing immediately after four minutes of vigorous exercise.

a) With reference to Traces **A** and **B**, calculate and insert the values in the following table.

	At rest	After exercise
Average tidal volume		
Rate of breathing		
Ventilation rate (the total amount of oxygen breathed per minute)		

b) **(i)** Explain why the spirometer trace shows an overall decrease in both Trace **A** and Trace **B** in Fig 17.Q2(b).
  (ii) Describe the physiological reasons for the steeper decrease shown in Trace **B**.

c) List **three** variables that you would need to control if this investigation was repeated to compare the effects of exercise on the breathing of an athlete and a non-athlete.

[UCLES 1993 Modular Sciences: Human Health and Disease Paper, q. 4]

3 Variations in the rate of water vapour loss from the leaves of different species can be linked to differences in epidermal structure.

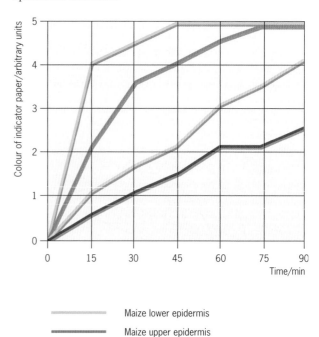

Maize lower epidermis
Maize upper epidermis
Daffodil upper epidermis
Daffodil lower epidermis

Adapted from Freeland, Problems in Practical and Advanced Level Biology (1985)
Fig 17.Q3

An experiment was carried out to compare the water vapour loss from daffodil leaves and from maize leaves. Pieces of blue (dry) cobalt chloride paper, 1 cm², were attached to the upper and lower surfaces of the leaves by means of clear adhesive tape. Colour changes were observed at 15-minute intervals over a period of 90 minutes. As the cobalt chloride paper became moist, it changed from blue to pink.

The colour changes during each 15-minute period were matched against a standard range of colours from blue (0) to pink (5). The colour changes were plotted against time and are shown in the graph of Fig 17.Q3.

a)   **(i)** Comment on the differences in the rate of water vapour loss from the leaves of the two plants.

      (ii) Give *two* reasons why this method of estimating rate of water vapour loss might not be very accurate.

b) The numbers of stomata on each surface of a daffodil leaf and of a maize leaf were estimated by counting the stomata in several 4 mm² areas of epidermis. From these, the mean number of stomata per cm² was then calculated. The mean numbers were given in the table below.

Species	Mean number of stomata/cm⁻²	
	Upper epidermis	Lower epidermis
Daffodil	3550	1850
Maize	5100	6300

Explain the relationship between rate of water vapour loss and number of stomata.

c) State *two* other structural features of a leaf, apart from the number of stomata, which might influence the rate of water vapour loss. In each case give a reason for your answer.

[ULEAC June 1993 Biology: Paper 3, q. 3]

4 The graph in Fig 17.Q4 shows changes in the mean diameter of the stomatal apertures when bean plants were moved from a chamber maintained at 25 °C to cooler conditions.

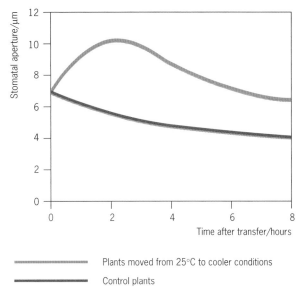

Fig 17.Q4

a) Giving a reason for your answer, describe how the control plants should have been treated.

b) What effect would the changes shown on the graph have on the transpiration rate of the experimental plants?

[AEB: June 1992, Biology: Common Paper 1, q. 12]

5 Discuss the importance of transpiration to plants.

[UCLES 1994 Biology: Paper 2, q. 7(b)]

6 Describe the features of gaseous exchange that are **common** to both mammals and plant leaves.

[UCLES 1995 Biology: Paper 2, q. 6(a)]

Assignment

COPING WITH ASTHMA

Asthma

Asthma is one of the commonest childhood ailments, affecting at least one in ten children, and many adults. The number of asthma cases has risen dramatically in recent years, and air pollution is thought to be one of the main causes.

To understand fully why asthma happens, you need to be familiar with the fine structure of the lungs – see page 258.

1

a) What is the difference in structure between the bronchi and bronchioles?

b) What type of muscle lines the bronchi and bronchioles?

What is asthma and what causes it?

Put simply, asthma is a difficulty in breathing caused by constriction of the smooth muscles of the bronchioles. Asthma sufferers find it hard to breathe because the airways leading to the alveoli become narrower, so requiring far more effort to deliver a normal amount of air to the lungs.

There are many factors that can cause asthma: an allergic reaction, exercise (especially in cold air) and lung infections. An allergy is an 'over-reaction' of the immune system to a substance that, in most other people, has no effect. The commonest allergens are house dust, fur, feathers and pollen.

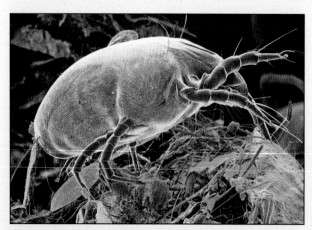

Fig 17.A1 **The house dust mite, *Dermatophagoides farinae*, is a normal inhabitant of our mattresses. They feed on the large deposits of skin that accumulate there, and the faeces they produce can cause an allergic reaction which leads to asthma in some people**

2 Suggest what measures could be taken to reduce the problems caused by the house dust mite shown in Fig 17.A1.

3 What advice would you give to a patient suffering from asthma caused by pollen?

Treating asthma

Several approaches are used together to help control asthma. A doctor's first priority with all but the youngest of children is to explain what is happening inside the lungs. This reduces fear and helps sufferers to come to terms with their problem. The worst physical effects of the condition are then controlled by a combination of drugs and prevention.

Generally, two types of drugs are prescribed to asthma sufferers: **bronchodilators** and **steroids**. Bronchodilators give instant relief from chest tightness and wheezing because they contain chemicals that relax the bronchiole walls. Many asthma sufferers carry an inhaler around with them. A common bronchodilator is the drug Salbutamol™, which is modified adrenaline. The hormone adrenaline relieves asthma, and is very useful in the treatment of severe attacks, but it is unsuitable for regular use.

4

a) What are the effects of the hormone adrenaline on the body? (You may need to look at page 369.)

b) Suggest what happens when an asthma sufferer takes an adrenaline-based drug?

c) Suggest why adrenaline is not suitable for use in long-term asthma treatment.

Steroids (see page 365) such as **Becotide™** act by reducing the degree of inflammation of the bronchioles. Steroids are preventative: asthma sufferers take regular doses morning and night to reduce the problem of over-reaction. The amount of steroids taken is minimal but they are effective if taken according to instructions.

5 Suggest why it is an advantage to inhale steroids rather than inject them.

6 The girl in Fig 17.A2 is using an inhaler. To get the full benefit from each dose – and not just coat the inside of her mouth – she must learn the correct technique. Find out what this is.

Fig 17.A2 **Using an inhaler is easy once you have mastered the right technique**

7 How would you investigate the hypothesis that air pollution is linked with the measured increase in asthma cases?

Assignment

POTASSIUM IN PLANTS

The potometer

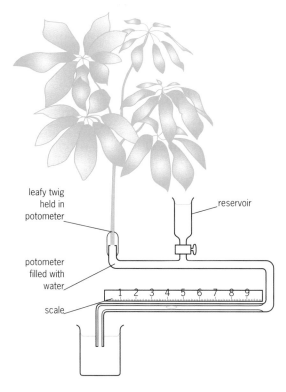

Fig 17.A3 **A potometer is a useful apparatus for measuring relative rates of transpiration in the laboratory. The rate of movement of the bubble is proportional to the rate of transpiration**

The following table includes some data collected from a simple potometer experiment.

Relative rate of bubble movement	Conditions	Explanation
5	light, still air	'Normal' rate of transpiration
7	Light, shoot placed in current of air	
1		Stomata closed
	light, still air, plant sprayed with a solution of abscissic acid	

1 Complete the table by suggesting conditions, providing an explanation, or suggesting a likely bubble rate, as appropriate.

Stomatal changes

The diagram in Fig 17.A4 shows the potassium ion molar concentrations in guard cells and other adjacent cells, the **subsidiary cells**, in a tropical plant in the dark and in the light. (Note: Stomata are open in the light and closed in the dark.)

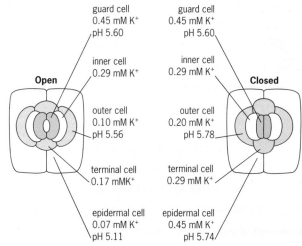

Fig 17.A4 **Potassium ion concentrations and pH values in vacuoles of guard and other cells in *Commelina***

2 What evidence is there from the data to suggest that potassium is involved in stomatal movement?

Researchers wanted to know more about the pumping of potassium ions. If potassium ions were the only ions being pumped, this would lead to a huge electrical imbalance (there would be a net positive charge inside the guard cell).

But the researchers found no such electrical difference. They knew that guard cells tend to produce organic acids in the light, which leads to the formation of H^+ ions. Using this and other information, they formed the two hypotheses shown in Fig 17.A5.

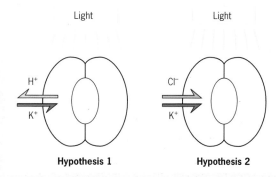

Fig 17.A5 **Two hypotheses about pumping potassium ions: Hypothesis 1. Potassium ions are exchanged for hydrogen ions Hypothesis 2. As potassium ions enter, another negative ion (chloride) enters too, to keep positive and negative charges balanced**

3 Using the pH data in Fig 17.A4, suggest which of these two hypotheses is more likely, and say why.

18 Transport systems

Despite a high cholesterol diet, the Masai people have low rates of coronary heart disease

CORONARY HEART DISEASE is one of the commonest causes of death in the western world. It is a condition in which the arteries supplying the heart muscle become narrowed by a fatty deposit.

You might expect someone who eats meat and drinks blood and milk almost exclusively to have a high cholesterol level and a high risk of developing coronary heart disease. However, despite eating exactly this diet, the Masai people of Kenya and Tanzania have *low* blood cholesterol. The Masai can eat a huge amount of cholesterol per day – up to 2000 milligrams – and still maintain a healthy blood cholesterol level of around 3.5 millimoles per litre. How do they do it?

Investigations into the Masai diet have suggested that the answer may lie in some of the tree-barks that the Masai add to their food. These have been found to contain cholesterol-lowering chemicals called saponins. These chemicals have been found in more than 100 different plant species including legumes such as chick peas and soya. It is thought that the saponins work by binding to cholesterol in the gut, so preventing its absorption into the bloodstream.

1 WHY DO ANIMALS NEED A CIRCULATORY SYSTEM?

> In larger animals, diffusion is simply too slow for all the cells to exchange materials quickly enough or in sufficient quantities to stay alive and healthy. Their bodies need a **transport system**. The mammalian circulatory system is an example of a **mass flow system**, in which large volumes of fluid are carried to all parts of the organism.

Evolution has produced large animals with specialised organs such as intestines, gills and lungs to increase the surface area available for exchange of materials such as food and oxygen (see Chapters 5, 16 and 17). These organs would be relatively useless without a **circulatory system**, a network of tubes filled with fluid that can deliver vital materials to all the cells of the body, and then take away their waste. In fact, in most animals, almost every cell is within a few micrometres of a branch of the circulatory system.

A circulatory system has three components:

- a *fluid* that flows in the system, carrying materials around the body;
- a *system of tubes* that carries the fluid;
- a *pump* (or pumps) that keeps the fluid moving through the tubes.

In vertebrate circulatory systems, **blood** (the fluid) is driven through **arteries**, **veins** and **capillaries** (the tubes) by the **heart** (the pump). In this chapter, we look at these three components in detail.

Single and double circulations

Fish have a **single circulation**: blood passes from the heart to the gills and then slowly round the body and back to the heart in a single circuit, as shown on the left in Fig 18.2. This system is adequate for fish, but animals with a higher metabolic rate need to deliver oxygen to their tissues faster than a single circulation allows.

Mammals and birds have a **double circulation**: a **pulmonary circulation** from the heart to the lungs and a **systemic circulation** from the heart to the rest of the body, as shown on the right in Fig 18.2 and in Fig 18.3. A double circulation must have a four-chambered heart. The right side of the heart receives deoxygenated blood from the body and pumps it to the lungs. Here, blood gains oxygen, but loses pressure. The left side of the heart receives the oxygenated blood and gives it a boost so that it can reach all the body parts quickly.

Fig 18.1 **The heart of a blue whale weighs about the same as a small family car. A massive muscle, this heart pumps blood to organs over 15 metres away, down an artery large enough for a child to crawl through**

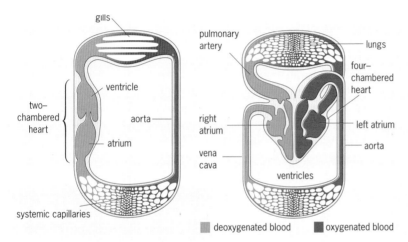

deoxygenated blood ■ oxygenated blood

Fig 18.2 **Far left: fish have a single circulation and a two-chambered heart. Blood first travels to the gills, where it passes through thin capillaries, picking up oxygen but losing pressure. It continues to travel around the body of the fish, back to the heart, but more slowly because of the low pressure**

Fig 18.2 **Near left: mammals have a double circulation and a four-chambered heart. Deoxygenated blood, coloured blue, passes into the right side of the heart and is pumped to the lungs where it picks up oxygen and releases carbon dioxide. Oxygenated blood, coloured red, returns to the heart to be pumped to all parts of the body except the lungs**

2 THE HUMAN CIRCULATORY SYSTEM

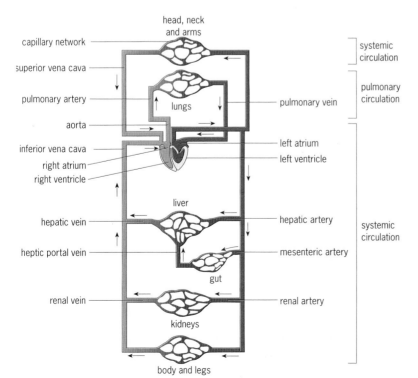

The human heart beats over 100 000 times a day, creating the pressure to force blood through more than 80 000 kilometres (50 000 miles) of arteries, veins and capillaries. Fig 18.3 shows the extent of the human circulatory system.

Fig 18.3 **An overview of mammalian circulation: this is the system for a human. Deoxygenated blood (coloured blue) passes into the right side of the heart through the superior and inferior vena cava.**

The right atrium pumps the blood into the pulmonary circulation and it picks up oxygen and releases carbon dioxide as it passes through the lungs. Now oxygenated (coloured red), the blood returns to the heart. The left ventricle then pumps it around the systemic circulation and it travels to all parts of the body (except the lungs) before returning to the heart

Structure of the heart

The mammalian heart is composed mainly of **cardiac muscle**: a specialised tissue that can contract automatically, powerfully and without fatigue, throughout the life of the organism (Fig 18.4).

The thickness of the walls in the different heart chambers reflects their function. The atria are thinly muscled because they only pump blood the short distance to the ventricles directly below them. The right ventricle is more heavily muscled than either of the atria because it has to force blood a much further distance to the lungs. The left ventricle has the thickest wall because it has to force blood all the way around the body.

It is important that blood flows through the heart in one direction only. Two sets of valves close to prevent backflow:

● The **atrio-ventricular valves (AV valves)**. These lie between the atria and the ventricles to prevent blood from returning to the atria when the ventricles contract. The **tricuspid valve** on the right side has three **cusps**, or flaps. The **bicuspid valve**, or **mitral valve**, on the left has two cusps. Both AV valves are subjected to great pressure and ultra-tough tendinous cords are needed to prevent them turning inside out – see Fig 18.4(b).

● The **semilunar valves** guard the openings to the pulmonary artery and the aorta to prevent backflow of blood into the ventricles. Both valves have three semilunar ('half moon'-shaped) cusps.

> The size of the heart is closely related to the organism's body size. Generally, the heart weighs 0.59 per cent of the total body weight. You can estimate the weight of your own heart using this formula.

> Valves are simply strong flaps of tissue: they cannot move on their own. A common mistake in examinations is to claim that valves can actively control blood flow. For example, the statement: 'the AV valves close, preventing blood flow', should read 'blood begins to flow back, forcing the valve shut and preventing any further flow'.

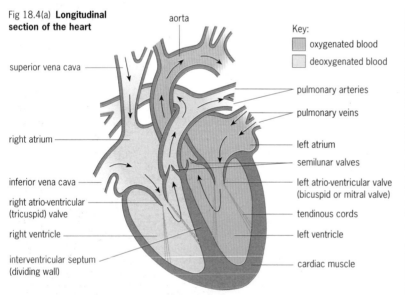

Fig 18.4(a) **Longitudinal section of the heart**

aorta

superior vena cava

right atrium

inferior vena cava

right atrio-ventricular (tricuspid) valve

right ventricle

interventricular septum (dividing wall)

Key:
- oxygenated blood
- deoxygenated blood

pulmonary arteries

pulmonary veins

left atrium

semilunar valves

left atrio-ventricular valve (bicuspid or mitral valve)

tendinous cords

left ventricle

cardiac muscle

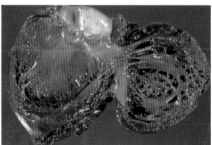

Fig 18.4(b) **The heart is a slightly twisted, asymmetrical organ: the best way to appreciate its three-dimensional structure is to dissect a pig or sheep heart or to study a model. This is a section through the left ventricle of a human heart. The string-like cords link heart muscle to valves**

Fig 18.4(d) **The fine structure of cardiac muscle. The rapid spread of impulses through cardiac muscle is possible because individual cells are connected by specialised junctions called** intercalated discs. **This system of communication ensures that all cells 'beat' at the same time**

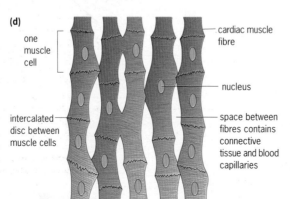

(d)

one muscle cell

intercalated disc between muscle cells

cardiac muscle fibre

nucleus

space between fibres contains connective tissue and blood capillaries

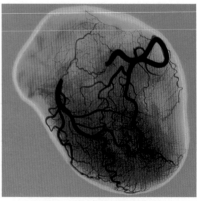

Fig 18.4(c) **The coronary circulation comprises all the blood vessels that supply the heart muscle. In this photograph, these blood vessels are easy to see because they have been filled with a dye that is opaque to X-rays**

HEART DISEASE

HEART DISEASE IS the single greatest cause of death in the western world, accounting for 26 per cent of all deaths. The underlying causes of heart disease are structural problems; **ischaemia**, (an inadequate coronary blood supply); or faulty electrical conduction.

Structural problems

These are defects in the anatomy of the heart, often present from birth. A fetus has a modified circulation because the placenta is doing the job of the fetal lungs, intestines and kidneys (Fig 18.5). Fetal circulation differs from that of an adult by having a **foramen ovale** and a **ductus arteriosus**.

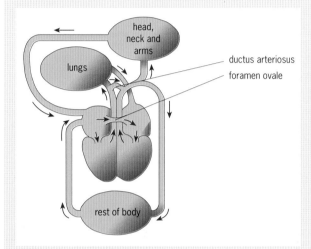

Fig 18.5 **The foramen ovale and the ductus arteriosus work together to greatly reduce the blood supply to the lungs (remember, after birth the amount of blood that passes through the lungs is as large as the amount that travels around the rest of the body)**

The foramen ovale is a hole between the right and left atria which closes immediately after birth. If it fails to close, the baby has a 'hole in the heart'. This allows oxygenated and deoxygenated blood to mix, reducing the efficiency of the circulation. A person can lead a normal, healthy life with a small hole in the heart, but larger ones need surgery.

Faulty valves allow blood to flow backwards and so reduce the efficiency of the heart. The result is a **heart murmur** that is heard with a stethoscope as a 'lub-swish' instead of lub-dup (see page 281). Artificial valves are used to correct very bad heart murmurs, but many do not need treatment.

Inadequate coronary blood supply

By far the most widespread heart condition is **coronary heart disease (CHD)**, also known as **ischaemic heart disease**. The coronary arteries narrow when a fatty deposit called an **atheroma** builds up in them. Atheromas can reduce, or even block, the blood supply to heart muscle, causing symptoms ranging from mild chest pain to a full heart attack.

Research shows that people who develop heart disease usually have one or more of the following: high blood cholesterol, high blood pressure, a history of cigarette smoking, diabetes mellitus, or an inherited (therefore genetic) tendency. They may also be obese (fat) and may not have exercised regularly.

CHD leads to two common conditions: angina and heart attacks. **Angina pectoris** literally means 'chest pain'. Blood flow to the heart muscle is sufficient at rest, but the coronary arteries are so blocked up that they cannot deliver the extra oxygen needed to cope with exercise. Any physical effort leads to chest pain, so the lifestyle of the sufferer is significantly restricted.

In a heart attack, also called a **myocardial infarction**, the blood flow to a particular part of the heart is blocked and that region of **myocardium** (heart muscle) dies. The body can recover if only a small area of muscle dies, but extensive infarctions are often fatal.

Faults in the conducting system

Contrary to popular belief, the heart doesn't have to stop beating to cause death, it just has to stop pumping effectively. Many deaths are due to **arrhythmias**: breaks in the normal rhythm of the heart. For example, **ventricular fibrillation** (rapid and uncontrolled contraction of the ventricles) is fatal if untreated. Ventricular fibrillation is caused by disease or traumas such as electrocution. It is treated with a **defibrillator**, a machine that delivers a strong current across the heart for a short time. This depolarises all the heart cells at the same time. All activity stops for 3 to 5 seconds, allowing time for the heart to re-establish normal electrical activity.

If a heart attack damages the conducting system of the heart, the coordination of the heartbeat can be disrupted. This condition is often called a **heart block**. A patient with a permanent heart block can be fitted with an artificial pacemaker (Fig 18.6).

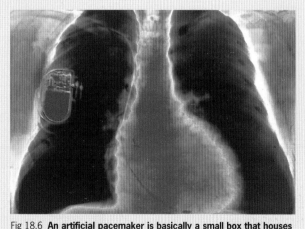

Fig 18.6 **An artificial pacemaker is basically a small box that houses small batteries and complex electronics. The device is fitted under a muscle in the upper thorax, and a wire leads down a vein into the heart. The wire ends in an electrode, seen at the base of the heart. The electrode touches the heart muscle, initiating (setting off) the cardiac cycle. New, sophisticated pacemakers can sense changes in breathing, movement and body temperature and can adjust heart rate accordingly**

How the heart beats

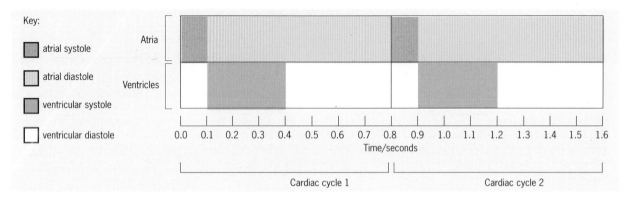

Fig 18.7 **The timescale of the cardiac cycle in a human at rest. The heart rate is 75 beats per minute. During exercise the heart rate can increase to over 150 beats per minute, so the cycle only takes half the time**

The sequence of events in a single heartbeat is known as the **cardiac cycle** (Fig 18.7). The cycle involves **systole**, or contraction, and also **diastole**, or relaxation, of the atria and ventricles. The cycle has four overlapping stages:

● **Atrial systole**. Both atria contract, forcing blood into the ventricles. This stage lasts 0.1 seconds.

● **Ventricular systole**. Both ventricles contract, forcing blood through the pulmonary artery to the lungs and through the aorta to the rest of the body. This takes 0.3 seconds.

● **Atrial diastole**. The atria relax, although the ventricles are still contracted. Blood enters the atria from the large veins coming from the body. This takes about 0.7 seconds.

● **Ventricular diastole**. The ventricles relax, and become ready to fill with blood from the atria as the next cycle begins. This takes about 0.5 seconds.

Given an average heart rate of 75 beats per minute, each cycle takes 0.8 seconds.

Control of the cardiac cycle

Heartbeat must be carefully controlled so that each chamber contracts only when full of blood. To achieve this, the events of the cardiac cycle are carefully coordinated (Fig 18.8).

A single heartbeat starts with an electrical signal from a region of specialised tissue called the **sino-atrial node (SAN)**, on the wall of the right atrium. This is the 'pacemaker', or heartbeat regulator. The electrical signal spreads out over the walls of the atria, causing them to contract.

From there, the signal does not pass directly to the ventricles. If it did, the ventricles would begin to contract before they had filled with blood. Instead, the impulse is delayed slightly. A second node, the **atrio-ventricular node (AVN)** picks up the signal and channels it down the middle of the **ventricular septum** through a collection of specialised cardiac muscle fibres called the **bundle of His**. From here the signal spreads throughout the wall of the ventricles, through the **Purkinje (or purkyne) fibres**, and the ventricles contract *after* they have filled with blood.

The heart continues to beat when removed from the body. Individual heart muscle cells grown in culture beat on their own! Because of this, we say that the vertebrate heart is **myogenic**: the stimulus that drives it to beat originates in the muscle itself. Some invertebrate hearts are **neurogenic**: they beat only when connected to external nerves.

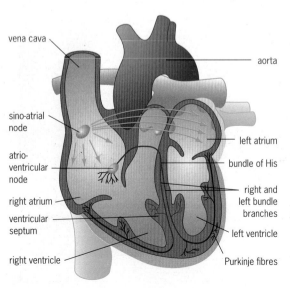

Fig 18.8 **The conduction system of the human heart initiates and controls a heartbeat. Cells in the sino-atrial node act as a pacemaker, initiating impulses that spread through the walls of the atria and, after a delay, through the walls of the ventricles via the atrio-ventricular node**

Electrical events and volume/pressure changes in the cardiac cycle

See question 1.

We have looked at what happens to the blood that passes through the heart during the cardiac cycle, and at the electrical events that coordinate the contraction and relaxation of the atria and ventricles. Fig 18.9 shows how these events correspond to pressure and volume changes in the heart and blood vessels, and also to the sounds that you hear when you listen to someone's heart beating. You will need to refer to this figure as you read on.

During **atrial systole**, the atria fill with blood from the vena cava and the pulmonary vein. Some of the blood that enters the atria flows immediately into the ventricles, without any need for contraction. Atrial systole is initiated when the SAN sends out an electrical signal that spreads out over both atria, causing them to contract and to force the remainder of the blood into the ventricles. There are no heart sounds. Atrial systole is very short, little more than a 'twitch'. Atrial relaxation, or **atrial diastole**, lasts for the remainder of the cycle. All the rest of the 'action' is in the ventricles.

Ventricular systole begins when the ventricles have filled with blood. The AVN picks up the signal from the SAN and then conducts impulses down through the bundle of His and on through the Purkinje fibres in the walls of the ventricles. This stimulates the ventricles to contract. Blood is forced upwards, forcing the semilunar valves open and the AV valves shut. The closing of the AV valves causes the 'lub' of the 'lub-dup' heart sound. Pressure in the arteries rises sharply as blood is forced into them.

Fig 18.9 **The cardiac cycle. Many exam questions use this diagram, without labels, to test understanding of the events of one heartbeat. It is important to remember what is cause and what is effect, ie:**

- **Pacemaker cells initiate systole (contraction).**

- **The squeezing of the muscle walls reduces the volume and so increases the pressure in the chamber(s), forcing blood in a particular direction.**

- **The direction of blood flow causes the valves to open or close. The valves ensure that blood flow through the heart is in one direction only**

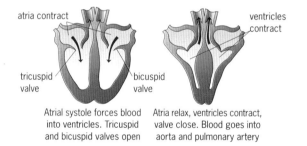
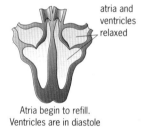

atria contract

ventricles contract

atria and ventricles relaxed

tricuspid valve

bicuspid valve

Atrial systole forces blood into ventricles. Tricuspid and bicuspid valves open

Atria relax, ventricles contract, valve close. Blood goes into aorta and pulmonary artery

Atria begin to refill. Ventricles are in diastole

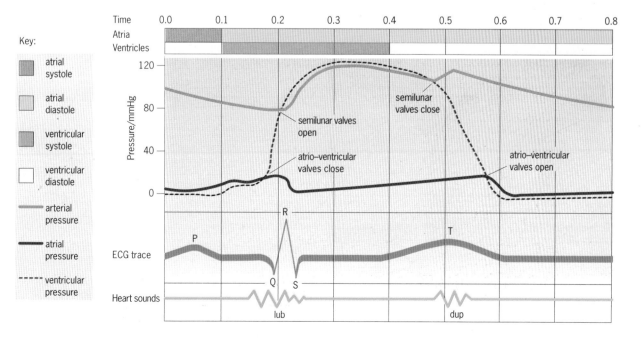

During **ventricular diastole**, the ventricle walls relax, arterial pressure falls and blood begins to flow back into the ventricles. This reversal of the flow causes the semilunar valves to shut, causing the second, 'dup', heart sound. Meanwhile, the atria have been filling with blood and, as the ventricles relax, blood flows from the atria into the ventricles, forcing the AV valves open again.

Stroke volume and cardiac output

The volume of blood pumped by the heart during one cardiac cycle is the **stroke volume**. A typical stroke volume in an adult is about 80 cm³: every time the heart beats, 80 cm³ of blood is forced through the pulmonary artery to the lungs and 80 cm³ is forced into the aorta to the body. So, the heart pumps over 8500 litres of blood per day. Stroke volume increases during exercise, and regular exercise brings about a permanent resting increase to 110 cm³, or more.

The volume of blood pumped in one minute is called the **cardiac output**. It is calculated by multiplying the stroke volume by the heart rate and is expressed in litres of blood per minute.

Control of heart rate

Heart rate is modified according to the needs of the body. It increases during physical exercise to deliver extra oxygen to the tissues and to take away excess carbon dioxide. At rest, a normal adult human heart beats about 75 times per minute: during very strenuous

---

Cardiac output = stroke volume × heart rate

This is measured in litres per minute. There are 1000 cm³ in a litre.

?

A Calculate the cardiac output:

(a) at rest when the stroke volume is 80 cm³ and the heart rate is 75,

(b) during vigorous exercise when the stroke volume is 100 cm³ and the heart rate is 150.

THE ELECTROCARDIOGRAM (ECG)

THE ELECTRICAL EVENTS that control the cardiac cycle are recorded by placing electrodes on certain parts of the body – Fig 18.10. A normal, healthy heartbeat produces a distinctive trace – right, in Fig 18.11. Certain heart defects produce a modified trace – below right, in Fig 18.11, and this makes the ECG a useful diagnostic tool.

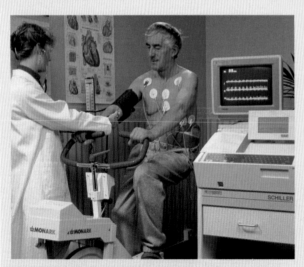

Fig 18.10 **The electrocardiogram: electrodes taped to the chest pick up the electrical events of the cardiac cycle as they pass through the body. People with heart problems can now transmit their ECG trace down the telephone to a specialist who can detect problems at a distance**

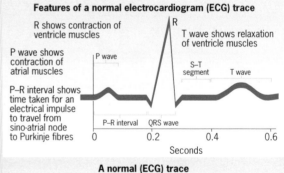

Features of a normal electrocardiogram (ECG) trace

R shows contraction of ventricle muscles

T wave shows relaxation of ventricle muscles

P wave shows contraction of atrial muscles

P–R interval shows time taken for an electrical impulse to travel from sino-atrial node to Purkinje fibres

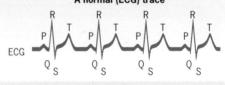

A normal (ECG) trace

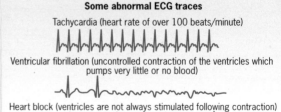

Some abnormal ECG traces

Tachycardia (heart rate of over 100 beats/minute)

Ventricular fibrillation (uncontrolled contraction of the ventricles which pumps very little or no blood)

Heart block (ventricles are not always stimulated following contraction)

Fig 18.11 **A healthy ECG trace and some abnormal traces, showing common heart defects that can be diagnosed using the ECG**

exercise it might beat 200 times per minute. Heart rate is controlled by the SAN. The rate goes up or down when it receives information via two autonomic nerves that link the SAN with the **cardiovascular centre** in the medulla of the brain (Fig 18.12):

- A **sympathetic** or **accelerator nerve** speeds up the heart. The synapses at the end of this nerve secrete **noradrenaline**.

- A **parasympathetic** or **decelerator nerve**, a branch of the **vagus nerve,** slows down the heart. The synapses at the end of this nerve secrete **acetylcholine**.

A **negative feedback system** operates to control the level of carbon dioxide and, indirectly, oxygen in the blood. During exercise, the level of carbon dioxide starts to rise. This is detected by **chemoreceptors** (cells sensitive to chemical change) situated in three places: the carotid artery, the aorta and the medulla. Impulses travel down the sympathetic nerve and heart rate increases. When the carbon dioxide level drops low enough, impulses pass down the parasympathetic nerve, and the heart rate returns to normal. The control of heart rate is closely related to the control of breathing (page 261).

Several factors affect heart rate:

- Secretion of **adrenaline** in response to stress, excitement or other emotions. Adrenaline is the hormone that prepares the body for action, and one of its effects is to increase heart rate (see page 373).

- Movement of the limbs, as in exercise. It is thought that stretch receptors in the muscles and tendons relay information to the brain, telling the cardiovascular centre that oxygen levels will soon fall and that carbon dioxide will soon build up. This initiates signals that increase heart rate and breathing rate.

- The level of respiratory gases in the blood, as described above.

- Blood pressure. When blood pressure gets too high, a fail-safe mechanism prevents any further increase in heart rate (see page 287).

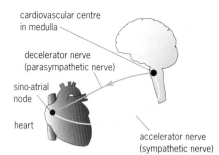

Fig 18.12 **The nerves that connect the cardiovascular centre in the brain to the heart. The accelerator and decelerator nerves are part of the autonomic nervous system and act** antagonistically **(they have opposite effects)**

As a general rule your maximum safe heart rate (in beats per minute) is 220 minus your age.

3 BLOOD VESSELS

There are three types of blood vessels: **arteries**, **veins** and **capillaries**. The structure of each is closely related to its function (see Fig 18.13 and Table 18.1).

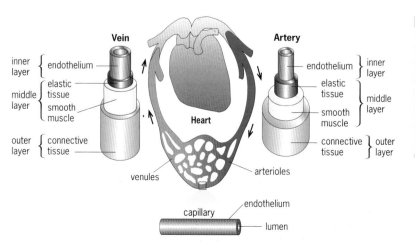

Fig 18.13 **The structure of blood vessels and their distribution in the circulatory system**

Arterial blood is under high pressure and surges occur every time the ventricles contract. The walls of the arteries are able to absorb and smooth out these pulse waves so that by the time blood reaches the arterioles it is flowing steadily.

Table 18.1 **A comparison of arteries, veins and capillaries**

■ tunica externa of collagen fibres

■ lumen

▨ tunica media of smooth muscle and elastic fibres

— tunica intima of endothelium

Arteries carry blood away from the heart towards other organs of the body. Arteries branch into smaller **arterioles**, which branch into tiny capillaries. These are **permeable** ('leaky') vessels whose walls are one cell thick to allow exchange of materials between blood and nearby cells. Blood flows from capillaries into **venules**, which drain into larger veins.

Vessel	Cross section	Direction of flow	Pressure	Oxygen content	Size of lumen	Presence of valves	Properties of wall
Arteries		away from heart	high	high*	relatively small	no	tough, powerful elastic recoil
Veins		back to heart†	low	low*	relatively large	yes	thin, distend easily
Capillaries	endothelial cell	through organs	medium	oxygen lost through wall	small, about the diameter of one red blood cell	no	one cell thick, permeable

* Except in pulmonary circulation, where oxygenated blood is carried in the veins and deoxygenated blood is carried in the arteries.

† Except portal veins, eg the hepatic portal vein, which carries blood **between** organs and not to or from the heart

As Table 18.1 shows, the walls of arteries and veins have the same three layers: the **tunica externa**, **tunica media** and **tunica intima** or **interna**. The relative thickness and composition of these layers varies according to the function of the vessel. The artery, which has to withstand pulses of high pressure, has an especially thick tunica media containing smooth muscle and elastic fibres. Veins have a thinner tunica media and a larger lumen to carry slower-flowing blood at low pressure. Valves prevent blood flowing backwards.

See question 3.

Capillary circulation

The circulatory system keeps all cells bathed in **tissue fluid**. The tissue fluid has a reasonably constant composition because permeable capillaries allow exchange of materials between the fluid and the blood.

> All living cells are surrounded by tissue fluid. This is also known as **interstitial** or **intercellular fluid**.

How tissue fluid is formed

Fig 18.14 shows what happens in living tissue. Blood from the arterioles is under high **hydrostatic pressure**. When blood enters the capillary, substances are forced out through the permeable capillary wall. The capillary walls act as filters and a proportion of all chemicals below a particular size are squeezed out, forming tissue fluid. The composition of tissue fluid is very similar to that of plasma, but tissue fluid lacks most of the large proteins that cannot pass through the capillary wall. Plasma proteins remain in the blood, where they play an important part in the drainage of tissue fluid.

> Hydrostatic pressure is physical pressure. The contraction of the ventricles of the heart creates hydrostatic pressure in the blood vessels.

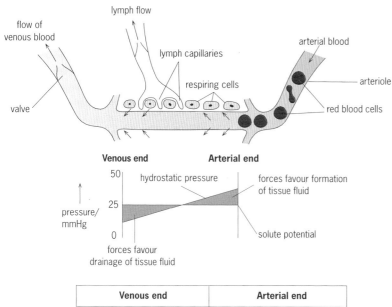

Venous end		Arterial end	
Hydrostatic pressure	= 16	Hydrostatic pressure	= 40
Solute potential	= 25	Solute potential	= 25
Net difference	= −9	Net difference	= +15
Tissue fluid drains into blood		Tissue fluid leaves blood	

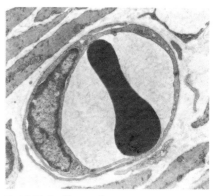

Fig 18.14 **The formation and drainage of tissue fluid. At the arterial end of the capillary, the high hydrostatic pressure forces water and small molecules out of the blood, providing the tissues with nutrients and oxygen. The hydrostatic pressure decreases as blood flows along the capillary. When it falls below the solute potential, fluid begins to drain back into the blood, taking with it wastes such as urea and carbon dioxide**

Fig 18.15 **Capillary walls are about 1 μm thick. Most capillaries have a lumen which is about the same diameter as the red cells that have to squeeze through them in single file as they unload their oxygen**

How tissue fluid returns to the blood

Two main forces act on the blood as is passes along the capillary:

- The hydrostatic pressure of the blood: this tends to force water and solutes *out* of the capillary.
- The solute potential of the blood: this tends to draw water *into* the capillary by osmosis.

As blood flows along a capillary, fluid passes to the tissues, and the volume and hydrostatic pressure of the blood in the capillary decreases. Since the solute potential created by the large plasma proteins is relatively constant, water begins to drain back into the blood when hydrostatic pressure falls below the solute potential (see Fig 18.14). Cell waste, such as urea, and substances that have been secreted into the tissue fluid, diffuse into the capillary (Fig 18.15), contributing to the composition of venous blood.

Blood flow in veins

Blood that drains into the venules is deoxygenated, under low pressure and contains many waste products and cell secretions. Several features allow blood to return to the heart:

- Valves in veins close to prevent backflow.
- Working muscles surrounding the veins squeeze blood along as they contract. The action of muscles, especially those in the legs, is so important that it has been called the 'secondary heart' or the 'venous pump' (Fig 18.16).
- Gravity helps blood to flow from organs 'above' the heart.
- The negative pressure created in the thorax during **inspiration** (breathing in) draws blood from surrounding veins.
- The action of the heart: after systole (contraction), the elastic walls of the chambers recoil, drawing blood in from the veins.

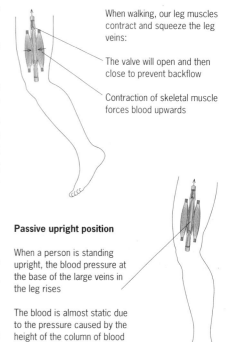

Walking

When walking, our leg muscles contract and squeeze the leg veins:

The valve will open and then close to prevent backflow

Contraction of skeletal muscle forces blood upwards

Passive upright position

When a person is standing upright, the blood pressure at the base of the large veins in the leg rises

The blood is almost static due to the pressure caused by the height of the column of blood above

Fig 18.16 **The action of the leg muscles squeezes blood along veins, and the valves ensure that flow can only be in one direction, This venous pump is very important to the circulation and regularly standing still for any length of time can lead to problems. Millions of people suffer from** varicose veins **or** haemorrhoids (piles): **damaged veins whose walls have been stretched by pools of accumulated blood**

Control of blood flow

The body often needs to alter blood flow to different areas, according to circumstances. Blood flow can be modified by **sphincters** and **shunt vessels**. A sphincter is a ring of muscle round a blood vessel that can contract, reducing the lumen size of the vessel and so reducing or preventing blood flow to a particular area. Sphincters can redirect blood into shunt vessels: these bypass a particular area. For instance, when we are cold, sphincters reduce blood flow to the skin surface, redirecting it along shunt vessels that keep the blood deeper in the body (see Chapter 25).

Blood flow can also be modified by altering the diameter of the vessel itself. Many blood vessels, particularly arterioles, contain smooth muscle fibres that can contract to reduce blood flow through the vessel: this is **vasoconstriction**. Blood flow increases when the muscle fibres relax: this is **vasodilation**. Blood pressure can be regulated by altering the degree of vasoconstriction or vasodilation.

?

B What effect do the following have on blood pressure?

(a) vasoconstriction, **(b)** vasodilation.

4 THE LYMPHATIC SYSTEM

The lymphatic system is part of the immune system (Chapter 26) and also part of the circulatory system: it returns to the heart the small amount of tissue fluid that cannot be returned by the veins.

We can think of the lymph system as an extra set of veins (look back at Fig 18.14 and see Fig 18.17). Flow through lymphatic vessels is slow, but important (Fig 18.18).

Lymphatic flow begins in the areas round blood capillaries (capillary beds), where small amounts of tissue fluid drain into tiny lymphatic capillaries (Fig 18.19). The walls of these vessels are more permeable than blood capillaries to lipids and large molecules such as proteins. This is why lymph contains a high proportion of these substances. Many cells secrete substances that are too large to enter the blood directly, and these substances pass into the general circulation via the lymphatics. The lymph capillaries drain into larger lymph vessels that look like thin, transparent veins. These vessels have valves to prevent backflow. Lymph contains no red blood cells, and so is pale and clear. All lymph vessels flow towards the upper thorax, passing through numerous lymph nodes that filter out bacteria and cell debris.

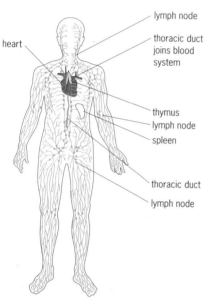

Fig 18.17 **The major lymphatic vessels in the human body. The lymph system drains into the blood system in the upper thorax. A large lymph vessel, the thoracic duct, connects with the subclavian vein (sub-clavicle = under collar bone). Here lymph mixes with blood before entering the vena cava on its way to the heart**

Fig 18.18 **Near right: This photograph shows someone suffering from** elephantiasis, **a condition caused by a nematode worm, *Wuchereria bancrofti*, which blocks lymphatic vessels. Lymphatic drainage takes about 120 cm³ of fluid out of the tissues every hour. When a lymphatic vessel is blocked, the limb quickly swells until the pressure interferes with normal blood flow.**

Fig 18.19 **Far right: Longitudinal section through a lymphatic capillary. The cells that form the capillary walls overlap slightly, forming tiny valves through which tissue fluid can flow. This picture shows such a valve.**

Lymph capillaries in the villi of the intestines are called lacteals, and are important in the drainage of lipids and high molecular weight substances (see page 249)

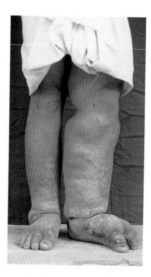

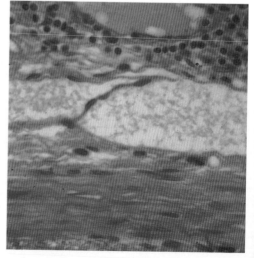

5 BLOOD PRESSURE

The term 'blood pressure' refers to the physical pressure of blood flowing through the arteries. This depends on the force created by the pumping of the heart, the volume of circulating blood and the size of the blood vessels. The body must keep blood pressure inside fairly strict limits. It must be high enough to force blood through all the capillaries so that all cells are well nourished, but not high enough to put the heart and blood vessels under strain: high blood pressure causes various health problems. Fig 18.20 shows how we measure blood pressure.

The diastolic pressure is used as an indicator for medical problems: a normal reading is between 60 and 80 mmHg, with anything over 90 regarded as high.

Have you ever stood up very quickly, and then felt faint or dizzy? This happens because the rapid change in posture has altered the body's blood distribution, causing a temporary lack of blood to the brain. Fortunately, body mechanisms quickly bring the blood flow back to normal.

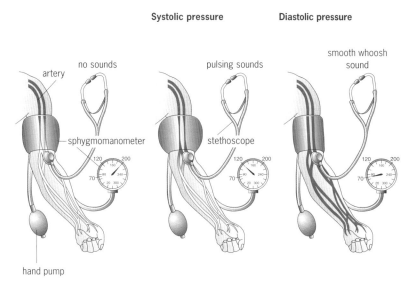

Fig 18.20 **Blood pressure is usually measured using a** sphygmomanometer.

An inflatable cuff is placed around the upper arm and inflated until all blood flow, in or out of the arm, stops. The blood flow in the brachial artery (at the elbow) is monitored using a stethoscope.

After inflation, there is no sound, but, as air escapes from the cuff, the pressure decreases until it falls just below that created by the heart as it contracts. At this point, blood is heard spurting through the constriction in the artery. This pressure is the systolic **value.**

Pressure in the cuff continues to drop until blood can be heard flowing constantly. This is the diastolic **value and represents the pressure to which arterial blood falls between beats**

See question 2.

Control of blood pressure

A negative feedback system keeps blood pressure inside safe limits. The pressure of the blood is detected by the **carotid sinus** (Fig 18.21), a small swelling in the carotid artery. When blood pressure gets too high, the walls of the sinus expand and **stretch receptors** in the carotid artery wall send information to the cardiovascular centre in the medulla. Signals from the medulla then lower heart rate and cause vasodilation, so lowering blood pressure.

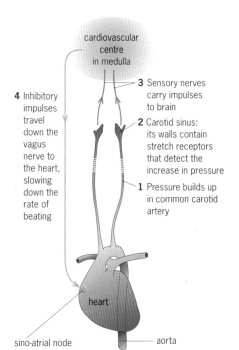

Fig 18.21 **Regulation of heart rate by the carotid sinus**

Many mechanisms work together to keep blood pressure constant. The size of blood vessels, the heart rate and the volume of circulating blood can all change to increase or decrease blood pressure, according to the needs of the body. The kidneys are particularly important because they control how much fluid we lose. A drop in blood volume and pressure causes more fluid to be retained in the blood and less to be lost in the urine (see Chapter 27).

C 'Blood pressure one twenty over seventy', shouts the doctor. What does this mean?

5 THE BLOOD

Blood is the fluid that flows through blood vessels, and the sight of blood is a sure sign that a vessel has ruptured. Losing large amounts of blood has serious consequences: death will result if the ruptured vessels are not sealed and blood volume is not rapidly returned to normal. In an emergency, when there is no time to find out an accident victim's blood group, he or she is given fluids called **plasma expanders**. These are isotonic fluids that contain no cells but which do increase blood volume.

Blood is a complex mixture containing cells, cell fragments and a range of dissolved molecules. It does the following:

- Transports materials. Blood transports food and oxygen to respiring tissues. It also takes carbon dioxide and waste products away from respiring cells to the various organs that remove them. It carries hormones from endocrine glands to target organs.

- Distributes heat. The blood helps to maintain a constant body temperature by distributing heat from metabolically active organs, such as the liver and working muscles, to the rest of the body (see Chapter 25).

- Provides pressure. Many organs of the body depend on the physical pressure of the blood to carry out their function. For example, filtration in the kidney, formation of tissue fluid and erection of the penis all depend on blood pressure.

- Acts as a buffer. The blood contains many proteins and ions that act as buffers, keeping the pH constant by 'mopping up' any excess acid or alkali. Haemoglobin in red blood cells is an important buffer (see Chapter 27).

- Defends the body against infection (see Chapter 26).

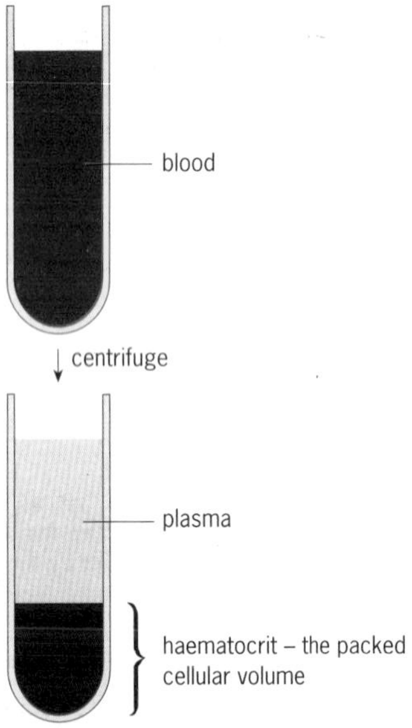

What's in blood?

Fig 18.22 shows that centrifuged blood separates out into two distinct layers: a **cellular portion** called the **haematocrit**, and the **plasma**. The cellular portion contains red cells, white cells and platelets.

Fig 18.22 **The percentage of blood taken up by cells is known as the haematocrit. A haematocrit of 39 would indicate that 39 per cent of the blood is composed of cells, mostly red cells. The average value for men is about 42 while women average 38. These values are affected by factors such as anaemia or high altitude**

D Suggest how **(a)** high altitude and **(b)** anaemia affect the haematocrit.

Plasma

Plasma is the fluid part of blood. The exact composition of the plasma varies greatly (Table 18.2). For instance, in the hepatic portal vein, which leads from the intestines to the liver, the plasma is much richer in dissolved foods such as sugars, vitamins and amino acids than the plasma in a vein in another part of the body.

Table 18.2 **The main constituents of plasma**

	Components	Function
Proteins:	albumin antibodies fibrinogen	osmotic balance immunity blood clotting
Salts:	sodium, potassium, chloride bicarbonate (hydrogencarbonate) calcium	osmotic balance, conduction of nerve impulses carriage of CO_2, buffering, blood clotting
Products of digestion:	glucose, amino acids, fatty acids, glycerol, vitamins	nourishment of cells
Hormones:	protein, eg insulin; lipid (steroids),eg testosterone	communication
Heat	distributed around body to maintain constant body temperature	
Oxygen	vital in aerobic cell respiration	
Waste products:	urea, CO_2	none: they must be removed by excretion (see Chapter 27)

Red blood cells

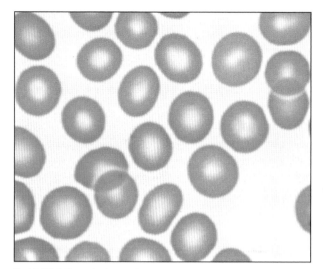

Fig 18.23(a) **If you look at a drop of blood under the microscope, the most obvious feature is the mass of red blood cells. Mammalian red cells are unique as they lack a nucleus**

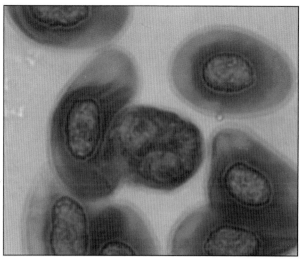

Fig 18.23(b) **Those of all other vertebrates, such as this example from a frog (an amphibian), are nucleated, oval in shape and generally much larger**

In this chapter we study red blood cells only (Fig 18.23): white cells and platelets are covered in Chapter 26.

Also known as **erythrocytes**, red cells are by far the most numerous in the blood. A single cubic millimetre of blood contains around 5 million, sometimes more. Mammalian red blood cells have no nucleus, but, as well as haemoglobin, they contain the enzymes and chemicals that allow them to function effectively. Red blood cells have one function: they carry the respiratory gases. The **haemoglobin (Hb)** they contain picks up oxygen in the lungs and swaps it for carbon dioxide in the respiring tissues.

Having the Hb inside red blood cells rather than in solution in the plasma gives several advantages:

- A much greater volume of Hb can be carried in cells than could be dissolved in plasma.

- Hb can be kept in a favourable chemical environment to allow faster loading and unloading of respiratory gases.

- Hb molecules of a particular age are kept together and can be replaced easily when old.

- Hb in cells does not increase the solute potential of the blood (free Hb would).

Red blood cells have a very regular shape, described as a biconvex disc (see Fig 18.23(c)). This shape is maintained by the cytoskeleton, a complex but flexible internal scaffolding made from protein fibres (see Chapter 1). When you think about what red blood cells do, it is easy to see why this is a good shape for them. They must have a large enough volume to carry useful amounts of oxygen but they also need a large enough surface area to load and unload it quickly. The biconvex shape is an efficient compromise between the maximum volume of a sphere and the maximum surface area of a flat disc.

Red blood cells are distorted when they have to squeeze through capillaries that often have lumens slightly smaller than the diameter of the red cell. Red cells rarely cause blockages because they have a flexible shape and a smooth membrane.

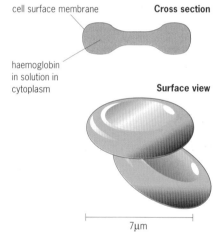

cell surface membrane   **Cross section**

haemoglobin in solution in cytoplasm

Surface view

7μm

Fig 18.23(c) **Mammalian blood cells are biconcave discs filled with haemoglobin**

Red blood cells are made in the bone marrow of the vertebrae, ribs and pelvis by specialised **stem cells**. The turnover of red cells is very rapid: we make about 20 million red cells per second! Red blood cells circulate for about 100 to 120 days before they are destroyed in the liver by the phagocytes called **Kupffer cells,** and also in the spleen.

Blood as a transport medium

Oxygen is carried in the blood in two ways: 98 per cent travels as **oxyhaemoglobin**, the oxygen-haemoglobin complex; the remaining 2 per cent is dissolved in the plasma. Carbon dioxide is carried in three ways: as HCO_3^- ions in the plasma (70 per cent), as **carboxyhaemoglobin** (23 per cent) and in simple solution in the plasma (7 per cent).

The structure of haemoglobin

The haemoglobin molecule is a conjugated protein with a relative molecular mass of about 64 500 kilodaltons. Each Hb molecule consists of four **globin** sub-units consisting of a polypeptide chain and a prosthetic group called **haem**. At the centre of each haem is an iron ion (Fe^{2+}), which combines with oxygen. There are four haem groups, so the overall equation for the reaction is:

$$Hb + 4O_2 \rightarrow HbO_8$$

haemoglobin + oxygen → oxyhaemoglobin

The polypeptide chains hold the haem groups in place and help to load oxygen. When the first haem group combines with an oxygen molecule, the haemoglobin molecule alters its shape so that the next haem group is exposed, making the loading of the subsequent oxygen molecules easier.

Haemoglobin in action

There are many substances that react readily with oxygen, but haemoglobin is one of the few that can combine with oxygen where it is abundant and then release it when the concentration falls. This property is illustrated in a graph called an **oxygen dissociation curve** (Fig 18.24). This is plotted by analysing the percentage of Hb saturated with oxygen at different concentrations of oxygen. The graph shows that Hb becomes fully saturated with oxygen at

Conjugated means 'joined'. A conjugated protein, such as haemoglobin, consists of a protein attached to a **prosthetic group** (a non-protein). The prosthetic group in haemoglobin contains iron, so we need a regular supply of iron in our diet to make this vital chemical.

The concentration of dissolved oxygen is often referred to as the **partial pressure**, or **tension**. The greater the concentration of dissolved oxygen, the higher the partial pressure or tension.

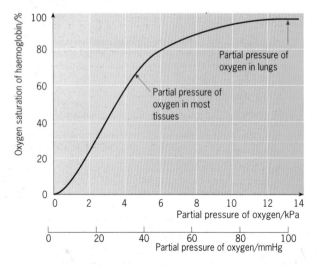

Fig 18.24 **The oxygen dissociation curve. At the oxygen tension found in the lungs, Hb becomes 97–99 per cent saturated with oxygen.**
Surprisingly, Hb releases only about 23 per cent of its oxygen in the respiring tissues, so the blood returning in veins is still about 75 per cent saturated. This suggests that three out of four haem groups are still bound to oxygen, which allows great flexibility: if a tissue such as a working muscle becomes particularly oxygen starved, the blood can release large amounts of extra oxygen

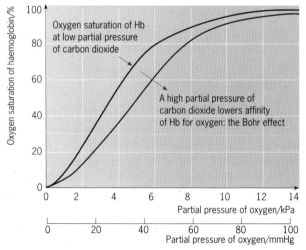

Fig 18.25 **If the oxygen dissociation curve is plotted at higher CO_2 concentrations, it moves to the right, showing that haemoglobin has a reduced affinity for oxygen: this is called the Bohr effect**

the concentrations of oxygen found in the lungs, and gives up a relatively large proportion of its oxygen in the lower oxygen concentrations that occur in the tissues.

The dissociation curve in Fig 18.24 was plotted using mixtures of gases in which the concentration of carbon dioxide was constant. Fig 18.25 shows that, at higher concentrations of CO_2, the curve moves to the right. This is a vital point: it shows that CO_2 lowers the **affinity** of Hb for oxygen. This means that when CO_2 concentration is higher, Hb does not hold on to its oxygen quite as well: Hb therefore tends to give up oxygen in areas of high CO_2 – such as in the respiring tissues that need it most. This lowering of affinity by CO_2 is called the **Bohr effect**, or the **Bohr shift**.

Fetal haemoglobin

Before birth, a mammalian fetus must obtain oxygen from its mother via the placenta. If fetal haemoglobin had the same affinity for oxygen as the mother's haemoglobin, no transfer of oxygen to the fetus would be possible. But, as Fig 18.26 shows, fetal haemoglobin has a higher affinity for oxygen than adult Hb. It can therefore pick up oxygen in the same conditions that cause the maternal blood to release it. After birth, the baby's body makes adult Hb, which gradually replaces the fetal version.

Unloading oxygen at the tissues

The series of events that lead to Hb unloading its oxygen in the respiring tissues is shown in Fig 18.27.

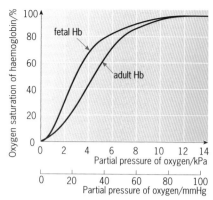

Fig 18.26 **The dissociation curve for fetal haemoglobin is to the left of the adult version. So, at the oxygen concentrations found at the placenta there is an efficient transfer of oxygen from mother to fetus**

See question 4.

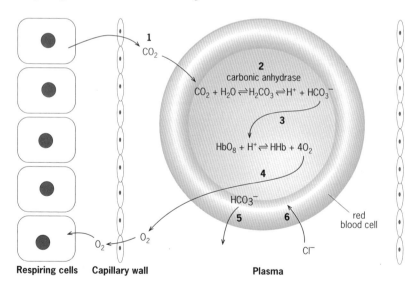

Respiring cells   Capillary wall   Plasma

Fig 18.27 **Unloading oxygen at the respiring tissues**

Step 1 **CO_2 diffuses from the respiring tissue, through the wall of the capillary and plasma, into the red blood cell**

Step 2 **Inside the cell, the enzyme** carbonic anhydrase **catalyses the conversion of CO_2 and water into** carbonic acid

Step 3 **The H^+ ions released by the carbonic acid destabilise the oxyhaemoglobin, causing each molecule to release its four O_2 molecules**

Step 4 **The oxygen is free to diffuse into the respiring tissues**

Step 5 **The accumulating HCO_3^- ions diffuse out into the plasma, leaving the inside of the red blood cell with a net positive charge**

Step 6 **To maintain a charge balance, Cl^- ions (the commonest negative ions in plasma) diffuse into the cell: this is the so-called** chloride shift

Myoglobin and muscles

Mammalian muscle contains the respiratory pigment, myoglobin (myo = muscle). The myoglobin molecule is one quarter of an Hb molecule: it consists of one polypeptide chain and one haem, and combines with one oxygen molecule. Myoglobin has a higher affinity for oxygen than haemoglobin and so it can pick up oxygen from haemoglobin and store it until the muscle becomes short of oxygen. When this happens (during exercise, for example), the oxygen tension in the muscle drops. Hb has no more oxygen to deliver and myoglobin then releases its oxygen.

The presence of myoglobin in muscles is associated with endurance. Chickens, not known for their flying ability, have flight muscles that contain little myoglobin: this is why chicken breast meat is pale. In contrast, ducks and geese are accomplished long-distance fliers, and so have dark meat because of the high levels of myoglobin.

PART II: PLANT TRANSPORT SYSTEMS

Materials move in living things in two main ways: by diffusion (along a concentration gradient), and by mass flow. (You can revise diffusion by turning to Chapter 5.) Large organisms need mass flow: examples of mass flow systems include blood systems, the breathing system of the lungs and the xylem and phloem transport system of plants.

6 WHY PLANTS NEED A TRANSPORT SYSTEM

In small water-living animals and plants, substances can move into cells by simple **diffusion**. The alga in Fig 18.28(a) is an example. Larger animals and land plants (Fig 18.28(b)) cannot rely on diffusion alone: they need a **mass flow system**. During the evolution of their transport system, plants have turned the presence of a cell wall to their advantage. The cellulose wall of a long cylindrical cell is, in effect, a ready-made reinforcing tube. Cells responsible for transport in plants have got rid of their cell contents and made holes in the end walls: columns of living cells have become long pipes like scaffolding poles.

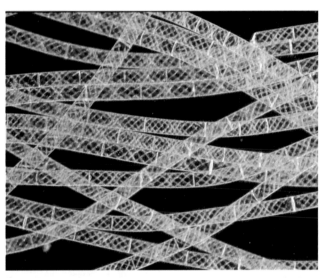

Fig 18.28(a) *Spirogyra* is a filamentous alga. All the cells photosynthesise and all take nutrients from the surrounding water, so this organisms does not need a transport system

Fig 18.28(b) Large plants need to transport substances over huge distances. Water absorbed by roots deep below ground has to rise over 80 metres to reach leaves at the top of this tree. Sugars made in the leaves have to move just as far to feed the growing tips of the roots

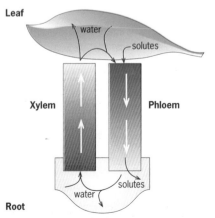

Fig 18.29 **This diagram shows the role of xylem and phloem in a land plant: materials move up xylem and down phloem**

7 TISSUES RESPONSIBLE FOR TRANSPORT IN PLANTS

Land plants have two transport systems to fulfil the need of all their cells for water and food (Fig 18.29):

- The **xylem** system moves water and dissolved minerals upwards from the roots.

- The **phloem** system moves food from photosynthetic tissues and food stores, to where it is needed. The majority of food moves downwards through the phloem, but it can also move upwards to feed the young growing tips of a plant.

In young shoots, the first or **primary** xylem and phloem are usually close to each other in a **vascular bundle**, as in Fig 18.30(a). A thin layer of **meristem tissue** (tissue that contains actively dividing cells) lies between the xylem and phloem. This is called the

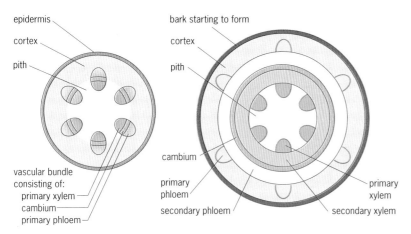

epidermis
cortex
pith

vascular bundle
consisting of:
  primary xylem
  cambium
  primary phloem

bark starting to form
cortex
pith

cambium
primary phloem
secondary phloem

primary xylem
secondary xylem

Fig 18.30 **The arrangement of xylem and phloem in: far left, a young green shoot, and in a slightly older woody shoot, near left**

F Look at the distribution of vascular tissue in Fig 18.30. Suggest why ringing a tree (removing the bark and the layers down to the xylem) can kill it.

cambium (see page 321). As the shoot gets older, the cambium in the bundle spreads out to form a complete ring, as shown in Fig. 18.30(b). **Secondary** xylem and phloem then form from the ring of cambium.

Xylem: the water-conducting tissue

Xylem tissue has evolved to carry water up the plant. Leaves can lose water quickly so, at times, the xylem must transport water very rapidly. The most efficient water-conducting tissue is the **xylem vessel** (Fig 18.31).

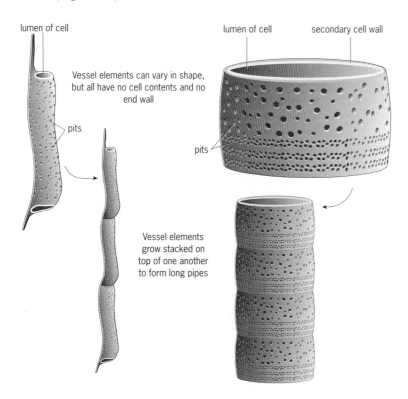

lumen of cell

Vessel elements can vary in shape, but all have no cell contents and no end wall

pits

lumen of cell

secondary cell wall

pits

Vessel elements grow stacked on top of one another to form long pipes

Fig 18.31 **Each cell in a xylem vessel is a** xylem vessel element. **Vessel elements grow stacked on top of each other to form long pipes**

Xylem vessels

Xylem vessels have two major adaptations. First, they have no cytoplasm, so there is nothing inside the cell to slow the passage of water. Second, they lose their end walls to allow water to flow easily from one cell to the next.

G Are xylem vessels living or dead?

Xylem vessels have other important features, and most of them are common to other types of xylem cell. A normal cellulose cell wall would be too thin to act as a water-conducting pipe. So xylem vessels have a **secondary cell wall** which is formed inside the primary wall. This secondary wall becomes impregnated with a variety of substances, especially **lignin**, which help to preserve the cell. The secondary cell wall never covers the cell completely: if it did, water could not pass from one cell to another. Microscopic connections called **pits**, seen in Fig 18.32, allow water to leak through. The properties of xylem that make it a good water conductor (strong, flexible, resistant to decay) also make it an ideal material for support (see the Extension box on this page).

Fig 18.32 **Pits develop in xylem vessel walls only where there is no secondary thickening. Pits allow water and dissolved substances to travel sideways to adjacent cells**

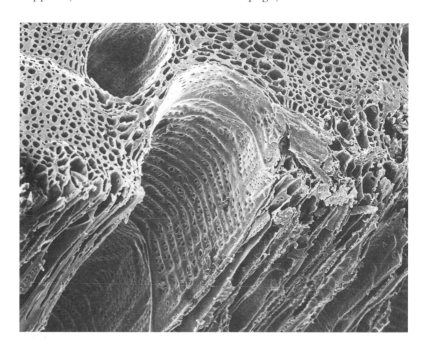

Conducting tissue and its role in support

Fig 18.33(a) **Left: Each year a new ring of xylem is created to transport water. Conducting cells made in spring are larger than those formed in summer. This seasonal variation in texture shows up as annual rings**

Fig 18.33(b) **Right: Individual vessels of different sizes can be seen more clearly in the micrograph**

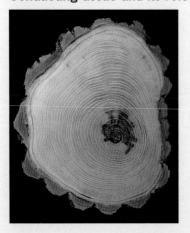

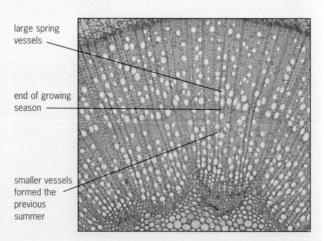

large spring vessels

end of growing season

smaller vessels formed the previous summer

In autumn, the conducting system of **deciduous** trees is affected in two ways: the trees lose their leaves and their roots grow. In spring, new sets of xylem and phloem cells must be created to connect the new leaves to the new root tips. In trees, last year's old 'piping' conducts very little but accumulates as, year by year, another layer forms (Fig 18.33). The accumulating xylem is known by its more common name of **wood**. Wood provides vital support to the growing tree.

A column of xylem vessel elements cannot have an open-ended vessel at either end or water would constantly be lost through the roots and air would enter the top of the vessels in the leaves (Fig 18.34). The finest branches of the xylem at the top and bottom of the plant end with a closed vessel called a **tracheid**. This is tightly surrounded by a small group of **parenchyma cells**. Together, these cells from a **bundle end** (Fig 18.35).

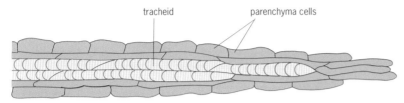

Fig 18.35 **Structure of a typical bundle end**

Other xylem cells

In among the xylem vessels are other cell types. These include:

- **Tracheids**. These also conduct water but are less well adapted for this function than xylem vessel elements. The principal difference between tracheids and xylem vessels is that tracheids do not have open ends: water must travel through the pits (see Fig 18.36).

- **Fibres**. These resemble heavily thickened tracheids. Fibres do not transport water, but help to support the plant.

- **Xylem parenchyma**. This acts as packing tissue and removes some minerals flowing in the vessels, enabling the plant to control the level of some minerals that reach the leaves (see Fig 18.37).

Fig 18.37 **Salt crystals on the outside of a plant that grows in the Florida mangroves. Living in salty conditions, it cannot prevent salt from getting into its roots. The xylem parenchyma removes some of the salt, which is then deposited in parts of the root where it does no harm, and then excreted. Water with a much lower concentration of salt continues up the plant to the leaves**

Phloem: the tissue that transports food

Organic substances like sugars cannot be pumped through dead, empty cells; they must be kept in living cytoplasm. **Sieve tube elements**, the parts of the phloem that transport sugars, are a design compromise. They have cytoplasm, so they are still alive, but they have lost their central vacuoles and most of their organelles, including the nucleus. These modifications allow material

Since xylem cells are dead they cannot divide. New xylem tissue must be produced from **meristem tissue** called the **cambium**. As a plant grows and thickens, more xylem is made from the cambium by division and differentiation.

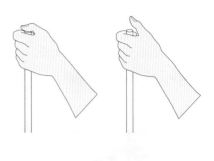

Fig 18.34 **If some liquid is in a thin open-ended tube and you put your thumb over the top, the liquid stays in the tube. If you take your thumb off, the liquid falls out of the bottom. This illustrates why the water-conducting vessels are closed by bundle ends at the top as well as the bottom**

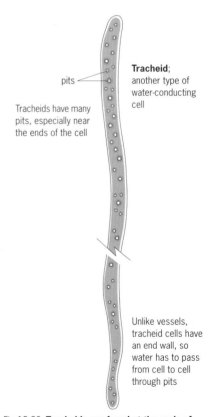

Tracheid; another type of water-conducting cell

pits

Tracheids have many pits, especially near the ends of the cell

Unlike vessels, tracheid cells have an end wall, so water has to pass from cell to cell through pits

Fig 18.36 **Tracheids are found at the ends of xylem vessels in flowering plants, the most advanced and numerous of the land pants. Simple land plants depend entirely on tracheids for water transport**

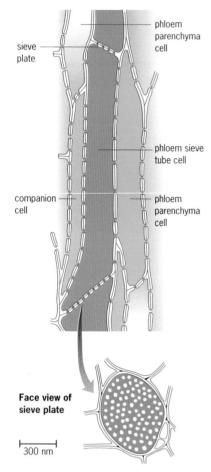

Fig 18.38 **The conducting cells of phloem need a companion cell to maintain their metabolism**

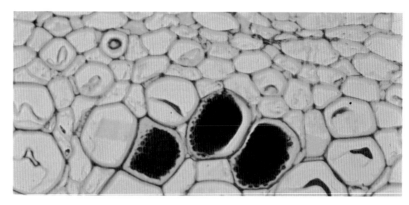

Fig 18.39 **Section through a stem showing three sieve plates. The pores in these sieve plates are large enough to allow fluid to move fairly freely from one cell to another**

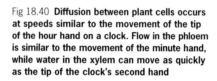

Fig 18.40 **Diffusion between plant cells occurs at speeds similar to the movement of the tip of the hour hand on a clock. Flow in the phloem is similar to the movement of the minute hand, while water in the xylem can move as quickly as the tip of the clock's second hand**

to flow more easily through the cell. Because their cytoplasm has degenerated, sieve cells cannot function without help. Each sieve cell needs its own **companion cell** to keep it alive (Fig 18.38).

The end walls of sieve tube elements are perforated by small holes, forming a **sieve plate** (Figs 18.38 and 18.39). The presence of cytoplasm and sieve plates means that flow through phloem is much slower than flow through the xylem (Fig 18.40). Fortunately, plants do not usually need food as urgently as they need water.

8 WHAT DRIVES PLANT TRANSPORT SYSTEMS?

Animals have a heart that pumps the blood round their circulatory systems. It may surprise you to learn that researchers are still not completely sure how plant transport systems work. In this section, we consider their favourite theories.

Water seems to travel upwards in the xylem by three different mechanisms:

● Capillary rise

● Transpiration pull

● Root pressure

Capillary rise in the xylem

Water tends to rise up inside narrow tubes by capillary action (Fig 18.41). This happens because water molecules are attracted to one another by a weak electrostatic force known as **hydrogen bonding**. The water molecules are also attracted to the wall of a glass capillary tube and, similarly, to the inside of a xylem cell. So, water rises up a capillary tube and up a xylem vessel, dragging up the other water molecules below it.

> The force of attraction between water molecules is called **cohesion**. The force of attraction between water molecules and the wall of a glass capillary or a xylem vessel is called **adhesion**.

Fig 18.41 **The principle of capillary rise**

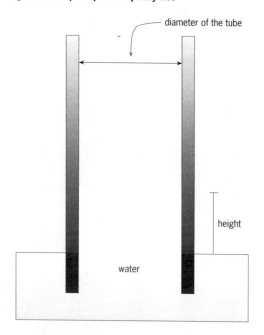

diameter of the tube

height

water

Factors that determine how far a liquid rises by capillarity include:

the surface tension
the density of the liquid
and g, the acceleration due to gravity

Water in glass behaves much the same way as it does in cellulose (the material of cell walls).

Examples:

Diameter of the tube μm (10^{-6} m)	Height water rises mm
(a) 0.01	3×10^6 (3 km)
2	15 300
20	153
(b) 80	38.3
(c) 200	15.3

(a) eg spaces within the cell wall
(b) eg typical tracheid
(c) eg typical vessel

These figures help to show that:

1 Water can travel long distances in the spaces within the cell wall (though the resistance is too great to supply the plant's needs in this way)

2 Water cannot travel very high just by capillarity inside the tracheids and vessels

Capillary rise may be important in very small plants but it could not possibly get water to the top of a tree. Two other mechanisms, transpiration pull and root pressure, make water move up the xylem of larger plants (see below).

Transpiration pull

We can imagine that xylem vessels form an unbroken tube running from the root right up to the leaf. Leaf cells are rarely more than three or four cells away from a vascular bundle, so every cell is close to a supply of water. As the water evaporates from the leaf (see Chapter 17), water from the xylem in the leaf replaces it. As water leaves the xylem, more is sucked up from below. Technically, a *negative gradient* in **hydrostatic pressure** forms, causing a column of water to rise up the plant. This is known as **transpiration pull**.

So, a column of water can withstand the tremendous strain caused by transpiration because of:

● the adhesion (which causes the water to 'stick' to the inside surfaces of the cells),

● the cohesion (which keeps the water molecules together).

If you measure the force needed to break the water column, you find that a column of water has a tensile strength nearly one tenth that of copper or aluminium wire of the same thickness.

Without the secondary wall in the xylem vessels, the huge tension in the column of water would make the cell collapse inwards. The faster the water evaporates in transpiration, the more rapidly it is drawn up the stem and the greater the tension inside the xylem. When it is warm or very windy, tension in the xylem can lead to a measurable decrease in the diameter of the shoot or branch. If the loss of water by evaporation through the leaf exceeds the rate at which water can be drawn up the xylem, the plant wilts.

?

H Why would water not pass up a tree if water had very little cohesion?

?

I Sometimes air can get into the vessels because of frost damage. Why might this harm a tree?

Root pressure

If a pot plant is cut just above the root, water seeps out of the cut stem (Fig 18.42). This is due neither to capillary rise nor to transpiration pull but to **root pressure**. Root pressure is a positive pressure caused by water entering the root from the soil.

How water enters the root

Near the root tip, see Fig 18.43, hairs grow from the outer layer of the **cortex**. Between the cells of the cortex are large air spaces that allow diffusion of gases and movement of water. The spaces (together with the gaps between the microfibrils of the cell walls) allow water to get in to the root by capillary action. So, water and dissolved minerals penetrate the root without actually entering any of the cells.

Fig 18.42 **The maximum height reached by the water in this tube is a measure of the force of the plant's root pressure**

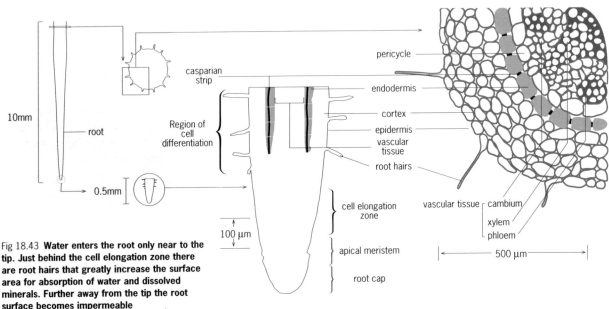

Fig 18.43 **Water enters the root only near to the tip. Just behind the cell elongation zone there are root hairs that greatly increase the surface area for absorption of water and dissolved minerals. Further away from the tip the root surface becomes impermeable**

Fig 18.44 **The whole of the root can be seen as two separate systems. The apoplast consists of the intercellular spaces and the spaces between the fibrils in the wall. The cytoplasm of all the living cells is interconnected by plasmodesmata forming the symplast**

These spaces between the cells and between the cellulose microfibrils are linked and form a continuous network of spaces called the **apoplast** (Fig 18.44). The cytoplasm of all the cells is also linked, via the plasmodesmata: this network is called the **symplast**. Water can move up through a root by both the **apoplastic pathway** and the **symplastic pathway**.

In the apoplast a single layer of cells, the **endodermis**, separates the **outer apoplast** from the **inner apoplast**. The cells of the

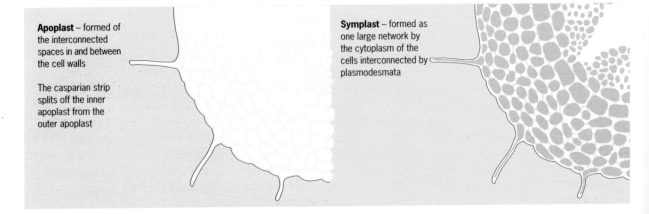

Apoplast – formed of the interconnected spaces in and between the cell walls

The casparian strip splits off the inner apoplast from the outer apoplast

Symplast – formed as one large network by the cytoplasm of the cells interconnected by plasmodesmata

endodermis have a band of wax round them called the **casparian strip**. This band stops water in the cortex getting past the endodermis by capillary action through the cellulose walls. Inside the endodermis is a layer of thin-walled cells called the pericycle. Water can only reach the pericycle by travelling across the endodermis by the symplast pathway. This means that the water goes through the endodermal cytoplasm, giving the plant some control over which substances enter the root. In this way, the endodermis with the casparian strip acts as a sort of 'passport point'.

From the cytoplasm of the pericycle, water and dissolved substances leak into the apoplast in the core of the root. Water passes easily from the apoplast into the transpiration stream of the xylem and then up the plant.

The mechanism that creates root pressure

Some dissolved substances move with the water that passes out of the cells of the pericycle and into the xylem, creating a solute potential in the xylem. This means that the xylem tends to absorb water from the surrounding tissues by osmosis. At night, when the rate of evaporation from the leaf is lower than in the day, transpiration often ceases. The osmotic pressure of the xylem sap remains the same, however, and continues to pull in water from neighbouring cells. This establishes a positive hydrostatic pressure in the xylem, and water is forced upwards. On humid nights, root pressure in some plants forces water out of the leaves, forming droplets at the edges. This rare effect is called **guttation**.

How food moves in the phloem

The phloem is less well understood than the xylem. The high pressures inside living cytoplasm make it difficult to study. What is obvious is that food moves up or down in the phloem, sometimes at the same time in the same plant. It is therefore convenient to think in terms of a **source**, where food is **loaded** into the phloem, and a **sink**, where food is **unloaded**. If we can suggest a mechanism to explain the flow between source and sink, it doesn't matter which particular organs are involved.

The cause of flow in the phloem

As long ago as 1930, Münch suggested that the movement in the phloem was caused by a gradient in hydrostatic pressure. His idea is simple and it is possible to build a laboratory model (Fig 18.45) based on the principles he suggests. However, Münch's hypothesis remains unproved.

See question 6.

J (a) Leaves sometimes drip water at night. Suggest a cause.

(b) Guttation usually happens at night when the stomata are closed causing water evaporation to slow down. What environmental conditions might cause guttation to happen during the day?

K For each of the following, say whether you think the food will be travelling up or down: **(a)** leaf to shoot tip, **(b)** leaf to root, **(c)** potato to new shoot, **(d)** seed to new shoot, **(e)** leaf to bulb.

L The energy for transpiration through the xylem comes from the Sun (warmth and wind causing evaporation). What is the energy source that drives movement in the phloem?

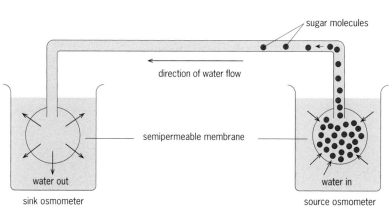

Fig 18.45 **This is a model of the Münch pressure-flow hypothesis to explain the movement of sugar in phloem. The concentrated (high osmotic pressure) sugar solution in the 'source' draws in water by osmosis. This causes a rise in hydrostatic pressure, forcing sugary water across to B. This model will only work for a little while, as there is no way of getting sugar out of the 'sink'**

✓

The osmotic pressure of a solution depends on the number of molecules, not the weight of material dissolved. When sugar is converted into starch, many molecules combine to make a few.

?

M Motorist who park under certain trees sometimes find their cars covered in sticky droplets. What might this be?

Fig 18.46 **An aphid inserts its stylet into a leaf in search of sugar-containing cells**

At the source (such as a leaf), the food is loaded into the phloem by **active transport**. There is evidence that the cell membranes of phloem cells have a sucrose **carrier** that can pump sugar into the sieve tube elements. As the sugar accumulates in the phloem, the osmotic pressure increases. This draws in water by osmosis from the surrounding less concentrated cells. The incoming water increases the hydrostatic pressure, forcing fluid (water, sugar and other dissolved substances) along the phloem in the direction of the sink. In the sink tissues, sugar is actively pumped out of the phloem to be used or converted into starch. When sugar becomes starch, the osmotic pressure goes down. As this happens, water is forced out under pressure into the surrounding tissues.

If Münch's pressure-flow theory is correct, the concentration in the phloem at the source should be higher than in the phloem near the sink. But this is hard to test. If you try to put a needle in the phloem to suck out the contents then the cells quickly clog up with protein. This seems to be a natural damage defence reaction, similar to a blood clot. Researchers have studied the contents of individual phloem using aphids. Aphids feed direct from the phloem using narrow **stylets** (mouthparts that form a tiny tube).

An aphid inserts its stylet into a sieve element and the pressure in the phloem forces the sugary water up the stylet into the aphid: it doesn't even have to suck. Researchers remove the body of an aphid, but leave in place the tiny stylet tube (Fig 18.46). The phloem contents continue to leak out and are collected and analysed. The concentration at source areas does indeed seem to be higher than at the sink areas.

FOOD AND DRUGS ON TAP

Aphids are not the only creatures that can plug into plants and steal their food. Many other insects do it, and now humans have learnt the same trick. Although it is possible to collect the sap from all sorts of plants, only a few species can withstand regular tapping: some of these are now economically important.

The sugar palm, for example, is tapped just as the flowers appear (Fig 18.47). Another well-known commercial plant *Hevea brasiliensis* produces rubber. Originally from Brazil, this plant is now cultivated all over the world. The white liquid latex that is used to make rubber comes not from phloem or xylem but from elongated cells called **laticifers**. Laticifers are stacked end to end and often lack end walls. They look as though they are adapted for transport, but their main purpose seems to be for defence against potential herbivores. Latex is rich in a wide variety of alkaloids that animals find unpalatable. Many of these, including morphine and codeine, are valuable in medicine.

Fig 18.47 **When the inflorescence at the top of this sugar palm is cut, over 10 litres of sugary sap may drip out of the sieve tubes of one plant. This is a graphic example of flow under pressure. The fluid collected may contain as much as 10 per cent sugar and is boiled to make molasses or refined sugar**

SUMMARY

Part I: Transport systems in animals

When you have read this part of the chapter, you should know and understand the following:

- Large organisms need a **circulatory system** to connect all living cells to the organs that exchange vital materials. Diffusion alone is not enough.

- Mammals have a double circulation: the **pulmonary circulation** to the lungs and the **systemic circulation** to the rest of the body.

- The mammalian **heart** is a four-chambered muscular pump made of **cardiac muscle**. This can contract powerfully and without fatigue.

- A single heartbeat, the **cardiac cycle**, consists of **atrial systole** (contraction) followed by **ventricular systole**, then **atrial** and **ventricular diastole** (relaxation).

- The heart is **myogenic**: the electrical impulses that control the cardiac cycle arise from the muscle itself.

- From the heart, blood passes into vessels in the following order: **arteries**, **arterioles**, **capillaries**, **venules**, **veins**.

- The **lymphatic system** acts as an extra set of veins, helping to drain fluid from the tissues back into the bloodstream.

- Blood consists of **red cells**, **white cells**, **platelets** and **plasma**.

- Red cells, **erythrocytes**, contain **haemoglobin**, a conjugated protein that can pick up oxygen where it is abundant (lungs) and release it where it is needed (the respiring tissues).

Part II: Transport systems in plants

When you have read this part of the chapter, you should know and understand the following:

- Over short distances, water and solutes pass between cells through **plasmodesmata**. Transport over longer distances requires **xylem** and **phloem**.

- Xylem conducts water and low concentrations of solutes, principally minerals, upwards from root to shoot.

- Xylem vessels are completely dead, have no cytoplasmic contents and lack end walls: they are ideally suited to the very rapid flow rates they have to sustain.

- Xylem flow is mainly due to **transpiration pull** and is driven by water evaporating from the leaf. A column of water in the plant is pulled upwards.

- Water can reach great heights because of the forces of **cohesion** and **adhesion**, coupled with the narrowness of the xylem vessels.

- Phloem carries smaller volumes of water containing high concentrations of solute (mainly sugar) from source to sink.

- Phloem transport, **translocation**, is due to loading and unloading solute and so requires metabolic energy.

- Phloem sieve elements have degenerated cytoplasmic contents and pores in the end walls. Their continuing metabolism depends upon the action of the adjacent **companion cells**. Sieve elements have pores in their end walls and transport of dissolved food occurs at moderate flow rates.

QUESTIONS

Part I questions

1 A mammalian heart continues to beat after it is deprived of nerve connections. The sino-atrial node (SAN) in a mammalian heart acts as the pacemaker. Cardiac muscle fibres serve as the conducting channels for a wave of excitation which spreads from the SAN to both atria. The wave eventually arrives at the base of the right atrium where there is a second node, the atrio-ventricular node (AVN). The wave reaches the AVN 0.045 seconds after leaving the SAN. A bundle of large muscle fibres (purkyne or purkinje fibres) arises from the AVN. This bundle of fibres (the bundle of His) forks into right and left branches which pass into the walls of the respective ventricles where fibres can be traced to all parts. A time delay of 0.12 seconds occurs at the AVN before the wave of excitation passes into the purkyne fibres. The right and left branches conduct the wave of excitation either side of the septum to the base of the ventricles, before spreading up the lateral walls to reach the top of the ventricles. The wave reaches the base of the ventricles 0.04 seconds after leaving the AVN and the top of the ventricles 0.08 seconds after leaving the AVN.

a) Complete the table below to summarise the timing of the passage of a single wave of excitation from origination to its arrival at the top of the ventricles.

Position	Time from origination of wave/s
SAN	0.00
AVN	
beginning of bundle of His	
base of ventricles	
top of ventricles	

b) State **one** piece of evidence that suggests that the origin of heart rhythm is an inherent property of the heart muscle itself.

c) Suggest, giving a reason, what the effect would be on heart rate of specifically cooling the SAN.

d) Suggest why it is important that there is a 0.12 second delay before the wave passes from the AVN into the Purkyne fibres.
  Experimental clamping of purkyne fibres squashes them and prevents them from conducting excitation waves.

e) Predict the effect of clamping the right branch of the purkyne fibres on ventricular systole.

f) Ventricular contraction starts at the base of the ventricles and not at the top. Suggest how this benefits the flow of blood through the heart.

[UCLES March 1995 Modular Sciences: Biology, Paper 2, q. 3]

2 A student cycled strenuously on an exercise bicycle for ten minutes. During this time, blood pressure measurements were taken. These measurements continued for a further five minutes after the exercise had finished. After the ten minutes cycling, the student was near exhaustion and felt very faint. The changes in the student's blood pressure are shown in Fig 18.Q2.

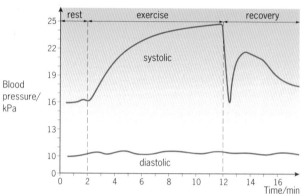

© David R Lamb, *Physiology of Exercise, Responses and Adaptations*, adapted by permission of MacMillan.

Fig 18.Q2

a) Distinguish between *systolic* and *diastolic* blood pressure.

b) With reference to Fig 18.Q2, describe the changes in systolic blood pressure between: **(i)** 2 and 12 minutes; **(ii)** 12 and 14 minutes.

c) Explain the change in systolic blood pressure during exercise.

d) With reference to blood pressure, suggest why the student felt faint immediately after exercise.
  Suggest how the student might avoid feeling faint at the end of strenuous exercise.

[UCLES Spring 1995 Modular Sciences: Biology, Paper 2, q. 4]

3 Fig 18.Q3 shows cross-sections of three different types of blood vessel. They are not drawn to the same scale.

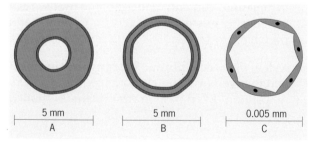

Fig 18.Q3

a) Identify blood vessels **A**, **B** and **C**.

b) State *two* ways in which vessel **A** is adapted for its functions.

[ULEAC 1996 Biology Specimen paper: Module Test B3, q. 1]

4 Fig 18.Q4 shows the effect of variations in the partial pressure of oxygen on the percentage saturation with oxygen of the haemoglobin from three different mammals. The pH and the partial pressure of carbon dioxide were the same in each case.

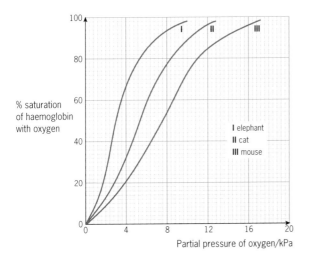

Fig 18.Q4

a) Why was it necessary for the pH to be the same in each case?

b) **(i)** Which type of haemoglobin has the greatest affinity for oxygen?
  (ii) Give evidence from the graph to support your answer.

c) How is the haemoglobin represented by curve III adapted to the high metabolic rate of an animal such as a mouse?

[AEB 1996 Biology: Specimen paper 1, q. 14]

Part II questions

5 Write an essay on 'The differences and similarities between transport systems in animals and plants'.

[AEB 1994 Biology: Specimen paper 2, q. 5]

6 The diameter of a branch from a small tree was measured over a 24-hour period. The results are shown in **Graph 1** of Fig 18.Q6.

a) Copy **Graphs 1** and **2** and sketch a curve on the pair of axes of **Graph 2** to show the rate of transpiration for the same 24-hour period as **Graph 1.**

b) The following have been used to explain the movement of water in xylem:

  A cohesion/tension
  B root pressure
  C capillarity

  (i) Which of these is best supported by the evidence in **Graph 1**?
  (ii) Explain your answer.

[AEB 1994 Biology: Specimen paper 1, q. 11.]

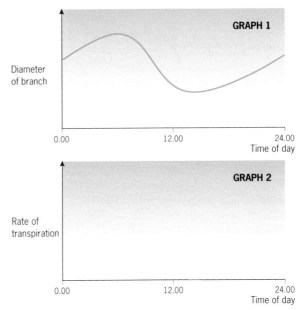

Fig 18.Q6

7 The roots of twelve bean plants were immersed in a nutrient solution containing radioactive phosphate (^{32}P) for 24 hours. The plants were then transferred to a non-radioactive nutrient solution and the leaves of six of the plants were covered with aluminium foil to exclude light.

  The graph of Fig 18.Q7 below shows the daily measurements of radioactivity in the leaves in counts per minute (cpm) of unit area.

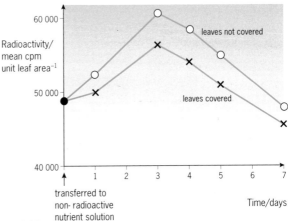

Fig 18.Q7

a) Explain how the ^{32}P passed from the nutrient solution to the leaves.

b) How would the amount of ^{32}P in the uncovered leaves have been affected if the plants had been exposed to moving air. Give a reason for your answer.

c) Suggest *one* reason for the difference in amount of radioactivity in the two sets of leaves.

d) **(i)** In which tissue would compounds containing ^{32}P have been transported out of the leaves?
  (ii) Outline an experiment which could be carried out to support your answer.

[ULEAC 1994 Specimen Biology Module Test B3, q. 7]

Assignment

PLANTS GET VIRUSES TOO

When a virus infects a plant it usually spreads slowly from cell to cell until it reaches the vascular bundles. Once in the vascular tissue, viruses spread rapidly to other parts of the plant. Tobacco mosaic virus, or TMV (Fig 18.A1), is a serious problem to commercial growers. Understanding how TMV infects plants may help scientists work out how to prevent the disease. In this Assignment we investigate whether TMV spreads through plants in the phloem or in the xylem.

1

a) Which type of vessel – xylem or phloem – would enable viruses to spread most easily through an infected plant and why?

b) Calculate the diameter of the TMV virus shown in Fig 18.A1(b).

c) Is TMV small enough to pass through the pores of the sieve plates in phloem? (You will need to refer to the diagram of the sieve plate in Fig 18.38.)

(a)

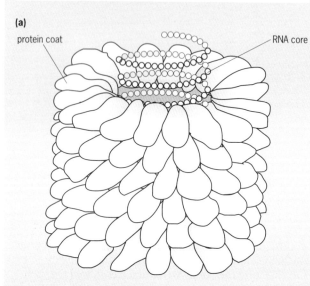

protein coat

RNA core

(b)

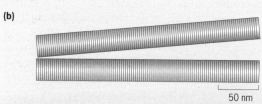

50 nm

Fig 18.A1 **Tobacco mosaic virus is transmitted by worm and insect pests. The virus consists of a long strand of RNA wrapped in a protein coat. It is small and infects plants easily, causing leaves to shrivel and die, as the leaves of the tomato plant have begun to do**

(a) **Diagram showing the detailed structure of TMV**

(b) **Diagram from an electron micrograph showing entire TMV particles**

(c) **Tomato plants showing symptoms of TMV infection**

(c)

Experimenting with the tobacco plant

So, theoretically, TMV virus could travel through a plant in either the xylem or phloem. Researchers used several approaches to learn more about long-distance movement of TMV in **Nicotiana**. Let's look at the experiments they did and the data they came up with.

The first thing the researchers did was to remove all the leaves from 8-week-old disease-free tobacco plants in two ways:

Twenty-four 8-week-old plants were used for group A. As Fig 18.A2 shows, the top of the shoot and all leaves were removed except the first leaf below the shoot.

In group B, 20 8-week-old plants were used. The top of the shoot was left on and all the leaves were removed except one vigorous leaf about 4 cm from the base of the stem (see also Fig 18.A2).

2

What is the main source and sink for sugars in the two different groups of plant after preparation? (Ignore the minor effect of the stem itself and of any axillary buds.)

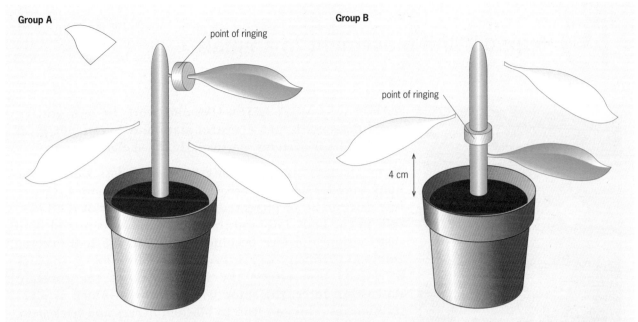

Fig 18.A2 **A summary of how the two groups of tobacco plants were prepared at the start of the investigation. The two groups were ringed differently. Ringing kills the outer cells. Half those from group A were ringed on the petiole of the upper leaf. The other 12 plants were left unringed. Half those from group B were ringed on the young part of the stem, above the leaf. The other 10 plants were left unringed**

The plants were then ringed by passing steam through a capillary tube held close to the tissues. The heat treatment lasted long enough to kill a ring of living cells, leaving only the xylem unaffected.

3

a) Why would the xylem cells be unaffected by heat treatment?
b) Identify which source–sink pathways are still working after the heat treatment.
c) Why do you think only half the plants were ringed in **(i)** group A, **(ii)** group B?

The remaining leaf on each of the plants was then inoculated with the virus by injection and kept in bright light and warm conditions. After a suitable period the plants were tested to see where the virus had spread. The results are shown in Table 18.A1.

Table 18.A1 **Number and parts of plants infected by TMV in each group**

Group A		Group B	
Number of plants infected by virus in leaf and stem		Number of plants becoming infected in their apical leaves	
	Leaf Stem		Apical leaf
ringed	12 0	ringed	0
not ringed	12 8	not ringed	10

4 What do these results indicate about the roles of the xylem and phloem in TMV movement?

If virus particles move freely in the phloem, then you might expect them to move at the same rate as sugars. To test this theory, researchers used radioactive sugars to study the rate of flow in the phloem. They found that sugar moved at about 1 cm per minute. Then they had to study how quickly TMV moved inside the plant.

Nicotiana plants were injected with the TMV virus and analysed after intervals of 18 and 24 hours. Researchers could not detect any movement in the first 18 hours.

5 What do you think the virus might be doing in this time? After 24 hours (ie 6 hours later) the virus could be detected 21 cm away from the point of infection.

6

a) Calculate the likely mean speed of movement after the virus has entered the phloem.
b) How does the speed of sugar translocation compare with the speed of movement of the virus?
c) Using your knowledge of sieve elements and their contents, suggest why the virus and the sugar might not move at the same speed.
d) As the virus spreads through the phloem, its numbers increase but it travels more slowly. Can you suggest why?

7 In nature, how do you think plants might become infected with viruses such as TMV?

19 Support and movement in animals

It may look fun, but in zero gravity the body is under a lot of stress

ASTRONAUTS MUST OFTEN LIVE and work in zero gravity. Weightlessness is not a restful state and it can put great stress on body systems, particularly the skeletal system.

Many changes occur in zero gravity. With less work to do, body muscles begin to break down and the bones start to lose calcium at an increased rate. The number of red blood cells in the body falls and there are dramatic shifts in fluid distribution: the face becomes puffy and the legs become thinner (astronauts call this condition 'bird's legs'). The lack of gravity also affects the spine. Without the constant downward force, the spine lengthens by as much as eight millimetres. This can lead to blocked nerves and back pain, and astronauts often lose their touch sensitivity.

NASA scientists are busy looking at forms of treatment and exercise that might prevent bone breakdown. Such information might also have practical benefits closer to home. By studying and finding ways to slow the accelerated changes produced in space, it may be possible to develop new treatments for bone diseases such as osteoporosis which occurs when bones become brittle due to a loss of calcium.

1 SUPPORT, MOVEMENT AND LOCOMOTION IN LIVING ORGANISMS

The bodies of most multicellular organisms need some form of support: body tissue is soft and collapsible and so needs to be held in a rigid frame. Land-living animals need more support than those that live in water.

Support in animals is usually provided by a **skeleton**. The skeletons of animals are of two basic types. **Fluid** or **hydrostatic skeletons** are used by a wide range of soft-bodied invertebrates such as earthworms, flatworms and jellyfish (Fig 19.1(a)). Arthropods, echinoderms and vertebrates have **rigid skeletons** (Fig 19.1(b) and (c)). (See Chapter 12 for more information about animal types.)

Fig 19.1(a) **Jellyfish have a hydrostatic skeleton**

Fig 19.1(b) **Many invertebrates including these lobsters have an exoskeleton**

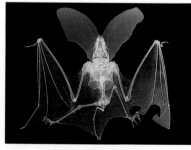

Fig 19.1(c) **This X-ray shows the bony endoskeleton of a long-eared bat**

A rigid skeleton supports important body organs, enabling them to work efficiently (they cannot operate properly when squashed), it protects the internal structures from damage and it allows the animal to **locomote**, or move from place to place.

Locomotion is different from movement. It refers not to a simple shifting of the body parts, but to the movement of the *whole* organism from place to place. Bacteria, for example, move from place to place using flagellae – Fig 19.2(b). Many protoctists (single-celled organisms) move around by altering their cell shape – Fig 19.2(c). But the most dramatic and most efficient forms of movement result from muscular contraction and occur in higher animals – Fig 19.2(a).

Movement is a feature of all living things and occurs at all levels. Atoms move within molecules and cell contents move inside cytoplasm. Plants move when they tilt their leaves towards the Sun and when their flowers open and close. The human rib cage moves up and down as we breathe.

Fig 19.2(a) **An adult cheetah can run at speeds approaching 120 kilometres per hour**

Fig 19.2(b) **The bacterium *Rhizobium* fixes nitrogen in the root nodules of plants such as peas and beans (see Chapter 37). It has a rigid flagellum which turns like a corkscrew, propelling the organism through its environment**

Getting enough food and finding a mate are major priorities for most animals. Animals need to locomote to do both. Animals may also move to escape from predators or harmful stimuli, to disperse themselves through a particular environment, to avoid competition from other individuals, or to move to conditions that are more favourable (into the shade on a hot day, for example).

In this chapter we first describe the features of different types of animal skeletons and then look at how each type of skeleton is used in locomotion.

Fig 19.2(c) **Amoebae locomote by forming pseudopodia**

2 HYDROSTATIC SKELETONS

A hydrostatic skeleton consists of an enclosed, fluid-filled cavity which provides support. Water is difficult to compress and so, when it fills body cavities such as the **coelom** of the earthworm (Fig 19.3), it provides a fluid mass against which muscles can contract.

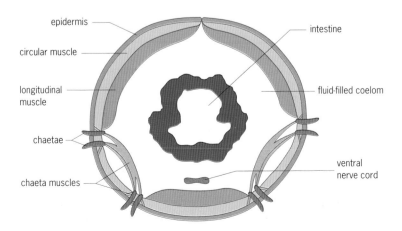

epidermis

circular muscle

longitudinal muscle

chaetae

chaeta muscles

intestine

fluid-filled coelom

ventral nerve cord

The supporting properties of water are also seen in animals that have other types of skeleton. In mammals, **amniotic fluid**, the fluid inside the uterus, surrounds and supports the developing fetus; and the **vitreous humour**, the jelly-like material inside the eyeball, supports the structures inside the eye.

Fig 19.3 **This cross-section through an earthworm's body clearly shows the coelom**

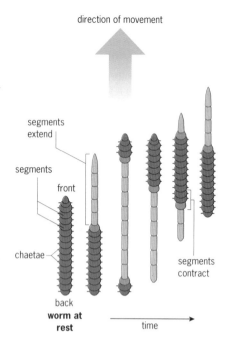

direction of movement

segments extend

segments

front

chaetae

back

worm at rest

segments contract

time

Fig 19.4 **Earthworm locomotion. Contraction of circular muscle causes the extension of groups of individual segments. Tough bristles (chaetae) then anchor the body and longitudinal muscle contraction shortens the body segments at the front, pulling the rest of the body forwards. Waves of contraction flow backwards away from the direction of movement**

Fig 19.5 **The typical form of locomotion seen in leeches and caterpillars. In the leech, anterior and posterior suckers anchor the ends of the animal. Contraction and relaxation of body muscles compresses fluids in the body spaces**

Locomotion in animals that have a hydrostatic skeleton

Animals such as earthworms, leeches, caterpillars, snails and slugs locomote using a hydrostatic skeleton. Earthworms are burrowing animals that need sustained and powerful locomotion. They can locomote because their muscles, attached to the body wall, pull against the fixed volume of fluid in the coelom.

The muscles are arranged in blocks and each enclosed segment of the earthworm's body is controlled separately. Contraction of the **longitudinal muscles,** which run along the length of each segment, shortens that segment. Contraction of the **circular muscles,** which run around the circumference of each segment, extends it.

To move forward, the earthworm extends the segments at the front of its body (Fig 19.4). This part of the body moves forwards, the front segments anchor themselves to the ground using bristles, or **chaetae**, and then these segments contract. The following segments 'catch up' with the front section by relaxing, anchoring and then contracting in the same way.

Other animals that have hydrostatic skeletons move in a very similar way. In the non-segmented animals, such as leeches, the whole body acts as a single hydrostatic system (Fig 19.5).

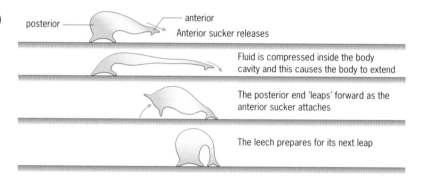

posterior — anterior
Anterior sucker releases

Fluid is compressed inside the body cavity and this causes the body to extend

The posterior end 'leaps' forward as the anterior sucker attaches

The leech prepares for its next leap

3 EXOSKELETONS

Exoskeletons, literally 'skeletons on the outside', occur mainly in arthropods such as crustaceans. Arthropods have a hardened outer 'skin' that protects and supports the body (Fig 19.6). This outer covering, the **cuticle**, is made up of a rigid polysaccharide compound

Fig 19.6 **An arthropod exoskeleton is a tough, protective covering. The outer cuticle of an arthropod skeleton consists of chitin and proteins arranged as a series of plates joined by thin membranes, rather like a medieval suit of armour protected by chain mail at the joints**

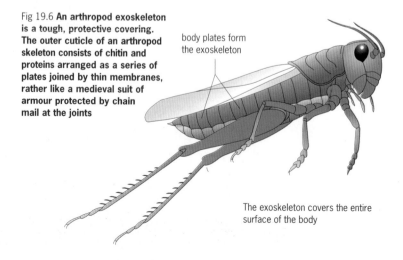

body plates form the exoskeleton

The exoskeleton covers the entire surface of the body

called **chitin** (see page 45). In land-living arthropod groups such as insects, spiders, centipedes and millipedes, this cuticle is covered by a layer of wax which provides a waterproof outer coat. The cuticle of crustaceans such as crabs and lobsters does not have a waxy layer (understandable in these mainly aquatic animals) and contains calcium salts for greater strength. The cuticle is not completely smooth. Internal folds add to its strength and ridges provide sites to which muscles attach.

How insects locomote using their exoskeleton

The arthropod leg (Fig 19.7) is basically a jointed hollow cylinder. Tendons attach muscles to projections on the inside of the exoskeleton and transmit the pulling force of the muscles to the hardened, chitinous cuticle. Areas of soft cuticle form flexible joints that allow the limbs to bend.

Insects walk like us. They bend and straighten each of their six legs in turn, while moving their body weight forwards. We can stand on one leg because we've got big feet, but an insect, which has small feet, needs to keep at least three of them on the ground. Muscles act across a joint working in pairs, one to **flex**, or bend the limb, the other to **extend**, or straighten it. Two sets of muscles that oppose each other in this way are called **antagonistic muscles**. Vertebrate muscles also work antagonistically (see page 312).

Most of the propulsive force needed for forward movement is generated by large muscles at the top of the leg.

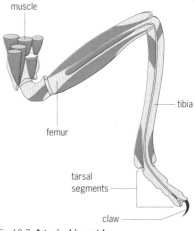
Fig 19.7 **A typical insect leg**

4 ENDOSKELETONS

An endoskeleton is a rigid structure that is inside the body, enclosed by soft tissues. Although we are more familiar with vertebrate endoskeletons, some invertebrates also have internal skeletons. The soft bodies of sponges are supported by sharp, mineral rods called **spicules**, and starfish and sea urchins are supported by small internal plates called **ossicles**.

CORAL: A SUBSTITUTE FOR BONE?

MENDING DAMAGED BONES is a major part of a surgeon's work, yet acceptable substitutes for bone are hard to find. It is possible to use bone from another part of the patient's body but only small amounts can be used. Artificial substitutes run the risk of being rejected by the body's immune system (see Chapter 26).

A few species of coral have a porous structure similar to that of bone. When grafted into the body, the honeycomb texture of coral provides the conditions necessary for new blood vessels to grow into it, and this promotes new bone growth. In addition, coral is tough, carries no risk of infection and is unlikely to be rejected by the body.

'Liquid bone' can also be made from coral. This is based on a calcium-rich solution and a phosphate-rich solution. The surgeon mixes the compounds in a little acid before applying it, rather like toothpaste, to the site of a fracture. Within 12 minutes it has solidified, and after one hour the 'new' bone is as hard as real bone.

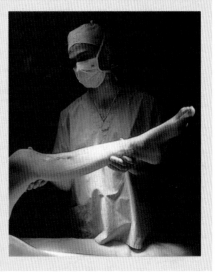

Fig 19.8 **Coral grafts have been used to repair shattered limbs, backbones and jaws. This broken leg is being examined before surgery**

Vertebrate skeletons are made up of **bone** and **cartilage**, which are specialised connective tissues (see Chapter 2). Some vertebrate skeletons, such as those of the cartilaginous fish (see Chapter 12), are made almost entirely of cartilage. In a developing vertebrate which has a bony skeleton, cartilage appears first and is then replaced by bone. We call this process **ossification**.

The human skeleton

The human skeleton has several functions:

● It acts as a framework that supports soft tissues.

● It allows free movement through the action of muscles across joints.

● It protects delicate organs and structures such as brain and lungs.

● It forms red blood cells in the red bone marrow.

● It stores and releases minerals from bone tissue.

Fig 19.9 shows the structure and features of the human skeleton.

?

A Look at Fig 19.9. Why are the bones of the legs thicker than the corresponding bones in the arms?

B Which structure is protected by the vertebral column or backbone?

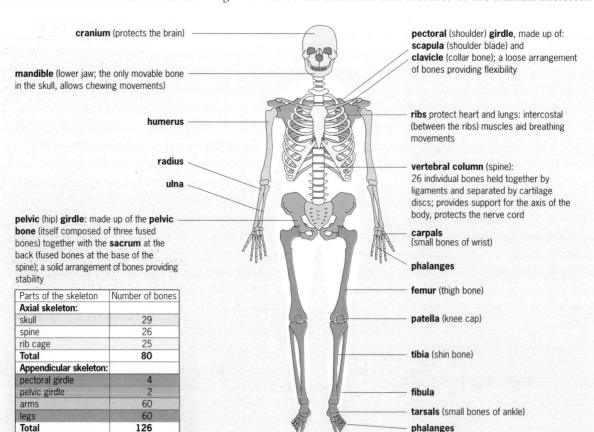

cranium (protects the brain)

mandible (lower jaw; the only movable bone in the skull, allows chewing movements)

humerus

radius

ulna

pelvic (hip) girdle: made up of the pelvic bone (itself composed of three fused bones) together with the sacrum at the back (fused bones at the base of the spine); a solid arrangement of bones providing stability

pectoral (shoulder) girdle, made up of: scapula (shoulder blade) and clavicle (collar bone); a loose arrangement of bones providing flexibility

ribs protect heart and lungs: intercostal (between the ribs) muscles aid breathing movements

vertebral column (spine): 26 individual bones held together by ligaments and separated by cartilage discs; provides support for the axis of the body, protects the nerve cord

carpals (small bones of wrist)

phalanges

femur (thigh bone)

patella (knee cap)

tibia (shin bone)

fibula

tarsals (small bones of ankle)

phalanges

Parts of the skeleton	Number of bones
Axial skeleton:	
skull	29
spine	26
rib cage	25
Total	**80**
Appendicular skeleton:	
pectoral girdle	4
pelvic girdle	2
arms	60
legs	60
Total	**126**

Fig 19.9 **The human skeleton contains a total of 206 bones. It is divided into two parts: an** axial skeleton **which comprises the** skull and vertebral column **and an** appendicular skeleton **which is made up of the** limbs and limb girdles.

All vertebrates show skeletal modifications related to their lifestyle. Since humans are bipedal (they walk on two legs), the hips and lower spine take most of the weight of the body and so the pelvic girdle (the hip) is larger and less flexible than the pectoral girdle (the shoulder)

The structure and properties of human bone

Bone is one of the hardest tissues in the human body and is second only to cartilage in its ability to absorb stress. It is made up of bone cells called **osteocytes** which sit in a bone **matrix**. The matrix consists of inorganic matter (mainly compounds of calcium and phosphorus) interwoven with **collagen fibres** (see page 54).

Bone needs to be tough and resilient. These properties are provided by two types of bone tissue: **compact bone** and **spongy bone** (Fig 19.10). **Compact bone** is deposited in sheets called **lamellae**

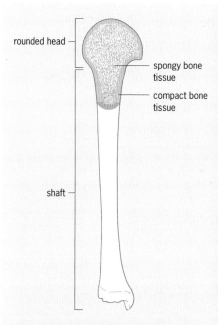

Fig 19.10(a) **A section through a human long bone which shows the distribution of compact and spongy bone**

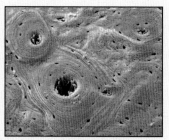

Fig 19.10(b) **Compact bone makes up the shaft of the long bone where strength and rigidity are important**

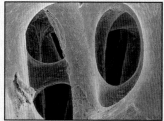

Fig 19.10(c) **Spongy bone forms the rounded head of long bones which absorbs the shocks and jolts of movement**

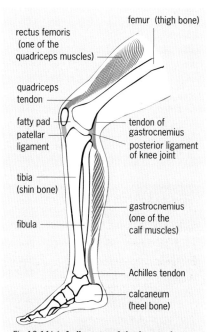

Fig 19.11(a) **A diagram of the human leg showing the position of the main ligaments (green) and tendons (blue)**

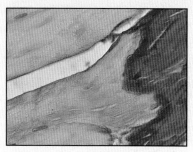

Fig 19.11(b) **This micrograph shows a slice of tendon (yellow) connected to a piece of muscle (red). Tendons and ligaments look similar but their composition is different: tendons contain mainly** collagen fibres **(see page 54); ligaments are made up of fibres formed from the protein** elastin

arranged as cylinders inside cylinders. Nerves and blood vessels run in a central canal. The compacted structure of the lamellae gives immense strength. The lamellae of **spongy bone** are arranged in a criss-cross pattern, forming a spongy honeycomb. This structure has excellent 'shock-absorbing' properties.

Joints

As in other vertebrates, the human skeleton is jointed. A joint is simply a point in the body where two bones meet. Most joints are **movable** but some, such as the bones making up the **skull**, the **sacrum** (base of the spine) and the **pelvis**, are **immovable**, or **fused**. Elastic **ligaments** bind bones together while tough inelastic **tendons** attach muscles to bone (Fig 19.11). The human knee joint is illustrated in Fig 19.12. Internally, the joint is lubricated by a viscous fluid, **synovial fluid**, secreted by the **synovial membrane**, the membrane that lines the joint.

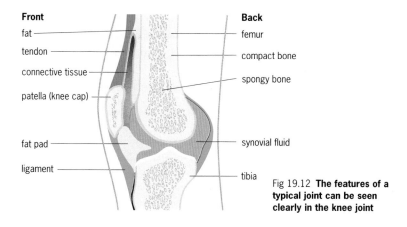

Fig 19.12 **The features of a typical joint can be seen clearly in the knee joint**

C Why are ligaments elastic?

D Why does the tendon need to be inelastic?

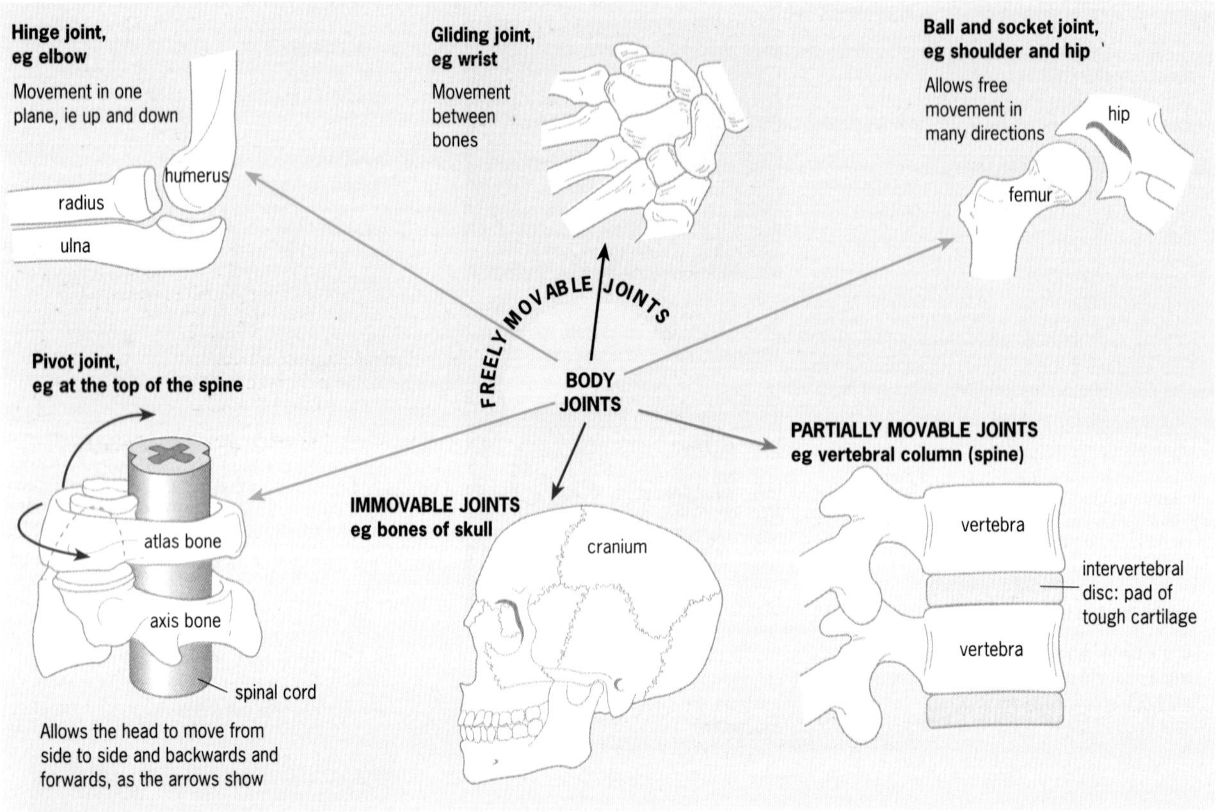

**Hinge joint,
eg elbow**

Movement in one
plane, ie up and down

humerus
radius
ulna

**Gliding joint,
eg wrist**

Movement
between
bones

**Ball and socket joint,
eg shoulder and hip**

Allows free
movement in
many directions

hip
femur

FREELY MOVABLE JOINTS

**BODY
JOINTS**

**Pivot joint,
eg at the top of the spine**

atlas bone

axis bone

spinal cord

Allows the head to move from
side to side and backwards and
forwards, as the arrows show

**IMMOVABLE JOINTS
eg bones of skull**

cranium

**PARTIALLY MOVABLE JOINTS
eg vertebral column (spine)**

vertebra

intervertebral
disc: pad of
tough cartilage

vertebra

Fig 19.13 **There are many different types of
joint in the vertebrate skeleton. They are
classified both by their shape and by their
mobility**

The ligaments and the synovial membrane together form a **joint
capsule** which surrounds the end of the bones. Different types of
joint allow for different kinds of movement. Some of these are
described in Fig 19.13.

How the human body moves

A simple action, like tapping your finger on the table, involves the
skeletal system, the muscular system and the nervous system. This
voluntary action begins as a stimulus in the cerebrum of the brain.
Motor nerves carry impulses to muscles and cause them to contract,
pulling on bones through the tough inelastic tendons.

Muscles can only pull, or contract; they cannot push. This means
that muscles rarely act alone: most of the time they work in groups.
Contraction of a muscle moves a bone at a joint, but a second
(**antagonistic**) muscle returns the bone to its original position. Take
movement in the arm, for example. The biceps muscle bends or
flexes the arm; the triceps muscle straightens or **extends** the arm.

We all know that machines can help us to lift heavy loads using
levers. The bones in our bodies also act as **levers**. The principle of
leverage allows a muscle to overcome the **resistance** supplied by a
weight. This happens in three ways:

- An arm bends because a force is applied between the **pivot** (the
  elbow) and the resistance (the weight to be lifted) (Fig 19.14(a)).

- We stand on tiptoes by pivoting the toes on the ground and
  using our calf muscles to raise the weight at the ankles. This is
  the wheelbarrow principle (Fig 19.14(b)).

- When standing or sitting, we can raise our head because muscles
  at the back of the neck pull the skull over the pivot at the top of
  the spine. This is the seesaw principle (Fig 19.14(c)).

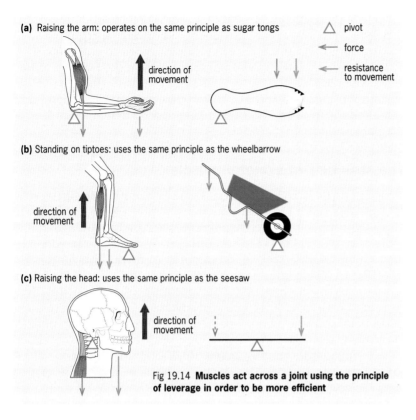

(a) Raising the arm: operates on the same principle as sugar tongs

△ pivot

← force

← resistance to movement

direction of movement

(b) Standing on tiptoes: uses the same principle as the wheelbarrow

direction of movement

(c) Raising the head: uses the same principle as the seesaw

direction of movement

Fig 19.14 **Muscles act across a joint using the principle of leverage in order to be more efficient**

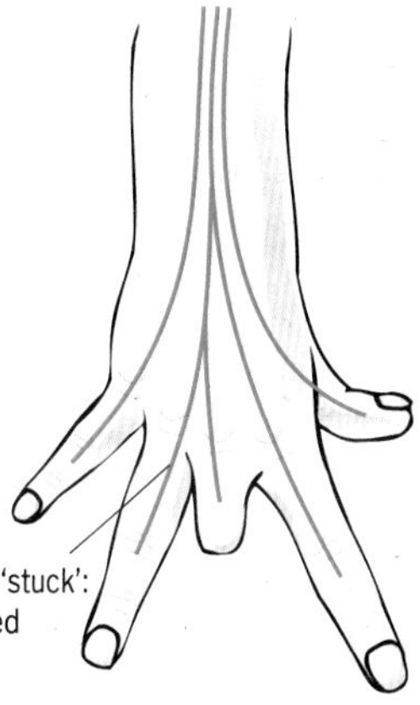

The ring finger feels 'stuck': this is due to a shared tendon connection

Fig 19.15 **The stuck finger investigation**

The effect of muscles working in groups can be demonstrated using the 'stuck finger' test (Fig 19.15). Each of your fingers has its own tendon connecting it to muscles in the forearm. To carry out this investigation curl up your middle finger and place the other four fingers on a hard surface as shown. Now lift up the other fingers one by one. You will find that you can lift all of your fingers except the ring finger. This is because these two fingers share the same tendon connection.

See question 1.

5 MUSCLES AND MOVEMENT

The bones of the human skeleton provide a basic system of levers and joints that makes the skeleton potentially movable, but neither levers nor joints can move without muscles. Skeletal muscle (also called striped or striated muscle) provides the main source of power for human locomotion. In this section we look at the structure and function of skeletal muscle.

The properties of skeletal muscle

Like other sorts of muscle, skeletal muscle has three basic properties:

- **Excitability:** it can receive and respond to a stimulus.
- **Extensibility:** it can be stretched or can contract.
- **Elasticity:** it can return to its original shape after it has contracted.

Skeletal muscle is also described as **voluntary muscle** because we can contract it when we want to. Smooth muscle and cardiac muscle is **involuntary muscle**: we cannot consciously control its contraction.

There are three types of muscle. **Skeletal muscles** are responsible for whole body movement. **Smooth muscle** is responsible for automatic movements such as those involved in peristalsis. **Cardiac muscle** is found only in the heart. The properties of the different muscle types are described in Chapter 2 and cardiac muscle is covered in detail in Chapter 18.

E What type of muscle fibre (skeletal, cardiac or smooth) is found in the following structures?

(a) stomach

(b) aorta

(c) biceps muscle

(d) left ventricle of heart

(e) face

Skeletal muscles contract and relax, moving bones at joints in the skeleton. This allows for movements like walking, running and waving, and fine movements like those needed to use tools. The action of skeletal muscles also helps to maintain the position of the body when we are sitting or standing, even though there is no obvious movement. The contractions of skeletal muscle also produce heat, and much of this is used to keep the body temperature at a normal level (see Chapter 25).

The structure of skeletal muscle

Skeletal muscle is made of specialised cells, the **muscle fibres** (Fig 19.16). Each cell is long and thin and contains several nuclei. Each fibre is surrounded by a cell membrane called the **sarcolemma**; the cytoplasm of the fibre is the **sarcoplasm**. The sarcoplasm contains many mitochondria and a large number of thread-like fibres, the **myofibrils**. These run along the length of the muscle fibre and are parallel to one another. The **sarcoplasmic reticulum** that surrounds each myofibril consists of a network of tubes that contain calcium ions. These play a major role in bringing about muscle movement (see page 315).

Each myofibril contains many thick and thin threads called **myofilaments** (Fig 19.17). Thick myofilaments are made of the large-molecule protein **myosin**, the thin myofilaments are composed of a smaller protein, **actin**. Actin myofilaments also contain two other proteins: **troponin** and **tropomyosin**.

See question 2.

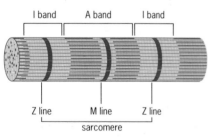

I band   A band   I band

Z line   M line   Z line
sarcomere

Fig 19.17 **The ultrastructure of the myofibril.** Each myofibril contains myosin and actin strands. Where the two overlap, the myofibril looks dark: this region is called the A band. In the regions where only actin is present, the myofibril looks lighter: this region is called the I band. The centre of each I band is the dark Z line. The centre of each A band is the dark M line. A bands and I bands alternate and so give skeletal muscle its striped appearance.

The sarcomere, the basic unit of muscle contraction, is the section of muscle fibre between one Z line and the next

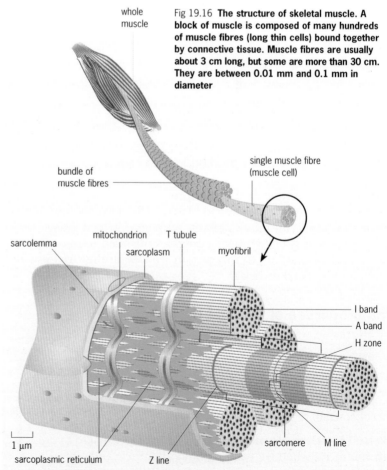

whole muscle

Fig 19.16 **The structure of skeletal muscle. A block of muscle is composed of many hundreds of muscle fibres (long thin cells) bound together by connective tissue. Muscle fibres are usually about 3 cm long, but some are more than 30 cm. They are between 0.01 mm and 0.1 mm in diameter**

bundle of muscle fibres

single muscle fibre (muscle cell)

sarcolemma   mitochondrion   T tubule
sarcoplasm   myofibril

I band
A band
H zone

1 µm
sarcoplasmic reticulum   Z line   sarcomere   M line

The sliding filament theory of muscle contraction

The relationship between the ultra-structure of skeletal muscle and its ability to bring about movement is explained by the **sliding filament theory of muscle contraction** (Fig 19.18).

As we have seen, the actin and myosin filaments lie parallel to each other. When the muscle is at rest, the 'heads' that stick out from the myosin filament cannot attach to the actin filaments that are alongside them. Movement occurs when the heads move towards the actin filament, attach to it and act as hooks to pull the actin myofilaments past them. As the actin myofilaments from opposite ends of the sarcomere are pulled towards each other, the sarcomere becomes shorter. This happens along the whole length of each myofibril and causes the muscle fibre to contract. Later, as the actin myofilaments become detached from the myosin heads, both filaments return to their original position and the muscle relaxes.

The role of calcium ions and ATP in muscle contraction

Contraction of skeletal muscle does not simply occur 'out of the blue'. It must be initiated by a stimulus in the form of a nerve impulse (see Chapter 21) which arrives at the muscle fibre from the nervous system. So when a muscle contracts and then relaxes, the following events take place (refer again to Fig 19.18):

● The brain makes a decision that the body is to move. Nerve impulses travel from the brain, along the spinal cord and through the motor nerves, towards the muscle.

● The electrical signal reaches the **motor end plate** at a specialised **synapse** (the junction between the nerve and the muscle). The arriving signal causes the release of acetylcholine (see Chapter 21, page 345). This chemical is a neuro-transmitter: it travels across the gap between the end of the nerve and the muscle, and binds to receptors on the muscle fibre membrane.

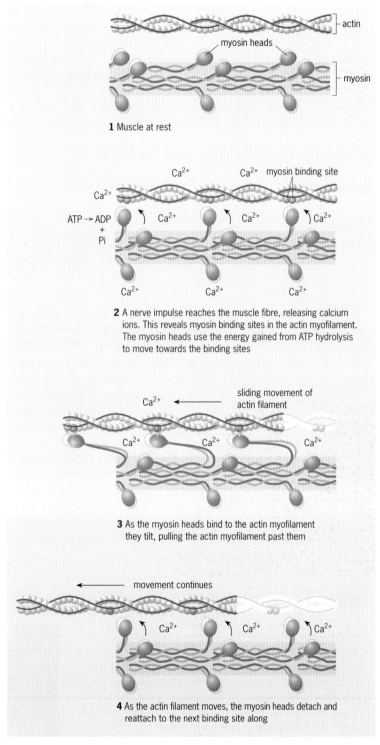

1 Muscle at rest

2 A nerve impulse reaches the muscle fibre, releasing calcium ions. This reveals myosin binding sites in the actin myofilament. The myosin heads use the energy gained from ATP hydrolysis to move towards the binding sites

3 As the myosin heads bind to the actin myofilament they tilt, pulling the actin myofilament past them

4 As the actin filament moves, the myosin heads detach and reattach to the next binding site along

Fig 19.18 **The sliding filament theory to explain muscle contraction**

See question 3.

- This binding causes an electrical change in the muscle fibre which triggers the release of calcium ions from the sarcoplasmic reticulum into the myofilament.

- Calcium ions bind to troponin and tropomyosin, the proteins that are closely associated with actin filaments. This changes the three-dimensional shape of the troponin–tropomyosin–actin complex, revealing parts of the actin filament that were previously hidden. These are the active sites to which the heads of the myosin filaments will attach.

- The calcium ions also act directly on myosin, activating it so that it splits ATP into ADP and inorganic phosphate, releasing energy. This energy is used to move the heads of the myosin filaments towards the newly exposed binding sites on the actin filaments.

- As cross-bridges form between the actin and myosin filaments, the myosin heads tilt, pulling the actin myofilaments past them. As the myofilaments slide, the heads detach from one site and attach to the next. As many as 100 such attachments can occur every second.

- As the actin myofilaments from opposite ends of the sarcomere move towards each other, the muscle fibre contracts.

- When the stimulation of the muscle fibre stops (when the brain decides the body should stop moving), the calcium ions are pumped back into the sarcoplasmic reticulum.

- As the level of calcium ions falls, troponin and tropomyosin move back to their original positions, blocking once again the active sites available for myosin attachment. The myosin heads can no longer attach to the actin filament, and both sets of filament return to their original position. The sarcomeres return to their normal length, and the muscle relaxes.

Fast and slow muscle fibres

A nerve impulse is the trigger for muscle contraction, but the length of time a contraction lasts depends on how long calcium ions remain in the sarcoplasm. This time is different in different types of skeletal muscle fibres. **Fast twitch fibres** and **slow twitch fibres** are classified on the basis of their contraction times.

TIME OF DEATH

WHEN A DEAD body is discovered, a pathologist investigates and tries to estimate how long the person has been dead. This information can be very important in a murder enquiry. The pathologist assesses the time of death partly by taking the internal temperature, to see how much the body has cooled from the normal body temperature of 37 °C, and partly by looking at the state of the muscles.

When death occurs, muscle protein starts to coagulate. Muscle tissue loses its elasticity and the body stiffens. As there is no ATP, cross-bridges that formed between actin and myosin filaments before death remain in place, stopping the muscles from relaxing. This condition, **rigor mortis**, happens in all body muscles. It appears about four hours after death and lasts about 24 hours. After this time, muscle proteins are destroyed by enzymes within the cells and so rigor mortis disappears.

Fig 19.19 **All suspicious deaths are investigated by a police pathologist: one of the first things he or she does is estimate the time of death**

There are three important differences between fast and slow fibres:

- Slow twitch fibres have less sarcoplasmic reticulum than fast twitch fibres. This means that calcium ions remain in their sarcoplasm longer.

- Slow twitch fibres have more mitochondria which provide ATP for *sustained* contraction.

- Slow twitch fibres have significantly more myoglobin than fast twitch fibres. Myoglobin has a higher affinity for oxygen than the haemoglobin in blood and so is particularly efficient at extracting oxygen from the blood.

The overall result of these differences is that slow twitch fibres are responsible for sustained muscle contraction, such as that which maintains body posture, while fast twitch fibres are responsible for the shorter-acting but more powerful contractions important in locomotion (see the Assignment on page 319).

SUMMARY

After completing this chapter you should know :

▓ There are three basic types of skeleton: **hydrostatic skeletons**, **endoskeletons** and **exoskeletons**. Invertebrates such as earthworms have hydrostatic skeletons, arthropods and crustaceans have exoskeletons, and vertebrates, including humans, have endoskeletons.

▓ Skeletons support and protect the body and allow animals to **locomote** (move from place to place).

▓ The human skeleton is made of **bones**, **tendons and ligaments**. The human skeleton is movable because different bones meet at **joints**, across which **muscles** are attached. Because muscles can pull but not push, muscles often work in pairs called **antagonistic pairs**.

▓ Muscles are **excitable**, **extensible** and **elastic**. **Skeletal muscle**, also called **striped**, **striated** or **voluntary muscle**, is responsible for human locomotion, maintaining body posture and generating body heat.

▓ Skeletal muscle is made of **muscle fibres** which contain bunches of **myofibrils** which in turn contain **myofilaments**. Thick **myosin** filaments lie alongside thin **actin** filaments.

▓ Muscle contraction can be explained at the molecular level by the **sliding filament hypothesis**.

QUESTIONS

1 Fig 19.Q1 shows the skeleton of a front leg from an armadillo.

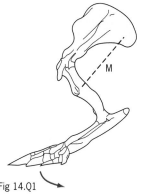

Fig 14.Q1

a) Draw the diagram and add a straight line to represent a muscle which would cause the lower leg to move in the direction shown by the arrow.

b) Suggest **two** features of the muscle **M** which would indicate that this mammal has powerful limbs suited to digging.

[AEB summer 1996 Biology: Paper 1, q.4]

2 Study the electron micrograph of striated muscle tissue shown below and then answer the following questions.

Magnification: × 13500

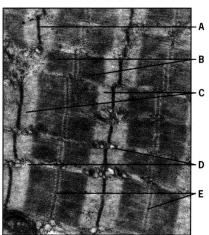

a) What name or term is given to the structures labelled

(i) **A**,

(ii) **C**,

(iii) **D**.

b)　**(i)** Measure, in millimetres, the distance between two successive **Z** lines. Repeat this twice at **two** other positions on the micrograph and note your three answers. Then calculate the mean value.

　(ii) Calculate the **actual** mean length of a sarcomere unit in this muscle tissue. (Show your working.)

c) Make an approximate measurement (in millimetres) of the (relaxed) length of the biceps muscle in your arm and note the result.

If a single striated muscle fibre (cell) stretches the complete length of **this** muscle, how many sarcomere units are required to reach from end to end of the muscle cell? (Show your working.)

d)　**(i)** What name is given to the structures that attach your biceps muscle to the bones?

　(ii) Name the two bones onto which the biceps is **inserted**.

e) Using only the letters given on the micrograph, identify the following:

　(i) a region where cross bridges are situated;

　(ii) a region that will decrease markedly in length on contraction of the muscle;

　(iii) a region composed primarily of actin protein;

　(iv) a region composed mainly of actin and myosin proteins.

f) Name an inorganic ion you would expect to be present in relatively high concentrations in regions **B** and **D**.

[Oxford June 1993 Biology: Paper 1, q. 3]

3 The diagram below represents a longitudinal section through striated muscle.

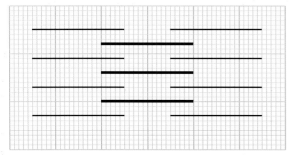

Fig 19.Q3(a)

a) Draw a similar diagram [on 2mm graph paper] to represent the appearance of the same section when the muscle has contracted.

b) Fig 19.Q3(b) represents a cross-section through this muscle.

Draw a vertical line on Fig 19.Q3(a) to show from where the cross-section might have been taken.

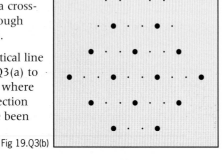

Fig 19.Q3(b)

c) Describe the part played in the contraction of striated muscle by: **(i)** calcium ions; **(ii)** the energy released from the breakdown of ATP.

[AEB 1996 Biology: Specimen Paper, q. 12]

4 The drawing below has been made from a slide of skeletal muscle tissue seen with a light microscope at a magnification of 800 times. It shows parts of two motor units.

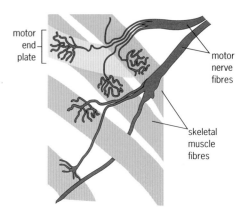

a) Use evidence from the drawing to suggest:

　(i) a meaning for the term *motor unit*;

　(ii) why all the muscle fibres shown will not necessarily contract at the same time.

b) Briefly describe the sequence of events at the muscle end plate which leads to an action potential passing along the muscle fibre.

The diagram below shows the pathways by which energy is produced for muscle contraction. Numbers **1** to **3** indicate the order in which the various pathways are called on to supply ATP as muscular effort increases.

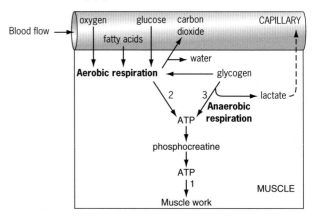

c)　**(i)** What happens to the lactate produced in pathway **3**?

　(ii) Explain the part played by phosphocreatine in supplying energy to the muscle.

　(ii) might account for the difference in speed of contraction of the two types of fibre.

[AEB 1996 Human Biology: Specimen Paper, q. 1]

Assignment

MUSCLE FIBRES AND ATHLETICS

This assignment follows on from the Chapter 8 Assignment **Energy and sport**. You may need to revise this to answer all the questions here.

The different types of muscle fibre

You many have noticed that athletes are highly specialised for their particular event. Marathon runners can run great distances without fatigue, but they lack the power to run as fast as the sprinters. In contrast, sprinters fatigue far too quickly to run long distances. These features are in part due to the type of muscle fibres they have inherited, and partly to the training they have done. Research has shown that there are two basic types of muscle fibre: **fast twitch** and **slow twitch**. Fast twitch fibres can be further subdivided, giving us the three different muscle types:

Slow twitch oxidative (type I) fibres. These contain large amount of myoglobin, many mitochondria and a dense capillary network. They split ATP slowly and therefore can only contract slowly. These fibres are red and have a high resistance to fatigue.

Fast twitch oxidative (type IIA) fibres. These show similar features to slow twitch fibres: a lot of myoglobin, many mitochondria and blood capillaries; but they are able to hydrolyse ATP far more quickly and therefore to contract rapidly. They are relatively resistant to fatigue, but not as resistant as type I fibres.

Fast twitch glycolytic (type IIB fibres). These fibres have a relatively low myoglobin content, few mitochondria and few capillaries. They contain large amounts of glycogen which provides fuel for anaerobic ATP production via the process of glycolysis. They contract rapidly but fatigue quickly, due to the build-up of lactic acid. Type IIB fibres appear whiter than the other two types.

1 Summarise the above information in the form of a table.

2 Movement of muscles requires energy which is provided by one or more of three energy systems: **(i)** the ATP/CP system, **(ii)** glycolysis and **(iii)** the aerobic system. Outline the essential features of these systems.

3
a) What is the function of myoglobin? (See page 291 for a reminder.)
b) Suggest why type IIB fibres have very little myoglobin.

Muscles are made up of motor units: a motor unit is a bundle of fibres controlled by a single motor nerve. Analysis of muscle tissue by biopsy shows that all the muscle fibres in a single motor unit are of the same muscle type, but there can be different muscle types in different units within the same muscle. The various muscles of the body have different proportions of the three types, according to their function.

4 Why is it an advantage to have all three types of muscle fibre in the same muscle?

The distribution of the different muscle types in the body

For the rest of this Assignment we will consider the two main types of muscle fibre: slow and fast twitch.

A central principle in physiology is that structure is closely related to function and this is illustrated clearly by the distribution of the muscle types in the body. For instance, the postural muscles – such as those in the neck and back – have a high proportion of slow twitch fibres. In contrast, the muscles which bring about fast, explosive movements like running, jumping and throwing are packed with fast twitch fibres.

5
a) Where in the body would you expect the muscles to have a high proportion of fast twitch fibres?
b) Explain why slow twitch fibres are suitable for postural muscles.

The distribution of different fibre types in different athletes

The muscle fibre types of a selection of age-matched athletes were analysed, and the results are summarised in Table 19.A1. These are average values: there is considerable variation between different athletes.

Table 19.A1 **Analysis of the average proportions muscle fibre types in selected muscles in different male athletes**

Type of athlete	Approx % fast twitch	Approx % slow twitch
Marathon runner	18	82
Swimmer	25	75
Cyclist	40	60
800 m runner	52	48
Untrained person	55	45
Sprinter and jumper	62	38

6 What does 'age matched' mean?

7 Explain in general terms how the proportion of fast and slow twitch fibres is related to the nature of an athlete's chosen activity.

8 What do you think the information about untrained people tells us?

20 Growth and coordination in plants

These seedlings are typical of those grown in microgravity on board Skylab

IT MAY BE NEWS to you that astronauts on the 1984 flight of the space shuttle Columbia took 'Teddy bears' with them up into space. Not the soft, cuddly kind, however: 'Teddy Bear' is the name of a variety of sunflower plants.

Sunflowers were the first type of plant ever to be grown away from the Earth's gravity. Scientists wanted to find out how seedling plants would develop in conditions of near weightlessness (a condition called microgravity). Their first worry was that the plants would not even start to grow. Since the water in the soil can go in all directions in weightless conditions, there was a possibility that the young seedlings might 'drown'.

However, the experiments showed that the seedlings grew very much as they do back on Earth but with one major difference: they did not grow straight. Instead of being neat upright plants, with roots pointing downwards and shoots growing upwards, they grew in all directions, eventually looking like hopelessly curled and knotted wire.

Since 1984, more Skylab experiments have shown that growing plants in space is an ideal way of trying to understand why roots grow down and shoots grow up. Scientists are looking in particular for the 'gravity genes', the genes that control the plant's response to gravity.

1 GROWTH IN PLANTS

Like most animals, human beings grow to adult size and then stop. Plants are different. They can keep on elongating and branching throughout their lives. New plant cells form in actively dividing regions called **meristems** (Fig 20.1). The main meristems occur at the tip of the root and the tip of the shoot (the **apical meristems**) and between the xylem and phloem (the **vascular cambium**).

Growth at the root and shoot apices

The newly formed cells just behind the **apices** (tips: singular = apex) of the shoot and root meristems are small at first, but they elongate rapidly. As a result, the shoot of a plant grows taller and its roots push more deeply into the soil. As Fig 20.2 shows, cell elongation is due to two events:

Fig 20.1 **In meristems, there are many newly produced cells with tightly packed cytoplasm. The cells have thin cell walls and small vacuoles**

- The cell forms a **vacuole**. Since the vacuole contains mainly water it can enlarge easily.
- The cell stretches. The cell wall is made of **cellulose microfibrils** (see Chapter 1), which slip over each other as the cell enlarges.

Fig 20.2 **As the cell grows, tiny vacuoles fuse into one large vacuole that occupies most of the cell. As new material is added to the cell wall the orientation of the microfibrils changes. When microfibrils run in all directions, the cell cannot grow any more**

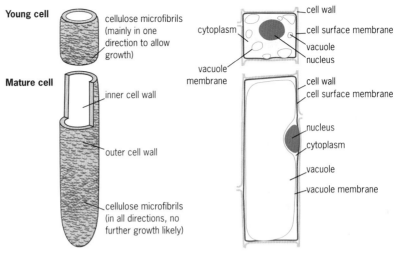

Fig 20.3 **This section through a bud (above) reveals next season's growth in miniature. Chemicals produced by the plant prevent it developing until the time is right. Under certain environmental conditions chemicals are produced by the plant which cause the apex to switch and develop into a flower, rather than a leafy shoot**

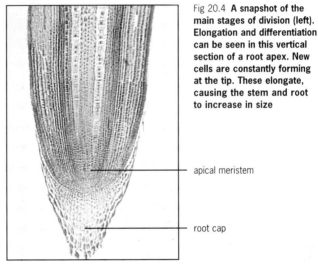

Fig 20.4 **A snapshot of the main stages of division (left). Elongation and differentiation can be seen in this vertical section of a root apex. New cells are constantly forming at the tip. These elongate, causing the stem and root to increase in size**

Under the microscope, a **shoot apex** is like a miniature shoot. Tiny bumps or **primordia** show where the new leaves will develop. Behind the apex the cells elongate and the first vascular tissue develops. During the winter many trees have no leaves – just a series of buds at intervals. Each bud contains a shoot apex wrapped in scales (Fig 20.3). The scales protect the meristem tissue from drying out and insect damage. In spring the bud scales open and the new shoot elongates.

As the **root apex** grows down into the soil, its meristem tissue is protected by a small layer of actively dividing cells, the **root cap**. Cap cells are constantly replaced as they are worn away. The root cap is also involved in detecting gravity (see page 324). Unlike shoots, roots do not form buds and so never carry flowers.

When a plant cell is fully grown it normally specialises. It may **differentiate**, for instance into a xylem vessel or a phloem sieve cell.

The pattern of growth of cells is controlled by the arrangement of the cellulose microfibrils (see Fig 20.2). The primary cell walls of young cells contain lots of hemicellulose between the microfibrils. This forms a gel that keeps the cell walls flexible. As the cell ages, it lays down more cellulose inside to form a secondary cell wall. The secondary cell wall of cells such as xylem and sclerenchyma are also **lignified** (they contain the polysaccharide lignin). Lignin strengthens the wall and makes it more rigid.

Growth in mature shoots

In the trunks and branches of trees, growth occurs mainly as an increase in diameter as the vascular cambium (the ring of meristem tissue between the xylem and phloem) divides by mitosis. Cells on the outer surface of the cambium (see Fig 20.5) differentiate into

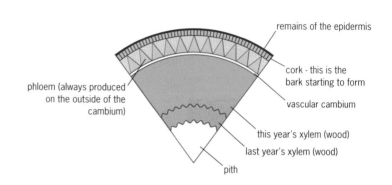

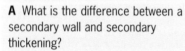

Fig 20.5 **This cross-section shows a two-year-old twig, changing from a green shoot into a woody branch. The thin layer of cork is the start of the layer of bark**

?

A What is the difference between a secondary wall and secondary thickening?

B Why is new bark formed each year?

phloem, while cells on the inner surface become xylem. This process is known as **secondary thickening**. **Wood** is xylem made during secondary thickening.

Trees also have another layer of actively dividing cells, the **cork cambium**, just below the surface. These form a protective layer of **bark**.

Plants do not live for ever

Plants may grow throughout their lives, but often their lives are short. Some plants, like cereals and peas, flower, fruit and then die in a single season. We do not yet fully understand why this happens – perhaps when the apex switches from growth to flower production, the shoot cannot recover.

In plants that live for several years, the constant production of new parts means that old parts (leaves, fruits, bud scales) die and must be shed, a process known as **abscission**. This usually follows a period of **senescence** ('old age') during which the leaves change colour as the chlorophyll is broken down.

2 GROWTH AND SUPPORT IN PLANTS

Young shoots support themselves by the turgor pressure of individual cells (see Chapter 5). If short of water, green shoots wilt, just like leaves. Away from the shoot tip, especially in plants that are long-lived, the weight of growth becomes too much for unspecialised turgid cells to support. Plants have specialised tissues to improve their support: the **collenchyma**, **sclerenchyma** and **xylem**:

- **Collenchyma cells** – Fig 20.6(a) – are normal plant cells with thickened walls that are usually not lignified (woody). They are found in the angles of young stems and in leaf midribs.

- **Sclerenchyma cells** – Fig 20.6(b) – can be different shapes but all are lignified and have lost their cell contents. These strengthening cells occur individually or in groups. (See also Chapter 17, page 269.)

- **Xylem tissue** is the major supporting structure in long-lived woody plants – Fig 20.6(c). The characteristics of xylem vessels and tracheids that make them suitable for water transport also make them ideal supporting tissues (see page 292 in Chapter 18).

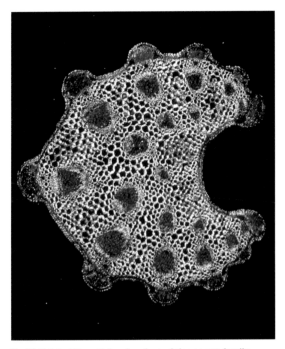

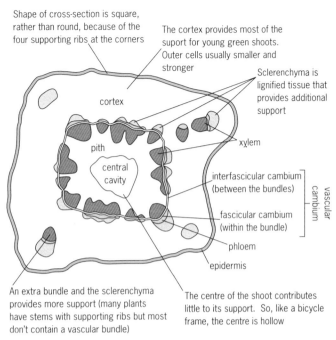

Shape of cross-section is square, rather than round, because of the four supporting ribs at the corners

The cortex provides most of the suport for young green shoots. Outer cells usually smaller and stronger

Sclerenchyma is lignified tissue that provides additional support

cortex

xylem

pith

central cavity

interfascicular cambium (between the bundles)

fascicular cambium (within the bundle)

vascular cambium

phloem

epidermis

An extra bundle and the sclerenchyma provides more support (many plants have stems with supporting ribs but most don't contain a vascular bundle)

The centre of the shoot contributes little to its support. So, like a bicycle frame, the centre is hollow

Fig 20.6(a) **In this petiole of a plant of the carrot family, collenchyma in the outer cortex forms ribs that provide support. You can see similar ribs very clearly on celery stalks**

Fig 20.6(b) **This is a section through a broad bean stem. Broad beans plants are often very tall. Sclerenchyma ribs help to provide support in the same way that collenchyma ribs support nettles**

Supporting tissue is necessary for most plants, but making it uses valuable resources (food and energy) that could be better spent on reproduction. A few plants have worked out how to 'cheat' to avoid this problem. Climbing plants like ivy are rooted in the soil but use other plants to support them as they grow towards the light.

3 PLANTS AND MOVEMENT

Plants may be rooted to the spot, but their individual parts can move, sometimes surprisingly quickly. The 'jaws' of a Venus fly trap close in a couple of seconds to trap an unsuspecting fly, and the leaves of the 'sensitive plant' *Mimosa pudica* droop immediately when touched.

Plant movements are caused by changes in the size of individual cells or groups of cells. Some size changes are permanent and result from growth. Others are temporary, happening as the key cell or cells swell and shrink repeatedly. Scientists recognise two main types of plant movement: **tropisms** and **nastic movements**. In practice, most tropisms are irreversible growth movements and most nastic movements are reversible. However, the underlying difference between tropisms and nastic movements is subtle.

Fig 20.6(c) **Some plants, usually trees, can live for hundreds of years. They can live a long time because their cambium remains active, producing new xylem and phloem each year**

Tropisms

Both tropisms and nastic movements occur in response to an outside influence. But in a tropic response, the direction of the stimulus determines the direction of the response movement, for example:

- **Phototropism**: plants bend towards the light.
- **Gravitropism**: plants grow down in the same direction as the force of gravity, or upwards, against the force of gravity. Strictly, growing upwards is *negative* gravitropism.

- **Thigmotropism**: part of a plant, such as a tendril, grows towards an object that is touching it. Often, tendrils can only curl in one direction, no matter where they are touched. This response is called **thigmonasty**. Both thigmotropism and thigmonasty are growth movements, but only the first is a tropism. A response is only a tropism if the *direction of the stimulus* determines the *direction of the response*.

When a plant shows a tropism, three things happen:

- **Perception**: the plant detects the stimulus (for example light, gravity or touch).

- **Transduction**: the plant converts the stimulus into a message which brings about a response.

- **Response**: the plant responds to the stimulus by growing unevenly.

Gravitropism – how plants detect and respond to gravity

Many plant parts respond to gravity and the way they detect it, known as the **perception mechanism**, seems always to be the same. Most research has been done on roots because their movements are not complicated by responses to light.

Many plants store food in the form of starch grains in organelles called **amyloplasts** (potato cells are full of them). In the root cap, some cells contain large amyloplasts called **statoliths** (Fig 20.7),

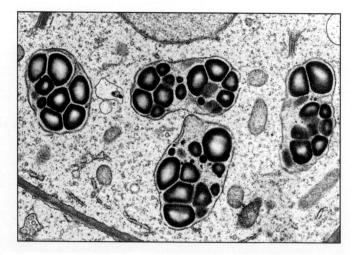

Fig 20.7 **The diagram on the right shows the position of statoliths in a root cap cell. The electron micrograph, far right, shows four statoliths at the base of a root cap cell. The diagram below summarises how statoliths are thought to enable the plant to detect and respond to gravity**

cell surface membrane

cell wall

statolith

1 The statolith might exert pressure on the endoplasmic reticulum

2 This could lead to calcium being pumped into the cytoplasm

stored calcium ions (not to scale)

statolith

cell surface membrane

3 This, in turn, could increase the export of calcium and auxin from the nearby parts of the cell surface membrane

endoplasmic reticulum

calcium ions **auxin**

cell wall

but these are not a food reserve. If the root cap is removed, or if it is treated so that the amyloplasts disappear, the root loses its ability to respond to gravity. When the statoliths re-form, the response to gravity is restored. This suggests that the plant uses statoliths to detect gravity.

Under a microscope, it is possible to see the statoliths slowly tumbling through the cytoplasm if the cells that contain them are turned upside down. The statoliths fall against the internal membranes of the cytoskeleton. This sets off a chain of events (a **transducer mechanism**) that results in uneven growth. (Exactly how this happens is unclear: the transducer stage is the least well understood in all tropisms.) As one side of the root grows faster than the other, the root curves.

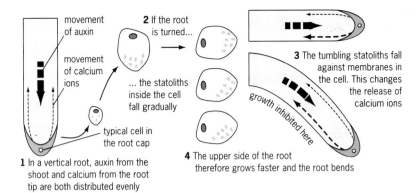

1 In a vertical root, auxin from the shoot and calcium from the root tip are both distributed evenly

2 If the root is turned... ... the statoliths inside the cell fall gradually

typical cell in the root cap

3 The tumbling statoliths fall against membranes in the cell. This changes the release of calcium ions

4 The upper side of the root therefore grows faster and the root bends

growth inhibited here

movement of auxin

movement of calcium ions

Fig 20.8 **Suggested explanation of how the movement of statoliths could change the pattern of growth**

The role of auxin in phototropism

Plants respond to light using a mechanism that involves the chemical **auxin,** a plant growth substance.

Most research into phototropism has been done on **oat coleoptiles**. A coleoptile is a sheath that surrounds the first true leaf. It is a simple structure, easily grown, responsive to light (and gravity) and so a favourite research subject. Also, since researchers can obtain established crop varieties, they can use genetically similar seeds – a useful element of control in experimental design. Experiments using oat coleoptiles have shown that light is detected by the coleoptile tip, but the growth response occurs further down the coleoptile (Fig 20.9). The coleoptile bends as one side grows faster than the other.

This effect is brought about by auxin. Auxin is constantly produced by cells at the tip of the coleoptile. Uneven lighting does not change the amount produced, but it does cause a redistribution of auxin. Auxin moves away from the light and becomes more concentrated in the side of the coleoptile tip that is furthest from the light source. More auxin travels down one side of the coleoptile than the other: because of the uneven growth that the auxin causes, the coleoptile bends towards the light (Fig 20.10 and Fig 20.11).

We know that the cells that respond to auxin have a **receptor** (just as animal cells do for hormones). But exactly how auxin speeds up growth is not fully understood. Different tissues respond differently – root growth may be inhibited by concentrations of auxin that speed up shoot growth, for example. Researchers have identified two underlying processes that seem to be involved.

Fig 20.9 **Although most of the coleoptile bends when unevenly lit, only the tip actually detects light**

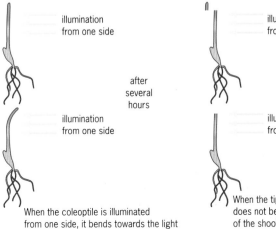

illumination from one side

after several hours

illumination from one side

When the coleoptile is illuminated from one side, it bends towards the light

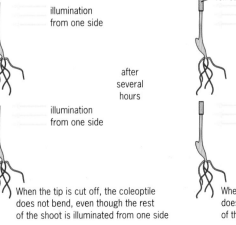

illumination from one side

after several hours

illumination from one side

When the tip is cut off, the coleoptile does not bend, even though the rest of the shoot is illuminated from one side

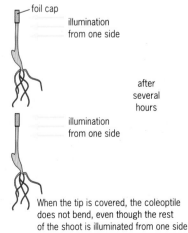

foil cap

illumination from one side

after several hours

illumination from one side

When the tip is covered, the coleoptile does not bend, even though the rest of the shoot is illuminated from one side

Fig 20.10 **This experiment shows that a chemical must be diffusing into the gel from the plant tip and causing the bending response**

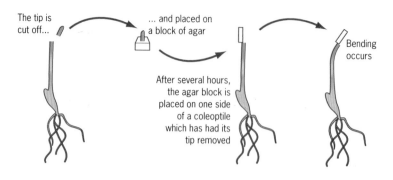

Fig 20.11 **This experiment suggests that the total amount of auxin produced by the tip remains constant but is distributed unevenly in conditions of uneven lighting**

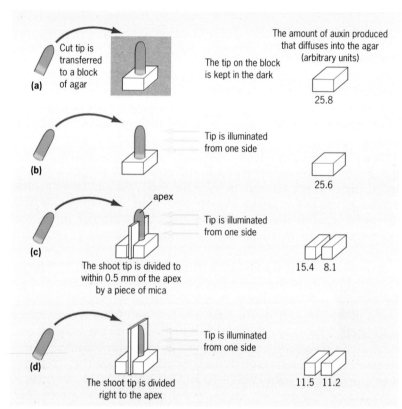

?

C Study the experiment outlined in Fig 20.10. Why does the agar block cause the shoot to bend?

D Study the experiment outlined in Fig 20.11. Which two treatments taken together prove that uneven lighting does not change the total amount of auxin produced?

E Which part of the experiment shown in Fig 20.11 is necessary to show that auxin is redistributed only within the very tip?

- First, auxin causes receptive cells to secrete hydrogen ions into the primary cell walls that surround them. The increased acidity weakens bonds between the cellulose microfibrils and these start to slip against each other. Turgor pressure then makes the cell swell. This can be a very quick process.

- Second, auxin appears to trigger a change in the rate of **transcription** (the process by which genes are read to produce messenger RNA). This has a more general effect on the rate of growth by increasing the rate of protein synthesis.

In other plant parts, phototropism is similar but not identical. In tissues that do not seem to work like coleoptiles, there must be another mechanism. It has been suggested that in such tissues, light might cause production of a growth inhibitor.

Nastic movements

Nastic movements are plant movements that do not involve growth in a particular direction in response to a directional stimulus. Closure of the Venus fly trap and the spectacular collapse of Mimosa

AUXIN AS HERBICIDE

SOME COMMERCIAL weedkillers are chemically similar to auxin and have similar effects. They cause affected plant cells to elongate and the plant grows. The exact mechanism is not yet understood, but the auxin seems to interfere with DNA transcription and RNA translation. The amount of auxin applied in a weedkiller is far greater than the amount produced within the plant, and so the rate of growth produced is much greater than normal. The plant cannot sustain this rate of growth: it becomes weakened, unable to reproduce, and then it dies.

Fig 20.12 **Auxin-like herbicides have a much greater effect on dicots than on monocots such as grasses. (This is partly because dicot leaves have a larger surface area than monocot leaves and so absorb more herbicide.) This means that the herbicides are good at getting rid of weeds like dock**

are both triggered by stimuli, but the direction of the response is predetermined – it does not depend on the direction of the stimulus. Some other nastic movements, such as the opening and closing of flowers, show a regular daily or **circadian** rhythm.

Nastic movements are usually caused by clusters of cells located at 'hinge' points, such as at the bases of petals and leaves. When these cells pump ions inwards, water follows by osmosis and the cells swell. Later, the ions are allowed to leak out, water again follows and the cells shrink.

> ✓ Both tropisms and nastic movements occur in response to an outside influence. But, in a tropic response, the *direction of the stimulus* determines the *direction of the response movement*. In a nastic movement this is not the case.

4 COORDINATION OF PLANT GROWTH AND DEVELOPMENT

Chemicals other than auxin are involved in plant development: some are plant growth substances, others are not. The rest of this chapter looks at how some other plant growth substances work. **Phytochrome** (a chemical that is not a plant growth substance, but that plays an important role in plant development) is covered in Chapter 29.

Gibberellins

In the nineteenth century, Japanese rice growers occasionally noticed plants that were larger than usual but rarely produced any rice. They became known as 'foolish seedlings'. The cause turned out to be a fungus *Gibberella fujikuroi*, which produces a growth-promoting chemical (Fig 20.13). Since this time, a whole family of similar chemicals called **gibberellins** have been discovered. Gibberellins also play a role in flowering and overcoming dormancy of buds and seeds (Fig 20.14).

Gibberellins are extracted from the fungus *Gibberella*. This is an expensive procedure, but worth while because of two valuable commercial uses: beer and wine making.

To make beer you need partly germinated (or **malted**) barley. As the seeds germinate their starch is changed into simpler sugars such as maltose. This sugar is then fermented by the yeast to produce the alcohol. Gibberellin helps the germination of barley by speeding up the conversion of starch to sugar. Adding extra gibberellin gives more sugar more quickly.

> ✓ Animal hormones are typically synthesised in one organ and released into the bloodstream to have an effect on a target organ somewhere else. Plant growth substances are often created in the very organ where they have their effect. For this reason some authors argue they should not be called hormones, reserving this term for animal hormones.

> ✓ Maltose is a disaccharide made up of two glucose molecules joined together.

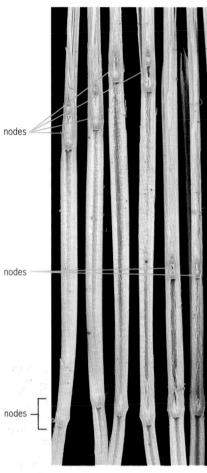

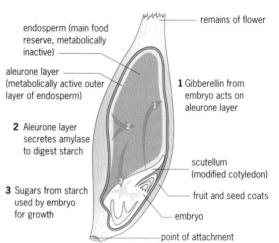

endosperm (main food reserve, metabolically inactive)

aleurone layer (metabolically active outer layer of endosperm)

remains of flower

1 Gibberellin from embryo acts on aleurone layer

2 Aleurone layer secretes amylase to digest starch

3 Sugars from starch used by embryo for growth

scutellum (modified cotyledon)

fruit and seed coats

embryo

point of attachment

Fig 20.14 **Gibberellin secreted from the embryo helps mobilise the food reserve during seed germination. Adding more gibberellin can speed up this process**

See question 1.

Vine growers benefit from gibberellin too. Gibberellin sprayed on developing grapes causes the stalks to elongate. This gives more room for the grapes to swell as they ripen, and the grower gets larger and less damaged fruit.

Cytokinins

A group of chemicals called **cytokinins** increase cell division. If a mixture of cytokinin and auxin is added to unspecialised cells (Fig 20.15), they will begin to differentiate. A high cytokinin to auxin ratio will lead to the formation of shoots, buds and leaves while a low cytokinin to auxin ratio will lead to root formation. The one treatment followed by the other provides a means of forming small plantlets. Such *in vitro* culture methods have become widely adopted for the rapid propagation of new plant varieties. The technique allows growers to produce very large numbers of plants, quite rapidly and in a small space. The alternative conventional propagation methods can take several years and a large area of land.

Fig 20.13 **Gibberellins make most plants grow taller.**
Here, wheat stems are cut lengthwise to show the nodes. The four plants on the left, treated with gibberellic acid, have internode lengths much greater than the internodes of the two untreated on the right

Fig 20.15 **If a small piece of pith from a shoot is placed on agar in aseptic conditions it will grow into a mass of unspecialised cells called callus (below). Plants can be grown from these cells (right)**

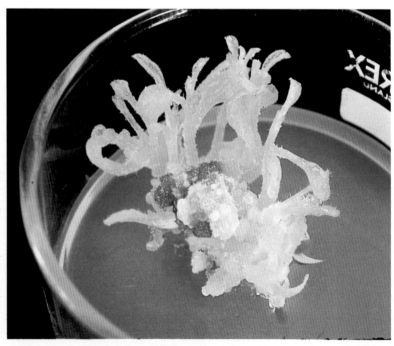

Abscisic acid

Abscisic acid (ABA) was thought to be the substance that caused abscission (leaf fall). We now know that it is a growth inhibitor and that leaf fall is probably one of the few processes that does *not* involve ABA.

ABA prevents some seeds from germinating – it keeps them dormant. It takes time to break down and, by the time this happens, the seeds have survived winter and can germinate in the more mild conditions of spring.

Sometimes, plants use two growth substances together. Using two growth substances with opposite effects can give better control. ABA causes dormancy in buds and gibberellin breaks it. The two substances work together to ensure buds grow at just the right season. Growth substances with *opposite* effects like this are said to be **antagonistic**.

Ethene

Ethene is a gas that seems to be involved in seed dormancy, fruit ripening and leaf abscission.

Early this century, growers found that their greenhouse plants were turning yellow and dying prematurely. They tracked the cause to the ethene that was produced because of incomplete combustion of the gas used to heat the greenhouse. Ethene is produced naturally by ripening fruit, and this can be a commercial problem today. It is common for fruit such as bananas to be picked and shipped under-ripe, but if the odd ripe banana produces ethene, the whole batch can ripen before they reach port. On board ship the air in the hold is flushed regularly to prevent ethene build-up, or the fruit is sealed with air that has a high carbon dioxide concentration (5 to 10 per cent is usually enough to prevent ripening).

Fig 20.16 **We say that 'one rotten apple spoils the rest of the barrel' because pieces of fruit near to a rotten one start to go bad quickly. This is because a damaged or fungus-infected fruit starts to produce ethene. In a closed barrel the concentration of ethene is high enough to quickly trigger the ripening process in neighbouring fruits making them more vulnerable to infection. These oranges are infected with the fungus *Penicillium digitatum*, which itself produces large amounts of ethene**

See questions 2 and 3.

SUMMARY

When you have read this chapter, you should know and understand the following:

▪ Growth continues throughout a plant's life because of groups of dividing cells, the **meristems**. The most important of these are the **apical meristems** (in the shoot and root tips) and the **vascular cambium** (between the phloem and xylem).

▪ Cell division, elongation and differentiation are separated in time and occur in different parts of the meristem.

▪ Cambium produces new xylem and phloem in a process called **secondary thickening**. This causes an increase in diameter.

▪ As plants grow, special supporting tissues are needed. The three main types of supporting cell are the **collenchyma**, the **sclerenchyma** and the **xylem**.

▪ **Tropisms** are plant movements that occur in a particular direction, as governed by a directional stimulus such as light or gravity.

▪ **Nastic movements** are responses independent of the direction of the stimulus. They often do not involve growth, and are usually due to changes in cell turgor pressure.

▪ Many different chemical substances help to coordinate the growth of a plant. Some of the most important are **auxin**, **gibberellins**, **cytokinins**, **ethene** and **abscisic acid**.

1 The embryo of a germinating barley seed produces gibberellins which stimulate the synthesis of enzymes in the aleurone layer.

The diagram below shows a vertical section through a barley seed. The positions of the embryo and aleurone layer are indicated.

embryo

aleurone layer

Fig 20.Q1

An experiment was carried out to investigate the effect of gibberellins on the synthesis of amylase by the aleurone layer. Gibberellins were added to aleurone layers isolated from barley seeds and the quantity of amylase present was measured over a period of 15 hours. The results are shown in the table below.

Time/hours	Amylase synthesised/arbitrary units	
	Gibberellins added	No gibberellins added
0.0	0.0	0.00
2.5	0.7	0.20
5.0	1.9	0.45
7.5	5.2	0.80
10.0	9.5	1.00
12.5	17.6	1.15
15.0	17.9	1.38

a) Plot these data in a suitable graphical form using a single pair of axes.

b) How much amylase is present after exposure to gibberellin for 9 hours?

c) State **two** precautions that would need to be taken when carrying out this experiment.

d) What conclusion can be drawn from these results.

e) Describe the function of amylase in germinating seeds.

f) Suggest **one** way in which these findings may be applied commercially.

g) State **three** functions of gibberellins in a mature plant.

[ULEAC January 1993 Biology: Paper 3, q. 8]

2 Write an essay on plant growth substances.

[ULEAC June 1995 Biology: Paper 3, q. 9a]

3 Read the following passage and then answer the questions.

One of the important growth-controlling systems in plants is provided by the so-called 'plant growth substances' or 'plant hormones'. A plant growth substance is an organic substance produced within a plant that will, at low concentrations, promote, inhibit or modify growth, usually at a site other than its place of origin. Its effect does not depend on its energy content nor on its content of essential elements.

One of the difficulties in the study of plant growth substances is that much of it is based on circumstantial evidence derived from experiments in which the chemical, or a closely related substance, is applied to the appropriate plant from the outside. The reasoning in such a case is usually as follows.

A We know that a substance X, or one very like it, occurs in a certain plant.

B We have a supply of substance Y, which is very similar to substance X.

C When applied to the relevant plant, substance Y causes a specific response (for example, stem elongation).

D Therefore it is likely or possibly that substance X has a role in controlling stem elongation in this plant.
(Adapted from Hill, *Endogenous Plant Growth Substances*, Arnold (1973))

a) The author of this passage used the term 'plant hormone' as an alternative to 'plant growth substance'. Other authorities reject the term 'plant hormone' on the grounds that the comparison with animal hormones is not valid.
  (i) State **two** features that plant growth substances and animal hormones have in common.
  (ii) State **two** differences between plant growth substances and animal hormones.

b) Using information in the passage, give **two** reasons why each of the following is not regarded as a plant growth substance: **(i)** carbon dioxide, **(ii)** glucose.

c) **(i)** Explain why the experimental evidence referred to in the passage (statements A, B, C and D) does not demonstrate conclusively that substance X has a role in controlling stem elongation in the plant concerned.
  (ii) Suggest **two** other pieces of evidence that might be sought to support the conclusion that substance X has a role in controlling stem elongation.

d) Assume that substance Y can be manufactured cheaply in large quantities. Suggest **two** possible commercial applications that might be investigated for substance Y.

[ULEAC January 1995 Biology: Paper 3, q. 7]

Assignment

AUXIN AND GROWTH IN PLANTS

Auxin is an important growth substance that is not just involved in phototropism. It can have many effects in the plant, sometimes acting as a growth promoter, sometimes as a growth inhibitor.

How does auxin travel?

We know that auxin is produced at the shoot apex and that it travels down the plant. It may be that auxin moves by simple diffusion (see Chapter 5). Fig 20.A1 shows two experiments that investigate this possibility.

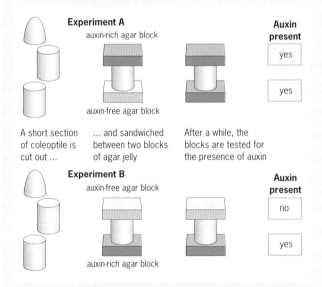

Fig 20.A1 **In Experiment A, an auxin-rich block is placed at the top end of the coleoptile section and an auxin-free block is placed at the bottom end. In Experiment B the position of the two blocks is reversed**

1

a) In which direction(s) would you expect the auxin to move within the short piece of stem shown?

b) Which way does auxin seem to travel in the two experiments A and B?

c) Do you think that diffusion can account for the results of these two experiments? Explain your answer.

What does auxin do?

Auxin travels from the tip (the main bud) of the plant downwards towards the root. This leads to a concentration gradient of auxin from shoot tip to root tip. This gradient can affect the growth and development of buds and the appearance of the whole plant. In the next section, we investigate the effect of auxin on buds.

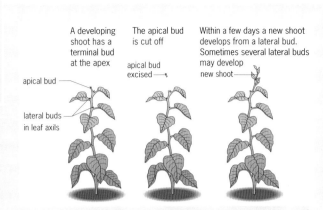

Fig 20.A2 **An experiment to investigate what happens when the apical bud of a plant is removed**

2

a) Look at Fig 20.A2. What effect does the apical bud have on other (lateral) buds?

b) What happens when the apical bud is removed?

c) Think of a situation where this may arise in nature, and describe it.

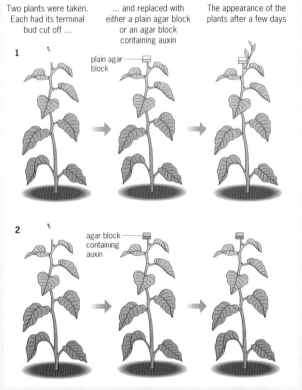

Fig 20.A3 **A further investigation into the effect of the terminal bud**

3

a) Look at the experiment shown in Fig 20.A3. What does this experiment suggest about the way the terminal bud works in preventing lateral bud development?

b) What further investigations might you do to confirm your hypothesis?

21 Nervous systems

There is still a long way to go to develop computer controlled robots that can move with the same ease as the human body

COMPUTER-OPERATED ROBOTS can now do many of the repetitive and dangerous jobs that people used to do. Jointed robotic arms with operating hands can drill holes, spray paint and move radioactive materials. The most recent mobile robots, developed for the planned exploration of Mars, shamble around on crude limbs, taking equipment from place to place. But their complex electronic circuits cannot match up to a nervous system: designing a robot that can walk on two legs with the skill that most people take for granted has, so far, proved impossible.

All parts of a nervous system – the brain, spinal cord and nerves – work together to coordinate body activity. The brain uses constantly updated information from body sensors, together with memory, to initiate and control body movements such as walking. We only become aware of the coordinating role of this highly specialised system when it stops working. In multiple sclerosis the nerves of affected people cannot carry signals properly. Sufferers gradually become weaker and less able to control movements or feel sensations. This loss of function is caused by damage to the insulating sheaths that surround nerves in the brain and spinal cord. Understanding more about diseases like multiple sclerosis helps scientists to develop better treatments and also often gives them more insight into how healthy nervous systems work.

1 PROFILE OF A NERVOUS SYSTEM

A **nervous system** is an intricate network of specialised nerve cells called **neurones**. It coordinates the activities of all but the simplest of animals. The nervous system receives sensory information, interprets it and generates appropriate responses using structures such as muscles. Environmental changes, **stimuli**, alter the electrical properties of neurone membranes.

Most complex nervous systems can be divided into two parts. The **central nervous system** receives all incoming information and is the site of decision-making. In humans, it consists of the brain and spinal cord. (See also pages 352–356.) All the other parts of the nervous system make up the **peripheral nervous system**, which relays information between the sense organs receiving stimuli, the central nervous system, and the structures which bring about responses. (See also page 357.)

Information passes along neurones in the form of electrical impulses called **nerve impulses**. Communication between neurones, at **synapses**, is via **chemical transmitters**. A mixture of electrical and chemical signals allows information to travel around an organism with far greater speed than if only chemical signals were

Fig 21.1 **Eating with chopsticks requires complex nervous coordination**

A Study the labels in Fig 21.2. For a person using chopsticks in Fig 21.1, write a similar succession of events and processes for each stage in a sequence from 'stimulus' to 'response'.

Fig 21.2 **The person in this diagram is carrying out a typical *voluntary response* to an outside stimulus. Later in the chapter, we will look at an *involuntary response*, or *reflex***

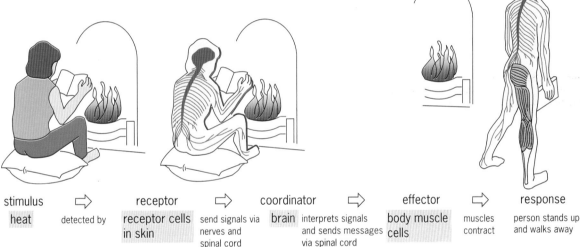

stimulus ⇨	receptor ⇨	coordinator ⇨	effector ⇨	response
heat	detected by	receptor cells in skin	send signals via nerves and spinal cord	brain	interprets signals and sends messages via spinal cord	body muscle cells	muscles contract	person stands up and walks away

used. Fig 21.2 shows the sequence of events in the nervous system of someone who is sitting too close to a fire.

In humans, neurones start to appear in the embryo when it is 10 to 14 days old. The neurones we are born with are already mature and are incapable of reproducing themselves by dividing. Those that die cannot be replaced, but it is possible for their most immediate neighbours to take over the roles of damaged or dead neurones.

The nervous system continues to be modified throughout our lives, as connections between neurones are changed and rearranged as a result of our experiences. This flexibility is built into the system and is called **behavioural plasticity**.

2 THE NEURONE: BASIC UNIT OF A NERVOUS SYSTEM

The neurone is the functional unit of a nervous system. It connects with many other neurones and receives, conducts and transmits information. Like other types of cell, the neurone is not easily seen by the naked eye. Neurones vary in shape and size depending on their position and function (those extending down your leg may be over one metre long) but all neurones have a similar basic structure.

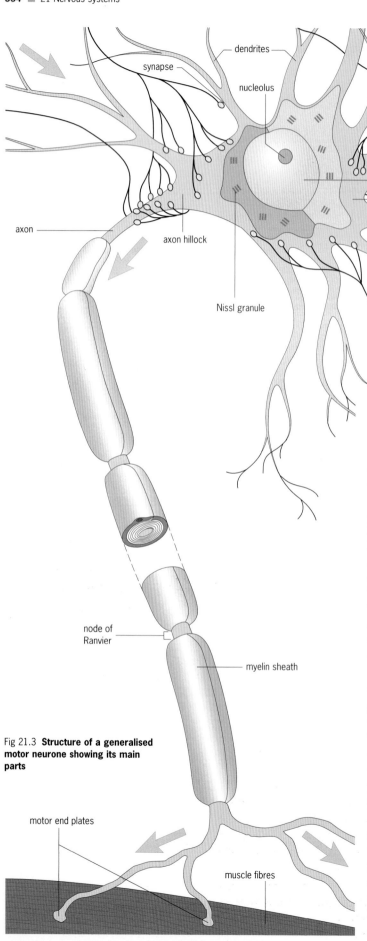

dendrites

synapse

nucleolus

direction of
information transfer

nucleus

cell body

axon

axon hillock

synaptic bulbs at
axon terminals

Nissl granule

node of
Ranvier

myelin sheath

Fig 21.3 **Structure of a generalised
motor neurone showing its main
parts**

motor end plates

muscle fibres

Basic structure of a neurone

Fig 21.3 shows the structure of a typical motor neurone which transmits signals to muscle fibres.

The **cell body** contains a large nucleus, a nucleolus and other organelles. It may have as many as 200 **dendrites**, thread-like branches that increase the surface area of the cell body, allowing many connections to be made with **synaptic bulbs** at the **axon terminals** of neighbouring neurones. Signals cross tiny gaps called **synapses** at these connections, and travel along the cell membrane towards the main part of the cell body.

When the signal reaches an area called the **axon hillock**, the nerve impulse is generated and the process of signal transmission to the next cell begins. The membrane of the cylindrical **axon** conducts the nerve impulse away from the axon hillock towards the **motor end plate**, a collection of synapses that pass signals on to a muscle cell.

Nissl granules are the site of protein synthesis in the neurone and are thought to be part of a maintenance system which monitors the state of the cell. The cytoplasm, or **axoplasm,** is continuous in the dendrites, cell body, axon and synaptic bulbs, and materials reach different parts of the cell by **axoplasmic transport**.

Types of neurone

Neurones can be classified according to how many processes they have, as shown in Fig 21.4. They can also be classified according to their role in the body.

Unipolar neurones have a *single* process coming out of the cell body which then divides into two branches, a dendrite and an axon. Unipolar neurones are very common in the peripheral nervous system of vertebrates. The sensory neurone, for example, has its incoming dendrite in the outer parts of the body, while the other branch, the axon, extends into the spinal cord. Unipolar neurones are also common in invertebrates.

Bipolar neurones have *two* processes which come out of the cell body. One process is a dendrite and the other is an axon. All neurones in the nervous system of a human embryo are bipolar, but most of them then develop into the other two types, so there are few bipolar neurones in the nervous system of an adult. They occur in the retina of the eye, the cochlea in the inner ear and in the nerves serving cells in the nose which sense smell.

Multipolar neurones have *many* processes coming out of the cell body – several dendrites and one axon. Multipolar neurones are the most common type of neurone in the brain and spinal cord of mammals.

Fig 21.4 **Three types of neurones named according to the way they branch**

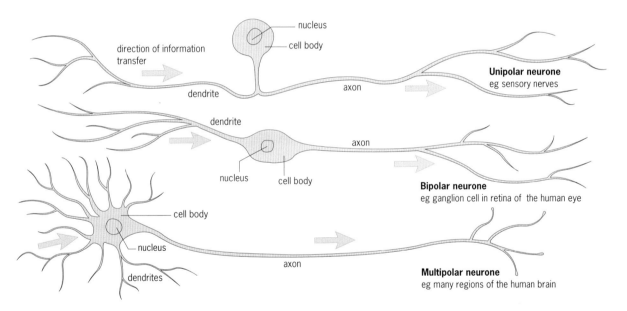

Cells associated with neurones

Glial cells are packed between neurones, making up tissue called the **neuroglia**. Glial cells:

- give mechanical support to the neurone network,
- provide electrical insulation by coming between the neurones.
- form myelin sheaths around myelinated neurones,
- have a metabolic role: they control the nutrient and ionic balance round neurones, and break down neurotransmitter substances after signals are transmitted (see page 344).

Glial cells make up about half the weight of the human brain, out-numbering neurones 50 to 1. In other parts of the nervous system, the proportion is much lower, at about 10 to 1.

Glial cells are also found in invertebrates.

Myelinated and non-myelinated neurones

In vertebrates, specialised glial cells called **Schwann cells** wrap themselves round the axons of some neurones in the peripheral nervous system. The Schwann cells form a thick, lipid-rich insulating layer called the **myelin sheath** (Fig 21.5). This electrically insulates the axon, rather like the plastic layer round a copper wire in an electrical flex. Neurones with myelin sheaths are said to be **myelinated**. They make up **myelinated nerves**.

Schwann cells cover most of the axon, but leave bare sections between the cells called **nodes of Ranvier,** as shown in Fig 21.3. Nerve impulses which travel along a myelinated axon 'jump' between these gaps. This is called **saltatory conduction** and is described on page 342. Myelinated neurones can conduct impulses at twice the speed of unmyelinated neurones (see page 342).

Fig 21.5 **Development of a myelin sheath round an axon**

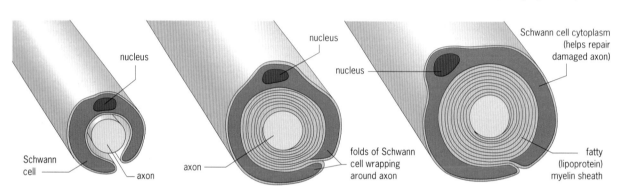

Fig 21.6 **The mouse on the right is called a 'shiverer'. Because of a genetic mutation, none of its nerves are myelinated, so that nerve impulses along one axon can generate impulses in adjacent neurones. Muscles receiving signals from neurones activated in this way are stimulated continuously, and so they contract uncontrollably, and the mouse shivers. People with multiple sclerosis have damaged myelin sheaths and have a similar problem: they cannot control muscle movements.**

EARLY THEORIES ABOUT HOW SIGNALS TRAVEL

ABOUT FOUR HUNDRED YEARS ago, people began to study how information travels around the body. Early scientists confidently believed that a fluid, the *succus nerveus*, was the message carrier. However, when dissection showed the absence of cavities in nerves and the lack of free-flowing nerve 'juice' from the cut ends, this theory was abandoned.

The link between electricity and the action of nerve and muscle in living organisms was discovered by the Italian anatomist Luigi Galvani (1737–1798). Galvani found that he could use static electricity to stimulate nerves in dissected frogs and produce muscle contractions. By 1849, Emil du Bois Reymond (1818–1896) had developed a machine that could detect a difference in electrical potential between the inside and outside of muscle fibres. He also showed the existence of electrical currents in nerves.

Fig 21.7 **Luigi Galvani used a small electrical current to stimulate the muscles of frogs and other animals, and so produced muscle contraction**

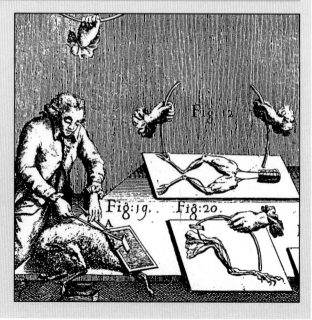

DISCOVERING HOW NEURONES CARRY INFORMATION

IN THE 1950s, two British physiologists, Alan Hodgkin (born 1914) and Andrew Huxley (born 1917) showed that there is a negative electrical potential between the inside of a neurone and the outside. With the equipment available at the time, they were unable to work with the very tiny axons from mammals. Instead they used squid axons with a diameter of up to 1 mm (several hundred times wider than a mammalian axon), inserting a tiny **microelectrode** into a length of a single axon.

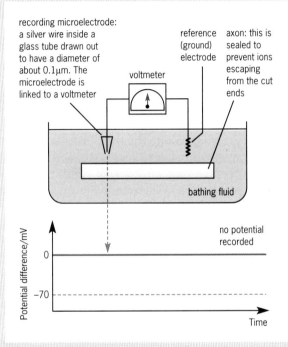

Fig 21.8 **The experimental apparatus which Hodgkin and Huxley used to investigate how nerve impulses are transmitted**

Fig 21.8 shows one of their key experiments. Part of an axon is bathed in a saline solution similar to the fluid that surrounded it in the living squid. A reference electrode – the **ground** electrode – is suspended in the bathing fluid. The **voltmeter** measures the electrical potential difference as a voltage between the two electrodes. When the microelectrode is inserted into the axoplasm, as shown in Fig 21.9, the voltage falls to –70 mV. This indicates that the inside of the squid axon is electrically negative with respect to the fluid surrounding it.

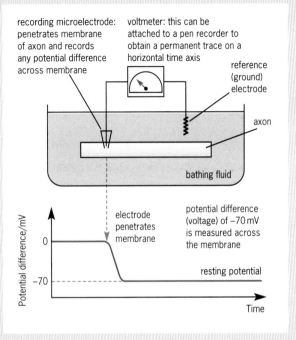

Fig 21.9 **Inserting a microelectrode into a resting axon to measure its membrane potential**

This experiment was later modified to find out what happened to the electrical potential when a signal was transmitted along the squid axon. Two electrodes were placed inside the axon, one at each end (as in Fig 21.10), and a brief pulse of electrical current was applied through the stimulating electrode to mimic a nerve impulse. The recording electrode's voltmeter registered a change of about 90 mV, from –70 mV to +20 mV. Then almost immediately, the reading reverted back to –70 mV.

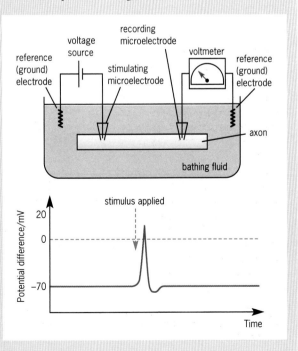

Fig 21.10 **Recording the membrane potential when an electrical stimulus is applied to the axon**

3 HOW DO NEURONES CARRY INFORMATION?

Neurones have two important properties that enable them to carry information. They are **excitable** (they can detect and respond to stimuli) and they are **conductive** (they can transmit a signal from one end of the axon to the other).

We are about to look at how a neurone transmits information but, first, we need to find out what is going on in the neurone before the information arrives.

The neurone at rest

When a neurone is not conducting a nerve impulse we say that it is at rest. But it would be a mistake to think that a resting neurone is completely inactive. At any given moment it needs to be ready to conduct impulses that might arrive from its neighbouring neurones. To do this, the membrane of a resting neurone must always be **polarised**. Fluid on the inside must be negatively charged with respect to the outside. This difference in charge of about −70 mV is called the **membrane potential** or **resting potential**. It sets up the *potential* for electrical activity along the axon membrane.

What causes the resting potential, and how is it maintained? Like most cells, neurones can maintain an internal fluid composition different from that outside the cell. As shown in Fig 21.11, there are *more* sodium ions (Na^+) and chloride ions (Cl^-) outside the axon than inside, and *more* charged proteins and *more* potassium ions (K^+) inside the axon than outside. This imbalance gives the **electrical gradient** needed to create a resting potential across the membrane.

Table 21.1 shows the distribution of ions inside and outside the squid axon. We can see that there is a very big imbalance of charged ions. How is this established and maintained? Let us look more closely at the different ions.

Na^+, K^+ and Cl^- ions can diffuse to and fro across the membrane, but the larger, negatively charged protein ions are confined to the inside of the membrane by their size: we describe the axon membrane as **selectively permeable**. Because of their negative charge, the protein ions establish an electrical potential across the membrane. So, as you might expect, the ions that can move tend to

Table 21.1 **The main ions inside the squid axon and in the surrounding tissue fluid: concentrations in millimoles per kilogram of liquid**

Ion	In tissue fluid	Inside axon
Na^+	460	50
K^+	10	400
Cl^-	540	100
Proteins⁻	0	360

Fig 21.11 **Concentrations of the main ions inside and outside the squid axon membrane**

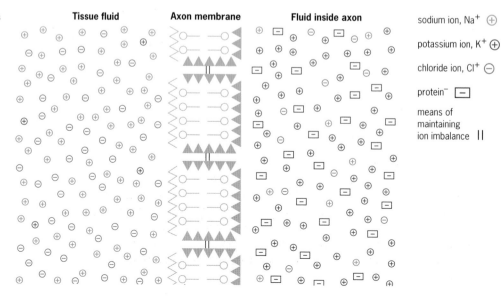

Tissue fluid   Axon membrane   Fluid inside axon

sodium ion, Na^+ ⊕

potassium ion, K^+ ⊕

chloride ion, Cl^+ ⊖

protein⁻ ☐

means of maintaining ion imbalance ||

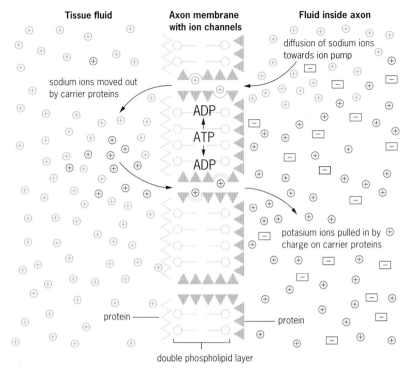

Fig 21.12 **Schematic diagram to show the sodium–potassium ion pumps. Sodium and potassium ions are moved against their concentration gradient in a process that requires energy, This is provided by ATP. Each type of ion has its own channel molecule, whose structure excludes all other ions**

redistribute themselves either side of the membrane to balance this charge. With diffusion only, a net movement of ions would continue until there was a balance across the membrane between the concentration gradients of ions and the electrical potential.

We would therefore expect different concentrations of ions inside and outside the membrane. But we would not expect such large concentration differences as we see in the table. Also, why do the K^+ ions inside the axon outnumber the Na^+ ions ten to one, and the Na^+ ions outside outnumber K^+ ions ten to one?

These ratios arise because of the process of **active transport**. The neurone actively transports Na^+ and K^+ ions across the membrane against their concentration gradients, using structures called **sodium–potassium pumps**. They are large protein molecules that form channels across the membrane, and they harness energy derived from ATP (see page 94) to move Na^+ out and K^+ in, to maintain the correct ratios of ions between the inside and outside of the cell (see Fig 21.12).

In each square micrometre of axon membrane surface, there are between 100 and 200 sodium–potassium pumps. Each pump moves about 200 Na^+ ions *out* and about 130 K^+ ions *in*, every second.

The neurone in action

When stimuli reach a neurone, subtle electrical changes take place at the membrane of the cell body. The resting potential of the membrane is changed, becoming slightly less negative. These small changes in potential spread to the axon hillock and, if they combine to exceed the **threshold level**, the membrane potential becomes positive and an action potential is triggered. An **action potential** is the new electrical potential that can be detected between the inside and the outside of the axon, just after it has been stimulated.

?

B Read the account in the box on page 337 of the work done by Hodgkin and Huxley, and look at Fig 21.8.

(a) What is the voltage when both electrodes are in the saline solution? Explain your answer.

(b) What is the purpose of the ground electrode?

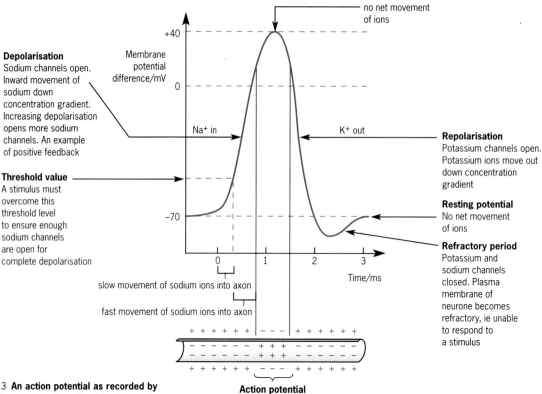

Depolarisation
Sodium channels open. Inward movement of sodium down concentration gradient. Increasing depolarisation opens more sodium channels. An example of positive feedback

Threshold value
A stimulus must overcome this threshold level to ensure enough sodium channels are open for complete depolarisation

no net movement of ions

Repolarisation
Potassium channels open. Potassium ions move out down concentration gradient

Resting potential
No net movement of ions

Refractory period
Potassium and sodium channels closed. Plasma membrane of neurone becomes refractory, ie unable to respond to a stimulus

Na+ in

K+ out

slow movement of sodium ions into axon

fast movement of sodium ions into axon

Action potential

Direction of nerve impulse

Fig 21.13 **An action potential as recorded by an oscilloscope, and how it relates to events in the axon**

Fig 21.13 shows the oscilloscope record of an action potential. It is also referred to as a **spike** because of the shape of the trace.

From the point at which the action potential starts, the wave of electrical activity travels, or is **propagated** along the axon – like fire travelling along a burning fuse. This process is described more fully below. Although it is similar in myelinated and unmyelinated nerves, there are some important differences. Let's look first at what happens in an unmyelinated neurone.

See question 1.

Depolarisation

As a stimulus reaches the resting neurone, its membrane at the site of stimulation becomes more permeable to Na+ ions. They move in relatively slowly at first, making the membrane potential less negative. Then, at a threshold of about –50 mV, a much more rapid inrush of Na+ ions is triggered, and the inside suddenly becomes positive (+40 mV) with respect to the outside. As the electrical charge across the membrane is reversed, we say that the membrane becomes **depolarised**.

Depolarisation initiates a nerve impulse. This is an **all or nothing** signal: if the threshold is not reached, depolarisation does not occur and no signal can be propagated.

The action potential

When a patch of the axon membrane depolarises, a flow of current is produced as (electrically charged) ions move. This stimulates the next patch of membrane, and the action potential begins to move along the axon (in the direction away from the cell body), as shown in Fig 21.13. The change in voltage (the amplitude) of the action potential is the same at all points along the axon. It does not increase or fall off as the signal travels.

C (a) Why do you think it is important to have a threshold?

(b) What would be the effect of having no threshold?

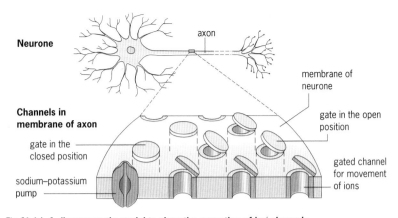

Neurone

Channels in membrane of axon

Fig 21.14 **A diagrammatic model to show the operation of ion channels**

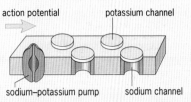

(a) The membrane of a neurone at rest. The inside is negative with respect to the outside. Most of the ion channels are closed.

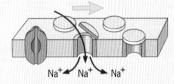

(b) An action potential is initiated. Gates of sodium channels open and sodium ions rush into the axon. The inside of the membrane becomes positive with respect to the outside: the membrane is depolarised.

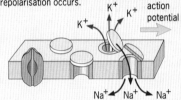

(c) The action potential travels to the next patch of membrane along the axon. At the original site, the sodium gates close and the refractory period begins. Potassium gates open and potassium rushes out of the axon, down its diffusion gradient. As the inside of the axon once again becomes negative with respect to the outside, repolarisation occurs.

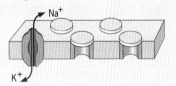

(d) The sodium–potassium pump uses energy to restore the sodium and potassium gradients. As only a tiny proportion of sodium and potassium ions move in and out of the axon when an action potential passes, several nerve impulses can be transmitted before the sodium–potassium pump is really needed.

Fig 21.15 **An action potential is propagated by ions moving through the axon membrane**

The start of the action potential has been shown to coincide with the opening of special **ion channels** in the axon membrane. Ion channels are large protein molecules straddling the membrane which are thought to change shape to allow or prevent the movement of ions. They are described as 'gated' channels because they can be opened and closed. When they are open, they allow specific ions to pass through, as shown in Fig 21.14. In the membrane of a neurone with a resting potential, most are closed.

Fig 21.15 shows an action potential travelling along an axon. As the **gates** of the **sodium channels** in the membrane open, sodium ions move rapidly into the axon. This inflow depolarises a small part of the axon. Only about one in ten million Na^+ ions in the fluid outside the axon crosses the membrane, but this is enough to give the inside of the membrane at that point a positive charge of approximately $+40$ mV relative to the outside. Depolarisation of the membrane lasts only a few milliseconds (thousandths of a second). Then, the sodium gates rapidly close. They cannot reopen until the resting potential has been restored.

Repolarisation

As the action potential travels away along the axon, **potassium channel gates** open, allowing K^+ ions to pass out of the cell. The inside of the membrane immediately 'behind' the action potential becomes positively charged, regaining its resting potential of -70 mV. Then, the membrane is said to be **repolarised**.

The refractory period

Nerves conduct messages by 'firing' repeated action potentials along the nerve fibres. The time delay between the conduction of one action potential and the next is called the **refractory period** (Fig 20.13). It has two phases. Immediately after the sodium channels close, no further impulse can be conducted at all: this short period of time is called the **absolute refractory period**. As potassium channels open, the membrane goes into a recovery stage and becomes increasingly responsive to a strong stimulus, even though it is not yet fully repolarised. This period of time is known as the **relative refractory period**.

The refractory period imposes a limit on the frequency of nerve firing. Large nerve fibres recover in 1 millisecond and could theoretically propagate 1000 impulses per second. Small fibres take longer to recover – about 4 milliseconds – and so could propagate about 250 impulses per second.

?

D (a) In the resting state, is the ion channel open or closed?

(b) In that position, what is its function?

(c) When sodium gates open, two forces act on Na+ ions outside the axon, causing them to move into the axon from the extracellular fluid. What are these forces, and how do they operate?

E An action potential normally travels in one direction along an axon. However, if it is artificially stimulated in the middle, two action potentials arise, each moving away from the point of stimulation. Explain this and include the idea of the refractory period in your answer.

See question 2.

F Why would a squid need giant axons up to 1 mm in diameter? (Those of humans, for example, average only a few thousandths of a millimetre across.)

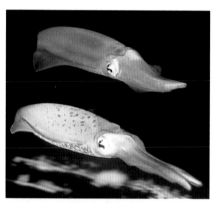

Fig 21.16 **The squid, which provides experimenters with axons of 1 mm diameter from the nerves serving its muscles. The squid uses strong muscles all at the same time to force a jet of water very rapidly through its siphon (visible under the body), and so propel it to safety**

G What effect does a myelin sheath have on conduction velocity?

The refractory period ensures that each action potential is separated from the next, with no overlapping of signals. We can think of the information the signal conveys as coded information. The refractory period also ensures that a nerve impulse flows in one direction only: the wave of depolarisation can only move away from the refractory region, towards the axon terminal, and therefore onwards to the next neurone in the communication chain.

The speed of an action potential

The speed at which a nerve impulse or action potential travels is known as its **conduction velocity**. In human nerve fibres, values range from 1 to 3 metres per second in unmyelinated fibres, and between 3 and 120 metres per second in myelinated fibres. In general, conduction velocity depends on the following factors.

- **Axon diameter**. The larger the axon, the faster it conducts.
- **Myelination of the neurone**. A nerve impulse travels faster in a **myelinated nerve** than an **unmyelinated nerve**. Myelin, a fatty material, acts as a barrier to the movement of ions across the membrane. Depolarisation occurs at the small gaps (nodes of Ranvier) between the Schwann cells. The action potential therefore 'jumps' from one gap to the next, a process known as **saltatory conduction** (see page 336).
- **Number of synapses involved**. Communication between neurones across the minute gaps at the synapses involves chemical release and a brief time delay. Therefore, the greater the number of synapses (or number of neurones) in a series of neurones, the slower the conduction velocity.

Saltatory conduction in myelinated neurones

Saltatory nerve conduction takes place in electrically insulated (myelinated) nerve fibres, which are found only in vertebrates. The insulating Schwann cells round the axon allow ions to cross the membrane *only* at the nodes – the gaps between the Schwann cells. As a result, action potentials arise only at the nodes, and conduction occurs in a series of saltatory 'jumps' from node to node, as in Fig 21.17. (*Saltare* is Latin for 'to leap'.)

Saltatory conduction has two advantages:

- The conduction of a nerve impulse is fast. In humans unmyelinated fibres nerve impulses travel at 1 to 3 metres per second, while myelinated fibres conduct at speeds of up to 120 metres per second.
- Metabolically, saltatory conduction is economical, because fewer ions move across the membrane, so the ion pumps need less energy to restore the ionic balance.

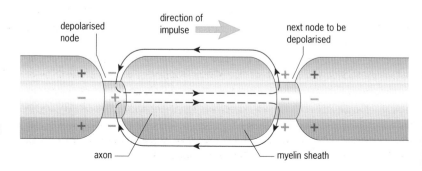

Fig 21.17 **Saltatory conduction in a myelinated nerve: the depolarisation 'jumps' from node to node**

Transmission through a nerve fibre

Understanding how information is carried along individual neurones helps us to explain how the nervous system functions as a whole. But it is important to remember that neurones do not act independently; they interact with other neurones. This interaction is responsible for the properties of **nerves** (bundles of many neurones) and of complete nervous systems.

For example, the size of an action potential in an individual axon is the same, no matter where along the axon the measurement is taken. But, if a nerve is stimulated, action potentials with differing peak values are recorded. These show the activity of individual axons, each of which has a slightly different depolarisation voltage.

The interaction between neurones also explains how we can tell if a touch is light or heavy, or if a sound is soft or loud. You might think that this property of the nervous system means that the all-or-none law of conduction must be wrong. In fact, this law still applies to *individual neurones* but a *nerve* can show a graded response. It can distinguish between a strong and a weak stimulus because of **impulse frequency**. A strong stimulus causes many individual neurones within a nerve to 'fire' often, setting up one action potential after another. A weak stimulus causes only a few neurones in the nerve to 'fire', and they do so only now and then.

Simple chemicals transfer information between neurones

The **electrochemical theory** of nerve action was confirmed by the German pharmacologist Otto Loewi (1873–1961), working with an English physiologist Henry Hallett Dale (1875–1968). They discovered the link between electrical conduction along nerves and the release of chemicals at nerve endings. One of these chemicals was identified as acetylcholine, a **neurotransmitter** which transmits information from one nerve cell to the next.

4 COMMUNICATION BETWEEN NEURONES: THE SYNAPSE

H A synaptic cleft is very narrow. Why should this be?

When an action potential reaches the end of an axon it is passed on to the next neurone, or on to an effector cell such as a muscle or gland. The axon of one neurone does not usually make direct contact with the cell body of the next; the two cells are separated by a gap called a **synapse**. On page 345, an electronmicrograph of a synapse is shown in Fig 21.20, and its structure is shown in Fig 21.18.

The cell which carries a signal towards a synapse is described as a **presynaptic** cell; the cell carrying the signal away from the synapse is the **postsynaptic** cell. Presynaptic cells are always neurones but postsynaptic cells can be either neurones or effector cells such as muscle cells or glands.

Neurones usually transmit information across synapses using transmitter chemicals also called **neurotransmitters**. Molecules of these chemicals cross the synaptic cleft and cause electrical changes in the membrane of the postsynaptic cell. If an action potential is produced, this then passes along the axon to the next neurone in the series. Direct communication between neurones is rare, but electrical synapses do exist. (See also page 346.)

See question 3. ■

?

I When a neurone is over-stimulated, it becomes fatigued. Suggest a physical basis for this fatigue.

Why have synapses?

Synapses are important because they allow the transfer of information in nerve networks to be controlled. Synapses:

- allow information to pass from one neurone to another.
- help ensure that a nerve impulse travels in one direction only.
- allow the next neurone to be excited or inhibited.
- can amplify a signal (make it stronger).
- protect nerve networks by not firing when over-stimulated. When this happens the synapse is said to be **fatigued**. Over-stimulation might damage muscle or glandular tissue.
- can filter out low-level stimuli. For example, you fail to notice the sound of a clock ticking because synapses are 'filtering out' the signal of sound.
- aid information processing by the action of **summation** (adding together the effect of all impulses received).
- are modifiable and can form a physical basis for memory.

Structure of the synapse

Fig 21.18 shows the main features of a chemical synapse. The axon terminal of a presynaptic neurone has a bulb-like appearance, and is often called the **synaptic bulb**. It meets the cell body or dendrite of the next axon, leaving a gap or **synaptic cleft** of about 20 nm between. This is less than one five-hundredth the width of a human hair.

The synaptic bulb contains many mitochondria which provide energy for the manufacture of chemical transmitters. **Synaptic vesicles** are temporary vacuoles which store **neurotransmitter** chemicals. Neurotransmitters are small molecules that can diffuse easily across the synaptic cleft.

Fig 21.18 **Main structures at a synapse**

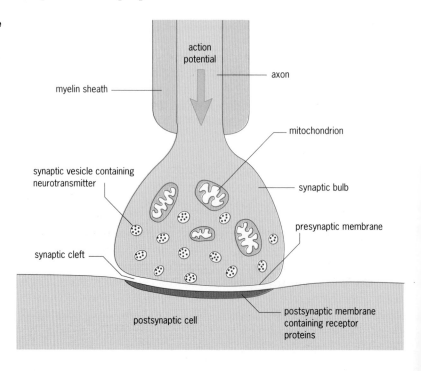

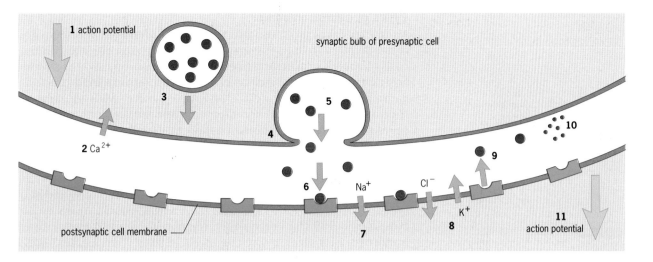

1 action potential

synaptic bulb of presynaptic cell

3

5

4

2 Ca^{2+}

6

Na$^+$

Cl$^-$

9

10

K$^+$

8

11
action potential

7

postsynaptic cell membrane

Fig 21.19 **The sequence of events in chemical transmission at a synapse**

What happens at the synapse?

As you read the next section, follow the stages of chemical transmission at a synapse shown in Fig 21.19.

At the presynaptic cell

An action potential arrives at the synaptic bulb. This opens calcium channels in the presynaptic membrane. As the Ca^{2+} ion concentration inside the membrane is lower than outside, Ca^{2+} ions rush in. As their concentration increases, synaptic vesicles move towards the membrane, fuse with it, and release neurotransmitter into the synaptic cleft. The short journey across the synapse takes about a millisecond, longer than an electrical signal takes to travel the same distance. This time is therefore called the **synaptic delay**.

At the postsynaptic cell

The neurotransmitter binds to receptors on the postsynaptic cell membrane, some neurotransmitters open sodium channels in the postsynaptic membrane, causing an inflow of Na$^+$ ions. This creates an **excitatory postsynaptic potential** (EPSP) in the membrane. This potential lasts for only a few milliseconds and can travel only a short distance, but it makes the membrane *more* receptive to other incoming signals.

Other neurotransmitters open chloride and potassium channels, causing Cl$^-$ ions to flow into the cell and K$^+$ ions to flow out. This creates the **inhibitory postsynaptic potential** (IPSP) which makes the postsynaptic membrane *less* receptive to incoming signals. If more EPSPs than IPSPs are produced in the postsynaptic membrane, the change in potential can exceed the threshold potential needed to create a new action potential.

Receptor binding can also lead to the formation a **second messenger** (a transmitter substance) such as **cyclic AMP** (cAMP). This also changes the ionic permeability of the membrane, but it has a longer lasting metabolic effect on the ion channels. Such long-term changes at receptor surfaces of neurones in the brain are thought to underlie **memory**.

Once the neurotransmitter has acted on the postsynaptic membrane, it is immediately broken down (hydrolysed) by an enzyme found on the postsynaptic membrane. If acetylcholine remained, it could continue to stimulate the neurone, even without new impulses coming from the presynaptic cell.

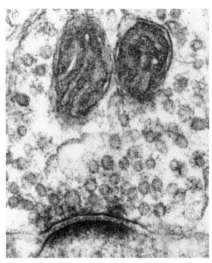

Fig 21.20 **A synapse showing the mitochondria (top) and vesicles (centre) in the presynaptic neurone, and neurotransmitter passing across the synaptic cleft**

?

J If the excitatory effect (EPSP) of stimulation is greater than the inhibitory effect (IPSP) but less than the threshold level, what is the response of the postsynaptic neurone?

Types of neurotransmitter

More than 50 neurotransmitters have been identified and there are certainly more to find. There are four main groups, classified on the basis of chemical structure: **acetylcholine** (neurones which release acetylcholine are described as **cholinergic**); **amino acids** such as gamma-aminobutyric acid (GABA), glycine and glutamate; **monoamines** such as noradrenaline, dopamine and serotonin (neurones which release noradrenaline are described as **adrenergic**); and **neuropeptides** (chains of amino acids) such as endorphins.

Acetylcholine also acts throughout the brain, modifying the activity of other neurotransmitters. Nerve pathways in which acetylcholine is a neurotransmitter seem to be involved in motivation and memory. **Acetylcholinesterase**, the enzyme which breaks down acetylcholine on the postsynaptic membrane, does so in one five-hundredth of a second. The chemicals in some 'nerve gases' work by inhibiting acetylcholinesterase.

NICOTINE – THE DRUG THAT MIMICS ACETYLCHOLINE

DO YOU WONDER why people continue to smoke, even though they know the increased risk of lung cancer and heart disease associated with their habit? The answer lies with one of the components of tobacco: nicotine.

Nicotine is addictive. It affects the brains of smokers, making them feel less stressed, better able to concentrate and less likely to eat sweet foods. Smokers become tolerant to nicotine over time, needing to smoke more to achieve the same effects. But how does nicotine cause addiction?

Studies carried out in the early 1980s using nicotine labelled with a radioactive tracer showed that it is taken into the brain very rapidly. Once there, it binds tightly to acetylcholine receptors, fooling postsynaptic cells into 'thinking' they are being stimulated. It also binds to other receptors which normally accept another neurotransmitter, dopamine.

The action of nicotine on both types of receptor in specific areas of the brain causes long-lasting changes to cell connections and may explain why it is addictive. Dopamine receptors, in particular, are known to be involved in addictions to other substances such as amphetamines and cocaine.

Because nicotine is addictive but not carcinogenic (that is the responsibility of the chemicals in tobacco tar), smokers keen to kick the habit can get help. They can buy skin patches and gum which deliver nicotine to the brain, but not tar to the lungs.

Although patches and gum do help smokers to cut down or stop smoking altogether, they are only a partial solution. Nicotine affects the acetylcholine receptors in the parasympathetic nervous system that are involved with the constriction of blood vessels. Over time, circulatory problems and heart disease can result, and it is best to avoid these effects altogether.

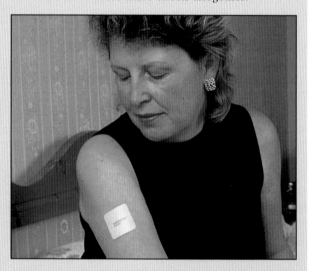

K Nicotine, in tobacco smoke, mimics acetylcholine. Explain its short-term effects, such as irritability, increased heart rate and blood pressure.

Electrical synapses

A few nerve cells communicate directly, by electrical synapses. Such cells are connected by hollow protein channels which allow an electrical impulse to travel directly from one cell to the other. Electrical synapses were first found in the crayfish, and later in vertebrates. They occur in human retinal ganglion cells. Because there is no synaptic delay at electrical synapses, signals pass almost instantaneously from one cell to the next.

5 COMMUNICATION BETWEEN NEURONES: PROCESSING INFORMATION

Neurones do not exist in isolation; they form complex networks which behave like the integrated circuits in electronics. (An integrated circuit is defined as 'an assembly of electronic components which cannot be subdivided without destroying its function'.)

Facilitation and summation

Synapses are sites of information processing. The action potential is an all-or-nothing event, but transmission of information across the synapse is *graded*.

A neurone may have both excitatory and inhibitory synapses with the different neurones that send messages to its cell body. The potential across the postsynaptic membrane of the receiving cell changes, but whether the cell develops an action potential at its axon hillock is determined by:

- the sum of all the excitatory and inhibitory synapses at any particular moment,
- the distance of the synapses from the axon hillock, since changes in potential decrease in magnitude the further they travel over the cell body.

The first wave of neurotransmitter to bind at a postsynaptic membrane sets up **excitatory postsynaptic potentials (EPSPs)**. Though they may not be strong enough to push the postsynaptic cell to its threshold potential, they make the postsynaptic membrane more excitable. As more neurotransmitter molecules bind and further EPSPs are set up, it is more likely that an action potential will be generated. This effect is called **facilitation** (see Fig 21.21(a)).

Most synaptic cells have receptors for many different neurotransmitters, and not all potentials that are set up are excitatory. **Inhibitory postsynaptic potentials (IPSPs)** make the membrane potential more negative, so that an action potential is *less* likely to occur (see Fig 21.21(b)). The effects of all the EPSPs and the IPSPs reaching a postsynaptic cell are combined in a process called **summation**. If the number and frequency of EPSPs is greater than the number and frequency of IPSPs, an action potential may result.

L A brain cell can send a repeat signal at the rate of about 10 per second (for example, to cause a muscle fibre to contract). There are about ten thousand million neurones in the brain and each can form synapses with ten thousand other neurones. Assuming that only one in a thousand brain cells is firing at one time, how many interconnections does the brain make per second?

Fig 21.21 (a) **Facilitation** and (b) **summation**

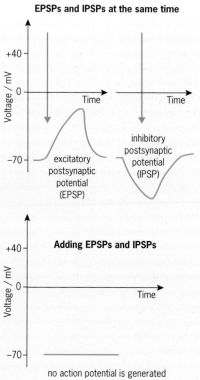

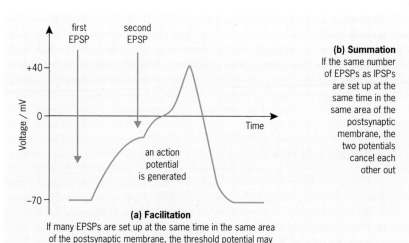

(a) Facilitation
If many EPSPs are set up at the same time in the same area of the postsynaptic membrane, the threshold potential may be reached and an action potential may result

(b) Summation
If the same number of EPSPs as IPSPs are set up at the same time in the same area of the postsynaptic membrane, the two potentials cancel each other out

Nerve networks

Nervous systems may contain millions of neurones. As well as transmitting information, they process it by being organised into groups, or nerve networks. Some groups simply relay information, others produce predetermined and often complex nerve responses.

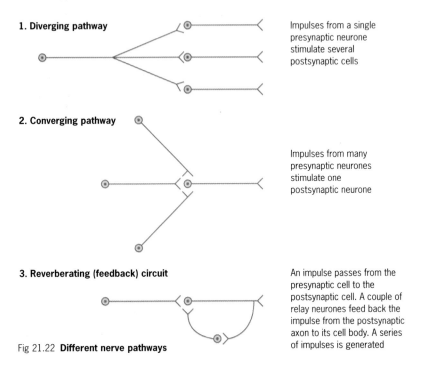

1. Diverging pathway

Impulses from a single presynaptic neurone stimulate several postsynaptic cells

2. Converging pathway

Impulses from many presynaptic neurones stimulate one postsynaptic neurone

3. Reverberating (feedback) circuit

An impulse passes from the presynaptic cell to the postsynaptic cell. A couple of relay neurones feed back the impulse from the postsynaptic axon to its cell body. A series of impulses is generated

Fig 21.22 **Different nerve pathways**

Nerve cell pathways vary in their complexity, as shown in Fig 21.22. Simple pathways merely stimulate the next neurone in the chain. But most sequences are more complex. **Diverging** and **converging pathways** allow neurones to interact in specific ways, while a **reverberating circuit** uses feedback to deliver a repeated series of impulses.

6 COMMUNICATION BETWEEN NEURONES: STRUCTURAL ORGANISATION

Neurones are grouped together within a nervous system to form **nerve fibres**. A bundle of such fibres enclosed in a sheath is called a nerve. In vertebrates, the correct term for a bundle of fibres and its sheath in the central nervous system is a **tract**, while a bundle and sheath in the peripheral nervous system is called a **nerve**. Nerve fibres and the neurones they contain are named according to the direction in which they transmit nerve impulses.

Sensory nerves

Sensory, or **afferent nerves** contain neurones which carry messages *from* **sensory receptors** *towards* the central nervous system. Sensory receptors can respond to stimuli that come from inside, or from outside the body. Sensory nerves carry impulses from these receptors to the spinal cord, or to the brain via the spinal cord. (Afferent is from the Latin for *carry towards*.)

Motor nerves

Motor or **efferent nerves** contain neurones which take messages *away from* the central nervous system – the spinal cord or brain via the spinal cord – towards **effectors** such as muscles and glands. (Efferent is from the Latin for *carry away*.)

Most cranial and spinal nerves are called **mixed nerves** because they are composed of many afferent *and* efferent nerve fibres.

Intermediate nerves

Interneurones in intermediate nerves carry impulses from sensory neurones to motor neurones within the central nervous system. Interneurones with long axons are called **relay neurones** and convey signals over long distances. Interneurones with short axons are called **local circuit neurones** and convey signals over short distances. Complex functions such as learning and memory depend on the interaction of thousands of local circuit neurones. Simpler reflexes involve only a few local circuit neurones, in some reflex arcs sensory and motor neurones are directly connected.

7 COMMUNICATION BETWEEN NEURONES: THE REFLEX ARC

A **reflex** response is a rapid, automatic response to a stimulus; pulling your hand away from a hot pan handle is a reflex action. A simple reflex involves communication between neurones in the peripheral nervous system and the spinal cord. The brain may be informed, but does not take part in the actual response. Although many reflexes have a protective function, they can also help the body to coordinate complex muscular events, such as swallowing. The nerve pathway involved in a reflex action is called a **reflex arc**.

> ?
> **M** Write down other responses that you might make without deliberately deciding to do so.

The knee-jerk reflex

The basic components of the knee-jerk reflex arc are shown in Fig 21.23. They include a receptor, a sensory neurone, a motor neurone and an effector. In everyday life, the knee jerk reflex helps the body to control the movement of the legs during running. As it starts with stretching of the muscle, it is known as a **stretch reflex**.

See question 4.

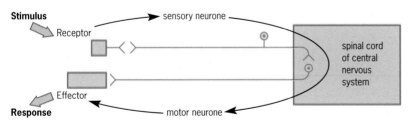

Fig 21.23 **Components of the knee-jerk reflex arc**

A tap on the patellar tendon with a small percussion hammer makes the front thigh muscles stretch very slightly. Receptors in the muscles called **spindles** respond to the change in length by generating nerve impulses, which pass along sensory neurones to the spinal cord. Here, the sensory neurones make synapses with motor neurones, and a message is sent straight back down the leg to the thigh muscles. The muscles contract, causing the lower leg to swing forward and giving the familiar knee jerk.

The knee jerk involves only a sensory and a motor neurone (no interneurone), and there is no direct communication with the brain. Because it has only a single synapse, this type of reflex is called **monosynaptic**.

In medicine, testing reflexes can help to detect disorders of the nervous system, allowing injured tissue along a nerve pathway to be located.

Other spinal reflexes

Polysynaptic reflexes (Fig 21.24) have more than one synapse and are more common than monosynaptic reflexes.

An example of a polysynaptic reflex is the withdrawal reflex, shown in Fig 21.25. Pulling your hand away from a hot pan involves a circuit containing sensory receptors, sensory neurones, spinal relay neurones (interneurones), motor neurones and effector muscles. The principle is the same: receptors are stimulated by the heat of the pan and impulses pass along sensory neurones towards the spinal cord. Here, instead of making synapses with motor neurones, they pass their signals on to relay neurones. These connect with motor neurones which cause muscles to contract, moving your hand away from the pan.

You know that you have touched the hot pan because some impulses do travel to the brain but the movement which is part of the reflex is **involuntary**, not under your conscious control.

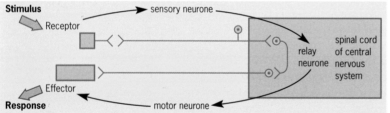

Fig 21.24 **A polysynaptic reflex arc**

Fig 21.25 **The polysynaptic reflexarc in more detail**

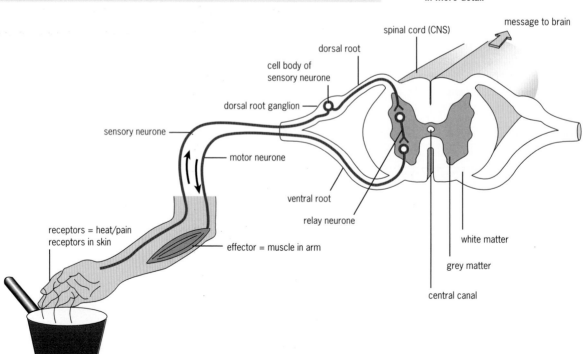

Nerve nets

The simplest type of nervous system is the **nerve net,** found in jellyfish and sea anemones (Phylum Cnidaria), simple, two-layered (diploblastic) animals which use diffusion to obtain oxygen and to transport food and waste around their bodies. Fig 21.26 shows the structure of a nerve net in a sea anenome. Nerve nets may be simple, but their individual neurones are well developed and organised into a complete conducting system.

It is possible to stain the entire nerve net in *Hydra* (Fig 21.26), a small freshwater relative of the sea anemone, and view it under a light microscope. There are no large tracts of nerve fibres, or any structure resembling a brain. This simple nerve net allows only slow conduction, yet is adequate for the *Hydra*'s way of life. *Hydra* has simple light receptors to detect light and dark, chemoreceptors to detect food chemicals, and mechanoreceptors that sense touch and can trigger the action of stinging cells. Its tentacles are mobile and, overall, *Hydra* has a high level of integration for an organism with such a simple nervous system.

Two-part nervous systems

The behaviour of jellyfish, sea anemone, and *Hydra* shows wide differences, but is very limited compared to most invertebrates. The body systems of annelid worms, molluscs and arthropods have a more complex organisation, and so need a greater level of coordination. This is provided by a more elaborate nervous system which is divided into two parts, a **central system** consisting of a **ventral nerve cord** and **ganglia** (collections of nerve cell bodies) and a **peripheral system** of nerves and receptors. The central system is the destination for all sensory information coming into the animal via receptors. The peripheral nerves act as conducting pathways between the two.

Fig 21.27 shows the nervous systems of the arthropods woodlouse, caterpillar, honey bee, water bug and mite, showing how ganglia have become fused during evolution.

Sea anemone

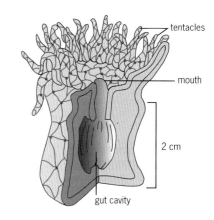

- tentacles
- mouth
- 2 cm
- gut cavity

Hydra

receptors mainly around tentacles and mouth

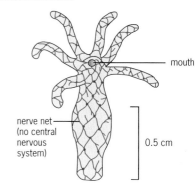

- mouth
- nerve net (no central nervous system)
- 0.5 cm

Fig 21.26 **Nerve nets in sea anemone (above) and *Hydra* (below)**

Fig 21.27 **Nervous systems of woodlouse, caterpillar, honey bee, water bug and mite**

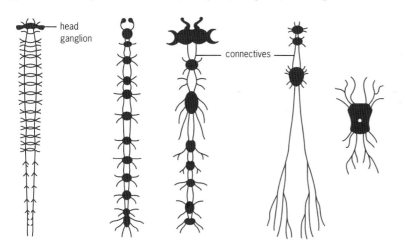

- head ganglion
- connectives

Cephalisation

Most multicellular animals are segmented. In the organisation of their nervous systems, a small group of nerve cells (a **ganglion**) controls each segment. This arrangement is seen in the nervous systems of the earthworm, leech and lobster shown in Fig 21.28. The more work a segment has to do, the larger the amount of nerve tissue that is needed to control it.

During the course of evolution, animals have become more complex. Feeding and sensory organs have become located at the front (anterior) end of the body. In more advanced organisms, a definite **head** has evolved. This trend is called **cephalisation**. Nerve tissue has tended to concentrate in the head, forming a **brain**. The human brain, it may be argued, is the most evolved example.

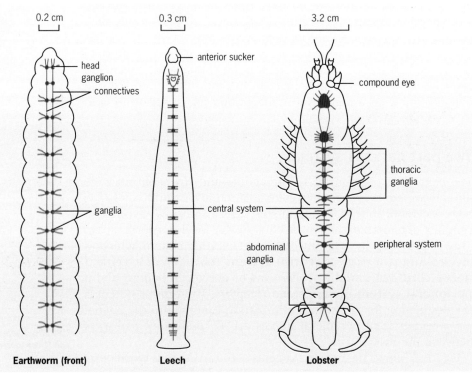

Fig 21.28 **Nervous systems of earthworm, leech and lobster**

9 ORGANISATION IN THE VERTEBRATE NERVOUS SYSTEM

Fig 21.29 shows how the vertebrate nervous system is organised.

The vertebrate nervous system can be divided into two main components. The **central nervous system** (CNS) is the site of information processing. The **peripheral nervous system** (PNS) is a highway conducting information to and from the CNS.

The central nervous system

The central nervous system is the central control and coordinating system of a vertebrate. It receives information from receptors, processes it and then passes instructions to effectors to allow the organism to respond effectively to its environment. The CNS consists of the brain and spinal cord.

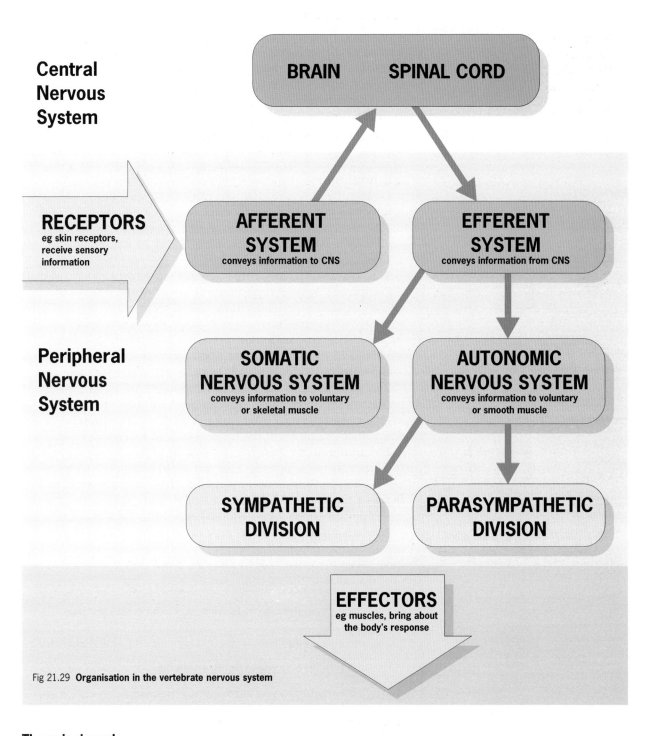

BRAIN   SPINAL CORD

RECEPTORS
eg skin receptors,
receive sensory
information

AFFERENT
SYSTEM
conveys information to CNS

EFFERENT
SYSTEM
conveys information from CNS

Peripheral
Nervous
System

SOMATIC
NERVOUS SYSTEM
conveys information to voluntary
or skeletal muscle

AUTONOMIC
NERVOUS SYSTEM
conveys information to voluntary
or smooth muscle

SYMPATHETIC
DIVISION

PARASYMPATHETIC
DIVISION

EFFECTORS
eg muscles, bring about
the body's response

Fig 21.29 **Organisation in the vertebrate nervous system**

The spinal cord

The spinal cord starts at the base of the brain and ends at the first
lumbar vertebra (this is roughly at waist level). It is enclosed
within the body cage of the **vertebral column** and has a diameter
of about 5 millimetres. Further protection is provided by three
layers of tough membranes called **spinal meninges** and by
the **cerebrospinal fluid** which cushions the cord, acting as a
'shock-absorber'.

Each vertebra has an opening on its right and left sides to let
spinal nerves pass through. These extend into the body, forming
the peripheral nervous system. In the human body there are 31
pairs of spinal nerves.

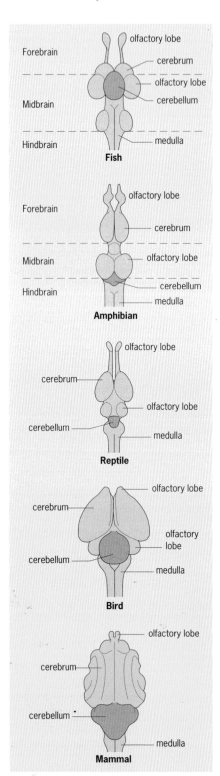

Fig 21.30 **Comparing vertebrate brains (viewed from below). Their size is adjusted on the assumption that all animals weigh the same. Fish and amphibians show the primitive three-part structure that is obscured as brains become more complex.**

The cerebral hemispheres of the cerebrum, originally the centre for olfaction (smelling) are particularly well developed in birds and mammals, as the centre for integrating sensory information.

The size of the cerebellum relates to the intricacy of body movements – large in fish, birds and mammals

The brain

The brain receives stimuli from inside and outside the body. It maintains basic involuntary body functions such as heart rate, breathing rate and temperature control. It also coordinates the semi-automatic muscular actions of the body such as swallowing, and it initiates and controls voluntary activities such as walking and running. The human brain is the site of higher mental functions such as reasoning, emotion and personality.

In primitive vertebrates, the brain is just a swelling of the front end of the spinal cord. It consists of three parts: a **hindbrain**, **midbrain** and **forebrain** which correspond to basic functional regions.

In fish, this three-part pattern can still be seen but in humans, it is only noticeable in the early stages of development of the embryo. Fig 21.30 shows the organisation of the brain in fish, amphibian, reptile, bird and mammal (human). Refer to this drawing as you read on.

The **hindbrain** has three distinct structures. The **medulla oblongata** is a swollen portion at the top of the **brain stem** that houses vital centres controlling heart beat, breathing and blood supply. The **cerebellum** controls body movement and maintains balance, and the **reticular activating system** (a collection of neurones in the centre of the brain stem) filters incoming stimuli and controls wakefulness and sleep.

The **midbrain** is particularly important in vision: studies of predatory fish have shown that those with good vision which live near the water surface have larger midbrains than those with poor vision that live down in the mud.

The midbrain in humans, as well as other animals, links the forebrain to the hindbrain (see Fig 21.31). Our emotions, which are located in the forebrain can affect basic functions of the hind brain such as control of blood vessel diameter, heart rate and sweating. We blush when embarrassed, turn red when angry, and pale when frightened. When very nervous, we sweat, take shallow breaths, and our pulse rate rises. All these survival mechanisms have taken on a new role in humans, since they also communicate our feelings.

Fig 21.31 **Vertical section through the human brain, showing its main structures**

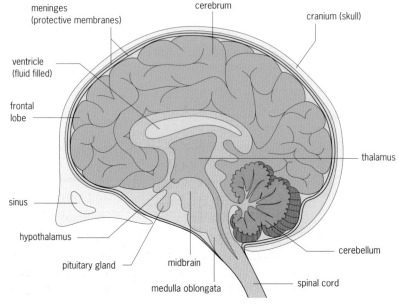

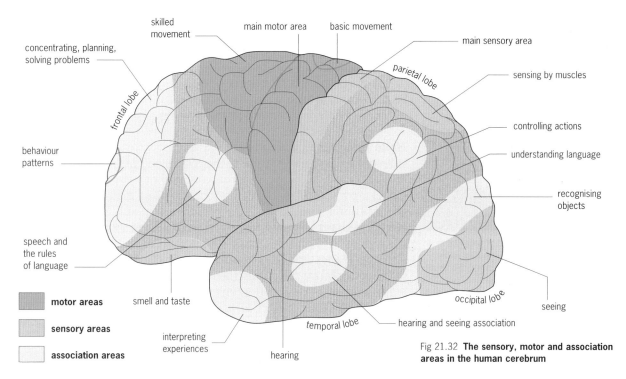

concentrating, planning, solving problems

skilled movement

main motor area

basic movement

main sensory area

parietal lobe

sensing by muscles

frontal lobe

controlling actions

understanding language

behaviour patterns

recognising objects

speech and the rules of language

■ motor areas

□ sensory areas

□ association areas

smell and taste

interpreting experiences

temporal lobe

hearing

hearing and seeing association

occipital lobe

seeing

Fig 21.32 **The sensory, motor and association areas in the human cerebrum**

The **forebrain** has two main parts, the cerebrum and a region containing the thalamus and the hypothalamus. The **cerebrum** is made up of two cerebral hemispheres. They have a thin outer layer, the **cortex** which is thrown into many folds, with **fissures** (grooves) between. The cerebral cortex has three main roles:

- It controls and initiates voluntary muscle contraction.
- It receives and processes information from the senses (see also Chapter 23) and all the body's receptors.
- It carries out the 'higher' mental activities of reasoning, and is regarded as the site of personality and emotion.

Different areas of the cerebral hemispheres are associated with different sensory and motor functions. **Association areas** link sensory and motor information and store memory. The areas are mapped out in Fig 21.32. The cerebrum of mammals is very large compared to the forebrains of other vertebrates. It is the dominant feature of the human brain where it surrounds many other structural features. Folds in the surface of the cerebrum increase the surface area for **centres of control**, where incoming nerve impulses are interpreted, or **integrated,** in the light of information already stored in the brain.

The **hypothalamus** is concerned with the emotions, and the drives of an animal such as hunger and thirst. It also helps control the autonomic nervous system in the control of body temperature – **thermoregulation** – and the water-salts balance in the blood – **osmoregulation**. It is linked to the pituitary gland and is involved in the release of hormones, including the antidiuretic hormone (which controls water reabsorption in the kidneys). The **thalamus** (see Fig 21.31) directs sensory information from the sense organs to the correct part of the cerebral cortex (see Chapter 23).

The brain of a human weighs about one-fiftieth of body weight. Being very delicate, about as rigid as thick blancmange, the brain is protected by the skull or **cranium** and by the **cranial meninges**, membraneswhich are continuous with the spinal meninges.

?

O Where in the human brain would you find centres controlling **(a)** temperature control, **(b)** heart beat rate, **(c)** personality and emotions?

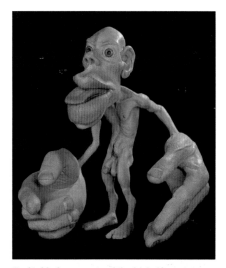

Fig 21.33 **Some parts of the body have more sensory receptors in a given surface area than others, and similarly, more tissue in the brain is allotted to processing information from them. This model of a human is distorted to reflect the different brain space allocated to receiving sensations from different parts of the body (except for the eyes, which have more brain space than all the rest of the body)**

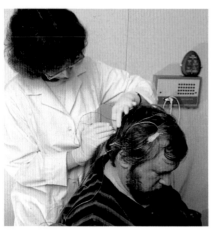

Fig 21.34 **Electrical activity in the subject's brain is investigated by attaching electrodes to his scalp and recording the electrical signals as an electroencephalograph or EEG**

Cerebrospinal fluid bathes the outside of the brain and fills the chambers (or ventricles). Twelve pairs of cranial nerves **innervate** (supply nerves to) various regions of the head. The human brain is thought to contain ten thousand million (10^{10}) neurones – about the same number as the stars in our Galaxy. Each neurone may be in contact with a thousand other cells, providing an immense number of different communication routes.

White matter and grey matter

Different areas of nerve tissue in the central nervous system are either greyish or white. **Grey matter** contains nerve cell bodies; their nuclei are responsible for the grey colour. Grey matter also contains glial cells and some nerve fibres (these are mainly unmyelinated). **White matter** consists largely of myelinated fibres (their fatty sheaths are white), some unmyelinated fibres and glial cells.

White matter and grey matter are distributed in specific areas. Grey matter forms the surface layer of the brain, regions deep in the brain known as the basal ganglia, and the central column of the spinal cord. White matter forms a layer between the two areas of grey matter in the brain and encloses the column of grey matter in the spinal cord: look back at the spinal cord in Fig 21.25.

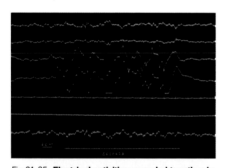

Fig 21.35 **Electrical activities recorded together in a sleeping subject. The top two traces are EEGs of brain activity. The third and fourth show movements of the right and left eyes. The fifth is a trace for heart activity (an electrocardiogram). The sixth and seventh show activity in the muscles of the larynx and neck**

The peripheral nervous system

The nerves of the peripheral nervous system behave like major road systems, carrying traffic in and out of the central nervous system. **Afferent** nerves, also called **sensory** nerves, carry information from sensory receptors into the CNS.

Efferent nerves, also called **motor** nerves, carry information from the CNS out to effector organs. The efferent system can be further subdivided into the **somatic** and **autonomic** systems. These differ in their functions rather than their structure or position in the body.

The somatic nervous system

The somatic nervous system contains both afferent and efferent nerves. It receives and processes information from receptors in the skin, voluntary muscles, tendons, joints, eyes, tongue, nose and ears, giving an organism the sensations of touch, pain, heat, cold, balance, sight, taste, smell and sound. It also controls voluntary actions such as the movement of arms and legs.

There is a parallel here between the nervous and muscular systems of the human nervous system. In the same way that we have voluntary and involuntary components of our nervous system, we also have **skeletal muscle** to deal with voluntary muscle contraction and **visceral muscle** to deal with involuntary muscle contraction.

See question 5. ■

The autonomic nervous system

The **autonomic nervous system** (ANS) includes a chain of ganglia which lie close to the spinal cord and are associated with their own neurones. The system is entirely motor, made up of efferent nerves only. It does not carry sensory information (feedback from muscles and glands travels via the somatic system).

The ANS controls basic 'housekeeping' functions such as heart beat, digestion, breathing and blood flow. Heart beat, for example, can speed up, or it can slow down. These two involuntary actions are **antagonistic** (opposite) actions, and two separate parts of the ANS control these antagonistic actions – the **sympathetic** and the

P What, in the autonomic nervous system, controls breathing?

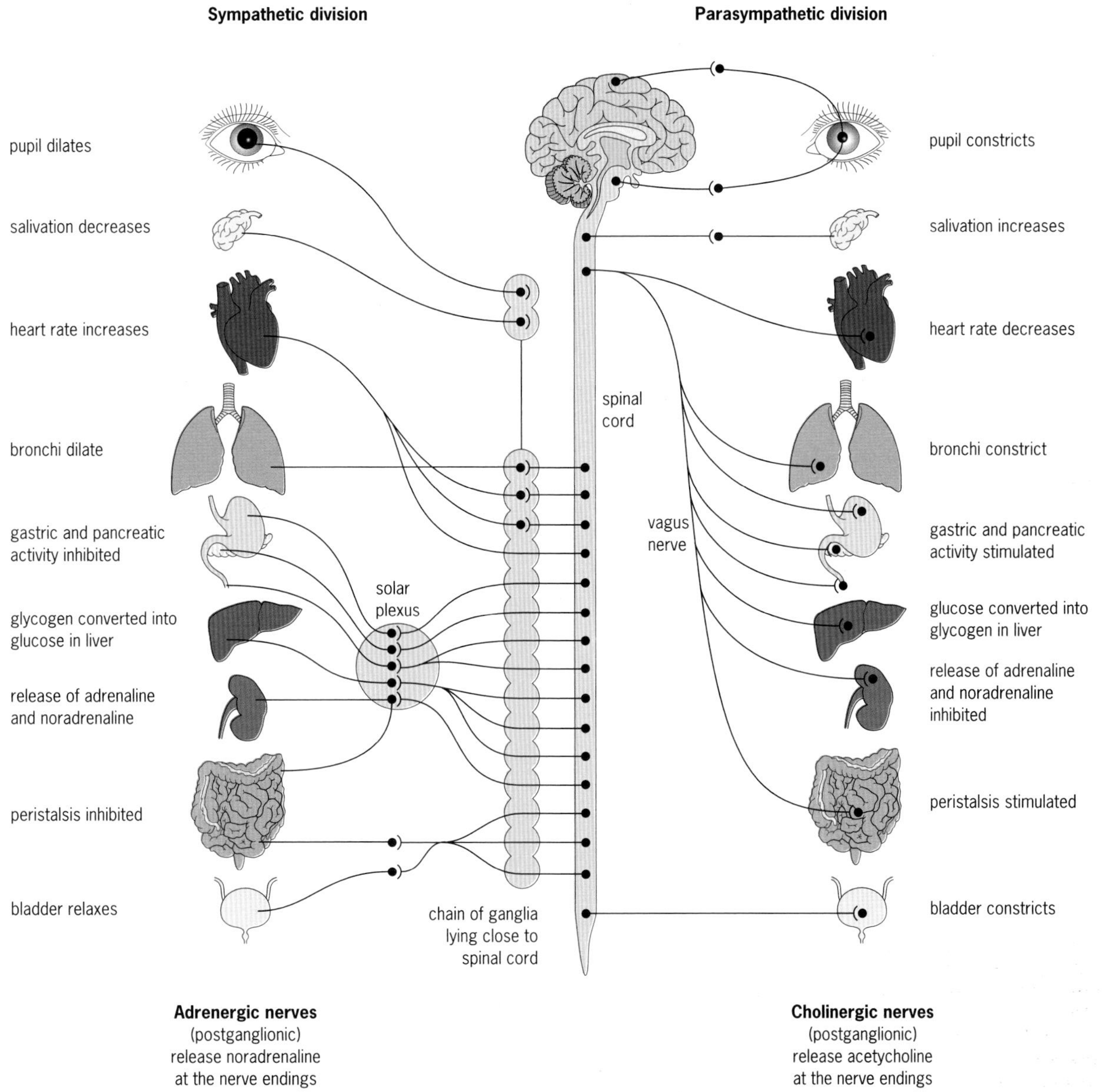

Sympathetic division **Parasympathetic division**

pupil dilates — pupil constricts

salivation decreases — salivation increases

heart rate increases — heart rate decreases

bronchi dilate — bronchi constrict

gastric and pancreatic activity inhibited — gastric and pancreatic activity stimulated

glycogen converted into glucose in liver — glucose converted into glycogen in liver

release of adrenaline and noradrenaline — release of adrenaline and noradrenaline inhibited

peristalsis inhibited — peristalsis stimulated

bladder relaxes — bladder constricts

spinal cord

vagus nerve

solar plexus

chain of ganglia lying close to spinal cord

Adrenergic nerves
(postganglionic)
release noradrenaline
at the nerve endings

Cholinergic nerves
(postganglionic)
release acetycholine
at the nerve endings

parasympathetic divisions (Fig 21.35). Most organs receive impulses from both divisions. In general, the sympathetic system tends to *stimulate* a particular function, while the parasympathetic has a *calming* effect.

Normally, the activity of both divisions is balanced. But if the body is stressed, then the 'fear, flight and fight' reactions of the sympathetic nervous system take over, causing an increase in heart rate, faster breathing, an increase in blood pressure and an increase in blood sugar level. This makes the body ready for sudden, strenuous activity. When the emergency is over, the parasympathetic system takes over. It decreases the heart and breathing rates and diverts blood supply back to 'housekeeping' activities such as digestion and food absorption.

The autonomic system was originally thought to be independent of the rest of the nervous system, hence the term autonomic, meaning 'on its own'. Now we appreciate that it is not autonomous or self-governing, but is regulated by areas within the central nervous system, including the hypothalamus, cerebral cortex and the medulla oblongata.

Fig 21.36 **The autonomic nervous system, showing the antagonistic (opposing) effects of the sympathetic and parasympathetic division.**
Autonomic neurones, like others in the nervous system, release chemical transmitter substances where they communicate with other cells. Autonomic neurones can synapse with other neurones within the ANS or with effectors such as muscles and glands. Neurones are classified as cholinergic or adrenergic on the basis of the transmitter they release

Q What body responses, normally controlled by the ANS, are sometimes (unconsciously) triggered by higher centres of the brain?

SUMMARY

When you have finished this chapter you should know and understand the following:

■ The **neurone** is the basic unit of any nervous system. Neurones are classified according to their structure and to their role in the body.

■ Some neurones are **myelinated**, some are **unmyelinated**. You should understand the difference between the two and be able to describe how both types carry information.

■ Information passes along the axon of a neurone as a wave of electrical disturbance on the axonal membrane, called an **action potential**.

■ You should be able to describe, in detail, the events that occur before, during and just after an action potential has passed along an axon.

■ Communication between neurones happens at **synapses**. Most of these are chemical synapses but a few nerve cells communicate directly, at electrical synapses.

■ Know the structure of a synapse and be able to describe how **neurotransmitters** carry information between one neurone and the next.

■ Neurones form complex nerve networks that can process information. You should understand the processes of **facilitation** and **summation** and should be able to describe some simple **nerve pathways**.

■ Nerves contain many neurones. **Sensory nerves** carry messages from sensory receptors towards the **central nervous system** (CNS). **Motor nerves** carry messages away from the CNS. **Intermediate nerves** link motor and sensory nerves, inside the CNS.

■ You should understand the structure and function of a simple **reflex arc**, and of a more complex **polysynaptic reflex**.

■ Nervous systems in different organisms show different levels of complexity. The **nerve net** of *Hydra* is one of the simplest. Most complex invertebrates need more complex nervous systems. You should understand the trend towards a two-part nervous system with the development of a brain (**cephalisation**) that fulfils this need.

■ You should be able to describe organisation within the vertebrate nervous system, particularly the human nervous system. You should understand the structure and function of the **central nervous system** (the brain and spinal cord) and of the **peripheral nervous system**.

■ The peripheral nervous system consists of the **somatic** and **autonomic** nervous systems. The latter consists of the **sympathetic** and **parasympathetic** nervous systems. You should know the basic role of each in the human body.

QUESTIONS

1 The graph shows the changes in the permeability of an axon to sodium ions and to potassium ions during an action potential.

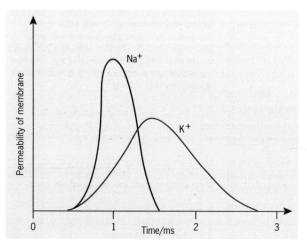

Fig 21.Q1

a) Describe how sodium ions enter the axon during the passage of the action potential.

b) Explain how the events shown in the graph:
    (i) lead to the inside of the axon becoming positive with respect to the outside during the first stage of an action potential;
    (ii) restore the resting potential.

[AEB: November 1993, Biology paper 1]

2 The table shows the speed of conduction of a nerve impulse along axons from three different species of animal.

Species	Diameter of axon/mm	Speed of conduction/m s-1
A	7	1.2
B	500	33.0
C	15	90.0

Which one of these three species is most likely to possess myelinated nerve fibres? Give a reason for your answer.

[AEB: June 1991, Biology common paper 1]

3 The diagram shows a synaptic knob of a neurone, next to a postsynaptic membrane.

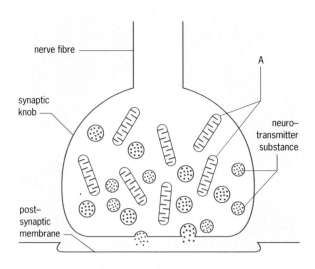

Fig 21.Q3

a) Name the process by which substances, such as the neurotransmitter molecules shown in the diagram, are released from cells.

b) Explain why a synaptic knob contains a large number of the structures labelled A.

c) With reference to the information in the diagram, explain why prolonged and frequent stimulation of the nerve fibre would make it temporarily unable to generate any more nerve impulses in the postsynaptic cell.

d) If the neurotransmitter were acetylcholine, what effect would the presence of acetylcholinesterase have on synaptic transmission? Explain why this effect would occur.

[AEB: June 1994, Biology paper 1]

4 The diagrams A to F represent reflexes and control mechanisms found in the human body.

Select the diagram that represents each of the following. You may use each letter once, more than once or not at all.

(i) A withdrawal reflex;
(ii) a knee-jerk reflex;
(iii) the control of plasma calcium levels;
(iv) the control of the iris diaphragm of the eye;
(v) the control of plasma thyroxine levels.

[AEB: 1990, Human Biology paper I]

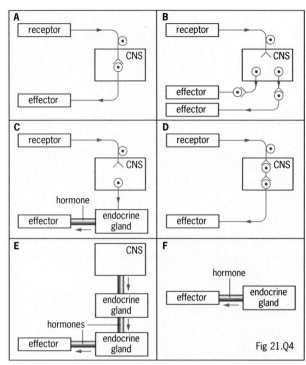

Fig 21.Q4

5 The table compares some effects of the sympathetic and parasympathetic systems.

Feature	Sympathetic	Parasympathetic
pupil of eye	dilates	constricts
salivary gland	inhibits secretion of saliva	stimulates secretion of saliva
lungs	dilates bronchi and and bronchioles	constricts bronchi and bronchioles
arterioles to gut and smooth muscle	constricts	no effect
arterioles to brain	dilates	no effect
heart rate		
stroke volume		

a) Complete the table by filling in the spaces to suggest the effects of these two systems on heart rate and stroke volume.
Use information in the table and your own knowledge to answer the following questions.

b) In giving dental treatment, it is important that any local anaesthetic stays close to the site of its injection. Explain why dental anaesthetics usually contain adrenaline.

c) Many people suffer from motion sickness when travelling in cars. A number of drugs are used to control this and some work by inhibiting the parasympathetic system.
Explain why side-effects of such drugs may include:
(i) dryness of the mouth; **(ii)** blurred vision.

[AEB: November 1992, paper 1]

Assignment

HEAD INJURIES

One in seven of children in hospital surgical wards are there because they have head injuries. For the best chance of recovery, their condition must be speedily diagnosed and treated fast and effectively.

 Anne was not wearing a cycling helmet when she fell backwards from her bike and was found unconscious. Her breathing was poor until someone released her tongue from obstructing the airway.

a) How would you give mouth-to-mouth resuscitation to an unconscious person?

b) How is the brain protected against physical damage? What went wrong in Anne's situation?

c) What happens when the oxygen supply to the brain is reduced?

d) When food and oxygen are both in short supply, how does the brain maintain a sufficient supply of oxygen? Does the brain store food to use in emergencies?

e) Materials from the blood pass to the brain at different rates because of a blood–brain barrier. What is it, and what is its importance? A head injury can damage it. Suggest the effects of this.

2

a) Anne was taken to hospital. How would damage to her skull be checked?

b) She was found to have a skull fracture and her condition began to worsen. She had a slow pulse, rising blood pressure and slow respiration. Which control centre of the brain may have been affected?

c) Reflexes often help to diagnose disorders of the nervous system. Anne remained unconscious and did not respond to sounds, but her basic body reflexes (eg pupil reflex) still operated. What are the nerve pathways for the pupil reflex? Is the knee-jerk reflex useful in determining brain function?

3

To see if blood was collecting in or around Anne's brain, CAT scans (computerised axial tomography scans) were taken. The scanner detects density differences between brain, blood and bone and produces pictures of 'slices' through the brain.

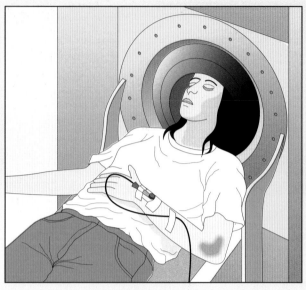

Fig 21.A1 **Anne has an urgent CT scan**

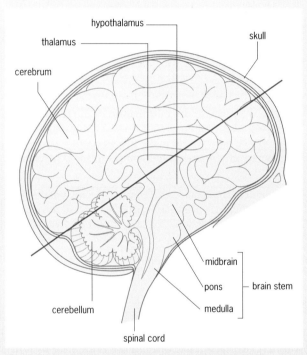

Fig 21.A2 **This diagram shows the brain 'slice' which is shown in the CT scans in Figs 21.A3 and 21.A4**

a) Look at Fig 21.A3, a scan of a normal brain. What are **A**, **B** and **C** on the scan?

b) The light region **X** in Fig 21.A2 is a **haematoma**, blood pressing on Anne's brain. Which region of the cerebrum is this? Check with Fig 21.32. How may the brain's function be affected?

c) If the haematoma were not treated, Anne's condition would rapidly worsen because the brain stem (and its structures) would be under pressure. Suggest the effects of damage.

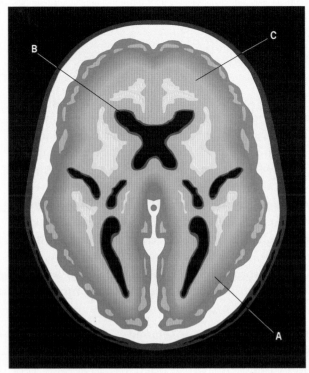

Fig 21.A3 **A CT scan of a normal brain**

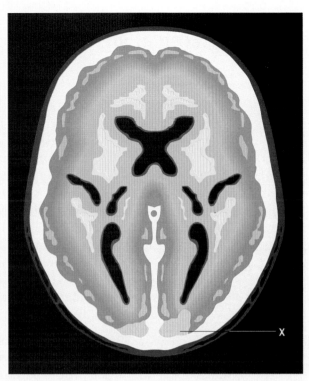

Fig 21.A4 **A CT scan of Anne's brain after the accident**

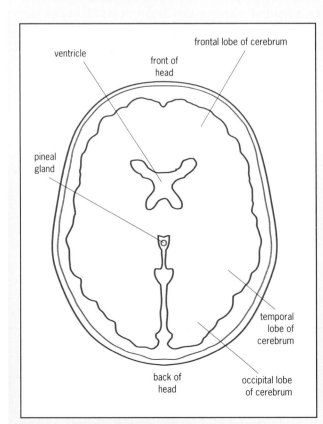

Fig 21.A5 **A simplified diagram showing the brain structures which appear in the CT scans shown in Figs 21.A3 and 21.A4**

4 The hospital takes an electroencephalogram (EEG) of Anne's brain.

a) What is an EEG?

b) The doctor diagnoses 'petit mal epilepsy'. What is **epilepsy**, and how does 'petit mal' differ from 'grand mal'? Find out and explain how an EEG helps to diagnose epilepsy.

5 The 'monoamine oxidase (MAO) inhibitor' group of antidepressant drugs can cause epileptic seizures. MAO is an enzyme produced by the postsynaptic neurone. It is secreted from the nerve endings of postganglionic fibres of the sympathetic nervous system. It breaks down the major neurotransmitter noradrenaline.

a) What is the sympathetic nervous system's effect on the body?

b) Find out the similarities between noradrenaline and adrenaline.

c) What effect would an MAO inhibitor have on the *quantities* of noradrenaline at the synapse?

d) Suggest how MAO inhibitor acts as an antidepressant.

The European fire bug, *Pyrrhocoris apterus*. Fire bugs grow through five nymph stages. Normally, as the fifth nymph sheds its skin, it emerges as a fully mature, reproductive adult

WE USUALLY THINK of hormones as complex chemicals that occur in the human body. But hormones control all sorts of processes in all sorts of organisms. The vital part that one hormone plays in the life cycle of an insect, the European Fire Bug, was discovered by accident. In the 1950s, a young Czech scientist, Karel Slàma travelled from Europe to Harvard University, carrying a small number of European Fire Bugs. This insect develops in five nymph stages, shedding its skin each time it grows. After the last moulting, it becomes an adult, capable of reproduction. When he arrived, Slàma was astonished to find that his bugs had not changed into mature adults: they had moulted more than five times and had grown into abnormally large nymphs.

The paper lining the insects' cages seemed to be responsible. Cages lined with *The Times* or the science journal *Nature* produced normal insects but those lined with *The New York Times* or *Scientific American* produced large nymphs. Slàma his colleagues later showed that wood pulp from balsam fir, used to make the paper for some North American publications, contains a chemical that mimics insect juvenile hormone. This hormone prevents development from nymph to adult. But, why should a tree mimic an insect hormone? Probably as a defence mechanism against aphids. Aphids which attack the balsam fir absorb the hormone and are then unable to mature into reproductive forms. Their colonisation attempt fails, and the tree remains free from aphids.

1 CHEMICAL SIGNALS: CONTROLLING THE PROCESSES OF LIFE

Chemical signals are produced by most living organisms. In plants they control processes such as growth and flowering (see Chapter 20). Animals use a wide variety of chemical signals to control and co-ordinate many aspects of their everyday life (Fig 22.1). In mammals, as Fig 22.2 shows, there are three main types of chemical signals:

- Locally acting chemicals such as **endorphins**, **prostaglandins** and **histamines** (Fig 22.3). These act on cells which are close to their site of production and are chemicals of *internal* communication: they act only on tissues *inside* the organism.

- **Hormones**. These are produced by special tissues called **endocrine glands**. They pass into the blood and act away from their site of production. Hormones affect **target cells** and **target organs** in the body, either stimulating or inhibiting them. The action of hormones is closely linked to the activity of the nervous system.

- **Pheromones** are chemical messengers which allow simple exchanges of information *between* organisms.

In this chapter we look at how chemical signals control many life processes in animals, concentrating on the systems that occur in humans. The main part of the chapter describes the human endocrine system and we finish with a brief overview of prostaglandins and endorphins and the effect of pheromones on animal behaviour.

The definition of a hormone in mammals is:

A chemical messenger which is produced by cells or tissues of the endocrine system and which travels through the blood to act on target cells or target organs, before being broken down.

Fig 22.1 **Chemical signals occur in many organisms**

Sexual reproduction in most organisms is controlled by several hormones. In humans, hormones control gamete production, puberty, pregnancy, birth and lactation

This dog is searching for olfactory clues: airborne chemicals produced by frightened prey or given off by a female dog to let potential mates know that she is 'on heat'

This dragonfly began life as a water-living nymph. Because its cuticle was not flexible, it shed its outer skin to grow bigger. The timing of these moultings is controlled by hormones. One, juvenile hormone, prevents metamorphosis into the adult form. Another, ecdysone, is a moulting hormone which helps the old cuticle to split (see page 196)

Local signals eg prostaglandins

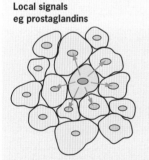

External signals eg pheromones

Distance signals eg hormones

pituitary gland

Pituitary produces the hormone prolactin which stimulates the breasts to produce milk in a woman who has recently given birth

Prolactin travels from the pituitary to the breasts in the blood

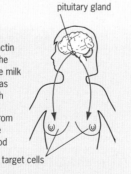

target cells

Fig 22.2 **A summary of the different types of chemical signalling that exist within and between organisms. Local signals include neurotransmitters such as acetylcholine (see Chapter 21) and local messengers such as prostaglandins, endorphins and histamines. Hormones allow communication between different organs, via the blood. Externally-produced messengers allow communication between organisms**

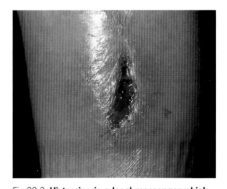

Fig 22.3 **Histamine is a local messenger which is released by injured cells. It acts on the cells around an injury like the deep cut shown above, causing blood vessels near to the injury to dilate, bringing more blood into the area. This is an important part of the inflammation response (see Chapter 26)**

2 THE HUMAN ENDOCRINE SYSTEM

In humans, more than a dozen tissues and organs produce hormones. Some, including the **pituitary**, the **thyroid**, the **parathyroid glands** and the **adrenal glands** are endocrine specialists: their *major* function is to secrete one or more hormones. These **glands** make up the human **endocrine system**, shown in Fig 22.4. This diagram provides a useful reference point for the rest of this chapter, and also for the hormones described in Chapters 25, 27 and 28.

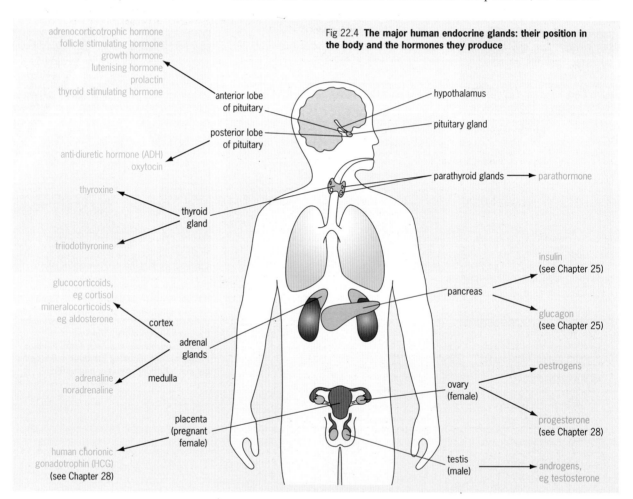

Fig 22.4 **The major human endocrine glands: their position in the body and the hormones they produce**

The endocrine system has four main functions:

- It maintains **homeostasis**, the balance of the body, by making sure the concentration of many different substances in body fluids are kept at the correct level. The control of blood sugar level, blood pH and water balance are all examples of homeostasis (see Chapter 25).

- It works with the nervous system to help the body respond to stress. The release of adrenaline in the **fight or flight** response is an example of this (see pages 370 and 373).

- It controls the body's rate of growth.

- It controls sexual development and reproduction (see Chapter 28).

We know of more than 50 human hormones, and we divide them into two groups according to their origin: those made from *fatty acids* and those made from *amino acids*. The first group, which include

steroids such as oestrogen and progesterone, are *lipid-soluble*. **Amino acid-based hormones** are either **modified amino acids**, **peptides** or **glycoproteins**: they are *water-soluble* and include hormones such as insulin and adrenaline. Table 22.1 lists most of the hormones produced by the endocrine system and shows where they act and what they do.

Gland	Hormone	Chemical structure	Effect
Pituitary 1 Posterior lobe	Anti-diuretic hormone (ADH)	Peptide	Reduces amount of water lost in urine. Raises blood pressure by constricting arterioles
	Oxytocin	Peptide	Contraction of smooth muscle during child birth. Stimulates secretion of milk from mammary glands
2 Anterior lobe	Adrenocorticotrophic hormone (ACTH)	Peptide	Stimulates production and release of hormones from adrenal cortex
	Follicle stimulating hormone (FSH)	Glycoprotein	Controls the development of follicles in the ovary and sperm cells in the testis (see ICSH, page 450)
	Growth hormone	Protein	Promotes growth (especially of skeleton and muscles). Affects body metabolism
	Luteinizing hormone (LH)	Glycoprotein	Stimulates ovulation and formation of the corpus luteum (stimulates testosterone production in males)
	Prolactin	Protein	Stimulates milk production and release during pregnancy
	Thyroid stimulating hormone (TSH)	Glycoprotein	Stimulates growth of thyroid gland; synthesis and production of thyroid hormones
Thyroid	Thyroxine + Triiodothyronine	Iodine-containing amino acids	Increases rate of cell metabolism. Controls aspects of growth and development
Parathyroid	Parathormone	Peptide	Raises blood calcium levels by stimulating release of calcium from bone
Pancreas (Islets of Langerhans)	Insulin (produced by the β cells	Protein	Lowers blood sugar levels, increases glycogen storage in liver
	Glucagon (produced by the α cells)	Peptide	Raises blood sugar levels by stimulating glycogen breakdown in the liver
Adrenal 1 Cortex	Glucocorticoids eg cortisol	Steroids	In response to stress, raises blood glucose through actions of the liver. Causes increase of glucagon from liver
	Mineralocorticoids eg aldosterone	Steroids	Concerned with water retention. Increases reabsorption of sodium and chloride ions in kidneys
2 Medulla	Adrenaline	Modified amino	'Fear, flight and fight reactions'. Prepares the body for heightened activity. Mimics effects of the autonomic nervous system
	Noradrenaline	Modified amino	As adrenaline
Gonads 1 Ovary (follicle)	Oestrogens	Steroids	Inhibits production of FSH by pituitary. Begins the rebuilding of the uterus wall
2 Ovary (corpus)	Progesterone	Steroids	Inhibits FSH. Maintains development of uterus wall (particularly during pregnancy)
3 Testis	Androgens (eg testosterone)	Steroids	Support sperm production. Important in the development of male secondary sexual characteristics
Placenta	Human chorionic gonadotrophin (HCG)	Steroid	Maintains the activity of the corpus luteum with regard to continuous progesterone

Table 22.1 **Human endocrine glands and their principal secretions. You do not need to memorise this, but you should find it useful for reference as you read the rest of the chapter.**

There are two basic types of gland in the body: **exocrine glands** (eg salivary gland), whose secretions are released through a tube or **duct**, and **endocrine glands** (eg thyroid gland), whose secretions are released into the bloodstream for transport to their eventual destination.

adrenal gland

main blood vessels

kidney

ureter

capsule

adrenal medulla

This secretes adrenaline

The adrenal cortex is made up of 3 layers of cells

These secrete glucocorticoids such as cortisone, mineralocorticoids such as aldosterone, and gonadocorticoids such as the sex hormones

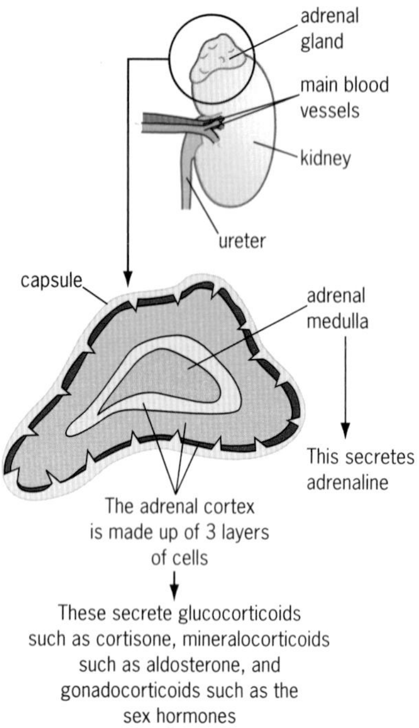

Fig 22.5 **The adrenal glands sit on top of the kidneys. They consist of the adrenal medulla, the tissue in the centre of the gland which secretes adrenaline, and an outer layer called the adrenal cortex**

See question 2. ■

Fig 22.6 **Control of metabolic rate involves a negative feedback loop**

Making and releasing hormones

Endocrine glands are tissues and organs that produce and release hormones. These hormones influence target organs elsewhere in the body. Each endocrine gland usually secretes more than one hormone. The **parathyroid glands** in the neck are an exception, releasing only a single hormone, **parathormone**. Fig 22.5 shows the basic structure of an endocrine gland, the adrenal gland.

Hormones are formed inside endocrine cells. Some amino acid-based hormones, such as insulin and glucagon, are produced directly from a copy of a gene by the processes of transcription and translation (see Chapter 3 and Chapter 31). Most hormones, including those based on fatty acids, are made as an end product of a series of chemical reactions. Each reaction makes minor changes to **precursor** molecules, until, eventually, the active hormone is produced.

All endocrine glands contain some of the hormone they produce, at any given time. Although the hormone thyroxine is released continuously, most are not. **Negative feedback loops** control the release of hormones from many endocrine glands so that homeostasis is maintained. This concept is explained in detail below and is also covered in Chapter 25, on page 398.

Thyroxine controls metabolic rate through a negative feedback loop

Metabolic rate is the rate at which all cells in the body carry out their biochemical reactions. It is a vital whole body function that must be controlled within very strict limits. The hypothalamus in the brain detects even a small decrease in metabolic rate and responds by releasing more **thyrotropin releasing hormone** (Fig 22.6). This acts on the pituitary gland, causing it to release more **thyroid stimulating hormone**. This passes to the thyroid gland which responds by secreting more **thyroxine**, the hormone which acts on individual cells to increase metabolic rate. As soon as metabolic rate gets back to normal levels, the hypothalamus responds by releasing less thyrotropin releasing hormone and, in a healthy person, homeostasis is maintained.

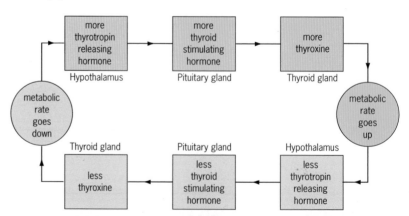

How hormones affect target cells

Hormones cause their effects in two different ways. Peptide hormones travel in the blood to all parts of the body. They do not, however, affect all cells in the body. Target cells or organs have specific proteins on their surface which act as receptor sites. The hormone fits into these sites with the same accuracy as an enzyme fits its substrate (see Chapter 4). Once in place, the hormone-receptor complex brings about changes inside the cell. Steroid hormones pass easily into cells. They bind to a specific receptor molecule inside the cytoplasm: this complex then causes biochemical changes inside the cell.

Peptide hormones and second messengers

Amino-acid-based hormones are water-soluble and vary in size from **thyroxine** (just two modified amino acids) to **growth hormone**, a protein made up of 190 amino acids. Because they are not lipid-soluble, they cannot get into cells through the lipid cell membrane. Instead, they act as **first messengers**, binding to target receptors on the outside of the cell membrane.

As Fig 22.7(a) shows, this binding event activates an enzyme, **adenylate cyclase** which is located on the inside surface of the membrane. The activated enzyme converts ATP into a substance called **cyclic AMP**. This is the **second messenger** which moves about inside the cell, causing biochemical changes. The final effect on cell function depends on the type of hormone and the type of target cell involved.

Steroid hormones

Steroids are small lipid-soluble molecules that can get through cell membranes easily. They enter cells and bind with **target receptors** inside the cytoplasm (Fig 22.7(b)). Cortisol, progesterone, oestrogen and testosterone all act in this way.

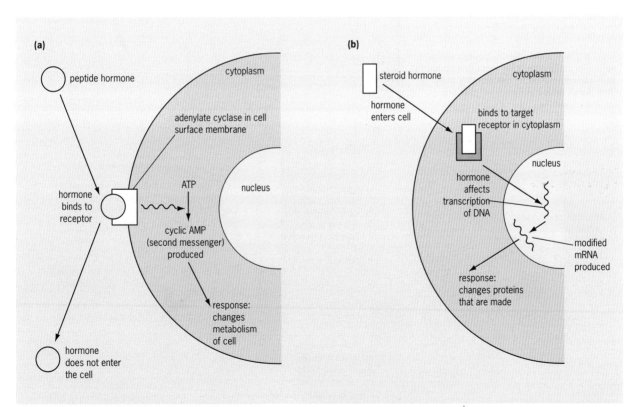

Fig 22.7(a) **Water-soluble hormones work via a second messenger which is formed inside the cell when they bind to receptors on the cell surface**

Fig 22.7(b) **When steroid hormones bind to a receptor inside the cytoplasm, they form a complex similar to an enzyme–substrate complex (see page 67). This enters the nucleus of the cell and binds directly to the DNA, interfering with the cells ability to read some of its genes. A steroid hormone can either switch on or switch off protein synthesis of particular genes. The overall effect of hormone action is different for each steroid hormone (see Table 22.1, page 365)**

3 COMPARING THE NERVOUS AND ENDOCRINE SYSTEMS

Hormones do not act independently. They work together with other hormones and also with the nervous system. The nervous system and the endocrine system both control and coordinate the function of different parts of the body and they both rely on chemical messengers, but they are obviously different (Table 22.2). Despite these differences, we now realise that the link between the endocrine system and the nervous system in humans is more definite than once thought. The main physical link between the two systems is between the **hypothalamus**, part of the base of the brain, and the **pituitary gland**, an endocrine gland just beneath it.

Table 22.2 **Differences between the nervous system and the endocrine system in humans**

Property	System	
	Nervous system	Endocrine system
Nature of signal	Although chemical signalling occurs at synapses, most communication within the nervous system is via **electrical** signals called nerve impulses (see Chapter 21)	All hormones are **chemical** signals
Size of signal	**Frequency modulated** – determined by the frequency of nerve impulses sent along a nerve fibre, and the number of nerve fibres being stimulated	**Amplitude modulated** – determined by the concentration of hormone
Speed of signal	**Rapid**. Human nerves conduct nerve impulses at speeds ranging between 0.7 metres per second and 120 metres per second	**Usually slower**, by comparison. The release of insulin from the pancreas in response to a rise in blood sugar level takes several minutes
Effect in the body	**Localised** effect – each individual neurone links with only one or a few cells	More **general** effect – hormones can influence cells in many different parts of the body
Capacity for modification	**Can be modified** by learning from previous experience	**Cannot be modified** by learning from previous experience

Fig 22.8 **The human pituitary gland is a small outgrowth at the base of the brain. It is connected to the hypothalamus by the pituitary stalk. Both the posterior and anterior pituitary release hormones**

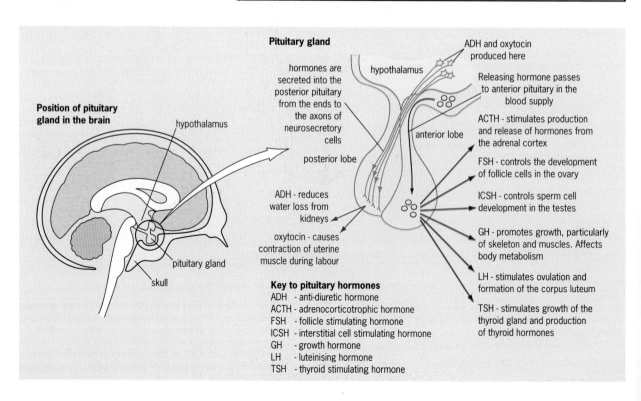

Position of pituitary gland in the brain

hypothalamus

pituitary gland

skull

Pituitary gland

hormones are secreted into the posterior pituitary from the ends to the axons of neurosecretory cells

posterior lobe

ADH - reduces water loss from kidneys

oxytocin - causes contraction of uterine muscle during labour

hypothalamus

anterior lobe

ADH and oxytocin produced here

Releasing hormone passes to anterior pituitary in the blood supply

ACTH - stimulates production and release of hormones from the adrenal cortex

FSH - controls the development of follicle cells in the ovary

ICSH - controls sperm cell development in the testes

GH - promotes growth, particularly of skeleton and muscles. Affects body metabolism

LH - stimulates ovulation and formation of the corpus luteum

TSH - stimulates growth of the thyroid gland and production of thyroid hormones

Key to pituitary hormones
ADH - anti-diuretic hormone
ACTH - adrenocorticotrophic hormone
FSH - follicle stimulating hormone
ICSH - interstitial cell stimulating hormone
GH   - growth hormone
LH   - luteinising hormone
TSH - thyroid stimulating hormone

The hypothalamus and pituitary: partners in communication and control

The pituitary has been called the 'master gland' because it controls the activity of most of the other endocrine glands, but it is itself actually under control from the hypothalamus. The range of hormones produced by the pituitary and the way the hypothalamus controls pituitary function is shown in Fig 22.8 on the facing page.

The human pituitary gland has a **posterior lobe** and an **anterior lobe**. The **pituitary stalk** connects the posterior lobe to the brain. This direct physical link allows nervous communication between the pituitary and the hypothalamus. A rich network of blood vessels also link the hypothalamus with the anterior lobe, allowing hormonal signals to pass from the brain to the pituitary.

Neurosecretory cells and the posterior pituitary

Neurosecretory cells carry nerve impulses and they also secrete substances which act directly on the cells. The neurosecretory cells which arise in the hypothalamus and end in the posterior pituitary, make two hormones; **anti-diuretic hormone (ADH)** and **oxytocin.** When the hypothalamus is stimulated by appropriate nerve impulses, it releases these hormones into the posterior pituitary. From here they pass into the blood and then travel around the body. ADH controls the reabsorption of water by the kidneys (see page 437), oxytocin causes contraction of the uterus during childbirth (see page 460).

See questions 1 and 3.

Hormonal control of the anterior pituitary

Blood vessels that connect the hypothalamus with the anterior pituitary carry hormones. These either stimulate or inhibit the release of pituitary hormones, including growth hormone, prolactin, the gonadotrophins, adrenocorticotrophic hormone and thyroid stimulating hormone.

Fig 22.4 shows which endocrine glands produce these hormones. Table 22.1 shows what each hormone does.

Nervous and hormonal control of digestive secretions

The body uses a combination of nerve communication and hormones to control the secretion of digestive juice to coincide with the presence of food in particular areas of the gut. Below is a brief summary of how this happens: for more detail see Chapter 16.

The taste or smell of food encourages the brain to send signals to the salivary glands and stomach to release saliva and stomach secretions. As food leaves the stomach, this stimulates the vagus nerve which then triggers the release of bile and pancreatic juice. Digestive secretions can also be controlled by hormones. The hormone gastrin, for example, is produced by the stomach wall. It travels in the bloodstream but exerts its effect locally, stimulating the production of both pepsinogen and hydrochloric acid.

Secretin and cholecystokinin control pancreatic and liver secretions. Both are formed by cells in the duodenal wall. Secretin causes the release of sodium hydrogencarbonate in the pancreas. In the liver, it increases the rate of bile formation. Cholecystokinin triggers the release of pancreatic enzymes such as lipase and trypsinogen. In the liver, it increases the rate of bile release.

A What would happen if the production of digestive juices was not timed to coincide with the presence of food in the gut?

The action of adrenaline

As the nervous system and the endocrine system are linked closely together, we should not be surprised that hormones can affect the way that we feel emotionally. The classic example of this is the **fear, fight or flight response** (see the assignment on page 373).

At the molecular level, adrenaline binds to the outside of a target cell and activates the enzyme adenylate cyclase. This converts ATP to cyclic AMP. Cyclic AMP, is a second messenger which sets off a **cascade**, a chain of reactions that convert glycogen to glucose.

As adrenaline binds to receptor molecules on the surface of a cell, changes in the membrane cause the production of **G proteins**. For every one molecule of adrenaline that binds, 10 molecules of G protein are produced. Each of those 10 molecules of G protein catalyses the production of 10 molecules of **adenyl cyclase**. Each molecule of adenyl cyclase stimulates the production of 10 molecules of cAMP. This **amplification** effect continues along a chain of reactions that eventually breaks down glycogen into glucose.

> ✔ The process of amplification seen in the enzyme cascade which results in the fear, fight or flight response, allows *one* molecule of adrenaline to stimulate the production of *millions* of glucose molecules.

4 PROSTAGLANDINS

Prostaglandins are a group of hormone-like compounds. They were originally thought to be produced by the prostate gland – hence the term prostaglandins. They are not true hormones as they are not produced by endocrine glands. Most mammalian cells synthesise prostaglandins which act locally on surrounding cells.

Prostaglandins control cell metabolism, probably by modifying levels of cyclic AMP inside the cell, and are involved in a wide range of activities including blood clotting, inflammation and smooth muscle activity. Prostaglandins are extremely powerful: one billionth of a gram produces measurable effects. The prostaglandins in human semen cause muscles of the uterus and oviduct to contract during female orgasm, helping the sperm on their journey towards the ovum. Synthetic prostaglandins can be used to stimulate labour in a pregnant woman whose baby is overdue.

5 ENDORPHINS

Endorphins are chemical messengers found in the brain. These polypeptide molecules mimic the effects of drugs such as morphine and heroin (endorphin = 'endogenous morphine'). Like prostaglandins, they are involved in a wide range of activities. Their most important role seems to be in the management of pain. It is thought that endorphins bind to pain receptors and so block the sensation of pain. Apparently long-distance runners who run until they collapse have abnormally high levels of endorphin.

Endorphins also seem to affect the 'pleasure centres' of the brain. Stimulation of these areas provide the intense feelings associated with orgasm. Scientists looking into the chemistry of pleasure have discovered that pleasurable sensations begin with the release of serotonin from the hypothalamus. This stimulates the release of endorphins which turns on the supply of dopamine and simultaneously turns off the supply of GABA (an amino acid which suppresses dopamine). Dopamine directly stimulates the pleasure centre.

Drugs like heroin work by mimicking endorphins. The brain of an addict stops making endorphins and so withdrawal from heroin results in a sudden lack of endorphin and a build-up of GABA, producing unpleasant withdrawal symptoms.

7 PHEROMONES AND BEHAVIOUR

Pheromones, sometimes called **ecto-hormones**, are substances which organisms release into the environment to communicate with organisms of the same species. Pheromones are small volatile molecules that spread easily into the environment. They are active in very small amounts: the pheromone of the female Gypsy moth causes a response in the antennae of the male moth at concentrations as low as one in a thousand million million molecules (Fig 22.8).

Pheromones are usually classified according to the type of response they produce:

- **Alarm pheromones** are produced by bees and ants when attacked. They excite other insects of the same species to swarm around the attacker.

- **Sex attractants** are released by moths, rats and possibly humans to attract members of the opposite sex. Humans do not have a particularly good sense of smell but there is some evidence that very young babies can recognise their mother by her characteristic smell. Some people also think that sexual partners can also recognise each other using their sense of smell, but there is less evidence to back this up.

- **Trail substances** are produced by ants during to show other ants where to find sources of food.

- The **queen substance** is produced by the queen bee within a hive to suppress the production of other queens.

Fig 22.8 **Sex attractants such as the pheromone bombykol are released into the air in minute quantities by female moths. The male is extremely sensitive to these pheromones, because his antennae are large and crammed with many specialist receptors. This male has about 17 000 receptors in each antenna that respond only to bombykol**

SUMMARY

- Animals have nervous and chemical control systems, plants, bacteria fungi and protoctists have only chemical control.

- There are three basic types of chemical control: **local signals** such as histamine, prostaglandins and endorphins; **hormones** such as insulin and adrenaline; external signals or **pheromones**.

- The human **endocrine system** has four main functions. It controls growth, sexual development and fear, flight and fight reactions. It maintains body homeostasis.

- Endocrine glands lack a duct: their secretions are delivered directly into the bloodstream.

- Hormone levels are generally controlled by **negative feedback loops**.

- There are two basic types of hormone: **peptide hormones** which are derived from amino acids (eg insulin) and **steroid hormones** which are made from fatty acids (eg oestrogen).

- Peptide hormones bind to receptors on the outer membrane of target cells and achieve their effect via a **second messenger** (eg cAMP). Steroid hormones pass through the cell membranes and bind to target receptors inside the cytoplasm.

- There is a close link between the nervous and endocrine systems, shown by the way in which the **pituitary gland** interacts with the **hypothalamus** of the brain.

- Adrenaline, the hormone which causes the classic **fear, flight and fight** response acts at the molecular level to bring about the production of millions of glucose molecules.

- **Prostaglandins** and **endorphins** are local messengers which affect many different types of cell. Prostaglandins are best known for their effects on the female reproductive system: endorphins are chemicals which influence our perception of pleasure and pain.

QUESTIONS

1 The anterior lobe of the pituitary produces and secretes six different hormones including follicle stimulating hormone (FSH), which initiates development of gametes in ovaries and testes, and adrenocorticotrophic hormone (ACTH) which controls the secretion of some hormones by the adrenal cortex. The anterior lobe also produces thyroid stimulating hormone (TSH) which controls the formation of thyroxine by the thyroid gland. The secretion of TSH itself is controlled by thyrotrophin releasing factor (TRF) in a negative feedback loop. TRF is produced by the hypothalamus.

a) Account for the different specificity of FSH and ACTH.

A neuroendocrine mechanism is involved in the control of body temperature in mammals. Some of the components of this mechanism are listed below in alphabetical order.

    anterior lobe of pituitary
    central nervous system
    cold (exteriostimulus)
    hypothalamus
    increase in metabolic rate
    thermoreceptor
    thyroid gland
    thyroxine
    TRF
    TSH

b) Complete the flow diagram below. Place each component in a box as shown, joining them with arrows as shown in the key.

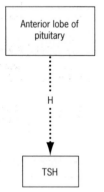

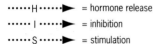

Keyed arrows:

······H······▶ = hormone release
······I······▶ = inhibition
······S······▶ = stimulation

[UCLES June 1995 Modular Biology: Transport, Regulation and Control Paper, q.1]

2
a) **(i)** Name an exocrine gland.
    (ii) State **one** of the features which distinguishes an exocrine gland from an endocrine gland.

Negative feedback is involved in the functioning of many endocrine glands and different glands may interact with each other in a coordinated manner.

b) **(i)** What is the role of negative feedback mechanisms?
    (ii) Explain briefly how the pituitary and thyroid glands interact with each other.

[UCLES 1994 Modular Biology: Specimen Paper, q.1]

3
a) The table shows some of the hormones produced by the pituitary gland, the organ or gland which they affect, and the effect they produce on this organ or gland. Complete the table.

Pituitary hormone	Target organ/gland	Effect on organ/gland
	Adrenal cortex	Secrets cortisol
		Increased secretion of thyroxin
Luteinising hormone (LH)	Ovary	

b) Somatotropic hormone is also a pituitary hormone. It plays an important part in normal growth. Suggest:
    (i) how somatotropic hormone may increase growth;
    (ii) how it acts to oppose the action of insulin.

[AEB 1995 Human Biology: Paper 1, q.13]

Assignment

ADRENALINE AND STRESS

When we are nervous, familiar symptoms appear. We get butterflies in the stomach, our pulse increases, our hands become cold and clammy. This happens because our body is preparing for action by secreting **adrenaline**. This hormone:

- increases heartbeat,
- increases ventilation rate,
- causes vasoconstriction, resulting in redirection of blood from intestines and skin to brain and muscles,
- increases metabolic rate,
- dilates pupils.

a) Which organs secrete adrenaline?
b) What is the overall effect of adrenaline on the body?
c) Use your answer to 1b to explain why adrenaline leads to an increased blood pressure.

The effects of adrenaline are very similar to the effects of the sympathetic nervous system (see Chapter 21). For example, secretion of adrenaline speeds up the heart, as does direct stimulation by a sympathetic nerve.

The difference is that adrenaline produces a general state of readiness for action, while the sympathetic nervous system is used for fine control. If your body needs to increase the heart rate a little, but does not need the other effects of adrenaline, it does so via the sympathetic nerves.

Fig 22.A1 **Adrenaline helps this mountain biker to concentrate and compete to the best of her abilities. Her heart is pounding, blood pressure is raised and all of the capillaries in the muscles are open to maximise performance**

Fig 22.A2 **There are many stressful lifestyles: being a busy executive is just one. Stress happens when adrenaline is continually released, ready for action that is not taken. This leads to raised blood pressure which can increase the risk of heart disease or stroke**

The effects of stress

It is difficult to define stress. What is a stressful situation to some people may be sheer enjoyment to others. Certainly, some stimulation is desirable – it makes us feel more alive.

In the short term, the responses of the body to stress have a beneficial effect and leave the individual better able to cope with the crisis if and when it comes. In the long term, however, the effects can be far from beneficial. Continued secretion of adrenaline and general sympathetic stimulation bombard the body with stress chemicals. Table 22.A1 outlines some of the common symptoms of stress.

Table 22.A1 **The effects of stress on bodily functions**

Normal state	Adrenaline response	Short-term effect	Long-term effect (stress)
Brain: normal blood flow	blood flow increases	think more clearly	headaches and migraines
Muscles: normal blood flow	blood flow increases	improved performance	muscular tension and pain
Heart: normal pulse and blood pressure	output and pressure increase	improved performance	hypertension and chest pain
Intestines: normal blood flow and peristalsis	blood flow decreases, peristalsis increases	slower digestion	abdominal pain and diarrhoea
General biochemistry: normal rate of oxygen use. Glucose and lipids liberated	oxygen, glucose and lipid use increases	more energy available quickly	rapid tiredness

a) List some activities/sports that people enjoy because they stimulate release of adrenaline.
b) List some occupations or lifestyles that are commonly associated with stress.

a) Discuss or write down some ideas that might explain why there is more stress-related illness today than in previous generations.
b) Discuss or write down some strategies that people might use to minimise stress.

Sensory systems in animals

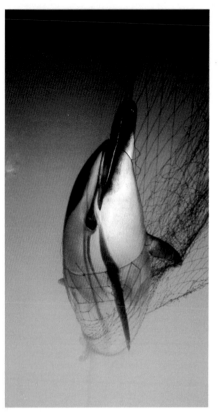

Many dolphins and porpoises die entangled in fishing nets each year

DOLPHINS, PORPOISES and toothed whales make sense of their environment by producing a variety of squeaks, whistles and clicks at high frequencies and then detecting echoes from nearby objects. Though sophisticated, this system can be fooled by human inventions like fishing nets. These are too fine to reflect the sonar signals that dolphins send out and are therefore 'invisible'.

Understanding more about how dolphins' sensory systems work has enabled scientists to devise a way to allow fishermen to carry on their activities without being a threat. Apparently, dolphins can detect shoals of fish by picking up the sonar reflections from their swim bladders. When fishing nets are fitted with plastic devices that look like small rugby balls, dolphins come to an abrupt halt some way from the net, firing rapid and repeated sonar at it. The plastic balls seem to 'look' a bit like swim bladders, but are not similar enough to attract the dolphins.

When trialled in the open sea, the initial curiosity of the dolphins turned to suspicion: they avoided the nets completely and, after turning back briefly to bombard the back of the net with more sonar, they then resumed their normal activities. Scientists hope to get more funding to do larger-scale tests. If they are successful, dolphin deaths from fishing net accidents could become a thing of the past.

1 RESPONDING TO THE ENVIRONMENT

All living organisms are **sensitive**. Whole organisms and individual cells respond to physical and chemical changes in their environment. Bacteria, fungi and plants respond to their environment. Plants, for example, grow towards the light. Single-celled protoctists such as amoeba, paramecium and euglena (Fig 23.1) react to touch, heat and light by simply moving closer or further away from the source. But it is in the animal kingdom that we find the widest range of body senses and the most sophisticated behavioural responses.

In this chapter we look at sensory systems in animals, concentrating on the human body senses, as illustrated by the human eye. We go on to look at animal behaviour in Chapter 24.

Sensory systems in animals

Animals have a **nervous system** (see Chapter 21) which receives information from the environment, decides on the best course of action and then signals the body to respond. Humans sense the environment through the body senses of touch, taste, sight, hearing and smell.

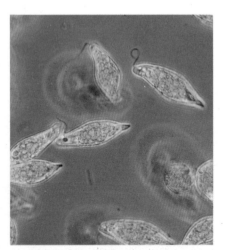

Fig 23.1 *Euglena gracilis*, a single-celled protoctist, is capable of hunting for food and eating prey, and it can also use light to build up sugars by the process of photosynthesis. It usually swims towards light in its environment

Before going on to look at sensory systems in more detail, you will find it useful to become familiar with some important terms:

- A **stimulus** is a physical event, usually some form of energy change, that affects an organism and excites its receptors. Examples include **mechanical stimuli** such as pressure, touch, and movement of air, **thermal stimuli** such as heat and cold, **light** and **chemical stimuli** such as taste, smell, and the concentration of carbon dioxide or oxygen in the blood.

- A **sensation** refers to a general state of awareness of a stimulus. Aristotle first identified the five senses of hearing, sight, taste, smell and touch, but we now know that there are actually many more senses than this. The skin can experience the sensations of light touch or deep pressure, heat, cold and pain. There are also the general senses of balance and body movement. Internally, the body can sense changes in blood pressure and blood levels of chemicals; and the body can generate the sensations of hunger and thirst when it needs food or fluids.

- **Perception** is a feature of more complex animals as it involves higher brain functions. Perception of a stimulus is different from just sensing that it is there – it allows us to assess the *significance* of the stimulus. If you see that a housefly has landed on your arm, you might just brush it away, but if you see a wasp, you would probably decide to get up and make sure it goes out of the window, to prevent it stinging you.

> ✔ Perception involves reception *and* interpretation of stimulus information

2 FEATURES OF SENSORY SYSTEMS

A sensory system must allow the following processes to take place:

- **Transduction**. The body detects a physical stimulus: **sensory cells** inside specialised **receptors** are **depolarised**: the *physical* stimulus is transduced, or changed, into an *electrochemical* event. In the skin, for example, touch receptors contain cells that become depolarised when pressed.

- **Transmission**. A nerve impulse is transmitted along a sensory neurone into the central nervous system (CNS) (see Chapter 21). The body can recognise the strength of a stimulus: the *stronger* the stimulus, the *more frequently* impulses are transmitted (the faster the rate of 'firing').

- **Information processing**. The CNS processes information that it receives from several parts of the body and makes a decision about the best course of action to take.

Receptors

A receptor is part of the nervous system that is adapted to receive stimuli. Sensory receptors are vital for the well-being of the organism: they sense the external environment and they also provide information to the CNS about the body's internal environment. Receptors that detect changes in the outside world are positioned at or near the body surface, usually at the front end (Fig 23.2). These include the specialised sense organs, the ear and eye, the smell receptors of the nose and the taste receptors of the tongue. Receptors that respond to internal stimulation, such as stretch receptors in blood vessels, are found deep inside the body.

Fig 23.2 **The head of this ragworm, an active hunter, contains many receptors. It has well developed eyes, chemoreceptor cells and large palps** to enable it to find its prey. **The ragworm withdraws into its burrow for safety, guiding itself into position using** tactile (touch) **cells and sensitive tentacles called** cirri

Types of receptor

Receptors can be classified in three ways: by the type of stimulus they respond to, by their location within the body, or by their level of complexity (Table 23.1).

Table 23.1 **Classification of receptors**

Receptor name	Action	Example
Classified by stimulus		
Photoreceptor	responds to light	eye
Chemoreceptor	responds to chemicals	taste bud
Thermoreceptor	responds to changes in temperature	temperature receptors in skin
Mechanoreceptor	responds to mechanical (touch) stimuli	nerve fibres around hairs
Baroreceptor	responds to changes in pressure	baroreceptors in carotid artery
Classified by location		
Interoceptor	responds to stimuli within the body	baroreceptors in carotid artery
Exteroceptor	responds to stimuli outside the body	eye and ear
Proprioceptor	responds to mechanical stimuli conveying information about body position	muscle spindle (see Fig 23.4)
Classified by complexity		
General senses	single cells or small groups of cells	temperature receptors in skin
Special senses	complex sense organs	eye and ear

?

A Look at the cells in Fig 23.3 (a), (b) and (c). What types of receptor cell are they? Classify each according to stimulus and location, as in Table 23.1.

There are also two basic types of receptor, and both can be any of the classes of receptor described in Table 23.1. Simple or **primary receptors** consist of a single neurone whose axon carries a nerve impulse into the CNS. The temperature and pressure receptors in the skin are both examples of primary receptors. **Secondary receptors** receive information from a cell outside the nervous system and then pass the information on to a nerve cell for transmission to the CNS. Taste buds in the tongue are secondary receptors.

Fig 23.3 **Some common examples of body receptors**

Fig 23.3 (a) **Below: the tall cells in the central band are the rod and cone cells in the retina of the eye**

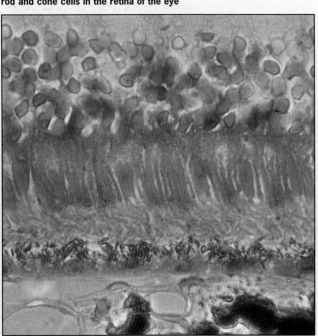

Fig 23.3 (b) **Left: sensory hairs on the mouth-parts of a blowfly are sensitive to chemicals, particularly sugars**

Fig 23.3 (c) **Below: taste buds in the human tongue. You can see the faint sensory hairs projecting from the sense cells**

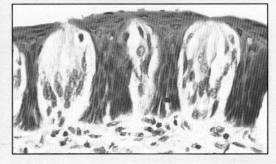

How receptors work at the molecular level

In 1950, Bernard Katz showed how stretching a **muscle spindle** (Fig 23.4) produces a local membrane depolarisation, called a **generator potential**, in an associated receptor cell. If the depolarisation goes beyond the **threshold level** for that cell, it triggers an **action potential** (see page 339). All receptors therefore act as **biological transducers**: they create action potentials in neurones in response to physical and chemical stimuli in the environment.

In their normal resting state, all sense cells maintain a **resting potential**. When the cell is affected by a stimulus, ion channels in its cell surface membrane open, allowing sodium ions (Na^+) and potassium ions (K^+) to flow down their respective electrical and chemical gradients. This sets up a **generator potential**: a local depolarisation of the receptor cell surface membrane. The greater the stimulation of the sense cell, the greater the depolarisation. Once the generator potential exceeds a threshold level, an action potential is produced and is transmitted along the axon.

Receptors are very sensitive – they have to be in order to detect small changes in the environment. Apparently the human nose can detect ethanoic acid (acetic acid, the acid in vinegar that makes it smell) at a concentration of 5×10^{11} molecules per litre of air; a dog's nose, though, can detect the same smell at a concentration of about 2 million times less, 2×10^5 molecules per litre.

Body senses also **amplify** the incoming signal. A stimulus is often quite weak – just a few photons of light or a few molecules of a chemical. But the action potential running from the eye to the brain, for example, has about 100 000 times as much energy as the few photons of light that triggered it.

A stimulus alters the electrical properties of membrane of the receptor cell. The charge on the membrane is reversed: it goes from –70 mV to +40 mV). We say that it has depolarised. When a membrane depolarises there is the formation of a nerve impulse, or moving action potential that is transmitted along a nerve fibre. Find out more about action potentials see Chapter 21.

The endocrine system also shows amplification. One molecule of the hormone adrenaline, for example, can release millions of glucose molecules in the **cascade effect** (see page 369).

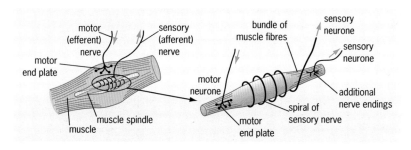

Fig 23.4 Muscle spindles are receptors in human skeletal muscle. Each one consists of a bundle of small, thin muscle fibres which are enclosed by a spiral of sensory neurones.

Muscle spindles are proprioceptors. They respond to an increase in muscle length and tell the brain the position of rest of the body. The motor nerve shown brings nerve impulses from the central nervous system to the muscle

3 THE EYE AND THE SENSE OF SIGHT

The next section looks at the eye as an example of a sensory structure. We look at the ear on page 388: you should use this information to compare these two major sense organs. The chemical senses of taste and smell are covered briefly in Chapter 16.

Vision is one of the most important body senses: almost all animals respond to light stimuli. Simple eye spots in single-celled organisms detect light: they can tell how bright the light is and the direction it is coming from. Complex eyes form an image. They usually include a lens which concentrates light onto photoreceptor cells, so increasing sensitivity when the light level is low. Because they focus light onto only a few receptor cells, lenses also increase the eye's ability to detect direction and movement. Sensory information from the eye travels to the brain where perception occurs. We therefore 'see' when nerve signals from the eyes enter the brain and are processed, not when light enters our eyes.

B Look at Table 23.1. What type of receptors are the photoreceptors of the eye: are they exteroceptors, proprioceptors or interoceptors?

4 STRUCTURE OF THE HUMAN EYE

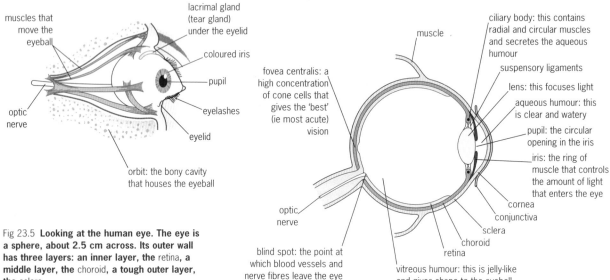

Fig 23.5 **Looking at the human eye. The eye is a sphere, about 2.5 cm across. Its outer wall has three layers: an inner layer, the** retina, **a middle layer, the** choroid, **a tough outer layer, the** sclera

The retina

Fig 23.6 **Detailed structure of the retina showing photoreceptors and their connections**

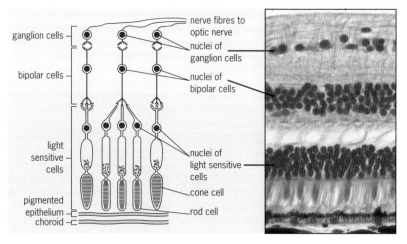

Fig 23.7 **The tapetum is a light-reflecting layer in the choroid of many vertebrates. It is particularly well developed in nocturnal animals, and in fish that live in deep water. In hoofed mammals like elk, the tapetum contains glistening fibres of connective tissue. In cats and other carnivores, it contains shiny crystals of guanine**

C The eye is often compared with a camera. Which part of the camera corresponds to the retina of the eye?

The retina is the light-sensitive layer at the back of the eye. It is a complex structure with a deep layer of light-sensitive cells called **rods** and **cones,** together with a middle layer of **bipolar neurones** which connect the rods and cones to a surface layer of **ganglion cells** (Fig 23.6). Fibres from the ganglion cells join up to form the **optic nerve.** Only the back of the retina is photosensitive. The front surface extends over the choroid and forms the inner lining of the **iris** and **ciliary body.**

The retina is about the area of a postage stamp and acts as a projection screen onto which images are directed. The retinal surface is covered with blood vessels and nerve cells but the brain ignores these obstructions: we do not see them as part of our image of the world.

Nocturnal animals such as cats have a shiny backing to the retina, a layer of cells called the **tapetum** (Fig 23.7). Their eyes seem to glow if a light shines into them. The tapetum reflects light back onto the receptors, making vision more effective in low light conditions.

The choroid

The choroid contains many blood vessels. Blood supplies the cells of the eye with nutrients and oxygen and removes waste. The choroid is dark-coloured due to a high concentration of the pigment **melanin** in its cells. These pigmented layers prevent internal reflection of light rays and so prevent us seeing a confused and blurred image.

At the front of the eye, the choroid expands to form the **ciliary body**. The smooth muscles in the ciliary body alter the shape of the **lens** (Fig 23.8). The lining of the ciliary body secretes aqueous humour, the fluid which fills the front of the eye.

The **iris** is also an extension of the choroid, partially covering the lens and leaving a round opening in the centre, the **pupil**. Its function is to control the amount of light entering the eye. The iris contains radial and circular muscles which can alter pupil size. Iris muscles are controlled by the **autonomic nervous system** (see page 357).

The sclera

The sclera forms the 'white' of the eye. It is a tough external coat mainly of collagen fibres. Six exterior muscles attached to the sclera enable the eye to look up, down and side-to-side. One-sixth of the sclera is colourless and transparent, forming a clear window, the **cornea**, through which light gets into the eye.

The main body of the eye

The main body of the eye is divided into two parts by a **biconvex lens**. This lens contains transparent **lens fibres** (long thin cells which have lost their nuclei) together with an elastic **lens capsule** made of glycoprotein. The lens is flatter at the front than at the back and is soft and slightly yellow. With age it gets flatter, yellower and harder and becomes less elastic. There is no blood in the lens. Substances diffuse in and out of it from the surrounding fluid.

The eyeball has three cavities: the region between the cornea and the iris, the region between the iris and the lens and the cavity that fills all the space behind the lens. The first two cavities are filled with a fluid called the **aqueous humour** which is thin and watery. The third cavity, the largest (it takes up about 80 per cent of the volume of the eyeball) is filled with a thicker fluid called the **vitreous humour**. This is a transparent gel which has the consistency of egg white and which contains 99 per cent water and 1 per cent **collagen fibrils** and **hyaluronic acid**. The vitreous humour preserves the spherical shape of the eyeball and helps to support the retina.

Eye protection

The eye is an important and delicate structure. In mammals, the orbit, or bony socket of the skull, protects the eye from physical damage, and a pad of fat behind the eyeball helps to cushion it from shocks. The other protective features of the eye are shown in Fig 23.9.

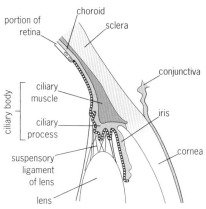

Fig 23.8 **The position of the ciliary body and the suspensory ligaments in the eye. When we focus on objects that are close to us, the ciliary muscles relax, increasing the tension on the suspensory ligaments. This pulls the lens into a thinner and flatter shape. When we look at objects further away, the ciliary muscles contract, the tension on the ciliary ligaments decreases and the lens becomes fatter (see also Fig 23.12)**

Fig 23.9 **Mammalian eyes are contained in the** orbit, **a bony socket of the skull. They are protected by the bone and also by a pad of fat at the rear of each eyeball. Other features of the eye are also protective:**

The eyebrows prevent sweat running from the forehead into the eye

The eyelashes prevent the entry of airborne particles such as flying insects, often adding to the efficiency of the eye blink reflex

Tears constantly bathe the surface of the eye removing the surface film of dust and dirt and killing bacteria

The conjunctiva, **a thin, transparent membrane which covers the cornea, also protects the eye from airborne debris**

?

D Fish lack eyelids and lacrimal glands but these are developed in amphibians following transformation from tadpole into adult frog. Suggest an explanation for this observation.

Bright light
Relaxation of radial muscles and contraction of circular muscles cause pupil to constrict

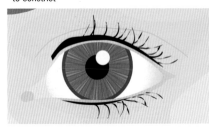

Dim light
Contraction of radial muscles and relaxation of circular muscles cause pupil to dilate

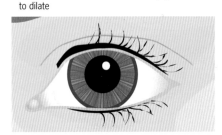

Fig 23.10 **The eye is prevented from being dazzled by a speedy reflex that makes the pupil smaller. To achieve this** constriction **of the pupil, radial muscles in the iris relax and circular muscles contract. In dim light, the opposite happens: radial muscles contract and circular muscles relax, causing the pupil to enlarge**

5 HOW THE HUMAN EYE WORKS

The process of seeing is complex and involves five different stages:

● Light enters the eye.

● An image is focused on the retina.

● The light that makes up the image is transduced into an electrical signal.

● Nerve impulses carry information about the image into the brain.

● The brain decodes the information and perceives the image.

Let's look at each of the stages in more detail.

How light enters the eye

Light enters through the cornea and then passes through the pupil, the aqueous humour, the lens and the vitreous humour, before it reaches the retina.

To operate well in conditions of different light intensity, the eye must control how much light reaches the retina. It does this by changing the diameter of the pupil. In bright light, the pupil **constricts** (gets smaller) to prevent overstimulation of the retina and the perception of 'dazzle'. In dim light, the pupil **dilates** (gets wider), letting in more light.

The size of the pupil is controlled by the muscles in the iris (Fig 23.10). These are themselves controlled by the autonomic nervous system and so pupil adjustment is a **reflex response**, not under the conscious control of the brain.

Focusing light on the retina

If you do not focus a camera correctly, the picture you take is blurred. Similarly, the eye must focus light to produce a high resolution image on the retina. The eye focuses an image by **refracting**, or bending light using the cornea and the lens, forming an upside down or **inverted image** on the retina.

Because of the composition and curvature of the cornea, most of the refraction of light occurs in this structure. The lens is important in the fine focusing of light onto the retina. The cornea is fixed but the lens is adjustable, allowing **accommodation**, a reflex that allows the eye to focus on objects that are at different distances from the eye.

Problems with the lens system

In some individuals the lens and cornea do not focus correctly (Fig 23.11). If you are short sighted, a condition known as **myopia**, you focus light from an object *in front of* the retina. This produces a blurred image. Short sight can result from an elongated eyeball or a thickened lens. It is corrected by using a diverging or concave lens.

If you are long sighted you are said to have **hypermetropia**, and you focus light *behind* the retina. Again, objects appear blurred. Long sightedness results from a shortened eyeball or a lens that is too thin, and is corrected using a converging or convex lens.

A more complicated visual defect is **astigmatism**. In astigmatism, the surface of the cornea is irregular, the object appears blurred because some of the light rays are focused and others are not (this

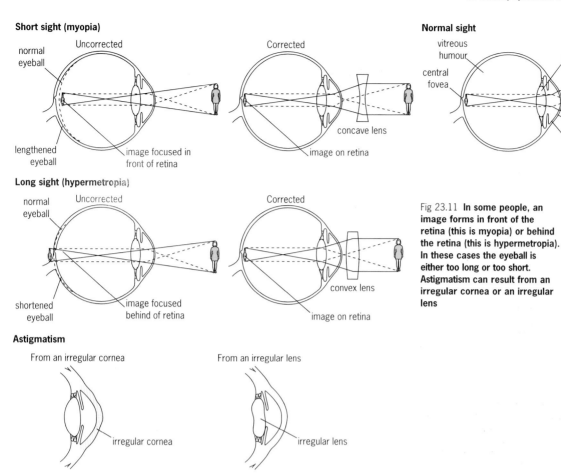

Fig 23.11 **In some people, an image forms in front of the retina (this is myopia) or behind the retina (this is hypermetropia). In these cases the eye ball is either too long or too short. Astigmatism can result from an irregular cornea or an irregular lens**

is similar to the distortion produced by a wavy pane of glass). Astigmatism is corrected using a cylindrical lens that bends light rays in one plane only.

Accommodation

In a normal eye, assuming that the lens is not adjustable, an object 6 metres away would be in perfect focus on the retina. The process of **accommodation** (Fig 23.12) allows us to focus on objects that are nearer and further away than this.

Fig 23.12 **The eye can focus on near and distant objects by altering the shape of the lens. This process is called accommodation**

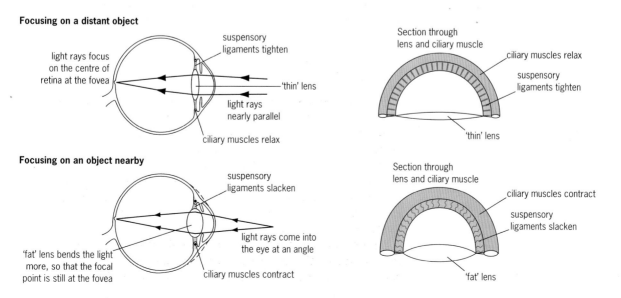

When we focus on an object that is only a metre or so away, the ciliary muscles of the eye contract involuntarily. This reduces the tension on the ligaments that hold the lens in place, and the lens becomes 'fatter'. The focal length changes, and the image is focused. When we then look at an object much further away, the ciliary muscles relax, again automatically. The tension on the ligaments supporting the lens increases, and the lens becomes 'thinner'.

See question 1.

?

E Imagine you are standing at a bus stop reading a book. What happens **(a)** to the ciliary muscle, **(b)** to the lens and **(c)** to the pupil of your eye as you look up to see the bus approaching in the distance?

Convergence and binocular vision

Humans and other mammals have **binocular vision**: although we have two eyes, we see only one image.

Normally, seeing an object that is in the middle distance or far distance with both eyes requires no particular effort. When we focus on a near object, however, the two eyeballs turn slightly inwards, so that both images fall on the corresponding points of both retinas at the same time. This action is called **convergence**.

The importance of binocular vision

All vertebrates and many invertebrates have two eyes, giving them binocular vision. This has many advantages over monocular vision. It provides a larger field of view, it reduces the risk of the animal becoming disabled following damage to one of its eyes and it allows **stereoscopic vision**.

We said above that animals with binocular vision see one image. In fact, each eye picks up the same image from a slightly different angle and we actually experience two images, very close together. When these are decoded by the brain, we get the impression of distance, depth and objects being three-dimensional. Stereoscopic vision allows us and other animals to judge distance and depth accurately.

Predators often have their two eyes placed centrally to maximise this overlap of retinal images (Fig 23.13). It gives them excellent stereoscopic vision to judge distances accurately and locate their prey. Prey species, on the other hand, tend to locate their eyes to the side of the head. They have reduced stereoscopic vision but an all-round vision that allows them to spot nearby predators, even if they try to sneak up from behind.

Fig 23.13 **Look at these photographs of predators and prey. Notice how the eyes of the predators look straight ahead while those of the prey are placed at the side of the head to give wrap-around vision**

Transduction: the action of photoreceptors

An image on the retina stimulates **photoreceptors**, cells that are specialised to detect light. This is a photochemical event: pigment molecules in the photoreceptors are altered by light and this results in a generator potential. When threshold is reached, action potentials are created and nerve impulses begin to travel towards the brain.

The two photoreceptors in the human eye are the **rods** and **cones** (Fig 23.14 and Table 23.2).

Table 23.2 **Comparing rods and cones**

Feature	Rod cells	Cone cells
Shape	rod-shaped outer segment	cone-shaped outer segment
Connections	many rods connect with one bipolar neurone	only a single cone cell per bipolar cell
Visual acuity	low	high
Visual pigment(s)	contain rhodopsin (no colour vision)	contain three types of iodopsin responding to red, blue and green light
Frequency	120 million per retina – twenty times more common than cones	7 million per retina – twenty times less common than rods
Distribution	found evenly all over retina	found all over retina but much more concentrated in the centre, particularly the yellow spot or fovea centralis
Sensitivity	sensitive to low light intensities (used in dim light)	sensitive to high light intensities (used in bright light)
Overall function	vision in poor light	colour vision and detailed vision in bright light

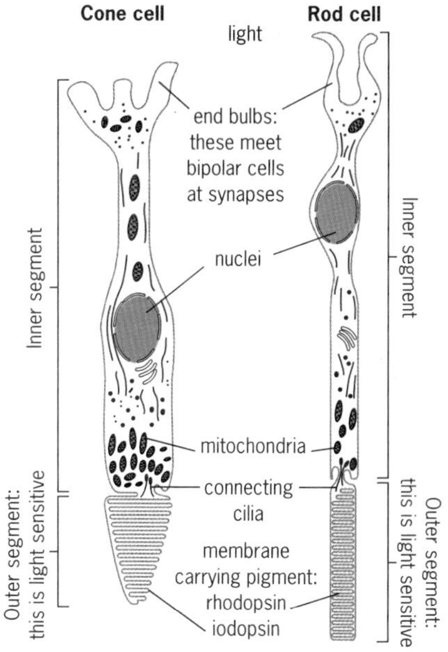

Fig 23.14 **Rods and cones: the two types of light-sensitive cell in the human eye. Look back at Fig 23.3a to see a photograph of rods and cones**

Photoreception in rod cells

Rods contain a reddish-purple compound called **rhodopsin**, or **visual purple**. This consists of two joined compounds – **opsin**, a protein, and **retinal**, a light-absorbing compound derived from vitamin A. When retinal is exposed to light and absorbs light energy, it changes shape (it is actually converted into an **isomer** and this causes the two compounds of rhodopsin to break apart. The free opsin acts as an enzyme and, through a series of reactions, activates a neurotransmitter called **cyclic GMP**.

Cyclic GMP closes membrane pores in the rod cell and the negative charge inside the cell increases. This causes the membrane of the cell to **hyperpolarise** (become more negative) and sets up signals in nearby nerve cells. Nerve impulses then pass the to brain for decoding. In the absence of further light stimulation, the retinal molecule goes back to its original shape and then recombines with opsin to form rhodopsin.

Rhodopsin is very sensitive to light and so rods are mainly used in dim light. In strong light the rhodopsin is broken down quicker than it can be reformed, but in dim light, production is able to keep pace with the slower rate of breakdown.

> ?
> **F** It has long been known that a deficiency of vitamin A causes the condition of night blindness. Give an explanation for this.

Photoreception in cone cells

Cone cells contain the photosensitive pigment **iodopsin**, a photosensitive pigment made up of a **photopsin**, a different protein from that found in rods, and retinal. The events that occur in cone cells stimulated by light are basically the same as those that occur in rod cells. There are, however, three different types of cone. Each one contains a slightly different pigment and responds to a different wavelength of light. One responds to red light, one to green light and one to blue light, so allowing us to see in colour (Fig 23.15).

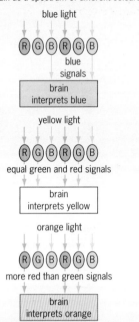

Sensory information from the cones is perceived by the brain as a spectrum of different colours

blue light

R G B R G B
blue
signals

brain
interprets blue

yellow light

R G B R G B
equal green and red signals

brain
interprets yellow

orange light

R G B R G B
more red than green signals

brain
interprets orange

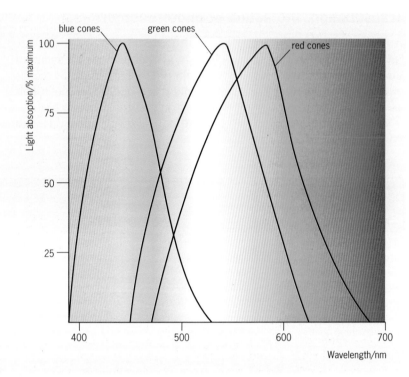

Fig 23.15 There are three types of cone cell responding (optimally) to three wavelengths of light. If a single type cones are missing from your retina, you cannot distinguish some colours and you are said to be colour-blind

The colour that we 'see', or more accurately *perceive* (interpret in our brain), depends on which cones are stimulated. When all the cones are fully stimulated we see 'white'. When very few are stimulated we see black. Stimulation of separate types produces red, green or blue, and all colours in between are produced by combinations of different levels of stimulation of the three types together. This model of how three different types of cone can produce the range of colour vision that we experience is known as the **trichromatic theory** of colour vision.

Fig 23.16 There are several theories that might explain why trichromatic vision evolved. Some people think it developed to aid communication between animals while others think it may have given animals, an advantage in seeing their prey. The photograph above illustrates another theory – that, in primates, colour vision evolved because it was an advantage that helped them distinguish between edible and inedible (and also possibly poisonous) food

See question 2.

How our eyes adapt to light and dark

If you are in the dark, or in very dim light, for longer than half an hour, the rate that your photopigments are formed is much faster than the rate at which they are broken down. Your eyes are **dark adapted**.

If your eyes are exposed to bright light for a long period of time, much of the light-sensitive pigment in the rods and cones is broken down and the sensitivity of your eyes to light is much reduced. In this situation, your eyes are said to be **light adapted**.

Visual acuity

When rods and cones are stimulated, they pass information on to bipolar cells. From here, the information passes on to ganglion cells and then leaves the retina through the optic nerve. The number of ganglion cells that connect with each rod cell determines how well different portions of the retina pick up fine details of an image.

The ability of the eye to see fine detail is known as its **visual acuity**. If we wish to look at an object closely we 'fix' that object in the centre of our field of view. We have learned to do this because

?

G Which cones are stimulated by **(a)** blue light at 490 nm, **(b)**, violet light below 440 nm? (Check with Fig 23.15.)

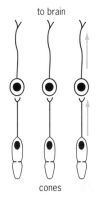
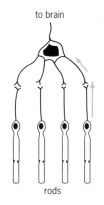
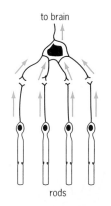

Fig 23.17 **Responses of rods and cones to light. Note that cone cells give a sharper image because they all have 1:1 connections with bipolar cells**

cones

When a narrow beam of light strikes a cone cell, action potentials occur in the connecting bipolar cells

rods

A narrow beam of light which strikes a rod cell does not produce an action potential in a bipolar cell

rods

A much broader beam of light does produce an action potential in a bipolar cell because it strikes several rod cells at once

the central portion of the retina, the **fovea**, is able to pass very finely detailed images to the brain. This is because the fovea has equal numbers of cones, bipolar cells and ganglion cells. Each cone cell therefore communicates directly with a single bipolar cell. Because of this 1:1 relationship, a much more detailed image passes to the brain.

Towards the edge of the retina, bipolar cells are over a hundred times less numerous than rods. In these areas, many rod cells therefore communicate with a single bipolar cell (Fig 23.17).

Transmission of nerve impulses to the brain

As described above, the photoreceptor cells are linked to a set of nerve cells in the retina called bipolar cells. These link with a second type of nerve cell called ganglion cells. It is the axons of ganglion cells that take information from the eyes to the brain. Ganglion cells fibres are bundled together to form the **optic nerve.**

We cannot see an object whose image falls on the retina at the point where axons of ganglion cells leave to form the optic nerve. Since it contains no receptor cells, any light striking this small area is not sensed; hence its name, the **blind spot** (Fig 23.18).

How the brain perceives an image

Action potentials do not differ very much from one another and so there are only two possible ways in which stimuli going into the brain can be coded. One is the *rate* at which action potentials arrive and the other is their *destination* in the brain.

The rate at which sensory neurones fire tells the brain the strength of the stimulus (see page 342): specific neurones tell the brain which part of the retina has been stimulated. The brain analyses this information to assess the nature of the original stimulus. This process, called **visual processing**, is immensely complex.

Specific areas of the cerebral cortex are associated with different sensory functions (see page 355). The **visual cortex** in the **occipital lobes** deal with visual information (Fig 23.19). However, body senses must not be considered in isolation. Sensory and motor systems interact within the control regions of the CNS and sensory information is processed at various points along the nerve pathway before it reaches the brain. In the brain it is processed further with information from other senses and stored memories.

Fig 23.18 **To demonstrate the blind spot, turn the book clockwise through 90° and hold it at arm's length with the two symbols straight in front of your eyes. Close your left eye and concentrate on the cross with your right eye. Bring the book slowly towards you. Keep looking at the cross on the left: eventually, as the dot falls on the blind spot of your right retina, the image of the dot on the right will disappear**

See question 3.

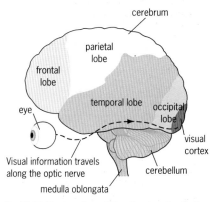

Fig 23.19 **The location of the visual cortex in the human brain**

The human ear

The human ear has two main functions: it allows us to hear sounds and it allows us to maintain balance and posture. The ear is a miniature receiver, amplifier and signal-processing system. It is divided into three parts – an **outer ear**, **middle ear** and **inner ear** – and is connected to the brain by the auditory nerve (see Fig 23.20).

The structure and function of the ear are shown below.

Fig 23.20 **Sound waves pass through the outer and middle ear and into the inner ear through the oval window. Pressure on the oval window squashes the fluid in the inner ear and compression waves are pushed through the canals of the cochlea. The round window eventually receives these movements of fluid and dampens them down (this prevents them from being reflected back into the cochlea) by bulging into the middle ear. Sensitive hair cells in the cochlea fire off and send action potentials along the auditory nerve to the brain, which analyses and processes the incoming information**

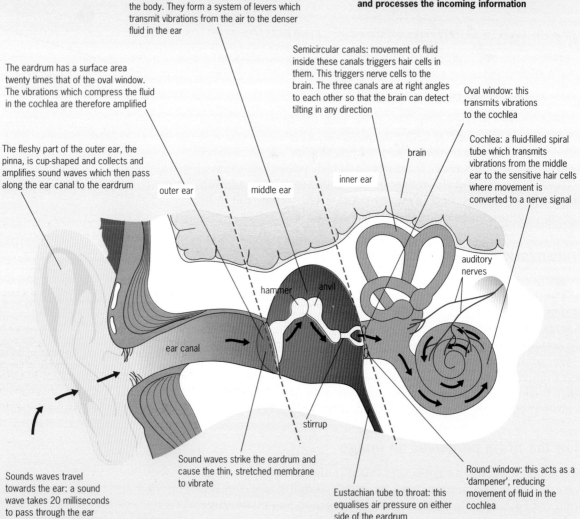

Ear ossicles: these are the smallest bones in the body. They form a system of levers which transmit vibrations from the air to the denser fluid in the ear

The eardrum has a surface area twenty times that of the oval window. The vibrations which compress the fluid in the cochlea are therefore amplified

Semicircular canals: movement of fluid inside these canals triggers hair cells in them. This triggers nerve cells to the brain. The three canals are at right angles to each other so that the brain can detect tilting in any direction

Oval window: this transmits vibrations to the cochlea

The fleshy part of the outer ear, the pinna, is cup-shaped and collects and amplifies sound waves which then pass along the ear canal to the eardrum

Cochlea: a fluid-filled spiral tube which transmits vibrations from the middle ear to the sensitive hair cells where movement is converted to a nerve signal

brain

inner ear

outer ear

middle ear

auditory nerves

hammer   anvil

ear canal

stirrup

Sounds waves travel towards the ear: a sound wave takes 20 milliseconds to pass through the ear

Sound waves strike the eardrum and cause the thin, stretched membrane to vibrate

Eustachian tube to throat: this equalises air pressure on either side of the eardrum

Round window: this acts as a 'dampener', reducing movement of fluid in the cochlea

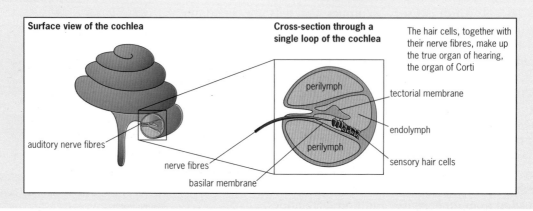

Surface view of the cochlea

Cross-section through a single loop of the cochlea

The hair cells, together with their nerve fibres, make up the true organ of hearing, the organ of Corti

perilymph

tectorial membrane

endolymph

perilymph

auditory nerve fibres

sensory hair cells

nerve fibres

basilar membrane

After completing this chapter you should know that:

▨ A **sensation** is a general state of awareness of a stimulus, **perception** involves the interpretation of sensory information.

▨ A **receptor** is a **biological transducer**: it converts energy from one type of system (eg chemical) to another system (eg electrical). A receptor is classified according to the type of stimulus it receives (eg photoreceptor), its location (eg interoceptor) or its level of complexity.

▨ The eyeball contains three outer layers, the **sclera**, **choroid** and **retina**.

▨ Incoming light rays are **refracted** (bent) by the cornea and by the lens of the eye.

▨ **Accommodation** is the ability to focus the eye automatically, to see near and distant objects.

▨ Rod and cone cells are stimulated by the effects of light on two pigments, **rhodopsin** and **iodopsin**. Breakdown of these compounds alters the electrical properties of the cell membranes.

▨ There are three types of cone cell pigment. Each responds to a different range of wavelengths of light. Humans have **trichromatic vision**: our cone cells detect three colours, blue, green, red.

▨ **Stereoscopic vision** enables us to perceive depth and distance.

▨ The human ear has two main functions, to detect sound and to maintain balance.

1 A number of changes occur in a human eye when it re-focuses from a distant object to a nearby object.

a) What is the term used to describe these changes?

b) Complete the table to describe the state of each of the structures when looking at an object at different differences from the eye.

Structure	Looking at a distant object	Looking at a nearby object
Ciliary muscle Suspensory ligament Lens		

[AEB 1994 Biology: Specimen Paper, q. 6]

2 The diagram of Fig 23.Q2 shows a single rod from a mammalian retina.

Fig 23.Q2

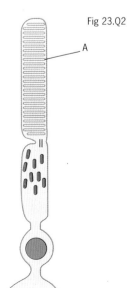

a) Name the parts labelled A and B and give *one* function of each. Copy the table and write in your answers.

Part	Name	Function
A		
B		

b) Draw an arrow next to the diagram to indicate the direction in which light passes through this cell.

c) State *two* ways in which vision using cones differs from vision using rods.

[ULEAC 1996 Biology Module Test B3: Specimen Paper, q. 4]

3 A person was instructed to close the left eye and stare with the right eye at a cross drawn on a plain white board. Different objects were then moved into the field of view from one side. The points where the object were first seen and where their colours were first identified were recorded. The results are shown in Fig 23.Q3.

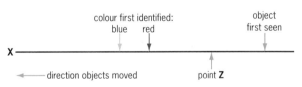

Fig 23.Q3

a) On what part of the right eye would the cross be focused.

b) Explain why an object might be seen at point **Z** but its colour would not be identifiable.

c) **(i)** Draw an arrow below the line on the diagram showing where you would expect the colour purple to be first identified.
(ii) Explain your answer to part **(i)**.

[AEB 1995 Biology: Specimen Paper, q. 7]

24 Animal behaviour

The tiny male Australian redback spider makes every effort to persuade his much larger partner to eat him after mating

SOME INVERTEBRATE ANIMALS are sexual cannibals. The black widow spider gets her name from her habit of eating her partner after mating with him. To us humans, this seems a strange way to end a relationship.

Another male spider, the Australian Redback spider also gets eaten, but he actively backflips into the female's jaws after mating. The female spider takes on a more passive role in eating him, and spits him out half the time. The male spider has no more than two percent of the female's mass and so cannot offer much in the way of nutrition, so why does the male sacrifice himself in this bizarre way?

One theory might explain the male behaviour in terms of competition with other males. Apparently, the longer the male can remain near the female, the longer the pair of spiders copulate. His leap into the female jaws prolongs his overall contact with her, keeping her busy, even if she is only trying to spit him out. By doing this, the male ensures that the female does not have chance to copulate with any other males before his own sperm have succeeded in fertilising her eggs. He is then the most likely father of her offspring. The male does not live very long (two to four months as an adult) and females are hard to find so it makes evolutionary sense to put all of his resources into one reproductive attempt.

1 INTRODUCING ANIMAL BEHAVIOUR

The biological study of animal behaviour is called **ethology**. When we study animal behaviour we look at what animals do in different circumstances and we try to find explanations for particular

Fig 24.1 **By studying the behaviour of newts, we know that courtship between smooth newts begins during April. Males develop crests and bright breeding colours to identify themselves to females. When he sees a female, the male dashes ahead and begins to move his tail from side to side to show off his crest and body patterns**

behaviour patterns (Fig 24.1). There are different branches of ethology: **neuroethologists**, for example, look at how behaviour patterns can be related to the function of the animal's nervous system while **behavioural ecologists** look at how animals adapt their behaviour to different environmental conditions.

The study of animal behaviour includes detailed **observation** and **experiment** and, historically, has involved two main approaches:

- Studying animal behaviour in **laboratory conditions**. Animal learning, motivation and the development of behaviour have all been examined in the laboratory. Animal behaviour specialists keep the animals under carefully controlled conditions and study individual aspects of their behaviour.

- Observing how animals behave under natural conditions, in **field studies**. The field biologist looks for reasons, often evolutionary reasons, that might explain why an animal behaves in a specific way. This may give a more realistic picture, but it is difficult to achieve exactly the same conditions on several different occasions, and so experimental repeatability and the reliability of results can be a problem.

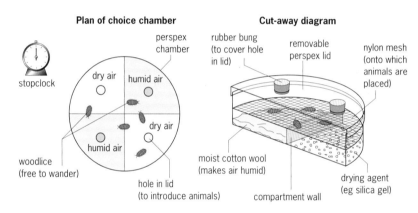

Fig 24.2 **A choice chamber is useful for observing many small invertebrates. This chamber allows the organisms to choose between two different environments. In an experiment, the conditions of differing humidity would be set up and left to stabilise for 20 minutes. Twelve healthy woodlice would then be put into the chamber and the experimenter would note their distribution and activity for 5 to 10 minutes**

The division between the laboratory and field methods is artificial and today's ethologists try to use a much more integrated approach which can involve:

- Observation of animals in their natural habitats.

- Experimental work in the field and the laboratory (Fig 24.2).

- Looking at how hormones and nerve and muscle physiology explain behaviour patters.

- Studying how behaviour changes over time.

- Studying how of behaviour patterns are passed from generation to generation.

Studying animal behaviour

Behavioural research tries to explain behaviour patterns. The starting point for any study is deciding what questions we need to ask. Look at Fig 24.3. An ethologist might ask: 'Why are those two stag beetles fighting?' 'Why does the peacock put on such as spectacular display?'. These sorts of 'why' questions can have lots of different answers. They might be answered by looking at the cellular basis of the behaviour (which nerves and chemicals are involved, for example) or by looking at the biological significance of that piece of behaviour.

Fig 24.3 **Rigid behaviour patterns in animals. The two male stag beetles (above) fight to the death. The peacock (below) displays the full glory of his feathers to attract nearby females. He struts around peahens shimmering and arching his tail feathers. If she is sufficiently impressed she will stand in front of him, ready to mate. The peacock's tail acts as a signal called a** releaser **that triggers sexual behaviour in the female**

A The (male) peacock is brightly coloured; the (female) peahen though is very drab. Why is this do you think?

One of the best ways to answer a question is to suggest a **hypothesis** that can then be tested. A hypothesis is a tentative proposition or a 'hunch' about the relationship between events. It can also be a suggested explanation for a set of observations. A hypothesis needs to be testable, plausible and have explanatory power (AEB Practical Assessment Criteria, 1997). It is the starting point from which research proceeds.

A reflex is a rapid, automatic response to a specific stimulus. It is a built-in response: it does not have to be learned (see Chapter 21).

Niko Tinbergen, one of the founders of ethology, suggested four important lines of investigation. Taking the example of the peacock behaviour shown in Fig 24.3 these would be:

● What *causes* the behaviour? Which hormones and nerve pathways are involved?

● What is the *function* of the behaviour?

● How does the behaviour *develop*?

● How did this behaviour *evolve*? Can we see examples of similar behaviours in related birds?

2 SIMPLE BEHAVIOURS

One of the simplest types of behaviour is **orientation.** This enables animals to move in order to find favourable environments and to avoid unfavourable ones.

Animals show two types of simple orientation:

● **Taxes**: movements in a specific direction which are directed by a stimulus such as light or food.

● **Kineses**: more random movements which are not directed by a stimulus.

Taxes and kineses are usually regarded as **reflexes**.

Taxes

In many orientation behaviours the animal moves directly towards or directly away from a stimulus. Maggots (fly larvae) have very simple eyes and can respond to the direction of light (Fig 24.4). Honey bees move towards a scent which is associated with food. These types of directional response are called taxes (singular: taxis).

A **positive taxis** involves movement towards a stimulus, a **negative taxis** involves movement away from the stimulus. As Table 24.1 shows, taxes are also classified according to the type of stimulus involved (light, water, gravity) or to the type of behaviour shown (straight-line responses, turning movements).

B How would you describe the behaviour of the maggot in Fig 24.4?

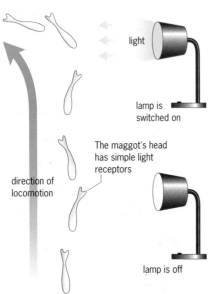

Maggot shows avoidance reaction when light is turned on

light

lamp is switched on

The maggot's head has simple light receptors

direction of locomotion

lamp is off

The maggot has to move its head from side to side to compare light levels on either side of its body

Table 24.1 **Summary of some common taxes. These are normally classified *either* by the type of stimulus that triggers them *or* by the type of response they elicit**

Name of taxis	Type of stimulus	Type of response
telotaxis		fixing on a stimulus without scanning from side to side
klinotaxis		moving head repeatedly from left to right to judge stimulus intensity
geotaxis	gravity	
thermotaxis	heat	
phototaxis	light	
anemotaxis	air currents	
rheotaxis	water currents	
chemotaxis	chemicals, eg salts	

Fig 24.4 **The maggot has simple eyes that can distinguish changes in light intensity but it cannot tell from which direction the light is coming. It moves its head from side to side to judge on which side the light is brightest**

Kineses

Imagine a ladybird, inactive during a cold night. It begins to stir at dawn. As the light level and temperature increase, the animal becomes more active and begins to move around and locate food. Conversely, woodlice are active in the dark and damp but become less active in brighter, drier conditions. These responses, which are not directional, are called kineses. Some common types of kineses are shown in Table 24.2.

Name of kinesis	Type of stimulus	Type of response
Photokinesis	light intensity	
Chemokinesis	chemical gradients	
Hygrokinesis	humidity levels	
Klinokinesis		Turns faster when stimulus is more intense
Orthokinesis		Moves with greater speed when stimulus is more intense

Table 24.2 **Some common kineses. Like taxes, these are normally classified** *either* **by the type of stimulus that triggers them** *or* **by the type of response they elicit**

3 MORE COMPLEX BEHAVIOURS

We can divide all animal behaviour into two categories: **instinctive**, or **innate behaviour** that is inherited and **learned behaviour** which is modified by experience.

Instinct

If we study behaviour closely we can find many examples where a piece of behaviour appears without any previous experience. A herring gull chick automatically pecks at the red spot on its parents bill (Fig 24.5). The digger wasp, *Ammophila*, builds a nest in the sand and immobilises a caterpillar using precisely the right amount of poison injected into exactly the right spot, without any prior knowledge. We describe this sort of behaviour as instinctive.

?

C Look at Fig 24.5. What is the benefit to the herring gull chick of having an inbuilt (innate) response to a contrasting red spot on a yellow bill?

(a)

(b)

Fig 24.5(a) **A hungry herring gull chick gets its parent to regurgitate food by pecking at the red spot on the parent's bill. But what is the precise stimulus for this behaviour?**

Fig 24.5(b) and (c) **In 1950, Tinbergen and Perdeck looked at this question. Using cardboard cut-outs to represent the parent bird, they found that it was not just the colour of the spot that elicited a pecking response. Herring gull chicks were also responding to the colour of the bill (the colour and shape of the head behind the bill did not seem to matter).**

Interestingly the model of a red bill gave an even greater response (more pecks) than a red spot. (The length of the grey bars on the diagram indicates the number of pecks.) But this feature is not found in nature. So why is there an increased response? Ethologists call these exaggerated stimuli (in this case a bright red bill) supernormal stimuli. Maybe this is why humans use make-up and body paints

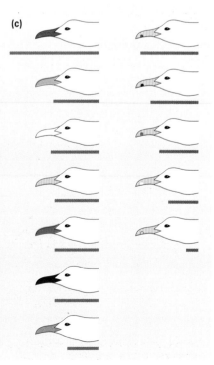

(c)

Fig 24.6 **Egg rolling behaviour in the greylag goose is a good example of a fixed action pattern. The parent bird retrieves an egg that has been moved out of the nest in a set sequence of movements. If the egg is taken away in the middle of the retrieval, the goose carries on moving an 'imaginary' egg, until the set sequence is complete**

Charles Darwin was the first to propose a definition of instinct in terms of animal behaviour. Natural selection, he argued, applies just as much to behaviour as it does to body parts and so behaviour can be inherited. Those animals that do not behave in the right way (eg finding the right food or looking for a mate) do not pass on their genes to the next generation.

But the classical, ethological view of instinct was put forward by Konrad Lorenz and Niko Tinbergen. They said that **motivational energy**: the more commonly-called drives and urges such as hunger, thirst and sex, controlled complex sequences of behaviour which they called **fixed action patterns** (Fig 24.6). These behaviours, which include nest-building, courtship and parental behaviour, are automatic and occur in response to specific stimuli called **releasers**.

The concept of instinct has changed several times since these early ideas. Modern research has shown that instinct is not 'driven from within'. Feedback from the body's receptors continuously modifies behaviour associated with feeding, sex and grooming. Behaviours are often described as *either* instinctive or learned. But, in reality, the majority of behaviours rely on both inheritance and the effects of the environment.

Learned behaviour

Learning is behaviour which is modified by experience. A three day old chick eventually learns to peck at food, not at stones on the ground and a bee learns to visit those flowers with most nectar. Such behaviours are useful because they allow animals to respond to changing conditions.

Some features of the environment, such as gravity and seasonal changes, are highly predictable. An animal does not need to learn how to respond to these situations: to do so would waste energy which could be used for other things. Animals learn:

● to avoid predators,

● to avoid harmful environments,

● to find food,

● to avoid harmful food,

● to find a suitable mate,

● to recognise important individuals within the group,

● to find their way home.

Fig 24.7 gives a simple classification of learned responses. The following section looks in more detail at some of these responses.

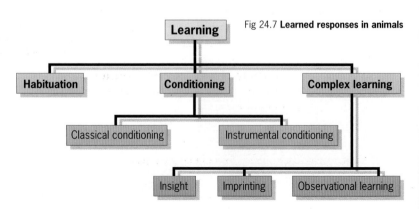

Fig 24.7 **Learned responses in animals**

Habituation

Habituation is the simplest type of learning. It is the reduced response which follows repeated stimulation and has been described as *learning not to respond*. If the sea slug, *Aplysia*, is touched at the side of its body, it immediately draws its gills in. Presumably this is a protective response. However, this response declines if the stimulus is applied over and over again (Fig 24.8). At this point the sea slug can be made to respond again by applying a strong stimulus elsewhere on the body and then re-applying the original (touch) stimulus. Because the animal can recover so quickly, biologists conclude that habituation in Aplysia is not due to muscle fatigue nor to any change in sensory information. Rather, it seems that there are changes in the motor nerves that control gill withdrawal.

Why should any animal habituate to, what is after all, potentially threatening stimuli? The answer seems to lie in energy savings. Take the common shore ragworm *Nereis*. This animal lives in a sandy tube which it leaves only to forage (Fig 24.9). If a shadow passes overhead it immediately retreats to the safety of its burrow. But, if there are repeated shadows, this response quickly subsides. A single shadow may well be a predator but if the shadows continue it is more likely to be seaweed, or some other object, floating above. A different type of stimulus, such as a vibration, results in an immediate response once again.

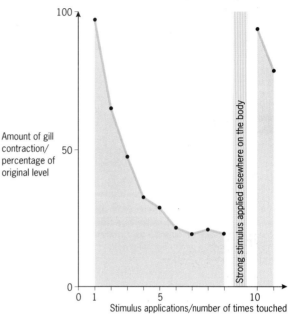

Fig 24.8 **The gill withdrawal reflex in the sea slug rapidly habituates with repeated touch**

Conditioning

Conditioning is another type of learning. An animal is able to detect a stimulus and then to predict what is likely to happen, simply because it has happened several times before. The animal associates two events which occur together. For example, cattle respond to the farmer entering the field in his tractor, but ignore other people with other vehicles. They associate the farmer and his tractor with food.

In his original experiments on conditioning, the Russian physiologist, Ivan Pavlov (1849–1936) recorded saliva produced by hungry dogs which were given food. Pavlov delivered a neutral stimulus, the sound of a bell, at the same time as the food. The dogs began to connect the neutral stimulus with food and, after a short time, the animals would salivate at the sound of the bell, whether food appeared or not. We now call this type of learning **classical conditioning** and the response it produces a **conditioned reflex**. It is found in all complex animals.

Fig 24.9 **The ragworm is an active carnivore. It emerges slowly from its sandy tube in order to feed. A sudden stimulus such as a shadow causes the animal to quickly withdraw into its burrow**

?

D Write down any ways you can think of in which farm animals show habituation?

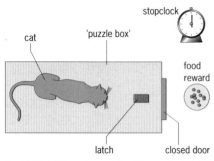

Fig 24.10(a) **The cat can solve this puzzle by trial and error learning (one of Thorndike's puzzle boxes)**

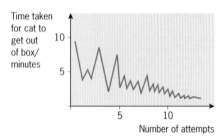

Fig 24.10(b) **A learning curve for a cat in one of Thorndike's puzzle boxes. Note the time taken to complete the task (ie obtain a food reward) decreases over time as the animal practises this procedure**

Classical conditioning occurs when an animal learns to associate a 'neutral' stimulus with an important one: the dog associates the bell ringing with the arrival of food.

Instrumental conditioning occurs when an animal learns to associate an action with a 'punishment' or a 'reward': the bird learns that eating a particular butterfly makes it sick.

E A cow touches an electric fence and gets a shock. Thereafter it avoids the fence. Is this an example of classical or instrumental conditioning?

While Pavlov was busy training his dogs to salivate in response to a bell, the American physiologist, Edward Thorndike (1874–1949), was looking at how cats solve problems. He placed a cat in a closed box and put its food outside (Fig 24.10(a)). If the cat operated the latch it could escape and obtain the food. Thorndike measured the time it took the cat to escape from the box and plotted these times over successive trials to get a learning curve (Fig 24.10(b)).

This example involves **positive reinforcement**: an action is reinforced by being rewarded immediately. An unpleasant stimulus, like a nasty taste, will discourage an action and so is called **negative reinforcement** (Fig 24.11). The process of learning in which a random action is either encouraged or discouraged is called trial and error learning, **instrumental** or **operant conditioning**.

Fig 24.11 **The blue jay soon learns to avoid distasteful food through trial and error learning. After eating a monarch butterfly for the first time, the bird vomits almost immediately (the butterflies contain chemicals that are toxic and irritant). The next time the jay sees a monarch butterfly it heads off in the opposite direction**

Complex learning

Not all learning is the result of conditioning or habituation. When a banana is left just out of reach outside its cage, a chimp might use a nearby bamboo pole to pull in the fruit. Or, if a banana is tied to the branch of a tree, again out of reach, the chimp may eventually stack boxes one on top of another and climb up to reach it. This behaviour appears to involve reasoning and we call it **insight learning**.

It is well known that birds and mammals form an attachment to people or animals that they come into contact with shortly after hatching or birth. This behaviour is called **imprinting** and it allows young animals to learn the characteristics of its parents. This is particularly important where young animals need to learn quickly who their parents are. Young ducks, for example, follow their parents to the water soon after hatching.

4 SOCIAL BEHAVIOUR

There is safety in numbers. Schools of fish are less vulnerable to predators than single fish: large numbers tend to confuse a predator. Living in a group may also give benefits in terms of feeding success – for example, flocks of birds eat better than single individuals because group members can 'share' tasks such as keeping a lookout when feeding. Birds, like geese, are also more likely to find food when there are many pairs of eyes to look for it. We call this type of group behaviour **social behaviour**.

In a society individuals cooperate (Fig 24.12). Animals take on individual tasks, a practice called **division of labour** and there are complex systems of communication involving recognition of either family or other species members. Social organisation can range from the simple cooperation between a male and a female during mating to the complex societies of insects and primates.

Group living also has its costs. In a large bird colony, for example, males run the risk of females mating with other males. Females run the risk of male desertion. In mixed-species flocks such as Australian finches, there can be an increased competition for food and in any large group there is always the risk of contracting disease or parasites.

F In which flock of geese would individuals probably spend more time feeding, a large flock of thirty birds or a small flock of six birds? Explain your answer.

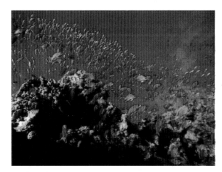

Fig 24.12 **Animals often find it beneficial to congregate in large groups**

Costs and benefits of social behaviour

Prairie dogs are rabbit-sized rodents with small ears and short legs. They live in large groups in the plains of North America digging a complex of tunnels up to 30 metres long (Fig 24.13). There are two types of prairie dog, the black-tailed and white-tailed forms and each community or 'town' may contain up to a thousand animals.

Prairie dogs graze above ground during the day when there are many predators around such as coyotes, badgers and hawks. They have 'sentries' sitting upright outside the burrows keeping a watchful eye for predators. The appearance of a predator provokes a series of whistling barks from the sentries.

A cost-benefit analysis has been carried out on the social behaviour of the prairie dog. An anti-predator response was tested by dragging a stuffed badger through a colony. In the larger groups of prairie dog, the predator was detected sooner. Black-tailed prairie dogs live in larger groups than their white-tailed cousins. Consequently they can afford to spend significantly less time scanning their surroundings for predators (35% of their average daily time budget compared with 45% in the white-tails). Less time being vigilant means more time to feed.

So why don't white-tailed prairie dogs live in large groups as well? One answer is that living in large groups has costs as well as benefits. Within either species, individuals in large groups fight more often than those in small groups. Also disease-spreading fleas are more common in large groups (3.3 fleas per black-tailed burrow, 0.5 fleas per white tailed burrow).

The evolution of group behaviour therefore is dependent upon balancing advantages against disadvantages. Other factors may also be involved, eg black-tailed prairie dogs live in open plains. There are many predators around therefore it is safer for them to live in larger groups.

Fig 24.13 **A prairie dog keeping guard as sentry**

Fig 24.14 **Woodlice are not social animals. They form aggregations because of the limited space available in the damp and dark places they need**

Fig 24.15 **Japanese macaques learn how to wash potatoes in the sea to remove soil. This behaviour is passed on to other members of the troop by observational learning**

Fig 24.16 **A great tit removes the top from a milk bottle before drinking the cream**

Not all group behaviour is social. There are many features of the physical environment that act to bring animals together on a more casual basis. Such **aggregations** can be found under stones and logs where animals such as woodlice are huddled together in the dark and damp conditions (Fig. 24.14). Nesting birds gather in large aggregations where nest sites are at a premium (eg kittiwakes on cliff ledges, sand martins on sandy river banks).

Cultural transmission of behaviour

One advantage of living in groups is that animals gain information from one another. Young animals, like mammals and birds, learn skills from their parents by watching and observing. On the Japanese island of Koshima, macaque monkeys learned to wash sweet potatoes originally provided for them by researchers in the early 1950s (Fig 24.15). This behaviour was passed on from parents to offspring and was observed by other members of the troop. After a few months, all the monkeys were washing their potatoes in the sea. Nearly fifty years later, this colony of monkeys still wash their potatoes even though they are not given dirty potatoes any more. Scientists think they like the salty taste.

Closer to home, blue tits and great tits have learned the 'trick' of opening milk bottle tops to get at the cream inside (Fig 24.16). These birds already had the habit of hammering and pecking at nuts with their beak. It was therefore a relatively small step to use the same technique on attractive, shiny bottle tops. This behaviour did not spread randomly but seemed to spread from certain focal points, suggesting that birds learned by watching each other.

In both these examples, **observational learning** occurred. The behaviour spread rapidly through the population until it became the norm. We call this phenomenon **cultural transmission**.

Human ethology

If children see pictures of small children's faces and adult faces they prefer the adult face. A change in preference occurs at 12 to 14 years of age in girls when they express a preference for baby pictures. Boys display the same trend two years later. If we sit too close to a stranger in a library then we tend to use physical 'barriers' such as books or bags to keep the stranger at a distance. If this person gets too close we often move seats.

These are examples of the types of behaviour studied in human ethology – the biology of human behaviour. It asks the question 'why do we behave the way we do?'

Ethologists view human behaviour in the same way as that of other animals: they think it is adaptive and has evolved to suit its purpose. In the two examples above, there is strong evidence to link children's behaviour with child rearing and adult behaviour with protection.

One goal of human ethologists is to look for patterns of behaviour shared by all peoples of the world. These are seen in examples of behaviour which involve body movements, rather than speech. Gestures such as smiling, raising the eyebrows, and behaviours such as grooming and hugging seem to be common to all human societies.

We all think that we have a free choice in our behaviour, but do we really? We may be under the control of our genes or be shaped by our surroundings. There is evidence for both points of view.

SUMMARY

After reading this chapter you should know and understand the following:

■ **Ethology** is the study of animal behaviour under natural conditions.

■ Patterns of behaviour are controlled by the nervous system: the **reflex** is the simplest form of behaviour.

■ A **taxis** is an orientation response related to the direction of the stimulus. A **positive taxis** describes movement *towards* a stimulus; a **negative taxis**, movement *away* from a stimulus.

■ A **kinesis** is an orientation behaviour where the response relates not to the direction of the stimulus, but to its intensity.

■ **Instinct** or **innate behaviour** is unlearned. It has survival value.

■ There are three main types of learning: **habituation** (learning not to respond), **conditioning** (associating a particular stimulus with a particular response) and **complex learning.**

■ Animals live in social groups for a number of reasons including protection, food capture, reproduction, shelter. They show cooperative behaviour.

■ Behaviour can be passed on from parents to off-spring genetically (innate behaviour eg fixed action patterns) and culturally (eg observational learning).

QUESTIONS

1 In order to investigate the effect of variation in temperature on the rate of activity of blowfly larvae (maggots), apparatus was set up as shown in Fig 24.Q1(a).

Fig 24.Q1(a)

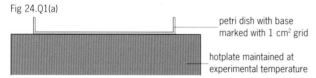

petri dish with base marked with 1 cm² grid

hotplate maintained at experimental temperature

A larva was placed in the Petri dish and its track marked by following it with a felt pen for two minutes.

a) **(i)** Name **one** factor that could also affect the activity of a larva which must be kept constant throughout the investigation.
  (ii) How would you treat the larva before placing it in the Petri dish?

The drawings in Fig 24.Q1(b) show the tracks of two larvae, one at 10 °C, the other at 20 °C.

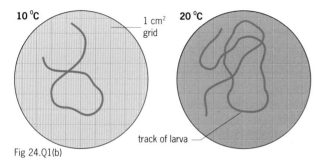

10 °C   1 cm² grid   20 °C

track of larva

Fig 24.Q1(b)

b) **(i)** Describe precisely how you would measure the rate of activity of the larvae from these results.
  (ii) Draw a table suitable for recording the results of this investigation. You are not required to enter any figures.

[AEB June 1994 Biology: Paper 3, q.3]

2 The graph shows the percentage of time spent by three infant gorillas within five metres of their mothers.

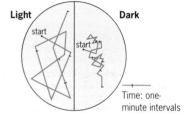

Light   Dark

start   start

Time: one-minute intervals

Fig 24.Q2

a) Suggest **two** advantages that the pattern of behaviour shown over the first 20 months might have to the infant.

Many of the adult gorillas in zoos were captured as very young animals in the wild and brought up on their own in captivity. When allowed to breed, the captive females showed different degrees of competence as mothers. The table compares various aspects of behaviour towards their offspring in first-time mothers and in those who have already had young.

Pattern of behaviour	Number of instances shown in:	
Pattern of behaviour	first-time mothers (*n* = 30 animals)	mothers who have already had young (*n* = 34 animals)
Cleaning infant	80	82
Suckling infant	40	68
Successful care of infant over the first week	33	59

b) In view of the fact that wild gorillas live in family groups which usually include several adult females explain the difference in the figures relating to the suckling of the infant.

c) Suggest why the figures for cleaning the infant are very similar.

[AEB 1996 Human Biology: Specimen Paper I, q.17]

25 Homeostasis

The emperor penguin (*Aptenodytes forsteri*) is over one metre tall and weighs about 35 kilograms, roughly half the weight of an average adult man. The males fast for two months while they incubate their egg and lose as much as 15 kg of body weight. Like their fathers, the chicks instinctively huddle together to reduce their heat loss

THE EMPEROR PENGUIN lives in the Antarctic, one of the coldest places on Earth. Surprisingly, it is a warm blooded animal and, even more surprising, it breeds in the middle of winter, when temperatures are at their lowest. At the beginning of winter, the penguins leave their feeding grounds in the sea and walk to their breeding grounds, up to 100 kilometres away. There, the female lays one egg, gives it to the male and then promptly returns to the open sea to feed.

The temperature plummets to −40 °C but the males have no choice but to stand on the ice for the next two months, incubating the egg between their feet and a fold of fat. If the egg accidentally rolls onto the ice, the male must gather it up again immediately, or the chick will die. Despite these dreadful conditions, the core body temperature of the male penguin remains at 38 °C, about a degree higher than that of most mammals.

The penguins maintain their body heat by respiring stored fat and by huddling together. Penguins are roughly cylindrical and snuggling up to their neighbours greatly reduces the surface area that is directly exposed to the freezing air. In an impressive example of social cooperation, the birds take it in turns to stand on the outside of the crowd. Scientists estimate that without this huddling behaviour, the males would not have enough fat to power the process of homeostasis that keeps them warm, let alone make the long return journey back to the open sea.

1 INTRODUCING THE CONCEPT OF HOMEOSTASIS

The word **homeostasis** means 'steady state'. When we study homeostasis we look at how organisms strive constantly to keep conditions inside their body as stable as possible. Organisms use many different homeostatic mechanisms to do this. Table 25.1 summarises some of them and shows where they are covered in this book.

Table 25.1 **A summary of the homeostatic mechanisms that are covered in this book**

Homeostatic mechanism	Covered in
Temperature	Chapter 25
Blood glucose	Chapter 25
Blood pressure	Chapter 18
Solute potential of blood	Chapter 27
Blood pH	Chapter 27
Blood volume	Chapter 27
Blood hormone levels	Chapters 23 and 28

Homeostasis is necessary because cells, especially those of higher animals such as ourselves, are efficient but demanding. To function properly they need to be bathed in tissue fluid which provides them with their optimum conditions. This means that higher animals must maintain the composition of their tissue fluid within very narrow limits.

The mechanism of homeostasis

When you start to study how the body manages to control a physiological factor such as temperature, blood glucose or blood pressure, it is important to organise your thoughts by asking the following questions:

- What conditions bring about change?
- What detects the change?
- How is the change reversed?

You will soon notice a pattern: whenever a physiological factor changes, the body detects the change and then, by using nervous or hormonal signals, or both, it reverses the change. The extent of the correction is monitored by a system called **negative feedback**. This makes sure that, as levels return to normal, the corrective mechanisms are scaled down.

Control of body temperature illustrates this mechanism well (Fig 25.1). When we are in a very hot environment or have been doing strenuous exercise, our body temperature rises. The brain detects this and sends signals to the body to bring the temperature down using various corrective mechanisms such as sweating and increased blood flow to the skin. As the body cools, the drop in temperature is monitored by the brain which then sends out less of the signals which lead to sweating. Sweating decreases and body temperature reaches its normal level and stays there until some other factor causes a further change.

A Why are homeostatic mechanisms often described as detection–correction systems?

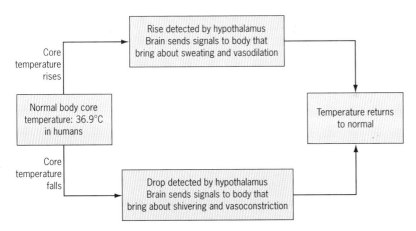

Fig 25.1 **The mechanism of homeostasis as illustrated by temperature control**

The opposite of negative feedback is **positive feedback**. In this situation a change is amplified rather than returned to normal. Positive feedback in living systems is rare, but there are a few examples. When the blood clots, for instance, damage to a blood vessel causes a cascade reaction: a few molecules of a substance become activated, and each one then activates many more (see page 417). Another example of positive feedback, the action of oxytocin in labour, is covered on page 460.

Positive feedbacks also occur in abnormal situations when normal homeostatic mechanisms get out of control. Elderly people whose sensory systems have deteriorated can suffer from **hypothermia**. When they start to become cold, their body systems fail to respond and the drop in temperature goes uncorrected. As it drops, their metabolic rate decreases and they produce less heat and they cool down even more quickly. Death occurs when their core temperature reaches about 25 °C.

Homeostatic mechanisms at the molecular level

We can also look at homeostasis at the level of molecules. In Chapter 4 we saw that enzymes control metabolic pathways, and that particular products are made, step by step, in a series of carefully controlled reactions. In a metabolic pathway the product is made only as fast as it is needed. Excess product inhibits one of the enzymes in the pathway, and the product is no longer made. When the excess has been used up, the inhibition is lifted and production continues. The process is self regulating and so is a homeostatic mechanism.

2 TEMPERATURE CONTROL AND HOMEOSTASIS

Life can exist at almost all the temperatures encountered on the Earth's surface, from the extremes of cold that occur at the polar regions to the extremes of heat that occur in hydrothermal vents (see page 176). This is possible because it is the *internal*, not the *external* temperature that is important to an organism. Animals that can withstand extremes of external cold, such as Arctic foxes and the emperor penguins, do so only because they maintain their body temperature inside this 'operating range', despite the outside temperature.

When the temperature gets too high

Why do animals die when the external temperature exceeds a certain limit? A common theory says that some vital enzymes become denatured, but it is probably not the whole story. It is more likely that death from excess heat is due to a metabolic imbalance, caused by enzymes working at different rates. As the temperature increases some enzymes work faster than others: some intermediate chemicals build up while some vital end-products become scarce.

When the temperature gets too low

Life can survive in a metabolically inactive **dormant state** at sub-zero temperatures. Sperm, eggs and even embryos kept in liquid nitrogen at −196 °C, can be thawed out and used successfully in infertility treatment (see Chapter 28). However, only a few organisms

> With the exception of some bacteria, organisms are active only at temperatures ranging from a few degrees below zero to about 50 °C.

> In Chapter 4 we saw that enzymes, the chemicals which control metabolism, are temperature sensitive. Enzyme activity increases with temperature and this increases the metabolic rate of the animal. The Q10 is a value which describes the effect of a 10 °C rise in temperature on the rate of reaction. Most enzymes have Q10 values of between two and three. This means that they double or triple their activity with every 10 °C rise in temperature, up to the temperature when the enzymes are denatured by heat.

Fig 25.2 **This Antarctic ice-fish lives in waters where the temperature is a constant −1.8 °C. The fish has an 'antifreeze' in its body fluids. This is actually a high concentration of a glycoprotein which prevents small ice crystals from developing into large ones. If the ice-fish swims into water with a temperature greater than 6 °C, it dies**

remain active when their body temperature drops below freezing (Fig 25.2). In most organisms, very low temperatures cause damaging ice crystals to form inside and also between cells.

If ice crystals damage the cellular structure, how do some cells survive being frozen in liquid nitrogen? The answer lies in the timescale involved: ice crystals take time to develop. When samples are immersed in liquid nitrogen the freezing process happens so rapidly that all the molecules simply 'lock' in position and ice crystals do not form. The cells and organelles are not damaged, and, when thawed, they function normally.

3 MECHANISMS OF TEMPERATURE CONTROL IN ANIMALS

Different animals control their body temperature in different ways. Many animals (fish, insects, amphibians etc) have a body temperature which is more or less the same as their surroundings. Mammals and birds usually maintain a constant body temperature, despite the external temperature.

These two categories of animal are commonly described as cold blooded and warm blooded. But these terms are not really satisfactory: a cold blooded animal, for example, is often not cold. In bright sunshine on a hot day, the core temperature of a reptile or amphibian may be higher than that of a mammal (Fig 25.3). Nor are the terms **poikilothermic** (cold blooded) and **homeothermic** (warm blooded) strictly accurate. The word *poikilos* means changeable, while *homeo* means constant. However, many cold blooded animals, especially aquatic ones, have a remarkably constant body temperature, due to the stability of their surroundings. Hibernating mammals, on the other hand, show a dramatic drop in their core temperature during their dormancy.

Fig 25.3 **These crocodiles, are gaping to keep cool. The external temperature is over 45 °C and so their blood is far from cold. Evaporation from their large mouths helps them to cool down**

To overcome these problems, we now use two other terms: **ectotherm** ('heat from outside') and **endotherm** ('heat from within'):

- **Ectothermic organisms** can control their body temperature only by changing their behaviour (Fig 25.4). All animals except mammals and birds are ectotherms.
- **Endothermic organisms** maintain a stable core body temperature using physiological and behavioural means (Fig 25.4). Mammals and birds are the only endotherms.

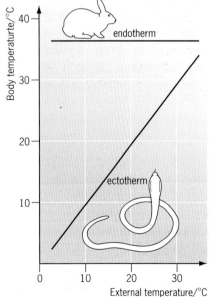

Fig 25.4 **The temperature of an ectotherm tends to be the same as the surroundings. An endotherm maintains a more or less stable body temperature over the whole range of external temperatures normally encountered**

Thermoregulation in ectotherms

Ectothermic animals **thermoregulate** (control their body temperature) with difficulty. Some, such as the aquatic ectotherms, always have the same temperature as their surroundings. Their gills, which have a large surface area to collect oxygen, are a liability in terms of heat loss because they carry large amounts of blood to within a few micrometres of the water.

Air-breathing ectotherms such as reptiles thermoregulate more successfully. These organisms do not have the fine physiological control over their temperature practised by birds and mammals, but they manage to control their body temperature to a large extent by their behaviour.

The Peruvian mountain lizard shown in Fig 25.5 is a classic example. This reptile lives in an environment where the days are hot but the nights are very cool. After a cold night, the lizard's body temperature is well below the optimum required for activity. Its movement is sluggish and it cannot hunt or avoid predators. So, early in the morning, the lizard emerges from its burrow and basks in the sun until its body temperature reaches about 35 °C. It is then fully active and goes off in search of food. As the day wears on and gets hotter, the lizard seeks shade to avoid overheating.

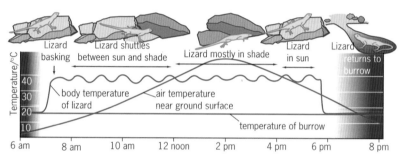

Thermoregulation in endotherms

Endotherms regulate their body temperature so that it stays at a constant level. When we refer to body temperature, we actually mean **core body temperature** (Fig 25.6). In a human, the limbs and other extremities may be substantially cooler than 37 °C, but the core usually remains constant. If the core temperature does fall, the person develops hypothermia (see page 399).

In the section that follows, we look at the mechanisms of thermoregulation in the mammal.

Fig 25.5(a) **The Peruvian mountain lizard, sunning itself in the morning**

Fig 25.5(b) **The Peruvian mountain lizard comes out in the morning when the air temperature is still below freezing. By positioning itself at right angles to the Sun's rays, the body temperature of the lizard rises quickly and is soon 30 °C higher than the surrounding air. Many other species also show this behaviour**

B Where would you find the greatest diversity of reptile species: in temperate countries (like the UK), in the tropics or the polar regions? Explain your answer.

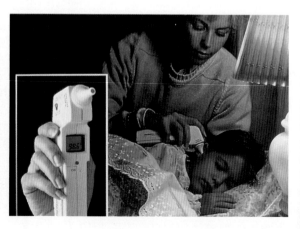

Fig 25.6(a) **Near right: a thermal image of a man taken with a camera sensitive to infra-red light. Areas of higher temperature look white, red or orange, cooler areas look green, blue or purple**

Fig 25.6(b) **Measuring core body temperature; a clinical thermometer provides a good estimate of core temperature when it is placed in the mouth, armpit or rectum. The thermal probe in the photographs (far right) allows an instant reading of the temperature of the blood near to the brain, without damaging the ear**

The role of behaviour

The mammal has complex body mechanisms which detect and deal with changes in core body temperature. However, just as ectotherms modify their behaviour to stay comfortable, so do endotherms.

In all mammals, including humans, body temperature is monitored by the **hypothalamus** (see page 355). This part of the brain is the body's thermostat. When we encounter a particularly warm or cold environment, temperature receptors in the skin inform the hypothalamus and also stimulate the higher, **voluntary**, centres of the brain. So, before physiology has a chance to act, we 'feel' hot or cold and decide to do something about it like changing position (Fig 25.7), changing our clothing or turning the heating up or down.

If behaviour cannot deal with the problem, blood temperature continues to change, rather than returning to normal. The hypothalamus then acts via the autonomic nervous system (see page 356) to initiate the appropriate corrective mechanism.

Physiological responses to cold

There are four main physiological responses that occur when the hypothalamus detects a drop in blood temperature:

- **Shivering** occurs when muscles contract and relax rapidly. Shivering muscles give out four or five times as much heat as resting muscles.

- **Vasoconstriction** occurs when the arterioles which lead to the capillaries in the surface layers of the skin **constrict** (narrow), so reducing blood flow to the skin (Fig 25.8 and see also page 286). This cuts down the amount of heat lost through the skin. Vasoconstriction is controlled by sympathetic nerves which pass from the **vasomotor centre** in the brain. This, in turn, receives information from the hypothalamus.

- **Piloerection** means 'erection of hairs' and refers to a reflex which obviously evolved in hairier animals than humans. In most mammals, piloerection makes the fur 'thicker', so that it traps more air to provide extra insulation. In humans, the **erector pili** muscles in the skin (see Fig 25.10) pull our tiny hairs upright, but only succeed in creating 'goose pimples'.

- **Increased metabolic rate**. The body secretes the hormone adrenaline in response to cold. This raises metabolic rate and therefore increases heat production. Mammals which live in cold conditions for a period of several weeks or months show a more permanent increase in metabolic rate due to the secretion of thyroxine (see Chapter 23).

Fig 25.7 **In a cold room, we may automatically assume a fetal position in bed to keep warm: when the arms and legs are drawn into the body this reduces the surface area that can lose heat. Conversely, in a hot room we spread out as much as possible**

?

C Which of the responses to cold will be the least effective in humans? Explain your answer.

Fig 25.8 **Blood flow to the skin can be altered by controlling the flow of blood in the arterioles leading to the surface capillaries. When tiny sphincter muscles in the arteriole walls contract to prevent blood flow to the surface, blood is forced through shunt vessels and away from the skin, so less heat is lost. When vasodilation occurs, the skin looks red; when vasoconstriction occurs, the skin looks pale**

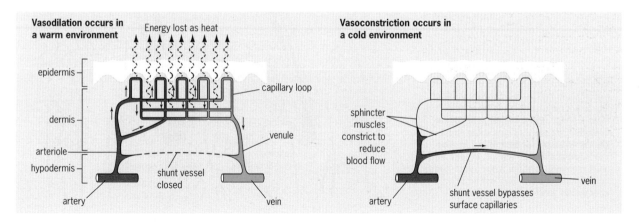

Physiological responses to heat

There are two main physiological responses to heat:

- **Vasodilation** occurs when arterioles which lead to skin capillaries dilate and shunt vessels are closed off, resulting in a greatly increased blood flow to the skin (Fig 25.9). As a result, more heat can be lost to the environment. Coupled with sweating, vasodilation is a very effective cooling mechanism.

- Sweating: sweat is a salty solution made by sweat glands, also known as **sudorific glands**. Evaporation of sweat from the skin's surface leads to cooling. The efficiency of sweating depends on the humidity. In dry air, humans can tolerate temperatures of 65 °C for several hours. In humid air, however, when the sweat cannot evaporate, and temperatures of only 35 °C (lower than the core body temperature) cause overheating.

> A common misconception is that blood vessels move to the surface of the skin. Blood vessels are fixed; it is the blood flow through them which can be changed.

> **D** Why do you think that people in the tropics are recommended to take salt tablets regularly?

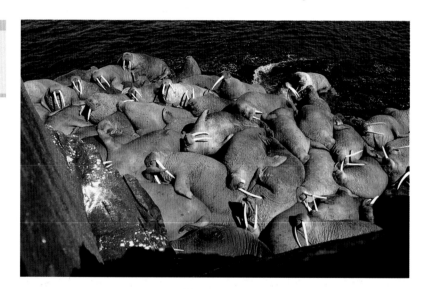

Fig 25.9 **The walrus is a large well-insulated mammal, well adapted to life in the cold sea. When on land, however, walruses are prone to overheating, especially in mild weather. This colony have gone pink because of the extra blood flow in the surface of the skin. Note the different colour of the walrus emerging from the sea**

The role of the skin in thermoregulation

In terms of area, the skin is the largest organ in the body. It is in direct contact with the external environment and so plays a central role in maximising or minimising heat loss. Fig 25.10 shows the structure of mammalian skin.

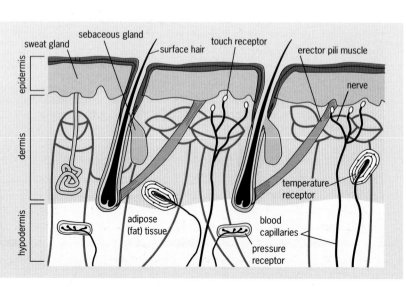

Fig 25.10 **The structure of human skin. Human skin differs from that of most other mammals in that we are effectively naked. We do not necessarily have fewer hairs, but most are tiny and no use for protection or insulation. We are also covered in sweat glands, and can lose heat by evaporation over most of our body surface. In contrast, most mammals have far fewer sweat glands, because evaporation from wet fur is not effective as a cooling mechanism**

Fig 25.11 **Above right: Black skin colour is due to the pigment melanin, a substance which protects skin from the damage by ultraviolet light. Skin cancer is rare in dark skinned races. Fair skinned people are not adapted to cope with strong sunlight.**
Below right: This is not a tan, but a first degree burn! When the redness has died down the skin will begin to make melanin, giving the desired tan. However, this is short lived and the individual runs the risk of developing skin cancer because ultraviolet light can cause genetic damage

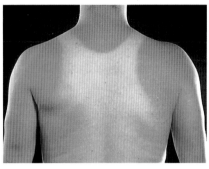

Structurally, the skin is divided into two layers, the **outer epidermis** and the **dermis** underneath. Forming the boundary between the two is the **Malpighian layer**, whose function is to produce new epidermal cells by mitosis. These cells make the protein keratin and once they are pushed up into the epidermis they flatten, die and dry up because they have no blood supply. We are constantly losing epidermal cells in a process called **desquamation**. House dust is mainly desquamated skin cells.

The dermis is much thicker than the epidermis, and contains many different structures such as nerve endings, hair follicles and blood vessels, held together with elastic connective tissue. Beneath the dermis is the **hypodermis** which usually contains at least some **subcutaneous fat**. This fat storage tissue, called **adipose tissue**, provides vital insulation in mammals which live in cold climates.

Mammalian skin has several functions:

- It detects stimuli using cells that are sensitive to heat, cold, light pressure, heavy pressure, and pain.
- It prevents excessive water loss or gain.
- It plays a role in thermoregulation by adjusting heat loss according to the circumstances.
- It prevents entry of microorganisms.
- Secretions such as hair, nails, claws and horns provide camouflage, defence and a variety of other functions.
- Communication. Most mammals live in a world where smell is very important, as anyone who has taken a dog for a walk will know. Most mammalian species communicate by means of chemicals called pheromones made by modified sweat glands (see Chapter 22).

See question 7.

?

E Consider the structure of human skin. Do you think that we are primarily adapted to a warm or a cold environment? Explain.

4 THE CONTROL OF BLOOD GLUCOSE

In normal circumstances, we obtain most of our energy by respiring glucose. Cells therefore need a regular supply of glucose. Some vital organs, notably the brain, cannot do without it for even a short time, and a lack of glucose can cause brain damage. Other cells can turn to other fuels such as protein or lipid for a short time, when glucose is unavailable.

There are three sources of glucose:

- Digestion of carbohydrates in the diet.
- Breakdown of **glycogen** (see Chapter 3). This storage polysaccharide is made from excess glucose in a process called **glycogenesis**. Glycogen is particularly abundant in the cells of the liver and muscles. When needed, glycogen can be quickly broken down to release glucose. This breakdown process is called **glycogenolysis**.

Glucose is one of the most abundant substances in our diet. Plant material contains **starch** and **cellulose**, meat contains **glycogen**. All three carbohydrates are broken down into glucose, which passes into the blood in large amounts following digestion.

✓

The terms used to describe glycogen metabolism are complex. Try remembering the origins of the words that make them up:

glyco, gluco = sugar
lysis = splitting
neo = new
genesis = generation, or formation

✓

A common mistake in examinations is to write 'Insulin converts glucose to glycogen.' Insulin has a more indirect connection; it opens gates in membranes, allowing glucose to pass into cells where enzymes may convert it into glycogen.

Fig 25.12 **The major part of the pancreas makes a juice containing digestive enzymes, but small patches of cells, called Islets of Langerhans, produce the hormones insulin and glucagon**

● Conversion of non-carbohydrate compounds. Following deamination, the organic acid part of the amino acid is converted into glucose. Pyruvate and lactate (see Chapter 8) can also be converted to glucose. The conversion process in either case is called **gluconeogenesis**, this means literally 'the generation of new glucose'. During prolonged fasting, blood glucose is maintained by conversion of the body's protein and lipid stores.

The blood of a normal person contains between 80 and 90 mg of glucose per 100 cm^3. This normal value is maintained even during prolonged fasting but rises to around 120 to 140 mg per 100 cm^3 shortly after a meal, when carbohydrate digestion is in full swing. Feedback mechanisms bring the levels back to normal within two hours.

The mechanism of blood glucose control

The control of blood glucose level is a classic example of homeostasis. A negative feedback mechanism operates to detect and correct the level of blood glucose, maintaining it within 'safe' limits.

The pancreas plays a central role in the control of blood glucose. The digestive functions of this organ are covered in Chapter 16 but in this section we are concerned with its **endocrine role**: how it produces the hormones **insulin** and **glucagon** to control blood glucose (Fig 25.12).

The pancreas detects any change in the level of glucose in the blood. If the levels get too high, β-**cells** in the **islets of Langerhans** respond by releasing insulin. This travels to all parts of the body in the blood , but mainly affects cells in muscles, liver and adipose (fat storage) tissue. Insulin lowers blood sugar by making cell membranes more permeable to glucose. It activates transport proteins, allowing glucose to pass into cells. Although the exact mode of action of insulin is still unclear, it also seems to activate enzymes inside the cells. These enzymes convert the glucose into glycogen, and also increase protein and fat synthesis.

If the levels of blood glucose get too low, α-**cells** in the islets of Langerhans secrete glucagon. This hormone fits into receptor sites on cell membranes, and activates the enzymes inside the cell which convert glycogen into glucose. The glucose then passes out of the cells and into the blood, raising blood sugar levels.

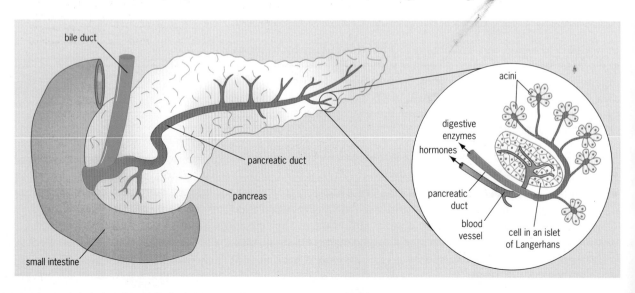

When control of blood glucose fails

Some people are unable to control the level of glucose in their blood. If levels get too high the solute potential of the blood increases, interfering with effective circulation and making the individual very thirsty. Low blood glucose levels mean that the cells are starved of their main fuel, and may have to respire other fuels. Both of these situations have a variety of causes and varying degrees of severity but both are described as the disease **diabetes mellitus**. In the UK, about 25 people in every thousand have diabetes: that means that there are over a million sufferers.

Diabetes produces a range of symptoms. Blood sugar levels get too high because glucose cannot pass into cells which are therefore starved of fuel. In the absence of glucose, cells must respire lipids and proteins, leading to weight loss and eventual starvation. Glucose appears in the urine because blood sugar levels are so high that the kidney cannot reabsorb it all. Before the Clinistix ™ test became available to test for sugar in the urine (see page 77), doctors used to taste the urine of patients they suspected of being diabetic.

Diabetes mellitus

There are two main types of diabetes: **type I** and **type II**.

Type I diabetes, also known as 'early onset diabetes', occurs when the body cannot make insulin. This is often caused by an auto-immune reaction which attacks and destroys cells in the islets of Langerhans. This form of the disease usually appears before the age of 20, and the onset is sudden. Sufferers have this condition for the rest of their lives, but they can be treated by regular injections of insulin (Fig 25.13), matched to their glucose intake (in diet) and expenditure (in exercise). Most diabetics lead normal lives. This is in great contrast to what used to happen before insulin was available: untreated type I diabetes was usually fatal within a year of diagnosis.

Type II diabetes, also called 'late onset diabetes' is more common than type I, accounting for more than 90 per cent of the cases in the UK. It tends to begin after the age of 40, especially in overweight women. 'Forty, fat and female' is the blunt way some doctors describe the classic circumstances for the onset of type II diabetes. Fortunately, it is not as serious as type I, and can usually be controlled by regulating the sugar content of the diet. The cause of this condition may be due to a drastic fall in the efficiency of islet cells, or of a failure of the cell membranes to respond to insulin.

> **Diabetes mellitus** is the name given to the inability to control blood glucose levels. It is not the same as **diabetes insipidus**, a disease which results from the inability to make anti-diuretic hormone (see page 437).

See questions 2, 9 and 10.

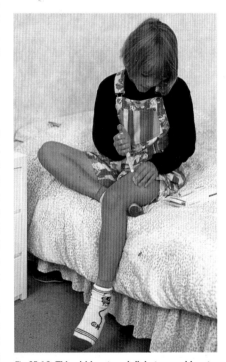

Fig 25.13 **This girl has type I diabetes, and has to inject herself with insulin every day. For many years the insulin came from cows or pigs, but both differ slightly from human insulin and gradually cause an immune reaction, reducing their effectiveness. Nowadays, diabetics use human insulin, made by genetic engineering**

5 THE LIVER AND ITS ROLE IN HOMEOSTASIS

The liver has the largest volume of any organ in the body and it plays a central role in homeostasis. It performs many hundreds, possibly thousands of different chemical functions that contribute to control of many different processes. We can condense these into a few basic headings and we deal with some of the main ones in this section.

The main functions of the liver are:

- Control of blood glucose levels.
- Control of amino acid levels.
- Control of lipid levels.

Although control of blood sugar is a 'whole-body' mechanism, the liver plays a central role. It is the first organ to receive blood from the intestines that contains high levels of sugar following a meal. Insulin therefore affects liver cells more than any other cells in the body. Its cells are also particularly rich in the enzymes involved in glucose metabolism.

- Detoxification.
- Synthesis of fetal red blood cells.
- Production of bile.
- Synthesis of plasma proteins.
- Destruction of red blood cells.
- Formation of cholesterol.
- Storage of vitamins.

The structure of the liver

The structure of the liver is shown in Fig 25.14. Follow parts (a) to (e) round in a clockwise direction.

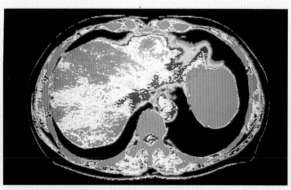

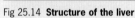

Fig 25.14 **Structure of the liver**

(a) This CAT (computerised axial tomography) scan through the abdomen of someone lying on their back shows just how much space the liver takes up: the liver is yellow and orange and the stomach is pink

(e) The fine structure of the liver. Blood is delivered to the liver cells in branches of the hepatic artery and the hepatic portal vein. As blood flows along the sinusoids, some chemicals are removed while others are added by secretion. The liver secretes bile into the canaliculi. These drain into small branches of the bile duct

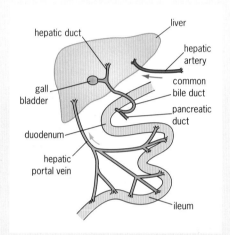

(b) **The blood supply to the liver and associated organs. The liver weighs about 1.5 kg in a normal adult**

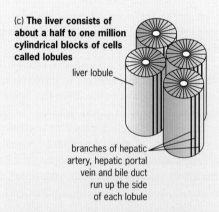

(c) **The liver consists of about a half to one million cylindrical blocks of cells called lobules**

(d) **A micrograph of liver lobules. Radiating sinusoids drain into the central veins**

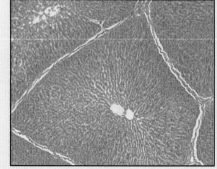

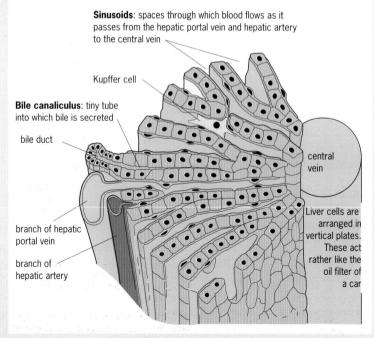

Blood flows into the liver in two blood vessels: about 30 per cent arrives in the **hepatic artery** which contains oxygenated blood, while 70 per cent is delivered by the **hepatic portal vein** which brings absorbed food from the intestines. Having already been through the intestinal capillaries, blood in hepatic portal vein contains relatively little oxygen. Blood leaves the liver in the **hepatic vein**.

The liver consists of thousands of **lobules**, about 1 mm in diameter, which surround branches of the hepatic vein. Channels called **sinusoids** radiate outwards from each central vein. These are surrounded by rows of liver cells, called **hepatocytes**. These apparently unspecialised cells perform the majority of the liver's functions, and the composition of the blood changes as it flows along the sinusoids towards the central vein.

Dotted along the sinusoids are numerous white cells called **Kupffer cells** (see Fig 25.14(e)). These are phagocytes (see Chapter 26) whose main function is to 'eat' bacteria and other debris. Parallel to the sinusoids are fine channels called **bile canaliculi** (singular: canaliculus). Hepatocytes secrete the constituents of bile into the canaliculi, which drain into the gall bladder. The contents and function of bile are discussed in Chapter 16.

The control of amino acid levels

The mammalian body cannot store proteins. Every day we need a minimum amount (40 to 60 grams, about the weight of one egg) to provide the amino acids that the body needs to repair and grow new cells. Most people take in more than this in their diet and the liver breaks down any amino acids that are not used.

Amino acids, like many other digested foods, reach the liver via the hepatic portal vein. They can be:

- **Deaminated**. This process removes the amino group ($-NH_2$) of an amino acid and forms ammonia (NH_3). The organic acid residue is usually respired while the toxic ammonia is quickly converted into a more harmless substance, **urea**, via the **ornithine cycle** (Fig 25.15). Urea is removed from the body by the kidneys (see Chapter 27).

- **Transaminated**. In Chapter 15 we saw that there are eight essential amino acids which must be present in the diet. The remaining 12 are termed non-essential amino acids because they can be made in the liver by transamination. Transamination

F Hepatocytes have many mitochondria and microvilli. What does this suggest about their function?

G Why can't we take in a week's supply of protein in a single meal by eating one large steak or omelette?

(a) **Deamination**
Amino group and hydrogen removed from amino acid as ammonia

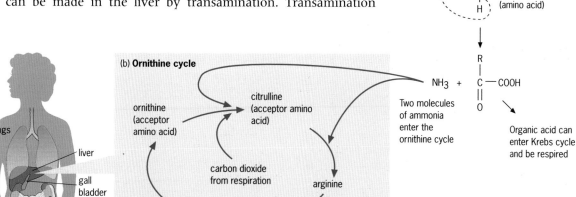

(b) **Ornithine cycle**

ornithine (acceptor amino acid)

citrulline (acceptor amino acid)

carbon dioxide from respiration

arginine

ornithine regenerated

arginine is hydrolysed to release urea

one molecule of urea produced

urea (contains two amino groups)

Two molecules of ammonia enter the ornithine cycle

Organic acid can enter Krebs cycle and be respired

lungs
liver
gall bladder
kidney
colon
ureters
bladder
urethra

Fig 25.15(a) **When an amino acid is deaminated, two molecules of ammonia (NH_3) are produced. These enter the ornithine cycle**

(b) **For every two molecules of NH_3 that enter the ornithine cycle, one molecule of urea (NH_2CONH_2) is produced**

involves the transfer of an amino group from an amino acid to an organic acid (derived from carbohydrate metabolism), thereby making a new amino acid.

● Used to synthesise plasma proteins such as fibrinogen.

● Released unchanged into the general circulation (most cells need a supply of amino acids for protein synthesis).

Detoxification and hormone breakdown

The liver concentrates and **detoxifies** (breaks down) many harmful chemicals. Some, hydrogen peroxide for example, are produced by the body. Others, such as alcohol and food additives, come from outside. Drinking large amounts of alcohol over a long period of time can cause the death of liver cells, followed by replacement with scar tissue. This is known as **cirrhosis** of the liver (Fig 25.16).

The liver also breaks down many circulating hormones. This removal of hormones from the blood is an important aspect of the control process as it ensures that hormones act only for as long as they are needed. Insulin, for instance is rapidly **metabolised** by the liver and has a half-life of about 10 to 15 minutes.

Storage

The liver stores relatively large amounts of vitamins A, D and B_{12} – enough to supply the body for several months. Other vitamins, such as most of the B complex, and the minerals copper and iron are also stored but in smaller amounts. This high vitamin and mineral storage capacity explains why eating liver is good for you, even though many people don't find it a pleasant experience.

Manufacture of blood proteins and blood cells

The liver makes many important blood proteins, including the most abundant plasma protein, **albumin**, and fibrinogen and other substances involved in blood clotting (see page 417). Given a supply of vitamin K, the liver can make prothrombin and factors VII, IX and X. Other plasma proteins such as globulins are also made mostly in the liver.

The liver of a fetus produces red blood cells but this complex manufacturing process is taken over by the bone marrow after birth.

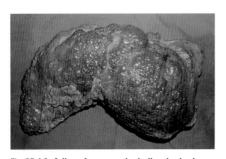

Fig 25.16 **A liver from an alcoholic who had cirrhosis. In the liver, alcohol (ethanol) is converted into ethanal by the enzyme alcohol dehydrogenase. Ethanal is toxic and may be responsible for much of the long term alcoholic liver damage, although the exact mechanisms are unclear**

See questions 3, 5 and 8. ■

SUMMARY

■ The word **homeostasis** means 'steady state'. Homeostatic processes keep conditions in the body within narrow limits.

■ Homeostasis is usually maintained by **negative feedback mechanisms**. The body detects a change in a particular internal factor, such as core body temperature, and then activates a corrective mechanism to reinstate the normal level.

■ Animals are divided into two groups according to their ability to thermoregulate: **endotherms** ('heat from within') maintain a stable core temperature by behavioural and physiological means; **ectotherms** maintain a relatively steady body temperature using behaviour only.

■ Blood glucose levels are monitored by the islets of Langerhans in the pancreas. If blood sugar levels get too high, cells in the islets secrete **insulin**. If they get too low, the cells secrete **glucagon**.

■ **Insulin**, a peptide hormone, increases the permeability of cell membranes to glucose. It also activates enzyme systems which **metabolise** glucose to glycogen, fat and protein.

■ **Glucagon** promotes the breakdown of **glycogen**, releasing more free glucose into the bloodstream.

■ The **liver** is a large organ which plays a major role in controlling the composition of the blood.

QUESTIONS

1

a) Explain:
  (i) what is meant by homeostasis;
  (ii) why homeostasis is important in living systems.

b) Hill walkers can encounter extreme changes in environmental conditions. Describe the processes involved in thermoregulation when a walker responds to a rapid fall in external temperature.

[NEAB June 1995 Modular Biology: Processes of Life Paper, q.9]

2
The diagram shows part of a muscle cell membrane, including a membrane receptor for insulin.

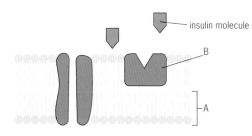

a) Name the molecules labelled A and B

b) Explain how glucose enters a muscle cell.

c) How is the rate of glucose uptake by a muscle cell affected by insulin?

[NEAB February 1995 Modular Biology: Physiology Test q. 5].

3

a) Name a polymer that is formed from amino acids by condensation reactions.

b) The diagram shows part of the structural formula of an amino acid.

$$H_2N - \overset{\displaystyle R}{\underset{\displaystyle H}{\overset{|}{\underset{|}{C}}}} -$$

  (i) Complete the structural formula of the amino acid.
  (ii) With reference to the diagram, explain what is meant by deamination.
  (iii) Describe what happens, in the liver, to the products of the deamination process.

[NEAB February 1995 Modular Biology: Physiology Test q. 6]

4
An investigation was carried out to find the effect of diet on the rate at which muscle glycogen was replenished after exercise. Three athletes each exercised for two hours. During the subsequent recovery period each athlete received a different diet. The results are shown on the graph.

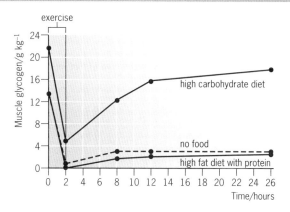

a) What features do the curves for all three athletes have in common?

b) Suggest **two** improvements that could be made to the design of this investigation to give more reliable results.

c) Explain how insulin is involved in the recovery of muscle glycogen levels in the athlete who was given the high carbohydrate diet.

d) In the athlete given the high-fat plus protein diet, much of the protein and fat was not converted into glycogen. Describe briefly what happens in the body to excess amino acids from the protein.

[NEAB June 1995 Modular Biology: Physiology Module Test q. 4]

5
The diagram shows part of a liver lobule.

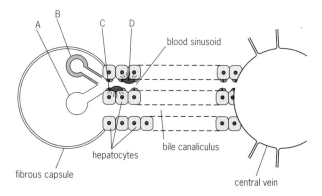

a) **(i)** Name vessels **A** and **B**.
  (ii) Describe the part played by vessels **A** and **B** in liver function.

b) **(i)** Draw an arrow in the sinusoid to show the direction of blood flow in the lobule.
  (ii) Explain the function of cells **C** and **D**.

[AEB 1994 Biology: Specimen paper 1, q. 4]

6
Write an essay on homeostasis (40 mins max – 24 marks)

[AEB 1991 Human Biology Paper 2, q. 4]

7 Describe the structure and functions of human skin
(40 mins – 24 marks)

[AEB 1994 Human Biology Paper 2, q. 4]

8 The diagram shows the main blood vessels going to
and coming from the liver.

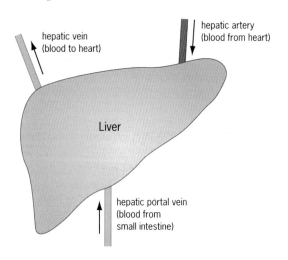

a) In a healthy person the blood glucose level in the
hepatic vein fluctuates much less than that in the
hepatic portal vein. Explain why this is so.

b) Blood sugar level is more or less constant, even if a
person has not eaten for several days. How does
gluconeogenesis help to maintain this constant blood
sugar level?

c) Suggest why people suffering from diabetes are advised
to eat their carbohydrate in the form of starch rather
than as sugars.

[NEAB June 1997 Biology: Physiology Module Test,
q.2]

9 The extract below is an entry from a dictionary of
biological terms.

blood glucose pool: the total amount of glucose in
the blood at any one time. Accurate control of blood
glucose is very important. If the concentration falls
too low, the central nervous system ceases to function
correctly. If it rises too high, then there will be a loss
of glucose from the body in the urine. Hormones
including *insulin* and *glucagon* play an important part
in maintaining a constant level of glucose in the blood.
Although the overall level stays within narrow limits,
it is important to realise that glucose is always being
added to and removed from the blood glucose pool.
Digestion and absorption of carbohydrates and
conversion from the body's stores of glycogen and
fats tend to increase the blood glucose level while such
processes as respiration decrease it. This is summarised
in the diagram.

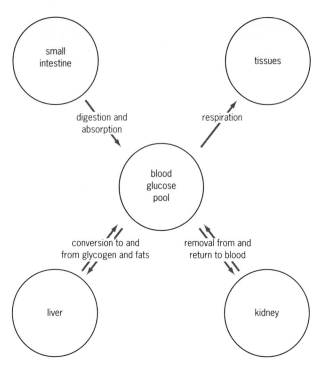

a) Describe the part played by insulin and glucagon in
maintaining a constant level of glucose in the blood.

b) In the kidney, glucose is removed from and returned
to the blood glucose pool. Explain how this occurs.

[NEAB June 1997 Biology: Human Systems Module
Test, q.8]

10 The diagrams in Fig 25.Q10 show the difference
between negative feedback and positive feedback.

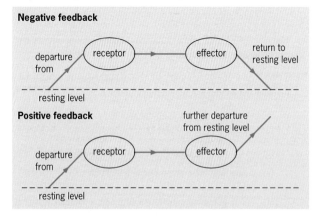

a) Suggest why negative feedback is frequently involved
in homeostasis.

b) Explain how negative feedback enables the carotid
and aortic bodies and the medulla to maintain a
constant blood carbon dioxide level.

c) Explain why the mechanism involved in the initiation
of an action potential is an example of positive
feedback.

[NEAB June 1997 Biology: Human Systems Module
Test, q.7]

Assignment

THE SECRET LIFE OF THE CAMEL

The popular image of the camel is not flattering. Described as, proud, arrogant, bad-tempered and 'a horse designed by a committee', the reputation of the camel does not do justice to its magnificent adaptations to its harsh desert environment. In this Assignment we consider two central aspects of homeostasis in the camel: the control of water balance and of body temperature.

Control of water balance

1 List a few of the main problems of desert living.

2 The most obvious feature of the camel is the hump (Fig 25.A1). Despite popular belief, the hump stores lipid, not water. Why is it an advantage for the camel to store its fat in one place rather than evenly distributed under the skin?

Fig 25.A1 **The dromedary (single humped) camel has been domesticated for at least three thousand years, and has played a vital role in the survival of desert dwelling people. Most 'working' camels are semi-wild, foraging for vegetation but relying on humans for water**

3 The camel obtains metabolic water from the respiration of stored lipid. What is meant by metabolic water?

4 List four ways in which mammals lose water.

5 Which of the methods that you have listed in your answer to question 4 increases in hot weather?

Although the camel can take some water from its food, it often becomes dehydrated. If a human dehydrates and loses more than 10 per cent of their body weight, death is usually inevitable. Amazingly, the camel can survive dehydration so severe that it loses 30% of its body weight. When it gets the chance to drink, the camel can drink up to 136 litres of water in a few minutes!

6 During dehydration, the solute potential of the blood increases. What effect would this have on human red blood cells?

Temperature tolerance

In addition to coping with dehydration, the tissues of the camel are also very tolerant to temperature fluctuations. Fig 25.A2 shows the daily fluctuations in core temperature of a dehydrated camel and a well watered camel.

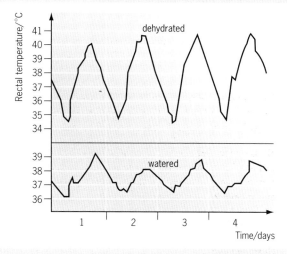

Fig 25.A2 **The daily temperature fluctuation (from midnight to midnight) in a well watered camel is about 2 °C. When dehydrated, however, the daily fluctuation is much greater: when the camel cannot 'afford' to lose heat by evaporation, it suffers much larger rises in body temperature**

7 What is the camel's maximum core temperature?

8 At what time of day is this peak temperature reached?

9 Explain why the temperature of the well-watered camel stays relatively constant.

10 A camel has two other adaptations: it can open and close its nostrils and it has wide feet with splayed toes. What the advantages do you think the camel gains from these two features?

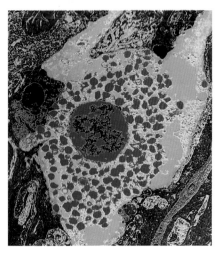

A false colour electron micrograph of a mast cell, a type of white cell involved in allergic reactions. The small red vesicles contain histamine, the chemical which produces many of the symptoms of allergy. Many people take antihistamines to dampen down their allergy to such things as pollen, house dust or animals

Fig 26.1(a) **Below:** this child has smallpox, a viral disease. Once a widespread killer disease, the last recorded case of smallpox occurred in October 1977. In 1980 the World Health Organisation declared that the disease had officially been eradicated – stamped out completely. This is the first time humans have managed to do this. It is possible that ongoing vaccination programmes will eradicate polio by the year 2000

IN 1995, A MAJOR MEDICAL JOURNAL reported the tragic case of a teenage girl who died within 20 minutes of taking a health supplement containing 'royal jelly', the substance made by queen bees. Sudden death in young people usually makes headlines, particularly when it seems to be related to a trivial event such as taking a normally harmless tablet. But similar deaths have also occurred as a result of a bee sting.

The victims all have in common a very rare and severe allergy. Just as some of us suffer from hay fever in the summer because we overreact to pollen in the air, some people overreact to the 'foreign' proteins produced by bees.

Normally, when someone is stung by a bee, the result is a painful red swelling which disappears again in a couple of days. In someone who is allergic to bee protein, the effects can be catastrophic. As soon as the poison gets into their bloodstream, many of their white cells release large amounts of histamine and other chemicals which cause severe inflammation. Fluid builds up in the tissues and all of the smooth muscles contract. The whole body is affected and both blood volume and blood pressure quickly plummet. Sometimes, the airways in the lungs narrow so much that the affected person can no longer breathe. These symptoms, collectively known as anaphylactic shock, can happen very quickly and can be fatal unless immediate treatment with adrenaline and antihistamines is available.

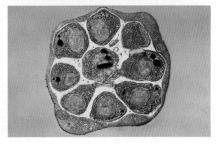

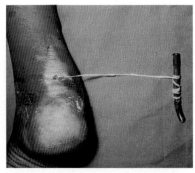

Fig 26.1(b) **Above: a guinea worm being removed from an ankle joint.** This parasitic worm can reach a length of 120 cm (4 feet) and is caught by bathing in infected water

Fig 26.1(c) **Left: a single-celled protoctistan called *Plasmodium* causes malaria.** After being injected into the bloodstream by mosquito bite, the parasites invade red blood cells where they multiply, hidden from the host's immune system. Nine malaria parasites are seen in this red blood cell

Fig 26 1(d) **This man has leprosy, a bacterial disease** that affects the skin, nerves and mucous membranes. With modern treatment, the deformities seen here should not develop

1 AN INTRODUCTION TO DISEASE AND IMMUNITY

We are all surrounded by bacteria, viruses, fungi and other organisms that are capable of invading our bodies and causing disease (Fig 26.1). We are able to overcome infections by these **pathogens** (disease-causing organisms) because we have an **immune system**. This is a complex system involving many different cells and tissues which allows us to develop **immunity** – resistance to infections.

Common pathogens of humans include bacteria, fungi, viruses, protoctists, such as *Plasmodium* which causes malaria, and parasitic animals such as tapeworms. Only a few types of these organisms are pathogenic: others are harmless. A pathogenic organism is able to:

- breach the physical barriers of the body and enter tissues or cells;
- resist the efforts of the immune system to destroy it, long enough to multiply inside the host's body;
- get out of one host and into another;
- damage the host's tissues – either directly, or indirectly by means of **toxins** (poisons) that it releases. Some bacterial **exotoxins**, such as the one produced by *E. coli* 507 (which caused an epidemic of gastroenteritis in Scotland in 1996), are very powerful and can cause death.

The easiest way to start understanding the immune system is to look at its overall functions, rather than concentrating on the individual parts. Fig 26.2 summarises the main lines of defence that an organism comes up against when it tries to infect a healthy person. We look at each of these in more detail in the following section.

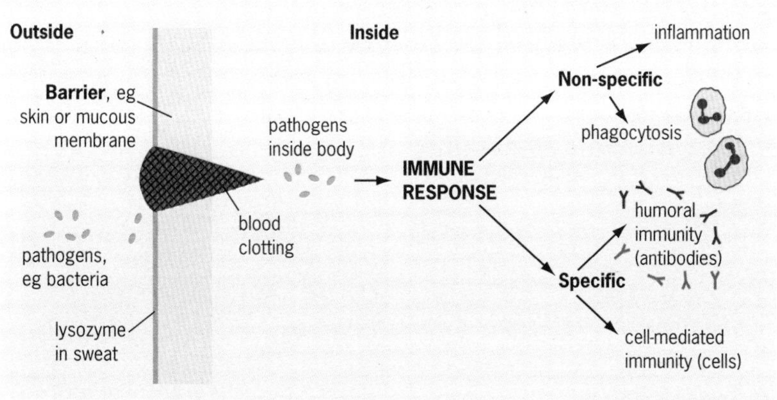

A What conditions inside our bodies make it ideal for the growth of microorganisms?

Pathogens are disease-causing organisms. Such diseases are **communicable** or **infectious**: they can pass from one person to another. Other diseases, such as diabetes and cancer are non-communicable.

Fig 26.2 **An overview of the body's defences: non-specific responses are general responses to damage. They include inflammation and phagocytosis of debris. Specific responses are targeted against individual types of microorganism**

Our study of the immune system covers:

- The ability of the immune system to distinguish between invading organisms and the tissues of its own body. This concept of **self** and **non-self** is important in determining whether the immune system keeps the body healthy, or whether it is overcome by infection.

- Problems of the immune system. In the Feature box on page 420, we see how the body can react against its own tissues, causing **autoimmune disease**.

- How medicine today can manipulate the immune system to provide life-saving treatments such as vaccines, blood transfusions, organ and tissue transplants and specific cancer therapy.

2 THE BODY'S BARRIERS TO INFECTION

B Smoking leads to paralysis of the cilia which line the airways. Suggest why this causes problems.

One of the most obvious ways to avoid infection is to stop potential pathogens getting into the body in the first place. The four main strategies that the body uses are shown in Fig 26.3 and are summarised below:

- **Mechanical** defence. **Nasal hairs** filter the air that is drawn into the upper airways. **Cilia**, which line the airways, sweep bacteria and other particles away from the lungs (Fig 26.3(a)).

- **Physical** defence. The skin (Fig 26.3(b)), made from **stratified squamous epithelium** (see page 30 in Chapter 2), forms a tough, impermeable barrier which normally keeps out bacteria and viruses. The mucous membranes that line the entry points to the body such as the nose, eyes (Fig 26.3(c)), mouth, airways, genital openings and anus produce fluids (see below) and/or sticky mucus. This traps microorganisms and stops them attacking the cells underneath.

- **Chemical** defence. Fluids such as sweat, saliva and tears contain chemicals that create harsh environments for microorganisms. Sweat contains lactic acid and the enzyme lysozyme, both of which slow down bacterial growth. Stomach acid kills many microorganisms that manage to get this far. When we are injured, blood clots at the injury site, sealing the breach to prevent entry of bacteria (see the next section).

- **Biological** defence. Normally, a vast number of non-pathogenic bacteria live on the skin and mucous membranes (Fig 26.3(d)). These do not harm the body but they 'crowd out' pathogenic bacteria, preventing them from gaining a foothold from which to launch a full-scale infection.

Fig 26.3 **Some of the body's barriers to infection**

Fig 26.3(a) **Light micrograph of a section through the human trachea. Household dust contains human skin cells, pollen, bacteria and fungal spores. Cilia in the respiratory system are able to filter out much of this before it reaches the alveoli, sweeping it upwards to minimise the chances of infection**

Fig 26.3(b) **The outer layer of (red) skin consists of flat, dead cells that consist mainly of the protein keratin. This forms a tough barrier which is impermeable to micro-organisms**

Fig 26.3(c) **You can see the pink mucous membranes of the eye when the eyelid is pulled down. These membranes also line the intestines and the respiratory, reproductive and urinary tracts. Cells and glands at these sites produce secretions that contain chemicals which kill bacteria**

Fig 26.3(d) **The skin and mucous membranes are covered in microorganisms such as these hyphae of *Trichophyton*, a fungus which causes athlete's foot. Normally there is a balance of harmless organisms which keep the pathogenic ones in check**

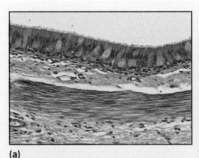
(a)

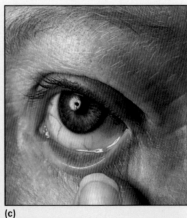
(c)

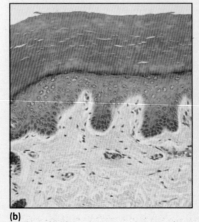
(b)

(d)

3 BLOOD CELLS AND DEFENCE AGAINST DISEASE

Blood clotting

Whenever blood vessels break, blood leaks out and clots in a process called **haemostasis** (Figs 26.4 and 26.5). We can see this happening when we cut ourselves, but blood can also clot deep inside the body. Clotting enables the body to avoid blood loss and, at the surface, to prevent infection. It is important that blood clots only when it should, because when a blood clot, a **thrombus**, blocks a vital blood vessel, the consequences can be fatal.

There are two separate pathways for blood clotting:

- the **extrinsic pathway**
- the **intrinsic pathway**

Both systems turn the *inactive* plasma protein prothrombin into *active* thrombin. In turn, this converts the soluble plasma protein fibrinogen into insoluble fibrin, producing a mesh which traps blood cells and so forms a clot.

Fig 26.5 outlines the major steps in blood clotting. Once initiated, blood clotting is a **cascade reaction** in which the activation of one molecule leads to the activation of many more. If any of the factors in the cascade are missing, the blood cannot clot. This occurs in the condition haemophilia, a sex-linked genetic disease in which the sufferer (usually a male) cannot make factor VIII. The genetics of haemophilia are discussed in Chapter 32.

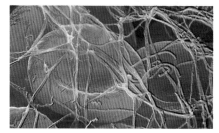

Fig 26.4(a) **A scanning electron micrograph of a blood clot. You can see red blood cells and platelets trapped in a fine mesh of fibrin. Blood clotting is a complex process that is set off when blood vessel walls are damaged and finishes when soluble fibrinogen is converted into an insoluble fibrin mesh**

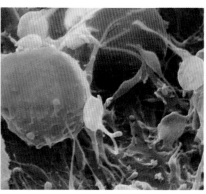

Fig 26.4(b) **A platelet is a tiny fragment of a cell which 'buds off' stem cells in the bone marrow. Each platelet is packed with the enzymes and chemicals needed for blood clotting. When they come into contact with damaged blood vessels, platelets (yellow) become sticky, attracting and activating other platelets, thus initiating the clotting process**

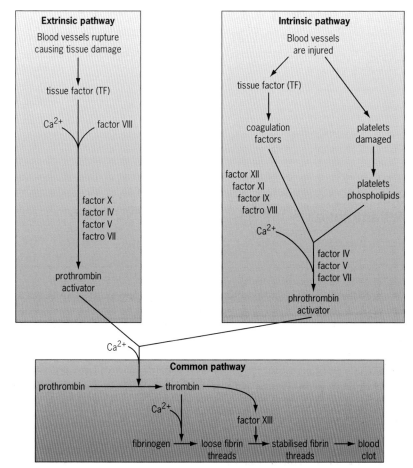

Fig 26.5 **Blood clotting is a cascade reaction: a complex set of chemical reactions which occur one after the other**

The process of thrombosis causes blood to clot inside vessels. The clot itself – a thrombus – can be fatal if it prevents blood from reaching vital tissues such as heart muscle. A coronary thrombosis is a common cause of heart attack while a thrombosis in the brain can cause a stroke.

C The saliva of several species of leech contains an anticoagulant. What are anticoagulants and why does the leech need them? Could 'leech spit' have any medicinal uses?

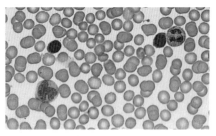

Fig 26.6 **In this blood smear, the white cells stand out because their nuclei are stained. The numerous red cells, which have no nuclei, do not take up the purple stain**

Classification of white cells

Fig 26.6 is a blood smear showing some white cells – **leucocytes** – along with many red blood cells. (The red cells outnumber the white by about 700 to 1). White cells are made in the bone marrow and are found throughout the body. They show amoeboid movement and are able to squeeze between cells, passing freely in and out of the circulation. Individual types of white cell are classified according to their appearance, origin or function (Table 26.1).

Table 26.1 **Different types of white cell. Neutrophils, eosinophils and basophils have a granular cytoplasm and are therefore called granulocytes. The other types are called agranulocytes**

Cell type	Diagram	How to recognise them	Relative abundance %	Function
Neutrophils		Lobed nucleus	57	Phagocytosis
Lymphocytes		Large round nucleus; little cytoplasm	33	Specific immunity. B-cells make antibodies. T-cells involved in cell-mediated immunity
Monocytes		Large, kidney-shaped nucleus	6	Phagocytosis. Monocytes develop into macrophages: general 'rubbish collecting' cells
Eosinophils		Stain red with eosin	3.5	Associated with allergy
Basophils		Stain with basic dye	0.5	Release chemicals such as histamine that are responsible for inflammation

?

D Use Table 26.11 to identify the white cells shown in Fig 26.6.

✔

White cells spend only about 10 per cent of their time in the blood, and so cannot really be called white *blood* cells. Leucocyte is the general name for all white cells.

✔

The *non-specific* immune response occurs in response to tissue damage. The processes are always the same, even though the damage can have a variety of causes. The response occurs immediately and either solves the problem or limits it, until the specific immune response can be activated.

The *specific* immune response is a highly targeted response to individual pathogens. It involves the production of **antibodies** and cells that are specific for individual molecules called **antigens**.

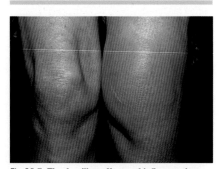

Fig 26.7 **The familiar effects of inflammation: reddened skin, swollen tissues and tenderness**

4 THE NON-SPECIFIC IMMUNE RESPONSE

When the body is damaged by cuts, scratches or burns, or is attacked by a pathogenic organism that manages to breach its defences, the body puts up a **non-specific immune response**. It is called a *non-specific* response because it occurs in response to tissue damage itself, not to the *cause* of the damage.

Inflammation

Inflammation is a rapid reaction to tissue damage. Whether it is in response to a cut, insect bite or a heavy blow such as a sport injury, the classic signs of inflammation are always the same:

● **Redness**: blood vessels dilate to increase blood flow to the area.

● **Heat**: also caused by the extra blood flow.

● **Swelling**: the extra blood forces more tissue fluid into damaged tissues.

● **Pain**: swollen tissues press on receptors and nerves. Also, chemicals produced by cells in the area irritate the nerve endings.

Inflammation (Fig 26.7) is triggered by damaged cells. Ruptured cells and some white cells (**mast cells** and **basophils**) release 'alarm' chemicals such as **histamine** and **kinins**. These substances dilate blood vessels and the increased blood flow leads to the classic signs of inflammation. The 'alarm' chemicals also attract white cells which clear bacteria and debris by **phagocytosis**.

Why inflammation is useful

Inflammation prevents the spread of infection and speeds up the healing process. It also provides a way of telling the rest of the immune system what is going on. When microorganisms are phagocytosed, fragments of their cells (particularly molecules that were originally on their surface) are processed by the phagocytes. Some of these surface molecules, which we call **antigens**, allow the specific immune system to recognise and remember the type of microorganism that has tried to invade the body.

Antigens stimulate the specific immune system to produce cells and chemicals that bind specifically to that antigen, and to no others. Find out more about specific immunity in the next section.

> An **antigen** is a structure (usually part of a molecule, for example, a protein) which is detected by the specific immune system as 'foreign': not part of the host's body. The immune system responds to an antigen by producing a very specific protein, an **antibody**, which reacts with that antigen, and that antigen only.

Non-specific chemical defences

Two main chemical systems help to protect the body against infection. The **complement** system is active mainly against bacteria; the **interferons** act primarily against viruses.

Complement

This is a collection of several proteins found in blood plasma. It is so called because it 'complements' the action of the body's specific immune mechanisms. It makes two important contributions to the non-specific immune response.

Firstly, it helps phagocytes clear up bacteria (see page 423). One of the complement proteins is an **opsonin**. This means it surrounds a bacterium, a bit like the sugar coating on a vitamin pill, making it much more appetising to the average phagocyte. **Opsonisation** ensures that bacteria are 'eaten' more quickly and any infection is kept in check.

Secondly, the proteins work together to kill bacteria directly. The whole complement system is activated by lipid–sugar molecules on the surface of some bacteria. When one of the complement proteins binds to this structure, a **cascade**, similar to that seen in the blood clotting process, is set off. About 11 different proteins react in a fixed sequence, forming a sort of protein 'hole punch' which jabs into the wall of the bacterium. Any bacterium with several of these holes cannot survive for long.

Interferons

There are several sorts of interferon. All are small protein molecules produced by body cells infected by viruses. This response happens very quickly. The interferons produced pass to other cells nearby, telling them to block protein synthesis. When viruses get into those cells, they have reached a dead end, because the cell cannot make the proteins the viruses need to reproduce. Interferons also activate white cells that kill cells infected by viruses. Such activated white cells can also kill cancer cells and so may have potential in medicine.

Phagocytosis

Neutrophils are the commonest type of white cell. Together with monocytes, they are known as **phagocytes** because of their ability to 'eat' pathogens by phagocytosis (Fig 26.8). In this process, the white cell engulfs the pathogen, takes it into a vacuole inside its cytoplasm and then digests it with **lytic** enzymes (see Chapter 5).

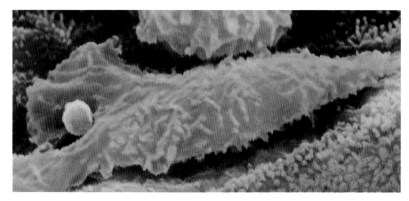

Fig 26 8 **This phagocyte (a neutrophil) is engulfing a speck of dust. These white cells occur in the bloodstream and throughout the tissues. Each phagocyte (which is spherical when not active) can ingest between five and 25 bacteria before the toxic products of breakdown kill the cell. At sites of infection there may be many dead, liquefied white cells which, together with dead bacteria and other debris, form** pus

AUTOIMMUNE DISEASE: WHEN THE BODY ATTACKS ITS OWN TISSUES

IN PEOPLE SUFFERING from autoimmune disease, the mechanism that enables the immune system to tell what is *self* and what is *non-self* breaks down. T- and B-cells (types of white cell – see text below) begin to attack the body's own cells and tissues. This is the underlying cause of diseases such as **multiple sclerosis**, **myasthenia gravis** and **rheumatoid arthritis**.

In multiple sclerosis, T-cells attack the myelin sheath around nerves. This severely limits nerve function, resulting in loss of movement and sometimes blindness.

In myasthenia gravis, the body makes antibodies that attack the motor end plates, the specialised synapses that connect motor nerves to muscles. If the motor end plates become damaged, the muscles cannot contract. The first symptom of this disease, which affects one in 30 000 people (mainly female), is rapid tiring during exertion. The muscles become progressively more unresponsive and the affected person can have difficulty breathing.

People with rheumatoid arthritis suffer from swollen and deformed joints (Fig 26.9).

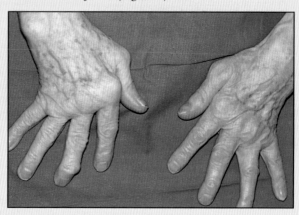

Fig 26.9 **This person suffers from rheumatoid arthritis. The cartilage at the joints has been attacked by the immune system, causing swollen and painful joints**

5 SPECIFIC IMMUNITY

The specific immune system protects the body from 'invasion' from microorganisms and parasites and also makes sure that the body's defences do not turn on its own tissues. The specific immune response is made up from two different systems which cooperate closely:

- **Humoral immunity**, also called antibody-mediated immunity, involves only chemicals: no cells are directly involved. The chemicals, called antibodies, attack bacteria and viruses before they get inside body cells. They also react with toxins and other soluble 'foreign' proteins. Antibodies are produced by white cells called **B-lymphocytes**, or **B-cells**.

- **Cell-mediated immunity**, as the name suggests, involves cells which attack 'foreign' organisms directly. Activated **T-lymphocytes**, or **T-cells**, kill some microorganisms, but they mostly attack infected body cells. Cell-mediated immunity is used by the body to deal with multicellular parasites, fungi, cancer cells and rather unhelpfully, tissue transplants.

Fig 26.10 summarises the differences between these two systems and Fig 26.11 shows a T-cell.

B-cells and humoral immunity

B-cells are produced in the **bone marrow** and are distributed throughout the body in the **lymph nodes**. B-cells respond to the 'foreign' antigens of a pathogen by producing specific antibodies. Antibodies are complex proteins that are released into the blood and carried to the site of infection. B-cells do not fight pathogens directly.

An antibody, or **immunoglobulin**, is a Y-shaped protein molecule that is made by a B-lymphocyte in response to a particular antigen. Fig 26.12 shows the overall structure of an antibody molecule. Antibodies interact with the antigen and render it harmless.

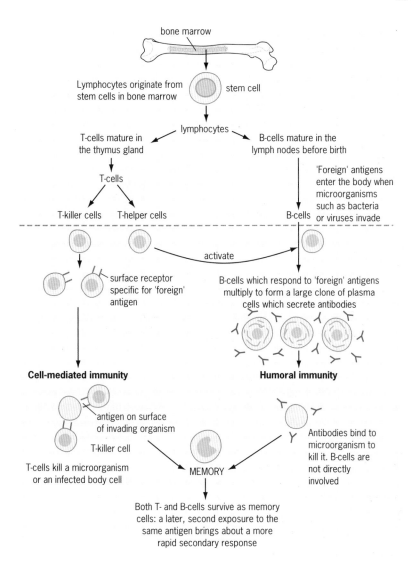

bone marrow

Lymphocytes originate from stem cells in bone marrow — stem cell

lymphocytes

T-cells mature in the thymus gland

T-cells

T-killer cells   T-helper cells

B-cells mature in the lymph nodes before birth

'Foreign' antigens enter the body when microorganisms such as bacteria or viruses invade

B-cells

activate

surface receptor specific for 'foreign' antigen

B-cells which respond to 'foreign' antigens multiply to form a large clone of plasma cells which secrete antibodies

Cell-mediated immunity

Humoral immunity

antigen on surface of invading organism

T-killer cell

T-cells kill a microorganism or an infected body cell

MEMORY

Antibodies bind to microorganism to kill it. B-cells are not directly involved

Both T- and B-cells survive as memory cells: a later, second exposure to the same antigen brings about a more rapid secondary response

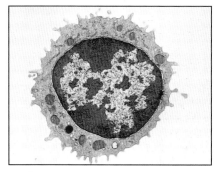

Fig 26.10 **Lymphocytes are made in the bone marrow by stem cells. Two distinct types of lymphocytes develop, T-lymphocytes and B-lymphocytes (T-cells and B-cells). Both T- and B-cells migrate from the bone marrow to the lymph glands, but the T-cells go via the Thymus while the B-cells go straight from the Bone marrow**

Fig 26.11 **A T-lymphocyte. T- and B-cells are identical: they have a characteristically large round nucleus and relatively little cytoplasm. The difference between them is their function: B-cells secrete antibodies while T-cells are involved in cell mediated immunity**

Fig 26.12(a) **Antibodies are Y-shaped molecules consisting of four polypeptide chains, two heavy and two light. Much of the molecule is constant but the tips of the Y are variable and match precisely part of a particular antigen molecule**

Fig 26.12(b) **A computer generated 3-D image of an antibody molecule (green) bound to an antigen (red)**

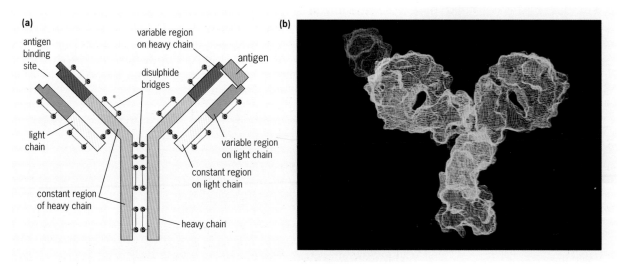

(a)

variable region on heavy chain

antigen binding site

disulphide bridges

antigen

light chain

variable region on light chain

constant region on light chain

constant region of heavy chain

heavy chain

(b)

When a pathogen tries to invade the body for the first time, each of its antigens activates one B-cell, which divides rapidly to produce a large population of cells. All the new cells are identical: we say they are **clones**, and they all secrete antibodies specific for the invading pathogen. When the infection is over, most of the newly

> A clone is a set of genetically identical individuals. In immunology, a clone refers to a population of B-cells, all of which can produce the same antibody because they are genetically identical.

made B-cells die: their job is done. We describe this sequence of events as a **primary immune response.**

So that the body can respond more quickly next time, some of the activated B-cells persist in the body for several years. These **memory cells** 'remember' what the pathogen is like and, if it tries to invade again, they all divide rapidly to produce an even greater number of active B-cells, all capable of secreting specific antibody. This response is called a **secondary immune response** and is very much quicker and more effective than the primary response (Fig 26.13).

Fig 26.13 **The immune system can remember antigens**

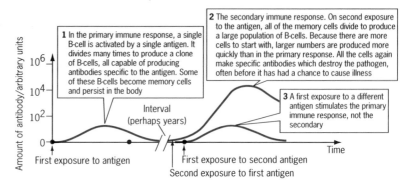

1 In the primary immune response, a single B-cell is activated by a single antigen. It divides many times to produce a clone of B-cells, all capable of producing antibodies specific to the antigen. Some of these B-cells become memory cells and persist in the body

2 The secondary immune response. On second exposure to the antigen, all of the memory cells divide to produce a large population of B-cells. Because there are more cells to start with, larger numbers are produced more quickly than in the primary response. All the cells again make specific antibodies which destroy the pathogen, often before it has had a chance to cause illness

3 A first exposure to a different antigen stimulates the primary immune response, not the secondary

This ability of the immune system is central to **vaccination** (Fig 26.14). A vaccine stimulates the body to produce a primary immune response to a particular pathogen, without becoming infected by it. A subsequent booster produces a secondary response. Later, if the pathogen tries to invade, the body can mount a very fast response and we can avoid becoming ill.

See question 2.

Vaccinations

Several infectious diseases overwhelm the normal primary immune response and so are fatal on first exposure. Thankfully we are able to speed up the specific immune response by giving vaccines against the pathogens that cause them. The basic idea behind a vaccine is that it contains some form of the pathogen, so that it stimulates memory cells to develop, ready to fight off the real pathogen should it be encountered. Obviously, the vaccine can't simply be the pathogen itself, or the toxins it makes. Somehow, the vaccine must be made less **virulent** – less able to produce disease. Examples of this are shown in Table 26.2.

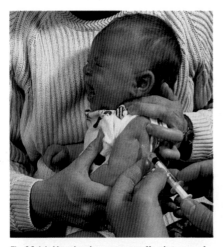

Fig 26.14 **Vaccinations are an effective way of stimulating the body's own defences so that we need not suffer the infectious diseases, such as measles, mumps and whooping cough, that used to be a common feature of childhood**

See question 3.

Table 26.2 **Recommended vaccination schedule for children in the UK**

Vaccination	Type of vaccine	Age due (UK vaccination schedule)
Diphtheria	Killed organism	2, 3 and 4 months
Tetanus	Modified toxin	2, 3 and 4 months
Whooping cough	Killed organism	2, 3 and 4 months
Polio	Live, non-virulent	2, 3 and 4 months
Haemophilus influenzae type B (Hib)	Purified bacterial capsule	2,3 and 4 months
Measles	Live, non-virulent	12–18 months
Mumps	Live, non-virulent	12–18 months
Rubella	Live non-virulent	12–18 months
Diphtheria	Modified toxin	3–5 years
BCG (bacille Calmette-Guérin – for tuberculosis)	Live, non-virulent	10–14 years
Hepatitis B	Genetically engineered antigens	For people at risk, eg health professionals

How do antibodies work?

On their own, antibodies are not usually capable of damaging pathogens: their role is to work with other parts of the immune response to fight the infection. They do this in four main ways:

● When antibodies combine with the antigens that stimulated their production in the first place, they act as opsonins. As we saw in the section on complement, this means adding a coating to the pathogen which acts as an 'eat me' sign to phagocytes (Fig 26.15).

● Antibodies also work with the complement system. When an antigen and antibody combine, this sets off the complement cascade. As many antibodies are made against the surface antigens of the pathogen, this means that the 'hole punch' effect of the complement system can take place at just the right place and can pepper the pathogen with holes (see page 419).

● Antibodies also combine directly with toxins released by bacteria, to neutralise their effect.

● In some cases, antibodies can act as **agglutinins**, rendering pathogens helpless by making them stick together.

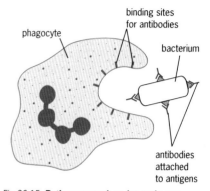

Fig 26.15 **Pathogens such as bacteria are covered in antigens which stimulate B-cells to produce antibodies. Many antibodies coat the pathogen, labelling it as foreign to stimulate attack by phagocytes**

MONOCLONAL ANTIBODIES

FOR SEVERAL YEARS now, scientists have been trying to harness the power of the immune system to provide successful cancer treatments. One line of research has been the development of antibodies that can hunt for cancer cells and help to destroy them, leaving normal cells untouched.

In 1975, Georges Kohler and Cesar Milstein came up with a way of making large quantities of pure antibodies against particular antigens. They called these antibodies **monoclonal antibodies (MABS)**. Fig 26.16 shows how monoclonal antibodies are made.

Possible uses of monoclonal antibodies in medicine

Monoclonal antibodies are used as tools in different fields of medicine:

● **Medical diagnosis**. MABS can be used to detect cancer cells, pathogens such as viruses, and chemicals such as hormones. The use of MABS is pregnancy tests is described on page 459, in Chapter 28.

● **Cancer treatment**. MABS can be attached to toxin molecules which kill cancer cells by binding to antigens on their surface. The antigens chosen are not present on normal cells and so these are unaffected by the treatment.

● **Transplant surgery**. MABS directed against the T-cells that lead to the rejection of some transplants are a valuable alternative to powerful immuno-suppressive drugs (see page 427).

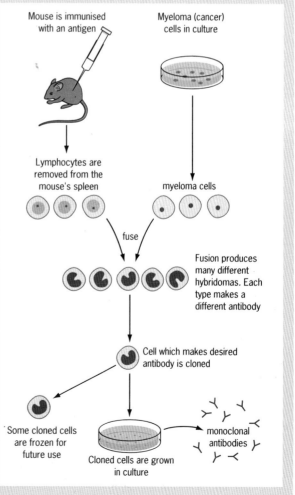

Fig 26.16 **The production of monoclonal antibodies (MABS)**

How mothers 'lend' immunity to their babies

Fig 26.17 **Babies borrow immunity from their mothers while their own immune system has a chance to develop. This baby is receiving antibodies in its mother's milk**

When babies are born, they emerge from the protective environment of their mother's womb. They are exposed to many potential pathogens. Normal babies have a fully functional immune system and can mount primary immune responses to many different antigens straight away. However, babies also get a bit of help from their mother, who 'lends' them some of her own immunity.

Antibodies pass across the placenta and, after birth, the supply continues through breast milk (Fig 26.17). Babies have very porous intestines that can absorb these large proteins directly into the bloodstream without digesting them. These large pores close by the age of one year. We call this kind of immunity – passed from one person to another – **passive immunity**. It does not last long, because the antibodies are broken down within a few days, but it can help a baby to fight off common pathogens.

Passive immunity is also used to treat some types of poisoning, such as snake bites. **Antiserum**, blood that contains antibodies specific to a particular snake venom, is raised in horses, purified and then given to people who have been bitten by a snake.

T-cells and cell-mediated immunity

Like B-cells, T-cells respond to specific antigens. When a pathogen first infects the body, each individual antigen stimulates a single T-cell. This divides to form a clone, in the same way that B-cells do. Some of the activated T-cells become memory cells and persist in the body, ready to mount a secondary response if the pathogen attacks again. The others, however, do not produce antibodies. They develop further to become one of three types of T-cell:

● **Helper T-cells** are so called because they help with, or rather orchestrate, the rest of the specific immune response. They tell B-cells to divide and then to produce antibodies, they activate the two other sorts of T-cell (see below), and they also activate macrophages, telling them to get ready to phagocytose pathogens and debris.

● **Killer T-cells** attack infected body cells and the cells of some larger pathogens (eg parasites) directly. The two cells face each other, membrane-to-membrane, and the killer T-cell punches holes in its opponent. The infected cell, or parasite, loses cytoplasm and dies.

See question 1.

- **Suppressor T-cells** are a sort of safety cut-out mechanism. When the immune response becomes excessive, or when the infection has been dealt with successfully, these T-cells damp down the immune response. Obviously, this is a good idea: if the body continued to make antibody and stimulate more and more T- and B-cells to divide, even when there was no need, this could damage the body and would be, at best, a waste of resources.

Allergies

An **allergy** or **hypersensitivity** is an overreaction to the presence of a harmless substance called an **allergen**. Some common allergens are pollen, house dust mites, animal fur and feathers, fungal spores, insect bites and penicillin (Fig 26.18).

The commonest symptoms of an allergy are sore eyes, runny nose, sneezing and asthma. Many of these symptoms are cause by an inflammation of the mucous membranes, caused by mast cells which release chemicals such as histamine. Many anti-allergy treatments suppress mast cells or neutralise histamine – chemists sell many **antihistamines** in the pollen season. For more information about asthma see the Assignment on page 274.

E What do you think it means when products such as make-up are labelled hypo-allergenic?

Fig 26.18 **A doctor performs skin tests to investigate the cause of this child's asthma. A small amount of an allergen such as pollen is placed under the skin with a sterile needle. If the patient responds to one allergen in particular, it is probably this substance that is causing the allergic reaction that is producing the asthma symptoms. The child in this photograph is allergic to house dust, as this produces inflammation**

6 BLOOD TRANSFUSIONS AND ORGAN TRANSPLANTS

In the last 100 years, advances in modern medicine have led to the technology and knowledge that allows doctors to give one person's cells, tissues and organs to another person. One of the main objectives in developing these life-saving treatments has been to overcome the natural reaction of the immune system to destroy transplanted cells and tissue, which it 'sees' as non-self.

Blood grouping and transfusions

Towards the end of the last century, medical scientists realised that accidents were often fatal simply due to blood loss. Losing large volumes of blood sent victims into shock and they died, even though their injuries were otherwise not too severe. Many women, in particular, often died when they lost blood when giving birth. Different doctors tried transfusing blood from a healthy person into the injured person, to try to restore their blood volume.

Sometimes this worked and sometimes it didn't. When it failed, the results were disastrous. We now know that, if the **blood groups** of the **donor** (the person giving the blood) and the **recipient** (the person receiving it) are different, the recipient's immune system reacts against the donated blood, producing a massive and deadly immune response.

In the early 1900s, an Austrian scientist, Karl Landsteiner, discovered that red blood cells from different people had different sets of antigens on their surface. The entire human population can be placed into one of four main blood groups: **A**, **B**, **AB** and **O**, according to the antigens on their red blood cells.

From Fig 26.19(a), you can see that people with blood group O have no antigens on the surface of their red blood cells. Blood from these people cannot cause an adverse immune response if it is transfused into people from the other three blood groups. We say that people

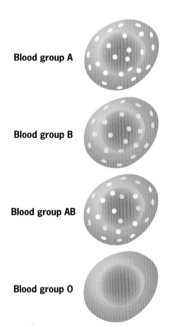

Blood group A

Blood group B

Blood group AB

Blood group O

Fig 26.19(a) **People can be placed in four groups according to which red blood cell antigens they have**

Fig 26.19(b) **We can find out what blood group someone has by mixing a sample of their blood with anti-A and anti-B antibodies. Blood from people with group A antigens agglutinates (forms clumps) with anti-A, blood from people with group B antigens agglutinates with anti-B. AB blood agglutinates with both, O with neither. In the diagram, the agglutination reaction produces a granular appearance as red cells clump together. In the smooth, dark samples, agglutination has not taken place**

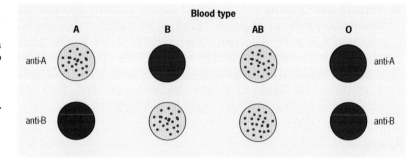

F Why do you think some of the early transfusions worked, even though they were done before we knew about the importance of blood groups?

See question 4.

with blood group O are **universal donors**. By the same reasoning, people with blood group AB cannot have any antibodies to either of the blood group antigens and so cannot mount an immune response if they are given blood by people in any of the other three groups. We say that people with AB blood are **universal recipients**.

Blood transfusions are dangerous when the recipient has antibodies to antigens present on the surface of red blood cells in the donor blood (Fig 26.19(b)). So, for example, a person of blood group A cannot donate blood to someone with group B. The recipient in this case has anti-A antibodies which react with the A antigens on the red blood cells of the donor blood. Table 26.3 gives a full list of possible transfusions.

Table 26.3 **Which transfusions are possible?**

Blood group	Antigens present on red blood cells	Antibodies present in blood	Can donate blood to	Can receive blood from
A	A	anti-B	A, AB	A, O
B	B	anti-A	B, AB	B, O
AB	AB	none	AB	A, B, AB, O
O	none	anti-A, anti-B	A, B, AB, O	O

G Look at Table 26.3. Why can someone with blood group AB donate blood only to someone else with the same blood group?

For examination purposes, a quick way to remember safe blood transfusions is as follows: O is the universal donor, so can be given to anyone. AB is the universal recipient, so can receive from anyone. Apart from that, only like-to-like transfusions (eg A to A) are safe. All others cause problems.

Transplants and grafts

Transplantation is the practice of taking cells, tissues or organs from one individual and placing them into another individual. Since the first organ transplant operations in the 1950s, patients have received hearts, lungs, skin, corneas, kidneys and various other organs. The success rate of these operations has improved steadily.

The biggest problem facing most transplant recipients is that of **rejection**. Transplanted tissue is covered in antigens which stimulate the specific immune response of the recipient, notably the cell-mediated response brought about by T-cells. Symptoms of rejection include the degeneration of blood vessels in the transplanted organ and the destruction of whole cells followed by their replacement with 'scar' tissue. A patient whose transplant is rejected becomes seriously ill: they lose the function of the organ concerned and the massive immune response puts an incredible amount of stress on their already weakened body.

Rejection can be minimised by tissue typing, by using immuno-suppressive drugs and by using monoclonal antibodies.

Tissue typing

Like red blood cells, other body cells also have surface antigens that are different in different people. Body cells, such as cells in the kidney or liver, have many antigens that differ, and so matching the tissue type of the donor and recipient is much more complex than matching blood groups. In practice, the cells from two people are never identical, so the aim is to match people who have as few differences as possible.

Immunosuppression

If the recipient of a transplant has a fully functional immune system, even a closely matched organ will be rejected to some extent. To counter this, transplant patients are given immunosuppressive drugs. The most effective is currently **cyclosporine**. This chemical, isolated from a fungus, inhibits the action of T-cells and so has a profound effect on cell-mediated immunity. Since the introduction of cyclosporine in the early 1980s, the success rates of some transplants has risen to over 90 per cent.

The problem with immunosuppressive drugs is, of course, that they reduce the body's ability to fight off infection. Transplant patients are more prone to infection from normally harmless microorganisms, particularly viruses. Although the dose of immunosuppressive drugs can be reduced after a few months, if there is little sign of rejection, transplantees must continue to take them for life.

See question 5.

Monoclonal antibody therapy

In the mid-1980s, a monoclonal antibody called OKT3, which attacks the T-cells responsible for transplant rejection, was developed for use in kidney transplant patients. It is more effective than cyclosporine in reversing the acute rejection that occurs very soon after transplantation. Another monoclonal, Campath-IG, was developed in the mid-1990s. This also attacks T-cells but is being used to treat bone marrow, after it has been taken from the donor, but before being put into the recipient. This technique seems to prevent rejection of the bone marrow, even when the patient and donor are not closely matched by tissue typing. It also allows them to avoid immunosuppressive drugs altogether.

SUMMARY

- We are surrounded by potential **pathogens** (disease-causing organisms) such as bacteria and viruses which can cause disease if allowed to enter and multiply inside the body.

- The body's defence system consists of barriers to keep microorganisms out, and mechanisms to detect and destroy those that do enter. These mechanisms are collectively called the **immune response**.

- The immune system is capable of non-specific responses and specific responses. Non-specific mechanisms (**inflammation** and **phagocytosis**) occur in response to any invading microorganism. Specific mechanisms allow the body to recognise and fight individual types of microorganism. There are two specific responses: **cell-mediated immunity** and **humoral immunity**.

- White cells called **lymphocytes** are responsible for the specific immune response. **T-cells** mature in the thymus gland and **B-cells** come directly from bone marrow. Both types are able to recognise foreign chemicals – **antigens** – which come from pathogens.

- B-cells are responsible for humoral immunity: they secrete specific proteins called **antibodies**.

- T-cells are responsible for cell-mediated immunity and help to control the overall specific response. **Killer T-cells** attack pathogens or infected cells. **Helper T-cells** activate B-cells, telling them to secrete antibodies. **Suppressor T-cells** damp down the immune response when the infection is over.

- **Auto-immune diseases** occur when the body's immune system attacks its own tissues. Examples include rheumatoid arthritis, multiple sclerosis and myasthenia gravis. **Allergies** occur when the body 'overreacts' to a normally harmless substance such as pollen.

- When cells or tissues are taken from one individual and given to another, they are often recognised as 'foreign' and destroyed. This process if often called **graft rejection**. To minimise the chance of rejection in blood transfusions and organ transplants, it is important to match blood and tissue types.

1 The diagrams in Fig 26.Q1 show the first stages in the development of humoral and cell-mediated immunity.

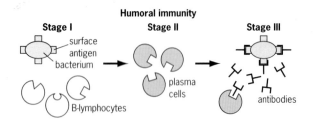

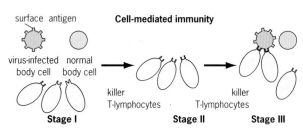

Fig 26.Q1

Source: adapted from *Family Health Encyclopedia*, 1990.

a) Describe **two** differences, shown in the diagrams, between the two types of immunity.

b) Describe **one** similarity at each of the stages, **I**, **II** and **III**, shown in the diagrams, between the two types of immunity.

[AEB 1995 Human Biology: Paper 1, q. 17]

2 In Fig 26.Q2, **A** shows the structure of an antibody molecule, and **B** and **C** show antibody molecules in contact with bacterial cells and virus particles.

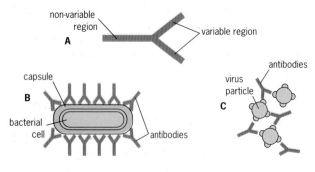

Fig 26.Q2

a) What do **B** and **C** suggest about the way in which antibodies combine with their respective antigens?

b) With reference to **B** and **C**, describe how antibodies help to bring about the destruction of: **(i)** bacteria, **(ii)** viruses.

c) Give **one** difference in antibody production after a second exposure to antigen compared to the first exposure.

[AEB June 1989 Human Biology: Paper 1, q. 12]

3 Explain how a vaccine derived from a pathogenic organism may protect an individual against the disease that the organism causes.

Explain why it is not necessary to vaccinate the whole of a population in order to prevent an epidemic.

[NEAB June 1995 Modular Biology: Health and Disease Paper, q. 8]

4

a) In a hospital pathology laboratory, two different samples of blood were mixed separately, on microscope slides, with three types of serum containing antibodies against red blood cell antigens (agglutinogens). A, B and Rhesus respectively. The slides were gently rocked and then visually examined. Fig 26.Q4 shows the results.

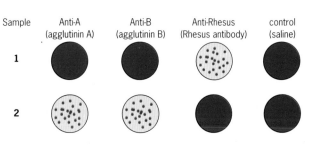

Fig 26.Q4

What is the blood group of **(i)** Sample 1, **(ii)** Sample 2?

b) Explain why it would not be safe to transfuse blood of Sample 1 into a patient who has the same blood group as Sample 2.

[AEB June 1989 Human Biology: Paper 1, q. 13]

5 A diseased kidney may be replaced with a healthy one transplanted from another person.

a) Explain how the action of T-lymphocytes would normally result in rejection of the transplanted kidney.

b) Describe **two** ways in which the chance of rejection may be reduced.

c) Explain why a kidney may be transplanted from one identical twin to another without risk of rejection.

d) Suggest why some illnesses, normally rare, occur more frequently in both AIDS patients and in those who have received transplants.

[NEAB June 1995 Modular Biology: Health and Disease Paper, q. 4]

Assignment

AIDS AND THE HIV VIRUS

What is AIDS?
Since its discovery in the early 1980s, few scientific topics have been as controversial as AIDS. Short for Acquired Immune Deficiency Syndrome, AIDS arises when the immune system ceases to function effectively. It is caused by the human immunodeficiency virus (HIV), shown in Fig 26.A1, which infects and destroys vital cells of the immune system, notably the helper T-lymphocytes.

1 HIV and AIDS are the same thing. True or false? Explain.

When the HIV virus enters the body, it can take many years before AIDS develops. Some people take well over ten years to develop full-blown AIDS and some people have been infected since the mid-1980s and have not as yet developed the syndrome. Perhaps they never will.

Tell-tale signs that the immune system is no longer effective are the appearance of diseases that would normally cause no trouble. These include Kaposi's sarcoma (a type of skin cancer), fungal infections (eg thrush), bacterial infections (eg tuberculosis) and viral infections (eg herpes).

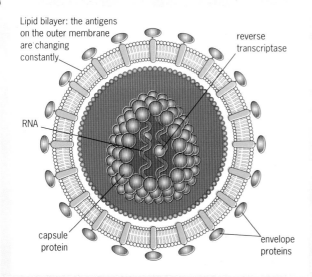

Lipid bilayer: the antigens on the outer membrane are changing constantly

reverse transcriptase

RNA

capsule protein

envelope proteins

Fig 26.A1 **The structure of the human immunodeficiency virus**

2 Outline briefly how viruses replicate.

HIV is a retrovirus, and contains the enzyme **reverse transcriptase**. Once the virus infects a cell, this enzyme allows the cell to make viral DNA from viral RNA. The foreign DNA is then inserted into the cell's own DNA where it acts as a gene, causing the manufacture of more viral RNA and, subsequently, many thousands of new viruses which eventually burst out of the cell and infect others.

3 Outline the essential difference between retroviruses and other viruses.

HIV viruses primarily affect a group of lymphocytes called helper T-cells. These cells normally activate other lymphocytes, including B-cells.

4
a) What is the function of B-cells?
b) What will happen if B-cells cannot become activated?

The transmission of HIV
Accurate knowledge of how the virus is transmitted has become a matter of basic survival for many people. Infection occurs when the virus is transferred from person to person, via body fluids. This usually occurs during sexual intercourse. The risk of someone becoming infected through sexual activity depends on the number of partners they have, the proportion of infected partners and the nature of their sexual activity.

5 What is the easiest and most usual way that a sexually active person can avoid HIV infection?

Prevention or cure?
Treating HIV infection is difficult. The viral DNA finds its way into cells in the brain, intestine and retina as well as white cells, and every time the cells divide, the viral DNA is copied. Much research is under way to develop an AIDS vaccine, but efforts have been thwarted so far by the virus's remarkable ability to vary the shape of the chemicals in its outer coat.

6 Suggest why it is difficult to make a vaccine for HIV.

Scientists agree that the most hopeful line of attack is to control the replication of HIV in the body and therefore delay the development of AIDS, perhaps indefinitely. However, this is not yet possible. The best we can do at the moment is to delay the onset of full-blown AIDS by a combination of treatments.

7 Discuss or write down details of any AIDS awareness campaigns you have seen. Which approaches do you feel are the most effective? What approach would you take if you were designing a campaign for young people?

Fig 26.A2 **An AIDS information poster in Kenya**

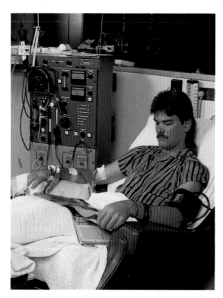

This man is undergoing haemodialysis. For several hours, blood flows from his forearm into the machine. This removes much of the waste, together with excess salt and water. Find out more about haemodialysis in the Assignment at the end of this chapter

IF ONE OF YOUR kidneys were injured or diseased and stopped working, you could probably still lead a normal life. But people whose kidneys both fail are faced with a crisis: water, urea and potassium build up rapidly in their body. They may continue to pass some urine, but they cannot get rid of all the waste produced by normal cell processes.

Most people suffering from kidney failure hope for a transplant, but there is a shortage of donors. Until a suitable organ becomes available, patients must rely on dialysis to filter their blood and balance their fluid intake. As the photograph on the left shows, dialysis is uncomfortable and inconvenient. To reduce the time they need to spend in dialysis, kidney patients must stick to a strict diet called the Giovanetti diet. This comes as something of a shock. They must limit their fluid intake to only one pint a day, about a quarter of the normal daily intake of a human adult. They must also keep their protein intake down to about 30 to 40 grams per day, the amount of protein in a small egg.

But perhaps the biggest problem is the need to regulate potassium. This ion is a normal constituent of the body, but large amounts cause serious problems, including heart failure. Potassium-rich foods include citrus fruits, bananas, instant coffee, peanuts, treacle and – a big blow for many people – chocolate. In this strange diet, carbohydrates and calories are not a priority. Few patients feel like eating much anyway, and this restricted diet makes the task of finding appetising food even more difficult, but it does keep them alive.

1 WASTE AND WATER CONTROL

All animals carry out the process of **excretion** and higher animals have a sophisticated **excretory system**. In mammals and all other vertebrates the **kidneys** are the main organs of excretion. As well as removing metabolic waste, the mammalian excretory system keeps the volume of fluid in the body at the correct level and maintains the proper concentrations of **solutes** such as sodium and potassium ions.

Most excreted waste leaves the body in the **urine**, but some waste is also lost in **sweating** and **breathing out**. Sweat contains mainly salt and water, while exhaled air includes carbon dioxide and water vapour.

In this chapter we concentrate on **nitrogenous excretion**, the removal of waste compounds that contain nitrogen. This type of excretion deals with the waste products of protein breakdown (Fig 27.1). In mammals, the commonest excreted nitrogenous

Excretion is the removal of chemical waste from the body, waste that is produced by the metabolic processes within cells and which would be toxic if allowed to accumulate. You should not confuse excretion with **egestion** or **defaecation**, the removal of undigested food and other debris from the intestine.

Fig 27.1 **Adult humans need about 40 to 50 grams of protein per day, and so there is enough here for over a week! The body cannot store protein and so the daily excess must be broken down. The urea produced by the liver as a result of protein breakdown is excreted in the urine**

compound is **urea**. This is made in the liver from the breakdown of excess amino acids. Enzymes in the liver remove the amine ($-NH_2$) group from amino acids and form ammonia (NH_3) in a process known as **deamination**. Ammonia is a highly toxic compound that must not build up. It immediately enters a series of reactions called the **ornithine cycle** (see Chapter 25), which produces the relatively harmless compound urea:

$$2NH_3 \ + \ CO_2 \ \rightarrow NH_2CONH_2 + H_2O$$

ammonia + carbon $\rightarrow$ urea
dioxide

We also look at the role that the kidneys play in **homeostasis**, the maintenance of constant conditions in the body. The kidneys selectively eliminate water and solutes, such as sodium, potassium and chloride ions, so that the water and solute balance of the body is kept at the correct level. The need to balance the solute concentration of body fluids is called **osmotic regulation**, or **osmoregulation**.

Osmoregulation is the maintenance of a constant solute concentration within the body. In Chapter 5 we saw why this is important: if we place animal cells in a *hypertonic* solution, they lose water by osmosis and shrivel. If we put them in a *hypotonic* solution, they gain water and may burst. Both situations are harmful to cells, and so it is important that animals maintain the solute concentration of their body fluids within narrow limits.

Different fluids are often compared on the basis of their solute concentrations. A solution that has a higher solute concentration than another is said to be **hypertonic**. A solution that has a lower solute concentration than another is **hypotonic**. A solution that has exactly the same solute concentration is **isotonic**. So, for example, we can say that sea water is hypertonic to human blood plasma, because sea water has a much higher salt concentration.

2 AN OVERVIEW OF THE HUMAN URINARY SYSTEM

The principal organs of the human urinary system are the two kidneys (Fig 27.2(a)). They lie at the back of the abdominal cavity, at about waist level, where they are protected to some extent by the spine and the lower part of the ribcage. Usually, the left kidney is slightly above the right.

The kidneys receive blood from the two **renal arteries** that branch off the aorta (see Chapter 18). Per gram of tissue, the kidneys receive the largest blood supply of any organ. About 1200 cm³ of blood flows to each of them every minute. High pressure blood is essential for proper kidney function: they cannot filter the blood effectively if the pressure drops. Blood leaves the kidneys through the **renal veins**.

Urine made by the kidneys is pushed down muscular tubes, the ureters, by the rhythmic muscle contractions of **peristalsis**. The ureters empty into the **bladder**, a muscular bag that stores **urine** until it is convenient to release it. The capacity of the human bladder varies from 400 to 700 cm³ or more. When it begins to get full, stretch receptors in the walls inform the brain. Urination, or **micturition**, happens when the **sphincter muscles** relax, allowing urine to pass out of the body through the **urethra**.

The urinary system is also know as the **renal system**.

A If a person has 5 litres of blood, and the kidneys receive 1.2 litres per minute to filter, how many times on average does the total volume of blood in the body pass though the kidneys every hour?

3 STRUCTURE AND FUNCTION OF THE MAMMALIAN KIDNEY

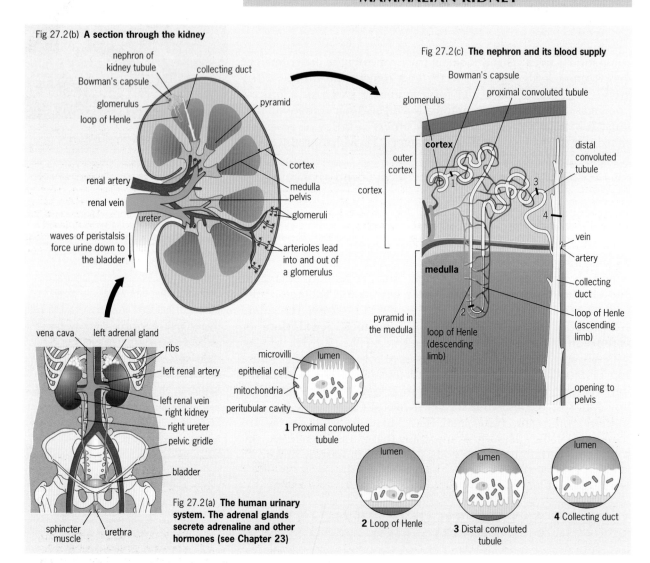

Fig 27.2(b) **A section through the kidney**

Fig 27.2(c) **The nephron and its blood supply**

Fig 27.2(a) **The human urinary system. The adrenal glands secrete adrenaline and other hormones (see Chapter 23)**

1 Proximal convoluted tubule

2 Loop of Henle

3 Distal convoluted tubule

4 Collecting duct

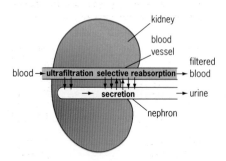

Fig 27.3 **A simple summary of kidney function. At the proximal (near) end of the nephron, blood is filtered to produce a fluid that is virtually identical to tissue fluid. The filtrate is then modified by active transport. This involves active secretion of substances into the filtrate and active reabsorption of substances into the blood. These processes are possible only because of the intimate association between the nephron and the blood system – see Fig 27.2(c)**

The structure of the kidney is shown in Fig 27.2(b). The functional unit of the kidney is the **nephron** or **kidney tubule**. It is important to know the position of the nephrons in relation to the overall plan of the kidney. As Fig 27.2(c) shows, the **outer cortex** of the kidney contains the **Bowman's capsules** and the **proximal** and **distal convoluted tubules**, while the **medulla** houses the **loops of Henle** and the **collecting ducts**. Bundles of collecting ducts form **pyramids**, which deliver urine into an open space called the **pelvis**. The ureters connect the kidneys to the bladder.

Kidney function involves two processes: **ultrafiltration** and **active transport** (Fig 27.3). Ultrafiltration is filtration under pressure: blood is 'squeezed' to separate a fluid called **glomerular filtrate** (often called simply the **filtrate**). Active transport then modifies the filtrate, secreting some substances and reabsorbing others according to the needs of the body. The result is that blood flows back into the body without much of its harmful waste. This waste, a solution containing urea, salts and various other chemicals, is called **urine.**

Let's now look at how individual parts of the kidney contribute to the overall process of excretion.

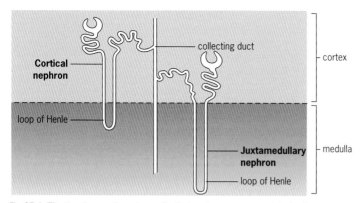

Fig 27.4 **The two types of nephron.** Cortical nephrons **occur mainly in the cortex. They have a short loop of Henle that extends only a short distance into the medulla. These are the main functional nephrons when water is plentiful.** Juxtamedullary nephrons **each have a long loop of Henle that extends deep into the medulla. These nephrons produce very hypertonic urine when water is in short supply. The kidneys of desert animals, such as the kangaroo rat, contain mainly juxtamedullary nephrons**

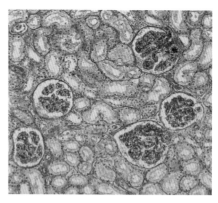

Fig 27.5 **The fine structure of the kidney as seen under the microscope. Five Bowman's capsules (containing glomeruli) are visible, surrounded by sections of proximal and distal convoluted tubules.**

An underwater dissection allows some of the fine nephrons to be teased out and viewed individually without the aid of a microscope. Each nephron, although only 60 μm in diameter, can be over 14 cm long when uncoiled

The nephron

Each human kidney contains about a million nephrons, together with a maze of blood vessels and some connective tissue (Fig 27.5). There are two different types of nephron, named after their position: **cortical** and **juxtamedullary** (Fig 27.4). In this section we deal with the function of each region of the nephron in sequence, but you must remember that the nephron functions as a whole. The activities of one region are essential to the effectiveness of others.

Ultrafiltration in the Bowman's capsule

Fig 27.6 shows the **Bowman's capsule** in detail. The capsule itself is shaped rather like a wine glass, containing a central knot of blood vessels called the **glomerulus**. This area of the nephron filters the blood by ultrafiltration – filtration under pressure. Obviously, this requires a *means of creating pressure* and a *filter*.

The kidneys receive blood from the first branch off the aorta, so the blood is already under relatively high pressure when it reaches the nephron. This pressure is maintained and enhanced because the **afferent arteriole**, the blood vessel that takes blood into the glomerulus, has a larger diameter than the **efferent arteriole**, which takes blood away.

B What type of nephrons would you expect to find in the kidneys of an otter? Explain.

The Bowman's capsule and the glomerulus together are often known as the **Malpighian body** or **renal corpuscle**.

Fig 27.6 **The fine structure of the Bowman's capsule, a region of the kidney adapted for ultrafiltration of the blood. Kidney filtrate is basically blood plasma without the larger proteins.**

Note the difference in size between the afferent (incoming) and efferent (outgoing) arteriole

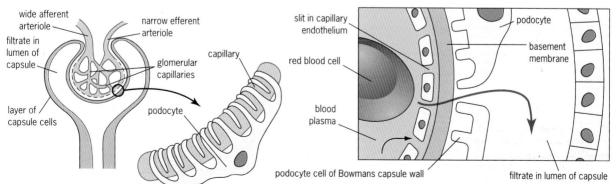

The walls of capillaries in the Bowmans capsule are much more permeable than those of normal capillaries: the cells do not fit tightly together, but have thin slits in between, through which all the constituents of the plasma can pass

The Bowman's capsule is lined with unique cells called **podocytes** ('foot-cells'). These cells, like those of the capillary, do not fit tightly together, but form a network of slits that fit over the capillary

Between these two relatively coarse filters is the continuous basement membrane. This finer filter prevents the passage of all molecules with a relative molecular mass greater than about 68 000 kilodaltons, so the larger molecules (mainly proteins) remain in the blood

This physical pressure, or **hydrostatic pressure**, forces blood against a filter that consists of three layers:

- the lining, or **endothelium** of the glomerular capillaries,
- the basement membrane,
- the cells of the Bowman's capsule itself.

Look again at Fig 27.6 to see how these three layers are arranged. The basement membrane in the centre acts as the fine filter and is therefore largely responsible for the chemical composition of the filtrate. At this stage, the filtrate is identical to tissue fluid (see Chapter 18).

The rate of filtrate production is high: about 125 cm³ per minute. Obviously, we don't produce anything like this volume of urine or we would dehydrate within minutes. On average we produce about 1 cm³ of urine per minute. The rest, over 99 per cent of the filtrate, is *reabsorbed*. In fact, after the Bowman's capsule, the rest of the nephron is concerned with adjusting the volume and composition of the filtrate. Necessary substances are reabsorbed, while toxic compounds, and water and any solutes present in the body in excess, are removed.

Several forces act on the fluid in the Bowman's capsule, opposing or encouraging filtration. The high hydrostatic pressure of blood is the main factor that promotes filtration but, as Table 27.1 shows, it is opposed by the hydrostatic and solute pressure of the filtrate.

The proximal convoluted tubule

Refer again to Fig 27.2(c) and see Fig 27.7. In the proximal convoluted tubule (or simply proximal tubule), the newly formed filtrate comes into close contact with blood vessels carrying blood away from the glomerulus, but now the conditions are very different from those in the Bowman's capsule. This blood has a *low* hydrostatic pressure but a relatively *high* solute potential, because there are plasma proteins remaining there that could not pass through the filter. This allows the blood to reabsorb a large percentage of the water from the filtrate by osmosis. In addition, many solutes, such as glucose and amino acids, are totally reabsorbed into the blood by active transport. Normal urine should not contain glucose. Only when blood glucose becomes very high, as in diabetes, for example, does the reabsorption mechanism fail (Fig 27.8).

As you can see from Fig 27.7, the cells that line the proximal tubule show the characteristic adaptations to active transport: a large surface area, provided by microvilli, and numerous mitochondria to provide the ATP needed to power the process.

Table 27.1 **Summary of the forces acting in the Bowman's capsule**

Force acting	Effect on filtrate formation	Approx value/ kPa
Hydrostatic pressure of blood	encourages	8.0
Hydrostatic pressure of filtrate in capsule	opposes	−2.4
Solute potential of blood	opposes	−4.3
Overall filtration pressure	encourages	1.3

?

C If an individual lost a large amount of blood in an accident, what effect would this have on kidney function?

Fig 27.7 **In cells of the proximal tubule, a sodium-potassium pump actively pumps sodium ions into the blood. Other protein pumps actively transport amino acids and glucose. Water follows passively by osmosis. Sodium, amino acids and glucose enter the tubule cells from the tubule lumen, down their concentration gradients**

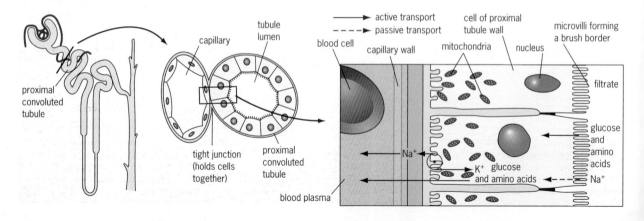

The loop of Henle

The **loop of Henle**, seen in Fig 27.9, is a long hairpin-shaped region of the nephron that descends into the medulla and then returns to the cortex. The loop creates a region of high solute concentration in the medulla. The collecting duct of each nephron passes through this region and the osmotic gradient between the inside of the collecting duct and the outside draws water out of the duct by osmosis. Consequently, the urine becomes more and more concentrated (compared to body fluids) as it passes down the duct. So, the loop is a vital adaptation to terrestrial life: it allows organisms to get rid of waste without losing too much water.

Fig 27.9 outlines how the loop of Henle works. Fluid in the two limbs flows in opposite directions. We describe this sort of arrangement as a **countercurrent system**. As fluid passes up the ascending limb, sodium chloride (NaCl) is transported actively out of the loop into the surrounding area. This causes water to pass out of the descending limb by osmosis. The net result is that the solute concentration at any one level of the loop is slightly lower in the ascending limb than in the descending limb. The longer the loop, the more chance there is for this mechanism to build up a high salt concentration. If the loop in Fig 27.9 were only half the length shown, the salt would only accumulate to about 600 units.

So the longer the loop, the greater the solute concentration, and the more concentrated is the urine that is eventually produced. It should therefore be no surprise that animals living in dry desert

The kidneys of fish, amphibians and reptiles do not possess loops of Henle. These animals cannot produce urine that is any more concentrated than blood plasma.

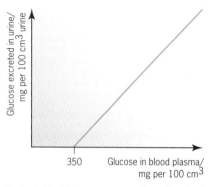

Fig 27.8 **The kidneys can normally reabsorb all of the glucose that passes through them and return it to the blood. Only when blood sugar exceeds a threshold of about 350 mg per 100 cm³ does glucose begin to appear in the urine**

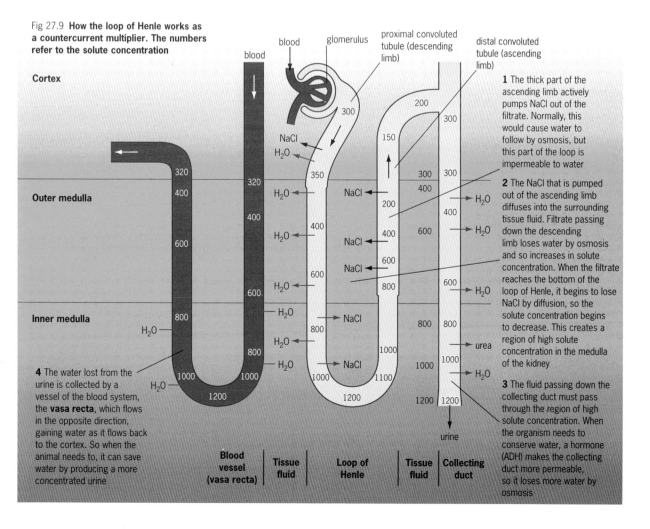

Fig 27.9 **How the loop of Henle works as a countercurrent multiplier. The numbers refer to the solute concentration**

D The loop of Henle is sometimes described as a **hairpin countercurrent multiplier**. Explain this description.

conditions have very long loops of Henle. To accommodate this extra length, the medulla of their kidneys is very thick. Table 27.2 shows the general rule: the more need an animal has to conserve water, the thicker the medulla and the more concentrated the urine. The kangaroo rat can produce urine that is fourteen times more concentrated than its blood plasma.

Table 27.2 The relationship between thickness of medulla (ie the length of loop of Henle) and the concentration of urine

Species	Habitat	Thickness of medulla (relative to beaver)	Temperature at which urine freezes °C*
Beaver	freshwater	1.0	−1.0
Human	land	2.6	−2.6
Cat	land	4.2	−5.8
Kangaroo rat	desert	7.8	−10.4

*The more concentrated the urine, the lower the temperature at which it will freeze. The measurement of freezing point is a relatively easy practical method for comparing the solute concentration of urine samples.

E People on survival courses are taught to assess their level of dehydration by looking at the colour of their urine. Explain how they would make an assessment.

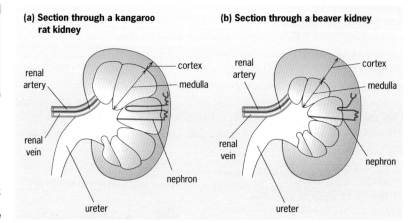

Fig 27.10 **The relative thickness of the cortex in two mammals from different environments.**
(a) **The kangaroo rat, a desert dwelling animal, has a thin cortex. The deep medulla contains mainly** juxtamedullary nephrons **with long loops of Henle**
(b) **The beaver, an aquatic animal, has a thick cortex. The relatively shallow medulla contains mainly** cortical nephrons **with short loops of Henle**

The distal tubule

While the proximal tubule reabsorbs most of the filtrate, the distal tubule 'fine tunes' the remaining fluid (see below), according to the immediate needs of the organism. As the cells that form the distal tubule also need to carry out active transport, they are very similar in structure to those of the proximal tubule, with microvilli and many mitochondria.

F Which areas of the nephron carry out active transport?

4 THE ROLE OF THE KIDNEY IN HOMEOSTASIS

See questions 1 and 2. ■

The kidney (particularly its distal tubules) contributes to several vital homeostatic mechanisms:

● regulation of water content and blood volume,

● maintenance of acid/base equilibria,

● maintenance of salt balance.

Water balance in humans

Table 27.3 shows a typical water balance sheet for an average person, assuming a normal level of activity and a comfortable external temperature.

On average, we get almost two-thirds of our water from drinks and the remaining third from food. We obtain a small but important proportion of water from metabolic reactions, such as cell respiration.

Table 27.3 **The water balance sheet for a 24-hour period**

Water gain	Amount/ cm³	Water loss	Amount/ cm³
Food and drink	2100	through skin	350
Metabolic water	200	sweat	100
		in breath	350
		urine	1400
		faeces	100
Total	2300	Total	2300

Some of the water loss shown in Table 27.3 is unavoidable. Metabolic waste must be removed in solution, and so some water loss in urine is inevitable. Similarly, lungs must maintain a moist lining and so exhaled air always contains a lot of water vapour. Research has shown that a significant amount of water is also lost by diffusion through our skin (this is not the same as sweating).

The mechanism of water balance

Like most homeostatic mechanisms, maintenance of water balance involves a negative feedback loop that has a **detector** and a **correction mechanism** (see Chapter 25).

A part of the brain, the **hypothalamus,** contains **osmoreceptor cells** that are sensitive to the solute concentration of the blood. When the concentration rises, indicating that water loss is greater than intake, the hypothalamus responds in two ways:

- It stimulates the thirst centre in the brain.
- It stimulates the pituitary gland to produce **anti-diuretic hormone (ADH)**.

Fig 27.11 summarises the mechanism of ADH action. ADH acts on the kidney to reduce the volume of urine produced. It achieves this by increasing the permeability to water of the distal tubule and the collecting duct. When ADH makes the collecting duct more permeable to water, the region of high solute concentration created by the loop of Henle has an even more dramatic effect than usual. The action of ADH causes more water to leave the tubule and re-enter the blood. A much more concentrated urine is produced and vital water is conserved.

Conversely, when fluid intake exceeds loss, the blood becomes more dilute. When this is detected by the hypothalamus, it reduces ADH production. The action of ADH on the kidneys lessens, less water is reabsorbed and larger volumes are produced of dilute, **insipid** urine.

People who cannot produce ADH because they have a faulty pituitary gland have the disease **diabetes insipidus**. Once known as the 'pissing evil', this condition results in the constant production of dilute urine, leaving the sufferer permanently thirsty and unable to venture very far from the toilet. Today, it can be treated by giving extracted or synthesised ADH.

Control of blood volume and pressure

Since it regulates water reabsorption, ADH also regulates blood volume. A drop in blood volume leads to a drop in blood pressure, which is detected by **stretch receptors** in the walls of the aorta and the carotid arteries. Impulses from these detectors pass to the hypothalamus, which then secretes more ADH. This acts on the kidneys to cause more water to be retained and so blood pressure increases.

Acid/base balance

The body must maintain a relatively constant pH. As we saw in Chapter 4, enzymes are very sensitive to changes in pH, and even a small change can inhibit their activity and so have serious consequences for an organism.

Under normal circumstances, the human body is engaged in a constant battle against accumulating acid, although this depends on

✔

When we sweat, we produce a salty liquid to keep our body temperature constant. On a hot day, or during exercise, we lose more water as sweat. Unless we drink more, our water balance is maintained by producing a smaller quantity of more concentrated urine.

?

G How would the water balance sheet change if the individual was suffering from a bowel infection such as dysentery?

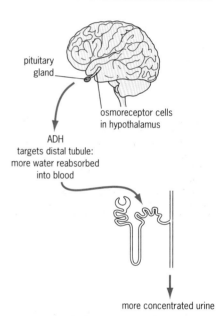

Fig 27.11 **When the solute potential of the blood rises (the blood becomes too concentrated), osmoreceptor cells in the hypothalamus initiate secretion of ADH from specialised nerve cells in the posterior lobe of the pituitary gland. ADH increases the permeability to water of the distal tubule and collecting duct, allowing more water to pass from the filtrate and into the blood. The blood becomes more dilute and blood volume increases**

?

H In desperation, a castaway drinks some sea water. Why would this not be a good idea?

See questions 3, 4 and 5.

HANGOVERS!

SOONER OR LATER, most people experience the unpleasant 'morning after' feeling that tends to follow after drinking too much alcohol. Many hangover symptoms are due to dehydration, rather than to the toxic effects of the alcohol or other ingredients. Research has shown that alcohol inhibits the production of ADH, so there is a greater water loss in the urine than the body needs. Many of the symptoms disappear when the body is rehydrated.

Fig 27.12 **Had he known about the dehydrating effects of alcohol, this man could have minimised his headache by having a long drink of water before he went to bed (assuming, of course, that he could find the tap)**

✔

pH is a measure of the relative abundance of hydrogen ions (H^+) and hydroxyl ions (OH^-) in solution. A pH of less than 7 is acidic and indicates an excess of H^+. A pH of more than 7 is alkaline and indicates an excess of OH^-.

---

several factors, including diet. All of the following activities tend to lower blood pH:

● Cell respiration: this produces CO_2, which dissolves in the plasma to form carbonic acid.

● Strenuous exercise: this produces lactic acid.

● Digestion: breakdown of certain foods, such as meat, eggs and cheese, result in the formation of acid products.

Generally, the body is able to regulate the pH of blood plasma and tissue fluid in the range 7.3–7.45 by using buffers and by secreting acids into the urine.

Buffers

A buffer is a chemical, or combination of chemicals, that can resist change in pH by mopping up any excess acid or base. The main buffers important in living organisms are:

● Blood proteins such as **haemoglobin** and **albumin.**

● Inorganic buffer systems in body fluids, such as the hydrogen-carbonate (bicarbonate) system:

$$CO_2 + H_2O \leftrightarrow H_2CO_3 \leftrightarrow H^+ + HCO_3^-$$

This system acts as a buffer because any extra acid tends to shift the equilibrium to the left, so the extra hydrogen ions are neutralised by the hydrogencarbonate ions.

The sodium hydrogen phosphate in plasma:

$$Na_2PO_4 + H^+ \rightarrow NaHPO_4 + Na^+$$

$$NaHPO_4 + H^+ \rightarrow H_2PO_4 + Na^+$$

This shows that, in the presence of extra hydrogen ions, the equilibrium changes so that hydrogen ions are swapped for sodium ions that do not affect pH.

Secretion of acid

Most people produce slightly acidic urine of between pH 5 and 6. Active transport mechanisms in the kidney enable the body to excrete excess acid but still retain the vital buffer chemicals mentioned above. If the blood becomes too alkaline, the kidney can reabsorb hydrogen ions and secrete basic ions such as ammonia.

Salt balance: the control of sodium chloride levels

The mechanism by which salt balance is regulated is shown in Fig 27.13. The concentration of sodium ions in body fluids is controlled by the hormone **aldosterone,** which is secreted by the **adrenal cortex** of the **adrenal glands** – see Fig 27.2(a). When sodium ions are actively transported, a negative ion, usually chloride, automatically follows to maintain **electrolytic balance** (balance of positively and negatively charged ions).

When the body loses sodium ions, it also loses water by osmosis, and blood volume and blood pressure fall. This is detected by a group of receptor cells, the **juxtaglomerular complex**, situated next to the Bowman's capsule. These cells respond to a fall in blood pressure by releasing an enzyme called **renin** (not to be confused with the digestive enzyme rennin). Renin converts a plasma protein into an active hormone called **angiotensin**. This stimulates the adrenal cortex to secrete aldosterone.

Aldosterone increases the reabsorption of sodium ions from the intestines and from the kidney. The increased salt concentration in the blood leads to a greater retention of water, so bringing blood volume and pressure back to normal.

Many western diets contain too much salt and this can pose a real problem for our salt balance. A lot of extra salt in food increases salt levels in our blood and so we retain more water to dilute it. As a result, our blood pressure increases. Many people suffering from high blood pressure, or **hypertension**, need a low salt diet.

```
┌─────────────────────────────┐
│ Loss of sodium and chloride │
│   ions (in urine or sweat)  │
└─────────────────────────────┘
              │
              ▼
┌─────────────────────────────┐
│        Loss of water        │
└─────────────────────────────┘
              │
              ▼
┌─────────────────────────────┐
│      Lowered blood volume   │
│         and pressure        │
└─────────────────────────────┘
              │
              ▼
┌─────────────────────────────┐
│         Detected by         │
│    juxtaglomerular complex  │
└─────────────────────────────┘
              │
              ▼
┌─────────────────────────────┐
│       Release of renin      │
└─────────────────────────────┘
              │
              ▼
┌─────────────────────────────┐
│   Activation of angiotensin │
└─────────────────────────────┘
              │
              ▼
┌─────────────────────────────┐
│    Release of aldosterone   │
└─────────────────────────────┘
              │
              ▼
┌─────────────────────────────┐
│    Reabsorption of sodium   │
└─────────────────────────────┘
```

Fig 27.13 **Summary flow chart to illustrate salt balance**

Osmoregulation in desert animals

Studies of kangaroo rats kept in captivity (Fig 27.14) have shown that they can survive on dried food and no drinking water for long periods. Their ability to control their water loss is remarkable and is a classic example of how animals can adapt to a harsh environment.

Table 27.4 **A study into the water balance for kangaroo rats over a four-week period. The rats ate nothing but 100 g of barley and were kept at a constant temperature of 25 °C and a humidity of 20 per cent. The kangaroo rat 'creates' most of its water (metabolic water) by metabolic reactions, notably cell respiration. The rest is absorbed from its food. So all the kangaroo rat's water comes either directly or indirectly from its apparently dry food. Metabolism of the protein in barley produces a substantial amount of urea which must be excreted in the urine. Faeces also need some water, but the most wasteful activity is breathing out**

Water gains	cm³	Water losses	cm³
Metabolic water	54.0	urine	13.5
Water absorbed from dried food	6.0	faeces	2.6
		evaporation (mainly in breath)	43.9
Total gain	60.0	Total water loss	60.0

This remarkable rodent is well suited to life in the desert. Not only does it avoid dehydration, but it avoids overheating without the cooling effects of evaporation, which it could not 'afford' in terms of water loss. The kangaroo rat shows the following adaptations to its dry environment:

● Its kidneys consist mainly of juxtamedullary nephrons (see Fig 27.4), which contain extra-long loops of Henle. These allow the animal to produce very concentrated urine and so it can lose its metabolic waste without losing significant amounts of water.

● It spends long periods underground, where the air is cooler and more humid. This reduces its water loss by evaporation.

● Its nasal passages cool the air before it is exhaled, so that much of the water vapour condenses within its nose instead of being breathed out.

Fig 27.14 **The kangaroo rat (Dipodomys deserti) lives in the Arizona Desert, where rainfall is scarce and unpredictable. This mammal is able to survive on dry plant material without drinking water at all**

SUMMARY

After reading this chapter, you should know and understand the following:

■ The kidneys remove metabolic waste and control water and solute levels in the body. As a result of these functions, the kidney also plays a vital role in the control of blood volume and pressure.

■ Each kidney is made from about 1 million **nephrons**, narrow tubules closely entwined with blood vessels.

■ At one end of the nephron is the **Bowman's capsule**. Here, the **glomerular filtrate** is formed by **ultrafiltration** (pressure filtration) of the blood.

■ The first filtrate has the same composition as tissue fluid. As it passes along the nephron, the composition of filtrate is altered by various active transport mechanisms that leave some substances in the filtrate and reabsorb others.

■ A large amount of filtrate is formed, but over 99 per cent of it is reabsorbed in the **proximal convoluted tubule**, mainly by active transport mechanisms and osmosis. Usually, all glucose molecules and amino acids are reabsorbed into the blood.

■ The movement of solutes from the **loop of Henle** creates a region in the medulla of high solute concentration. As filtrate (now **urine**) passes through the **collecting ducts**, water leaves the urine by osmosis, and so the urine becomes **hypertonic**.

■ The **distal convoluted tubule** regulates salt concentration, solute concentration and acid/base balance. It can alter the composition of urine according to the needs of the body.

QUESTIONS

1 The diagram of Fig 27.Q1 represents a nephron from a human kidney.

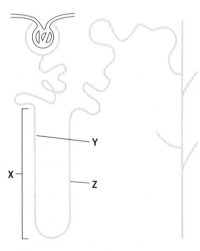

Fig 27.Q1

a) Name the part labelled **X**.

b) Sodium chloride is actively pumped out of **Z** into the medulla of the kidney. This sodium chloride moves back into **Y**.
   Explain the effect of the sodium chloride concentration in the medulla of the kidney on the reabsorption of water from the collecting duct.

c) Most of the sodium chloride filtered into the glomerular filtrate is reabsorbed.
   (i) From which parts of the nephron does this reabsorption take place?
   (ii) How is the reabsorption of sodium chloride controlled?

[AEB 1996: Biology: Specimen paper, q. 1]

2

a) If the glomerular filtration rate of the kidneys is 120 cm³ min⁻¹ and the tubular reabsorption rate is 114 cm³ min⁻¹, calculate the rate of urine formation per minute.

b) State **two** differences between tubular reabsorption of water in the first (proximal) convoluted tubule and the second (distal) convoluted tubule.

c) What might reduce the rate of tubular reabsorption of water?

d) The minimum rate of urine production is 300 cm3 per day. Explain why it is necessary for some urine to be produced each day.

[AEB 1995 Human Biology: Paper 1, q. 6]

3 The graph of Fig 27.Q3 shows the volume of urine collected from a subject before and after drinking 1000 cm³ of distilled water. The subject's urine was collected immediately before the water was drunk and then at intervals of 30 minutes for several hours.

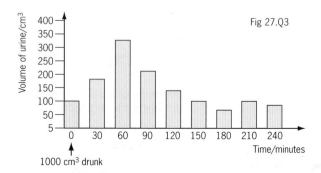

a) **(i)** Describe the changes in urine output during the period of the experiment.

   (ii) Name the process responsible for these events.

b) Explain the difference in the volume of urine collected at 60 minutes and at 90 minutes.

c) The experiment was repeated, but this time the subject exercised vigorously for 10 minutes before drinking the water. How would you expect the results of the second experiment to differ from those in the graph? Explain your answer.

d) If, in the original experiment, the subject had drunk 0.9 per cent NaCl solution instead of 1000 cm³ of distilled water, would the same volumes of urine have been collected? Explain your answer.

   (Note 0.9 per cent NaCl solution is isotonic with blood plasma, i.e. it has the same osmotic potential as blood plasma.)

[AEB 1995 Human Biology: Paper 1, q. 8]

4 Nephrosis is a kidney condition in which damage to the glomeruli results in large quantities of protein passing into the glomerular filtrate. This protein finally appears in the urine.

a) Suggest why this protein is not reabsorbed into the blood in the proximal convoluted tubule of the nephron.

b) As a result of nephrosis, large amounts of tissue fluid accumulate in the body, especially in the ankles and feet.

   (i) Explain why the loss of protein from the blood results in the accumulation of tissue fluid.

   (ii) Suggest why this fluid accumulates especially in the ankles and feet.

c) **(i)** Explain how the action in the kidney of the hormone aldosterone controls the sodium content of the blood.

   (ii) Suggest why little aldosterone is produced by a person suffering from nephrosis.

[NEAB 1995 Modular Biology: Physiology Test, q. 6]

5 Fig 27.Q5(a) shows a vertical section through a mammalian kidney.

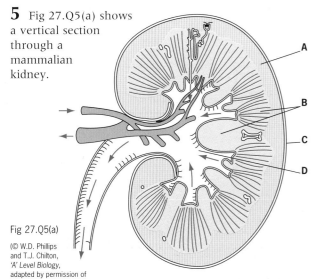

Fig 27.Q5(a)

(© W.D. Phillips and T.J. Chilton, 'A' Level Biology, adapted by permission of Oxford University Press)

a) Label **A** to **D**.

   The effective filtration pressure (EFP) in the glomerulus of the mammalian nephron depends on three factors. These are the three forces shown labelled on the left of Fig 27.Q5(b).

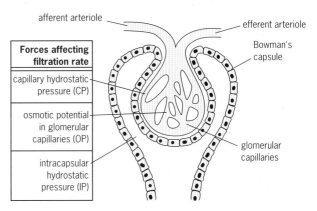

Fig 27.Q5(b)

(© M. Stewart (ed.) Animal Physiology, adapted by permission of Hodder & Stoughton/The Open University)

b) **(i)** Suggest which blood component is likely to provide the major contribution to the osmotic potential in the glomerular capillaries.

   (ii) Complete the table below, which summarises the effect of the three forces on glomerular filtration, by placing a tick in each appropriate space.

Force	Glomerular filtration	
	encouraged	opposed
CP		
OP		
IP		

   (iii) Construct a word equation that shows how the three forces interact to generate the effective filtration pressure.

   (iv) Suggest **two** situations, or conditions, where the blood pressure in the glomerulus will fall, leading to impaired renal function.

c) Explain **briefly** the significance of microvilli and mitochondria in the functioning of the proximal convoluted tubule.

d) Outline **briefly** the function of the loop of Henle.

[UCLES 1995 Modular Biology: Transport, regulation and control Paper, q. 3]

Assignment

TREATING KIDNEY FAILURE

In many cases of **chronic kidney failure**, there is a gradual decline in kidney function that gives the patient plenty of warning. In contrast **acute renal failure** is a crisis in which all kidney function effectively stops. There are many causes of acute renal failure, but generally they can be placed into the following categories:

  sudden loss of large amounts of fluid (blood or tissue fluid),
  inadequate blood flow to the kidneys,
  bacterial infection in the kidneys,
  effect of toxins,
  a blockage in the urinary tract.

1 Why would a reduced blood flow to the kidneys interfere with kidney function, even though enough blood may get through to provide the kidney cells with the nutrients and oxygen they need to stay alive?

2 As we saw at the start of this chapter, the immediate problems of kidney failure are a build-up of fluid, urea and potassium. To minimise these problems, patients must follow the Giovanetti diet.
a) What are the essential features of this diet?
b) Suggest why protein intake needs to be limited.

Haemodialysis and kidney machines

Dialysis is a method of separating small molecules from larger ones using a semipermeable membrane. Blood dialysis,

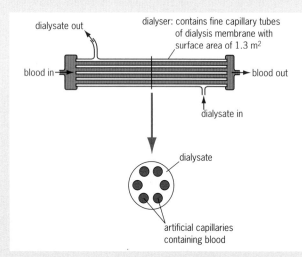

Fig 27.A1 **The fine tubes seen in this dialysis filter are artificial capillaries. Blood flows through the middle of these tubes, while the dialysing fluid flows along the outside in the opposite direction. Each filter is an expensive piece of precision engineering, but can only be used for a few dialysis sessions and must then be discarded**

or **haemodialysis**, separates the smaller constituents of plasma such as urea and solutes from the larger ones such as proteins.

Blood is taken from the patient, usually from a forearm vein, and passed into the machine, where it runs through minute artificial capillaries (Fig 27.A1). These are made from a semipermeable plastic that filters the blood. While blood flows inside the artificial capillaries, a special fluid, the **dialysate,** flows around the outside in the opposite direction. In dialysis, molecules are exchanged between the blood and the dialysate. The composition of the dialysate is carefully controlled so that there is a net movement of urea, water and salts *out* of the blood.

3
a) By what physical process do solutes enter or leave the blood during dialysis?
b) Why do the blood and dialysate flow in opposite directions?
c) Suggest two problems which might occur if the dialysate was pure water.
d) Why must the dialysate contain glucose, amino acids and salt?
e) Why is there no urea in the dialysate?

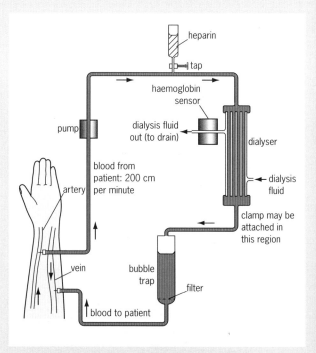

Fig 27.A2 **The main features of a haemodialysis machine**

4 Fig 27.A2 shows the circuit the blood takes as it passes through the dialysis machine.
a) Calculate the volume of blood that is processed by the dialyser in four hours.
b) Why is heparin, an anti-coagulant, added to the blood?
c) Why is heparin not given in the last hour of dialysis?
d) Why is a filter included in the blood circuit?

e) Suggest why the omission of the bubble trap could prove dangerous to the patient.

f) What would a positive reading on the haemoglobin sensor indicate?

g) The dialysis fluid is maintained at approximately 40 °C. Suggest two reasons for this.

h) Excess water can be removed from the blood in a number of ways. One way is to increase the amount of glucose in the dialysing fluid. Explain how this method would work.

i) Another method involves partially clamping the blood tube at the region shown. Explain the principle involved in this method of removing water.

CAPD

CAPD stands for **continuous ambulatory peritoneal dialysis**. In this fairly new treatment, individuals with kidney failure can use one of their own membranes, the **peritoneum**, as a dialysing membrane (Fig 27.A3). Over 3000 people in the UK are presently using this technique because of the advantages it offers compared to the conventional dialysis machine.

The basic principle is simple: patients have a hole, or **stoma**, made in their abdomen wall near the navel. Through this a large volume of dialysing fluid reaches the abdominal cavity through a tube or **catheter**. The patient is then free to walk around (hence the ambulatory part of the name) while dialysis occurs across the peritoneum between the blood and the dialysate. Every four to six hours the dialysate is replaced. This is a relatively simple exchange procedure that patients can carry out themselves after some basic training.

 5

a) Suggest the biggest problem with the exchange procedure.

b) Explain why it would be no use leaving the dialysate inside the body for longer than the recommended time.

Kidney transplants

When a donor kidney becomes available, it is a relatively simple operation to put it into the body. Surprisingly, the old kidneys are left in place: they are rather inaccessible and so difficult to remove, but do no harm. The new kidney is placed in the lower abdomen. Surgeons choose this site because the new kidney can easily be attached to a large artery (the **femoral artery** supplying the leg) and is usefully right next to the bladder.

 What is the main problem with a transplant once it has been carried out?

A major problem with any transplant is a shortage of suitable donors. Only a small minority of people carry donor cards. Permission to used body parts from recently deceased people has to be given by distressed relatives, who often say no. The problems of transplants are covered in more detail in Chapter 26, but you might want to find out more.

7

a) Discuss or jot down some ideas about what could be done to encourage more people to become organ donors.

b) It has been suggested that everyone should automatically be organ donors, unless they take the decision to opt out of the scheme. Discuss whether this idea would work or not.

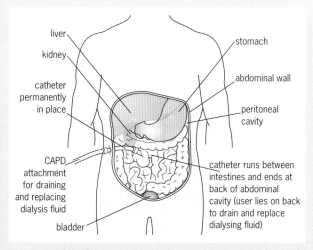

Fig 27.A3 **The peritoneum is a semipermeable membrane that lines the abdominal cavity and organs such as intestines. Dialysate introduced into this cavity draws waste and excess water out of the blood**

28 Reproduction in animals

HUMAN COUPLES CAN be interested in sex at any stage of the woman's menstrual cycle. In many other mammals, the female is not receptive to the male unless an egg is ready to be fertilised: when an egg is about to emerge from the ovary, a surge of the hormone oestrogen in the female's body ensures that she becomes interested in mating. We say that the female has entered a phase in her reproductive cycle called oestrus – this word comes from the Greek for 'frenzy'.

Anyone who has ever kept an unneutered female cat knows just how powerful oestrogen is. As soon as the female is 'in season' or 'on heat' she transforms from domestic pet into sex-hungry pest. Secretions from her reproductive system act as pheromones, chemical signals that attract males from all over the neighbourhood. The urge to respond is overwhelming and many a precious pedigree female has escaped while on heat and returned pregnant, courtesy of a local moggy.

In cats, mating usually leads to pregnancy because the male cat's penis has several backward-pointing barbs, like fish hooks. When the male withdraws, these barbs dig into the female, causing a sharp pain that actually triggers ovulation. This extra insurance makes it virtually impossible for the sperm to miss the eggs, and a litter of kittens make their inevitable appearance nine weeks later.

1 THE NEED TO REPRODUCE

The need to reproduce, so passing genes on to the next generation to ensure the survival of the population, is one of the basic features of living organisms. In many organisms, the urge to reproduce can over-ride the urge to live (Fig 28.1).

It has been suggested that individual organisms are merely disposable containers that exist to ensure the survival of the genes they contain. None of us is immortal, but many of our genes have persisted in the human population over hundreds of thousands of years. And, although our genes may mix with those from other people to form future generations, many will survive long after we are forgotten.

Reproduction is either **sexual** or **asexual** (non-sexual). All animals reproduce sexually, and some animals also reproduce asexually in particular circumstances.

Fig 28.1 **The overwhelming urge to reproduce is illustrated by these salmon. Born in fresh water, they swim downstream to the sea where they spend several years feeding and growing. Eventually, they return to the river of their birth, often using massive amounts of energy to complete the journey upstream. Finally, they lay and fertilise their eggs and then die**

2 ASEXUAL REPRODUCTION IN ANIMALS

Asexual reproduction involves no fertilisation: one individual organism produces one or more new individuals. Since there is only one parent, there is no mixing of genetic material and so the new organisms are **clones** – organisms that are genetically identical to the parent. Asexual reproduction can quickly increase the number of organisms in a population and it avoids the need to find a mate. This works well if the environment is favourable and does not change. The problem with asexual reproduction, though, is that the organisms produced are all identical and so the population cannot adapt to change.

The only variation that occurs in asexual reproduction is by mutation (see Chapter 31). This means that organisms that reproduce asexually tend to evolve more slowly, and hardly any animal species uses it as their only means of reproduction. Asexual and sexual reproduction are compared in Table 28.1. The processes of asexual and sexual reproduction in plants are covered in Chapter 29.

There are three main methods of asexual reproduction in animals:

- **Budding**. Some cnidarians, flatworms and annelids multiply by budding (Fig 28.2). A new individual simply grows out of the body of the parent.

- **Parthenogenesis**. This term means 'virgin birth' or reproduction without fertilisation. It occurs in a variety of animal types, including some insects, fish, amphibians and lizards, when unfertilised eggs develop into new individuals. See the Feature box on aphids on the opposite page.

- **Regeneration**. Many echinoderms such as the starfish (Fig 28.3) and some flatworms and annelids reproduce successfully by regeneration. If their bodies are broken into fragments, each one can develop into a new individual.

Fig 28.2 **Budding is one type of asexual reproduction. In this** *Hydra* **(a cnidarian), the new individuals simply grow out of the body wall of the parent**

Fig 28.3 **A memorable example of the power of some organisms to regenerate was when some fishermen in the USA launched a campaign to prevent starfish from destroying their oyster beds. They collected a large number of starfish, chopped them up and threw them back into the sea. Bad move. As the picture shows, a severed arm can regenerate into a whole new organism. The fishermen caused a population explosion**

Type of reproduction	Advantages	Disadvantages
Asexual	Fast – one individual can build up a population in favourable circumstances.	All offspring are identical. There is less genetic variation, so evolution is slower.
Sexual	Produces variation. Allows organisms to evolve and adapt to changes in their environment.	Slow – requires two individuals. Individuals are vulnerable during mating and pregnancy.

Table 28.1 **Comparing sexual and asexual reproduction**

APHIDS ARE BORN PREGNANT!

APHIDS ILLUSTRATE the pros and cons of sexual and asexual reproduction quite dramatically. These insects, which include greenfly, are pests: they cause millions of pounds of damage to crops world-wide. Although the exact details vary from species to species, a generalised aphid life cycle is as follows:

After surviving the winter on tree bark, aphid eggs hatch in spring, producing a population of wingless females. These females are born pregnant – aphid embryos produced by parthenogenesis are already developing inside them (Fig 28.4).

- The aphids build up their numbers rapidly, taking advantage of the very favourable conditions at the start of the summer – big healthy plants full of sap, no competition for food and no predators.

- The large numbers of insects – and their waste – damage host plants and so food becomes scarce. (No population can go on expanding indefinitely: there are always limiting factors.) Predators such as ladybirds become more common as they take advantage of the abundant food and, towards the end of summer, conditions generally become unfavourable to the aphids. This triggers a 'switch'

and the reproductive process produces winged aphids that can fly away and colonise new plants.

- Towards the end of summer, as the days get shorter, aphids of both sexes are born. Winged males and females then reproduce sexually, mating and laying eggs that can survive the winter.

Fig 28.4 **Aphids give birth to live young produced by the process of parthenogenesis. Each wingless female aphid is born with embryos of the next generation already growing inside her. This phenomenon is also known to occur in organisms as advanced as fish, amphibians and reptiles**

3 SEXUAL REPRODUCTION IN ANIMALS

Sexual reproduction involves the fusion of **gametes**, or sex cells. Female sex cells, called **egg cells** or **ova**, fuse with the male gametes, called **spermatozoa** or **sperm**, in a process known as **fertilisation**. The resulting cell, the **zygote**, develops into a new individual.

In sexual reproduction, the genetic material of two different individuals is mixed and combined to produce an individual that is genetically different from either parent. This produces **variation** within a population (see Chapter 30). All individuals, unless they have an identical twin, are **genetically unique**: different from all other individuals in the group.

Most species are **unisexual**: individuals are either male or female. But some organisms carry both male and female sex organs in the same body: they are described as **hermaphrodites**. Many familiar animals such as earthworms and snails are hermaphrodites.

Having both types of sex organs in the same body gives an organism greater reproductive capacity, or **fecundity**. When two snails mate, each female sex organ receives sperm and both snails become pregnant and can lay eggs. One mating therefore results in twice the number of offspring. In some circumstances a hermaphrodite can fertilise itself and so start up a new population.

?

A Why is self-fertilisation uncommon in hermaphrodite organisms?

Animal life cycles

Animals are diploid organisms that produce haploid gametes (Fig 28.5). These fuse to produce a new diploid individual, so there is no alternation of generations (in contrast to plants – see page 204).

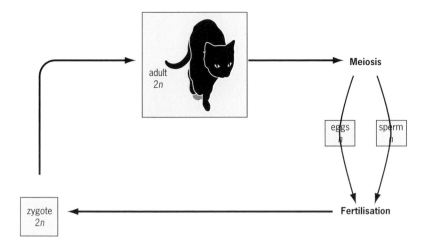

Diploid cells contain two sets of chromosomes, haploid cells contain only one set. In animals, **somatic** (body) cells are diploid while gametes (sex cells) are haploid.

Fig 28.5 **Animals have diploid life cycles. All of the body cells of this cat are diploid, except for its sex cells. These are produced by meiosis and are haploid. When a male gamete fertilises a female gamete, a diploid zygote is produced. This divides by mitosis many times. The new cells continue to divide and** differentiate **(cells become specialised to carry out particular functions), to produce a new individual**

Internal versus external fertilisation

The male gametes of all animals can swim. Animals that reproduce in water, such as fish and amphibians, fertilise egg cells outside the female's body simply by releasing sperm (Fig 28.6). These swim through the water to reach the egg cells. We now know that the egg cells of many species release chemicals that attract the sperm, greatly increasing the chances of successful fertilisation.

Fig 28.6 **Species that breed in water often fertilise externally. In these green pullers the female deposits her egg cells and the male follows behind, releasing millions of sperm**

As animals evolved and began to colonise the land, they had to overcome the problems of reproducing out of water. Internal fertilisation became a necessity. Animals developed specialised organs such as a **penis** and complex behaviour patterns to enable the male to introduce sperm directly into the female's body (Fig 28.7).

Animals have developed various strategies to protect the growing embryo(s) and prevent them from drying out. Some animals – birds and some reptiles, for example – produce waterproof eggs. The egg has a tough material round the fertilised ovum. Eggs are laid by the animal and provide the embryo with a self-contained, watery environment. Organisms that lay eggs are said to be **oviparous** (ovi = egg, parous = birth).

Even so, eggs are vulnerable to damage and can be eaten by predators. A much safer place for fertilised ova is inside the female, and it makes sense to lay eggs as late as possible in development. Many species have taken this to its logical conclusion: their eggs hatch inside the mother and the young are born as immature miniatures of their parents. This process is known as **ovoviviparity** (ovo = egg, vivi = live, parity = birth). Many species of fish (Fig 28.8), amphibians and reptiles reproduce in this way.

Fig 28.7 **In terrestrial animals such as reptiles, birds and mammals, fertilisation must be internal. Mating introduces sperm directly into the body of the female**

Fig 28.8 **Many sharks give birth to 'live' offspring: the eggs hatch inside the female. This process avoids the vulnerable egg stage of the life cycle**

B Why is it curious to say that the offspring born by viviparity and ovoviviparity are born 'alive'?

Mammals also give birth to live young, but these do not hatch from eggs. In most species of mammal, the embryo grows inside a specialised organ called the **uterus** and is fed by a **placenta**. The young are born live, but their state of maturity at birth depends on the species. This reproductive process is called **viviparity**.

Reproduction in mammals

Mammals show a great range of reproductive patterns that reflect their circumstances, but the basic life cycle of most mammals follows a general pattern. After mating, the fertilised egg cell(s) develop inside the uterus, nourished by the placenta. The **gestation period** of a mammal, the time between fertilisation and birth (Table 28.2), often reflects the metabolic rate of the organism. It is shorter in small, short-lived mammals such as mice and longer in large, long-lived mammals such as elephants.

Table 28.2 **The gestation period of some mammals**

Species	Approx. gestation period
Hamster	18 days
Stoat	28 days
Rabbit	30 days
Cat	63 days
Lion	110 days
Human	40 weeks
Horse	11.5 months
Elephant	22 months

Other factors complicate the length of gestation. For example, it is also limited by the size of the fetal skull (which must fit through the female's pelvis at birth) and the mobility of the mother. In species where the mother needs speed and agility to survive, the gestation period tends to be shorter. Birth is followed by a period when the mother gives close protection and suckles her young. The length of upbringing varies greatly: it depends on factors such as degree of maturity at birth and the amount of learning required to survive.

C The gestation period of the lion is short for such a big animal. Suggest a reason for this 'early' birth.

The onset of sexual maturity marks the change from **juvenile** to **adult**. Most female mammals conceive as soon as they become sexually mature and enter their first **oestrus** (season). From then on, throughout their reproductive life, they are either pregnant, feeding young or both. A female's reproductive life is usually brought to an end only by death. This has been the harsh fact for humans, too, until fairly recently. Even now, in societies where contraception is either unavailable or not used for cultural reasons, many women reproduce continuously.

✓

The period of time after birth during which a female mammal suckles her young is called the **lactation period**.

The role of hormones in mammalian reproduction

For many animals the timing of reproduction is vital. If there is a severe variation in the seasons, animals must ensure that their offspring are born when food is plentiful and conditions are mild. **Hormones** that control and synchronise reproductive events are produced by three different organs: the **hypothalamus**, the **pituitary gland** and the **gonads** (sex organs).

The hypothalamus is a major point of contact between the nervous and **endocrine** (hormone) systems (see Chapter 22). Internal and external stimuli reach the hypothalamus, which responds by releasing hormones. They, in turn, control the rest of the endocrine system (Fig 28.9).

Day length is a classic example of an external stimulus. When the hours of daylight reach a threshold level, perhaps indicating the arrival of spring, the hypothalamus responds by releasing a hormone called a **gonadotrophin releasing factor (GnRF)**. This controls the activity of the nearby pituitary gland, stimulating it to release hormones called **gonadotrophins**.

D The hypothalamus controls the level of sex hormones via a negative feed-back mechanism. Suggest how the hypothalamus would respond to a fall in oestrogen levels.

Gonadotrophins have a direct effect on the gonads (the ovaries and testes). These respond by releasing **steroid sex hormones** (Chapter 3) such as **oestrogen** and **progesterone** and **testosterone**.

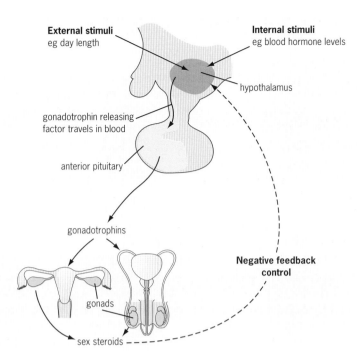

External stimuli
eg day length

Internal stimuli
eg blood hormone levels

hypothalamus

gonadotrophin releasing
factor travels in blood

anterior pituitary

gonadotrophins

**Negative feedback
control**

gonads

sex steroids

Fig 28.9 **The hormonal control of reproduction.
The hypothalamus can respond to external and
internal nervous stimuli and is also sensitive to
hormone levels in the blood. The hypothalamus
controls the levels of several different circulating
hormones using a negative feedback mechanism**

E Animals use day length as an
indicator of time of year. Which other
indicators could be used, and why are
they not suitable?

F Suggest why smaller mammals such
as mice and rats tend not to have
breeding seasons.

THE TIMING OF REPRODUCTION

SHEEP THAT LIVE in temperate climates are **seasonal
breeders**. They synchronise their reproductive cycle to
produce lambs in early spring (March/April), when
new grass is plentiful. This also gives the young as
much time as possible to mature before winter.
Ensuring that the young are born at the right time
needs careful timing on the part of the reproductive
process: the ewe needs to produce eggs cells, and then
mate, at just the right time. The gestation period in
sheep is 21 weeks, so mating takes place in
October/November.

For sheep, the most reliable indicator of the time of
year is day length. In the autumn the stimulus of
shortening days (or lengthening nights) affects the
hypothalamus, which responds by secreting GnRF.
This stimulates the pituitary gland to secrete a
gonadotrophin called **follicle simulating hormone**
(**FSH**, or **follitropin**). This in turn stimulates the
development of follicles in the ovaries. The follicles
secrete oestrogen, which causes the ewe to come into
season. As she begins to give off vaginal secretions that
contain **pheromones**, nearby rams are aroused and
mating occurs (Fig 28.10).

Seasonal breeding in other mammals
In sheep and many other seasonal breeders, the gestation
period exactly fills the length of time between the mating
season and the most favourable time for birth. Some
species, however, have an extra advantage: they can
mate and conceive, but then delay the time when the
embryo implants in the wall of the uterus. In bears and
kangaroos, for example, the embryo does not implant
immediately after fertilisation, but remains in a kind of

Fig 28.10 **When ewes come into season, the farmers release rams
into their field for a night. In the morning, all the females should
have been impregnated. Just to make sure none has been missed,
some farmers take the precaution of taping a block of dye to each
ram's chest. This rubs off during mating. Any ewes that do not
have dyed bottoms by morning have to stay with the rams for
another night**

limbo until the time is right. In the case of the kangaroo,
the embryo is kept on hold until the previous baby
(the joey) has been fully weaned and has left the pouch.
Then implantation takes place, and the pregnancy
proceeds as normal.

The mechanism of delaying implantation is of great
interest to scientists, because these animals are able
to 'freeze' the development of embryos at body
temperatures. A knowledge of how they achieve this
can have many uses: preserving cells, embryos and tissues
without risking damage from the ice crystals that can
result from actual freezing.

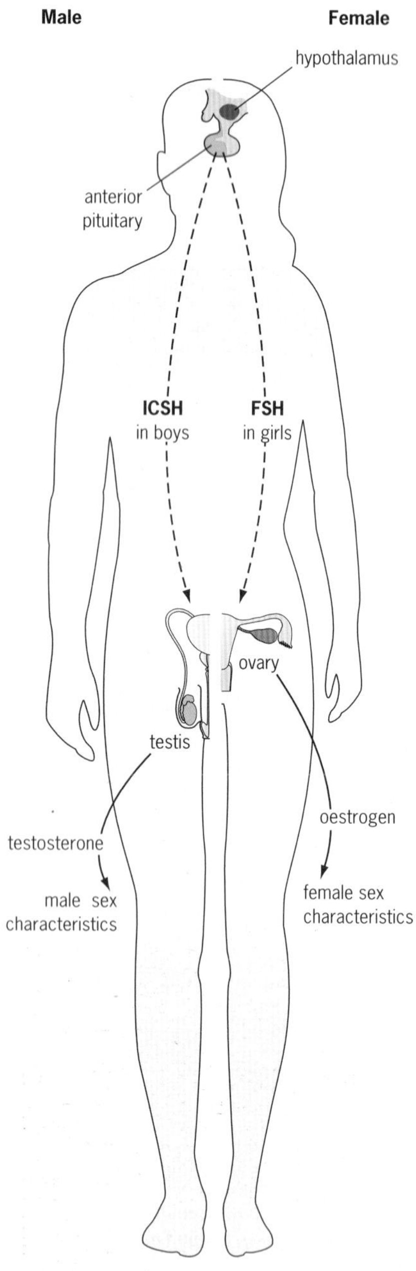

Male **Female**

hypothalamus

anterior
pituitary

ICSH FSH
in boys in girls

ovary

testis

testosterone

oestrogen

male sex
characteristics

female sex
characteristics

Fig 28.11 The onset of puberty is controlled by hormones.

Oestrogen causes the ovaries, oviducts, uterus and vagina to mature and brings about the development of secondary sex characteristics. Breasts grow, pubic and underarm hair grows, the shape of the pelvis changes and the body fat is redistributed so that a woman tends to have wider hips and a more curved shape than a man

Testosterone causes the penis, scrotum and testes to enlarge and mature. It also causes enlargement of the larynx (voice-box) which deepens the voice. Body hair is more extensive, with coarse hair appearing on the chest and face. Ironically, in later life, the same hormone leads to male-pattern baldness. A man's shoulders and chest tend to be large and there are often changes in facial structure, with the nose and chin becoming more prominent. The muscles tend to be better developed and generally the male has greater physical strength than the woman

4 HUMAN SEXUAL REPRODUCTION

Sexual activity in humans has evolved to achieve more than just fertilisation. Although humans become sexually mature in their early to mid-teens, many societies have cultural and legal controls aimed at restricting sexual activity before the late teens. Although beyond the scope this book, it is interesting to consider whether human sexual behaviour may have evolved to strengthen the emotional bond between a couple. This 'social cement' may make it more likely that couples will stay together and so provide a stable environment for raising children.

Puberty

Puberty is the time between childhood and adulthood: it marks the process of sexual maturing. We still don't know exactly what triggers the onset of puberty, but we know that, early on, GnRF is secreted from the hypothalamus (look back to Fig 28.9). GnRF travels the short distance to the anterior pituitary, where it stimulates the release of gonadotrophins, **follicle stimulating hormone** (**FSH**) in females and **interstitial cell stimulating hormone (ICSH) in** males (Fig 28.11). These two hormones are chemically identical, but have different names because they have different effects in the two sexes.

In girls, FSH targets the ovaries, which it stimulates to produce **oestrogen**. This steroid hormone is responsible for many of the female sex characteristics. In addition, oestrogen stimulates the ovaries to start producing egg cells (ovulation) and this leads to the first monthly period, or **menstruation**, an event known as the **menarche**.

To start with, periods tend to be irregular and unpredictable, and may sometimes occur without ovulation. Within about a year, hormone levels have increased to the point where they stimulate the regular development of follicles. This makes periods more regular and, unfortunately for many girls, painful. Period pain is mainly due to the hormone progesterone, which causes uterine cramps.

In boys, ICSH targets the **interstitial cells** of the testes. These are embedded in the connective tissue between the seminiferous tubules – see Fig 28.12(b). ICSH stimulates these testis cells to secrete **testosterone**, the steroid hormone that stimulates development of male sex characteristics.

The age of onset of puberty

The average age for the onset of puberty is 12 to 13 in girls, 13 to 15 in boys. Interestingly, this is much earlier than in previous centuries. Two hundred years ago the average age of the menarche in girls was 16, around four years later than it is now. The reason for this is almost certainly the improvement in diet. A better diet enables us to grow faster and so reach the same stage of maturity at an earlier age. In girls, the proportion of body fat appears to be important: a girl who is a dedicated athlete (a gymnast for example) and has a high muscle:fat ratio, may find that the menarche is delayed. Also, girls who have started their periods but who then crash diet and lose a lot of weight may suddenly find their periods stop for a while.

The male reproductive system

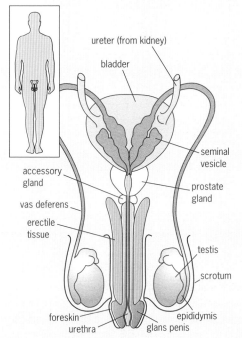

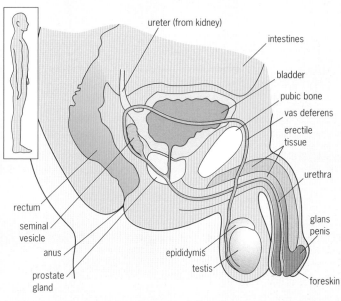

Fig 28.12(a) **The main parts of the male reproductive system, seen from the front and the side**

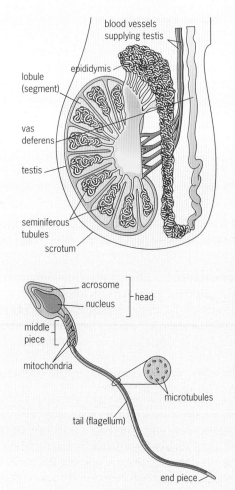

Fig 28.12(b) **The microscopic structure of the testis showing a section through a seminiferous tubule. The testis contains many seminiferous tubules. Inside each of these at any time, many thousands of sperm are maturing**

The overall structure of the male reproductive system is shown in Fig 28.12. The male system secretes testosterone, makes and stores sperm and delivers them into the female's body.

Spermatozoa are made in the testes, a pair of organs that are held in a pouch of skin called the **scrotum**. It may seem odd that such delicate and vital organs are relatively unprotected outside the body, but there is a reason for this. The process of sperm production, **spermatogenesis**, is most efficient at around 35 °C, two degrees cooler than the core of the body. Men whose testes do not descend into the scrotum cannot produce healthy sperm.

See questions 3(a) and 3(b).

The penis

The penis introduces sperm into the female's body so that internal fertilisation can occur. Some animals do not have a penis – most birds and reptiles, for example – and their attempts at fertilisation are a little more haphazard. The male produces semen from his genital opening and simply rubs it onto the female's genital opening.

The testes and spermatogenesis

The production of sperm in human males is a continuous process, beginning at puberty and continuing well into old age: men in their nineties have fathered children. Spermatogenesis centres around the process of meiosis, and occurs at a remarkable rate: over a thousand human sperm are made every second. Look again at Fig 28.12(b) and you will see that each testis is composed of a series of **lobules**. They contain **seminiferous tubules**, the structures in which sperm production takes place. This process, shown in Fig 28.13, has three main phases:

- **Multiplication**. As large numbers of sperm are needed, cells of the **germinal epithelium** divide by mitosis to produce many **spermatogonia** (sometimes called sperm mother cells).

- **Growth**. The spermatogonia grow into **primary spermatocytes**. At this stage the cells are still diploid ($2n$).

- **Maturation**. The diploid primary spermatocytes undergo meiosis. After the first division they become **secondary spermatocytes** and when meiosis is complete they have become haploid **spermatids**. In the final part of the maturation process, spermatids differentiate into the familiar spermatozoa (sperm).

Throughout their development, sperm cells are closely associated with **Sertoli** or **nurse cells**, from which they obtain nutrients. In the lumen of the seminiferous tubule – see Fig 28.12(b) – the tails of the spermatozoa are clearly visible: their heads are attached to Sertoli cells.

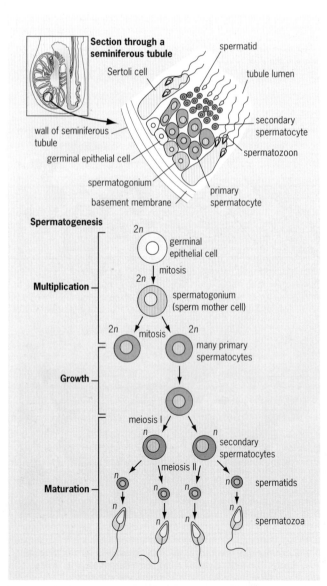

Fig 28.13 **The process of spermatogenesis. Sperm cells develop as they pass from the outer wall of the seminiferous tubule to the lumen**

Sexual arousal in males

Men become sexually aroused by thinking about sex, as a result of physical stimulation or a combination of both. Nerve impulses from the brain pass down parasympathetic nerves (see Chapter 21) and cause arterioles leading to the penis to dilate. The penis receives more blood than can drain away, spongy **erectile tissue** in the shaft becomes filled with blood, and an **erection** results.

Flaccid (non-erect) penises vary greatly in size, largely depending on how much blood is retained in the spongy tissue. When erect about 90 per cent are between 14 and 16 cm long. The end of the penis, the **glans**, is particularly sensitive, and continued stimulation from rhythmic thrusting eventually leads to a series of reflexes known as **ejaculation**. Stored spermatozoa are propelled along the **vas deferens** by powerful peristaltic waves. As they pass various accessory glands, different secretions are added to the sperm and the final ejaculate is a milky fluid called **semen** (Table 28.3).

Human males ejaculate, on average, about 5 cm^3 of semen, which contains between 50 to 200 million sperm. Most sperm never get anywhere near the egg, even after unprotected sex. For most of the monthly cycle, the cervix is blocked by a plug of mucus which sperm cannot penetrate. Only at around ovulation time does the mucus consistency change, allowing sperm to pass through easily.

Table 28.3 **The constituents of semen**

Gland	Secretion	Purpose
Seminal vesicles	fructose	energy for spermatozoa
	mucus	lubrication
	protein	forms clots, which alter consistency of semen
	prostaglandins	stimulate peristalsis in the female system
Prostate	alkaline chemicals	neutralise acid in vagina
	clotting agent	clots the protein from the prostate, forming a gelatinous mass
Cowper's gland	clear fluid	cleans urethra prior to ejaculation

The female reproductive system

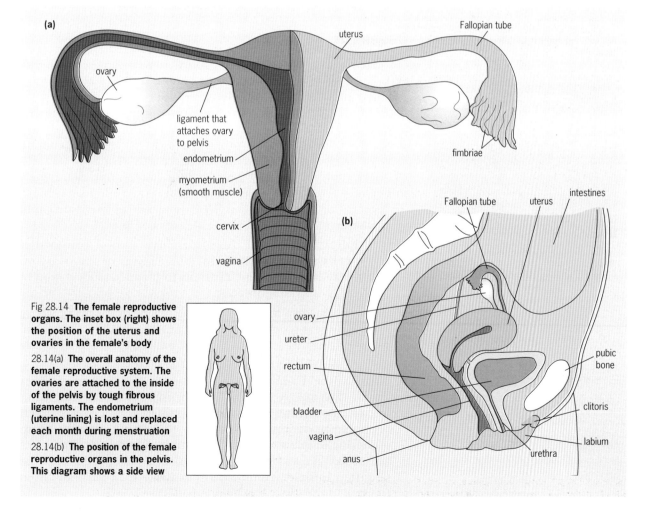

Fig 28.14 **The female reproductive organs. The inset box (right) shows the position of the uterus and ovaries in the female's body**

28.14(a) **The overall anatomy of the female reproductive system. The ovaries are attached to the inside of the pelvis by tough fibrous ligaments. The endometrium (uterine lining) is lost and replaced each month during menstruation**

28.14(b) **The position of the female reproductive organs in the pelvis. This diagram shows a side view**

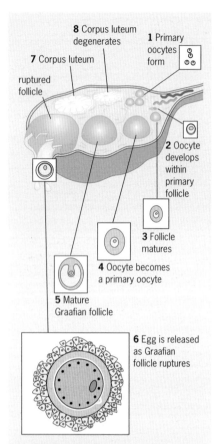

Fig 28.15(a) **Section through an ovary showing the egg cell at different stages of development**

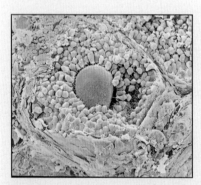

Fig 28.15(b) **Scanning electron micrograph through a primary oocyte showing an egg surrounded by follicle cells**

Meiosis produces four daughter cells. In oogenesis, one ovum is formed; the other three cells degenerate.

Fig 28.14 on the previous page shows the structure of the female reproductive system. This includes a pair of **ovaries** which are the primary female sex organs. These produce egg cells, or **ova**, and also secrete the hormones **oestrogen** and **progesterone**. Each ovary is about 3 to 4 cm across and is attached to the inside of the pelvis by a **ligament** (a tough band of connective tissue). The oviducts, or **Fallopian tubes**, connect the ovaries to the **uterus** (womb). Each tube ends in finger-like **fimbriae**, which move close to the ovary at the time of ovulation.

The uterus is a compact muscular organ that nourishes, protects and ultimately expels the fetus. The human uterus is able to expand from about the size of a small orange with a capacity of about 10 cm³ to accommodate a full-term baby. This is a five-hundred-fold increase in the capacity of the uterus. The bulk of the uterine wall is made from smooth muscle and is known as the **myometrium**. The lining of the uterus, the **endometrium**, consists of two layers. The underlying layer is a permanent **basement membrane** which produces the surface layer, the **decidua**. This layer is built up every month and shed during **menstruation**.

The **cervix**, or neck of the uterus, is a narrow muscular channel which is usually blocked by a plug of mucus. During sexual arousal the muscles of the **vagina** (a muscular tube which leads to the outside of the body) relax and glands in the vagina secrete lubricating mucus. This allows the male's penis to enter without discomfort. During childbirth, the cervix dilates to around 10 cm in diameter to allow the baby to pass through.

Oogenesis

Throughout this section we refer to the female gamete as an **egg cell**. The egg cell is surrounded by several layers of cells and the complete unit is called a **follicle**.

The production of egg cells, **oogenesis**, takes place within the ovaries of the developing female fetus. At birth, a girl already has about 2 million **primary oocytes**. Most of these degenerate during childhood and by puberty there are only about 200 000 left. Of these, only about 450 ever mature fully – one per month throughout the female's reproductive life. As in spermatogenesis, the process of oogenesis (Figs 28.15 and 28.16) is divided into three phases:

● **Multiplication**. As the female embryo grows, **primordial germ cells** in the **epithelium** (outer layer) of the ovary go through a series of mitotic divisions to produce a population of larger cells called **oogonia**.

● **Growth**. Oogonia move towards the middle of the ovary where they grow and go through further mitotic divisions to become **primary oocytes**. Each oocyte is surrounded by a layer of follicle cells. Together they form a **primary follicle**.

● **Maturation**. From puberty onwards, a few primary follicles mature each month. Usually, only one completes its development, the rest degenerate. The remaining primary follicle grows larger, becoming an **ovarian follicle**. Its cells secrete **follicular fluid**, producing droplets which join together to form a fluid that fills the space known as the **antrum**. The mature follicle, the **Graafian follicle**, is almost 1 cm in diameter. It protrudes from the wall of the ovary just before ovulation.

Inside the developing follicle, the oocyte begins its first meiotic division. There is no need for more than one egg cell, so the second set of chromosomes formed at meiosis I is discarded, passing into a small cell (with very little cytoplasm) known as the **first polar body**. This appears to have no function, but it often completes the meiotic division, producing two similar cells; both later break down. After meiosis I, the egg cell is known as a **secondary oocyte**. It then begins the second meiotic division but gets no further than metaphase. The division is completed only if the egg cell is fertilised.

When fertilisation occurs, meiosis II is completed and the egg cell becomes the mature **ovum**. This produces another 'spare' set of chromosomes, the **second polar body**, a cell that also degenerates.

Ovulation

On around day 14 of the human female's menstrual cycle, an egg cell (a secondary oocyte) is released from the ovary in a process called **ovulation**. Pressure in the antrum builds up and ruptures the Graafian follicle, forcing the egg cell out. The egg cell is released into the body cavity, but very few get lost since the fimbriae of the Fallopian tubes hover close to the ovaries, and the ciliated lining of the tubes creates a current that gently sucks in the egg cells.

Remarkably, a woman can become pregnant even if she has lost an ovary on one side and a Fallopian tube on the other. This suggests that the functioning fimbria actively seeks out the productive ovary by moving right across to the other side of the woman's body.

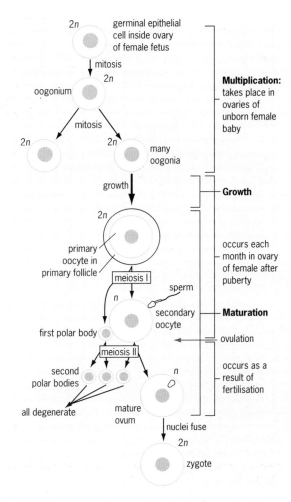

Fig 28.16 **The process of oogenesis begins in the ovaries of a developing fetus but is not completed unless the egg is fertilised**

THE MENOPAUSE AND HORMONE REPLACEMENT THERAPY

THE MENOPAUSE IS a natural event that occurs when the ovaries stop working. For most women, this happens between the ages of 45 and 54. This time is often difficult and traumatic. The lowered levels of oestrogen and progesterone are directly responsible for the unpleasant symptoms of the menopause:

- Circulatory problems such as hot flushes and night sweats.
- Psychological problems, such as depression, anxiety and insomnia.
- Skeletal problems. Oestrogen inhibits reabsorption of bone, and after the menopause, bone loss can be as great as 7 per cent per year. This condition, **osteoporosis**, affects the spongy bone particularly and the sufferer is more likely to break a bone.
- Oestrogen is also thought to give women some protection against some types of heart disease. Women below menopausal age are less likely than men to have heart disease, but afterwards they catch up.

Hormone replacement therapy can help to reverse many of these symptoms and effects. The basic idea behind HRT is simple: to restore hormone levels to those of the early follicular phase of the menstrual cycle using tablets, implants or transdermal patches.

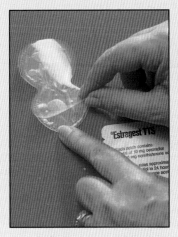

Fig 28.17 **One method of administering the hormones in HRT: a transdermal patch (a patch attached to the skin) that releases controlled amounts into the blood-stream. The hormones used in HRT are extracted from the urine of pregnant mares or made artificially**

See question 1(b).

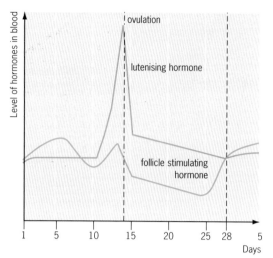

Fig 28.18(a) **How levels in luteinising hormone and follicle simulating hormone change during the menstrual cycle**

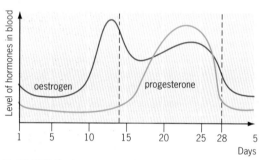

Fig 28.18(b) **How oestrogen and progesterone levels change during menstruation**

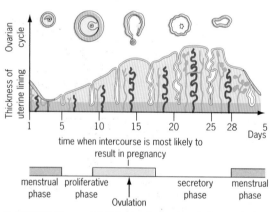

Fig 28.18(c) **Stages of follicle development and changes in the thickness of the endometrium. Note that pregnancy is more likely to occur after unprotected sex between days 9 and 17, in a cycle where ovulation occurs on day 14**

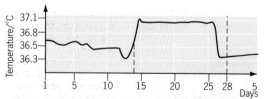

Fig 28.18(d) **A woman's body temperature rises just after ovulation. In couples finding it difficult to conceive, taking daily temperature measurements is a good way to pinpoint ovulation**

The human female menstrual cycle

The average length of the human menstrual cycle is 28 days, but variations from 24 to 35 days are normal, and greater variations are not uncommon. The cycle is divided into four phases:

● The **proliferative phase** in which the endometrium regenerates.

● The **ovulation phase** in which the ovum (egg cell) is released.

● The **secretory phase** in which the endometrium secretes nutrients in preparation for implantation.

● The **menstrual phase** in which the endometrium is shed from the body.

By convention, the first day of the period (the most obvious event) is called day 1 of the menstrual cycle. Fig 28.18 shows the timing of the phases listed above. It is important to relate the events of the menstrual cycle to the hormonal changes shown. You will need to refer to these diagrams as you read the following text.

The proliferative phase

On day 2 of the cycle, the pituitary gland releases **follicle stimulating hormone** (**FSH**). This stimulates the development of several ovarian follicles. At around day 6, one of the follicles dominates and begins to secrete oestrogen. The others degenerate. The remaining follicle develops into a **Graafian follicle** and continues to secrete oestrogen until day 14 of the cycle. Consequently, blood oestrogen levels rise. Oestrogen causes **proliferation** (growth) of the endometrium to replace the layer lost during the previous menstruation. After 14 days the repair is complete.

The ovulation phase

On day 12 to 13, the blood oestrogen level reaches a threshold level which triggers the release of **luteinising hormone** (**LH**), from the anterior pituitary gland. The rapid increase in LH levels triggers ovulation around day 14. The ovum lives for only 24 to 36 hours. During this time it moves only a few centimetres from the ovary.

The secretory phase

Luteinising hormone has a second effect: it causes the Graafian follicle to develop into a **corpus luteum** ('yellow body'). The name comes from the yellow appearance of the secretory cells that develop inside the 'remains' of the Graafian follicle. The corpus luteum secretes oestrogen and progesterone. Progesterone causes spiral-shaped blood vessels to grow into the endometrium. This thickened lining begins to secrete nutrients and mucus to prepare for an embryo to be implanted. During this phase the high levels of progesterone inhibit the production of FSH. As long as progesterone levels are high, the endometrium is maintained and no new follicles are stimulated. The 'contraceptive pill' takes advantage of this inhibition (see Feature box on page 461).

If the egg cell is not fertilised, the corpus luteum lasts for about 10 to 12 days and then degenerates, ceasing to secrete progesterone. This is a key event because the inhibition of FSH is lifted. The endometrium is no longer protected and the cycle can start again.

The menstrual phase
The drop in progesterone and oestrogen levels cause the uterine capillaries to rupture, and the endometrium is lost from the body through the cervix, together with some blood. Renewed secretion of FSH begins around day 2, and the cycle begins again.

The events of pregnancy
If the egg cell is fertilised, the menstrual cycle is interrupted and the female's body changes in response to the events of pregnancy.

Fertilisation
Fertilisation is a complex sequence of events that begins when the sperm reaches the egg cell (oocyte). The events of fertilisation are shown in Fig 28.19.

It may also help you to look back at Fig 28.16. The secondary oocyte is surrounded by several layers of follicle cells, the **corona radiata**, and a layer of glycoprotein, the **zona pellucida**. Before a sperm can penetrate the layers surrounding the oocyte, it undergoes a process called **capacitation**. The actual mechanism is poorly understood, but it seems to set off the **acrosome reaction**, which allows the sperm head to enter the oocyte. The **acrosome** is a bag of digestive enzymes on the tip of the sperm head (see Fig 28.19). During the acrosome reaction the bag splits, releasing the enzymes, which digest a pathway through any remaining follicle cells and the zona pellucida.

As soon as the outer membrane of the first sperm penetrates the cell surface membrane of the oocyte, a rapid reaction occurs. Many **cortical granules** fuse with the zona pellucida, forming a **fertilisation membrane**. This reaction starts at the point of entry of sperm head and spreads rapidly over the surface of the oocyte, preventing entry of other sperm. So, only one sperm enters the diploid secondary oocyte, even though many reach it at the same time. Entry of the sperm nucleus triggers the completion of meiosis II in the female nucleus, leading to the formation of the second polar body and a haploid mature ovum.

Almost immediately afterwards, a spindle forms and the paternal and maternal chromosomes come together, forming a diploid zygote. Within 12 hours, the first mitotic division takes place. Cell division is now rapid, forming a bundle of cells called a **morula** (Latin for the blackberry it resembles). As divisions continue, this becomes a **blastocyst** and moves slowly along the Fallopian tube through the action of cilia, which create a steady current of fluid towards the uterus (see Fig 28.20).

It takes around 6 or 7 days for the embryo to complete the journey down the Fallopian tube. When it arrives at the uterus on day 21 of the cycle, the lining must be in just the right condition to accept it. For a successful pregnancy there must be exact timing between the preparation of the endometrium and the development of the embryo.

G Many women do not have a 28 day cycle. Although the second half of the cycle (ovulation to menstruation) is usually 14 days, the first half can vary considerably. If a woman's cycle is 35 days, on which day is she likely to ovulate?

See question 1(a).

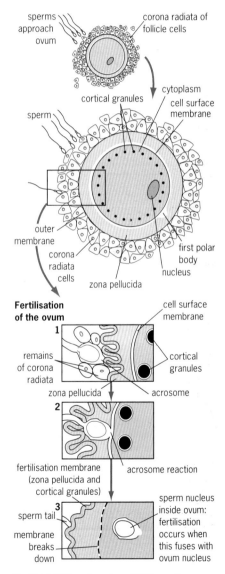

Fig 28.19 **With a diameter of 100 μm, the egg cell is the largest cell in the human body. Most of the volume is taken up by inert food material that will fuel the embryo during its first few cell divisions. A spermatozoon must pass through any remaining follicle cells (the corona radiata) and the zona pellucida, before it can fertilise the egg cell. The head of the sperm enters the body of the egg cell, but the tail remains outside**

See questions 3(b) and 3(c).

Implantation

Pregnancy begins not with fertilisation but when the embryo **implants** in the wall of the uterus. This happens about a week after fertilisation and is not always successfully completed. Many women trying to conceive may have a 'near miss', when an egg is fertilised but fails to implant.

Implantation begins when the blastocyst makes contact with the endometrium (Fig 28.20), usually on the back wall. The outer layer of the blastocyst, the **trophoblast**, causes an **inflammatory-type response** (normally a response to damage) and this causes an outgrowth of the endometrium at this point. The placenta develops where the trophoblast and the endometrium interact.

Fig 28.20 **Following fertilisation, the embryo undergoes several mitotic divisions as it passes slowly down the Fallopian tube. When the embryo reaches the uterus, it implants in the endometrium**

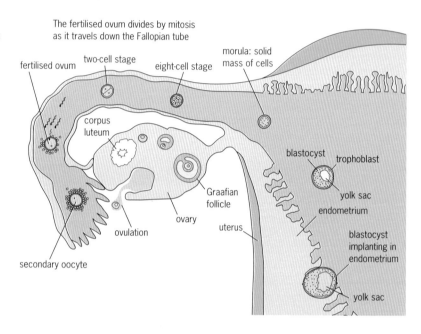

If implantation is successful, the embryo begins to secrete **human chorionic gonadotrophin** (**HCG**). This hormone forces the corpus luteum to continue to secrete progesterone, thereby maintaining the endometrium and inhibiting FSH production.

The **chorion**, one of the membranes that later grows and surrounds the embryo, develops **villi** (projections) that burrow into the endometrium. These are thought to break down the mother's blood vessels, causing the chorionic villi to become bathed in maternal blood.

See questions 3 and 4

The placenta

The placenta is a temporary organ that allows the blood systems of the fetus and the mother to come into close contact, without actually mixing (Fig 28.23). The placenta allows nutrients and oxygen to pass to the fetus from the mother, and allows metabolic waste back into the mother's blood (see Table 28.4).

The placenta is an organ adapted to maximise the exchange of materials so, as you would expect, it has a large surface area provided by the **chorionic villi**. A close look at the cells of the villi shows that the membranes are folded into microvilli and also contain many mitochondria: these two features maximise the processes of diffusion and active transport (see Chapter 5). There are many small vesicles in the cells of the villi, suggesting that substances are being absorbed by pinocytosis (see Chapter 5).

Table 28.4 **Exchange of materials across the placenta**

Mother to fetus	Fetus to mother
oxygen	carbon dioxide
glucose	urea
amino acids	other waste products
lipids, fatty acids and glycerol	
vitamins	
ions; Na, Cl, K, Ca, Fe	
alcohol, nicotine, many drugs	
viruses	
antibodies	

THE PREGNANCY TEST

BY THE TIME A PREGNANT woman would have been due for her next period, the embryo will have implanted and would be starting to secrete HCG. Enough of this hormone passes into the urine for it to be detected by a modern pregnancy testing kit. This kit makes use of monoclonal antibodies (Chapter 26), which are specific for HCG and which are attached to a coloured chemical.

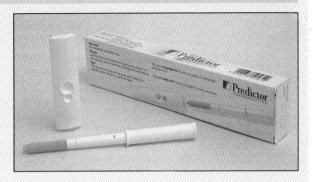

Fig 28.21 **A home pregnancy test kit**

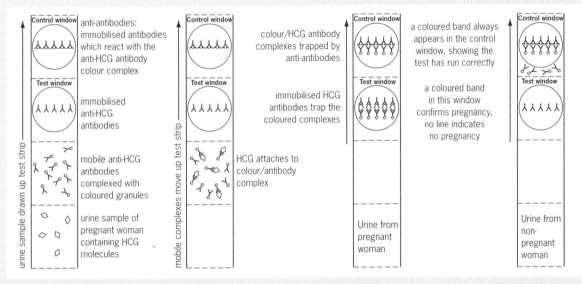

Fig 28.22 **How a home pregnancy test kit is used.**

(a) **Urine containing HCG is drawn up the test strip by capillary action**

(b) **HCG attached to the colour/antibody complex moves up the test strip**

(c) **In the urine of a pregnant woman, HCG attached to the colour/antibody complexes moves up to dock with the immobilised (fixed) HCG antibodies in the test window. All the complexes are stopped at one place, forming a visible coloured line and confirming pregnancy**

(d) **For the urine of a non-pregnant woman, the colour/antibody complexes have no attached HCG, so all the complexes move past the immobilised HCG antibodies in the test window and attach to the anti-antibodies in the control window showing the test has worked**

Labels in Fig 28.22:

(a) Control window — anti-antibodies: immobilised antibodies which react with the anti-HCG antibody colour complex

Test window — immobilised anti-HCG antibodies

mobile anti-HCG antibodies complexed with coloured granules

urine sample of pregnant woman containing HCG molecules

urine sample drawn up test strip

(b) Control window

Test window

mobile complexes move up test strip

HCG attaches to colour/antibody complex

(c) Control window — colour/HCG antibody complexes trapped by anti-antibodies

Test window — immobilised HCG antibodies trap the coloured complexes

Urine from pregnant woman

(d) Control window — a coloured band always appears in the control window, showing the test has run correctly

Test window — a coloured band in this window confirms pregnancy, no line indicates no pregnancy

Urine from non-pregnant woman

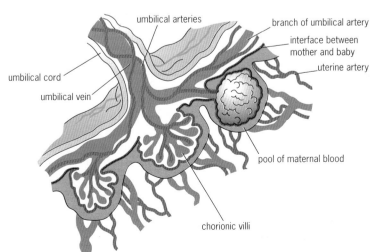

umbilical arteries

branch of umbilical artery

interface between mother and baby

uterine artery

umbilical cord

umbilical vein

pool of maternal blood

chorionic villi

Fig 28.23 **The fine structure of the placenta. The capillaries within the chorionic villi are bathed in pools (lacunae) of maternal blood. The placenta is a large, disc shaped organ which is usually attached to the back wall of the uterus. At birth, contraction of the uterine muscle causes the chorionic villi to split away from the endometrium and the placenta is delivered after the baby: hence its common name of afterbirth. Most mammals eat the placenta after the birth of their young. It is an important source of nourishment at a time of great need**

See question 4. ■

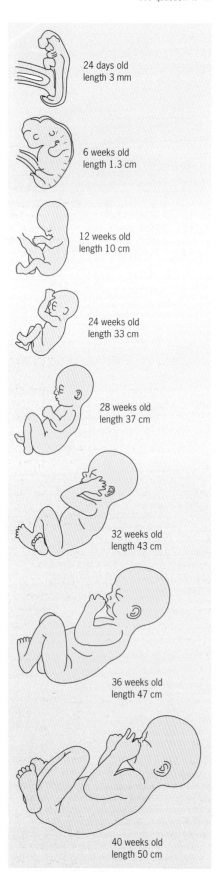

Fig 28.24 **Stages in the growth of a baby in the uterus**

The placenta is an important **endocrine organ**. It secretes the hormones that maintain pregnancy, taking over from the corpus luteum at about 12 weeks. The placenta secretes progesterone (which maintains the endometrium), oestrogen (which inhibits the ovulatory cycle), human chorionic gonadotrophin (see the Feature on the opposite page) and **human placental lactogen** (which stimulates the development of the mammary glands).

Pregnancy, labour and birth

In humans, pregnancy lasts for an average of 40 weeks. Some of the main stages of growth of a baby are shown in Fig 28.24.

From around the twelfth week of pregnancy, progesterone secreted by the placenta inhibits uterine contractions. The level of progesterone rises steadily until just before birth, usually around 38 weeks after conception, when it starts to fall dramatically. This lifts the inhibition of uterine contractions. The mother's anterior pituitary begins to secrete **oxytocin** and the placenta secretes prostaglandins, two hormones that actively promote contractions. Oxytocin stimulates uterine contractions at about 40 weeks. The resulting tension in the muscle and pressure on the cervix are stimuli that bring about further secretion of oxytocin, causing more powerful contractions.

There are three stages of labour:

● Stage 1. The cervix dilates (opens) to a diameter of 10 cm.

● Stage 2. The fetus is pushed out of the uterus.

● Stage 3. The placenta and umbilical cord are expelled.

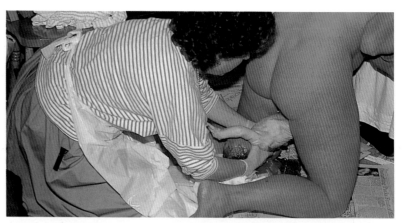

Fig 28.25 **Childbirth is one of the most intense human experiences. Childbirth used to be a time of great danger to both mother and child, but advances in medicine have greatly reduced the risk. Today, in the developed world, 99 out of every 100 babies born survive beyond their first birthday (two hundred years ago, only 54 out of every 100 lived)**

Lactation

All mammals produce milk from specialised **mammary glands**. Humans are unique in having permanent breasts but these are normally composed only of fatty tissue. Throughout pregnancy, however, oestrogen and progesterone stimulate the development of milk-producing tissue.

After birth, the first fluid released from the breasts is called **colostrum**. This contains no fat and very little sugar, but has important antibodies that 'lend' the baby immunity until it has time to develop its own. After about three to four days, normal milk is produced.

Milk is produced constantly and, although a certain amount of leaking from time to time is usual, milk flow only happens when the baby suckles. The stimulus of sucking at the nipple causes the posterior pituitary gland to secrete oxytocin. This hormone travels in the blood and causes contraction of the muscular **myo-epithelial cells** surrounding the milk glands or alveoli. This squeezes the milk out of the alveoli, through the milk ducts and into the infant's mouth. This mechanism, called the **let-down reflex**, takes several seconds to take effect but is quite powerful. If the unsuspecting infant lets go of the nipple, it can get a jet of milk in the eye.

Throughout the period of lactation the pituitary continues to secrete **prolactin**, a hormone that maintains the milk ducts and, to some extent, inhibits ovulation. This inhibition is called **lactational anoestrus** and reduces the risk of conception so soon after birth. However, it is not always effective and breastfeeding cannot be relied on as a contraceptive.

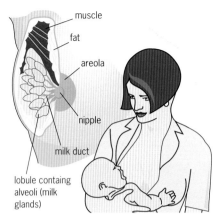

Fig 28.26 **During pregnancy the milk-producing ducts grow, substantially increasing the size of the breasts. When the new baby suckles, the physical stimulation leads to a series of hormonal events that increase the amount of milk that is made**

See question 6.

CONTRACEPTION

THE TWENTIETH CENTURY has seen a revolution in 'family planning'. People can regulate the number and spacing of their children by using one or more of the wide range of effective methods of contraception now available (Table 28.5).

Table 28.5 **The efficiency of the commoner methods of contraception**

Method	Mode of action	Failure rate (pregnancies per 100 women/year)
The pill	prevents ovulation	0–3
IUD (coil)	prevents implantation	0.5–6
Condom	prevents sperm from reaching cervix	3–20
Diaphragm + spermicidal gel	prevents sperm from reaching cervix, gel kills sperm	3–25
Spermicidal gel alone	kills sperm	3–30
Coitus interruptus	penis withdrawn before ejaculation	10–40
The rhythm method	abstain from sex near to time of ovulation	15–35
Vaginal douche	washes sperm out of vagina	80
Male sterilisation (vasectomy)	vas deferens cut so that semen contains no spermatozoa	0–0.15
Female sterilisation (tubal ligation)	prevents egg from passing into uterus	0–0.05
Nothing	no contraception at all – a control for the other methods	85

Hormones and contraception

'The pill'

The pill – or, to use its full name, the **combined oral contraceptive pill** – has been used widely since the 1960s. It contains a mixture of artificially produced oestrogen and progesterone. The pill is effective because it raises the blood hormone levels to those encountered during the secretion phase of the menstrual cycle, or those of early pregnancy. The high level of progesterone inhibits the secretion of FSH, preventing the development of new follicles and therefore eggs cells. The high progesterone levels also inhibit menstruation, so the pill is usually taken for 21 days and then not taken for 7 days. This allows menstruation to take place but does not leave enough time for any follicles to develop. Strictly speaking, this is not normal menstruation but a 'withdrawal bleed' caused by the drop in progesterone.

The 'mini-pill'

This variation of the pill contains low doses of hormones, not enough to prevent ovulation, but enough to change the consistency of the mucus at the cervix. Instead of becoming permeable to sperm at ovulation, the mini-pill causes continued secretion of the thick, non-ovulatory mucus which prevents sperm getting through.

Injections and implants

Injections and implants both contain progestogen, an artificial form of progesterone. Injections are intra-muscular – usually in the buttock – after which the hormone is slowly released into the bloodstream. Progestogen acts as a contraceptive by inhibiting FSH, although the cervical mucus is also thickened.

Implants, such as Norplant, come in the form of six small tubes that are surgically placed under the skin of the upper arm. The progestogen is released slowly and lasts for five years. In this case, the contraceptive effect is mainly due to an alteration of the thickness of the mucus at the cervix, although there may be some inhibition of FSH.

SUMMARY

After studying this chapter, you should understand and be able to answer questions on the following aspects of reproduction.

■ **Asexual reproduction** usually involves one individual. All offspring are genetically identical: they are **clones**.

■ **Sexual reproduction** involves two sexes which produce **haploid gametes**. Males produce **spermatozoa** by **spermatogenesis**, females make **ova** (egg cells) by **oogenesis**. Both processes involve a special type of cell division called **meiosis**.

■ **Fertilisation** can be either external or internal. Generally, aquatic animals can fertilise externally but terrestrial animals (insects, reptiles, birds and mammals) must fertilise internally by mating. Male mammals have a penis, which allows the sperm to be delivered inside the female's body.

■ At fertilisation, a new **diploid zygote** is formed. The zygote, which is genetically unique, grows and develops by a series of mitotic divisions.

■ The timing of reproductive events in humans is controlled by hormones. The **hypothalamus** controls the **pituitary gland**, which releases **gonadotrophins**. These stimulate the **gonads** to secrete steroids: **oestrogen**, **progesterone** and **testosterone**.

■ The human menstrual cycle lasts, on average, for 28 days. The cycle begins with **menstruation**, which occurs as the next ovum (egg cell) is prepared inside the ovary. By day 14 a new endometrium has grown. Ovulation takes place on day 14 and the egg cell is then available to be fertilised. If the egg cell is not fertilised the endometrium breaks down and the cycle repeats.

■ In humans, the egg cell is fertilised in the Fallopian tube. The **zygote** begins mitotic divisions, becoming a **morula** and then a **blastocyst** as it moves along the tube towards the uterus. When the blastocyst has implanted in the endometrium, it is known as an **embryo**.

■ The developing embryo (which is called a **fetus** after about 8 weeks) is nourished via the placenta, a temporary organ that does the job of the fetal lungs, intestines and excretory system.

■ Birth is brought about by a series of hormonal changes that begin uterine contractions. In mammals, the infant is nourished by milk produced by the mother.

QUESTIONS

1

a) (i) Describe the changes that occur in the uterus endometrium during a normal menstrual cycle.
(ii) Explain how progesterone is involved in the control of these changes in the endometrium.

b) (i) Explain what is meant by hormone replacement therapy (HRT), and
(ii) describe the circumstances in which it might be used.

[UCLES Autumn 1995 Modular Biology: Growth, Development and Reproduction Paper, q. 1]

2

a) Copy and complete the table which refers to hormones controlling the mammalian oestrous cycle.

Hormone	Site of production	Effect
FSH	anterior lobe of pituitary gland	
	ovary	repair of uterine lining
	anterior lobe of pituitary gland	ovulation
progesterone		maintenance of uterine lining

b) In some mammals changes in day length (amount of daylight) stimulate the release of reproductive hormones and the onset of a breeding season.

Outline how the nervous system is involved in detecting changes in daylength and co-ordinating the release of reproductive hormones.

[NEAB June 1995 Modular Biology: Processes of Life Paper, q. 7]

3

Fig 28.Q3 shows the structure of a human sperm.

a) (i) Name the parts **A** to **F**.
(ii) Apart from **B** to **E**, sperm contain few other cytoplasmic organelles. Suggest why this is so.

Fig 28.Q3

b) Outline the main events that take place from when a sperm first reaches an egg until a zygote is formed.

c) Suggest why very large numbers of sperm are produced yet only one is involved in fertilisation.

[UCLES 1996 Modular Biology: Growth, Development and Reproduction, Specimen Paper, q. 2]

4 The graph shows the concentration of the hormones progesterone and human chorionic gonadotrophin (HCG) in the blood during the early stages of a human pregnancy.

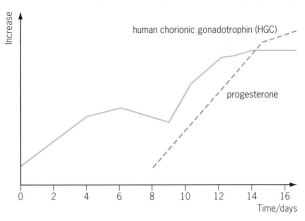

a) Describe how the progesterone curve would differ if pregnancy had not occurred.

b) Name the site of secretion of progesterone during **(i)** the period shown in the graph; **(ii)** the last 3 months of the pregnancy.

c) **(i)** Suggest the main function of human chorionic gonadotrophin in early pregnancy.
(ii) Give evidence from the graph to support your answer to (c)(i).

[AEB 1996 Biology: Specimen Paper 1, q. 13]

5 When the Pill gets under your skin

When news of a new contraceptive method reached British women in 1993, family planning organisations were flooded with enquiries. The contraceptive is an implant called Norplant. Not everyone, however, is thrilled by Norplant. Organisations around the world are concerned that it might be used as a method of social control.

Norplant delivers in a new way. It consists of six capsules, each 34 millimetres long. Each capsule contains 38 milligrams of a synthetic progesterone hormone. This hormone thickens the mucus produced by the cervix (neck) of the uterus. It also inhibits the production of LH (luteinising hormone).

A health worker inserts the contraceptive capsules through an incision on the inside of the upper arm and they remain under the skin for five years, steadily releasing progesterone into the bloodstream.

(Adapted from an article in *New Scientist*)

a) Suggest how the thickening of the mucus produced by the neck of the uterus might help to prevent contraception.

b) **(i)** Describe the role of LH in the menstrual cycle and then explain how the inhibition of LH production prevents contraception.
(ii) Hormone levels are often affected by negative feedback processes. Explain as full as you can what is meant by negative feedback.

c) Suggest and explain the advantages and disadvantages of using Norplant as a contraceptive.

[NEAB Feb 1995 Modular Biology: Life Processes Paper, q. 8]

Assignment

INFERTILITY TREATMENT

Nine out of ten couples who use no contraception and actively try to conceive, are successful within a year. But, one in ten couples may face the distressing problem of infertility. Many are 'sub-fertile' rather than infertile, and can be helped by the various methods described in this Assignment. However, for a few, nothing works.

In this exercise we look at the possible causes of infertility and at some of the treatments. When you have read both and worked through the questions, you should be in a position to prescribe the best course of treatment for a particular couple.

1 Causes of infertility

Doctors accept that a couple may need infertility treatment if they have been trying to conceive for at least a year without any success. The first objective is to establish whether either partner has an obvious problem which is preventing conception.

Female infertility may be caused by:

- Blocked Fallopian tubes.
- Altered hormone levels leading to a failure to ovulate or implant. Failure to ovulate, known as **anovulatory infertility**, is usually due to a failure to secrete the right balance of hormones.
- Cervical mucus that halts, repels or kills sperm.

1

a) How would a specialist find out if a woman's Fallopian tubes were blocked?
b) Why might blocked Fallopian tubes cause infertility?
c) Which hormones combine to cause ovulation?
d) What should happen to the cervical mucus around the time of ovulation?

Male infertility may be caused by:

- A low sperm count. Samples which are found to have fewer than 20 million sperm per cm^3 are said to be abnormally low.
- Production of large numbers of abnormal sperm (more than 4 per cent).
- Production of antibodies that make the sperm stick together.

2

Men who wear tight-fitting clothes, or who spend a lot of time in a hot bath, have sometimes been found to have a low sperm count. Suggest a reason for this.

Treatment methods: ovulation induction

A woman may not ovulate because the balance of hormones in her body is abnormal. To restore the hormone levels and initiate follicle development and ovulation, she can have treatment with artificial gonadotrophins or drugs that stimulate the natural secretion of gonadotrophins. One such drug, clomiphene, works by increasing FSH secretion.

After such treatment, the response of the ovaries can be followed by ultrasound, which shows how many follicles are developing in each ovary. A follicle is considered to be ready for ovulation when it reaches 17 mm in diameter. The endometrium is also checked – it should be at least 8 mm thick at this time.

When a ripe follicle is detected, ovulation can be stimulated artificially by injecting human chorionic gonadotrophin (HCG), a hormone that has a similar effect to LH. Ovulation should occur after about 36 to 48 hours, and so intercourse should be timed to coincide with this.

3

a) Why will increased FSH secretion help a woman who isn't ovulating regularly?
b) What is a potential problem with treatment which stimulates ovulation? Hint: this may not be a problem if the couple want a large family.

Treatment methods: in vitro fertilisation (IVF)

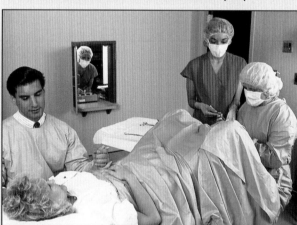

Fig 28.A1 **A woman undergoing IVF treatment must endure weeks of hormone treatments, followed by the uncomfortable process of egg cell collection. Up to about 20 egg cells may be recovered. Her egg cells are fertilised by her partner's sperm 'in vitro' and, when the tiny embryos have developed, two of these are put back into her uterus. IVF has, at best, about a 30 per cent success rate**

In vitro (= 'in glass') fertilisation is what most people think of as 'test-tube baby' treatment. In IVF, the egg cells and sperm are taken from the couple and fertilised in a dish. The process of fertilisation normally takes 12 to 15 hours and after this the new embryos start to develop. Cell division is taken as a sign that fertilisation has been successful. Then, the tiny embryos (no more than balls of 8 to 16 cells at this stage – see Fig 28.A2) are placed into the uterus.

4 How often does an ovary normally produce an ovum?

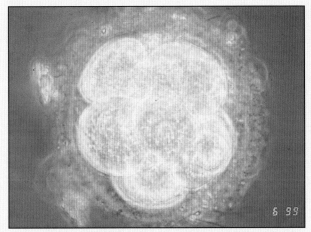

Fig 28.A2 **A human embryo at the 8-cell stage**

Treatment methods: gamete intra-Fallopian transfer (GIFT)

GIFT involves stimulating the ovaries and collecting the ova in much the same way as in IVF treatment. The important difference is that in GIFT the egg cells are mixed with sperm and immediately introduced into the Fallopian tubes (Fig 28.A3), without waiting to see if fertilisation occurs. The advantage of this procedure – which would seem to be less controlled than IVF – is a 5 per cent better success rate. This is possibly because the Fallopian tube is the natural site for fertilisation: it may secrete chemicals which stimulate the process.

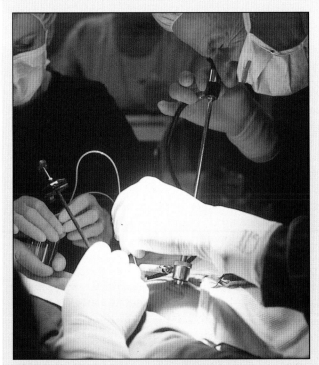

Fig 28.A3 **A mixture of egg cells and sperm is being inserted into the woman's Fallopian tube through a small insertion**

5 Many specialists feel that counselling couples is an important part of IVF and GIFT treatment. Think of two reasons why these treatments might cause anxiety.

Treatment methods: intra-uterine insemination (IUI)

In around 20 per cent of infertility cases, there seems to be no problem with either partner. Surprisingly, in some of these cases, intra-uterine insemination proves successful. The basic idea behind IUI is to introduce the partner's semen into the uterus.

The steps in IUI are as follows:
* Follicle development is stimulated and monitored.
* Treatment to induce ovulation is given.
* A fresh sperm sample is introduced into the uterus.

6 What advantage does IUI give over trying to conceive naturally?

After working through the Assignment so far, you should now be able to tackle the following question:

7 Copy and complete the following table, matching up the causes of infertility to suitable treatments. Remember that there may be more than one suitable treatment.

Your options are:

OI – ovulation induction

IVF – in vitro fertilisation

GIFT – gamete intra-Fallopian transfer

IUI – intra-uterine insemination

Problem	Possible treatments
Blocked Fallopian tubes	
Hostile cervical mucus	
Low sperm count	
Woman has antibodies against partner's sperm	
Many abnormal sperm	

8 A programme of infertility treatment can put a great strain on a relationship. Suggest reasons for this.

9 Hormone replacement therapy can be used to allow post-menopausal women to become pregnant. A donated egg can be fertilised by sperm from the woman's partner, and can lead to women as old as 60 bearing children. Discuss the advantages and disadvantages of this technology.

29 Reproduction in flowering plants

This looks like a clump of trees, but they are all joined together below ground level. There is a single root system and lots of tree trunks. Because all the trunks and roots are genetically identical and still connected, botanists argue that an aspen forest is a single organism

IF YOU WERE to ask people, 'What is the largest living organism in the world?', what answer would you expect? High on the list of suggestions would be the blue whale, weighing in at around 180 000 kilograms (far larger than any dinosaur ever was). But the 'California Redwood' or giant sequoia can weigh much more. The largest individual redwood, the 'General Sherman', was last estimated to be 83 metres high and 11 metres in diameter at its base, and to weigh just under 2 million kilograms: the equivalent of about 11 blue whales growing out of the ground.

It's hard to imagine that anything living could be bigger, but the quaking aspen *Populus tremuloides*, a flowering plant, might surprise us all. Although they grow vertically, aspens also spread out vegetatively, rather like grass and brambles. As an aspen spreads, it forms a clone of genetically identical shoots: but instead of sending out small plants, the aspen grows a collection of enormous tree trunks. In America, in the state of Utah, one individual aspen seems to have covered an area of 43 hectares forming over 47 000 trunks. Together, these have an estimated weight of over 6 million kilograms, making this particular aspen the largest known organism in the world.

1 SEXUAL REPRODUCTION IN FLOWERING PLANTS

Any organism that reproduces sexually transfers its genetic material – not an easy task for an organism that cannot move, such as a plant. Plants have evolved many wonderful strategies to make sure that their genes pass to the next generation. Sexual reproduction allows a plant species to increase in number and it also introduces genetic variation. Without this variation, plants could not evolve and adapt to changes in their environment.

In this chapter we start with the events of sexual reproduction in flowering plants (Fig 29.1 provides a useful overview). We then study individual stages in more detail; we look at reproductive strategies and at the way flowering plants control the timing of flower and seed production. The chapter ends with a brief description of asexual reproduction.

Sexual reproduction in flowering plants has many stages. In the first, flowers are formed that have male and female parts. Unlike the sex organs of animals, which straight away form gametes, plant sex organs first produce male and female **spores**. The male parts, the **stamens**, make male microspores, more usually called **pollen**. The female parts, the **ovules**, form female spores called **megaspores**.

Pollination occurs when pollen that is released from the stamens reaches the female plant parts. The pollen grain produces **sperm**

Like other organisms in the Kingdom Plantae (see Chapter 13), flowering plants show **alternation of generations** (page 204). The independent plant stage is the **sporophyte**, which produces **spores** by meiosis. These spores develop into the **gametophyte**. In flowering plants, the gametophyte stage is hardly noticeable: the germinated pollen grain and the female embryo sac produce the male and female gametes. At fertilisation, two gametes come together and form an embryo sporophyte plant.

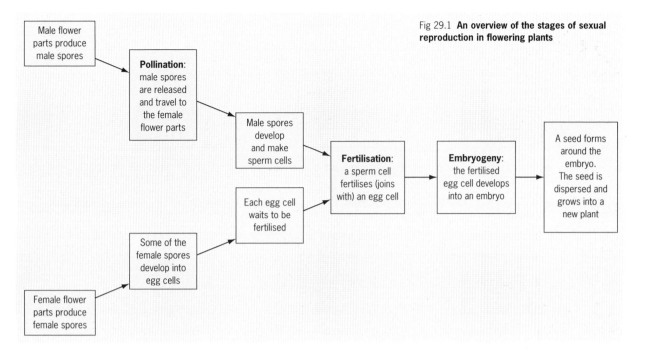

Fig 29.1 **An overview of the stages of sexual reproduction in flowering plants**

nuclei, which pass through a **pollen tube** to the **egg** that forms inside the female megaspore. When sperm meets egg, **fertilisation** occurs. A fertilised egg grows into an **embryo** plant, enclosed as a **seed**. The seed protects the embryo, supplies it with food and often carries it to a new site, where it can grow into a new plant.

Flower formation

The structure of a generalised flower is shown in Fig 29.2. All flowers are made up of a few basic parts, often arranged in rings or **whorls** at the end of a flower stalk, or **axis**. The centre of the flower contains the reproductive structures, sometimes either male or female, but usually both. These are surrounded by the protective envelope, the **perianth**, which consists of **petals** and **sepals**. In some plants, the petals and sepals look identical: in this case we call both structures **tepals**.

Some flowers have not changed much since flowering plants first evolved. Flowers of primitive species such as the protea (Fig 29.3) usually have many petals and sepals. The numbers of sepals and

Flowering plants make up a phylum of plants called the **Angiospermophyta**, commonly called the angiosperms (see Chapter 13). They have evolved to be the most advanced type of plant. The word angiosperm means 'hidden seed': the seeds of angiosperms are hidden inside the ovary, not exposed as in gymnosperms (eg conifers).

Fig 29.3 **Protea is a primitive flower. The petals and sepals, which are quite difficult to tell apart, have evolved from the leaves that surround the flower. This large South African flower is normally pollinated by sugarbirds**

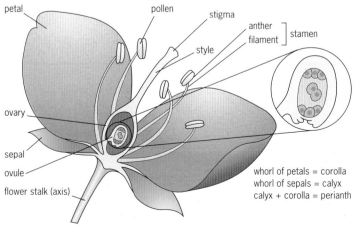

whorl of petals = corolla
whorl of sepals = calyx
calyx + corolla = perianth

Fig 29.2 **The main parts of a generalised flower**

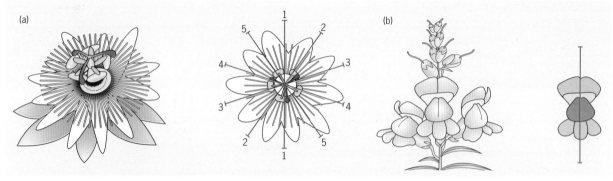

(a)

(b)

Fig 29.4(a) **A flower with more than one plane of symmetry is said to be actinomorphic. This passion flower has five main planes of symmetry**

Fig 29.4(b) **A flower, like this snapdragon, that has a single plane of symmetry is said to be zygomorphic**

A **diploid** cell has two of each chromosome. In mitosis, one diploid cell divides to give two diploid daughter cells. These are genetically identical to each other and to the parent. In meiosis, one diploid cell divides to give four **haploid** cells, cells with half the number of chromosomes of the parent. Meiosis also involves genetic 'mixing', so that each of the haploid cells produced are genetically different from each other and from the original parent. See Chapter 6 for more detailed information about mitosis and meiosis.

petals present varies from flower to flower. More primitive flowers are usually **radially symmetrical**, or **actinomorphic**, as shown in Fig 29.4(a).

Other species have evolved quite a lot. Clover flowers, for example have fewer petals, and the number is constant from flower to flower. Flowers that have evolved for insect or bird pollination (see page 470) are often **bilaterally symmetrical**, or **zygomorphic**, as shown in Fig 29.4(b). Although insect pollination is an important evolutionary step, many highly advanced plants, such as grasses, remain wind pollinated.

Reproductive structures: the stamen

The size and shape of the reproductive parts of a flower vary between species, but both male and female parts always show the same basic structures. Fig 29.5 shows the structure of the male part, the **stamen**.

Within the pollen sacs of the stamen, each *diploid* pollen mother cell divides by *meiosis* to form four *haploid* pollen cells. These mature into pollen grains (Fig 29.6) using the nutrients released as the surrounding tissue breaks down. When the pollen is ripe, the sacs split open and the pollen is released. By this point, the *haploid* pollen cell nucleus has usually divided once more, by *mitosis*, to form a haploid **generative nucleus** and a haploid **vegetative nucleus**. The generative nucleus forms the two **male gametes**, the **sperm cells**.

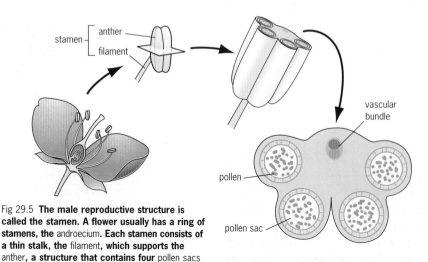

Fig 29.5 **The male reproductive structure is called the stamen. A flower usually has a ring of stamens, the** androecium. **Each stamen consists of a thin stalk, the** filament, **which supports the** anther, **a structure that contains four** pollen sacs

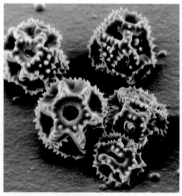

Fig 29.6 **Dandelion pollen grains. Most pollen has a highly sculptured surface. It is often possible to identify a plant from a single grain of its pollen**

Reproductive structures: the ovule

Fig 29.7 shows the female structure, the **carpel**. Part of the carpel is stretched out into a long **style**. At the end of the style is a special surface called the **stigma** which accepts pollen. The swollen part of the carpel is the **ovary**. Each ovary contains one or more **ovules**, depending on the plant.

The way the ovule develops is shown in Fig 29.8. Each ovule is made up of a rugby ball-shaped **nucellus**, surrounded by two layers of cells called **integuments**. Just one of the cells inside the nucellus becomes the **megaspore mother cell**. This divides by meiosis to give four haploid **megaspores**. Three do not develop. The surviving cell becomes the megaspore that develops into the **embryo sac**.

Inside the embryo sac there are further nuclear divisions but the cytoplasm does not divide to form separate cells (ie there is no cell wall formation). Eventually, the embryo sac contains eight nuclei, three of which play a key role in fertilisation. Two of them, called polar nuclei, fuse together to form a diploid **central cell**. The third nucleus forms the **egg cell**. This is the female gamete.

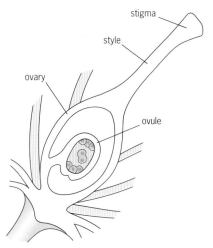

Fig 29.7 **The female reproductive part of the flower is the carpel which consists of one or more ovaries. This carpel has just one ovary which contains a mature ovule and is ready for fertilisation**

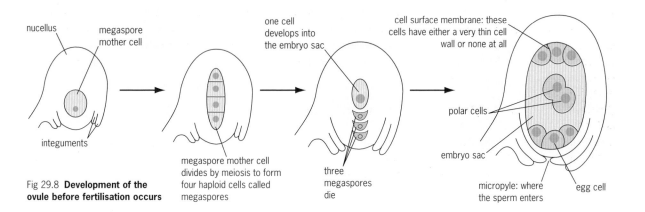

Fig 29.8 **Development of the ovule before fertilisation occurs**

Pollination

The reproductive structures of the flowering plant have now formed. As plants cannot move to find a mate, they must transport their pollen, often over large distances, to make contact with a female plant from the same species. Once the pollen has made it to the stigma, the male gamete must make a smaller but equally crucial journey to meet the female gamete within the ovule.

Most plants are pollinated either by wind or by insects, but there are other **vectors** (pollen carriers), including bats, mice, slugs, birds, small reptiles and water.

A The flowers of protea in Fig 29.3 are large and quite robust. Suggest why.

Wind pollination

Wind pollination, or **anemophily,** is the most straightforward method of pollination. The anthers of a plant ripen in dry weather and release pollen grains into the air. Pollen often spreads to a radius of 5000 metres around the parent plant, and some pollen grains are known to have travelled thousands of miles. The chance of any pollen grain landing on a stigma of the right plant is slim, so large amounts of pollen are produced.

Wind pollination works well for plants that exist in large numbers in a particular area. Grasses and many woodland trees, such as oak,

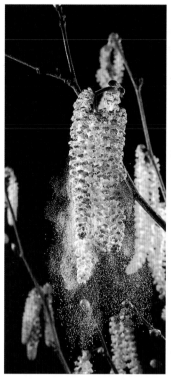

Fig 29.9(a) **Hazel produces clusters of tiny flowers called catkins early in the year, so pollen grains do not get trapped by the leaves which develop later**

Fig 29.9(b) **Maize is a very large grass: its huge tassels are amongst the largest stigmas in the plant kingdom**

Fig 29.10 **Some flowers have inflorescences which, at first glance, look like a single flower. Outer showy flowers attract pollinators to the mass of smaller flowers crowded onto the flat disc**

birch and hazel, are successful wind-pollinated plants. Their flowers are not 'showy': they do not need to be. Instead, they have large feathery stigmas that stick out from the flower to catch pollen, as seen in Figs 29.9(a) and (b).

Insect pollination

Insect-pollinated (or **entomophilous**) plants are pollinated by insect vectors (carriers), rather than the wind. The shape of their flowers has evolved to ensure that pollen brushes off onto the visitor and is later transferred to a stigma when the insect visits another flower. Such flowers need to be seen and so they advertise themselves with large white or brightly coloured petals. But, if the plant invests large amounts of energy in making large petals, it has fewer resources available to make gametes. So, many plants make 'cheaper', smaller flowers and crowd them together in clusters, called **inflorescences** (Fig 29.10).

Pollen grains are rich in oils and proteins, and producing large amounts drains resources from the plant. Insect-pollinated flowers tend to produce small numbers of larger and stickier pollen grains. Insect pollination is more efficient than wind pollination because less pollen gets wasted and so less is needed. Since pollen itself is nutritious, a plant may have to sacrifice some of it to the insect as a reward for successful pollination. But some plants have found that a more economical bribe of sugary water, **nectar**, can be just as effective. If insects get plenty of nectar, they are more likely to visit another flower of the same type. Both insect and plant profit. This mutual benefit is the basis of many plant–insect relationships (see the Feature box on page 472).

Fertilisation

Pollen lands on the stigma; this marks the completion of pollination and the start of fertilisation. The process of fertilisation is shown in Fig 29.11. The pollen grain draws water out of the stigma by osmosis and swells. This bulge develops into a **pollen tube**, which penetrates the style and grows towards the ovule.

Osmosis is the movement of water from a more dilute to a more concentrated solution, across a semipermeable membrane.

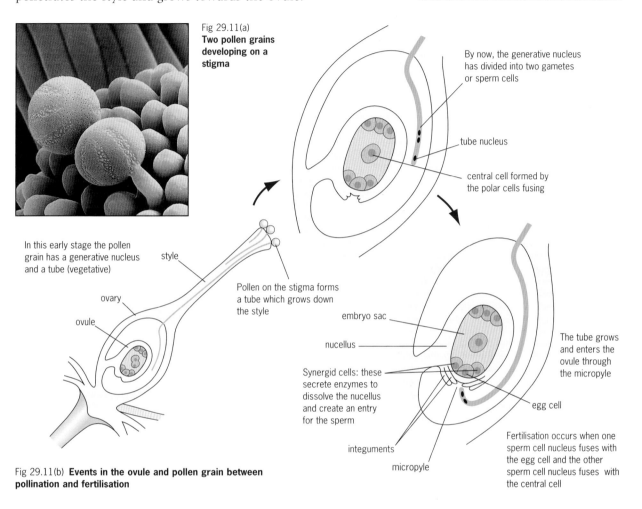

Fig 29.11(a)
Two pollen grains developing on a stigma

By now, the generative nucleus has divided into two gametes or sperm cells

tube nucleus

central cell formed by the polar cells fusing

In this early stage the pollen grain has a generative nucleus and a tube (vegetative)

style

ovary

ovule

Pollen on the stigma forms a tube which grows down the style

embryo sac

nucellus

Synergid cells: these secrete enzymes to dissolve the nucellus and create an entry for the sperm

The tube grows and enters the ovule through the micropyle

egg cell

Fertilisation occurs when one sperm cell nucleus fuses with the egg cell and the other sperm cell nucleus fuses with the central cell

integuments

micropyle

Fig 29.11(b) **Events in the ovule and pollen grain between pollination and fertilisation**

Each pollen grain contains two nuclei, the generative nucleus and the tube (or vegetative) nucleus. The tube nucleus directs the growth of the pollen tube. The generative nucleus forms two sperm cells. The nuclei are not separated from each other by cellulose walls: they simply share the cytoplasm, so they do not really occupy separate cells. The sperm and tube nuclei move down the pollen tube as it grows and they enter the ovule through a gap, the **micropyle,** to reach the embryo sac.

The embryo sac opens, the sperm nuclei enter, and then a **double fertilisation** occurs. In one fertilisation, one of the sperm nuclei fertilises the egg cell. The product of this fusion becomes the zygote, the diploid cell that develops into the embryo. In the second fertilisation, another sperm fuses with the central cell. This fusion produces a **triploid** cell (it has three copies of each chromosome: two from the diploid central cell and one from the haploid sperm). This triploid cell divides by meiosis to form the endosperm tissue. The endosperm contains the food supply that will nourish the growing embryo. Sexual reproduction in flowering plants always involves a double fertilisation.

D Most of the cells of a plant are diploid: they contain two sets of chromosomes. How many sets of chromosomes are in **(a)** the sperm cell, **(b)** the central cell and therefore **(c)** the cell produced when the sperm and central cell fuse to form the endosperm tissue?

Fig 29.12 is an overall summary of spore production, gamete production and fertilisation in a flowering plant. After fertilisation in most plants, the flower withers and the ovary swells to form the fruit containing the seeds. Inside each seed, the embryo develops.

Fig 29.12
Summary of the life cycle of a flowering plant

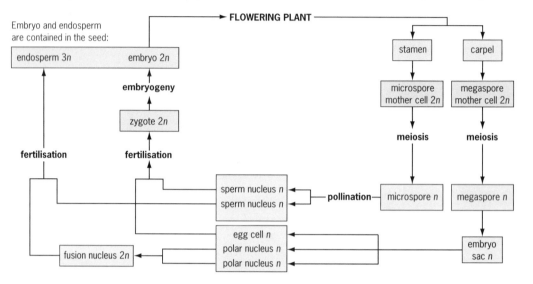

THE INTIMATE RELATIONSHIP BETWEEN PLANTS AND THEIR POLLINATORS

INSECTS AND FLOWERING PLANTS have evolved together. Some species are now completely dependent on each other: without the plant the insect doesn't get food, and without the insect the plant cannot reproduce. Most plant–insect relationships are not as narrow as this: most plants attract a range of potential pollinators. For each species the range is never very great, because particular colours, shapes, scents, opening times, flowering seasons and so on all tend to be attractive to particular groups of insects.

Red clover, for example, must be **cross-pollinated** (pollinated by pollen from another plant) before it can produce seeds. It is pollinated by only a handful of different insects, but bees are the most important.

Red clover is adapted to pollination by bees in several ways. As Fig 29.13(a) shows, the flowers develop as inflorescences. Each flower in the cluster has a long tube with a tiny hole at the top that gives access to the nectar, as shown in Fig 29.13(b). A visiting bumble bee inserts its proboscis and long tongue into the tube to feed on the nectar. As the bee does this, it brushes its underside against the anthers and picks up sticky pollen. When the bee visits another red clover flower, its underside brushes against the stigma, some of the pollen is transferred, and so pollination is accomplished.

The tiny access hole and the length of the tube make it hard for other insects to reach the nectar. Alternatively, the shape of the flower may prevent them from landing on it. Honey bees, which have slightly shorter tongues, can reach the nectar when the level is high enough. If not, they may adopt a 'smash and grab' approach, as in Fig 29.13(c), where the bee bites a hole in the flower near the base and sucks out the nectar.

Fig 29.13(a) **An inflorescence of red clover, *Trifolium pratense*. Individual parts of the cluster are small, but the inflorescence is easily spotted by a flying bee**

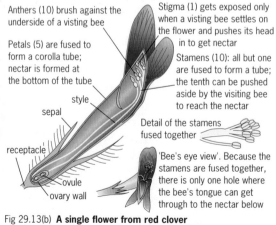

Anthers (10) brush against the underside of a visiting bee

Petals (5) are fused to form a corolla tube; nectar is formed at the bottom of the tube

style

sepal

receptacle

ovule

ovary wall

Stigma (1) gets exposed only when a visiting bee settles on the flower and pushes its head in to get nectar

Stamens (10): all but one are fused to form a tube; the tenth can be pushed aside by the visiting bee to reach the nectar

Detail of the stamens fused together

'Bee's eye view'. Because the stamens are fused together, there is only one hole where the bee's tongue can get through to the nectar below

Fig 29.13(b) **A single flower from red clover**

Fig 29.13(c) **A honey bee steals nectar from a field bean flower. The flower tube is longer than the bee's proboscis so it takes a short cut and pierces the bottom of the flower. Short-tongued bees rob red clover flowers in the same way**

Development of the embryo: embryogeny

At fertilisation, the egg and sperm nuclei fuse to form a diploid nucleus, the **zygote**, which develops into an embryo plant (Fig 29.14). This consists of a tiny root, or **radicle**, connected to the young shoot, the **plumule**, which later develops leaves. The first leaves, the **cotyledons**, are different from all later leaves. They are usually a different shape and have a simpler network of veins.

When the central cell is fertilised it divides to form a special triploid tissue called the **endosperm**, which nourishes the embryo. In some plants, such as the maize in Fig 29.15(a), endosperm accumulates and is stored inside a **seed**, ready to feed the young plant when the seed **germinates**. In other species, such as peas (Fig 29.15(b)) and peanuts, the endosperm food reserves are used by the embryo at a much earlier stage and any spare food is transferred and stored within the cotyledons.

Fruits and seeds

Once the seed is released from the parent it can develop into a new plant. But most seeds don't start growing, or germinating, straight away. If seeds are **dispersed**, the parent is not overcrowded by offspring competing for light, water and nutrients.

Seeds survive dispersal because the **testa**, a tough coating, prevents them from drying out or being digested by animals that eat them. The testa forms when the outer part of the ovule is strengthened by the polysaccharides **cutin** and **lignin**. The testa is sometimes so tough that the seed needs harsh treatment, such as freezing, soaking or a period of wear and tear in the soil, before it can germinate. This ensures that seeds do not germinate until the following year, long after the winter cold, or the summer drought.

Some seeds are light enough to be carried on the wind, like pollen. But most seeds need help to disperse. This help comes from the **fruit** that surrounds them. In some plants, the ovary wall develops into a fleshy, soft fruit, designed to entice animals to eat it. After digestion, the seeds pass out in the faeces, perhaps many miles away, still protected by the testa. Other fruits have explosive mechanisms to throw the seeds about, and others have hooks to catch on animals' feet or fur to 'hitch a lift'.

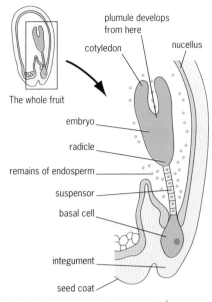

Fig 29.14 **The very early embryo often appears round (but never forms a hollow ball of cells like an animal blastula). Soon the first leaves, the cotyledons, and the young root, the radicle, start to take shape, as in this shepherd's purse embryo. The embryo grows as the seed coat develops and the fruit ripens**

?

E Are peas monocots or dicots? (See Fig 29.15 and page 212 for clues.)

F Why are many fruits brightly coloured?

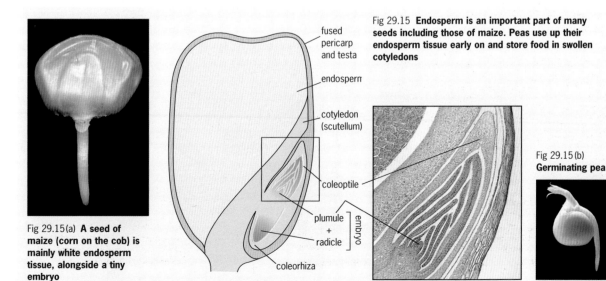

Fig 29.15(a) **A seed of maize (corn on the cob) is mainly white endosperm tissue, alongside a tiny embryo**

Fig 29.15 **Endosperm is an important part of many seeds including those of maize. Peas use up their endosperm tissue early on and store food in swollen cotyledons**

Fig 29.15(b) **Germinating pea**

2 PLANTS AND SEX: STRATEGIES IN REPRODUCTION

Sex in plants and animals is vital for two reasons: it creates new offspring and it shuffles the genes, giving variation. If plants remained genetically the same they would never evolve.

But, as we have seen, sex can be difficult for plants. Unlike animals, they can't chase after a mate. So a plant that depends on a 'mate' for breeding will lose out if it grows too far away from others of the same species. Because it is better to have some offspring instead of none at all, many plants in this situation pollinate and fertilise themselves.

Self-fertilisation leads to less variation among the offspring, so this creates a dilemma. If self-pollination is too easy, a plant might end up doing this all the time, and so miss out on the chance to shuffle its genes with others. To ensure cross-fertilisation and the advantages that this brings, many plants have developed strategies that make self-fertilisation less likely than 'mating' – but not completely impossible, just in case.

Fig 29.16 **In this plantain flower, the stigmas at the top are ready to receive pollen but the stamens round them are not fully ripened**

Careful timing

Many plants have a definite flowering season. Within that season, individual plants produce the stamens and carpels of their flowers at slightly different times to increase the chance of cross-pollination (Fig 29.16). In the system called **protandry**, the stamens develop first and pollen is shed from the anthers before the stigma of the flowers on the same plant can accept it. When the female parts develop before the male, this is **protogyny**.

Different flower shapes

G Why is it important that thrum and pin flowers are produced on different plants?

Primroses make two types of flower, the **thrum** and **pin** forms, though never together on the same plant: this is known as **heterostyly**. Thrum and pin flowers differ in the heights of the stamens and the stigma within the flower (Fig 29.17). Insect visitors to thrum flowers pick up pollen at a point on their proboscis that is level with the stigmas in pin flowers. Similarly, pollen from pin flowers is more likely to be transferred to stigmas in thrum flowers.

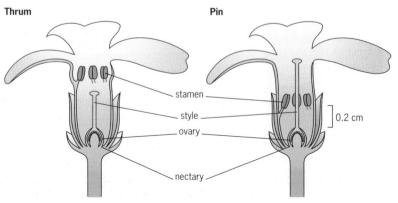

Fig 29.17 **Structure of the thrum and pin primrose flowers**

Keeping the sexes separate

Some species have separate male and female flowers, but both may grow on the same plant (Fig 29.18). Other species are **dioecious**: they have separate sexes, so self-pollination is impossible (Fig 29.19).

Plants can be choosy about the pollen they accept

Even if a plant species does pollinate itself, it can make self-fertilisation more difficult. In some plants, special genes slow up the rate at which the pollen tube can grow. Pollen from a genetically different parent may develop more rapidly and overtake a tube that is growing from its own pollen.

Recently, researchers have also found that some plants can recognise the pollen of *related* plants, as well as their own pollen, and also prevent it from growing properly. This **kin recognition** enables them to avoid self-fertilisation and fertilisation by close relatives.

Fig 29.18 **Cucumbers have separate (larger) male and (smaller) female flowers on the same plant. Insect visitors have to travel from flower to flower, so cross-pollination is made more likely**

Fig 29.19 **This common woodland plant, dog's mercury, has plants that are of separate sexes, so cross-pollination is assured. The photograph on the left shows male flowers; the photograph on the right shows female flowers**

3 HOW PLANTS KNOW WHEN TO FLOWER

Plants don't all flower at the same time. Each species has its own flowering season, sometimes quite brief. So how do plants know when to start flowering? The key is a light-sensing mechanism that involves a pigment called **phytochrome**. The phytochrome molecule is a protein that exists in two forms, called **photoisomers**, that are converted one to another by light (Fig 29.20).

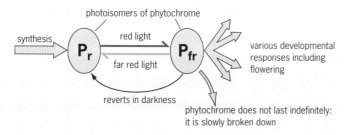

Fig 29.20 **Summary of changes in phytochrome brought about by light**

Phytochrome is created in the plant in one form, P_r. If the P_r form of the protein is exposed to red light, it is converted to the isomer P_{fr}. P_{fr} is changed by far red light back into P_r. (Far red light is light just bordering on the infra red.) Daylight contains more red light than far red. So plants exposed to sunlight contain only the P_{fr} form of phytochrome. In darkness, P_{fr} slowly changes back to P_r.

When the nights are long there is plenty of time for all the P_{fr} to decay to P_r. So in winter, at dawn, all the phytochrome would be in the P_r form. During summer, the nights are short, and so there is less time for the P_{fr} to decay. At dawn in summer there may still be phytochrome in the P_{fr} form left over from the previous day. So the relative amounts of the two forms of phytochrome found in a plant depend on how long the day is and therefore on what time of year it is. Phytochrome can switch on or off a wide variety of processes of the plant's development, including flowering (Fig 29.21). The mechanism by which it does so is beyond the scope of this book.

Fig 29.21 **Commercial growers can sell plants like these chrysanthemums in full bloom out of season. Growers can induce flowering by controlling the daylight in the growing houses with lamps. This is possible because of the way light interacts with the phytochrome system to control flowering**

4 ASEXUAL REPRODUCTION

Many flowering plants can reproduce asexually as well as sexually. While sexual reproduction adds variety to the species, asexual reproduction allows plants to increase their numbers or to survive between growing seasons, without going through the complex processes of gamete formation, pollination, fertilisation and seed production. It gives them extra flexibility: they can still reproduce and live, even outside the flowering season.

Vegetative propagation

Many flowering plants can produce new plants without forming flowers and seeds. A single parent gives rise to offspring that are genetically identical, both to it and to each other. We call this type of reproduction **vegetative propagation**.

Along much of our coastline there are extensive sand dunes. Many of the dunes nearest the sea are held in place by marram grass. Only marram grass succeeds on the dunes where other plants become smothered by the shifting sand. Marram grass forms underground stems, **rhizomes**, which grow very rapidly and produce new roots and leaves at intervals (Fig 29.22). A single individual will quickly spread and colonise a newly formed dune.

Other plants reproduce asexually because they can regenerate and repair themselves: fragments of stems and leaves broken off by passing animals may fall to the ground, root and grow into new plants. Gardeners and farmers take advantage of this ability in the well-known practice of 'taking cuttings'. Many modern varieties of apple are maintained by taking cuttings and then grafting the young plants onto a special rootstock that keeps the tree short. Vines too are often grafted onto special roots that are resistant to *Phylloxera*, a small weevil that nearly ruined the wine industry in France in the 19th century, and that threatens to do the same in California 100 years on.

Plants can reproduce asexually by the way that they overwinter. During the growing season they develop specialised underground stems, or **rhizomes**, which they pack with stored food. When cold winter temperatures make the plant die back, the rhizomes remain alive and then use the stored food to produce new shoots in the next growing season. **Corms, bulbs** and **tubers** (Fig 29.23) are other modified shoots adapted for food storage over winter. All these structures branch from **buds** (the 'eyes' on potatoes are buds). Each bud can form a new individual the following season.

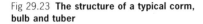

Fig 29.22 **Marram grass can grow in shifting sand. The lines of plants grow from a rhizome, a horizontal stem just below the surface**

Fig 29.23 **The structure of a typical corm, bulb and tuber**

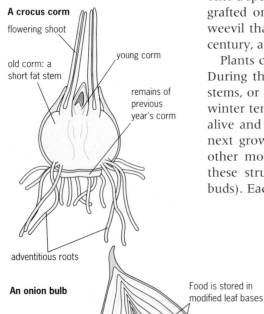

A crocus corm

flowering shoot

old corm: a short fat stem

young corm

remains of previous year's corm

adventitious roots

An onion bulb

apical bud

Lateral bud: this will grow into a new shoot which could form a separate bulb next season

Food is stored in modified leaf bases

outer papery protective leaves

short stem

roots

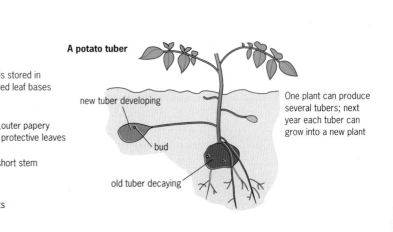

A potato tuber

new tuber developing

bud

old tuber decaying

One plant can produce several tubers; next year each tuber can grow into a new plant

SUMMARY

After reading this chapter, you should know and understand the following:

■ Sexual reproduction enables plants to pass their genes on, add to the number of the species, increase variation and evolve.

■ Flowering plants produce **spores** before they produce **gametes**. There are two types, **microspores** and **megaspores**.

■ Microspores or **pollen grains** are transferred by wind or an animal vector to the stigma of another flower, often on different plant of the same species.

■ The pollen grain grows into a tube that penetrates the female part of the flower. **Sperm nuclei** travel along this tube to meet the **egg**.

■ **Double fertilisation** occurs in plants: this leads to a **zygote** and a **triploid cell**. The zygote grows into an **embryo**, the triploid cell develops into a food store, the **endosperm**.

■ The **ovule** develops into a seed, which protects the embryo for a while. The **ovary** develops into a fruit.

■ There are a variety of means to promote cross-pollination.

■ Many plants can self-fertilise, but many have strategies to make sure that they cross-pollinate.

■ **Phytochrome** enables plants to determine the season and to flower at the correct time.

■ Many plants reproduce vegetatively, by **asexual reproduction**.

QUESTIONS

1 The diagram of Fig 29.Q1 shows a section through the ovary and pollen tube of a flowering plant just before fertilisation.

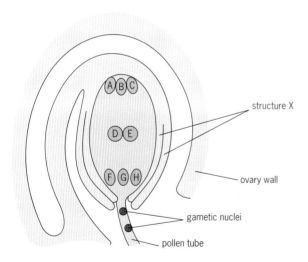

Fig 29.Q1

a) Which of the nuclei labelled **A** to **H** fuse with a gametic nucleus to form: **(i)** the endosperm; **(ii)** the zygote?

b) After fertilisation, into what structure does each of the following develop: **(i)** the ovary wall; **(ii)** the structure labelled **X**?

c) In this plant, the diploid number of chromosomes is 14. How many chromosomes would you expect to find in: **(i)** the nucleus labelled **C**; **(ii)** a nucleus in the endosperm?

[AEB 1996 Biology: Specimen Paper, q. 1]

2 The diagram of Fig 29.Q2 shows the structure of a cereal grain in section.

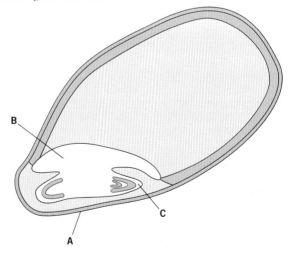

a) Name the parts labelled **A**, **B** and **C**.

b) State *one* function for each of the parts labelled **A**, **B** and **C**.

[ULEAC 1996 Biology/Human Biology: Specimen Paper, q. 2]

Assignment

POLLEN IN PEAT

Pollen grains have an outer wall that is resistant to decay – so resistant that pollen grains that are hundreds of years old can be found in peat and lake mud. Researchers can tell a plant species just from its pollen: by digging cores of peat or soil and studying the pollen, pollen analysts can build up a picture of the plants that were growing in that place hundreds or thousands of years ago. They have been able to piece together how vegetation in Europe has changed since the last ice age, 10 000 years ago.

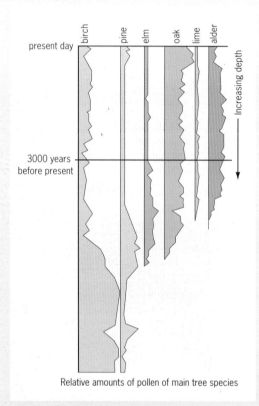

Relative amounts of pollen of main tree species

Fig 29.A1 **The pollen data given here is typical of many places in lowland Britain. It reflects the abundance of tree species since the ice retreated about 10,000 years ago**

a) What species of tree was most abundant during the coldest period immediately following the ice age?

b) Today, birch and pine trees can be found in exposed areas in some of the coldest parts of Britain, but trees such as elm, oak and lime cannot live in such conditions. Between 7500 and 5000 years ago the proportions of pollen changed considerably. So some plants must have done better and some fared worse.
Suggest a cause.

c) One tree species seems to have suffered a decline at a time when there was no evidence for a change in climate. It has been suggested that the tree might have been eaten by cattle, or perhaps the land where it grew was cleared for crops.
What tree species seems to have suffered a decline at the start of the bronze age?

Pollen grains can help archaeologists too

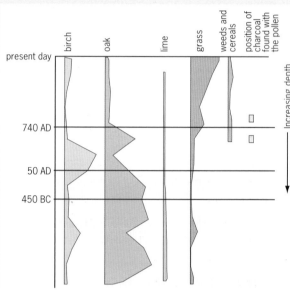

Relative amounts of pollen of selected species

Fig 29.A2 **This pollen diagram is from soil at a site where archaeologists were trying to understand the history of the area. From around 300 AD. there was a decline in the amount of oak pollen found at this site**

2

a) Look at Fig 29.A2. What does this suggest about the abundance of oak trees in the vicinity?

b) Do any other species seem to have the same pattern as oak?

3 Researchers had to consider several theories about the decline of oak, including:

1 It was due to a deterioration in the climate.

2 The land was cleared to make room for human settlement and cultivation.

3 Oak suffered from a disease.

a) What pollen evidence is there in favour of the second theory?

b) Is there any evidence against theories 1 or 3?

4 Is there anything to suggest that the land in the vicinity was cleared by humans? Explain your answer.

GENETICS

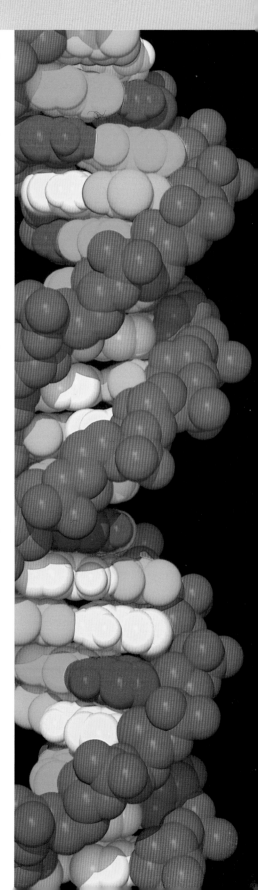

WHY ARE WE all similar? Why are we all different? These are two of the fundamental questions on which the science of genetics is based. When the sperm from your father met the egg from your mother, a unique individual was created, different from anyone who had gone before or will ever exist again.

But, would it surprise you to learn that you have quite a lot of genes in common with bacteria, carrots, pine trees and vampire bats? Certain essential genes - such as those which code for the enzymes of respiration - are very similar in all organisms as they are vital to this basic life process. Such genes must have been copied faithfully, generation after generation, for the last three billion years or so. It is also intriguing to realise that, although almost all organisms die young before reproducing, not a single one of our ancestors did. Their DNA must have had what it takes to be passed on.

In this section we begin by outlining some basic concepts in genetics in Chapter 30. In Chapter 31 we go on to look in more detail at the DNA molecule: how it codes for the manufacture of proteins, and how it can be copied faithfully, time after time. Chapter 32 focuses on inheritance and particularly on the work of Gregor Mendel. Then, in Chapter 33, we consider the mechanisms of evolution, before ending the section in Chapter 34 with an overview of modern genetics. We see, above all, how the DNA technology revolution is likely to have an increasing effect on all of our lives.

30 Basic concepts of genetics

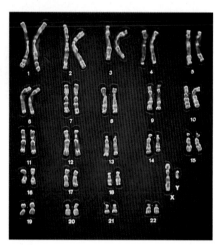

Fig 30.1 **A complete set of human chromosomes. Each body cell contains two sets of 23 chromosomes and each chromosome contains up to 4000 genes. The total amount of DNA in a cell is called the genome. The human genome consists of about 60 000 genes**

1 WHAT ARE GENES?

Genes are the 'instructions' that control everything that happens inside a cell. Physically, genes are different sized stretches of DNA which code for the manufacture of proteins such as enzymes. Enzymes organise the chemistry of an organism, often using other proteins such as collagen as structural materials. Throughout the life of an organism, the genes remain in the nuclei, directing the activities of the cell.

New individuals grow and develop according to the genes they have inherited from their parents. It's a harsh fact, but nothing lives forever: the inevitable fate of all organisms, often sooner rather than later, is death. In order to avoid extinction, species must reproduce and pass their genes on to the next generation.

?

A Some simple unicellular organisms such as bacteria and amoeba can reproduce by binary fission; they simply split in half. Is it correct to say that such organisms never die?

✔

In organisms which reproduce sexually, a set of genes from the male combines with one from the female to form a unique individual (see Chapters 28 and 29).

---

2 STUDYING GENETICS WITH THIS BOOK

Studying genetics means learning about DNA and what it does. In this section, we divide our study of this remarkable molecule into four areas:

- **Molecular biology** is the branch of science which deals with all aspects of the DNA molecule: its structure and function.

    How are genes used to make products? Why are some genes active when others aren't? How can DNA copy itself? How do mutations occur? This is covered in Chapter 31.

- **Genetics** is the study of inheritance: the way in which genetic information is passed on from one generation to the next.

    Why are only some genes passed on? Why are some characteristics always shown while others are hidden? This is covered in Chapter 32.

- **Evolution** looks at how species change and develop with time.

    What is the mechanism of this change? How long does the process take? Can we observe evolution in action? This is covered in Chapter 33.

- **Genetic engineering** is the popular name for the technology which manipulates the DNA molecule.

    How can we change DNA? How can we transfer DNA between organisms? What are the benefits of genetic engineering? Can we cure genetic disease? Should we pursue this technology at all? This is covered in Chapter 34.

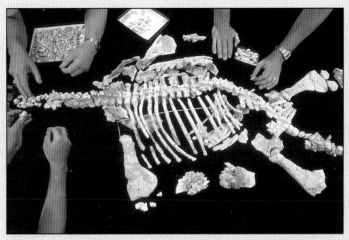

There has been much speculation about the possibility of extracting DNA from extinct species and using it to make new individuals. We have the technology to take tiny samples of DNA and copy them indefinitely (Chapter 34), but bringing long-dead species back from extinction looks unlikely

Our knowledge of genes tells us that the tortoiseshell kitten is almost certainly a female

Fig 30.2 **The applications of our knowledge of DNA**

Wheat has been bred selectively to produce a greater yield. Many new types are easier to harvest and are also resistant to disease

This fermenter houses genetically engineered bacteria which contain the human growth hormone gene. Inside the fermenter the bacteria multiply and produce large amounts of human growth hormone

Male and female sperm can be separated according to the amount of DNA they contain. The sperm are stored in liquid nitrogen until needed. Farm animals can be artificially inseminated to produce offspring of the required sex

3 THE POTENTIAL OF GENETICS

Few scientific subjects make the headlines as often as advances in genetics and DNA technology. We are witnessing an ever-accelerating revolution which is likely to have far-reaching effects. Soon, we will have mapped the complete human genome and perhaps that of other species too.

The implications of this are enormous:

- Two out of every three people die for reasons connected to their genes. Although only a small percentage inherit a lethal genetic disease, many more inherit a *tendency* to develop a condition such as heart disease, premature senility or cancer. An in-depth knowledge of the human genome will make it possible for doctors to screen babies before or soon after birth for genetic abnormalities and tendencies to develop such diseases. But, this raises ethical and moral questions, not just scientific ones: would you want to know just how likely you are to develop cancer in middle age? Would you want anyone else to know – life insurance companies, potential employers, friends or partners? Who would have access to your DNA files?

- We can isolate genes for useful products, such as insulin and growth hormone, and place those genes into another organism. This is **recombinant DNA technology**. Bacteria already produce human insulin and other products on a large industrial scale, and it is also possible to transfer such genes into higher animals such as sheep so that they produce milk containing the required protein.

- We can isolate faulty genes, such as those which cause cystic fibrosis, and replace them with multiple copies of healthy genes. Although still at the experimental stage, this **gene therapy** holds much promise for the future.

- We may be able to halt some of the genetic causes of ageing, and allow more people to live even longer.

Only the bravest scientists would dare to speculate about where our genetic knowledge might eventually take us. It could be to a greatly improved world where suffering is reduced while quality of life and life-expectancy are improved. Genetic engineering and agricultural techniques might produce food-producing species which can live in the harshest of conditions, so providing food in areas of most need.

But can we be this optimistic? There are still many things we don't know about living organisms and it is impossible to predict exactly the effects of interfering with an organism's DNA. Already, companies are trying to patent genetic material and its products, leading to hot debate about whether or not such natural products can be owned, like a brand name or an invention. Will genetic advances be used solely to make money, and therefore be available only to those who can pay?

The ethical implications of some very recent developments also need to be considered. The first mammal has now been cloned, but it is inconceivable that research in this country could ever move towards cloning humans. The controls and safeguards would not permit it. Such safeguards will need to be updated constantly. As you study genetics, keep an open mind.

3 WHAT GOES ON IN THE NUCLEUS?

The nucleus of every eukaryotic cell contains DNA (Fig 30.3). The vast majority of species have a set number of long, elaborately coiled DNA molecules which condense into chromosomes. Genes, the individual instructions, are dotted along the chromosomes.

In an adult organism most cells are not dividing: the chromosomes are uncoiled, and active genes are producing proteins to control the activities of the cells. Not all genes are active at the same time: controlling which genes are active and which are not enables different cells to carry out different functions. For instance, all cells in the human body contain two copies of the insulin gene. Only in certain specialised cells in the pancreas, however, are the genes switched on and used to make insulin.

Glossary of genetic terms

Like many branches of science, genetics comes with its own jargon, seemingly designed to make some very straightforward ideas inaccessible to the average person. You will need to know the following terms:

Diploid: cells or organisms containing two sets of genes. Human body cells, for instance, are all diploid.

Haploid: cells or organisms containing one set of genes. Human sex cells (egg cells and sperm) are haploid.

Gene: a length of DNA which codes for a particular polypeptide. Some proteins consist of more than one polypeptide; these are coded for by more than one gene.

Allele: one form of a gene. There may be two or more alleles of any particular gene. Diploid organisms contain two alleles, one on each pair of chromosomes. For example, in pea plants, the gene for flower colour may have a red allele and a white allele.

Genotype: the genetic make-up of an individual.

Phenotype: the physical expression of the genotype. Many alleles are expressed, or used, and these contribute to the characteristics of the individual. Other genes – recessive alleles in the presence of dominant version – are not expressed and so remain hidden.

Dominant: the allele which, if present, is expresssed. If the red allele in the pea plant is dominant, the pea has a red flower, even if it has a white allele as well.

Recessive: the allele which is expressed only in the absence of the dominant. In the pea, if the white allele is recessive, the plant only has white flowers if it has two white alleles.

Homozygous: when both alleles of a particular gene are the same. In the pea example, an individual would be homozygous if it has two red alleles or two white alleles.

Heterozygous: the two alleles of a gene are different. In a pea plant, a heterozygous individual could have a red allele and a white allele.

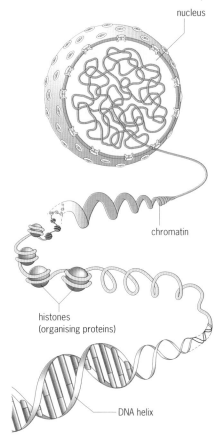

nucleus

chromatin

histones (organising proteins)

DNA helix

Fig 30.3 **Most of the DNA in the cell is locked away in the nucleus where, in effect, it forms a reference library. The genes, which act as instructions for all cell functions, must be passed on when the organism reproduces**

As we saw in Chapter 6, chromosomes condense and become visible during cell division.

Normal cell division – mitosis – duplicates the DNA and then divides it accurately into two so that the two new cells formed are genetically identical.

Meiosis is the type of cell division which produces haploid cells such as eggs and sperm. This process halves the amount of genetic material that goes into each new cell and shuffles it at the same time so that no two sex cells are the same.

4 VARIATION: WHY IS EVERYBODY DIFFERENT?

Fig 30.4 **Like most species, humans show great variation between different individuals**

The group of people in Fig 30.4 shows what we all know: that everyone is different. Height, weight, intelligence, personality, eye colour, size of feet – the list is endless. No two people have the same combination of features. Even identical twins show subtle differences. So what makes us the way we are?

Each human baby is born with a unique set of genes – we call this its **genotype**. However, only some of these genes affect the observable characteristics of the organism – its **phenotype**. Recessive genes, for example, remain hidden, masked by dominant versions.

The other factor which contributes to the phenotype is the environment – the circumstances in which the organism is brought up affects the way in which genes are expressed. For example, humans are born with a set of genes which give us the potential to grow to a certain height. We cannot realise this potential, however, without an adequate diet. Two hundred years ago the average height was several centimetres shorter than it is now: the difference is due to improvements in nutrition and general health care.

> The phenotype of an organism = its expressed genes + effects due to the environment that it is exposed to.

Why is variation necessary?

The simple answer is: survival. Variation in a population has great survival value because it greatly increases the chance that at least some individuals will adapt to changing conditions. To illustrate this, consider a **clone**, a population of genetically identical individuals. All is well when conditions are favourable, but when a serious disease comes along it is very likely that all of the organisms in the population will die. In a more varied population which has developed as a result of sexual reproduction and variation, there is a much greater chance that some of the individuals could be resistant to the disease. These individuals survive to breed, ensuring that the population, and perhaps the whole species, continues to exist.

This 'survival of the fittest' is more correctly called **natural selection**. In the long term, natural selection, acting on the variation in a population, is the main driving force behind the process of **evolution** – the subject of Chapter 33.

> A population is a group of interbreeding individuals of a particular species in a particular place (see Chapter 33). A species usually consists of many different populations. Some endangered species become particularly vulnerable because they consist of just one isolated population, restricted to a particular area.

How variation occurs

Variation arises because individual genes **mutate**. **Mutations** are changes in the genetic material of an organism, often caused by faults in the copying of the DNA (see Chapter 31). Once there is some variation to work on, the differences are maximised by the process of sexual reproduction. Using humans as an example we can see how this happens.

Firstly, gametes are produced by meiosis. This is a special type of cell division (see Chapter 6) which not only produces haploid cells, but shuffles the genes so that all eggs or sperm made by a particular individual are different. In meiosis, there are two key events which produce variation:

- 1 **Crossover during prophase I**. Blocks of gene are swapped between chromosome pairs, bringing some new genes together and separating others.

- 2 **Independent assortment during anaphase I**. Either chromosome from each pair can pass into the gamete. As a human cell has 23 pairs of chromosomes, 2^{23} different combinations are possible.

Secondly, following meiosis, **fertilisation** joins gametes from two separate individuals to form a genetically unique zygote from which the new diploid organism develops.

Continuous and discontinuous variation

Table 30.1 **Examples of continuous and discontinuous variation**

Continuous	Discontinuous
Most dimensions in animals and plants such as height, weight, length of bones, number of leaves	Ability to roll tongue in humans
Tolerance to adverse conditions, such as cold, heat, dehydration	Ability to taste phenlythiourea
Human fingerprints	Flower colour in peas ABO blood group in humans Coat colour in cats

There are two main types of variation in a population: **continuous variation** and **discontinuous variation** (Table 30.1):

- Continuous variation is so-called because the factors, or variables, can be any value inside a given range. Examples of continuously varying characteristics include height and weight in humans (Fig 30.5(a)), and the number of leaves in plants.

- Discontinuously varying features do not display a range: they are either one thing or the other. The ABO human blood group system is a classic example of discontinuous variation: people are either A, B, AB or O. They cannot be anything in between. Height in pea plants is another example: pea plants can be either tall or dwarf (Fig 30.5(b)).

Often, continuous variation is **quantitative** (you can put a numerical value on it) and it is **polygenic** (controlled by several genes). Discontinuous variation is more often qualitative (you can describe it, but not measure it) and is controlled by one or only a few genes. In Chapter 32 we look at the principles of inheritance using examples of discontinuous variation.

?

B Look at Fig 30.5(b) on the next page. A pea plant is 34 cm in height. Is it a tall or a dwarf plant?

Fig 30.5(a) **Height in human males is an example of continuous variation. Most men have values around the middle, giving a bell-shaped curve known as a** normal distribution

Fig 30.5(b) **Height in pea plants is an example of discontinuous variation. Individual plants are either dwarf or tall. Although there is some variation in both categories, there is no overlap: individuals fall into one category or the other**

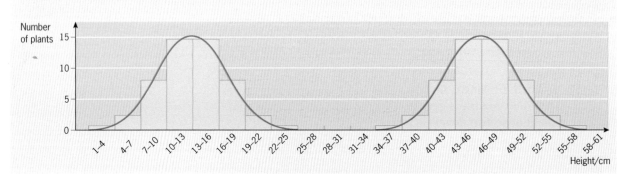

5 A BRIEF HISTORY OF GENETICS

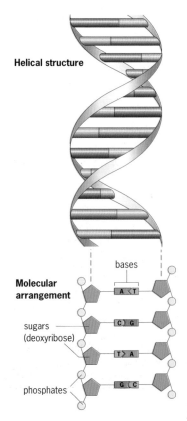

Helical structure

Molecular arrangement

bases

sugars (deoxyribose)

phosphates

The history of genetics gives an insight into the changing nature of scientific progress. Early scientific discoveries and breakthroughs were few and far between, often made by individuals who observed and studied nature and put forward radical theories to explain what they had seen. Mendel (page 505) and Darwin (page 524) are classic examples.

More recent discoveries have been made by teams of dedicated people working together, using the scientific method and latest technology to build on knowledge accumulated by others. Table 30.2 outlines some of the landmarks in this most dynamic of scientific areas.

Fig 30.6 **The discovery of the structure and function of DNA is one of the greatest scientific discoveries of the twentieth century. The DNA molecule stores genetic information and, with the help of enzymes, makes exact copies of itself, time after time**

Year	Scientist(s)	Discovery
1858	Charles Darwin, Alfred Russel Wallace	Jointly announced their theory of evolution by natural selection: 'Survival of the fittest'
1859	Charles Darwin	Published his book *The origin of species*
1866	Gregor Mendel	Published his laws of genetics following studies on pea plants. His findings were ignored
1900	Hugo de Vries, Carl Correns, Eric von Tshermak	Discovered meiosis, providing an explanation for Mendel's laws: his work is rediscovered by all three and gains recognition
1905	Nettie Stephens, Edmund Wilson	Independently discovered the principles of sex determination: XX = female, XY = male
1910	Thomas Hunt Morgan	Proposed the theories of sex linkage, mutation, linkage and chromosome maps following work on *Drosophila* (fruit fly)
1928	Fred Griffiths	Proposed that some 'transforming principle' had changed a harmless strain of bacteria into a lethal one. The hunt for DNA began
1941	Beadle, Tatum	Irradiated the bread mould *Neurospora*, producing mutations which suggested that genes code for enzymes
1944	Oswald Avery	Purified the 'transforming principle' in Griffiths' experiment, showing it to be nucleic acid (DNA)
1950	Erwin Chargaff	Discovered that the base pairing ratios in DNA were always the same, whatever the organism (Chargaff's principles: A=T, C≡G)
1951	Rosalind Franklin	Obtained high-quality X-ray diffraction studies of DNA, showing that it has a helical structure
1952	Hershey, Chase	Used bacteriophages to show that DNA, not protein, is the material of heredity
1953	Crick, Watson	Built on the work of Chargaff and Franklin to work out the three-dimensional structure of DNA
1958	Meselson, Stahl	Used radioisotopes of nitrogen to prove the semi-conservative mechanism of DNA replication
1958	Arthur Kornberg	Purified DNA polymerase from *E. coli* and used it to make DNA from nucleotides in a test tube
1961	Jacob, Monod	Propose a mechanism for switching genes 'on' or 'off': the operon hypothesis of metabolic control
1966	Marshall Nirenberg, Gobind Kharana	Cracked the genetic code: particular triplets of bases on DNA code – via mRNA – for the 20 amino acids
1970	Smith, Wilcox	Isolated the first restriction enzyme, *Hind*II, which can cut DNA
1972	Paul Berg, Herb Boyer	Produced the first recombinant DNA molecules
1973	Annie Chang, Stanley Cohen	Showed that DNA could be inserted and cloned inside a bacterium
1977	Fred Sanger	Developed a method for sequencing DNA
1977		The first Genetic engineering company (Genentech) is founded, using recombinant DNA to make pharmaceuticals
1983	James Gusella	Located the gene responsible for Huntington's chorea on chromosome 4 (see Fig 30.1)
1984	Cary Mullis	Developed the polymerase chain reaction (PCR) in which minute samples of DNA can be copied, prior to analysis
1984	Alec Jeffreys	Developed DNA profiling (or 'fingerprinting')
1988		The Human Genome Project began to map the human DNA sequence
1989	Francis Collins	Identified the gene (CFTR) responsible for cystic fibrosis on chromosome 7
1990		First attempts at gene therapy: T-cells of a 4-year-old girl were exposed to viruses containing working copies of her defective gene. After treatment, her immune system began working again
1994		Genetically engineered 'Flavr savr' tomatoes went on sale
1995		Transgenic sheep made to express human genes in their mammary glands, so that they produce milk containing valuable pharmaceuticals
1997		First mammal cloned. Dolly the sheep fuels controversy about genetic research

Table 30.2 **Some landmark discoveries in genetics. In many of these, the scientists mentioned by name led a much larger team of scientists who worked towards the discovery described**

31 How genes work

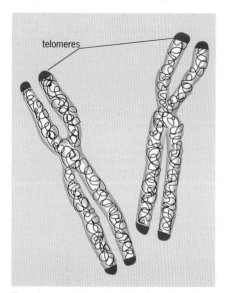

telomeres

The telomeres, shown here in red are the regions at the ends of the chromosomes which are thought to be responsible for control of ageing. They are long when we are young, and get shorter as we age. When they have gone, the cell can no longer divide

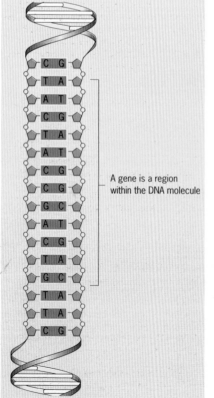

A gene is a region within the DNA molecule

WHY DO WE AGE? Until a few years ago, scientists and doctors thought that our bodies simply wore out like machines. They now think that many aspects of ageing are genetically determined. Organisms develop and grow by cell division, and the same process happens in adults to repair and maintain their bodies. It seems likely that each of our cells has a genetic timer which counts how many times a cell divides: after a maximum number of divisions, the cell dies. But what sets this limit?

The latest research is focusing on the ends of chromosomes, the telomeres. Every time a cell divides, the telomeres shorten, and when they disappear completely, the cell is unable to divide further and its repair systems break down. Researchers have been intrigued to find that eggs and sperm are 'immortal' – they have an enzyme, telomerase, which ensures that the embryo's telomeres are complete. This is necessary to make sure that the genetic clock that produces a baby is set to zero. It may seem obvious, but babies need to be born with 'new' cells so that they have their full life expectancy, regardless of the age of their parents.

The discovery of telomerase opens up the exciting, if ethically difficult possibility that we may be able to manipulate the ageing process. If we can use telomerase to repair the telomeres of normal body cells, it may be possible to extend potential life expectancy to 150 years, or more. The social and economic consequences of this are massive, and are bound to be the cause of much heated debate in years to come.

1 WHAT IS A GENE?

The nucleus of a eukaryotic cell has a set number of chromosomes – every human body cell has 46. Each chromosome is a single, elaborately-coiled DNA molecule which has individual genes dotted along its length (Fig. 31.1). A gene is a length of DNA which codes for the synthesis of one polypeptide chain.

We used to think that a gene contained all the information needed to build a complete protein, such as an enzyme. However,

Fig 31.1 **Genes are sequences in the DNA molecule. The sequence of bases shown here represents one very short gene. In the genetic code, a group of three bases codes for one amino acid, so a protein consisting of 500 amino acids requires a gene of at least 1500 base pairs long, probably more (see page 494). When unravelled, chromosomes are incredibly long – each one consists of up to 4000 genes and these make up only about 10 per cent of its total length (the rest is non-coding DNA)**

many proteins consist of more than one polypeptide chain, so these molecules are coded for by more than one gene. For instance, haemoglobin molecules consist of four polypeptide chains; two alpha and two beta, so it takes two genes to make a complete haemoglobin molecule.

Genes in action:
the central concept of molecular biology

Can we sum up the way genes work in one sentence? It is tricky, but the underlying theme, or **central dogma** of genetics is:

> **The DNA of a gene codes for the production of messenger RNA which, in turn, codes for the production of a polypeptide (Fig 31.2).**

In this chapter we look at how genes work, and at how they are copied to allow cell division, reproduction and growth.

After Crick and Watson worked out the structure of DNA in 1953, two vital questions remained:

- How does information get from the DNA in the nucleus to the site of protein synthesis in the cytoplasm?
- How does the base sequence of the DNA translate into the amino acid sequence in a polypeptide?

To answer the first question, Crick, Brener and Monod developed the **messenger hypothesis**. This states that a specific molecule, named **messenger RNA** (mRNA), is copied directly from the DNA sequence of the gene. mRNA is therefore a mobile copy of a gene. It can move from the nucleus to the cytoplasm. Here, ribosomes use mRNA as a template (pattern) to assemble amino acids in the correct order to make the required polypeptide.

Crick also answered the second question: he proposed that an 'adapter' molecule existed which had two specific binding sites. At one side, the molecule fitted a specific DNA base sequence and at the other, a specific amino acid. This molecule was later discovered and named **transfer RNA** (tRNA).

2 THE GENETIC CODE

Along the DNA molecule the base sequence is continuous, but it is read in blocks of three. Each three, or triplet, is called a **codon.** Each codon codes for a particular amino acid (see Chapter 3 for the actual codes). A particular sequence of bases therefore codes for a specific amino acid sequence. When the amino acids are assembled into a polypeptide, they interact to twist and bend the chain into its final shape. When the polypeptide chain is complete, it forms all or part of a protein which performs a vital role in the organism.

You will have noticed that there are two strands of bases in the DNA molecule. One strand, called the **template strand** or **sense strand**, contains the all-important genetic code for any particular gene. The corresponding part of the other strand is simply there to stabilise the molecule. However, in the same DNA molecule, different genes are present on different sides and the enzymes which transcribe DNA can use either side as the sense strand.

The genome is defined as 'all the DNA sequences in an organism'. It is a complete set of genes, together with the non-coding DNA in between. The human genome is thought to consist of over 3 billion base pairs. Remarkably, the entire genome is present in every one of our body cells.

The position of a gene on a chromosome is known as its **locus** (plural: loci).

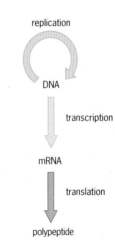

Fig 31.2 **The central concept of molecular biology. The base sequence on a particular gene is copied, or transcribed, to produce a messenger RNA template. This mRNA template passes out of the nucleus and moves to the ribosomes where it is read, or translated, into a polypeptide**

Retroviruses such as the Human Immunodeficiency Virus (HIV) are an exception to the central dogma. Retroviruses contain RNA from which they make a DNA copy using the enzyme **reverse transcriptase**.

A Name three structural differences between DNA and RNA molecules.

The base pairings are always the same. In DNA, T bonds with A and C with G. In RNA, T is replaced by U, so A bonds with U, and C with G (see page 57).

?

B The RNA polymerase enzyme moves along a DNA sequence reading ATACGCTAT.

(a) What is the corresponding sequence on the RNA molecule?

(b) How many amino acids are encoded in this sequence?

(c) Use Table 3.6 on page 58 to find out which amino acids are encoded.

Transcription

DNA can be thought of as a permanent reference library: it does not leave the nucleus and so cannot be used directly for protein synthesis. Instead, the genetic code for a gene is transferred, or **transcribed,** onto a smaller RNA molecule, **messenger RNA**, which can then pass to the ribosomes to act as a template for protein synthesis. In effect, mRNA is a mobile copy of a gene.

Before transcription can begin, the DNA helix must unwind and the two halves of the molecule must come apart, so exposing the base sequence. This process begins when the DNA is 'unzipped' by specific enzymes and the enzyme **RNA polymerase** attaches to the DNA molecule at the **initiation site**. This is a base sequence

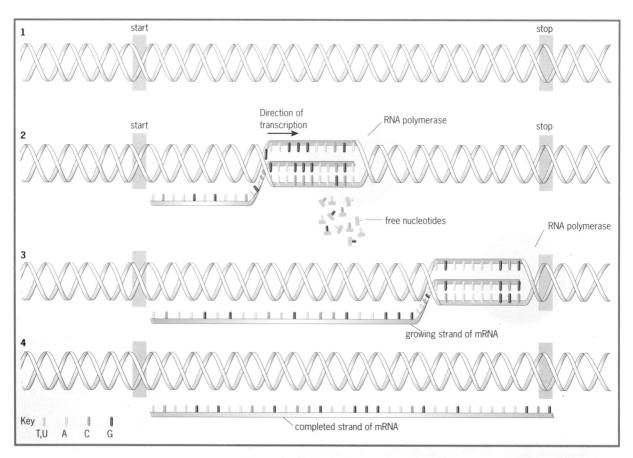

Fig 31.3(a) **This diagram shows the process of transcription. The enzyme RNA polymerase is a big molecule, (not shown to scale) and covers about 50 base pairs at once. Beginning at start the initiation site, the enzyme moves along the template or sense strand of DNA, using free nucleotides from the cytoplasm to make messenger RNA**

Fig 31.3(b) **This electronmicrograph shows part of two DNA strands. Each part contains a gene. The 'branches' are many lengths of mRNA that are being transcribed from the gene.**

Each length of mRNA is moving from left to right, attached to the DNA by an RNA polymerase molecule. These molecules are seen as dark dots on the DNA. RNA polymerase travels along the DNA with its length of mRNA, and adds nucleotides to the mRNA. When transcription of an mRNA molecule is complete, it detaches from the DNA strand and moves out of the nucleus

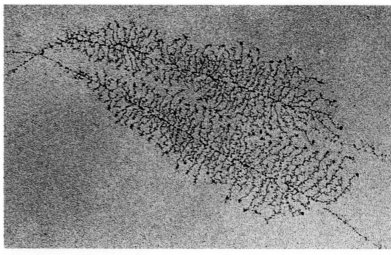

at one end of the gene which effectively says: Start here. Once in place, the enzyme moves along the gene (Fig 31.3(a)), assembling the messenger RNA molecule by adding the matching nucleotides, one at a time.

Once formed, the mRNA copy begins to peel away from the DNA (Fig 31.3(b)). When RNA polymerase has passed a particular region of DNA, the double helix rewinds. When the whole gene has been transcribed, the complete mRNA molecule leaves the nucleus.

Translation

Translation occurs when the base sequence on the mRNA is used to synthesise a polypeptide. The process is called translation because the information contained in the gene – and delivered by the mRNA – is *translated* into a polypeptide.

Protein synthesis is one of the major activities of cells, and it is highly likely that you are already familiar with many of the organelles and chemicals involved. The roles of individual organelles in the process are listed in Table 31.1. See also Chapter 1 and Chapter 3 for background information.

C How do mammals ensure that a regular supply of amino acids reaches all of their cells?

D Is protein synthesis an anabolic or a catabolic reaction (hint, it's a vital part of body building)?

Table 31.1 **The main organelles and chemicals involved in protein synthesis**

Organelle/molecule	Role in protein synthesis
Nucleus	Houses the DNA
Nucleolus	Manufactures the ribosomes
Ribosome	Site of protein synthesis – where amino acid assembly occurs
Endoplasmic reticulum	Isolates, stores and transports polypeptides
Golgi apparatus	Modifies and packages polypeptides
DNA	Stores the genetic information
Messenger RNA	Is a mobile copy of a gene on DNA
Transfer RNA	Each brings a specific amino acid to the ribosomes

Translation occurs on **ribosomes**, structures which hold all the components together as an amino acid chain is made. The mRNA strand passes out of the nucleus and attaches to the ribosome. At the point of attachment, two tRNA molecules deliver the required amino acids and hold them in position so that they can be added to the growing polypeptide.

The mechanism of translation is shown in detail in Fig 31.4. The process continues until an mRNA **stop codon** – which has no tRNA molecule – moves into the ribosome's first binding site. At this point all the components separate, releasing the completed polypeptide.

After translation, polypeptides are processed according to their final destination:

- Those which are to be exported from the cell, such as digestive enzymes, are threaded through pores in the endoplasmic reticulum to accumulate on the inside. Here they are processed and packaged before being secreted (see pages 13 and 15).

- Polypeptides that will form membrane proteins follow the same route as those for export, but they remain on the cell surface membrane rather than being released.

- Polypeptides that will be used inside the cell, such as haemoglobin, remain free in the cytoplasm.

Genes code for the manufacture of polypeptides. Fig 3.23 on page 49 shows the importance of protein in the human body.

Many people find protein synthesis a difficult topic to learn. A valuable exercise is to make the different components – either from modelling clay or just paper – and work through the events shown in Fig 31.4(b).

You will need to make a ribosome, an mRNA molecule with several codons written on it, and enough tRNA molecules and amino acids to match to all the codons on the mRNA.

Fig 31.4(a) **The ingredients needed for protein synthesis**

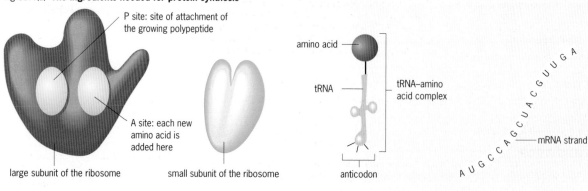

Fig 31.4(b) **The process of protein synthesis**

(1) The two subunits of the ribosome come together. The mRNA strand binds to the ribosome. Then an amino acid–tRNA complex with an anticodon complementary to the first codon on the mRNA, binds to the P site

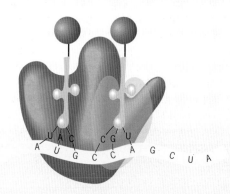

(2) In the next step, an amino acid–tRNA complex with an anticodon complementary to the next codon on the mRNA strand binds to the A site

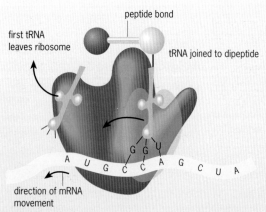

(3) A peptide bond forms between the two amino acids which have been brought together on the ribosome. The bond between the first amino acid and the first tRNA is broken and the tRNA molecule leaves the ribosome. The ribosome moves along the mRNA strand and the second tRNA molecule, now joined to a dipeptide, shifts across from the A site to the P site

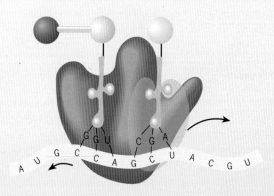

(4) A third amino acid–tRNA complex with an anticodon that matches the third codon on the mRNA strands comes into the A site

The sequence of events from **(2)** to **(3)** are repeated until the entire length of the mRNA strand has been translated. The polypeptide is then complete

The role of transfer RNA

Transfer RNA molecules are smaller than mRNA molecules, containing only about 75 to 80 nucleotides. Each consists of a single strand of nucleic acid folded back on itself to form a 'clover-leaf' shape (Fig 31.5). Transfer RNA molecules bring specific amino acids from the cytoplasm to the ribosome so that they attach to the growing polypeptide.

Fig 31.5 **Two ways of representing the tRNA molecule.** (a) **The 3D shape of the molecule, stabilised by hydrogen bonds as the single nucleotide chain folds back on itself.** (b) **A schematic diagram: each type of transfer RNA transports a specific amino acid to the ribosomes, the site of protein synthesis. At one end of the molecule is the anticodon which attaches to the corresponding codon on the messenger RNA. At the other end of the molecule is the amino acid attachment site**

(a) amino acid attachment site   amino acid

hydrogen
bonds between
paired bases

anticodon

(b)
amino acid attachment site   amino acid

hydrogen
bonds between
paired bases

anticodon

The process by which an amino acid binds to a tRNA molecule is controlled by an enzyme. The process also involves the splitting of ATP. This is important because ATP gives the tRNA–amino acid complex enough energy to form a peptide bond when the amino acid is added to the growing polypeptide.

At one end of the tRNA molecule is the **anticodon**, a three-letter base sequence which matches the codon on the mRNA molecule. At the other end is a particular amino acid. So, for instance, if the codon on the mRNA reads AUG, which codes for the amino acid methionine, it needs a tRNA molecule with an anticodon of UAC which arrives at the ribosome carrying a methionine molecule.

There are 64 different codons, but three of them code for 'Stop translating'. When such a codon arrives at the ribosome, no tRNA is needed. This means that tRNA molecules must match with 61 different codons. There are only 20 different amino acids, so some amino acids are translated from more than one codon. For example, the codons GGG, GGA, GGC and GGU all code for the amino acid glycine.

Like messenger RNA, transfer RNA is also made by transcription. In eukaryote DNA there are many genes whose sole function is to make tRNA. In any diploid cell, most genes occur only twice, but there are hundreds of genes which code for tRNA synthesis, because so many of these work-horse molecules are needed. The fact that tRNA genes do not code for proteins is another exception to the central dogma.

Polysomes and the rate of translation

We have seen that the mRNA molecule is a long single strand of nucleotides, and that the code is translated into a polypeptide at the point of contact with a ribosome. To achieve protein synthesis at a reasonable speed, ribosomes occur in clusters, called **polysomes** or **polyribosomes**, which all translate a different bit of the mRNA at the same time (Fig 31.6).

?

E Why must several different tRNA molecules carry the same type of amino acid?

▓ See questions 1 and 2.

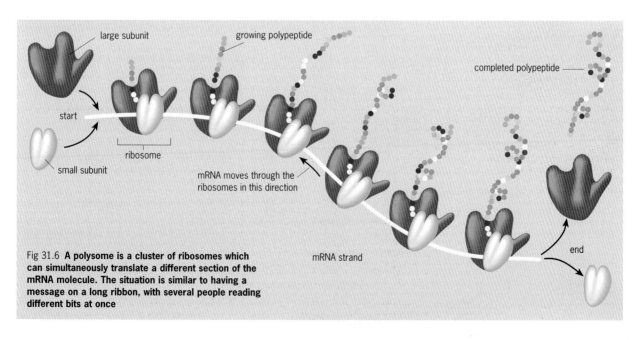

Fig 31.6 **A polysome is a cluster of ribosomes which can simultaneously translate a different section of the mRNA molecule. The situation is similar to having a message on a long ribbon, with several people reading different bits at once**

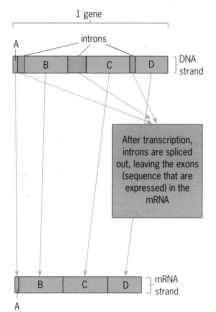

Fig 31.7 **Even in a gene, there are non-coding lengths of DNA called** introns. **Only the** exons **are encoded into the mRNA molecule, and therefore expressed**

3 A JOURNEY ALONG A CHROMOSOME

If one single chromosome were unravelled, like pulling the wool out of a piece of knitting, we would be left with one long DNA strand. Although this is one single molecule, it is several *centimetres* long and contains up to four thousand genes.

How do you go about investigating such a molecule? One way is to work backwards; if you can isolate the mRNA molecules which are in the cytoplasm, you can trace their origin in the DNA.

A clever technique called **nucleic acid hybridisation** is used to pinpoint the origin of the mRNA, giving the location of the gene itself. In this technique, DNA strands are denatured: when heated to 87 °C, hydrogen bonds which hold the two strands together are broken, and the individual strands separate. If messenger RNA strands from the cytoplasm are added to the mixture, they stick to the region of DNA on which they were originally made.

Studies using this technique have unexpectedly revealed that there is some non-coding DNA in genes (Fig 31.7). Most eukaryote genes contain regions, called **introns**, which do not find their way into the mRNA molecule at all. The parts of the gene which *are* expressed in the mRNA are called **exons** because they are the **ex**pressed regions of the gene.

Generally, a gene consists of a DNA sequence which is present only once. The non-coding sequences, however, tend to consist of repeated or 'stuttered' sequences. Humans all have the same repeated sequence, but in different individuals it is repeated a different number of times and in different places. This is the basis of *DNA profiling* – see the Assignment in Chapter 3.

The function of non-coding DNA is not known. It is probably not useless: much of the repeated DNA in a chromosome occurs around the region of its centromere and may have an important role in maintaining the physical stability of the chromosome as a whole.

The DNA could be like a page of writing in a book. If the letters were strung together continuously, we would find it difficult to read: the gaps between words help us to make sense of the sentences.

4 THE CONTROL OF GENE EXPRESSION

Most multicellular organisms start life as a zygote: a single fertilised cell. This cell has a complete set of genes. As the cell divides by mitosis, the DNA is copied faithfully, so each new cell also contains the same complete set of genes. How, then, do cells differentiate? How, for example, do some animal cells specialise into muscle, nerve or skin?

The answer lies in **gene expression** – how different genes are 'switched on', or **expressed**, in different cells (Fig 31.8).

At any one time, the average cell is probably using only 1 per cent of its available genes. Some of the expressed genes are used for 'housekeeping' – they code for the proteins needed by all cells, such as the enzymes involved in respiration. In contrast, other genes are expressed only in specialised cells. The genes for making insulin, for example, are expressed only in the beta cells of the islets of Langerhans in the pancreas.

Selective activation of genes is the key to the development of a complex multicellular organism and, not surprisingly, it is the subject of

Fig 31.8 **A spectacular example of the power of gene expression is seen in the life cycle of some insects such as this peacock butterfly. The genes present in the fertilised egg are still present in the adult, but a different set are activated to give rise to the caterpillar and to control its metamorphosis into a pupa and then an adult – which looks like a completely different organism**

intense research. If we could understand how and why some genes are activated while others remain dormant, we could unlock the mystery of embryonic development – how a single fertilised egg grows and differentiates into something as complex as a human being. Unfortunately, this interesting area is very complex and is beyond the scope of this book.

5 DNA REPLICATION

Whenever a cell divides, the DNA must be copied. Otherwise, there would be no reproduction and no growth. When cells divide, each new daughter cell must have a complete set of genes. **Replication**, the mechanism of DNA copying, is therefore of fundamental importance.

Just three years after Crick and Watson published their model of DNA structure in the journal *Nature*, their theory about DNA replication was confirmed by Arthur Kornberg. He put DNA into a mixture containing all four **nucleotides** (each of the four bases attached to a phosphate and a sugar), together with the enzyme DNA polymerase, and he showed that the DNA could replicate without any other factors present.

In the mixture, each DNA strand unwinds and acts as a template for the construction of a new strand. The exposed strand acts as a template on which the free nucleotides arrange themselves in exactly the same sequence as the intact strand they replace. This model is called **semi-conservative replication** because each of the resulting strands of DNA contains one strand from the original DNA and one newly synthesised strand: half has been conserved and half is new.

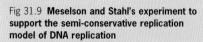

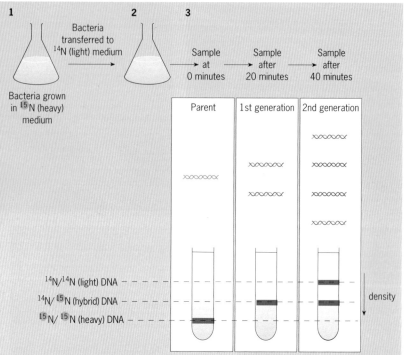

Fig 31.9 Meselson and Stahl's experiment to support the semi-conservative replication model of DNA replication

1 To start with, many generations of *E. coli* bacteria are grown in a medium containing nucleotides labelled with the heavy nitrogen radioisotope, ^{15}N . Eventually, almost all of the bacterial DNA contains ^{15}N, and so is denser than normal DNA. A control culture of the same bacterium is grown with 'normal' ^{14}N nucleotides for comparison

2 The bacteria grown in ^{15}N are transferred to ^{14}N medium. After 0, 20 and 40 minutes, samples of the DNA are extracted, placed in a solution of caesium chloride and centrifuged. This separates out the DNA according to its density and allowed Meselson and Stahl to distinguish between DNA containing ^{15}N and ^{14}N.

3 The results confirm that, initially, all the DNA contained ^{15}N. After 20 minutes the DNA had replicated, producing two strands of $^{15}N/^{14}N$ hybrid DNA. After 40 minutes, the second generation strands consisted of half $^{15}N/^{14}N$ hybrids and half all-new ^{14}N strands

?

F Predict the outcome of the third generation of DNA molecules in Meselson and Stahl's experiment: Of the eight DNA strands produced, how many would be hybrids and how many would be all 14N strands?

Fig 31.10(a) When DNA replicates, the double helix unwinds and each exposed strand acts as a template for the synthesis of a new strand. This is a simplified diagram showing how free nucleotides are added to produce two new DNA helices

Fig 31.10(b) DNA replication is actually more complicated and several enzymes other than DNA polymerase are involved. DNA helicase unwinds the parent helix and then this is 'held open' by DNA binding proteins so that the DNA polymerase can gain access to the parent strand

Strong support for the semi-conservative model of DNA replication came from the work of Meselson and Stahl (Fig 31.9). By growing bacteria with bases containing the radioisotope ^{15}N (sometimes called *heavy nitrogen*) and then following its progress, they showed that each new DNA strand contains half the original strand.

The mechanism of semi-conservative replication

The basic mechanism of DNA replication is shown in Fig 31.10(a). DNA is copied before the cell divides (see Chapter 6) and the basic idea is simple: the helix unwinds and each strand acts as a template for the manufacture of a new, identical strand.

In practice, DNA replication is rather more complex than it appears, and is a good illustration of the importance of enzymes (Fig 31.10(b)). Two **helicases** use energy from ATP to separate the DNA strands. Next, **DNA binding proteins** attach to keep the strands separate. **DNA polymerase** enzymes then move along

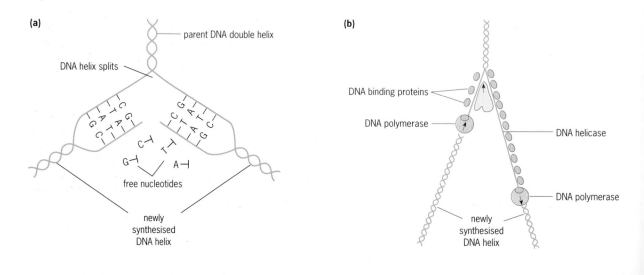

the exposed strands, synthesising the new strands, often in short segments. Finally, **DNA ligase** enzymes join the segments together, completing the new DNA strands.

See question 4.

DNA proofreading

Mistakes in DNA replication occur regularly. The chances of the wrong base being added are between 1 in 10^8 and 1 in 10^{12}. This may seem low, but so many nucleotides are added that this rate means approximately one mistake occurs for every ten genes that are copied. Clearly, such a rate would cause the death of most organisms. This disaster is prevented by the polymerase enzymes themselves. They can detect when the wrong base has been inserted, and pause to correct the mistake. In this way 99.9 per cent of mistakes are corrected. Uncorrected mistakes give rise to **mutations**.

There are many different DNA repair mechanisms in cells to put right the genetic damage which inevitably accumulates during the life of an organism. The condition xeroderma pigmentosum illustrates the importance of these mechanisms: an affected person is liable to get skin cancer because their cells cannot repair genetic damage caused by exposure to ultraviolet light (Fig 31.11).

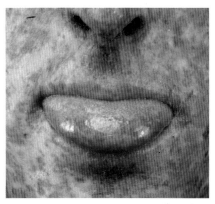

Fig 31.11 **Xeroderma pigmentosum is a rare genetic disease characterised by dry, freckled skin which is extremely sensitive to sunlight**

HOW DO YOU COMPARE THE DNA OF DIFFERENT ORGANISMS?

THE TECHNIQUE OF DNA hybridisation (Fig 31.12) can be used to compare the similarity of the DNA of different species, and is proving to be a useful tool in the search to piece together evolutionary trees.

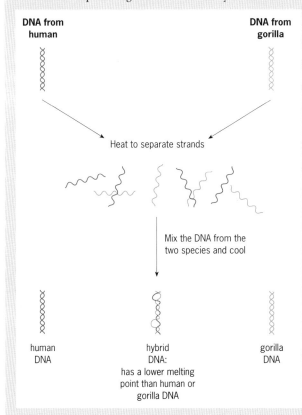

Fig 31.12 **The principle of DNA hybridisation**

DNA is much more stable than protein because it is held together by regular hydrogen bonds along the whole of its length. However, when DNA is heated to about 87 °C, known as the melting point of DNA, the hydrogen bonds break and the two halves of the molecule separate. When the temperature drops below 87 °C, the two strands join up: they **re-anneal**.

This is the basic process of DNA hybridisation:

DNA from two different species is cut into small fragments and heated to 87 °C so that the two strands separate.

The separated strands from each species are then mixed together. As the temperature drops the DNA strands re-anneal. Some of the molecules formed will be hybrids: they contain one strand from each species.

The temperature at which the two hybrid strands re-anneal tells us about the degree of similarity between them. When two strands from the *same* species re-anneal, they do so at 87 °C. Closely related DNA re-anneals at temperatures slightly lower but close to 87 °C. This indicates that they have many base sequences in common and so form many hydrogen bonds along the length of the molecule. In contrast, DNA from more distantly related species must cool to much lower temperatures, because there are fewer hydrogen bonds to hold the strands together.

For example, human DNA can be hybridised with gorilla DNA or rabbit DNA. The melting point for DNA from all three single species is 87 °C. The melting point for the human/rabbit DNA is lower than for the human/gorilla DNA. This shows that humans and gorillas have more DNA sequences in common than humans and rabbits.

See question 3.

Mutations

A mutation is a spontaneous change in the genetic material of an organism (Fig 31.13). It usually occurs when DNA is copied and cells divide. Generally, there are two types of mutation:

● a gene or point mutation,

● a chromosome mutation.

Both of these may occur in **somatic cells** (normal body cells) or **germ cells** (sex cells).

Somatic cell mutations are not passed on to the next generation. Generally, organisms accumulate somatic mutations as they get older, and this is thought to be one of the causes of cancer (see page 111). In contrast, when a mutation arises in an egg or sperm cell which then goes on to become a zygote, the mutation is passed on to all the cells of the new individual. Germ cell mutations alter the *genome* of an organism.

Fig 31.13(a) **To many people, this is the image conjured up by the word mutant. But in biology, a mutation is a fault in the copying of genetic material and is usually lethal or makes no difference to the organism. Only in rare cases will mutation give rise to new variation, and it is sure to be less spectacular than science fiction writers would have us believe**

Fig 31.13(b) **An example of a real-life mutation. A chance mutation in a single gene of the speckled moth Biston betularia produced a fortunate change from the normal speckled coloration, seen in the moth on the left, to the melanic form seen on the right. In sooty areas, this new form was better camouflaged and so had a better chance of avoiding predators. More of the black moths passed on their genes to the next generation, making the melanic form dominant in industrialised areas**

Gene mutations

A gene mutation is a spontaneous change in the base sequence of DNA. We have already seen that when DNA replicates, two identical strands are formed. Occasionally, however, a fault may occur. In a typical gene of, say, 1400 base pairs, a change in just one of the base pairs has the potential to make the whole gene useless, and may prove lethal to the organism. A change in a single base is known as a **point mutation**.

How can such a tiny change be so disastrous? It is because a change to any base will alter a codon so that it will probably code for a different amino acid. In turn, this 'wrong' amino acid can affect the way the polypeptide chain folds, so changing the shape of the whole protein molecule. It might then be unable to function. (Remember: many genes code for enzymes or parts of enzymes whose proper functioning relies on precise shape.) Table 31.2 shows the different types of gene mutation that can occur, using words instead of codons. In all cases, the original meaning of the message is lost.

From Table 31.2 we can see that the two least damaging types of mutation are substitutions and inversions, because these alter only one of the seven amino acids. In some cases, if the new amino acid occurs in a non-vital part of the chain, or has similar properties to the correct one, then the mutation may not matter at all.

In contrast, deletions and additions are potentially much more damaging because they cause **frame shifts**: only one base may be added or lost, but this causes the whole sequence to shift along, altering all the codes. In this case, the protein made is completely different from the original and is unlikely to function as it should. The Assignment at the end of this chapter deals with sickle cell anaemia, a disease caused by a fault in just one base.

How common are gene mutations?

The rate at which genes mutate varies between species and between genes at different loci. At a rough estimate, there is an average mutation rate of about 10^{-5} per locus per generation, meaning 1 in 100 000. So, in a population of one million individuals, you could expect ten to have inherited a mutation at any particular gene. The rate of gene mutation (and chromosome mutation) can be greatly increased by environmental factors, known as **mutagens**, such as ultraviolet, X-rays, alpha and beta radiation, and chemicals such as mustard gas and cigarette smoke.

Chromosome mutations

During mitosis (normal cell division) chromosomes are copied, condensed and pulled apart. The end result should be that each daughter cell receives a complete set of perfect chromosomes. But faults can occur. A chromosome mutation occurs when the structure of a whole chromosome – or set of chromosomes – is altered in some way.

Fig 31.14 shows the four basic types of chromosome mutation: deletion, duplication, inversion and reciprocal translation.

Table 31.2 **Different types of gene mutation**

Type of gene mutation	Effect on code (using words rather than codons)
Normal code	THE FAT OLD CAT SAW THE DOG
Addition (frame shift)	THE EFA TOL DCA TSA WTH EDO G
Deletion (frame shift)	THE ATO LDC ATS AWT HED OG
Substitution	THE FAT OLD BAT SAW THE DOG
Duplication	THE FAT TFA TOL CAT SAW HTH EDO G
Inversion	THE TAF OLD CAT SAW THE DOG

Fig 31.14 **The main types of chromosome mutations**

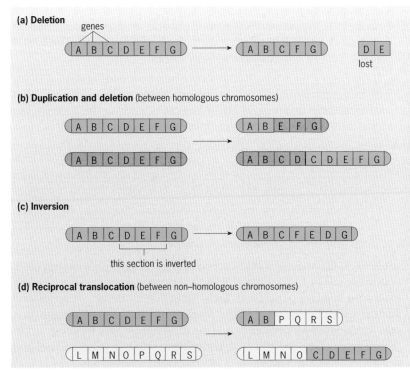

(a) Deletion

genes

(b) Duplication and deletion (between homologous chromosomes)

(c) Inversion

this section is inverted

(d) Reciprocal translocation (between non–homologous chromosomes)

(a) **Deletion occurs when a chromosome splits and a fragment is lost as the parts rejoin. In this case, a whole set of genes is lost, and so this is often fatal**

(b) **A duplication occurs when a section of chromosome is copied twice. This can occur when pairs of homologous chromosomes break and then reconnect to the wrong partners. Thus, one of the partners would have two copies of a particular sequence, but the other would have neither. This can have serious consequences**

(c) **An inversion results when a broken segment is reinserted in the wrong order. Any genes contained in this region will be transcribed and translated backwards, and will therefore not make the correct proteins**

(d) **A reciprocal translocation occurs when two non-homologous chromosomes exchange segments. This makes the chromosome pairs of unequal size, and this can cause problems in meiosis, with daughter cells receiving the wrong number of chromosomes**

While gene mutations normally affect just one gene, chromosome mutations involve the disruption of whole blocks of genes. For this reason, many chromosomal mutations give rise to **syndromes**, which are complex sets of symptoms with a single underlying cause.

In addition to the structural mutations outlined above, another common failure is **non-disjunction**. This occurs when pairs of chromosome fail to separate during anaphase I of meiosis. Some gametes get both chromosomes, while others receive neither. Downs syndrome, for example, occurs when a gamete containing two copies of chromosome 21 joins with a normal gamete containing one, giving the affected individual three copies of chromosome 21 (see page 111).

> ✓
>
> Meiosis is also known as reduction division. See Chapter 6 for more detailed information about this important process.

Polyploidy

Occasionally, a mutation causes a whole set of chromosomes to be changed. For instance, a fault in meiosis might result in some gametes having two sets of chromosomes while others receive none. When such a diploid gamete joins with a normal one, the result is a **triploid** zygote – one with three sets of chromosomes. In another case, a fault in mitosis after fertilisation could result in the chromosomes doubling but not separating, leaving four sets in a single **tetraploid** cell.

The possession of multiple sets of chromosomes is known as **polyploidy**. This phenomenon is very common in flowering plants and has played a vital role in their evolution. Botanists estimate that over half of the world's flowering plant species are polyploid, including some of the world's most economically important crops: sugar beet, tomatoes and tobacco.

> ?
>
> **G** Polyploid cells can be induced experimentally by adding the chemical colchicine, a substance extracted from the crocus. Colchicine inhibits the formation of the spindle during cell division. Suggest why colchicine results in polyploid cells.

Why mutations are important

A mutation can have one of several effects:

- It may be lethal: an organism with a lethal mutation cannot survive and may not even develop beyond the zygote. Genes code for vital proteins such as enzymes, so many gene mutations mean that the protein is the wrong shape, and so cannot function metabolically. Chromosome mutations which involve the loss of whole blocks of genes are usually lethal.

- It may have no effect. It may occur in a non-coding part of DNA, or in a gene which is not expressed. Alternatively, it may result in a different amino acid being incorporated into the protein, but one which does not alter its ability to function.

- It may be beneficial. Occasionally, a mutation produces an improvement in phenotype. Statistically, this is a very rare event, but given the number of DNA replications and the number of individual organisms, it is bound to happen sooner or later. Beneficial mutations are hugely important because, ultimately, they are the source of all variation. Natural selection favours these 'improved' mutants (see Chapter 33).

Fig 31.13(b) on page 498 shows an example of a beneficial mutation. Following the Industrial Revolution, a mutation in the peppered moth produced a black (melanic) form which was better camouflaged against sooty backgrounds. The black moth had an advantage as it was rarely seen by predators and so survived to pass its genes on to the next generation. Consequently, the mutated gene spread and the genome of the species was altered (see Chapter 33).

SUMMARY

■ A chromosome is a large, densely staining body consisting of a single, supercoiled DNA strand. The genes along a chromosome are found at particular points called **loci**.

■ A gene is a length of DNA which codes for the synthesis of a particular polypeptide or protein. Only about 10 per cent of a chromosome consists of genes; the rest is non-coding DNA.

■ The genetic code is the sequence of bases contained along one side of the DNA molecule. The code is read in groups of three bases, called **codons**. Each codon codes for a particular amino acid.

■ The central dogma (principle) of molecular biology is that genes are read by **messenger RNA** which in turn moves out into the cytoplasm where it codes for the assembly of a protein.

■ At the ribosome, messenger RNA attracts **transfer RNA** molecules with the right amino acids attached. The amino acids are held together so that peptide bonds form; in this way the protein is assembled.

■ DNA replication occurs when the two strands separate and the enzyme **DNA polymerase** assembles two new strands on the originals. This is called semi-conservative replication as each old strand is *conserved* as half of the new DNA.

■ A fault in DNA replication leads to a **gene mutation**. The base sequence is altered so that the gene no longer codes for the correct protein. Chromosome mutations can result in whole blocks of genes being lost, or translated backwards for example.

■ Most mutations are harmful, some have no effect, but a few can be beneficial. Beneficial mutations might give the organism an advantage and in this way mutations give rise to variation and hence drive evolution.

QUESTIONS

1 The DNA coding system contains the information for the production of polypeptides by a cell.

a) In the DNA coding system, describe:
  (i) the form in which the message is transmitted from the nucleus to the cytoplasm;
  (ii) precisely where in the cell the message is translated.

b) Explain why different alleles of the same gene produce similar, but not identical, polypeptides.

[AEB June 1996 Human Biology: Paper 1, q.5]

2 The diagram in Fig 31.Q2 shows part of a messenger RNA (mRNA) molecule.

U A C C G A C C U U A A

Fig 31.Q2

a) **(i)** How many codons are shown in this section of mRNA?
  (ii) What is specified by a sequence of codons in an mRNA molecule?

b) A tRNA molecule carries a complementary base sequence for a particular codon.
  (i) write the complementary sequence for the first codon in the mRNA sequence given above.
  (ii) Describe the role of tRNA molecules in the process of protein synthesis.

[ULEAC June 1996 Biology and Human Biology Module Test, q.5]

3 DNA hybridisation is a method of comparing DNA from different species. In this technique purified DNA samples from two different species are heated to separate the strands, then mixed and allowed to cool together. As the separated strands cool they re-combine into double helices. Some of the new double helices are hybrid: that is, they consist of one strand from each species. This process is shown in Fig 31.Q3.

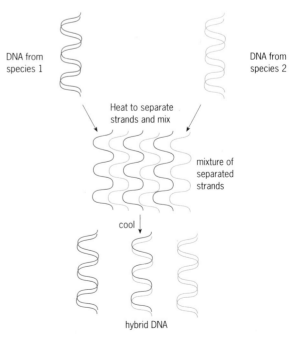

DNA from species 1

DNA from species 2

Heat to separate strands and mix

mixture of separated strands

cool

hybrid DNA

Fig 31.Q3(a)

a) The strands in hybrid DNA separate at a lower temperature than those in DNA from a single species. The more closely related the two species, the lower the difference between the separation temperatures of hybrid DNA and single-species DNA.

    (i) Suggest why the strands in hybrid DNA separate at a lower temperature than those in DNA from a single species.

    (ii) Suggest why the separation temperature of hybrid DNA from distantly related species is lower than that of hybrid DNA from closely related species.

b) The table below shows the difference between the separation temperatures of hybrid DNA and single-species DNA for a number of pairs of primate species or groups. Each figure is a mean of many tests using DNA from different individuals.

Sources of hybrid DNA	Difference in separation temperature/°C
Human/chimpanzee	1.8
Gorilla/chimpanzee	2.2
Human/gorilla	2.3
Human/orang-outan	3.6
Gorilla/orang-outan	3.6
Chimpanzee/orang-outan	3.6
Gibbon/other apes	4.8
Old World monkeys/apes	7.2

It is assumed that the difference in separation temperature is directly proportional to the time since the evolutionary lines of the two groups diverged.

    (i) The evolutionary lines of Old World monkeys and apes are thought to have diverged 30 million years ago. Use this figure to calculate the time represented by a difference of 1 °C in separation temperature of hybrid DNA. Show your working.

    (ii) Calculate how long ago the evolutionary lines of humans and chimpanzees diverged, according to DNA hybridisation studies. Show your working.

c) The diagram of Fig 31.Q3(b) is an incomplete family tree of primate evolution based on the table in part **b)**.

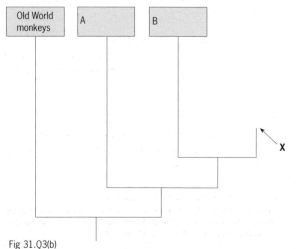

Fig 31.Q3(b)

    (i) Fill in the names of the missing species or groups in boxes A and B.

    (ii) Complete the family tree from the branching point marked X.

d) The relationship and divergence times calculated using DNA hybridisation data have been the subject of considerable debate and disagreement among students of primate evolution. In particular, it has been claimed that the difference in separation temperature between hybrid DNA and single-species DNA may not be directly proportional to the time since the evolutionary lines of the two groups diverged.

    (i) Give *two* possible reasons for doubting that the difference in separation temperature between hybrid DNA and single-species DNA is directly proportional to the time since the evolutionary lines of the two groups diverged.

    (ii) Give *two* alternative sources of evidence concerning relationships and divergence times in primate evolution.

[ULEAC June 1994 Human Biology: Paper 1, q.13]

4 In 1958, Meselson and Stahl published the results of an experiment which provided strong evidence that cells produce new DNA by a process of semi-conservative replication.

a) Why is replication of DNA described a semi-conservative?

Meselson and Stahl's experiment is outlined in Fig 31.Q4 (^{15}N is a heavy isotope of nitrogen).

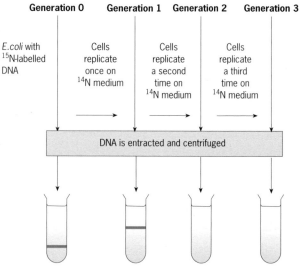

Fig 31.Q4

b) Which component of the DNA was labelled with the ^{15}N?

c) Explain why centrifugation separates the DNA labelled with different isotopes of nitrogen.

d) Copy 31.Q4 and draw in the results you would expect for generations 2 and 3.

[NEAB February 1996 Modular Biology: Continuity of Life Paper, q.4]

Assignment

SICKLE CELL ANAEMIA

In addition to the topics covered in this chapter, this Assignment assumes that you know some simple genetics (monohybrid inheritance) and evolution (natural selection).

A small minority of people, particularly those of African origin, inherit sickle cell anaemia (SCA), a condition in which their haemoglobin acts abnormally. SCA is a genetic disease: it results from a gene mutation which is inherited. Carriers of the sickle cell allele are usually perfectly healthy, and in some circumstances even have an advantage over people who possess two normal alleles. However, individuals with two sickle cell alleles have the disease.

The faulty allele makes haemoglobin S instead of the normal haemoglobin A. Problems arise when oxygen tensions are low, as they are in actively respiring tissues. At low oxygen tension, abnormal haemoglobin tends to come out of solution and crystallise, distorting the red blood cells into crescent shapes (Fig 13.A1). This has two main consequences. Firstly, sickle cells tend to form clots which can block blood vessels. Secondly, the spleen destroys abnormal sickle cells at a much greater rate than normal, causing anaemia.

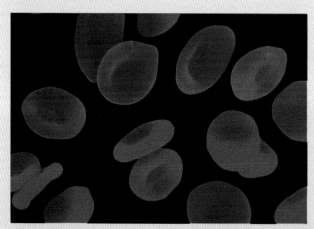

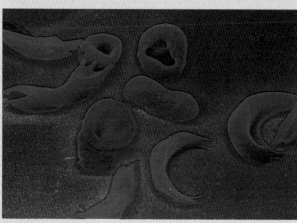

1 What is the function of haemoglobin?

Haemoglobin molecules consist of four polypeptide chains: two α-globins and two β-globins. Research shows that sickle cell anaemia results from a fault in just one base in the β-globin gene. Here is the base sequence of the first seven codons of this gene:

Normal base sequence GUA–CAU–UUA–ACU–CCU–GAA–GAG

Sickle cell sequence GUA–CAU–UUA–ACU–CCU–GUA–GAG

2
a) How can you tell that these codons are from mRNA and not DNA?
b) Which codon has mutated?
c) What is the name given to this type of mutation?
d) Work out the amino acid sequence for both base sequences (you will need to refer to the table on page 58).
e) Explain in general terms how this slight change can produce abnormal haemoglobin.
f) Explain why other types of gene mutation can have a much greater effect on the final protein than the example seen here.

The inheritance of sickle cell anaemia

Generally, people have two copies of each gene, and this is the case with sickle cell anaemia. The two alleles involved are:

Hb^A = coding for normal haemoglobin
Hb^S = coding for sickle cell haemoglobin

So the three possible genotypes are:
Hb^{A}Hb^A Normal
Hb^{A}Hb^S Carriers of the disease but with few, if any, symptoms
Hb^{S}Hb^S Have the disease

3 Sickle cell anaemia is described as an autosomal, recessive disease.
a) What do the terms **(i)** autosomal and **(ii)** recessive mean?
b) Explain why carriers are usually healthy.
c) If two carriers have children, what is the probability that the first child will have the disease?

The sickle-cell gene is far commoner than you would expect for one which is selected against. The reason for this is that the carriers usually show a higher resistance to malaria than normal.

4
a) What is meant by the phrase 'selected against'?
b) Explain why the frequency of the sickle cell allele is higher than you would expect in some areas.

Fig 31.A1 **Above left: Normal red blood cells.
Below left: Abnormal red blood cells from someone with sickle cell anaemia. Haemoglobin S tends to polymerise, forming crystal-like structures which distort the red blood cells**

32 How genes are inherited

Woody Guthrie, the famous American folk singer, suffered from Huntington's disease. His son, Arlo Guthrie was born before his father knew he had a disease that could be passed on to his children. Luckily, Arlo is unaffected: he has not inherited his father's faulty gene

ABOUT EIGHT PEOPLE in every 100 000 are born with the genetic disorder Huntington's disease. The condition is caused by a single faulty gene on chromosome four which somehow causes premature degeneration of the brain. This leads to a slow decline into dementia which, ultimately, is fatal.

Diagnosis is difficult because early symptoms are vague: sufferers can have bursts of temper, become more clumsy than usual or have difficulty remembering things. Also, the symptoms do not usually become apparent until the sufferer is in their thirties or forties, usually after they have passed the gene on to the next generation.

Until recently, people with a family history of the disease knew that they were at risk, but they could never be sure whether they were actually carrying the gene. This led to a dreadful dilemma: should they go ahead and have children and risk passing the gene on, or should they decide to remain childless, only later to experience the anguish of finding out that they were free of Huntington's disease when it was too late to try for a baby.

Today, tests are available which can tell people whether they are carriers of the Huntington's gene. A sample of DNA can even show if an unborn baby carries the disease. This does not make the decisions involved any easier, but it does give affected families accurate information on which to base those decisions.

> Some important terms and concepts relating to inheritance are covered in Chapter 30. If you are new to genetics, read Chapter 30 before starting this chapter.

> A gene is a length of DNA which acts as a template for the production of messenger RNA. mRNA forms a mobile copy of a gene, travelling from the cell's nucleus to the cytoplasm. Ribosomes use this template to synthesise a specific amino acid chain. Complete proteins are made from one or more of these chains.

1 MENDELIAN INHERITANCE

As Fig 32.1 shows, there is often a marked family resemblance between parents and their children and between siblings. However, all new-born babies, except identical twins, are genetically unique. Every new-born baby has around 60 000 genes and because humans – like most animals – are diploid organisms, a baby has two copies of each gene. It inherits one copy from its mother and one copy from its father.

It is difficult to imagine what happens when 60 000 different genes are mixed. Not surprisingly, although a vast amount of genetics research is underway, we are still a long way from completely understanding what happens when those mixed genes are passed on to the next generation. Some genes are activated while others remain hidden, some master genes switch on whole blocks of other genes.

How can we begin to make sense of it all? Amazingly, the man who first worked out the underlying rules did so without any

Fig 32.1 The tendency for certain features to be inherited is shown clearly in this family. However, no two people ever inherit exactly the same set of genes unless they are identical twins

Fig 32.2 Gregor Mendel (1822-1884). Mendel's discoveries, like those of many scientists, were due to a combination of hard work, good scientific method, a touch of genius and a significant amount of luck

knowledge of genes, chromosomes, DNA or cell division. Known as the father of genetics, **Gregor Mendel** (Fig 32.2) worked with pea plants. These are much simpler organisms than humans and provide a good starting point to understand some of the basic rules of **Mendelian inheritance**.

The work of Gregor Mendel

Gregor Mendel, a Czech monk, was the first to work out the basic laws which govern the inheritance of genes. Before Mendel's work became known, it was widely thought that inheritance was due to 'blending' of different features. So, for example, a tall person married to a short person would produce children who would eventually grow to an intermediate height. Mendel showed it was wrong to assume that characteristics blended and demonstrated how individual characteristics are passed down to the next generation.

For years, Mendel bred and studied the edible pea plant (Fig 32.3). He chose these plants because they had easily observable features, they were easy to cultivate, they had rapid life cycles and their pollination could be controlled. His breeding experiments centred on the inheritance of a few features such as flower colour, tallness and wrinkled/smooth seed texture.

Mendel was lucky because he chose features that were controlled by single genes. In addition, pea plants tend to fertilise themselves and produce pure breeding homozygous individuals. Had Mendel worked with more complex features or species, he may not have been able to work out the underlying laws of inheritance.

Mendel conducted his experiments using a good scientific method. He controlled his experimental conditions carefully, ensuring that plants were pollinated only by the pollen he transferred. He made careful observations and then repeated experiments as many times as was practicable, keeping meticulous records. Mendel was trained in mathematics and he applied statistical methods to his results in a way that was very advanced for the time.

His genius really showed in the way he interpreted his results. Mendel proposed the **particulate theory**, in which he stated that there are individual units of heredity which cannot be diluted, only

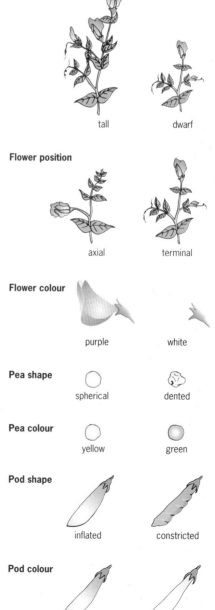

Plant height

tall   dwarf

Flower position

axial   terminal

Flower colour

purple   white

Pea shape

spherical   dented

Pea colour

yellow   green

Pod shape

inflated   constricted

Pod colour

green   yellow

Fig 32.3 The features of pea plants studied by Mendel. Pea plants tend to self fertilise, producing true-breeding strains. For example, white-flowered plants contain only white-flowered alleles: they are not 'hiding' any purple alleles and so always produce white flowers

hidden. Mendel came up with the idea that each pea has two factors for each character, one inherited from each parent. Only one unit can be present in the gamete and so be passed on to the next generation. Mendel's 'particles' are, of course, genes.

In 1866, Mendel published a detailed account of his findings, but their significance was not appreciated by the scientific community. In 1884, Mendel died without ever receiving true recognition for his work. In 1900, his work re-emerged thanks to the independent work of three other European scientists – see Table 30.2 on page 487. In the time between 1866 and 1900, the process of meiosis has been discovered and this provided a mechanism which would explain Mendel's observations. All three scientists cited Mendel's original work, which had suggested the basic mechanism of inheritance.

Meiosis and Mendel's findings

Mendel's findings can be summarised into two laws, both of which are explained by the process of meiosis (Fig 32.4).

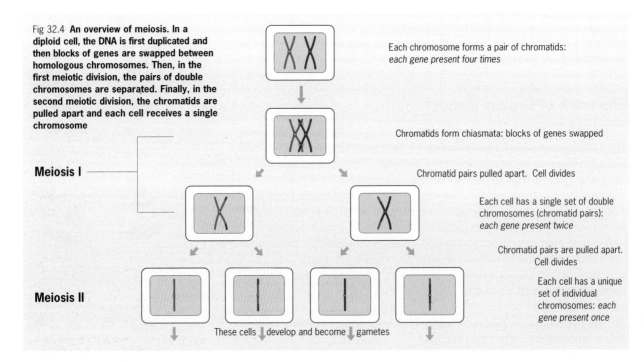

Fig 32.4 **An overview of meiosis. In a diploid cell, the DNA is first duplicated and then blocks of genes are swapped between homologous chromosomes. Then, in the first meiotic division, the pairs of double chromosomes are separated. Finally, in the second meiotic division, the chromatids are pulled apart and each cell receives a single chromosome**

Meiosis I

Meiosis II

Each chromosome forms a pair of chromatids: *each gene present four times*

Chromatids form chiasmata: blocks of genes swapped

Chromatid pairs pulled apart. Cell divides

Each cell has a single set of double chromosomes (chromatid pairs): *each gene present twice*

Chromatid pairs are pulled apart. Cell divides

Each cell has a unique set of individual chromosomes: *each gene present once*

These cells develop and become gametes

- Mendel's first law, the **law of segregation**, states:

  An organism's characteristics are determined by internal factors which occur in pairs, only one of which can be present in a single gamete.

  This is due to the fact that meiosis separates homologous pairs of chromosomes, so that only one of each pair passes into the gamete.

- Mendel's second law, the **law of independent assortment** states:

  Each of a pair of contrasted characteristics may be combined with each of another pair.

  Again, this is explained by meiosis; any one of a pair of chromosomes can pass into the gamete with any member of another pair. The significance of this is seen in dihybrid (two-gene) inheritance.

Monohybrid inheritance: how single genes are passed on

Monohybrid (single gene) inheritance concerns the inheritance of different alleles (usually two) of a single gene. Like Mendel, we start with the pea plant. Peas have several easily observable features which are controlled by single genes (see Fig 32.3).

For example, pea plants have one gene for height The height gene has two alleles: tall (T) and dwarf (t). Pea plants are diploid and so have two alleles. There are therefore three possible genotypes:

TT – homozygous for T (homo = same)

Tt – heterozygous (hetero = different)

tt – homozygous for t

Consider what happens when a homozygous tall plant is crossed with a homozygous dwarf plant (Fig 32.5). All the gametes from the tall plant contain a T allele and all those from the dwarf plant have a t allele. These combine at fertilisation to give offspring all with the genotype Tt. The **first generation**, known as the **F1**, are all tall, because tallness is dominant. However, although they look identical to the tall parent plant, they are different in one very important respect: they are heterozygous and not homozygous.

If two of these heterozygous plants are crossed, half of the gametes from each parent are T and half are t, giving us four possible genotypes in the **second generation**, the **F2**:

25% are   TT
25% are   tT
25% are   Tt
25% are   tt

The first three genotypes give tall plants (the T allele is present) but a quarter of the plants are dwarf (they are homozygous for the t allele).

It is important to realise that chance plays a very important part in genetics. If we took four F2 pea plants we might expect to get three tall plants and one dwarf, but it is highly likely that we would get something else, such as four and none or two and two. The greater the number of offspring, the higher the chance that the numbers reflect the expected ratio. This is why Mendel had to carry out *hundreds* of crosses before he became convinced of the underlying ratio – see the Assignment at the end of this chapter.

The test cross

It is often important to know whether an individual displaying a dominant trait is homozygous or heterozygous for a particular allele. For example, how can we tell if a tall pea plant is pure breeding, or if it is carrying a hidden dwarf allele? To find out we do a **test cross**. This involves crossing the tall plant with a homozygous dwarf plant (tt). If the tall plant is homozygous (TT), all the offspring produced are tall. However, if it is a heterozygote (Tt), about half the offspring are tall and half are dwarf.

In genetics, different alleles are often denoted by letters, such as Aa. Usually, the capital letter stands for the dominant allele. As a tip, if you are choosing letters to represent alleles, chose those whose capital cannot be confused with the lower case version. So, Rr is better than Ss. This is very important in exams, where your handwriting may be worse than usual.

Homozygous individuals are said to be **true breeding**. This means that they will always produce the same phenotype of offspring because they are not hiding a recessive allele.

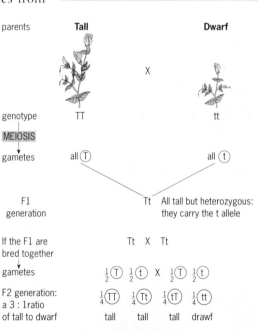

Fig 32.5 **Monohybrid inheritance in pea plants. Pure breeding tall and dwarf pea plants are homozygous (TT and tt). When crossed, the plants of the first (F1) generation are all tall but they are heterozygous (Tt), carrying the allele for dwarfness. If these heterozygous individuals are selfed (bred together) the dwarf form reappears in the next (F2) generation in a ratio of 3 : 1 (ie 1 in 4). We say that the heterozygotes do not breed true**

A In pea plants, yellow seeds are dominant to green. If a heterozygous yellow-seeded plant were crossed with a green-seeded plant, what ratio would you expect in the F1 generation?

B Why is a homozygous dwarf plant used in a test cross? Why not use a homozygous tall plant?

Monohybrid inheritance in humans

Clear-cut examples of monohybrid inheritance in humans are relatively rare, and often involve genetic disease where people inherit one or more faulty alleles. Genetic diseases are often recessive because faulty alleles which fail to make an important protein can be masked by normal ones which function properly. Table 32.1 shows examples of monohybrid inheritance in humans.

In contrast, some genetic disorders such as Huntington's disease (see the Opener of this chapter) are caused by dominant alleles. The alleles concerned code for a product which actively causes damage; the symptoms are not due to an allele not doing its job. Such alleles are dominant because the presence of a normal allele cannot mask the symptoms.

Table 32.1 **Examples of monohybrid inheritance in humans**

Traits	Features
Dominant traits	
Huntington's disease	See Opener
Freckles	
Dimple in chin	
Recessive traits	
Sickle cell anaemia	Haemoglobin polymerises, distorting red blood cells into a sickle shape. This leads to circulatory blockages and anaemia. See the Assignment on page 505.
Attached ear lobes	
Phenylketonurea (PKU)	Inability to process the amino acid phenylalanine. Protein intake must be monitored to prevent build-up of the amino acid to harmful concentrations. Can be fatal if not treated.
Haemophilia (also sex linked – see below)	Inability to produce a critical blood-clotting factor, carried on the X chromosome. Haemophilia is now treated by supplying the affected individuals with the clotting factor.
Albinism (inability to make pigment melanin)	
Rhesus blood group	
Cystic fibrosis	Excessive mucus production, especially in lungs and pancreas. Breathing is affected, and sufferers are very susceptible to infections.
Galactosemia	Inability to convert galactose to glucose in the liver. All lactose must be avoided to prevent fatal levels of galactose from accumulating.
Tay-Sachs' disease	Inability to produce critical lipases, which results in fatty accumulations in the brain.
Lactose intolerance	Inability to breakdown the disaccharide lactose to glucose and galactose. Leads to vomiting, diarrhoea, flatulence. Avoiding lactose prevents symptoms.

✔

The ability to roll the tongue is often cited as a simple example of inheritance in humans. However, although people seem to fall neatly into one of two categories – you can either roll your tongue or you can't – there is considerable debate about whether or not this skill can be learned, and whether it is controlled by one gene or more.

Inbreeding and outbreeding

Everyone agrees that brothers and sisters or other closely related people should not have children. In most human societies such **inbreeding** is illegal for social, moral and biological reasons.

Biologically, inbreeding is undesirable because genetic variation in the offspring is reduced. Inbreeding promotes homozygosity: inbred individuals are homozygous for many more alleles than individuals produced by outbreeding. So, faulty recessive alleles are more likely to be paired up, resulting in a much higher proportion of genetic defects and disease.

In practice, inbreeding tends to happen in small, isolated human and animal populations, or as a consequence of artificial selection. Pedigree dog and cat breeding is a good example of this. To maintain the most desirable features, champion animals are often bred with near relatives. Analysis of the pedigree certificates of the champions

CYSTIC FIBROSIS: AN EXAMPLE OF MONOHYBRID INHERITANCE IN HUMANS

THE DISEASE CYSTIC FIBROSIS, covered in detail in Chapter 34, is caused by the mutation of a single allele. A normal allele (C) makes a membrane protein essential for the proper functioning of certain epithelial cells. The mutated allele (c) does not code for a functional protein.

Most people have two healthy alleles: their genotype is CC. In the UK population, however, around one person in 25 carries a faulty allele. These carriers (Cc) are perfectly healthy because they also possess a normal allele which makes the working protein.

However, in one couple in 625 (1 in 25 × 1 in 25) both partners are carriers. If they have children, there is a one in four chance that the child will inherit both faulty alleles and therefore be a cystic fibrosis sufferer. As one CF child in four is born to one couple in 625, approximately one child in every 2500 is born with this condition.

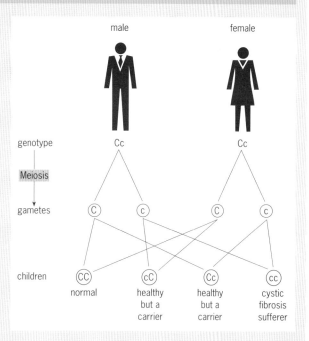

Fig 32.6 **The inheritance of cystic fibrosis. If both parents are carriers, there is a 25 per cent chance that any child will be a cystic fibrosis sufferer. There is also an equal chance that a child will inherit no faulty alleles and that his/her descendants will be completely free from the disease. There is a 50 per cent chance of the child being a carrier of cystic fibrosis**

in a particular breed will often show that the same individuals crop up again and again. The consequence of this artificial selection (see page 527) is a relatively high proportion of genetic defects.

Outbreeding (the breeding of unrelated individuals) is more desirable as it produces more new genetic combinations and greatly reduces the chance that faulty genes will be expressed. Many organisms have developed mechanisms which prevent inbreeding (Fig 32.7). If pedigree animals are cross-bred, rather than inbred (cross a Siamese cat with a Persian, for example), you often get a fitter, healthier animal. We say that such offspring show **hybrid vigour** and cross-breeding is a favourite tool of plant breeders who find that hybrids often grow faster and show better disease resistance.

Fig 32.7 **When woodlice lay a batch of eggs, the offspring are either all male or all female. This means that brothers and sisters cannot mate, thus reducing the chances of inbreeding**

Dihybrid inheritance: how two genes are passed on

What happens when a tall, purple-flowered pea plant is crossed with a dwarf, white-flowered individual? Do you always get tall plants with purple flowers, or can the features re-combine, producing tall, white-flowered plants, and dwarf, purple-flowered ones. The simple answer is yes, you can get these recombinants, thanks to the process of meiosis.

The inheritance of two separate genes is called **dihybrid inheritance**. If, as is usual, the two genes are on separate chromosomes, Mendel's second law (page 506) applies: Each of a pair of contrasted characteristics can be combined with each of another pair. So, an organism with a genotype of AaBb can produce gametes of AB, Ab, aB or ab. In other words, one allele from each pair passes into the gamete, and all four combinations are possible. This is due to **independent assortment** during meiosis (Fig 32.8).

C There are approximately 700 000 live births every year in the UK. How many of these babies would you expect to have cystic fibrosis?

D Suggest why there is less inbreeding in most human societies today than in previous centuries.

Fig 32.8 **Consider a diploid cell with the genotype AaBb. Meiosis produces four different types of gametes by independent assortment**

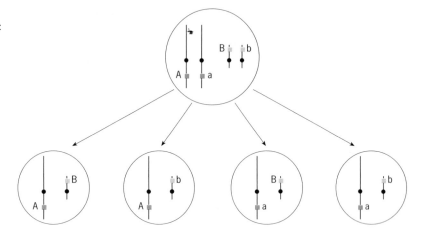

For an example, let's return to pea plants. In addition to tall and dwarf, we can consider a gene for seed texture. The allele R for round seeds is dominant to r, the allele for wrinkled seeds.

Fig 32.9 shows a cross between a pure breeding, tall, round-seeded plant, (TTRR), and a pure breeding dwarf, wrinkled-seeded plant (ttrr). One allele from each pair is passed into the gametes. All of the gametes from the tall, round-seeded plant are TR and all those from

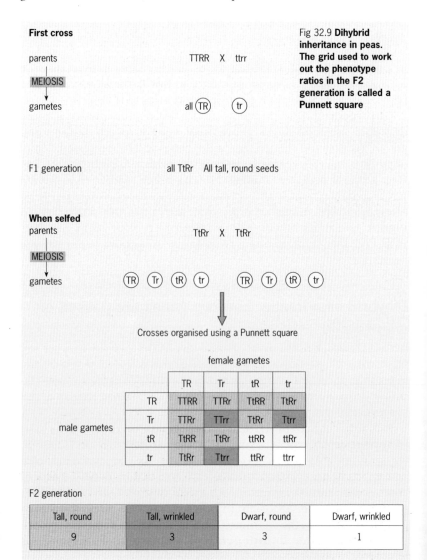

Fig 32.9 **Dihybrid inheritance in peas. The grid used to work out the phenotype ratios in the F2 generation is called a Punnett square**

the dwarf, wrinkled-seeded plant are tr. All of the F1 generation have the genotype TtRr and are tall with round seeds.

When we cross two TtRr individuals, things get more complex! Each plant produces four different gametes. Fertilisation is random, producing sixteen different combinations in the F2 generation. To organise our thoughts we can use a **Punnett square** (see Fig 32.9)

Dihybrid inheritance questions in examinations

Many examination questions require the candidate to work out the ratios of offspring resulting from a particular cross. If you have not practised these questions they can be very time consuming, leading to stress and silly mistakes.

The example given above (TtRr × TtRr) is the most complex dihybrid situation you can get. This is because both male and female are heterozygous for both alleles, and so produce four different gametes each, resulting in sixteen different combinations. All other genotypes will produce fewer combinations. If you practise these combinations you can learn to work out the gametes and resulting ratios in your head without resorting to Punnett squares and counting the results.

Consider the cross TtRR × Ttrr. The first individual produces the gametes TR and tR in equal amounts. The second one produces Tr and tr in equal amounts. So, there are four possible combinations: the same number as in a monohybrid cross.

You should see that the offspring will be one quarter TTRr, two quarters TtRr and one quarter ttRr, giving three tall plants with round seeds to one dwarf plant with round seeds. You would not get any dwarf plants with wrinkled seeds.

Some exam questions give the ratios of offspring phenotypes and ask you to work backwards to the original genotypes.

If you do not get the expected ratios, consider **linkage** as an explanation (see below).

2 WHEN MENDEL'S RULES DON'T APPLY

Linkage and crossing over

So far, we have considered the inheritance of genes on different chromosomes which could be separated at meiosis. We say that genes that are on the same chromosome are **linked**. Such genes cannot obey Mendel's laws because they do not undergo independent assortment. They are inherited together unless separated by crossing over during prophase 1 of meiosis (Fig 32.10).

To illustrate linkage, let's consider the sweet pea plant (*Lathyrus odoratus* – a different species from the edible pea we have been looking at). In this species, the allele for purple flowers is dominant to red flowers, and the allele for elongated pollen grains is dominant to round pollen. If pure breeding plants with purple flowers and elongated pollen grains are bred with plants with red flowers and round pollen grains, the F1 generation all have purple flowers and elongated pollen.

Humans have over 60 000 genes present on just 23 pairs of chromosomes. So, any single gene is linked with a few thousand others because they are all on the same chromosome.

Fig 32.10 **When two genes are close together on the same chromosome they are almost always inherited together: they are linked. The chance of being separated by crossover depends on the distance between them**

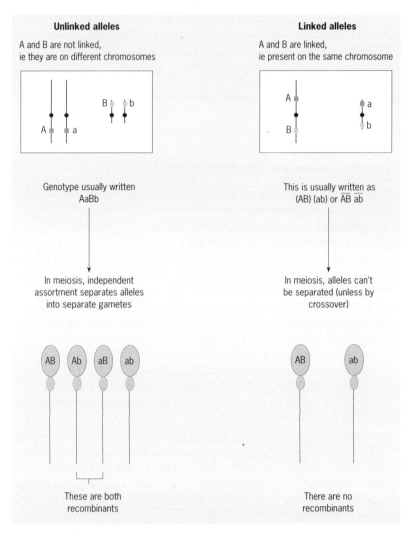

When the F1 are **selfed** (bred together) you would expect a 9:3:3:1 ratio, but this does not happen. The F2 generation are mainly like the parents (purple and elongated or red and round) but there are a few recombinants (red and elongated or purple and round). This is because the genes for flower colour and pollen shape are linked (present on the same chromosome) and so are inherited together. Any recombinants we get must therefore be due to crossing over.

Consider the set of results shown in Table 32.2.

Table 32.2 **An example of a phenotypic ratio which indicates linkage**

Phenotype	Approx. expected numbers if no linkage (9:3:3:1)	Observed numbers in F2 generation
Purple, elongated	405	336
Purple, rounded	135	31
Red, elongated	135	28
Red, rounded	45	325
Total	720	720

The total number of offspring is 720 of which 59 (31 + 28) are recombinants. Eight per cent of the offspring result from crossing over and so, by convention, we can say that the **loci** for flower colour and pollen shape are 8 units apart on the chromosome. If the percentage of recombinants was higher, we would know that the loci were further apart on the chromosome.

The term **locus** is used to describe the position of a gene on a chromosome.

When genes are not linked, we expect them to be separated 50 per cent of the time because there is a 50 per cent chance that alleles on separate chromosomes will be inherited together. When looking for linkage, geneticists look for a frequency of recombinants significantly less than 50 per cent.

See question 1.

Chromosome maps

The frequency with which linked genes are separated by crossover can tell us a lot about the relative positions, or loci, of these genes on the chromosome. Two alleles that are close together are rarely separated by crossing over, while those located on opposite ends of the chromosome are separated almost every time. Analysis of crossover frequency can be used to work out the relative positions of alleles and so make chromosome maps (Fig 32.11).

Genes Y and Z have a crossover frequency of 10 per cent and so are 10 units apart. A third cross between Y and X determines the relative positions of X, Y and Z. If the order of genes is XYZ (as shown), the crossover frequency between X and Y is 13 per cent. If the order is XZY, the crossover frequency is 33 per cent.

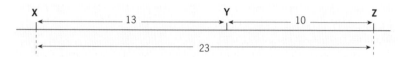

Fig 32.11 **The position of genes on chromosomes can be worked out by interpreting crossover frequencies. Consider three genes, X, Y and Z. Genes X and Z are found to have a crossover frequency of 23 per cent, so they are 23 units apart on the chromosome**

EXAMPLE

An A-level genetics question involving linkage

This example is a modified question from an UCLES Modular Biology examination in Applications of Genetics.

The mosquito, *Aedes aegypti*, has a spotted and a spotless form on a grey or yellow body. The dominant allele is spotless, S, and the recessive allele is spotted, s. The allele for grey body G, is dominant to the allele for yellow body, g. A cross was made between homozygous spotless, grey bodied mosquitoes and spotted, yellow-bodied mosquitoes. The F1 individuals were then crossed with the double recessive strain and the numbers of the resulting phenotypes were counted. The results are shown below.

Phenotype	Number
Spotless, grey bodied	442
Spotless, yellow bodied	458
Spotless, yellow bodied	46
Spotted, grey bodied	54

Q State the genotype and the phenotype of the F1 using the above symbols.

A The question is asking for the genotype of the first generation. You know that the parents are homozygous spotless, grey (genotype: SSGG) and spotted, yellow (genotype: ssgg). The genotype of the F1 must be SsGg, giving a phenotype of spotless, grey bodied.

Q Using a genetic diagram, explain fully the results shown in the table.

A The table shows the result of a cross between an F1 individual, SsGg, and a double recessive which must be ssgg.

So we have:

Parents with genotype	Spotless, grey SsGg	× Spotted, yellow ssgg		
Gametes	SG Sg sG sg	×	all sg	
Expected ratio Phenotypes	25% SsGg Spotless, grey 1	25% Ssgg Spotless, yellow : 1	25% ssGg Spotted, grey : 1	25 % ssgg Spotted, yellow : 1

You should see that the expected results are not the same as those given in the table, so there must be an extra complication. The key feature from the table is that there are more of the parental phenotypes than expected, and fewer recombinants. This should scream **LINKAGE** at you!

To produce a really thorough answer you should also work out the percentage of recombinants.

The total number of crosses in the table is 1000, of which 100 (46 + 54) are recombinants. This leaves you with the simple calculation to show that 10 per cent of the mosquitoes are recombinants, and so the two genes are 10 units apart on the chromosome.

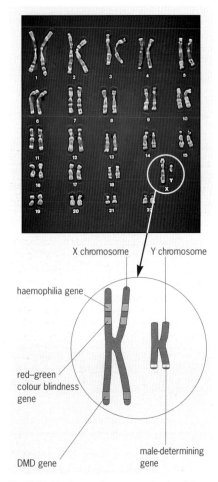

Fig 32.12 **The human karyotype showing 22 pairs of autosomes and one pair of sex chromosomes. This is a male: females have two identical X chromosomes. Note that the Y is much smaller than the X, and carries fewer genes. It does, however, carry the male determining gene**

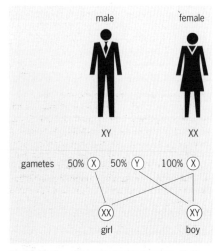

Fig 32.13 **Sex determination in humans. Half of the sperm carry an X chromosome, half carry a Y**

See question 2.

3 SEX DETERMINATION AND SEX LINKAGE

Living things have evolved a variety of mechanisms to determine whether a new organism will be male or female. In crocodiles, for instance, the temperature of egg incubation determines the sex of the embryo. However, in many animals, including humans, sex determination depends on the inheritance of special **sex chromosomes**.

Every human cell contains 23 pairs of chromosomes: 22 pairs of **autosomes** and one pair of sex chromosomes (Fig 32.12). Females have two large X chromosomes and so we say they are the **homogametic sex**. Males have one X chromosome and one smaller Y chromosome, and are therefore the **heterogametic sex**.

In the female parent, all eggs receive an X chromosome during meiosis (Fig 32.13). However, meiosis in the male parent produces sperm, half of which receive a Y chromosome and half an X. So it's a race: if a Y sperm fertilises the female's egg, the offspring is male. If X wins, its a female.

Interestingly, in humans, about 114 boys are conceived for every 100 girls. The reason for this is unknown. However, more male embryos die and so at birth the numbers are down to 106 to 100. By puberty the numbers are equal and in old age the females outnumber males by two to one.

In 1990, geneticists discovered that the Y chromosome carries a male determining gene which codes for a **testis determining factor** (**TDF**). All embryos are female unless the active TDF imposes maleness upon it. When a male and female embryo share the same blood supply, it is possible that the TDF can produce hormones which can make an XY *and* an XX embryo develop as a male. This is a very rare situation in humans but is common in cattle.

Sex-linked inheritance

Genes carried on the sex chromosomes are said to be **sex-linked**. Human females have two X chromosomes which, like the autosomes, carry two alleles of every gene. Females therefore have two sets of sex-linked alleles. In males, however, the Y chromosome is smaller and cannot mirror all of the genes on the X chromosome, so males have only one set of most sex-linked alleles. This is why males suffer from the effects of X-linked genetic diseases more often than females. There are no known Y-linked traits, probably because the Y chromosome carries so few genes.

A classic example of a sex-linked trait transmitted by the X chromosome is **haemophilia**. People suffering from this disease do not make factor VIII, an essential component in the complex chain reaction of blood clotting (see page 417). In addition to the problems caused if they injure themselves, haemophiliacs suffer from internal bleeding as a result of normal activity. Bleeding at the joints during even light exercise is a particular problem. However, haemophiliacs can usually live a full and active life by having regular injections of factor VIII.

Haemophilia is caused by a sex-linked recessive allele. Females have a pair of alleles but males possess only one. So, if a male inherits the haemophilia allele, he has the disease since he cannot possess another healthy allele to mask its effect.

Using the following notation:

X^h = the haemophilia allele on the X chromosome

X^H = the healthy allele on the X chromosome

Y = the Y chromosome which does not carry the allele

we can see that the possible genotypes are:

$X^H X^H$ = healthy female

$X^H X^h$ = healthy, carrier female

$X^H Y$ = healthy male

$X^h Y$ = haemophiliac male

The inheritance of haemophilia can be studied by using a **pedigree diagram**. The word pedigree means 'of known ancestry' and applies as much to humans as it does to domestic animals. The pedigree of the Royal family can be traced back to 1066 and shows that Queen Victoria was a carrier of the haemophilia allele and passed it on to four of her nine children (Fig 31.14). Other sex linked traits in humans include Duchenne muscular dystrophy (DMD, Fig 32.15) and red–green colour blindness (Fig 32.16).

The pedigree of haemophilia in Queen Victoria's family

Key: ◯ normal female ☐ normal male
     ● known female carrier ■ haemophiliac male

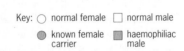

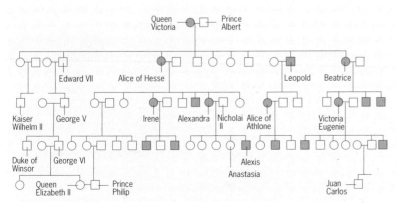

Fig 32.14 **Queen Victoria (1819–1901) had nine children and passed the haemophilia allele on to two daughters and one son. Her eldest son, Edward VII, did not inherit the faulty allele and so the disease is not carried by his descendants, the present-day British Royal Family**

E What would the notation be for a female haemophiliac? Suggest why they are much rarer than male haemophiliacs.

Fig 32.15 **Duchenne muscular dystrophy affects mainly boys. The body muscles are gradually replaced by fibrous tissue and become weak, usually confining sufferers to a wheelchair by the age of 10. Few live beyond 20. One male child in 4000 is born with this disease, sometimes caused by a spontaneous mutation, so the mother may not even be a carrier. Gene therapy holds new hope for the future**

Fig 32.16 **A person with red/green colour blindness cannot distinguish between red, green, orange and yellow. On these Ishihara test charts, a person with normal vision would see numbers but a sufferer would see only a random pattern of dots. This condition is caused by a defective allele which codes for the one of the three groups of light-sensitive cone cells in the retina. Eight per cent of males but only 0.4 per cent of females suffer from colour blindness**

F If a female carrier of the haemophilia allele marries a normal male (as was the case of Queen Victoria and Prince Albert) and they have a son, what is the probability that he will have haemophilia?

4 MULTIPLE ALLELES

Sometimes there are more than two alleles for a particular gene. A classic example which shows multiple alleles of the same gene is the ABO blood group in humans.

The entire human population can be classified into A, B AB or O depending on the proteins they carry on the membrane of their red blood cells. The ABO system is controlled by one gene (I) with three alleles, I^A, I^B and I^O. The I^A allele codes for A proteins, I^B for B proteins and I^O for no relevant proteins. I^A and I^B are codominant over I^O. Each individual inherits two alleles which combine to produce the blood group as shown in Table 32.3.

Blood group testing can sometimes be used to disprove (but not prove) parentage. Table 32.4 shows, for instance, that a man of blood group AB must pass on either an I^A or an I^B allele, and so cannot possibly be the father of a child with blood group O. DNA fingerprinting, of course, provides a much more accurate (if more expensive) test which can say with a great deal of certainty who **is** the father. The Assignment in Chapter 3 looks at the process of DNA fingerprinting.

Table 32.3 **The genetics of the ABO blood group system**

Genotype	Phenotype (blood group)
$I^A I^A$ or $I^A I^O$	A
$I^B I^B$ or $I^B I^O$	B
$I^O I^O$	O
$I^A I^B$	AB

G If a child has the blood group B and his mother is blood group O, can his father be a man with blood group A? Say why.

Table 32.4 **Possible blood groups of children born to parents of particular blood groups**

			Paternal blood group			
			A	**B**	**AB**	**O**
		Genotypes	$I^A I^A$ or $I^A I^O$	$I^B I^B$ or $I^B I^O$	$I^A I^B$	$I^O I^O$
	A	$I^A I^A$ or $I^A I^O$	A, O	A, B, AB; O	A, B, AB	A, O
Maternal blood group	**B**	$I^B I^B$ or $I^B I^O$	A, B, AB, O	B, O	A, B, AB	B, O
	AB	$I^A I^B$	A, B, AB	A, B, AB	A, B, AB	A, B
	O	$I^O I^O$	A, O	B, O	A, B	O

5 POLYGENIC INHERITANCE

So far in this chapter, we have been concerned with **discontinuous inheritance**: all-or-nothing features controlled by single genes. Many features, however, are **continuous** and are controlled by several genes which act together. We describe the inheritance of such characteristics as **polygenic inheritance**.

Skin colour in humans is an example of polygenic inheritance. The depth of skin colour depends on the amount of melanin present. This is determined by at least three genes. The alleles A, B and C contribute cumulatively to the overall amount of melanin but the alleles a, b and c do not. So, someone with the genotype AABBCC would have the darkest skin, while the genotype aabbcc would confer a very pale skin colour.

As Fig 32.17 shows, a cross between two purely homozygous individuals would result in an intermediate (heterozygous) F1 generation, but the F2 generation would show a wide variation in skin colour depth. The graph shows a **normal distribution**: most individuals have skin colours in the middle of the range. This range is one of the characteristic features of continuous variation (see page 485).

The inheritance of height in humans is polygenic.

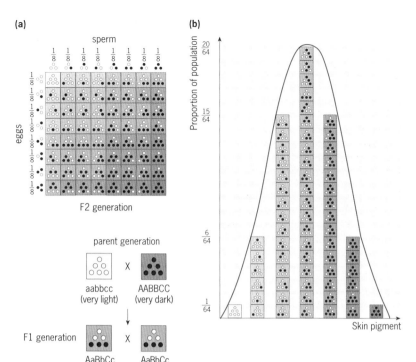

(a)

sperm

F2 generation

parent generation

aabbcc
(very light)   X   AABBCC
(very dark)

F1 generation

AaBbCc   X   AaBbCc

(b)

Proportion of population

Skin pigment

Fig 32.17 **Human skin colour is an example of polygenic inheritance which shows a normal distribution in the F2 generation**

6 EPISTASIS (WHEN GENES INTERACT)

Up to now, we have discussed the inheritance and effect of genes which act on their own. In practice, however, many genes interact with each other: the presence of one allele often affects the expression of others. We call this interaction **epistasis**.

A classic example of epistatic gene interaction is coat colour in mice. The natural coloration of wild mice is called **agouti** and is produced from banded hairs (Fig 32.18).

Two genes are involved, each with a dominant and a recessive allele. We will represent them by the notation Aa and Bb, where A and B are dominant over a and b.

The allele A codes for the ability to produce hair pigment: AA and Aa mice have pigmented hairs but all aa individuals are albinos. The B allele codes for the ability to make banded hair: BB and Bb mice have banded hair, bb mice have hair which is all one colour.

Only mice which have both dominant alleles A and B will show the agouti coloration. These can have the genotype AABB, AaBB, AABb or AaBb. Mice which are aa cannot make pigment, so, whether they are BB, Bb or bb, no bands are visible. Mice which can make pigments but not banded hair (genotypes AAbb or Aabb) are plain black.

See question 5.

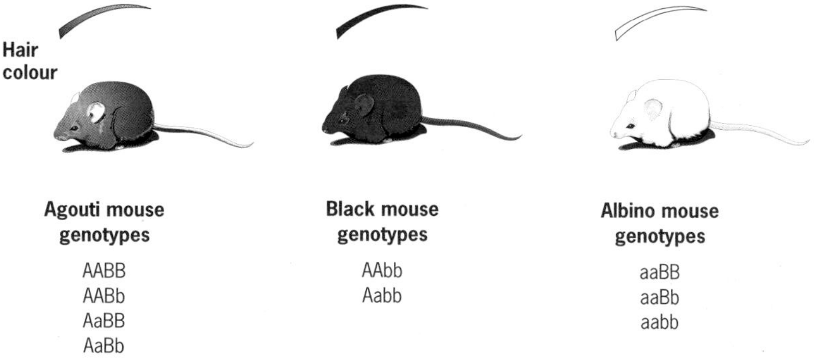

Hair colour		
Agouti mouse genotypes	**Black mouse genotypes**	**Albino mouse genotypes**
AABB	AAbb	aaBB
AABb	Aabb	aaBb
AaBB		aabb
AaBb		

Fig 32.18 **The base of the hair of a mouse with agouti coloration is dark brown/black fading to lighter brown towards the tip**

H Two agouti mice, genotypes AaBb, are bred together. What phenotypic ratio would you expect in the next generation?

SUMMARY

■ Most organisms are **diploid**; they carry two versions, or **alleles**, of each **gene**. If both alleles are the same, the organism is **homozygous** for that allele. If the pair are different, the organism is **heterozygous**.

■ During **meiosis**, only one of the alleles passes into the **gamete**. This is Mendel's first law.

■ **Monohybrid inheritance** is the study of one gene. A cross of AA × aa gives all Aa in the **first generation (F1)**. A cross of Aa × Aa gives a 3:1 **phenotypic ratio** in the **second generation (F2)**.

■ **Dihybrid inheritance** involves two separate genes, usually on separate chromosomes. Homozygous individuals AABB × aabb produce offspring which are all AaBb in the F1 generation. However, due to **independent assortment** during meiosis, AaBb individuals produce gametes of AB, Ab, aB and ab. If bred together, AaBb individuals produce sixteen different genotypes and four different phenotypes with a ratio of 9:3:3:1.

■ Genes present on the same chromosome are said to be **linked**. Such genes are inherited together unless separated by crossover in prophase 1 of meiosis. The greater the distance between the genes on the chromosome, the more often they are inherited separately.

■ Humans have 23 pairs of chromosomes: 22 pairs of **autosomes** and one pair of **sex chromosomes**. The X chromosome is larger than the Y and carries more genes.

■ The sex of a baby is determined at the moment of conception. Because gametes are produced by meiosis, half of the sperm carry a Y chromosome and half carry an X. If a Y sperm fertilises the egg, a male results; X gives a female.

■ Genes carried on the sex chromosomes are said to be **sex-linked**. Females have two complete sets of sex-linked genes, but males often have only one copy, because the Y chromosome carries far fewer genes.

■ Some diseases such as haemophilia are sex-linked; caused by defective genes on the X chromosome. Females have two alleles and so can be carriers. Males have only one allele and so cannot be carriers; they are either healthy or they have the disease.

■ Some genes have more than two alleles. The human ABO blood group is an example of such a **multiple allele**. In this case, one gene has three alleles, A, B and O.

■ Many characteristics, such as height in humans, are controlled by many different genes. We describe such characteristics as **polygenic**; they often produce a range of values. Human height is a good example of a polygenic characteristic.

■ The presence of one allele can affect the expression of an allele of a different gene. This type of gene interaction is called **epistasis**. For example, one allele may give an organism the ability to make pigment, while another allele determines the colours and/or the distribution of that pigment.

QUESTIONS

1 In maize, the allele for coloured grain, **C**, is dominant to the allele for colourless, **c**, and the allele for rounded grain, **R**, is dominant to the allele for wrinkled grain, **r**.

a) Explain what is meant by: **(i)** a gene; **(ii)** an allele.

b) Maize plants heterozygous for both characteristics were crossed. The following results were obtained:
83 coloured, rounded
33 coloured, wrinkled
29 colourless, rounded
8 colourless wrinkled
Suggest why the results are not an exact 9:3:3:1 ratio.

c) Some of the colourless rounded grains would have the genotype **ccRr** and some would be **ccRR**.

Explain the genetic crosses necessary to distinguish between these genotypes.

[NEAB Feb 1996 Modular Biology: Continuity of Life Paper, q. 2]

2 The diagram of Fig 32.Q2 shows the inheritance of one type of myopia in three generations of a human family.

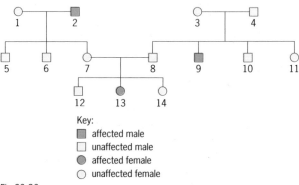

Key:
■ affected male
□ unaffected male
● affected female
○ unaffected female

Fig 32.Q2

a) Give **one** piece of evidence from the diagram which suggests:

(i) that the myopia is determined by a recessive allele;

(ii) why it is **not** likely that myopia is a sex-linked condition.

b) Using **A** for the dominant allele and **a** for the recessive allele, give the possible genotype or genotypes for:
(i) individual 7; **(ii)** individual 12.

[NEAB Feb 1996 Modular Biology: Continuity of Life Paper, q. 7]

3 Pure-breeding pea plants with grey seed coats and tall stems were crossed with plants with white seed coats and short stems. The offspring (F1 generation) all had grey seed coats and tall stems.

a) State suitable symbols for the alleles for colour of seed coat and height of stem.

b) Explain, by means of a genetic diagram, the ratios of genotypes and phenotypes you would expect if the F1 generation were self-fertilised.

c) Explain how you could find out the genotype of a plant with grey seed coats and tall stems.

[UCLES Spring 1996, Modular Biology: Central Concepts in Biology Specimen Paper, Section A, q. 1]

4 Fig 32.Q4 shows part of a family tree in which the inherited condition of phenylketonuria occurs.

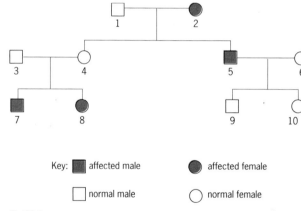

Key: ■ affected male   ● affected female
□ normal male   ○ normal female

Fig 32.4

a) Identify and explain **one** piece of evidence from this family tree to show that the allele for phenylketonuria is recessive to the allele for the normal condition.

b) Giving a reason for your answer in each case, identify **one** individual who must be: **(i)** heterozygous; **(ii)** homozygous.

c) If individual 10 married a man who was heterozygous for the gene, what is the probability that their first child would be affected?

[AEB 1996 Human Biology:Specimen Paper 1, q. 7]

5 Fig 32.Q5 shows four different comb shapes in chickens, resulting from the interaction of two unlinked gene loci, **R/r** and **P/p**. The genotypes of each comb shape

are given. A dash (–) in the genotype indicates that either the dominant or the recessive allele may be present.

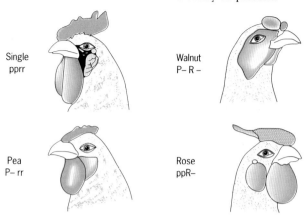

Single
pprr

Walnut
P– R –

Pea
P– rr

Rose
ppR–

Fig 32.Q5

a) Single-combed chickens were crossed with pure-breeding walnut-combed birds. The F1 were then bred together to produce the F2 generation. Draw genetic diagrams to show the genotypes and phenotypes of the F1 and F2 generations.

NB Take care that **P** and **p** cannot be confused in your answer.

b) Explain how varieties of chickens with very different plumage or comb shapes can be produced.

The interaction of these two unlinked gene loci produces four separate phenotypes and is an example of discontinuous variation.

c) **(i)** Suggest **one** phenotypic characteristic in chickens which shows **continuous** variation.
(ii) Explain the genetic basis of continuous variation.

[UCLES Spring 1996 Modular Biology: Genetics and its Applications Module Specimen Paper, Section A, q. 1]

6 Familial hypophosphataemia is a sex-linked condition caused by a dominant allele on the x chromosome. The pedigree of a family with this condition is shown in Fig 32.Q6.

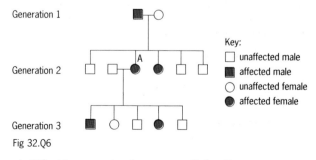

Generation 1

Generation 2

A

Generation 3

Key:
□ unaffected male
■ affected male
○ unaffected female
● affected female

Fig 32.Q6

a) What is meant by the term *sex linkage*?

b) Explain why none of the males in generation 2 suffers from hypophosphataemia.

c) If person A were to have another son, what is the probability that this son would suffer from hypophos-phataemia? Give a reason for your answer.

[ULEAC June 1994 Human Biology: Paper 3, q. 3]

THE CHI-SQUARED TEST

The chi-squared test is a simple statistical method that scientists use to tell if their results are significant or due to chance. For instance, if you were to roll a dice, you would have a one in six chance of getting any particular number. If you rolled the dice six times, you would probably not get each number once; you may get three twos or some other combination. However, the more times you roll the dice, the greater the probability that all six numbers will be represented evenly.

If you are playing a game with someone who gets a higher proportion of sixes than normal, you might suspect them of playing with a loaded dice. Without investigating the dice itself, how could you decide whether that person was just lucky or that the difference was due to cheating? You could use the chi-squared test.

The nature of science and the null hypothesis

Most people accept that there is a link between smoking and lung cancer but, in science, it is virtually impossible to prove anything absolutely. What you can do is gather support for your **hypothesis**.

You could do some research into lung cancer victims and divide them into smokers and non-smokers. You could then use the chi squared test to decide if there was a significant difference in the number of smokers who developed lung cancer compared with the non-smokers.

To use this statistical test you must approach the problem from a specific angle, and this involves developing a **null hypothesis**.

In the lung cancer example, you would use the hypothesis: 'There is **no** difference in the incidence of lung cancer between smokers and non-smokers'. You could then perform the chi-squared test (or some more sophisticated statistics) and come up with the conclusion that 'there is a very low probability that these results are due to chance' in which case you can reject the null hypothesis and support the idea that there *is* a link between smoking and lung cancer.

1 Write a null hypothesis that you might use to investigate the following questions:

a) What is the effect of lawn fertiliser on the size of the earthworm population in your garden?

b) What is the effect of radiation on the incidence of genetic mutation?

The test itself

The basic idea behind the chi-squared test is that you compare the observed results with those you would expect. The extent to which the results vary is sometimes described as 'goodness of fit'. This means do the actual results fit the results that we had expected?

The formula for the chi-squared test is:

$$X^2 = \sum \frac{(O - E)^2}{E}$$

In this formula
O = the observed result,
E = the expected result and
Σ = the sum of

In practice it is best to use a table like the one below to progressively work out all the steps needed by the formula. In this worked example, a geneticist performed one of Mendel's classic pea experiments and got the following results.

Yellow, round	318
Yellow, wrinkled	103
Green, round	106
Green, wrinkled	33
Total	560

The geneticist wanted to know whether these results did actually reflect the expected 9:3:3:1 ratio, or whether they deviated so significantly that other forces were at work, and that the 9:3:3:1 hypothesis was in doubt. She did the following calculations:

Category	Hypothesis	Observed	Expected	$O - E$	$(O-E)^2$	$\frac{(O-E)^2}{E}$
Yellow, round	9	318	315	3	9	0.028
Yellow, wrinkled	3	103	105	−2	4	0.038
Green, round	3	106	105	1	1	0.009
Green, wrinkled	1	33	35	−2	4	0.114
Total		560				= 0.189

So the value of chi squared is 0.189. What next? The hard work is now done and we only need one more piece of information; the number of **degrees of freedom**, which is a measure of the spread of the data. The number of degrees of freedom is always one less that the number of categories of information. So in this case, it is four categories minus one which equals three degrees of freedom.

We now look up the value of 0.189 with three degrees of freedom on a table of chi-squared values. The following is an extract from the chi-squared tables:

Table 32.A1 **Extract from chi-squared tables**

	Probability (eg 0.5 = 50%)									
	0.99	**0.98**	**0.95**	**0.9**	**0.7**	**0.5**	**0.3**	**0.2**	**0.1**	**0.02**
1 D of F	.00016	.0063	.0039	.0158	.148	.455	1.07	1.64	2.71	5.41
2 D of F	.0201	.0404	.103	.211	.713	1.39	2.41	3.22	4.60	7.82
3 D of F	.115	.185	.352	.584	1.42	2.37	3.66	4.64	6.25	9.84

From Table 32.A1 we can see that our value of 0.189 lies between the 0.98 and 0.95 column. From this we can say that the probability that our results are significant is between 95 and 98 per cent. Put another way, it tells us that the probability that these results are due to chance is between 2 and 5 per cent. By convention, scientists say take the cut-off point as 5 per cent. If the test shows that there is a less than 5 per cent probability that your results are due to chance, you can reject your null hypothesis.

2 Use the chi-squared formula to work out the following:

a) A student tosses a coin one hundred times. She gets 64 tails and 36 heads. Is the coin biased or are the results due to chance?

b) A geneticist managed to get his hands on a breeding pair of incredibly rare Matabili dung beetles. The male had green eyes and a blue abdomen. The female had pink eyes and an orange abdomen. When he crossed them he got the following results:

Table 32.A2

Phenotype	Numbers of offspring in F1
Green eyes, blue abdomen	563
Pink eyes, blue abdomen	190
Green eyes, orange abdomen	183
Pink eyes, orange abdomen	64
Total	1000

Do these results conform to the expected 9:3:3:1 ratio?

c) Apart from their diet, these beetles look like being ideal organisms for the study of genetics. What features make an organism suitable for the study of inheritance in the laboratory?

d) Much progress in the field of genetics has been made using the fruit fly **Drosophila**. Find out some of the features that can be studied using Drosophila.

Chi-squared in examinations

If you have worked your way through the above examples, you may have concluded that chi squared involves a lot of arithmetic and is time consuming. The examiners have come to the same conclusion and therefore it is unlikely that you will be set a full investigation to work through from scratch: they want to test your biology, not your maths. However, this test is a very useful tool and examiners will want you to appreciate its uses, and possibly to do some simple calculations. Another example of how to use the chi-squared test is provided on pages 598-599.

33 Evolution

Fifty species of bird are unique to New Zealand. Of these, fourteen have evolved to become poor flyers or are totally flightless. The kakapo, in particular, is close to extinction

ON A TINY REMOTE ISLAND off New Zealand's South Island the world's largest parrot – the kakapo – is just clinging to existence. Looking like a large green football with a beak, the kakapo was once an important part of New Zealand's bird-rich ecosystem. Like many island birds that have evolved in an environment free from predators, the kakapo is flightless. It has also developed a few other odd characteristics.

Locals knew when it was the kakapo's mating season because of its incredible mating call: a sort of giant heartbeat in the night. The male kakapo sits on a hillside and emits a huge 'whuumph' from his vocal sacs. The problem is that, like all bass sounds, it is non-directional. This is a big problem because the females cannot tell where the call is coming from. All is well when numbers are high, but when kakapos are few and far between, the females cannot easily find the males.

Like the dodo, the kakapo's idyllic life-style came to an end and their numbers started to plummet after human settlers arrived with their animals. The Polynesian colonists (the Maoris) settled in New Zealand 1000 years ago, bringing dogs and rats. These ate the eggs from the ground-level nests of the kakapo and other flightless birds. More recently, Europeans arrived, adding cats to the predators. Kakapos have not developed any new behaviour to deal with such enemies.

There are now less than one hundred kakapos left, and these occupy Little Barrier Island: a sanctuary free from humans and their pets. If the females can manage to find the males, they may yet be able to make something of a recovery.

1 WHERE DO LIVING THINGS COME FROM?

To most human eyes, the living world seems fairly static. Like produces like: elephants give birth to baby elephants, frogs make frogs. It seems that no new species appear. But, as we saw in the second section of this book (Chapters 10 to 13), there are literally millions of different species on Earth. Where did they come from?

Theories on the origin of species

Ideas about the origins of species – including ourselves – have been central themes in philosophical and religious thought for centuries. Until the nineteenth century, virtually everybody in the Christian world looked to the Bible. The book of Genesis provided the answers: all animals and plants were created at the same time, by

Fig 33.1 **The discovery in the nineteenth century of fossils such as this *Triceratops* fuelled the idea that the living world was not static. Different layers of rock held different sets of organisms and this led to the acceptance that the living world was not fixed: plants and animals could change with time**

the great Creator. Even the eminent taxonomist Linnaeus, who in 1742 published his system for classifying all living plant species known at the time, said 'there are as many species as God created in the beginning'.

The idea that species were not fixed, and that the infinite variety of living things had developed through a slow process of evolution, seems to have arisen first in early Greek philosophy and reappeared from time to time throughout the following centuries. It failed to gain lasting acceptance because no explanation could be found to show how the process might have come about. In addition, in many cultures, any 'non-religious' thought was strongly discouraged, often on penalty of death.

A catalyst in the development of the theory of evolution was the discovery of fossils (Fig 33.1). These were clearly the remains of extinct creatures. Why were they no longer alive?

One explanation said that extinct species were the victims of a great global catastrophe, and that a new set of organisms had been created to replace them. It was even suggested that only the passengers aboard Noah's Ark survived a great global flood. When it was pointed out that different rocks contained different types of plants and animals, it seemed that there would have had to be many such floods, and so the catastrophe theory gradually lost credibility.

An early attempt to explain the mechanism of evolution came from the French naturalist Lamarck, who had made extensive studies of different life forms. He was convinced of the process of evolution, but the mechanism he proposed in 1809 to explain it was wrong. Lamarck supposed that organisms adapt to their environment by developing new structures and losing old ones. Such differences, *acquired during the animal's life*, were then passed on to the next generation. As an example he cited the webbed feet of water birds. Lamarck suggested that, in an effort to swim, the birds extended their toes and so stretched the skin between them. The stretched condition was then inherited and this process was repeated until it produced a fully-webbed foot. With no knowledge of genes, chromosomes or DNA, this idea of 'the inheritance of acquired characteristics' was perfectly reasonable for the time.

?

A What was wrong with Lamarck's theory of evolution?

CHARLES DARWIN

Fig 33.2 **Charles Darwin, 1809 to 1892**

CHARLES DARWIN (Fig 33.2) was the son of a doctor. After studying medicine at Edinburgh he changed his mind and prepared for a career in the Church. However, despite going to Cambridge to study theology (religion), Darwin retained his interest in biology and geology.

Darwin was offered the post of naturalist on the HMS Beagle, embarking on a scientific survey of South American waters. It was observations he made during this trip that first stimulated the young Darwin into contemplating the origins of species. Not only did he become convinced of the process of evolution, but he developed an idea about how it might happen. This trip was not only the turning point in Darwin's life; it was also the start of the greatest ever revolution in our perception of the living world, and of our place within it.

On returning home, Darwin buried himself in his studies, determined to prove or disprove his ideas about the evolution of species. Like Lamarck, Darwin developed his theories with no knowledge of genes, chromosomes or DNA. Mendel's work with peas (see page 505) was going on during Darwin's lifetime but it remained hidden in an obscure journal.

Even though he managed to build up an overwhelming case for evolution, Darwin was reluctant to publish his theories, knowing the conflict they would cause with the Church. For more than 20 years Darwin carried on gathering evidence to support his theories. He was finally pushed into publication after exchanging letters with **Alfred Russel Wallace**: he too had come to the same conclusion about evolution. They agreed to make a joint announcement, and the Darwin–Wallace paper was presented to the Linnaean Society of London in July 1858.

In November 1859, Darwin published his revolutionary work, ***The Origin of Species by Means of Natural Selection***. Few books before or since have caused quite such a controversy. Although Church leaders largely remained quiet, many people bitterly attacked it because it questioned the theory of creation. Darwin was called 'the most dangerous man in England'.

A working definition of evolution is 'a change in the genetic composition of a population over time'. It has also been defined as 'a gradual process in which new species develop from pre-existing ones'. But as we shall see in this chapter, populations can evolve without changing into new species.

2 CHARLES DARWIN AND THE THEORY OF EVOLUTION

In a book whose title was shortened to *The origin of species*, Charles Darwin included these four ideas in his theory of evolution:

- **The living world is changing**, not static.
- **The evolutionary process is usually gradual**, not a series of jumps such as catastrophes and re-creations.

Fig 33.3 **This cartoon first appeared in Punch in 1861. It shows how poorly people of the time understood Darwin's theories**

THE LION OF THE SEASON.
ALARMED FLUNKEY. "MR. G-G-G-O-O-O-RILLA!"

HOW OLD IS THE EARTH?

ACCURATELY ESTIMATING THE AGE OF THE EARTH is central to the theory of evolution. In 1650 James Usher (then the Archbishop of Armagh) announced that his study of Hebrew literature had led him to the conclusion that the date of creation was the year 4004 BC. Dr Lightfoot, vicar of the University church at Cambridge, took it a step further and proclaimed that the date of creation was 9 a.m. on 23 October, 4004 BC. Baron George Cuvier later stated that the Earth must be much older, more like 70 000 years.

Faced with the large array of different fossils, nineteenth century geologists decided that the Earth was much older still. In 1830, Charles Lyell suggested that the true age of the Earth would turn out to be millions, rather than thousands of years. He was the first to propose that old rocks could be uncovered or brought to the Earth's surface (by volcanic activity, for example), so revealing the remains of long-dead species.

This century, ultra-sensitive dating techniques which measure the decay of radioactive isotopes have put the age of the Earth at about 4 600 000 000 (4.6 billion) years. This is an important breakthrough because, although our brains cannot comprehend such a time scale, it fits in well with the idea of progressive evolution. The Assignment at the end of this chapter should help you to put this geological time scale into perspective.

- **The common ancestor theory**. This suggested that closely related species evolved from one basic ancestor. Darwin said that man and apes evolved from a common ancestor, but the popular press took this to mean that humans had recently evolved from apes, leading to cartoons such as the one in Fig 33.3.

- **The theory of natural selection** as a mechanism to explain evolution.

Natural selection: 'survival of the fittest'

Darwin's observations of the living world led to three basic conclusions:

- Generally, organisms produce far more offspring than can possibly survive.

- Living things are locked in a struggle for survival. They compete for food, space, mates etc. In short, they are trying to eat and not be eaten, so that they can reproduce.

- Individuals of the same species are rarely identical: they show **variation**.

There was nothing particularly original about these observations, or in the idea that organisms evolved. What Darwin did, however, was to put all these factors together and suggest **natural selection** as the mechanism for evolutionary change.

Natural selection is commonly simplified to 'survival of the fittest'. Biologists define fitness as the ability of an organism to survive and reproduce. The fittest organisms survive to produce more offspring than less fit ones. In this way, the characteristic of a population can gradually change from generation to generation.

To illustrate the principle, imagine a population of blackbirds in a woodland. Like most organisms, they reproduce sexually and there is variation in the population. Some have better reflexes than others, forage for food more successfully, others have a better immune system. Normally, there is competition for resources such as food, mates and nesting sites. In this situation, the fittest birds survive and produce more offspring than others. The 'fitter' birds may gather more food and so rear larger families than less fit individuals (Fig 33.4). Or they may rear two clutches of eggs in the time it takes others to produce one. The vital point here is that the *fittest individuals pass their genes on to more of the next generation*.

The cornerstone of Darwin's theory is that the mechanism of evolution is natural selection – this is known as **Darwinism**. The application of twentieth century knowledge about chromosomes, genes and DNA to Darwin's theories is called **neo-Darwinism**.

Fig 33.4 **Blackbirds which can gather the most food tend to raise the greatest number of offspring, and so pass on their genes to more of the next generation**

Fitness in biology is a measure or reproductive success. So, a fit flowering plant produces and disperses more seeds than its competitors. Fit fungi are more successful because they produce more spores.

Fig 33.5 **When the European rabbit is introduced into countries where it is not endemic, it tends to out-compete the native herbivores. With few predators around, the rabbit population can reach plague proportions. When this has happened in the UK and Australia, farmers faced ruin. The deadly myxoma virus was introduced into the rabbit population, killing 99 per cent. The remaining 1 per cent were naturally resistant to the virus and are the ancestors of today's, largely resistant, rabbit population**

The severity of the situation is called the **selection pressure**. The greater the selection pressure, the faster the evolutionary process. When organisms have a short life cycle, the process can be very swift indeed. In the 1950s farmers in the UK and Australia took the drastic step of introducing a deadly disease, myxomatosis, into the ever-expanding rabbit population (Fig 33.5). This virus killed over 99 per cent of the rabbits: this is selection pressure at a massive level. Only the rabbits whose genes gave them resistance to myxomatosis were able to survive and pass on their genes to the next generation.

Natural selection in action

Many people tend to think that evolution is something that happened in the past and that the world of today is the finished product. But this is far from the truth: evolutionary forces are still at work and natural selection is continually changing the characteristics of populations, often with great speed. Consider the following examples:

● The widespread use of antibiotics such as penicillin rapidly led to the development of resistant strains of bacteria. Penicillin works by inhibiting one of the enzymes which the bacteria need in order to manufacture cell walls. However, many resistant strains have a slightly different enzyme, one which is unaffected by penicillin.

● The use of warfarin as a rat poison has led to the development of warfarin-resistant rats. Warfarin is an anti-coagulant which kills rats by inducing **haemorrhage** (internal bleeding). Resistant rats have a modified enzyme in their blood-clotting system which allows their blood to clot in the presence of warfarin.

● Copper tolerance in the grass *Agrostis tenuis*. Copper is a metabolic poison which prevents many plants from growing on copper-polluted land (Fig 33.6). However, resistant plants have evolved. These can transport copper out of their cells, so that it accumulates in the cell wall and does not interfere with the plant's metabolism. (In non-polluted areas, the normal type of grass out-competes the copper-resistant strain.)

The organisms in these examples have a short life cycle and high reproductive capacity, and so can respond quickly to the selection pressure placed on them. The effects of selection pressure on organisms which live longer and have a lower reproductive capacity take much longer to become apparent.

> **?**
>
> **B** Although we have waged war against species such as locusts, weevils and beetles, we have never succeeded in making any pest species extinct. On the other hand, many larger species such as rhinos, whales and giant pandas are facing extinction, despite our efforts to save them. Give two reasons for this difference.

Fig 33.6 **This disused copper mine at Parys Mountain in Anglesey (near right) is devoid of life. Copper is a metabolic poison which kills virtually all plants. The grass *Agrostis tenuis* (far right) is one of the few that has evolved copper-resistance**

ARTIFICIAL SELECTION

HUMANS HAVE PRACTISED artificial selection, or selective breeding, for thousands of years. We haven't needed a scientific understanding of genetics to do this: animals or plants with the most desirable features are simply crossed or mated. For example, if cows with the highest milk yield are impregnated by the healthiest and fastest-growing bulls, many of the next generation are big, healthy cows which produce large quantities of milk. Those which do not have the desired features are not used for breeding. In this way, *artificial selection* rather than natural selection is used to bring about great changes in the phenotype in a short time: animals and plants can be 'domesticated' in just a few generations.

Selective breeding is still important in many different areas of food production:

● Increasing the growth rate and the meat, milk or egg production of livestock.

● Increasing the disease resistance of crop plants (increasing the disease resistance of animals by selective breeding is not as reliable).

● Changing the muscle to fat ratio of livestock, so that more lean meat is produced.

● Increasing the yield and nutritive value of crop plants such as wheat.

● Increasing the tolerance of plants and animals to drought, heat or pollutants.

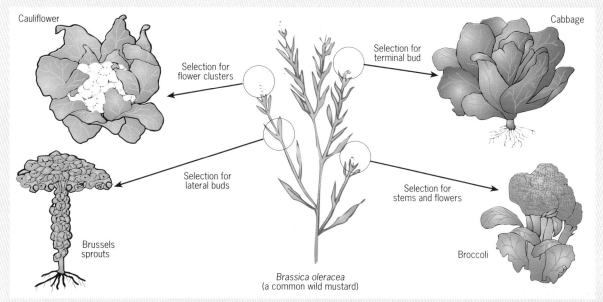

Fig 33.7 **Many familiar vegetable types have arisen from the artificial selection of the wild mustard, *Brassica oleracea*. In the original wild population, plants show variation in the size and shape of leaves, stems, terminal buds, lateral buds and flower clusters. Selection for individual features rapidly produces the familiar crops you see here**

Fig 33.8 **Like all pedigree animals, this champion Siamese cat is the product of many generations of artificial selection. Breeders consider the wedge-shaped face a desirable characteristic and so have concentrated on this feature. Breeding animals in this way can lead to problems, and some people strongly disagree with breeding animals for superficial 'perfection'**

Different types of natural selection

From what you have learned about natural selection, it would be tempting to assume that it is just a mechanism for change. However, this is not always the case: it can be a means of keeping things static. Many types, crocodiles and sharks for example, have not changed significantly for millions of years.

There are basically three different types of natural selection:

● **Stabilising selection** (Fig 33.9(a)). In a favourable and unchanging environment, stabilising selection can lead to a 'standardising' of organisms by selecting against the extremes. An example of this is human birth weight. Studies show that particularly large or small babies have a higher mortality than those in the mid-range.

- **Directional selection** (Fig 33.9(b)). This mechanism tends to occur following some kind of environmental change which causes selection pressure. The organisms with a particular extreme of phenotype may have an advantage. At some stage, directional selection must have operated to produce the long neck of the giraffe. Probably, in times of food shortage, only the tallest individuals could reach enough food to survive and pass on their genes to the next generation.

- **Disruptive selection** (Fig 33.9(c)). This type of selection acts at both ends of the distribution, favouring individuals with extreme rather than average characteristics. Rarer than the other two types, disruptive selection is important in the formation of new species, where natural selection acts differently on different populations. An example of this is seen in Darwin's finches (see page 529) where disruptive selection has acted on beak size. In some situations, those with longer, thinner beaks would have succeeded in exploiting insects as a food source, while those with shorter, stronger beaks would have succeeded in cracking seeds. Those with average beaks were adapted for nothing in particular and would have reproduced less successfully.

?

C From your knowledge of the living world, give three examples of directional selection.

Fig 33.9 **The different types of natural selection. Individuals in the shaded areas in the top row of graphs have a selective advantage**

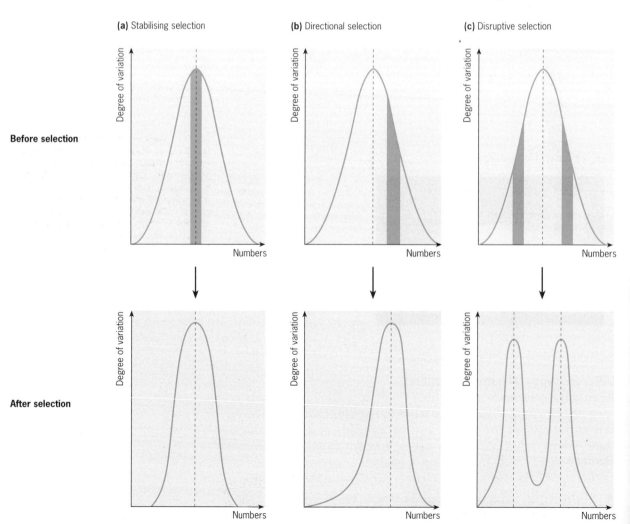

(a) Stabilising selection **(b)** Directional selection **(c)** Disruptive selection

Before selection

After selection

3 THE EVIDENCE FOR EVOLUTION

Today there are few scientists who seriously doubt the theory of evolution, or that it happens by natural selection. In this section we briefly consider some of the lines of evidence.

Fossils

Palaeontology, the study of fossils, supports the theory of evolution. Most fossils are found in sedimentary rocks, formed by layers of silt which preserve the bodies of organisms before they have a chance to decay in the normal way. In many places (see Fig 33.10) it is clear that different layers, or **strata**, represent different time periods. By studying the fossils in these strata we find obvious evidence for the gradual evolution of some species.

Geographical distribution

A study of the distribution of species presents strong evidence that organisms evolved 'into' their ecosystem rather than being placed there during a single act of creation.

During his time in South America, Darwin encountered a classic piece of evidence for evolution. On the South American mainland, finches have short, straight beaks to crush seeds. Other species have taken advantage of the other food sources. Woodpecker finches, for instance, have long beaks for extracting insects from holes in trees. On the Galapagos islands, however, things are different. For one reason or another, the finches managed to get to the islands, while many other species of birds did not. In the absence of competition, the finches have evolved to take advantage of the range of food sources on offer (Fig 33.11).

Fig 33.10 **The grand canyon in Arizona. The horizontal rock strata here can be clearly seen and span some 500 million years of geological history**

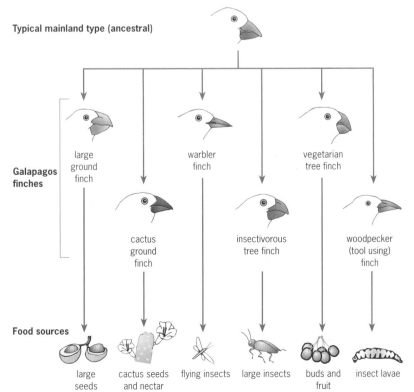

Fig 33.11 **Darwin's finches on the Galapagos islands. In a process called** adaptive radiation, **one common ancestor has evolved into many different species of finch, each exploiting a particular food source, such as seeds, insects and buds. Interestingly, Darwin did not immediately recognise the significance of all of these different species; he simply collected a large number of specimens in a bag and presented them to a bird specialist on his return home**

EVOLUTION ON ISLANDS

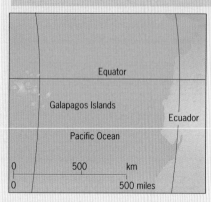

Fig 33.12 **The Galapagos Islands are an archipelago situated near the Equator in the Pacific Ocean, about 600 miles west of Ecuador**

AS DARWIN FOUND OUT ON HIS VOYAGE, islands are particularly interesting places to study evolution. Young, volcanic islands such as the Galapagos (Fig 33.12) are often colonised by plants whose seeds have been blown there by the wind, and by flying animals which have been driven off course by storms. In later years, of course, colonisation by non-flying animals (such as rats and cats) is likely to be a result of human activity.

As we saw in the Opener, the evolution of birds on islands can be very spectacular. Free from predators, birds often lose the power of flight and sometimes grow very large. Before the Maori settlers arrived from Polynesia, New Zealand was a bird-dominated ecosystem. Huge flightless birds, **moas**, some as tall as 3.5 metres, evolved to fill the niches normally occupied by mammals. They were easy prey for the Maoris, however, who hunted them to extinction within a few centuries. Ironically, the only survivor of the moa family, the kiwi, has become the country's national emblem.

Comparative anatomy

The anatomy (body plan) of plants and animals can reveal a lot about evolutionary relationships. Fig 33.13 shows one of the classic examples of comparative anatomy, the **pentadactyl limb** (five-fingered limb). This structure appears in various forms throughout the vertebrates: the feet of reptiles and amphibians, the human hand, the bat's wing, the seal's flipper, for example.

This basic five-fingered plan has been modified in a variety of ways to suit the animal's mode of locomotion. Structures with a common origin but with a different function are said to be **homologous**. So, the wing of a bat is homologous to a human hand. The possession of the pentadactyl limb in so many different organisms is persuasive evidence that these animals have evolved from a common ancestor.

Fig 33.13 **The vertebrate pentadactyl limb provides strong evidence for evolution. In all these examples, the basic bone layout is the same (in the diagram, the same type of bone is given the same colour in each animal) but the limbs have become modified in different ways according to the way the animal moves**

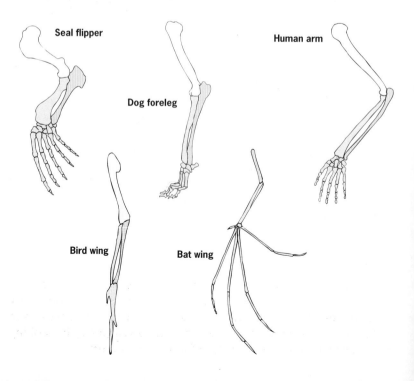

Seal flipper

Dog foreleg

Human arm

Bird wing

Bat wing

Fig 33.14 Convergent evolution occurs when unrelated species solve the same evolutionary problem in a similar way

(a) Cacti, which are native only to the Americas, have evolved to cope with desert living. The leaves have become spines which lose no water and protect the plant. The fleshy stem expands and stores water and has taken over the job of photosynthesis

(b) You might guess that this euphorbia from Africa was a cactus, but you would be wrong. The two plants have evolved in different continents and are unrelated; they look very similar because they have developed the same features to survive desert conditions

(c) and (d) The flipper of the whale and the fin of the shark are a classic example of convergent evolution. The two animals are completely unrelated and the internal structure of forelimbs is totally different. However, they are both the same shape because of the job they have to perform

In contrast, totally *unrelated* structures can become adapted for the same function. The wings of insects and the wings of birds, for example, have a totally different anatomy and origin in the embryo, but are similar in shape because they perform the same job. Such structures are said to be **analogous**. The process by which unrelated species evolve to resemble each other is called **convergent evolution** (Fig 33.14). A classic example of this is seen in seals (mammals), penguins (birds) and fish. Due to their need to swim efficiently, all have evolved the same streamlined shape. The forelimbs of the seal and the wings of the penguin have both evolved to resemble fins.

> **?**
>
> **D** Are the following pairs of structures analogous or homologous?
>
> **(a)** Trout's fin and dolphin's flipper.
>
> **(b)** Horse's leg and bat's wing.

Embryology

Studying the development of an embryo reveals some interesting evolutionary secrets. The scientist Ernst Haeckl said 'ontogeny recapitulates phylogeny' which means that you can 'revisit' the evolutionary development of a species (its **phylogeny**) by looking at the development of each individual (its **ontogeny**).

Fig 33.15 **Four developmental stages in embryos of fish, turtles, chicken and humans. You can see that the early stages at the top are very similar in all four. For example, in the drawings in the second row, all four embryos have branchial arches, the 'folded' areas just under the head at the front. Only in the later stages do differences become obvious**

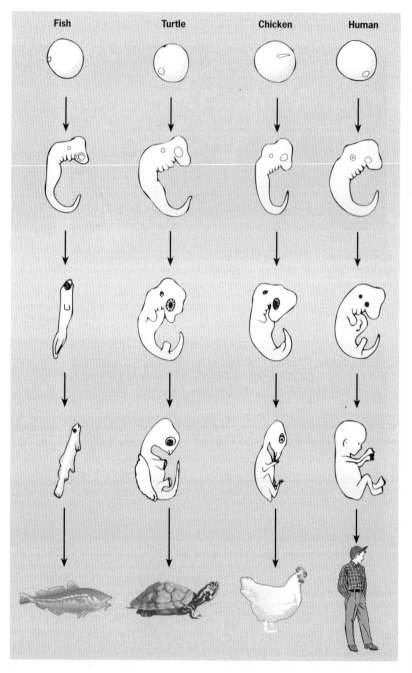

A number of vertebrate embryos follow a very similar path from fertilisation through several stages of early development – so similar that it is often difficult to tell them apart (Fig 33.15). For example, all vertebrate embryos develop branchial arches; these are relics of the gills possessed by aquatic ancestors. However, although the study of the developing embryo provides further evidence for evolution, it does not give us much information about the actual evolutionary process.

Comparative biochemistry

The biochemistry of different species shows some striking similarities. Many chemicals such as nucleic acids, ATP and cytochromes, organelles such as ribosomes, and pathways such as respiration, are almost universal. This is strong evidence that all living things have a common ancestry.

4 HOW A NEW SPECIES EVOLVES

We have looked at natural selection and seen that it can lead to stability or change. In this section we examine how natural selection can lead to the formation of a new species.

What is a species?

We instinctively think that we know what the word 'species' means, but an accurate definition is difficult, if not impossible. This is a common definition:

> **A species is a population or group of populations of similar organisms which are able to interbreed and produce fertile offspring.**

This would seem to be reasonable; a horse and a donkey can mate to produce a mule (Fig 33.16), which is sterile, so the horse and the donkey can be said to belong to different species. However, this definition runs into trouble. There are many closely-related species, wolves, coyotes and domestic dogs, for example, which can interbreed to produce perfectly fertile offspring. The only reason why the wolves and coyotes remain distinct is that the hybrids do not compete as successfully as the pure strains.

The difficulty we have defining the term species stems from the very nature of the process of evolution. New species evolve from pre-existing ones, and this is usually a slow transition which may take many generations and thousands of years. There are many living species which are still in the process of separating from each other. The wolf and the coyote almost certainly evolved from a common ancestor and for all practical purposes are separate species (Fig 33.17). If left to evolve naturally over the next few thousand years, the two species may well accumulate enough genetic differences to prevent interbreeding.

So, we can modify the definition to:

> **A species is a collection of recognisable organisms which shares a unique evolutionary history and is held together by cohesive forces of reproduction, development and ecology.**

Although more of a mouthful, it is a more satisfactory description because it recognises that all members of a species are usually (but not always) recognisable, and that they share the same 'biology', occupying the same niche in the ecosystem and having particular social structures, courtship, mating habits etc.

Fig 33.16 **This mule is the product of a mating between a donkey and a horse. Because of their stamina, mules have been used as beasts of burden for thousands of years**

Fig 33.17 **The coyote (*Canis latrans* – below left) and the wolf (*Canis lupus* – below right) are classed as different species, but they are capable of interbreeding and producing fertile offspring**

The development of new species

Speciation, the development of new species, happens when different populations of the same species evolve along different lines. **Populations** are groups of interbreeding individuals of the same species occupying a particular geographical region. Examples of populations may be the water fleas in a pond, all the oaks in a forest or the elephants in a game reserve.

It is thought that speciation usually happens in the following way:

- Part of a population becomes isolated in some way that prevents them from breeding with the rest of the population. Very often, the breakaway population will, by chance, have a different genetic constitution from the original (see genetic drift – page 537).

- The two populations experience different environmental conditions, so that natural selection acts in different ways. In this way the genetic make-up of the two populations changes.

- Eventually, genetic differences accumulate to a point where individuals from separate populations that come together again can no longer interbreed.

It is worth mentioning that new species can also arise from a hybridisation between two different species. This particularly applies to plants, and plant breeders take advantage of this ability to produce new crop varieties.

There are two basic methods of speciation:

- **Allopatric speciation** (allo = different, patris = country), where the two populations are physically separated.

- **Sympatric speciation**, where the two populations are reproductively isolated in the same environment.

How populations become isolated

Usually, species evolve when populations become isolated by physical barriers such as stretches of water or mountain ranges. This is known as **geographic isolation** (Fig 33.18).

Fig 33.18 **The principle of geographic isolation**

(a) The original population of mice has a 50:50 distribution of the alleles A and a, ie A is as common as a. A river changes its course and cuts off a few individuals. In the new, isolated population the allele frequency, by pure chance, is 4:10. This change in allele frequency is called the founder effect. The fewer the individuals in the breakaway population, the more dramatic the change in allele frequency is likely to be

(b) The founder population starts to grow, but they are in different conditions from the original population. The alleles A and a were neutral up to now, but then a water-borne disease spreads through the new population, which is confined to marsh land. The aa genotype confers resistance on those who carry it. By natural selection, the A allele is selected against, and so the a allele becomes even more frequent. In this way, the genetic make-up of the two populations diverges

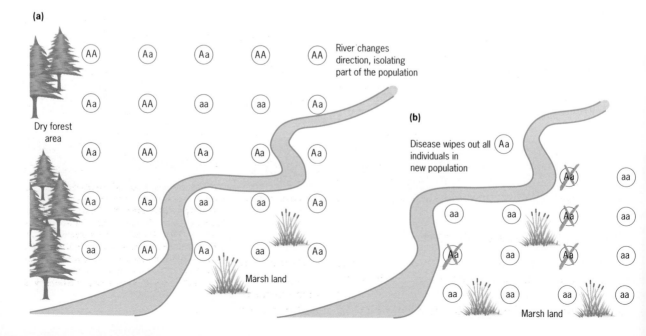

Evidence suggests that geographical isolation is the most common way in which populations become reproductively separated. However, speciation can happen even when two populations overlap. For instance, two populations of frogs may have a different mating call, and the females may only respond to calls of males in 'their' population. Table 33.1 summarises the different isolating mechanisms.

Mechanism	How it works
Prezygotic mechanisms, ie gametes don't come into contact	
Geographic	Two populations do not come into contact due to some physical barrier, eg mountain range, water
Temporal	Different timing, eg flowering times or mating seasons do not coincide
Behavioural	Individuals of other populations are not acceptable as mates, eg wrong courtship, mating call, or pheromones
Mechanical	Structural differences prevent mating, eg the shapes of the genitals in insects
Postzygotic mechanisms, ie even if mating takes place	
Hybrid non-viability	Hybrid zygotes fail to develop to sexual maturity
Hybrid sterility	Hybrids do not produce functional gametes
Hybrid weakness	Hybrids have reduced survival or reproductive rates

Table 33.1 **A summary of isolating mechanisms**

5 EVOLUTION IN ACTION

The process of evolution is generally slow, and it is no surprise that for centuries people thought that the living world was static. However, there are situations where you can see the process in action. Organisms with short life spans and high reproductive capacities, such as bacteria or insects, can evolve in a very short time indeed – see the examples of natural selection on page 526.

Also of great interest to evolutionary biologists is a **polymorphism**, a situation where two or more different forms, **morphs**, of a species exist in a population. Each morph confers a different survival advantage in different situations. Polymorphisms such as that of wing colour in the peppered moth (below) allow us to see how natural selection leads to changes in allele frequency and so observe evolution in action.

The peppered moth

Fig 33.19 shows the original speckled form of the peppered moth *Biston betularia*. This moth was common across the UK at the time of the industrial revolution. In areas of heavy industry, many buildings and trees became covered in soot. The speckled moth lost its camouflage in these areas and so was eaten more often by its predators.

Luckily for the moth, a mutation occurred around this time that produced a black moth known as the **melanic form**. In industrialised areas the melanic moth had a selective advantage. However, in 'cleaner' areas the reverse was true: the speckled moth thrived and the melanic form was rarely seen. Both morphs belong to the same species, but their survival depends largely on the environment.

How can we study evolution?

If evolution is so slow, how can we observe it happening? An important principle here is that evolution involves a *change in allele frequency*. For instance, when myxomatosis was introduced into the rabbit population, the alleles for resistance were selected for, and so

Fig 33.19 **The two forms of the peppered moth; normal (speckled) and melanic (black). Against the background of the bark of a tree shown in this photograph, the two speckled moths on the right are almost invisible – the melanic form on the left is far more likely to be seen and caught by a predator. However, it becomes invisible on a black surface such as a sooty wall, and so is more likely to survive in an industrial environment**

they increased in frequency. In any particular situation, if we can show that allele frequencies are changing, we know that evolutionary forces are at work.

Evolutionary studies are usually based on interbreeding populations, which are known as **demes** or **Mendelian populations**. The sum total of all the genes circulating within such a population is called the **gene pool**.

Populations and allele frequency

When the gene pool remains more or less the same, a population is not evolving. Natural selection may be acting, but it is often a force for stability rather than change. However, if the gene pool of a population is changing, natural selection is acting as a force for change and this is evolution in action.

To show how we measure allele frequency, imagine a population of 500 rats. In the population are two alleles for coat colour: black (B) is dominant over brown (b). The rats are diploid – each individual has two alleles – so the population contains 1000 coat colour alleles. If 600 of the alleles are B, then we can say that the frequency of B is 60 per cent, or 0.6 (statisticians prefer decimals). As there are only two alleles, it follows that the frequency of b must be 40 per cent or 0.4. (Remember that the sum of the allele frequencies must be 100%, or 1.)

If one gene is dominant to the other – as in this case – the symbol p is used for the frequency of the dominant allele B, and q is used for the frequency of b. It is clear that $p + q$ must equal 1.

The values of p and q change when a population evolves. To find out if this is happening we use the **Hardy–Weinberg principle**.

> ✓
>
> $p + q = 1$
>
> Frequency of dominant allele + frequency of recessive allele = total number of alleles of that gene.

The Hardy–Weinberg principle

The principle, named after the British mathematician G.H. Hardy and the German biologist W. Weinberg, states that there will be no change in the frequency of alleles in a population so long as the following conditions are met:

● The population is very large, diploid and reproduces sexually.

● Mating is totally random – there is no tendency for certain genotypes to mate together.

● All genotypes are equally fertile, so there is no natural selection.

● There is no emigration or immigration.

● No mutations occur.

We can use the Hardy–Weinberg rule to estimate allele frequencies using just the information about the frequency of a particular genotype. Consider the example of albinism in humans. One in 10 000 humans are born albinos – they have two copies of a recessive gene which means they cannot make the pigment which gives colour to hair, skin and eyes (Fig 33.20). The frequency of the allele for albinism in the population is difficult to estimate because carriers of the allele are not albinos. So, how can we find the frequency of the albinism allele?

If we call the normal allele A and the albinism allele a, most of the population are AA, some are carriers: Aa or aA, and a small number are albinos: aa. Using the Hardy–Weinberg formula we can estimate the percentage of the population who carry the a allele. Remember

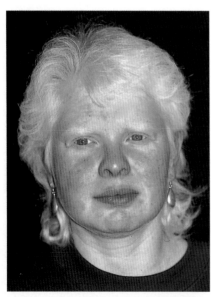

Fig 33.20 **This person is an albino. The white hair, pink eyes and pale skin are the result of an inability to make the pigment melanin**

that p and q represent the frequency of the A and a alleles respectively. The Hardy–Weinberg equation:

$$p^2 + 2pq + q^2 = 1$$

Put into words, it says:

> **The frequency of AA individuals, added to the frequencies of Aa, aA and aa, must equal 1.**

We also know that $p + q = 1$, and we can use these two equations to work out the proportion of heterozygotes in the population, and any other frequencies involved:

1 in 10 000 is a frequency of 0.01%, or 0.0001 expressed as a decimal.

1 in 10 000 have the genotype aa, so the frequency of aa; qq or q^2, has the value of 0.0001.

So $q = \sqrt{0.0001} = 0.01$.

If $q = 0.01$, then p must be $1 - q$, which is 0.99.

So, in other words, 99 per cent of the alleles in the human population are A.

From this we can calculate that the number of carriers of the albino allele $2pq$ is:

$$2pq = 2 \times 0.99 \times 0.01 = 0.0198 \approx 0.2.$$

So, almost 2 per cent of the population carry the albinism allele.

▨ For examples of exam questions involving Hardy–Weinberg calculations, see questions 4 and 5 at the end of the chapter.

When allele frequencies change

The Hardy–Weinberg principle states that allele frequencies remain constant as long as certain conditions apply. It follows that when these conditions are not met, the allele frequencies can change and the population can evolve. Consider what happens when the conditions listed on page 536 are not met.

- **The population is small**. If the population is small, chance plays a large part in determining which alleles pass to the next generation. The smaller the population, the greater the probability that the genes of one generation will not be accurately represented in the next. The name given to a change in allele frequency caused by chance is **genetic drift**. These are two important causes of genetic drift: **bottlenecks**, when a population declines and then recovers; and the **founder effect**, when a population starts from just a few individuals (see below). As Fig 33.21 shows, this can result in a drastic change in the allele frequency.

- **Mutations occur**. Mutations are ultimately the origin of all genetic variation (see Chapter 31). They are rare, and most are harmful or neutral. Occasionally, however, a mutation gives an organism a survival advantage, or the environment changes so that previously harmful/neutral mutations become advantageous. For example, a mutation in a bacterial gene which leads to the production of a modified enzyme could make the bacterium resistant to an antibiotic. That individual would survive, even in the presence of the antibiotic, and pass its resistance gene on to the next generation.

- **Individuals move in or out of populations**, that is, immigration or emigration. If the new individuals breed, this movement may lead to gene flow. When a few individuals move into a new environment and start a population, we see the founder effect. Like the beads in Fig 33.21, the individuals in the founder population are probably not representative of the original population. A group of six human survivors on a desert island, for example, might have two red-haired people and two blondes. A population founded by these people would have a much higher percentage of red-heads and blondes than the population in their country of origin. In the most extreme case, a new population can be founded by one individual, such as a self-fertilising plant or a pregnant female animal.

- **Mating is non-random**. In many cases, individuals choose mates with a particular genotype. In this case, mating is not random, it is **assortive**. Such assortive mating can change allele frequency.

- **Natural selection occurs**. When there is competition for resources – and this is usually the case in an ecosystem – natural selection acts and only the most successful individuals pass on their genes.

Fig 33.21 **An illustration of genetic drift. In the original 'population' there are 1000 beads; half red, half blue. If ten beads are extracted at random, there is a high probability that there won't be five of each. If these ten 'organisms' then become the founders of a new population, the proportion of colour (the allele frequency) will be significantly changed, in this example to 7:3**

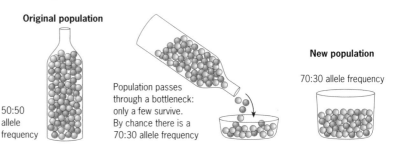

Original population

50:50 allele frequency

Population passes through a bottleneck: only a few survive. By chance there is a 70:30 allele frequency

New population

70:30 allele frequency

SUMMARY

- **Evolution** is defined as a change in the genetic composition of a population over time. New species develop from pre-existing species in this way.

- The basic mechanism of evolution is **natural selection**. This acts on the genetic variation that occurs in a population. Individuals whose genotype gives them an advantage over others are more likely to pass their genes on to the next generation.

- A **population** is a group of interbreeding individuals of the same species in the same geographical area.

- New species are formed when populations become isolated in some way so that they cannot interbreed. Natural selection then acts on the different populations, changing the frequency of alleles. With time, the differences between the two populations can become so great that an individual from one cannot interbreed with an individual from the other.

- Humans can speed up the evolutionary process by **artificial selection**. Breeding organisms

with desirable features, such as cattle with a high milk yield, changes the allele frequency rapidly. Within only a few generations, organisms are produced which are very different from the original wild stock.

- The evolutionary process is fastest in species which have large population size, large reproductive capacity (**fecundity**) and rapid life cycles. Many disease-causing and pest organisms fall into this category.

- Many endangered species have a low reproductive capacity and a long life cycle.

- Evolution happens when there is a change in the **allele frequency**. The **Hardy–Weinberg principle** states that, in a large population, the allele frequency remains constant unless some outside force acts to change it. These forces are chance (the smaller the population, the greater the effects of chance), migration of organisms, selective mating and natural selection.

- When the conditions for the Hardy–Weinberg equilibrium are not met, the allele frequency changes and the population evolves.

QUESTIONS

1 Read the following passage.

Shrimps Surface from Coal-dust Soup in Mine

Fifty shrimps have been rescued from a colony living in a dark, sulphur-filled pool at the bottom of a 300 m mine shaft at Wearmouth Colliery, Tyne and Wear.

Dr Phil Gates of Durham University believes that they may even be a new species and has installed them in a laboratory aquarium in the 'coal-dust' soup in which they were found.

Dr Gates said, 'They've been isolated from other shrimps for most of this century and that makes them very interesting. The pool is full of sulphur, sulphur dioxide and hydrogen sulphide and we thought that if we put them straight into freshwater it would probably kill them.'

The shrimps are thought to have been feeding on sulphur-eating bacteria. 'It is possible that we have found an entirely new species,' he said, 'but more likely they are a sub-species which has evolved independently of ancestors in water at the surface. They will have been evolving in this extreme environment – total darkness, very acid water and living on sulphur-eating bacteria. In terms of evolutionary genetics, the mine shrimps are going to be fascinating. They'll be totally different from shrimps in pools on the surface.'

Adapted from: The Daily Telegraph, 13 May 1994

a) Dr Gates said, 'In terms of evolutionary genetics, the mine shrimps are going to be fascinating. They'll be totally different from shrimps in pools on the surface. Explain how natural selection could account for the mine shrimps becoming genetically different from shrimps in pools on the surface.

b) **(i)** What do you understand by the term 'species'?
(ii) Outline an experiment that you could carry out in a laboratory that would enable you to determine whether the mine shrimps were a different species from those in pools on the surface. Explain how you would interpret the results.

[NEAB February 1995 Modular Biology: BY2 Continuity of Life, q.8]

2 Rats and mice are common pests. Warfarin was developed as a poison to control rats and was very effective when first used in 1950.

Resistance to warfarin was first reported in British rats in 1958 and is now extremely common. Warfarin resistance in rats is determined by a single gene with two alleles W^S and W^R. Rats with the genotypes listed below have the characteristics shown.

$W^S W^S$ Normal rats susceptible to warfarin
$W^S W^R$ Rats resistant to warfarin needing slightly more vitamin K than usual for full health
$W^R W^R$ Rats resistant to warfarin but requiring very large amounts of vitamin K. They rarely survive.

a) Explain why:
(i) there was a very high frequency of W^S alleles in the British population of wild rats before 1950;
(ii) the frequency of W^R alleles in the wild rat population rose rapidly from 1958.

b) Explain what would be likely to happen to the frequency of W^R alleles if warfarin were no longer used.

c) Mice show continuous variation in their resistance to warfarin. What does this suggest about the genetic basis of warfarin resistance in mice?

In humans, deaths from conditions where blood clots form inside blood vessels occur frequently in adults over the age of fifty. These conditions maybe treated successfully with warfarin. However, some people possess a dominant allele which gives resistance to warfarin.

d) Why would you not expect this allele to change in frequency?

[NEAB February 1995 Modular Biology: BY2 Continuity of life, q.6]

3 The dog whelk lives on rocky shores around Britain. The graphs show the variation in shell height in two different populations, one from a rocky shore exposed to strong wave action and the other from a sheltered rocky shore. Shell height was measured as shown in Fig 33.Q3.

Sheltered shore

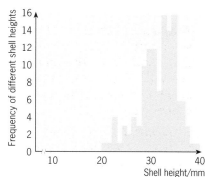

Exposed shore

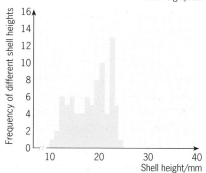

Fig 33.Q3

a) What type of variation is shown in the graphs?

b) Describe **two** differences shown in the graphs between dog whelks on the exposed and on the sheltered shore.

c) In a follow-up investigation, it was found that dog whelks on the sheltered shore had much thicker shells than those on the exposed shore. On the sheltered shore, there were more crabs, which are predators of dog whelks. Describe how natural selection could account for this difference in shell thickness.

[NEAB February 1996 Modular Biology: BY2 Continuity of life, q.3]

4

a) The following are types of selection:
directional
stabilising
disruptive

Identify the type of selection that would best describe each of the examples below.
  (i) Early farmers selected cattle to produce some breeds that were horned and some that were hornless.
  (ii) There is a higher death rate among very light and very heavy human babies than among those of average mass at birth.

b) The two-spot ladybird exists in the two different colour forms shown in Fig 33.Q4.

Fig 33.Q4

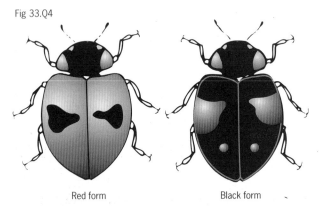

Red form   Black form

These forms are controlled by a single gene with the allele for black, B, dominant so that for red, b.

Use the Hardy–Weinberg formula to predict the frequency of red and the frequency of black ladybirds in a population if the frequency of alleles B and b are the same.

[AEB 1996 Biology: Specimen Papers, Paper 1, q.8]

5

Shells of the snail, Cepaea nemoralis, may have banded patterns or no bands at all. This is determined by a single gene, the allele for banded patterns, (b), being recessive to that for not bands, (B). Another gene determines the background colour of the shell, which may be dark (brown) or light (yellow) in colour. Examples of two contrasting types are drawn below.

Samples were taken from two neighbouring populations of snails, one living in the grassy verge of a country

Banded, light shell

Unbanded, dark shell

Fig 33.Q5

lane, and the other among the leaf litter in a beechwood. The proportions of different kinds of shell found are represented as percentages in the table below.

Table 33.Q5

Population	Habitat	Banded light	Banded dark	Unbanded light	Unbanded dark
X	Grass verge	70	11	13	6
Y	Woodland	1	3	11	85

a)   (i) Use the Hardy-Weinberg equation ($p2 + 2pq + q2 = 1$) to calculate the frequency of the b allele in Population X.
   (ii) If this population is in Hardy-Weinberg equilibrium, then this frequency should be exactly the same in the next generation of snails in this habitat. Give three factors which could lead to a change in allele frequency.

Thrushes are common predators of these snails, breaking open the shells by holding them in their beaks and striking them against a convenient stone known as an 'anvil'. An investigation was performed to discover how natural selection might be operating in these populations. Equal numbers of the banded, light and the unbanded, dark snails were collected and marked before being released into the centre of each habitat. A count was made at a later date of the number of broken marked shells found around anvils in each habitat.

b)   (i) Suggest a method of marking that would have minimal or no effect on the outcome of the investigation. Explain your suggestion.
   (ii) Why was it necessary to mark the snails in this investigation?

The table below shows the numbers of broken marked shells found at the end of this investigation.

Habitat	Banded light	Unbanded, dark
Grass verge	23	76
Woodland	68	30

c)   (i) From the information in this table, in which habitat is selection acting more strongly? Explain your answer.
   (ii) Explain how the process of natural selection operating in this habitat could account for these results.

[NEAB June 1994 Biology: Paper 2, Section A, q.3]

Assignment

THE HISTORY OF THE WORLD

Rocks, and the fossils they contain, can now be dated by **radiometric dating techniques**, which measure levels of certain isotopes. We can use this technique to construct a geological time scale – a biological history of the world. This Assignment outlines the process of dating using radioactive isotopes and puts the history of life on Earth into perspective.

Dating rocks and fossils

The principle behind radiometric dating is simple: some elements exist as **radioactive isotopes** which decay at a predictable rate. Measurement of this decay allows us to estimate the age of the rock.

For example, when volcanoes erupt, the lava contains radioactive potassium, ^{40}K. This decays steadily into argon (^{40}Ar). The half-life of ^{40}K is 1.3 billion (1.3×10^9) years. So, in 1.3 billion years, half of the potassium decays. Half of the remaining potassium will decay in the next 1.3 billion years, and so on. By measuring the ratio of radioactive potassium to argon, scientists can estimate the age of the rock strata to the nearest 50 000 years.

In many places, layers of volcanic lava alternate with sedimentary rocks which contain fossils. If we can date the lava, we can also estimate the age of the fossils. Some fossils of primitive bacterium-like organisms are thought to be around 3 500 000 000 years old.

1

a) What does half-life mean?
b) What is the half-life of ^{40}K?
c) What fraction of the original ^{40}K will be left after 3.9 billion years?

Carbon dating

This process can be used to date samples which still contain some organic material. Cosmic radiation is constantly bombarding the atmosphere, turning $^{12}CO_2$ into $^{14}CO_2$. This becomes fixed into organic molecules by the process of photosynthesis and then passes up the food chain. When an organism dies, the ^{14}C decays into ^{14}N, with a half-life of 5730 years. Carbon dating can be used to measure the age of any carbon-containing material up to 50 000 years old.

2

A human skull is found to have only one sixteenth of the ^{14}C it would have has when 'new'. Estimate its age.

The Earth is currently estimated to be about 4 600 000 000 years old. This time scale is impossible to comprehend, but if we condense this huge span into one year we can begin to put it into perspective.

The Earth calendar

Date	Event
1 Jan	Earth forms
26 Feb	Oldest rocks – nothing was permanent before this time
23 March	First life – bacteria like organisms obtain energy from organic molecules (food) by the process of glycolysis
April–May	Organic food runs out; photosynthesis evolves, makes food and releases oxygen as a by-product
30 June	Oxygen atmosphere fully developed. Aerobic respiration evolves some time after this
17 Sept	First eukaryotes
19 October	First coordinated multicellular colonies (like sponges)
5 Nov	First invertebrates (worms and jellyfish)
15 Nov	Rapid increase in invertebrate diversity
20 Nov	First fish
21 Nov	First coral reefs
29 Nov	First colonisation of land – vascular plants appear
4 Dec	First amphibians
6 Dec	First trees
7 Dec	First insects
8 Dec	First reptiles
17 Dec	First mammals
19 Dec	First birds and dinosaurs
21 Dec	First flowering plants
25 Dec	Last of the dinosaurs
26 Dec	First primates
31 Dec	
14.00	First upright bipedal hominids
21.00	Homo erectus
22.45	Recent Ice Age starts
23.00	Peking man appears – can cook food
23.48	Neanderthal man appears
23.54	Cro-Magnon man appears
23.55	Australian aborigines settle
23.58.49	Agriculture invented
23.59.28	Beginning of recorded history
23.59.29	Pyramids built
23.59.41	Buddha born
23.59.45	Christ born
23.59.53	Battle of Hastings
23.59.57	Start of the scientific method
23.59.58	Australia 'discovered' by Cook. The Industrial Revolution
23.59.59	Most of the knowledge contained in this book is discovered.

3

a) In our one-year scale, how many seconds represents 1000 years?
b) How long in actual time is represented by one minute, one hour and one day?
c) Use your answer to (a) to estimate how long the dinosaurs lived for.

4

The rise of humans as the dominant species has been meteoric.
a) What features of humans account for our success.
b) What dangers does this success bring?

5

Write down what you feel to be the four most significant advances made by humans. These could be inventions or social and cultural advances. If you are working in a group, use your own list to provide a discussion.

34 DNA technology

Tracey, a transgenic sheep at PPL Therapeutics in Edinburgh

THIS IS TRACEY, A TRANSGENIC SHEEP. Although Tracey is a perfectly normal, healthy animal, she has a remarkable talent: each litre of her milk contains about 35 grams of the protein human alpha-1-antitrypsin, a potential treatment for lung diseases such as emphysema and cystic fibrosis.

Transgenic organisms have DNA which has been modified by genetics scientists. In Tracey's case, the human gene which codes for alpha-1-antitrypsin was inserted into her DNA when she was an embryo. The gene was 'adopted' by her cells and is now expressed in her milk-producing glands.

Like many biological molecules, alpha-1-antitrypsin is far too complex to synthesise in a laboratory, and extracting the protein from human tissue is not a possibility – the tissue is just not available in the quantity required. Although all Tracey needs to make it is a diet of grass, her milk is worth several thousand pounds per litre! In the future we might have whole herds of transgenic organisms making life-saving medical treatments.

1 THE NEW GENETICS

Genetic engineers manipulate DNA to allow individual genes to be transferred from one organism to another. Often the two organisms are from completely different species.

The science of genetics seeks to explain the mechanisms of inheritance and evolution and, like many other 'pure' sciences, it can be applied to solve practical problems. In the 'new genetics' that has developed rapidly in the last few years, humans are taking a much more active role in determining how genes pass between individual organisms. Not only can we study DNA, we can manipulate it and use it to our advantage. This branch of genetics is called **genetic engineering**.

Transferring genes between organisms offers many exciting prospects. In agriculture, for example, genes from nitrogen-fixing bacteria could be put into plants which would then fix their own nitrogen and so make some of their own fertiliser. This could greatly improve the efficiency of food production. Crop plants given genes which code for disease resistance or tolerance to pollutants could be grown on previously inhospitable land, such as industrial land or desert margins. This sort of technology also has many potential applications in medicine.

The genome of an organism, its entire collection of DNA sequences, can now be explored and mapped. Theoretically, any individual gene can be located, cut out, replaced or inserted into another organism. Of course, this sort of research is regulated very carefully, and the ethical, moral and social consequences need to be taken into account (Fig 34.1).

We are in the process of mapping the entire genome of some species. The bacterium *Escherichia coli* – a prokaryote – has already been sequenced, and so has a simple eukaryote, *Saccharomyces cerevisiae* (baker's yeast). The ~~H~~human Genome Project, a far more ambitious project is due to finish in 2005 (see the Opener to Chapter 3).

Once we know what individual human genes do and where to find them, the possibilities are endless:

● We can take individual genes and make multiple copies: this is **gene cloning**.

● We can replace defective genes with healthy ones: this is **gene therapy**.

● We can extract genes which code for useful products and insert them into other organisms: this is **recombinant DNA technology.** Organisms produced in this way are called **transgenic organisms**. More genes which code for human products could be transferred into microorganisms for mass production. Products already made in this way include insulin, somatostatin, alpha-1-antitrypsin, human growth hormone and blood clotting substances such as factor VIII.

In this chapter we focus on the techniques which allow gene manipulation, and we look briefly at some of the possibilities for the future and at some of the ethical implications of genetic technology.

2 GENE HUNTING

The first problem that genetic engineers tackle is finding the gene they are interested in. Every human body cell contains 46 chromosomes each of which have several thousand genes. So, how, for example, do you go about finding the insulin gene that you want to insert into a bacterium?

The following are three different approaches.

A: Isolate the gene from the rest of the genome

This has been compared with looking for a straw needle in a haystack. However, there are several sophisticated techniques which can be used to map chromosomes. For instance, if you know part of the base sequence in the required gene you can make a **genetic probe**. This is a single-stranded piece of matching DNA labelled with a radioactive or fluorescent marker. When mixed with the genomic DNA – in the right conditions – the probe seeks out the complementary base sequence and binds to it, showing you exactly where the gene is.

In the long term, of course, having a complete library of human genes will make this a lot easier. The ultimate goal is to know where each gene is located, what it does, how it interacts with other genes and what goes wrong in cases of genetic disease.

B: Use messenger RNA

If you can isolate the mRNA molecule that is a copy of the gene that you are trying to find, you can use it as a genetic probe as described above. You can also use it to make a DNA copy using the enzyme reverse transcriptase. This means essentially that you make an 'artificial' gene. For example, you could find the mRNA which codes for insulin in the cells where the insulin gene is expressed – the

Fig 34.1 **'You were so keen to see what you could do, you didn't stop to think about whether or not you should.'** This is a scene and a quote from the film Jurassic Park. It is quite unlikely that we will ever be able to recreate dinosaurs, but the quote reflects an important point: all genetic research has the potential for abuse and we should think very carefully about its consequences

DNA is common to all organisms, and it always has the same basic structure. This allows us to combine genes from organisms as diverse as humans and bacteria.

?

A Suggest what the enzyme reverse transcriptase does.

cells in the islets of Langerhans. DNA made from mRNA is called **complementary DNA**, or **cDNA**. Artificial genes are shorter than those in the genome because they contain no introns (Chapter 31).

C: Work backwards from the protein

If you work out the amino acid sequence of the desired protein – a relatively easy process – you can make a piece of DNA which codes for it. It would obviously be a tedious process to make a large gene, but now there is equipment which can synthesise artificial DNA quickly and easily. The end product of this process is also a short, artificial gene made from cDNA.

3 MANIPULATING DNA USING ENZYMES

Once you have located the gene you want, the next step is to use it. Enzymes are the tools of the trade here.

Restriction enzymes

Restriction enzymes can be thought of as 'molecular scissors' – they cut DNA strands at specific points (Table 34.1). More properly called **restriction endonucleases**, they are made by bacteria in response to attack by viruses called bacteriophages. Their name reflects their function: they *restrict* damage by chopping bacteriophage DNA into smaller, non-infectious fragments, and they make their cuts *inside* the nucleic acid molecules.

> ✔
>
> The prefix *endo-* means *inside*; the prefix *exo-* means *outside*. An *endonuclease* cuts up a DNA strand by making cuts along the whole length of the molecule. In contrast, an exonuclease cuts DNA by chopping off the nucleotides at the ends of the molecule.

Fig 34.2 **Endonuclease enzymes are wonderful tools for the genetic engineer because not only do they cut DNA, but they make *staggered* cuts, leaving 'sticky' ends which make re-joining DNA strands much easier. Endonucleases cut at the sugar–phosphate bonds in the DNA molecule, leaving only the hydrogen bonds intact. At room temperature and above, these bonds are not enough to hold the molecule together, and so it splits, producing a 'sticky end'**

Action of restriction endonuclease:

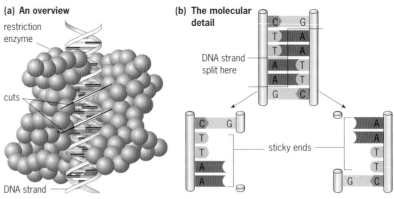

There are many different endonuclease enzymes, produced originally by different species of bacteria. Each enzyme cuts DNA at a different base sequence, a point known as the **recognition site**. The enzymes make staggered cuts in the DNA (Fig 34.2), commonly called **sticky ends**. These can combine with complementary sticky ends, and this allows lengths of DNA that have been removed from one organism to be **spliced** – inserted – into the DNA of another. EcoR1, for example, is a common endonuclease isolated from the bacterium *Escherichia coli*. This is a very popular enzyme with genetic engineers because it is readily available. It is also reliable: it cuts DNA at its recognition site with precise accuracy, showing very little **star activity** (cutting DNA at sites other than the recognition site).

Table 34.1 **The target sites of some restriction enzymes**

Enzyme	Bacterial origin	Recognition site
EcoRI	*E. coli*	G|AATTC CTTAA|G
HindIII	*H. influenzae*	A|AGCTT TTCGA|A
Bam H1	*B. amyloliquefaciens*	G|GATCC CCTAG|G

You may be asking, 'If restriction enzymes cut DNA, why is the bacterium's own DNA not damaged?' The answer lies in a process called **methylation**: the bacteria add a methyl ($-CH_3$) group to the sequences in their own DNA that could act as recognition sites. This stops the restriction enzymes doing any damage, presumably because the methylated region no longer fits the active site of the enzyme.

DNA ligases

The sticky ends of DNA strands produced by the action of restriction enzymes are joined together by enzymes called **DNA ligases** (ligate means *to join*). The normal function of DNA ligases is to join strands of DNA during replication (Chapter 31) and they are also involved in DNA repair. Genetic engineers use them as 'molecular glue' and they are essential partners for restriction enzymes. An example is **T4 DNA ligase** which is made by *E. coli*.

Enzymes which modify DNA

Restriction and ligase enzymes can cut and splice DNA, but there are many other enzymes which are useful tools for the genetic engineer. Some of these are shown in Table 34.2.

Enzyme	Function
Exonucleases	Remove terminal base from DNA – useful in analysis of base sequences.
DNase	Makes random cuts in DNA: chopping into varying-sized fragments.
Kinases	Add groups such as phosphates to DNA: this helps with labelling and analysis, particularly when phosphates are made with ^{32}P.
Polymerases	Synthesise nucleic acid molecules from nucleotides. This is essential for the polymerase chain reaction (PCR) – see Feature box. Reverse transcriptase is a polymerase which makes DNA from RNA. It occurs naturally in retroviruses such as the human immunodeficiency virus, HIV.

Table 34.2 **Some enzymes commonly used in genetic engineering**

B Different restriction enzymes have differently sized recognition sites. Those in Table 34.1 are described as 'six cutters' because their recognition sites are six bases long, but four or five cutters are also common. Why do you think four cutters produce a larger number of fragments than five cutters?

THE POLYMERASE CHAIN REACTION (PCR)

THE FIRST RECOMBINANT DNA molecules were produced in the early 1970s when bacteria were made to copy foreign DNA along with their own. However, in 1983 an American called Cary Mullis cloned DNA without bacteria using the relevant enzymes. This process, the **polymerase chain reaction** is gene cloning in a test-tube. It allows DNA to be **amplified** (copied many times).

The idea behind PCR is very simple (Fig 34.3). You mix together your original piece of DNA with the enzyme DNA polymerase in a solution of nucleotides. Next, add some **primers**, short pieces of DNA which act as signals to the enzymes, effectively saying 'start copying here'. You then heat the DNA so that it denatures into two strands. The thermostable DNA polymerase gets to work and produces two identical strands of DNA (see Chapter 31 for the detailed mechanism of DNA replication). In a second cycle of reactions the two strands become four, and so on. Typically, the cycle is repeated about 20 times and within an hour you have millions of copies of the original piece of DNA.

This technique has many applications. Forensic scientists use it to amplify tiny DNA samples from spots of blood or hair roots, to obtain enough material for forensic analysis by DNA profiling. PCR also allows archaeologists to study small samples of DNA from historical material such as the preserved bodies found in peat bogs or ice graves.

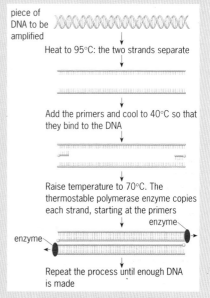

Fig 34.3 **The polymerase chain reaction**

4 CLONING GENES

Armed with restriction enzymes and ligases, genetic engineers can create recombinant DNA molecules. However, once a recombinant DNA fragment has been made in vitro ('in glass'), it usually needs to be copied repeatedly so that enough molecules are available for analysis or for commercial use.

Although DNA fragments can be multiplied in a test-tube (see the Feature box on PCR on page 545), it is often much easier to put the DNA into organisms such as bacteria which will then adopt the new DNA as their own, and copy it for you. This is called **gene cloning**. The basic procedure is outlined in Fig 34.4. As an added bonus, the bacteria can also express the gene, making large amounts of the product.

> ✔
> The word *protocol* is often used in genetic engineering. It means 'standard procedure' and so a cloning protocol would be a standard method for cloning a gene.

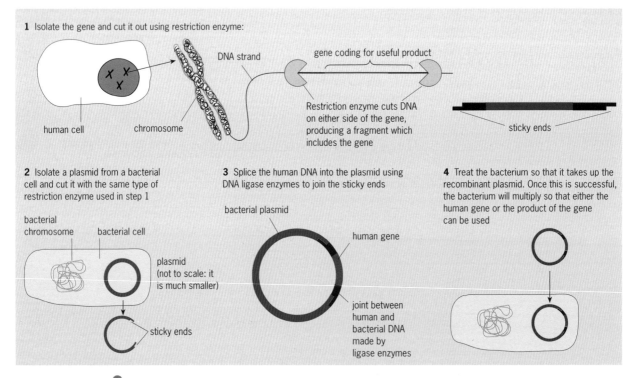

1 Isolate the gene and cut it out using restriction enzyme:

human cell • chromosome • DNA strand

gene coding for useful product

Restriction enzyme cuts DNA on either side of the gene, producing a fragment which includes the gene

sticky ends

2 Isolate a plasmid from a bacterial cell and cut it with the same type of restriction enzyme used in step 1

bacterial chromosome • bacterial cell

plasmid (not to scale: it is much smaller)

sticky ends

3 Splice the human DNA into the plasmid using DNA ligase enzymes to join the sticky ends

bacterial plasmid

human gene

joint between human and bacterial DNA made by ligase enzymes

4 Treat the bacterium so that it takes up the recombinant plasmid. Once this is successful, the bacterium will multiply so that either the human gene or the product of the gene can be used

Fig 34.4 **The principles of recombinant DNA technology. In this example, the human gene which codes for insulin is transferred into a bacterium which then multiplies to form a large population. The bacteria all express the human gene, and large amounts of human insulin can be recovered**

> ?
> **C** Suggest the ideal conditions that would need to be created in a fermenter to ensure maximal bacterial growth.

Using vectors to transfer genes

Human genes are often transferred into bacteria or yeast because these microorganisms multiply quickly to form a huge population which expresses the gene and makes large quantities of the gene product. This can then be extracted and purified. The microorganisms are grown in a **fermenter**, a giant vat with ideal conditions for growth (Fig 34.5).

Fig 34.5 **Transgenic bacteria or yeasts are cultured in large cylindrical sterile fermenters, (one is shown top left) although the process is not strictly speaking fermentation. These containers were originally used for making wine and the name has stuck**

Putting genes into bacteria

Putting a human gene into another organism is not an easy task.

The first step is to attach the gene to a **vector**, or carrier. One such vector is a **plasmid**, a tiny circular piece of DNA which occurs in bacteria (see Chapter 1). Like restriction enzymes and ligases, plasmids can be bought commercially.

The plasmid is cut using the same endonuclease enzyme that was used to cut the human DNA to obtain the gene. Using the same endonuclease is important – the DNA must be cut at the *same base sequence* to produce complementary sticky ends that will join up. The gene and the plasmid are mixed together and then a DNA ligase is added to join up the sticky ends. The DNA molecule produced is circular, like the original plasmid.

In the second step, the genetic engineer must induce the bacterium to accept the plasmid. One technique involves soaking the bacteria in ice-cold calcium chloride and then incubating them at 42 °C for 2 minutes. Nobody seems to know exactly why this works, but it does. Bacteria which have accepted the plasmid now contain recombinant DNA, and are, by definition, transgenic organisms.

The conversion process is not very reliable: for every bacterium that takes up the recombinant plasmid, about 40 000 do not. So how do you tell which ones are recombinant? A clever trick here is to insert two genes into the plasmid; the one you want to use and one which makes the recombinant bacteria easy to detect. For instance, you can add a gene which confers antibiotic resistance and then culture the bacteria on agar plates containing the antibiotic. Only the bacteria which have the modified plasmid will survive and grow.

An alternative is to add a second gene which codes for an enzyme that metabolises a coloured substrate. When the bacteria are grown on agar plates that have been made with the substrate, colonies of bacteria that have taken up the plasmid will be a different colour from colonies of non-recombinant bacteria.

Once you have identified the colonies of recombinant bacteria only, you can use these to start pure cultures. Transgenic bacteria are often described as 'sick' because they multiply more slowly then normal bacteria. Even so, they multiply at an enormous rate: the population doubles every 30 minutes instead of every 20 minutes. Within ten hours, one transgenic bacterium can produce over a million copies of itself, each one containing a working clone of the original gene.

Modified bacteriophages also make good vectors. These viruses reproduce by inserting their own DNA into the DNA of host bacterial cells. Bacteriophages are able to transfer larger amounts of DNA than plasmids, but, at present, their use is limited.

Putting genes into eukaryotic cells

Usually, a gene is transferred into a different organism so that it can be expressed, making greater volumes of product. This happens only when the gene is able to use the synthetic machinery of that organism's cells (the mRNA, ribosomes, and Golgi apparatus etc) for protein synthesis. Some eukaryotic genes are not expressed effectively by prokaryotic cells, and so must be transferred into another eukaryotic cell. Yeast, a single-celled fungus, is often used.

Producing recombinant eukaryotic cells is even more difficult than producing recombinant bacteria. The cell walls of fungal cells are a major barrier. One solution is to digest them away with suitable enzymes. A cell without a cell wall is called a **protoplast**. In this 'naked' state, the yeast cell can be 'persuaded' to accept plasmids.

Other methods of transferring DNA into eukaryotic cells include the following.

Microprojectiles. It is possible literally to shoot DNA into host cells. Tiny pellets of tungsten or gold are coated with DNA and fired at high speed at the target cells. Remarkably, some of the cells recover and accept the foreign DNA.

Electroporation. This technique involves exposing the host cells to rapid, brief bursts of electricity which create temporary gaps in the cell surface membrane, through which foreign DNA can enter.

Liposomes. This is a more subtle method of smuggling DNA into cells. DNA is inserted into liposomes – spheres formed from a lipid bilayer. The liposomes fuse with the surface membrane of the host cell and introduce the DNA into the cytoplasm. Liposomes might be useful in gene therapy (see the Assignment on page 552).

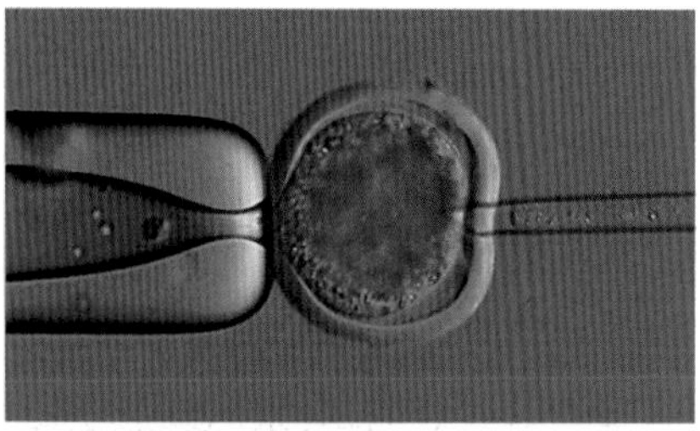

Calcium phosphate precipitation. Plasmids are mixed with calcium phosphate. As the calcium phosphate precipitates, the grains that form contain DNA. These grains can enter cells by endocytosis (see page 95).

DNA injection. Fig 34.6 shows this delicate technique. Using a very fine pipette, and a device called a micromanipulator which avoids the inevitable hand tremor, DNA can be inserted directly into a cell. The transgenic sheep featured in the Opener was created when the gene for alpha-1-antitrypsin was manually injected into the cells of a sheep embryo. This method leaves much to chance: many embryos have to be treated before one will correctly take up the foreign DNA.

Fig 34.6 **Injecting DNA directly into a sheep cell. The cell is held steady by gentle suction from the pipette on the left, while the even finer pipette on the right penetrates the cell surface membrane and introduces the foreign DNA**

CLONING WHOLE ORGANISMS

IN 1997, WHEN SCIENTISTS SUCCEEDED in cloning a sheep, there was a lot of fuss in the media. Was it ethically and morally right to clone animals as advanced as mammals? Would this lead to scientists attempting to clone humans?

The definition of a clone is 'a genetically identical copy' and the term can apply to individual strands of DNA, individual cells or whole organisms. Cloning is, in fact, a common natural process: all organisms made by asexual reproduction are clones of their parent, unless there is a mutation. Even humans make clones – they are more commonly called identical twins.

There is much commercial interest in the cloning process because it can produce exact copies of organisms with desirable characteristics. If, for example, you have a 'perfect' apple tree which produces large amounts of delicious, healthy, blemish-free apples, you can clone it rather than risk losing its valuable genotype in the lottery of sexual reproduction.

Cloning plants is easy, and can be done with relatively little equipment in schools and colleges. Cloning mammals, however, is a lot more difficult but technically feasible. Eggs and sperm or fertilised embryos are collected from prize specimens of cattle, for example. Cells can be pulled apart when the embryo is at the eight or sixteen cell stage and each fragment can then develop into a complete embryo. Each embryo is a clone of the original. The process can be repeated many times and then the cloned embryos can be introduced into the uteruses of normal cows. After the normal gestation period, several very average animals give birth to a herd of prize specimens.

There is no doubt that is it *possible* to clone humans, and this has led to some alarm. Much of the concern is justified and, in the UK, the Human Fertilisation and Embryology Authority has banned experiments on human cloning. However, some people have been worried unnecessarily by media reports based on ignorance and misunderstanding. A human clone would be, in every respect, a twin of the original person. He or she would still be a unique human being, with an individual personality. Your experiences and memories, as well as your genes, make you into the person you are. The misconception that people can have their cells frozen so that they can be cloned after their death to 'live again' is complete nonsense.

5 GENETIC ENGINEERING AND PLANTS

As we saw in the introduction to this chapter, genetic techniques have an enormous potential to improve plant crops to feed our ever-expanding population. Most of the world's important crop plants – eg maize, potatoes, wheat, oilseed rape, cotton and tobacco – are candidates for genetic improvement. So far, genetic engineers have concentrated on producing the following improvements:

- **Tolerance to herbicides**. Adding genes to make plants resistant to herbicides allows weeds to be killed without affecting the crop plants.

- **Resistance to insect pests**. Genes which code for products that are lethal to insects are added to the plant's genome.

- **Resistance to viral disease**. Genes for resistance to important diseases such as mosaic virus can now be built-in.

- **Improvements in crop quality**. Plants such as oilseed rape can be modified so that they make other, more useful oils that can be used as detergents, fuel oils and other non-food products. There is over-production of food in many areas of Europe and America, and new oil crops could provide a valuable alternative for land use. Other crops, such as tomatoes, are being modified to make them easier to store and process (Fig 34.7).

- **Manipulation of growth rate**. The growth rate of many plants appears to be genetically determined. One interesting line of research is manipulating the growth rate of trees such as oaks. Normally, oaks produce high quality wood but grow very slowly. If the genetic clock from a faster growing species, such as the poplar, can be transferred into the oak, we might obtain a renewable source of valuable hardwood.

Fig 34.7(a) **This tomato has been genetically modified to improve the quality of the fruit and its processing characteristics.**

The normal ripening process in a tomato is accompanied by softening of the tomato fruit making it susceptible to disease and damage during harvesting. By targeting and reducing the expression of the one of the tomato's ripening genes, polygalacturonase, a fruit has been produced which ripens normally but softens more slowly. The genetically modified fresh tomatoes delivered for processing are of greater quality and, because their 'softening' enzyme has been deactivated, they can be used to make a thicker tomato puree or a firmer tomato dice

Fig 34.7(b) and (c) **The tomatoes on the left are the same variety and the same age as those on the right. The deterioration and spoilage shown by the normal variety on the left is normal. The tomatoes on the right have been genetically modified and stay fresher for longer. They have been given a gene which blocks the normal production of ethene, a plant hormone which promotes ripening**

The advantages of genetic engineering

Modern genetic engineering techniques have several important advantages over traditional plant breeding:

- Selective breeding is slow, and requires large areas of land. In contrast, genetic engineering operates at the level of individual cells which multiply rapidly in the laboratory (Fig 34.8).

- Plant breeders must use closely related strains or species. Genetic engineers can insert genes from any organism into their plants.

- Selective breeding involves the mixing of whole genomes, so that unwanted genes can be bred in along with the desired ones. There is no chance of this happening when individual genes are intentionally transferred.

Fig 34.8 **Producing transgenic plants. Beneficial genes can be introduced into plant cells using the bacterium *Agrobacterium tumefaciens*, which causes tumours called crown galls on infected plants. This bacterium contains a plasmid called Ti which is able to transfer DNA directly into the genome of the plant cell. Normally, the effect is to cause abnormal plant growth, but the modified plasmids can be used to insert beneficial genes into plant cells. The transgenic cells can then be cultured to produce whole plants**

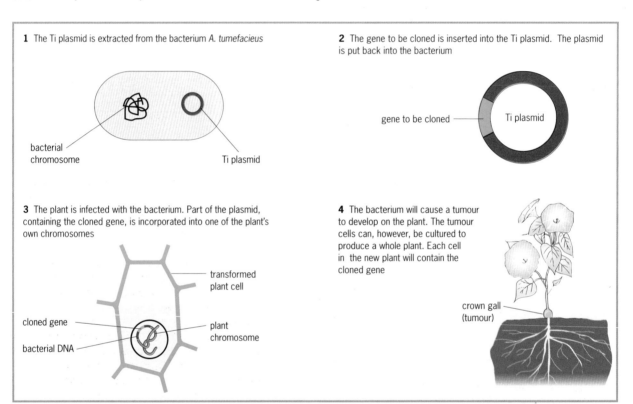

1 The Ti plasmid is extracted from the bacterium *A. tumefacieus*

bacterial chromosome

Ti plasmid

2 The gene to be cloned is inserted into the Ti plasmid. The plasmid is put back into the bacterium

gene to be cloned

Ti plasmid

3 The plant is infected with the bacterium. Part of the plasmid, containing the cloned gene, is incorporated into one of the plant's own chromosomes

transformed plant cell

cloned gene

bacterial DNA

plant chromosome

4 The bacterium will cause a tumour to develop on the plant. The tumour cells can, however, be cultured to produce a whole plant. Each cell in the new plant will contain the cloned gene

crown gall (tumour)

SUMMARY

- Genetic engineers isolate, cut out and transfer genes between organisms. Genetic engineering technology has many applications, including making human products such as insulin on a large scale.

- Strands of DNA are cut using **restriction endonuclease enzymes**, which always cut DNA at a particular base sequence. Strands of DNA can be joined by using **DNA ligase** enzymes.

- In order to transfer DNA into a living cell such as a bacterium, a **vector** (carrier) is needed. Common vectors are **plasmids** (circles of DNA found in bacteria) or viruses.

- Once the foreign DNA is inside the host cell, it is adopted by the host. Genes in the foreign DNA are expressed using the host cell's synthetic machinery (ribosomes, mRNA etc). In addition, the foreign DNA is **cloned** (copied) every time the host cell divides.

- **Gene therapy** is a technique which replaces faulty genes with healthy ones. It provides great hope for sufferers of genetic diseases such as cystic fibrosis.

- The **polymerase chain reaction** (PCR) allows samples of DNA to be cloned in a test-tube using the necessary enzymes, instead of inside a host cell.

QUESTIONS

1 Read the following passage on gene technology (genetic engineering), then replace the dashed lines with the most appropriate word or words to complete the text.

The isolation of specific genes during a genentic engineering parocess involves forming eukaryotic DNA fragments. These fragments are formed using enzymes which make staggered cuts in the DNA within specific base sequences. This leaves single-stranded 'sticky ends' at each end. The same enzyme is used to open up a circular loop of bacterial DNA which acts as a ___ for the eukaryotic DNA. The complementary sticky ends of the bacterial DNA are joined to the DNA fragment using another enzyme called ___. DNA fragments can also be made from ___ template. Reverse transcriptase is used to produce a single strand of DNA and the enzyme ___ catalyses the formation of a double helix. Finally new DNA is introduced into host ___ cells. These can then be cloned on an industrial scale and large amounts of protein harvested. An example of a protein currently manufactured using this technique is ___.

[ULEAC 1996 Biology/Human Biology, Specimen Module Paper, q.4]

2 Scientists have shown that kidney beans are resistant to cowpea weevils and adzuki bean weevils, two of the most serious pests of African and Asian pulses (vegetables related to peas and beans). This is because the beans produce a protein that inhibits one of the weevils' digestive enzymes. Weevils that eat the pulses soon starve to death. The researchers have identified the gene that produces the inhibitor and removed it from the kidney bean DNA. They inserted the gene that produces the inhibitor into the DNA of a bacterium called *Agrobacterium tumefaciens*. Using this bacterium they have been able to add the inhibitor gene to peas. They hope soon to be able to add the gene to African and Asian pulses.

a) Describe how scientists could:
   (i) remove the gene athat produces the inhibitor from kidney beans.
   (ii) insert this gene into the DNA of a bacterium.

b) The DNA in the bacterium is able to replicate to produce many copies of itself for insertion into pea cells. Describe the structure of a DNA molecule and explain how this structure enables the molecule to replicate itself.

[NEAB June 1995 Modular Biology: BY2 Continuity of Life, q.8]

3 The Colorado beetle is a pest of potato crops. A soil bacterium, *Bacillus thuringiensis*, produces a substance called Bt which kills Colorado beetles but is harmless to humans. Scientist have isolated the gene for Bt production from bacteria and inserted it into potato plants so that the plant produces Bt in its leaf tissues.

a) **(i)** What is a gene?
   (ii) Suggest how the gene for Bt production could be isolated from the bacteria and inserted into cells of the potato plant.

b) Bt can also be used as a spray. Colorado beetles may be killed if they ingest potato leaves which have been sprayed with Bt.

Suggest and explain **one** reason why using Bt-producing potato plants might increase the rate of evolution of Bt-resistance in the beetles compared with using Bt as a spray.

[NEAB February 1997 Modular Biology: BY2 Continuity of Life, q.3]

4
a) Suggest why it is important that there are many different types of restriction endonuclease enzymes available to genetic engineers.

People suffering from haemophilia need treatment with the blood-clotting protein, factor VIII, because they are genetically unable to produce their own.
   Factor VIII used to be extracted from human blood but a genetically engineered kind has just been made available. An artifical version of the human gene has been produced which codes for the same sequence of 2338 amino acids that is present in natural factor VIII. This is inserted into hamster kidney cells which are grown in large fermenters. The cells secrete large amounts of factor VIII into the surrounding fluid.

b) **(i)** Why was it necessary to know the amino acid sequence of factor VIII before an artificial version of the numan gene could be produced?
   (ii) What is the minimum number of nucleotides which must be present in the gene for factor VIII?
   (iii) Suggest **one** advantage of the use of genetically enegineered factor VIII over that extracted from blood.

[NEAB February 1995 Modular Biology: BY2 Continuity of Life, q.4]

Assignment

CYSTIC FIBROSIS AND GENE THERAPY

Cystic fibrosis (CF) is a genetic disease that is due to one defective allele (see Chapter 31). About one person in 25 carries the defective allele, but they also carry the normal allele and do not suffer from the disease.

1 Why do heterozygotes not suffer from the disease?

2 A couple are both carriers for cystic fibrosis.

a) What is the probability that their first child will have cystic fibrosis?

b) What is the probability that their second child will have cystic fibrosis?

The normal allele codes for a protein called **cystic fibrosis transmembrane regulator** (CFTR). This essential membrane protein in epithelial cells transports chloride ions out of the cells and into mucus. Normally, when chloride ions are secreted, sodium ions follow, and this decreases the water potential of the epithelial mucus. Water follows outwards by osmosis, making normal, watery mucus which can be moved by the cilia (tiny hairs) lining the airways.

Cystic fibrosis sufferers make a protein which differs in just one of its 1480 amino acids. Although slight, this fault prevents CFTR from functioning normally. Chloride ions and sodium ions cannot be secreted, and so the mucus becomes much thicker than normal. Dead epithelial cells also accumulate in the mucus, adding to the general congestion of the airways.

3 Explain how a fault in just one amino acid can have such a drastic effect.

Sticky mucus is also a real problem in the pancreas. The mucus blocks the pancreatic duct, preventing the secretion of pancreatic juice.

4 Suggest why cystic fibrosis sufferers have to take tablets containing digestive enzymes.

Treatment of CF includes regular physiotherapy in which the chest is patted to dislodge the mucus. Even so, infections are common and most CF sufferers have to take a variety of antibiotics according to the infection they have at the time.

Gene therapy for cystic fibrosis

Now, one exciting possibility is that the faulty genes can be replaced by healthy ones. In 1989 the cystic fibrosis allele was located on chromosome 7. The base sequence of the healthy gene was then compared to the defective allele, and the nature of the fault was narrowed down at the molecular level. This opened up the exciting possibility that if healthy genes could be somehow introduced into the epithelial cells, they might be expressed and so make the correct membrane protein, solving the problem for a time.

The basic steps are as follows:

1. The CFTR gene is isolated, and cut out.
2. The gene is cloned many times.
3. The genes are encapsulated, either by putting them into liposomes (spheres made from lipid) or viruses.
4. The gene particles are inhaled, so that they can pass into the epithelial cells of the lung to be incorporated into the DNA of the cells.

Once in place, if all goes well, the healthy genes are transcribed, making the correct protein.

5

a) What is used to cut the CFTR gene out of chromosome 7?

b) Name two different methods of cloning the CFTR gene.

c) Suggest why the pancreatic genes would be more difficult to replace.

d) Suggest why repeat treatments at regular intervals would be necessary.

Germ-line gene therapy

This area of research is an ethical minefield. So far in this Assignment we have looked at body cell gene therapy – the replacement of the faulty allele in certain cells of the body. However, the eggs or sperm of the sufferer would still carry the defective alleles. In **germ-line gene therapy,** the original alleles in the zygote would be changed. The individual would grow and develop with healthy alleles in all cells and all of the individual's children would inherit the healthy allele.

Germ-line gene therapy sounds attractive but there are problems, so much so that this area of research is banned in the UK. Ethically, there is the familiar 'where do you stop?' problem. A new treatment for cystic fibrosis or haemophilia would be desirable, but having the technology might allow for all sorts of temptations. Our knowledge of the human genome could reveal, for example, genes which code for intelligence, height or skin colour. Tampering with these characteristics would lead to obvious problems.

Biologically, germ-line gene therapy is potentially dangerous because we know very little about how genes function in the embryo. Tampering with genes in the zygote could have effects which might become apparent only in later life.

6 In discussion groups, prepare a case which you could present either for, or against, germ-line gene therapy.

THE ENVIRONMENT

IMAGINE IT'S SUMMER and you are a greenfly living on the underside of a leaf. That leaf is your whole world. The upper surface of the leaf is lashed by rain and wind, then scorched by the Sun. But, underneath, you are safe from the worst of the weather, and fairly well hidden from many of the predators which feast on aphids. You settle next to a leaf vein which provides you with food 'on tap'. You are female - at this time of the year all aphids are female - and you were born pregnant just a few days ago. Tomorrow you will give birth to ten more little aphids - all female, all pregnant. The whole of your short life will be spent giving birth to live, pregnant young.

Learning about the habitat of an aphid, the physical conditions around it and also the way it interacts with other living organisms, is part of the branch of biology known as ecology. Ecology is the study of living things and their surroundings. This branch of science can be thought of as 'whole-organism' biology. In contrast to much of this book which looks at what goes on inside organisms, ecology focuses on the study of complete individuals and whole communities.

Ecology is a vast topic. It can include a study of the ecosystem that exists in your mouth, but it also encompasses whole-planet issues such as global warming and ozone depletion. In this section, we begin in Chapter 35 with some of the basic concepts of ecology, before moving on to look ecosystems in more detail in Chapter 36. In Chapter 37 we concentrate on energy flow and nutrient cycling. Later in the section in Chapter 38, we look at the ecology of populations. Then we end the book by considering the effects of the most resource-depleting of population explosions – that of human beings.

35 Ecology: the basics

Fig 35.1 **Ecology is the study of whole organisms: we learn how an organism interacts with individuals of its own species and other species, and with the physical environment that surrounds it. The jigsaw pieces show just a few of the factors that influence the lives of these zebras in the savannah grasslands of Africa**

The word ecology is loosely used to mean environmental concern or environmental conservation. These topics are important, and they depend heavily on knowledge gained through ecology, but they are not the same thing. We define ecology as: **'the scientific investigation of living organisms in their natural surroundings'**.

1 THE SCIENCE OF ECOLOGY

Ecology is a branch of biology. The word ecology was first used by Ernst Haeckel in the 1860s. It comes from two Greek words: *oikos* meaning home and *logos* meaning understanding. His definition of ecology was 'the knowledge of the sum of the relations of organisms to the surrounding outer world, to organic and inorganic conditions of existence'. Put more simply, ecology is the study of organisms in their natural surroundings. It looks at how they are adapted to their environment and how they interact with both the living and non-living world around them.

Each organism and each physical feature of an environment are separate, but they all interact and interlock, forming a complex system that is a bit like a three-dimensional jigsaw puzzle: ecologists study different parts of the puzzle and then try to work out how it is put together (Fig 35.1).

Ecology and other branches of science

Ecology is closely related to two other branches of science: **environmental biology** and **environmental science**. The first of these looks more generally at living things in the environment. The second uses the physical, biological and earth sciences to help us to explore our environment.

It is also impossible to study ecology without bearing in mind the genetics, evolution and behaviour of the organisms at the centre of a study. Ecological genetics and behavioural ecology are beyond the scope of this book but it is important to remember that each organism you study in the context of its environment has developed to fit that environment through the process of natural selection (see Chapter 33). And, every time something in its environment changes, an organism may respond by showing an altered behaviour pattern. Some aspects of animal behaviour are covered in Chapter 24.

Practical ecology

Fig 35.2 **Safety and field trips. Ecologists conduct studies outside and must consider safety when working near water or in isolated areas.**
Here, a student is identifying organisms from a pond, and two students are measuring the rate of water flow in a stream

As in other branches of science, progress in ecology depends heavily on scientific investigation. Looking back at our definition of ecology, it should come as no surprise that many of these studies are done outside. Field studies involve a great deal of observation, but some experimental studies are also carried out in natural conditions. Others are carried out in the laboratory, where they can be more strictly controlled. Ecologists use these three basic approaches together to gain a complete picture of the organism or the environment under investigation.

The exciting thing about ecology is that it is still a relatively young science. In just over 100 years of ecological study, we have barely begun to even scratch the surface. Many environments of the world remain unexplored and there are many organisms that we know of, but whose habits and living conditions are known only sketchily, or are a complete mystery. Unlike biochemistry or physiology, ecology offers an A-level student the chance to do original practical work, and perhaps to make an important scientific discovery.

The Assignment in Chapter 36 is based on the work of real students on an ecology field trip.

Fig 35.3 **Caterpillars of the peacock butterfly feeding on a stinging nettle.**
About 100 species of invertebrate are associated with nettles: approximately 90 per cent of these are insects, but you are also likely to find woodlice, harvestmen, spiders and snails. You can use the plant and its animal partners to study a whole range of topics such as food chains, the effect of biotic and abiotic factors and population ecology

PLANNING A FIELD STUDY

BEFORE YOU CAN start taking measurements, you have to decide what kind of ecological investigation you want to do. There are two main types. **Synecology** is the study of an entire community of animals and plants: nettles, for example (Fig 35.3). **Autecology** is the study of a single animal or plant species. It includes an in-depth look at the habits of an organism together with its immediate environment (Fig 35.4).

Every study should be carefully planned. Like other scientists, ecologists follow a basic procedure called the **scientific method**. This involves five steps:

● They first put forward a **hypothesis**. A hypothesis is a statement that explains a particular problem or set of observations. The study sets out to test the hypothesis.

● The second step involves **data collection** (see pages 588 and 597).

● The third step is **data analysis**. Ecologists must summarise their data and put it into a form that allows conclusions to be made. Often, mathematical comparisons are made between different sets of data (see page598).

● In the fourth step, the data is used to **draw conclusions** (see also page598).

● The final step involves presenting the results in a **report**, so that others can examine the study (see page 607). Eventually, the results of investigations might be presented at conferences or written up in scientific journals. Some studies prove completely that a hypothesis is wrong. But even when a great deal of evidence exists to support a hypothesis and it has become widely accepted, nature is too full of surprises for scientists to say that anything is 100 per cent proven.

Fig 35.4 **Autecological studies have been carried out on species as diverse as ladybirds, vampire bats (seen here) and puffins**

2 TERMS AND CONCEPTS IN ECOLOGY

Fig 35.5 **The biosphere is the part of the Earth's surface that supports life. It extends from the bottom of the deepest oceans into the part of the Earth's atmosphere that contains breathable air: in total a vertical distance of between 30 and 40 kilometres.**

It covers most of the hydrosphere, the surface layers of the lithosphere and some of the atmosphere. The hydrosphere is the part of the Earth's surface that is covered by water. The lithosphere is the layer of soil and rock that forms the Earth's crust. The atmosphere is the blanket of gases which envelop the planet, maintaining a mean temperature of about 7 °C. This temperature is crucial to life because it allows most of the water on Earth to exist as a liquid (see Appendix 1)

The relationships between an organism and the physical features of its environment, and between all the other organisms that live with it, are incredibly complex. Any description of them uses terms that you have probably not come across before. So, before getting into detail about any particular aspect of ecology, the next section gives an overview of some basic terms and concepts.

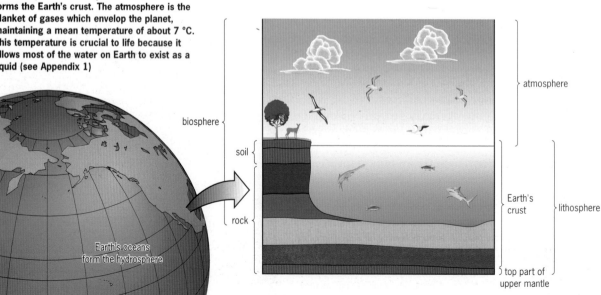

The biosphere and its division into biomes

All the living organisms that we currently know of live on Earth. But not all parts of the Earth support life. The part that does forms a sort of 'skin', the **biosphere,** which is shown in Fig 35.5.

The biosphere is made up of lots of different areas that have very different environmental conditions. We call each of these fairly broad areas **biomes**. The major biomes of the world are listed in Table 35.1 and some of these are shown in Fig 35.6. Many of the best-studied biomes are the **terrestrial biomes** (those that exist on land) but there are also **aquatic biomes** (those that exist in water).

> **?**
>
> **A** Look at Table 35.1. Which biome do you think exists in Britain?

Table 35.1 **The major biomes of the world**

Biome	Type	Major features
Savannah	terrestrial	Tropical grassland with few trees. Example: the Serengeti in Africa
Temperate grassland	terrestrial	Grassland with hot summers and cold winters. Some broad-leaved plants. Example: the prairies in the US
Desert	terrestrial	Hot and dry. A few highly specialised plants and animals. Example: the Kalahari desert in Africa
Tundra	terrestrial	Cold for most of the year: very short growing season. No trees: some specialised plants. Example: Northern Canada
Tropical rainforest	terrestrial	Tropical climate. Lush vegetation. Great diversity of animal and plant life. Example: forests of equatorial Africa
Temperate rainforest	terrestrial	Cool and wet. Tallest trees in the world. Example: redwood forests of North America
Temperate deciduous forest	terrestrial	Hot summers and cold winters. Dominated by broad-leaved deciduous trees. Example: oak woodlands of central and western Europe
Lakes and ponds	aquatic: freshwater	Large bodies of standing freshwater. Example: Great Lakes of North America
Streams and rivers	aquatic: freshwater	Flowing freshwater. Example: the Amazon
Marine rocky shore	aquatic: sea water	The border between the land and the sea. Most rocky shores show **zonation**: there are characteristic bands of different environmental conditions that lead to colonisation by very different plants and animals
Coral reef	aquatic: sea water	The tropical rainforests of the seas. Reefs form in warm, shallow waters and form a biome that contains an incredible diversity of living organisms

Fig 35.6 **The biomes of the world show great extremes of environmental conditions**

(a) **Arctic tundra**

(b) **Desert**

(c) **Coral reef**

(d) **Rainforest**

B The following components of its ecosystem all affect an aphid living on the underside of a rose leaf. Which are biotic and which are abiotic?

(a) the humidity of the air

(b) predation by ladybirds

(c) wind speed

(d) temperature

(e) competition for food

(f) disease

Fig 35.7 **A schematic diagram showing the essential features of an ecosystem**

The ecosystem concept

The **ecosystem** is the basic functional unit of ecology. It is a single working unit that consists of a group of interrelated organisms and their physical environment (Fig 35.7). We can divide any one of the biomes described above into many different ecosystems.

When we study a particular organism in its ecosystem, we look at the physical features that might affect it, such as rainfall, soil type, temperature and so on. These physical or **abiotic** features help to determine what range and type and numbers of living organisms live inside the ecosystem. Of course, we also look at the living or **biotic** part of the ecosystem, to see how our chosen organism relates to the other organisms present.

Later in this section, we look at the abiotic and biotic features that affect the distribution of organisms in an ecosystem in much more detail (see Chapter 36).

| Ecosystem | = | Producers (green plants) | + | Consumers (herbivores, omnivores, carnivores) | + | Decomposers (mainly bacteria and fungi) | + | Non-living components (eg weather, soil composition) |

Levels of organisation in an ecosystem

Individual living organisms may look randomly arranged in an ecosystem, but they are organised into recognisable units. Fig 35.8 provides an overview of organisation in an ecosystem.

Fig 35.8 **An ecosystem has two basic parts: the abiotic part and the biotic part. Within both of these, there are further levels of organisation**

habitats

Environment: physical or abiotic part

populations

Ecosystem

Community: living or biotic part

individual organisms

ecological niche

From ecosystem to individual

We can study the human body and narrow down our investigation from a body system to an important organ, to the level of the individual cell. In a similar way, we can focus in on the components of an ecosystem. Every ecosystem contains a **community** of organisms, which consist of many different **populations**. Each population is made up of many **individuals**. It is important to define these basic features of an ecosystem:

- A community is a collection of groups of organisms from different species that live in close association in the same ecosystem. Organisms in a community interrelate (see Chapter 37).

- A population is a group of individuals of the same species that live in a particular area at any one moment in time. We can look at a population of rabbits or a population of beech trees. Different populations interact in an ecosystem, forming a community.

- An individual is a single organism within a population. Some populations contain organisms that are genetically identical. Some plant populations, for example, are identical because they have developed as a result of asexual reproduction (see page 476). But most populations contain individuals that are the result of sexual reproduction: these individuals show genetic variation (Fig 35.9 and see Chapters 28 and 29).

Fig 35.9 **This individual stoat is part of a larger population. Like other members of the group, she is genetically unique. She has evolved to become well adapted to her woodland environment and, if her particular genes allow her to live a long life, she will pass her genes on to several new members of the next generation**

Environment, habitat and ecological niche

The divisions listed above relate primarily to the *living* components of an ecosystem. We can also sub-divide an ecosystem with reference to its *non-living* components. The term **environment** describes the overall physical surroundings that occur in a biome or an ecosystem (although it is often used more generally: people speak of the environment of the Earth).

Within the environment of a single ecosystem are **habitats**, individual areas in which particular groups or individual organisms live. The habitat of any organism is its normal home. Some organisms, the giant panda, for example, has only one habitat – bamboo forest. Others, like humans, live in many different ecosystems and biomes and so have many different habitats.

Another important term is **ecological niche**. To describe an organism's niche, we need to show how that organism relates to the physical and biological components of its surroundings. It is not just *where* it lives, it is also *how* it lives and *what* it does.

?

C What is the different between an organism's habitat and its ecological niche?

3 AN OVERVIEW OF THIS SECTION OF THE BOOK

In this section of the book, we provide an introduction to ecology at advanced level. In the next three chapters, 36, 37 and 38, we describe the characteristics of an ecosystem in greater detail. In Chapter 36, we look at the physical and biological components of an ecosystem. In Chapter 37, we see how energy is transferred through the living component, and how nutrients cycle through both the living and non-living component. In Chapter 38 we examine how organisms within communities and populations interact. In Chapter 39 we find out how human activity can influence the local and global environment and look at some of the important issues that concern ecologists today.

36 The biology of ecosystems

The loss of grassland and forest has caused extensive soil erosion, and much of Australia now has desert ecosystems rather than grassland. Seventy per cent of the native mammal species are endangered, because they have lost their habitat

THE ABIOTIC AND BIOTIC parts of any ecosystem interact in many different ways, some subtle and some not so subtle. Australia provides one of the most dramatic examples of how disrupting the delicate balance of an ecosystem can devastate the environment and the populations of living organisms that depend on it.

Two hundred years ago, Australia was hot, dry, and covered predominantly by grasslands, with some areas of forest. When European settlers arrived they started to clear the forests and brought in two alien species of mammal: sheep and rabbits. The rabbits in particular multiplied rapidly in conditions that suited them very well, and were soon eating the grass faster than it could grow.

Within ten years of the introduction of just 24 rabbits, the wild population had grown to millions and started destroying crops and wild vegetation as well as any grass that was left. They out-competed native mammal species, driving many of them to extinction. This plague continued until the late 1950s, when farmers introduced the myxomatosis virus. This killed over 99 per cent of the rabbits, with a few surviving because they were naturally resistant. Today, Australia is in danger of being overrun by another plague of rabbits.

1 THE BASIC FEATURES OF AN ECOSYSTEM

Before looking in detail at the components of an ecosystem, we shall consider an example of a real ecosystem. Fig 36.1 shows the main features of a familiar aquatic ecosystem, the garden pond. This is simplified (a real pond contains many more organisms), but you can see that an ecosystem is made up of physical or abiotic components such as water, mineral nutrients, soil and rock, and also living or biotic components such as fish, snails, insects and water plants. (See Chapter 35 for an introduction to the terms abiotic and biotic.)

Think about one of the insects in the pond. It can only survive and live there if the physical conditions in the pond suit it. So, the pH and the temperature of the water must be within a suitable range. The insect also needs to interact with other organisms: it needs to eat smaller organisms or to feed on the sap of plants; and it needs to be with other insects of the same species so that it can find a mate and reproduce. And there is a 'grey area': some of its physical needs, for instance dissolved oxygen, are only available if the insect lives with other organisms, such as a large population of phytoplankton which release oxygen as a product of photosynthesis.

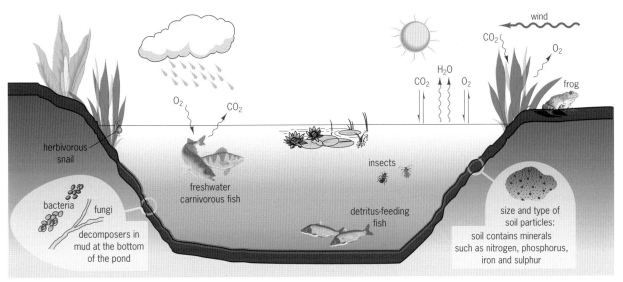

Fig 36.1 **A schematic diagram showing a garden pond as an ecosystem. Every living organism in the pond is affected by abiotic factors such as temperature, light levels and mineral availability, and by biotic components – other living organisms**

2 THE ABIOTIC COMPONENTS OF AN ECOSYSTEM

The abiotic components of an ecosystem give it a physical form, providing places for organisms to live. Physical conditions help to determine where each organism survives best: plants adapted to dry, almost desert conditions will not thrive if a river changes its course and the environment becomes boggy and wet. The nature of the physical environment in any ecosystem depends on:

- the rock and soil types present,
- the type of landscape: mountain range, flood plain etc,
- the position on Earth: altitude, latitude etc,
- the climate and weather, including light, temperature, water availability, wind and water movements,
- the potential for catastrophe: fire, flood etc.

Some abiotic factors, such as the position of an ecosystem on Earth, remain unchanged as time passes. Others, such as the landscape and soil type, change very slowly. Although the basic climate of an ecosystem is usually pretty constant, you can get daily and often hourly changes in weather. Catastrophes such as fires and floods are sudden and short-lived, but they can produce severe, long-lasting effects (Fig 36.2).

Fig 36.2 **In some ecosystems, catastrophic events are necessary. Some species of plant, such as *Banksia* from Western Australia (below) reproduce only after a serious fire (below left). *Banksia* seeds are protected by closed capsules that can be broken open only by the heat of the flames: without exposure to fire, the seeds cannot germinate**

Water stays in soil because it is stuck, or **adsorbed**, to the surface of the soil particles. Clay soils have a lot of very small particles and therefore have a large surface area that can adsorb water. Sandy soils have fewer, larger particles with less surface area and so adsorb less water. Soils of both types with a high humus content retain more water than soils with little organic matter.

Soil as an abiotic factor

The effects of subtle differences in soil type are important to individual species of plants and animals. To gain an overview of how soil type affects plant populations and communities, it is useful to look at two ecosystems with extreme soil compositions. Fig 36.3 shows a semi-desert in central Australia. The 'soil' here is mainly sand, with little organic matter and virtually no microorganisms. This soil cannot easily support plant life: it does not retain any water which falls on it, it warms up very quickly to high temperatures and it contains hardly any mineral nutrients.

Fig 36.3 **Soil erosion is a serious problem in many parts of Australia. Originally, the soil was held in place by grass on which only kangaroos grazed. With human settlers came huge numbers of domesticated sheep and a rabbit population that grew at a phenomenal rate. Both of these animals ate the grass faster than it could grow. With no plant roots to hold the soil together it blew away or was washed away by rain. The result is the semi-desert conditions that exist there today**

The plants that survive here have developed extreme strategies to cope with the extreme environment. Many have very short growing seasons, completing their whole life cycle in the 30 days that follow a decent spell of rain. Others survive long droughts as underground bulbs. Only a few, the succulents, manage to survive above ground for most of the year (Fig 36.4). Because so few plants live in a desert soil, few animals can live there either, and there is little material available for decomposition.

Fig 36.4 Fig 36.4 **Although these Australian succulents (the plants with the vivid magenta flowers, bottom left) are close to the sea, the sand dunes on which they grow retain little water, and are baked by the Sun all day long. We say that these and other succulents are** xeromorphic **(see page 269): they are adapted to living in a hot, dry environment**

Our second ecosystem example, a peat bog (Fig 36.5(a)), is the opposite of a desert. Peat bog soil is permanently wet and cold, it contains little air and, as a result, the organic matter in it cannot decompose. The peat soil is very deficient in nutrients which remain locked up in the undecomposed matter. Sphagnum mosses are amongst the few plants which achieve any success in a bog ecosystem (Fig 36.5(b)). A few other species manage to survive by obtaining minerals by 'alternative' methods: carnivorous plants, such as pitcher plants, catch insects, and plants such as the bog myrtle have nitrogen-fixing bacteria in root nodules (see page 583).

Fig 36.5(a) **Right: areas which contain peat bogs get a lot of rain. Since rain-water contains only small quantities of mineral ions, rain does not add to the mineral content of the peat. In addition, rain leaches any nutrients that are in the peat. Most plants fail to grow in this sort of soil: their roots become water-logged and they cannot absorb enough minerals**

Fig 36.5(b) **Left: as sphagnum mosses grow in a peat bog, their lower parts decay and add to the mounting pile of organic debris below them**

Soil type affects plants directly: they depend on soil to provide water and nutrients to manufacture food and to anchor them in place. The composition of soil may also directly affect some animals in the ecosystem, for example invertebrates such as worms and snails (Fig 36.6). However, more usually it affects animals indirectly: only those herbivores which can eat the plants that grow in a particular type of soil can live successfully in that area.

Soils at either extreme are not able to sustain many species of living organism, but soils which fall somewhere between these two extremes support more diverse ecosystems (see Chapter 10). Most of the soils in Britain, for example, are **loams**, soils which retain moisture year-round and contain a reasonable balance of different sized particles and decomposing organic matter.

The main soil types

The soil is the upper, weathered layer of the Earth's crust. The importance of soil cannot be overstated: it underpins the Earth's ecosystems and therefore the agriculture and economy of all countries in the world. All soils contain:

- Mineral particles of various sizes. These come from the rock of the area which, when weathered produces **sand**, **silt** and **clay**. Sand particles are the largest (60–2000 μm in diameter), silt particles are of intermediate size (2–60 μm in diameter) and clay particles are the smallest (less than 2 μm in diameter). We can classify and determine soil types by looking at how much of each type of particle is present in a soil sample (see Fig 36.9).

Fig 36.6 **The distribution of snails and slugs is sometimes determined by soil composition. Snails are usually found on chalky soils (they need the calcium to build their shells), slugs are more common on acid, non-chalky soils**

Fig 36.7 **A summary of the different soil types that can exist. All three extreme soil types can be improved: you can, for example, make a clay soil lighter and more fertile by digging coarse sand or fine gravel into it, and adding plenty of humus**

- Organic matter in various stages of decomposition. Organic matter that has been completely broken down by fungi and bacteria is called **humus**. Soils with a high humus content retain water much better and usually have a higher mineral content than soils with little organic matter.

- A collection of living organisms. Soil contains an enormous number of bacteria and fungi and some invertebrates such as earthworms, soil arthropods and nematodes (see Chapter 12).

These components are present in different proportions in different types of soil. As Fig 36.7 shows, there are many soil sub-types which are usually put into three main categories:

- **Sandy soils** are light, heat up quickly and allow water to drain away very rapidly. Sandy soils are made up of relatively large particles with relatively large air spaces. A soil made up of more than 90 per cent sand particles is called just 'sand'.

- **Clay soils** and **silty soils** tend to be cold, heavy and often waterlogged. They contain small particles with small air spaces. A clay soil is any soil with more than 40 per cent clay particles .

- **Loam**, an intermediate soil type, is dark and has a good crumb structure. Loam has a mixture of particles of different sizes and usually a fairly high humus content. Most of the agricultural and garden soils in Britain are loams (Fig 36.7(d)).

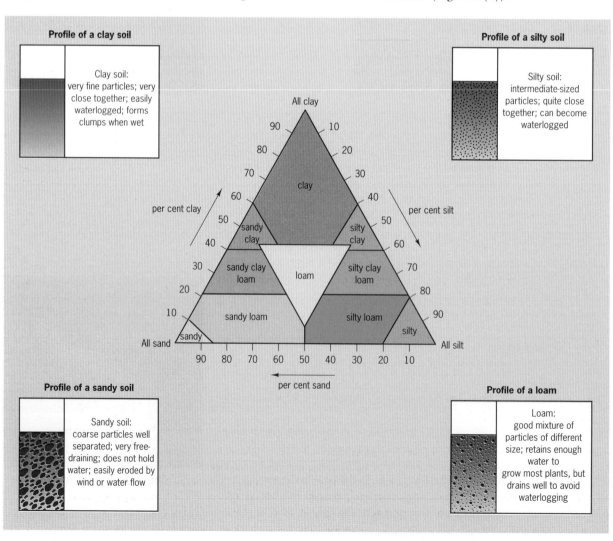

FIELD STUDY NOTES

Testing for soil type

You can find out what type of soil you have at a particular study site using a key devised by the Agricultural Development and Advisory Service of the Ministry of Agriculture, Fisheries and Food (Fig 36.8).

Fig 36.8 **A simple key you can use to identify soil types. To carry out the test, take a desert-spoonful of moist soil (wet it if necessary to obtain maximum stickiness) and knead it thoroughly between finger and thumb until it 'crumbs'. Then, go through the key**

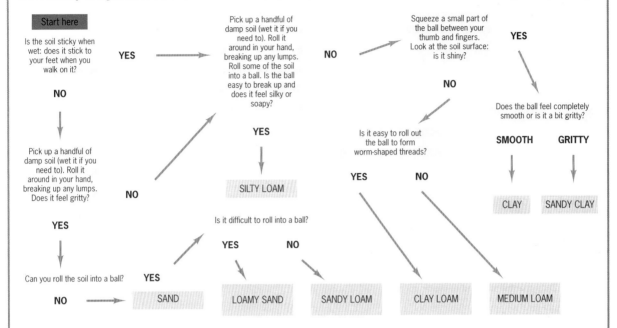

Studying soil composition

You can demonstrate the basic composition of a soil sample by the **sedimentation method**. Mix a small handful of soil with water, shake well and pour into a large measuring cylinder. The inorganic contents settle out in order of size, heaviest particles first. You can then work out the relative amounts of sand, silt, and clay in the sample (Fig 36.9).

You can also measure the **moisture content** of soil. Take another small handful and weigh it carefully. Then put it in an oven set at 105 °C and leave it overnight to dry out. Weigh again and then ensure the sample is fully dried by returning it to the oven and leaving it for another two hours, before weighing again. You may need to repeat this last step until the weight of your sample stops decreasing.

The value of knowing how much water is present in a sample is fairly limited because the amount of water in any soil varies according to what time of day it is, how deep you dig for your soil sample and whether it has just rained. A much more useful measure is the **field capacity** of the soil type: the amount of water that it can hold. Field capacity is a fixed value for each soil type.

To determine the field capacity of a soil sample, dry it thoroughly, weigh it, and then pack it into a perforated container such as a kitchen sieve. Gently pour a known volume of water onto the soil and measure the volume of water that comes through the bottom of the sieve. The amount of water left behind is the field capacity of the dry weight of soil you have used (units = $cm^3 g^{-1}$).

Soils of different types contain varying amounts of organic matter. To measure **organic content**, weigh and heat a sample of dried soil (you could use the one that you produced in the moisture content analysis) in an oven at 450 °C. This burns off all the organic material present. Weigh again after burning to work out how much organic matter your soil contains.

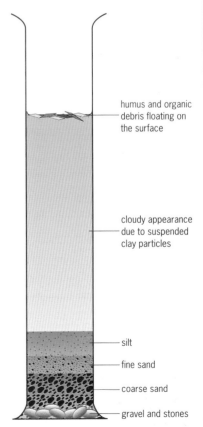

humus and organic debris floating on the surface

cloudy appearance due to suspended clay particles

silt

fine sand

coarse sand

gravel and stones

Fig 36.9 **The sedimentation method of soil analysis**

Fig 36.10(a) **Bacteria which live in deep sea ecosystems next to hydrothermal vents in the ocean floor can survive temperatures as high as 250 °C**

Fig 36.10(b) **Polar bears and other animals which live in polar regions can survive temperatures of – 40 °C. They are well adapted to their environment, having thick fur and a thick layer of body fat**

?

D Why is it an advantage for plants such as daffodils to survive the winter as underground bulbs rather than as seeds?

Fig 36.11 **In a forest ecosystem, different plants receive different amounts of light**

✔

The compensation point is the light intensity at which the rate of photosynthesis and the rate of respiration are equal.

The effect of temperature

The temperature range that living organisms can tolerate is narrow: most organisms live in temperatures between 0 °C and 40 °C, although some specialists manage to survive outside this range, between about – 40 °C to 250 °C (Fig 36.10). Although large swings in temperature can cause problems, such as when a pond freezes during a sudden period of cold, living organisms have evolved to cope with the usual range of temperatures that occur in their normal environments. The strategies used by ectotherms and endotherms to survive in hot and cold environments is covered in detail in Chapter 25.

The effect of light

Light often affects the distribution of animals in a particular ecosystem because, like soil type, it determines which plants grow there. As we saw in Chapter 9, light is *the* crucial environmental factor that affects photosynthesis. Light also affects and controls many other processes in plants, such as flowering, by interacting with the chemical phytochrome (see page 475). The length of daylight also determines the behavioural patterns of animals, whether they are **diurnal** (active in the day) or **nocturnal** (active at night).

The effects of light depend on its:

● intensity

● wavelength

● duration

● direction

A temperate deciduous forest ecosystem contains a large community of plants. As you can see from Fig 36.11, not every plant gets the same amount of light for the same length of time. Plants in such an ecosystem have life cycles that are adapted so that different plants grow and reproduce at different times of the year. Plants cannot grow at temperatures around freezing, but as soon as the slightly warmer temperatures of spring arrive, and the length of daylight increases, plants such as primroses and crocuses grow. They complete their life cycle quickly, flowering and setting seed, before the light and temperature increase to the levels at which the trees above them come into leaf.

Other plants show physiological adaptations which enable them to survive under the cover of overhead trees. Plants which grow in the shade, such as bluebells, ferns and mosses have a lower **compensation point** than plants which grow in bright light. Shade plants have slower rates of respiration and reach their compensation point much faster than a sun plant. Shade plants often show anatomical adaptations: they tend to have thick, broad leaves with a large surface area, to trap as much of the light that reaches them as possible.

The wavelength of light is also important. Land plants use mainly red, and some blue light, in photosynthesis. Plants which grow in aquatic ecosystems where they are submerged for long periods of time show a different absorption spectrum (see page 150): for example, lower shore seaweeds use mainly blue light as this is the wavelength of light that penetrates the water at this level.

Water availability

Water is an essential component of all living cells (see Appendix 1, page 616). In aquatic ecosystems, a long period of drought can cause a pond or stream to dry up completely, so killing the vast majority of living organisms that are part of the system. In terrestrial ecosystems, plants and animals are able to survive with smaller quantities of water and some specialised organisms, such as succulents, cacti, camels and desert rats, can live without water for extended periods. Plants that have become adapted to dry conditions are covered on page 269: some of the animals which live in hot, dry environments are described on pages 401, 402 and 413.

On a less extreme scale, the amount of water that a particular plant receives in a season can affect its root growth. If freshly planted shrubs are watered every day with relatively small amounts of water (as would happen in a drought when there is a hosepipe ban), the plant develops a shallow root system that spreads sideways in the soil (Fig 36.12(a)). If they are watered less frequently, but with larger quantities of water, the root system extends deeper into the soil, deep enough to reach the level of groundwater (Fig 36.12(b)). If the watering stops and the drought continues, as might happen when gardeners go away on holiday, the deep-rooted plants stand a much better chance of survival.

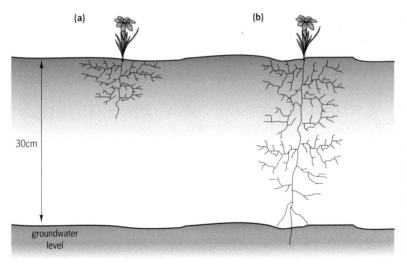

Fig 36.12(a) **An annual garden plant develops a mass of shallow roots when watered 'little and often'**

Fig 36.12(b) **The same plant develops a much deeper root system if it is given a large amount of water once or twice a week**

Fig 36.13 **Heat loss is more of a serious problem than water loss for animals in cold, windswept conditions**

Wind and water movements

Most organisms find it difficult to survive in an environment which is permanently windswept since, in windy conditions, plants are likely to lose a lot of water through transpiration. This is why plants that live in such conditions, for example plants adapted to life in the Swiss Alps, have developed fleshy leaves with thick cuticles.

Animals are also affected by air movement (Fig 36.13). Windswept ecosystems tend to support fewer species, and the populations of individual species are small. In general, aquatic ecosystems which have strong water currents are also less diverse. This is because the current sweeps away and dislodges organisms from their habitats. Both wind and water movements can increase suddenly, causing the catastrophe of a gale, hurricane, typhoon, a flood or tidal wave (Fig 36.14).

Fig 36.14 **Catastrophic events, such as the hurricane which devastated this mangrove, can wipe out whole populations and even whole communities**

E List two precautions you must take when measuring air temperature with a mercury thermometer.

F A male stickleback guards his nest by threatening and chasing other males from his territory. Is this an example of intraspecific aggression or interspecific aggression?

In a real ecosystem the abiotic and biotic factors all act together to determine the environmental conditions for each organism. Both interact to constantly change the conditions in the ecosystem. How this happens depends on the ecosystem and the physical conditions. The Opener to this chapter gives a thought-provoking example.

FIELD STUDY CASENOTES

Measuring abiotic factors related to climate

Temperature

You can use a standard laboratory **mercury thermometer** to measure air temperatures. (Take care not to place the bulb in direct sunlight or you will get unusually high readings.) To measure soil temperatures or temperatures of leaf litter, you should use a more accurate and robust instrument such as an **electronic thermometer** or **thermistor**.

Wind speed

You can measure wind speed using an **anemometer**: this has cup-shaped propellers attached to a small current generator (Fig 36.15(a)). When wind rotates the propellers, the generator provides a reading of wind speed. A flow meter (Fig 36.15(b)) uses the same principle to measure the speed of water flow in a river or stream.

(a)    **(b)**

Fig 36.15 **The anemometer and the flow meter work on the same principle**

Humidity

The amount of water that animals and plants lose from their surface depends on the level of **air humidity** (the amount of water held in the air). You can measure relative humidity using a **hygrometer**. If you don't have one, you can use blotting paper that has soaked in 5 per cent cobalt thiocyanate solution and been left to dry. The paper changes from red to blue when the relative humidity of air is between 50 and 90 per cent. It is not very accurate but it does allow you to find out, for example, which of two rock crevices is damper.

Light levels

To measure light intensity, you can use the kind of electronic light meter that is used by photographers.

Biotic components of an ecosystem

The biotic environment which surrounds an organism in an ecosystem results from the activities of all the other organisms living there. Complex relationships called **intraspecific interactions** occur between organisms of the same species; between organisms from different species, they are called **interspecific interactions**.

The main types of activity that make up intraspecific and interspecific interactions are summarised in Table 36.1, and described in more detail in Chapter 38.

Table 36.1 **Types of interactions that occur between organisms in an ecosystem**

Activity	Type of interaction	What happens
Reproduction	Intraspecific	Location of, selection of and competition for a mate
Caring for young	Intraspecific	Both parents, one parent or, more rarely, older siblings feed the young, keep them warm and sheltered and protect them from predators (Fig 36.16)
Social behaviour	Intraspecific	Animals cooperate to find food and to defend themselves against competitors or predators (Fig 36.17)
Competition	Intraspecific	Organisms compete for resources such as food, space, light, water and mineral nutrients
Reproduction	Interspecific	Animal vectors pollinate some species of plant and help to disperse the fruits and seeds of others (Fig 36.18)
Caring for young	Interspecific	Rare, but some species rear the young of another species. The cuckoo, for example, lays its eggs in the nests of smaller birds which then rear the fledglings
Mutualism	Interspecific	Two organisms both benefit from a long-term association
Parasitism	Interspecific	Individuals from one species use another species to provide food and shelter whilst giving nothing in return. Many hosts actually suffer from the interaction: many parasites cause disease
Predation	Interspecific	All animals obtain food by eating another organism. In its widest sense, predation also applies to herbivores feeding on plants
Protection	Interspecific	Many species make use of other species to hide from predators. Some copy the markings of other species to mimic warning signals that predators avoid
Competition	Interspecific	Organisms in a community compete for resources such as food, space, light, water and mineral nutrients
Defence	Interspecific	Organisms have developed various strategies to defend themselves against predators, parasites and competitors including mimicry (Fig 36.19)

Fig 36.16 **The female garden warbler makes regular trips back to her nest with food for the tiny chicks**

Fig 36.17 **Meercat sentries keep a look out for predators while other members of the group search for food and look after the young**

Fig 36.18 **Rufous humming birds drink nectar from the scarlet gilia, pollinating it at the same time**

Fig 36.19 **The hover fly (upper) mimics the markings of the much more dangerous wasp (lower) to fool other organisms into thinking that it, too, has a powerful sting**

SUMMARY

■ An ecosystem is made up of **abiotic** or physical components, and **biotic** or living components.

■ Abiotic factors that affect an ecosystem include **climate** and **weather** (including **humidity**, **temperature**, **air** and **water movements**), **soil type**, **rock type**, **natural catastrophes**, **water** and **mineral availability** and the position of the ecosystem on Earth.

■ Soil type helps to determine the type of plant that grows in an ecosystem: this affects which herbivores can live there and, in turn, which carnivores live there.

■ The size of rock particles and the amount of **humus** present determines soil type.

■ There are three main soil types: **clay/silt**, **loam** and **sand**. Loams are the most fertile soils because they contain a balance of particles and are rich in humus.

■ Temperature, light availability, humidity, water availability and wind and water movements are abiotic factors that you need to study in more detail. You can measure them in the field using simple techniques.

■ Biotic factors describe the interactions between living organisms. **Interspecific interactions** occur between organisms of different species; **intraspecific interactions** occur between organisms of the same species.

■ Biotic and abiotic factors affect each other: the physical environment determines the distribution of organisms in the ecosystem but living organisms can also change the physical properties of an ecosystem.

QUESTION

1 The drawings in Fig 36.Q1 show two plants which live on moorland.

In dry conditions, fires may spread rapidly across moorland.

Use evidence from the drawings to suggest explanations for each of the following observations:

a) cotton grass often becomes widespread in the years immediately following a fire;

b) in subsequent years heather becomes the dominant species.

[NEAB: Feb 1996, Ecology: Module Test, q. 7]

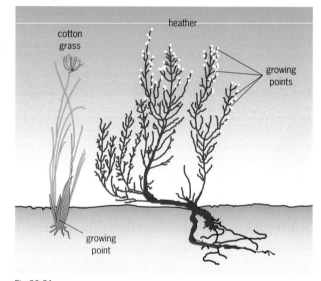

Fig 36.Q1

Assignment

OFF ON A FIELD TRIP...

James and Yasmin are part of a group of second-year A- level biology students attending a field studies week in May. They visit a hedgerow site to investigate some of the abiotic and biotic components of this ecosystem. Fig 36.A1 shows a rough plan of the site they choose to study.

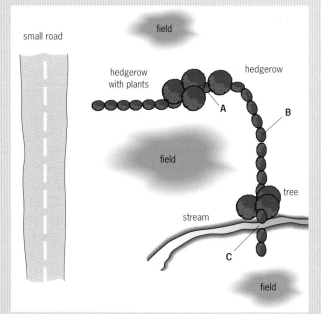

Fig 36.A1 **Map showing sites of study along the hedgerow**

James wants to investigate whether the distribution of different species of plant corresponds with a particular soil type. So he takes soil samples from the three points marked **A**, **B** and **C** in Fig 36.A1. He looks at the type of soil present, its field capacity and its humus content.

1

a) Describe the techniques that he might use to carry out this sort of analysis.
b) What other properties of the soil might James decide to test, if he has access to the right equipment?

2

Table 36.A1 show his results.
a) Which type of soil would you say is present at each point?
b) Give one factor that could explain why the soil at points **A** and **C** contain more organic matter than soil taken at point **B**.
c) James finds bluebells growing near **A** and **C**, but not at **B**. Do you think this distrubution could be due only to soil type? What other abiotic factor could be involved?

Table 36.A1 **Results of James' soil analysis**

Test	A	B	C
Properties of moist soil when rolled in the hand	Rolls into a ball: feels soapy	Feels gritty: won't roll into a ball	Feels gritty, easy to roll into a ball
Amount of water poured into 100 g dry soil	100 cm³	100 cm³	100 cm³
Amount of water passing through	75 cm³	90 cm³	82 cm³
Starting weight before heating at 450 °C for 4 hours	100 g	100 g	100 g
After heating	81.67 g	94.61 g	74.64 g
After further 30 min	81.59 g	92.71 g	73.98 g

While James is busy with the soil analysis, Yasmin decides to look at the plants in the hedgerow that are infested with aphids. She finds three elderberry shrubs, one at each of the points **A**, **B** and **C**. Fig 36.A3 shows the distribution of the aphids at the three different sites.

Stems with leaves:

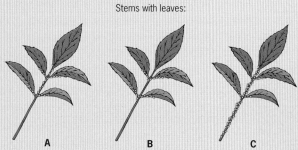

Stems with elderberries starting to form:

Fig 36.A2 **Diagram to show where aphids are found on elderberry bushes near points A, B and C**

3

a) Explain why aphids are distributed differently on the same plant at the three different sites. Which two abiotic factors do you think are responsible for this distribution?
b) What are the three main ways that the aphids can harm the plant?
c) Give an example of how interspecific and intraspecific competition can affect the aphid population.

4

Why are there fewer hedgerows in the British countryside today than, say, 40 years ago? Discuss the long-term environmental effects that result from the destruction of hedgerows.

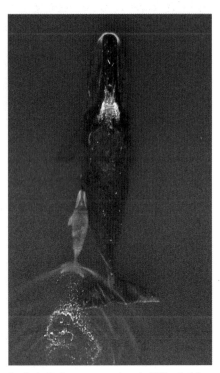

Bacteria that are a normal part of this whale's stomach contents can break down crude oil fractions and PCBs. If these bacteria can be introduced into the stomachs of other species such as dolphins and seals, this could guard them against the worst effects of some pollutants

ECOSYSTEMS ARE sophisticated recycling 'factories' in which elements and simple molecules are continually extracted from 'waste' chemicals and then used to synthesise new chemicals.

Toxic pollutants that are released into the environment by industrial processes cause problems if they cannot be broken down and recycled by living organisms. Pesticides, some oil components and PCBs (polychlorinated biphenyls) are particularly persistent and can build up to dangerous concentrations as they move through the trophic levels of food chains and webs.

One solution to toxic pollutants is to prevent the use or release of such chemicals. However, this is not practicable as many important industrial processes depend on them. Until recently, scientists had drawn a blank in the search for microorganisms that might have enzyme systems weird enough to break such chemicals down into harmless products. But progress is now being made. In the mid-1990s, American scientists discovered bacteria in the stomach of the bowhead whale which can break down naphthalene and anthracene, two persistent and cancer-causing fractions of oil. A couple of the other species found there prefer to feast on PCBs.

Not only could this finding explain why bowhead whales are particularly resistant to the effects of pollutants; it could lead to the identification and mass culture of bacteria which can help clear up the effects of oil spills and industrial contamination.

1 RECYCLING IS NOT NEW

In the last years of the twentieth century, recycling and reusing 'waste' has become a normal part of life. However, we humans cannot claim to have thought of it first. Every ecosystem that has ever existed has practised the same principle. Water molecules are constantly cycling through the environment, evaporating from the surface of the Earth, condensing as clouds, falling again as rain and then cycling through various living organisms. When you turn on the tap, many of the water molecules that come out have been through the kidneys of many people, via water treatment systems. Carbon, nitrogen and other minerals also cycle through the environment.

The supply of these vital chemical elements is limited, so they are used again and again. Autotrophs like plants build them into large,

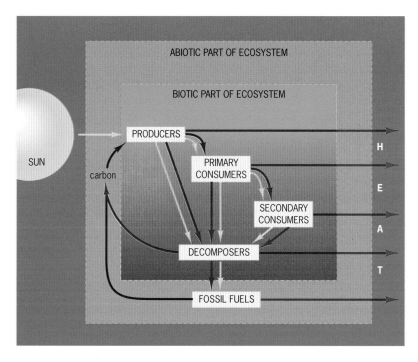

Fig 37.1 **The one-way transfer of energy (yellow arrows) and the cyclical movement of mineral nutrients (for example carbon) in an ecosystem**

✔
Autotrophs make their own food using an external energy source (usually sunlight) and a simple inorganic supply of carbon (usually carbon dioxide).

Photoautotrophs include green plants, algae and bacteria which make their own food by photosynthesis.

Chemoautotrophs are all bacteria: they also make their own food but they may use sources of carbon dioxide and energy that are different from those used by photosynthetic organisms. See Chapters 7 and 9.

Heterotrophs cannot make their own food: they must take in food molecules from their surroundings (see Fig 15.1). They use an **organic carbon source** (carbon that was once in living things) so they are totally dependent upon organisms like green plants that are capable of making their own food. See Chapters 7 and 15.

complex molecules, and heterotrophs break the molecules down again, respiring them to gain energy.

The very simple model of an ecosystem in Fig 37.1 illustrates how elements cycle through the compartments of an ecosystem. These processes are all driven by energy from the Sun. As you can see from the diagram, energy is transferred through different organisms in the ecosystem: it is never recycled.

2 ENERGY TRANSFER IN ECOSYSTEMS

Sunlight is the only significant source of energy for most ecosystems. Green plants, and some algae and bacteria, carry out photosynthesis. They use the energy from sunlight to power chemical reactions that combine simple inorganic molecules to form complex organic compounds. We say that the autotrophs in an ecosystem are **producers**: the products that they make form the food that ultimately feeds all other organisms in the system.

Animals, fungi and some protoctists are heterotrophs: they meet their need for energy by feeding on other organisms. We say that they are the **consumers** in an ecosystem. **Primary consumers** are herbivores: they obtain their energy by eating producers. **Secondary consumers** are carnivores or scavengers: they eat primary consumers. **Tertiary consumers** are carnivores that prey on other meat-eaters. This chain of dependence is usually called a **food chain** (Fig 37.2). The distinct levels of each chain are called **trophic levels**.

✔
Omnivores such as badgers or pigs or humans eat a mixture of plant and animal food. An omnivore can therefore be a primary, secondary and even a tertiary consumer all at the same time.

?
A Look at Fig 37.2. What do the arrows in a food chain represent?

Fig 37.2 **A simple flow chart summarising the overall structure of a food chain which has four trophic levels**

Food chain

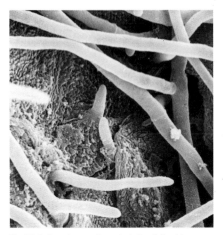

Fig 37.3 **Fungi grow into their food (in this case, human skin), secreting enzymes which digest the food externally. The soluble products of digestion are then absorbed through the walls of the hyphae**

?

B List the foods you ate at your last meal. Sketch out the food chains which produced them and say which trophic level you, the consumer, represent.

As one organism eats another, there is a transfer of energy from the bodies of the producers (or consumers) into the bodies of consumers through the trophic levels of a food chain. Energy transfer occurs in one direction only: it is not recycled, and much of the energy escapes from the ecosystem as heat (see Fig 37.1). Energy transfer in ecosystems is relatively inefficient (see Chapter 7).

Another major group of heterotrophs in an ecosystem are the **decomposers** (see Chapter 15). These are mainly bacteria and fungi. They break down the bodies of dead animals and plants, usually by extracellular digestion (Fig 37.3) and absorb the soluble products. Transfer of energy from producers and consumers to decomposers is also one-way, with a high proportion eventually lost as heat. The process of decomposition allows mineral nutrients to be recycled in the ecosystem (see page 579).

Food chains and food webs

Two examples of real food chains are shown in Fig 37.4. Most animals have a varied diet. A food chain is therefore only part of a much larger feeding picture: the **food web**. Take our top freshwater carnivore, the pike. It does not feed on stickleback alone but also roach and insects such as dragonfly nymphs and pond skaters. Most rivers and lakes also include a wide variety of producers ranging from the microscopic plant plankton to the larger pondweeds, rushes and flowering plants.

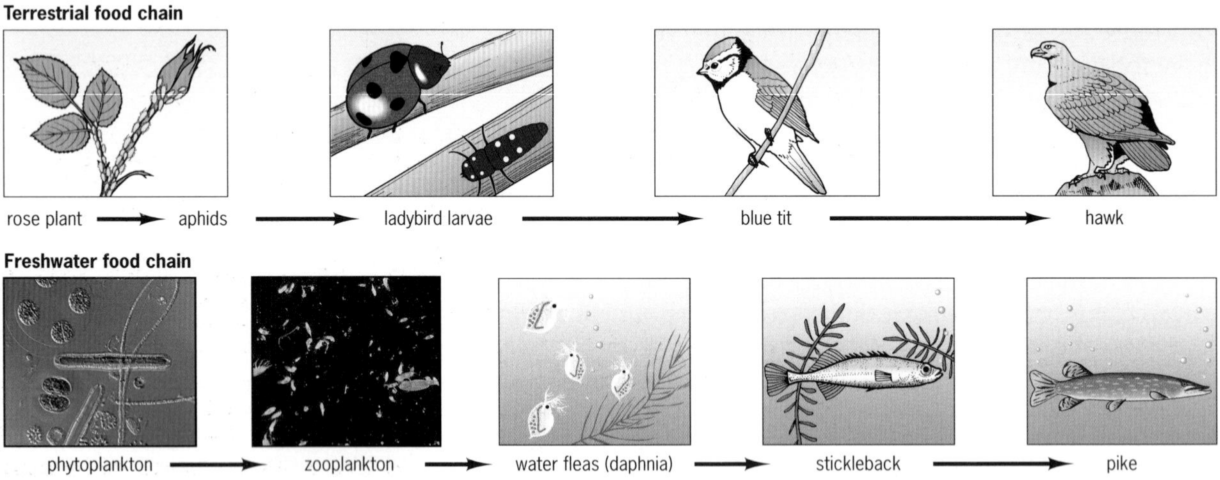

Terrestrial food chain

rose plant ⟶ aphids ⟶ ladybird larvae ⟶ blue tit ⟶ hawk

Freshwater food chain

phytoplankton ⟶ zooplankton ⟶ water fleas (daphnia) ⟶ stickleback ⟶ pike

Fig 37.4 **An example of a terrestrial and a fresh-water food chain. Each chain has five trophic levels (most food chains have fewer trophic levels than this)**

Fig 37.5 on the next page shows a simplified example of a food web in the Antarctic.

Pyramids of numbers

Food chains and food webs tell us a lot about the feeding relationships of organisms in an ecosystem, but they are **qualitative** rather than **quantitative**. That means that we know *which organisms* are part of the chain or web, but we have no idea of the exact *numbers of organisms* involved at each trophic level.

Clearly, numbers are very important: there are many more individual producers than primary consumers, and, as you carry on along a food chain, the numbers of organisms at each trophic level continues to decrease. At the same time, the size of individual organisms usually increases as you go up the trophic levels. It is difficult, if not impossible, to find out exactly how many producers

See question 1. ■

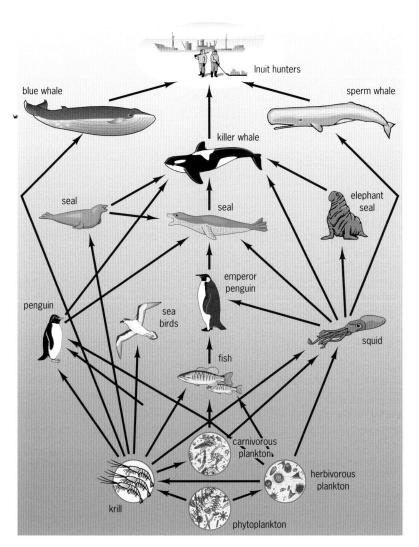

?

C How many trophic levels are there in the longest food chain in Fig 37.5? Is this usual or unusual, and why?

Fig 37.6 **A simple pyramid of numbers. The width of each box represents the relative number of organisms present in each trophic level at any one particular time. In this example, many grass plants are needed to sustain all the antelope that one individual lion needs to eat to survive**

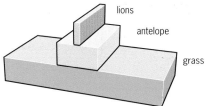

Fig 37.7 **Two examples of inverted pyramids of numbers. Neither is completely inverted; they simply do not conform to the expected pyramidal shape because one of the levels is distorted**

(a) **One sycamore tree can feed many thousands of aphids, making the first trophic level of this pyramid of numbers much larger than the second**

(b) **This pyramid is complicated by the top trophic level which involves a parasite. Many parasites can infect one host. _Wuchereria bancrofti_ is a nematode worm which causes the disease elephantiasis in humans (see page 286)**

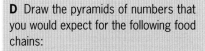

are eaten by a primary consumer and so on, but we use a pyramid of numbers to show the general idea (Fig 37.6).

Two odd things can happen when you use pyramids of numbers. Sometimes the base of the pyramid is very narrow, indicating that there are fewer producers than primary consumers. This seems not to make sense, but these examples arise when a few large plants such as trees produce food for thousands of tiny plant-feeders, such as aphids (Fig 37.7(a)). In pyramids of food chains which have a population of parasites at the top, the upper level can be much larger than the level below it. This is because many parasites can feed on one host (Fig 37.7(b)). We say that both of these pyramids are **inverted pyramids**.

?

D Draw the pyramids of numbers that you would expect for the following food chains:

(a) grass – rabbit – fox

(b) hazel trees – squirrel – fleas

How would you describe these two pyramids?

FOOD WEBS AND PESTICIDES

IN DEVELOPING a new pesticide, scientists look for a chemical which kills only the pest that we want to get rid of, harms no other species, breaks down into something harmless after a short time, does not lead to the development of resistance, and which is cheap. This is a pretty tall order: no pesticide yet developed is perfect, but modern pesticides are much safer than some of those developed and used in the 1950s and 1960s.

DDT (dichlorodiphenyltriflouroethane) was used as a pesticide in many countries after World War II. It breaks down slowly, remaining in the environment for between two and five years.

During the 1950s and 1960s it became obvious that DDT was affecting whole food webs. Fig 37.8 shows what happened when DDT was transferred up the trophic levels of a food web in an ecosystem in the US. DDT is not excreted and concentrates in the fatty tissues of organisms. So, although the concentration of DDT in zooplankton, the primary consumers in the chain, was only 0.04 parts per million, its concentration in the tissues of a top carnivore such as the osprey or bald eagle was 25 parts per million. The bodies of the large birds of prey converted the DDT into a substance that made their egg shells very fragile. As a result, very few birds managed to breed and their numbers fell quickly.

In 1972, DDT was banned and since then, the numbers of the large predatory birds has recovered. However, there are still problems because of the illegal use of DDT.

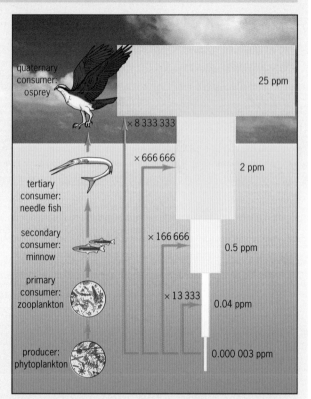

Fig 37.8 **Bio-amplification of DDT through an aquatic food chain. DDT accumulates in the fatty tissues and cannot be excreted. The fat of the osprey contains over 8 million times more DDT than the water at the bottom of the food chain in which the producers live**

Pyramids of biomass

Fig 37.9(a) **A generalised pyramid of biomass for a terrestrial ecosystem**

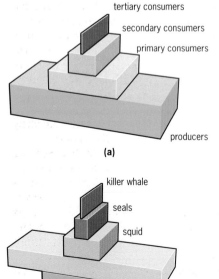

Ecologists often use pyramids of **biomass** (Fig 37.9(a)), to avoid the problem of inverted pyramids of numbers. Pyramids of biomass give a more accurate model of what happens in a food chain. They show the weight or mass of all the organisms at each trophic level – dry mass is the most useful measure. This is obviously extremely difficult to do: obtaining the dry weight of organisms means killing them, drying them and measuring the remains. Not surprisingly, very few pyramids of biomass have ever been determined in this way.

However, using samples of organisms and their wet weight allows ecologists to make estimates and devise models. These show that the pyramid of biomass provides a more realistic way of comparing trophic levels. But, even this type of pyramid can be inverted, as Fig 37.9(b) shows.

To obtain an even better model of an ecosystem which avoids inverted pyramids altogether, we must look at the way energy is transferred between the different trophic levels.

Fig 37.9(b) **In the waters of the Antarctic, the weight of zooplankton (animal plankton) is five times that of the phytoplankton (plant plankton). So, if we draw up a pyramid of biomass for one of the food chains that forms part of the more complex web shown in Fig 37.5, we get an inverted pyramid. This is because the zooplankton eat the phytoplankton so quickly that the plant plankton never gets the chance to attain a large biomass. Instead, phytoplankton has a high turn-over rate, reproducing very quickly. So the biomass of organisms produced each year is large, although the number alive at any one time is often less than the number of consumers**

Efficiency of energy transfer between different trophic levels

As we saw in Chapter 7 (page 120), the energy transfer between organisms at different trophic levels of a food chain is never 100 per cent efficient. In fact, less than 30 per cent of the energy available ever reaches the organisms in the next level. There are three main reasons for this:

■ See questions 2 and 3.

- Not all the organisms in one trophic level are eaten by organisms from the next level. Many die and provide food for decomposers.

- Not all food that is eaten is digested and not all digested food is absorbed – some is lost as faeces.

- Most of the energy gained from the food that is absorbed is lost through the processes of excretion (production of metabolic waste) and cell respiration.

> **?**
>
> **E** The proportion of energy that is lost as heat from farmed mammals such as cows and sheep is much higher than that lost from farmed trout. Suggest a reason why.

FEEDING THE WORLD: PRODUCTIVITY IN ECOSYSTEMS

THE RATE AT WHICH the producers in an ecosystem capture light energy to build biomass is known as the **gross primary productivity** of that ecosystem. This is a theoretical maximum. In reality, all producers use energy for respiration and other life processes. When we subtract the energy that they use to live, taking it away from the energy that they capture, we find the **net primary productivity** of the ecosystem. This is the amount of energy that is available to the consumers of the system.

net primary productivity = gross primary productivity – energy used by producers to live

Different ecosystems of the world have different primary productivities. As Fig 37.10 shows, an ecosystem which is part of tropical rainforest is one of the most productive in the world. A major reason for this is that tropical rainforests grow around the Equator of the Earth, where more light energy is available than further north or south. Ecosystems which have high levels of water and mineral nutrients, such as an estuary or a marsh swamp, are many times more productive than deserts which are dry and severely lacking in nutrients.

The primary productivity of an ecosystem limits the number and variety of consumers that can survive there. This explains why tropical rainforest ecosystems support more diverse animal populations than other parts of the world. As rainforests are cut down to make room for expanding human populations, the potential of the rainforest to produce enough food to support its populations decreases. As less energy is available in food webs and food chains, the numbers of animals at the higher levels are likely to fall. More species are likely to become endangered and, eventually, extinct.

Many scientists think that the conflict between the need to use the land for humans to live on and the need to leave enough land to produce food will reach a crisis in many areas of the world in the next 50 years. The famines that are tragically common in desert areas may well become more of a feature of life in many other parts of the world, as primary productivity cannot keep pace with the growth of human populations.

The scale of this problem is so large that it is difficult for any individual country to tackle alone. However, many countries world-wide are now working together to try to ensure that the worst outcome does not happen.

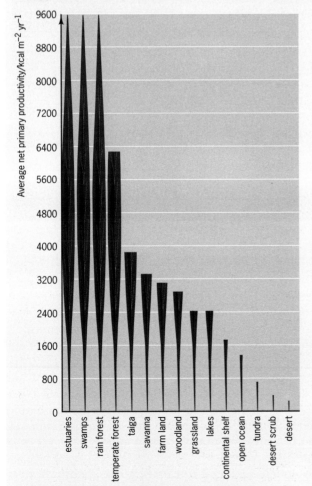

Fig 37.10 **Rainforests, estuaries and swamps and marshes are the most productive ecosystems in the world. Open oceans, arctic tundra and deserts are the least productive**

Pyramids of energy

Fig 37.11 shows a typical pyramid of energy. It shows the amount of energy that is transferred up the food chain and how much is lost as heat at each level. The more levels in the chain, the less energy there is available at the top level. Consequently, very few food chains have more than four or five levels. It also explains why the top carnivores such as the bit cats, tigers and large whales are the first to become endangered species when their ecosystems come under pressure from humans.

Fig 37.11 A pyramid of energy for an Antarctic food chain similar to the one shown as a pyramid of biomass in Fig 37.9(b). **As you go up the pyramid, there is a 90 per cent loss of energy at each trophic level. This energy, of course, does not disappear: it is 'lost' as heat. For a more detailed investigation of energy transfer, see the Assignment at the end of this chapter**

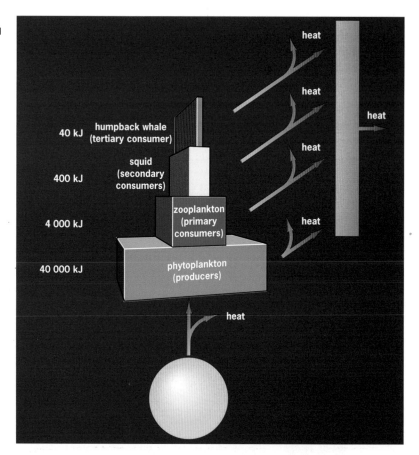

3 MINERAL NUTRIENT CYCLING

All organisms need a supply of energy to carry out the processes of life, but they also need to obtain the mineral nutrients (carbon, nitrogen, phosphorus etc) which make up the complex chemicals in their bodies. Unlike energy, mineral nutrients are recycled over and over again, passing between organisms in the same ecosystem, and also other ecosystems of the world. So the carbon, oxygen and nitrogen atoms in our bodies could have been part of the soil a year ago. They may have made up the proteins of some long extinct dinosaur. And one popular anecdote says that, during the course of our lives, each of us breathes out six carbon atoms that were once part of Napoleon Bonaparte.

Producers accumulate mineral nutrients. These then pass along the food chain and are finally released back into the abiotic part of an ecosystem by decomposers. Decomposers replenish supplies of these elements in the soil and in water, and they also help to return them to the atmosphere.

Why living organisms need mineral nutrients

As we saw in Chapter 3, the chemicals of life, carbohydrates, proteins, fats and nucleic acids, contain mainly carbon, hydrogen and oxygen. Proteins also always contain nitrogen and often sulphur, and nucleic acids contain phosphorus. In addition to these elements, living things need small amounts of calcium, sodium, potassium, iron, copper, chlorine, magnesium and other trace elements. (Table 15.3 on page 225 gives a list of minerals that the human body needs.)

In this chapter we concentrate on the cycling of carbon and nitrogen.

Cycling of individual mineral nutrients

Mineral nutrients occur in four basic compartments in an ecosystem:

- The organic compartment: living organisms and their debris (faeces etc).

- As available nutrients that are held on the surfaces of clay particles in soil or in solution, where they can be taken up by plants and microorganisms.

- As nutrients that are temporarily unavailable because they are bound up in soil and rocks.

- In the atmosphere, as gases that occur in the air.

The carbon cycle

Life on Earth is based on carbon compounds (see Appendix 2, page 617). Carbon circulates between the abiotic and biotic parts of the environment, as shown in Fig 37.12. Autotrophs capture carbon during photosynthesis, and all living organisms return it to the atmosphere as they respire and also when they die and decompose.

Fig 37.12 **The carbon cycle. Photosynthetic organisms fix about 80 billion tonnes of carbon each year. But this represents less than 1 per cent of the total carbon in the biosphere. Carbon can be locked up for long periods of time in limestone rocks, inorganic solutions (containing soluble carbonates) and fossil fuels such as coal: these reservoirs of carbon are often called carbon sinks. When volcanoes erupt, they return carbon dioxide to the carbon cycle**

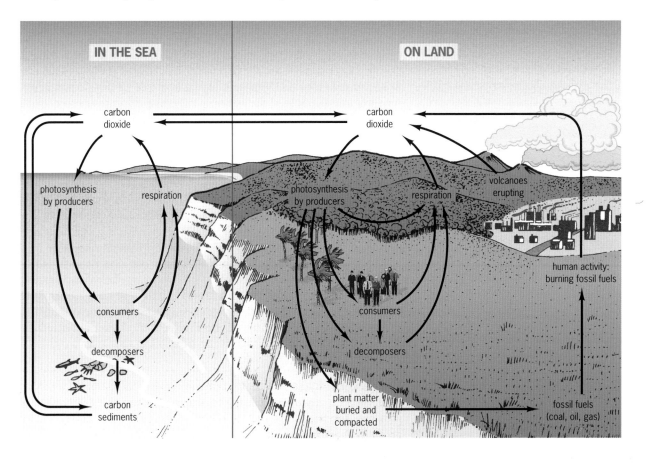

See question 4. ■

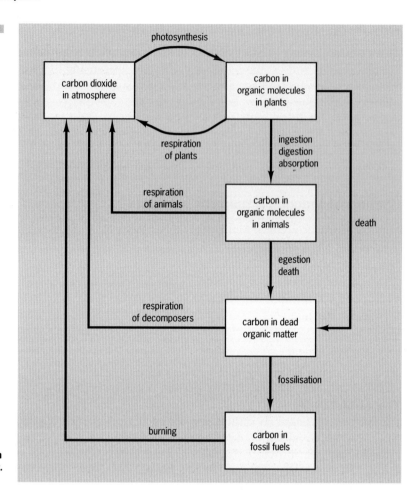

Fig 37.13 **An exam/revision version of the carbon cycle. You should find this diagram easy to draw. Learn it so that you can reproduce it in exams**

> ✔
>
> Decomposers stop organic material accumulating in an ecosystem. They also ensure that essential nutrients, such as nitrates, sulphates and phosphates, are released back into the soil.

The role of microorganisms in the carbon cycle

By looking at Fig 37.12 or Fig 37.13, you should be able to see that carbon would still cycle through the biosphere for some time if decomposers were removed. However, because of their general role in decomposition, decomposers play a key role in the carbon cycle. Bacteria and fungi feed by secreting digestive enzymes onto organic material to digest it. Some of the 'freed' mineral nutrients are absorbed by these microorganisms. Others escape into the soil.

How humans influence the carbon cycle

Humans influence the carbon cycle by two activities not practised by other species. Firstly, humans burn fossil fuels such as coal, oil and gas. This has been going on in a big way since the start of the Industrial Revolution in Europe. Burning such fuels is the only way that their 'locked' carbon is released back into the atmosphere.

The second activity is cutting down forests, particularly rainforests in recent times, to make room for farms, fields and homes. This increases the amount of carbon dioxide in the atmosphere because it reduces the total number of photosynthetic organisms, the 'trappers' of CO_2. The possible effects of these activities are discussed in the Feature box on the next page.

The nitrogen cycle

Like carbon, nitrogen is vital to organisms for the formation of proteins, nucleic acids and their products. Although nitrogen is all around us (it makes up 79 per cent of the air we breathe), most

See question 5. ■

CARBON DIOXIDE, THE GREENHOUSE EFFECT AND GLOBAL WARMING

CARBON DIOXIDE IS one of the main **greenhouse gases**. Together with other gases, it prevents heat from escaping from the atmosphere (Fig 37.14). This effect is often called the **greenhouse effect**. Although pollution may be increasing the greenhouse effect, this phenomenon is not *caused* by pollution: it is a natural physical effect caused by the presence of an atmosphere. In fact, having an atmosphere is one of the reasons that life developed on Earth: it allows surface temperatures to be much more stable than they would otherwise be.

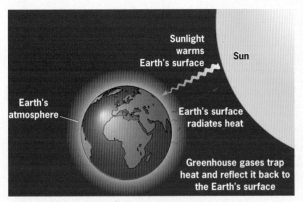

Fig 37.14 **The greenhouse effect. Energy from the Sun is either transmitted, reflected or trapped by the atmosphere. Energy radiated back into space is at longer wavelengths (in the infrared region) than the energy arriving from the Sun, and the consequence is that more heat enters than can escape.**
Greenhouse gases (CO_2, CH_4, nitrous oxides, CFCs), together with water vapour in clouds, contribute to the 'blanketing' effect of the atmosphere

Accurate measurements of atmospheric CO_2 concentrations began in the mid 1950s, in Hawaii and at the South Pole. These sites were chosen because they are far away from any major sources of pollution. The curve, shown in Fig 37.15, clearly shows an annual cycle of peaks and troughs corresponding to summer and winter. Superimposed on this is a small but steady rise in overall CO_2 levels. Levels of other greenhouse gases such as methane, nitrogen oxides and chlorofluorocarbons have also increased during the same period.

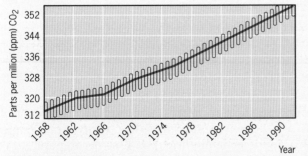

Fig 37.15 **The rising level of carbon dioxide in the atmosphere. Data was collected at the Mauna Loa observatory in Hawaii. There is good evidence to suggest that changes in the average temperature at the surface of the Earth are closely linked to changes in the levels of greenhouse gases**

What are the likely effects of global warming?

Many scientists think that an increase in carbon dioxide in the atmosphere will cause global warming. They believe that the average temperatures on the Earth's surface will increase, causing widespread disturbances in world climate.

Using all the evidence and data that is available at the moment, a computer model has been developed to predict what might happen in the future. This projection, shows that the mean surface temperature might increase. It is important to remember that any projection is an informed guess: we cannot say exactly what will happen. Let us assume that nothing is done to stop the amount of greenhouse gases that are being released into the atmosphere by human activity. The projection indicates that, at best, there may be a mean temperature increase of only 1.5 °C by the year 2050. At worst, there may be a temperature rise of more than 5 °C.

The effect of any global warming that does occur will not be felt evenly around the globe. There will be greater warming at the high latitudes (at the poles, for example) and at middle latitudes than at the equator. The Northern Hemisphere will warm more than the Southern Hemisphere, because it has more land mass (land heats up more quickly than water). More areas of the world would have extreme heatwaves and more forest and brush fires. The average sea level would rise, perhaps by two to four centimetres every 10 years. Many low-lying areas would become flooded and maybe even submerged.

What is being done about the problem of global warming?

Though scientists debate the extent and exact effects of global warming, they generally agree that we should be taking steps to cut the emissions of greenhouse gases. This means reducing the amount of fossil fuels that we burn: we could make a big difference by improving the efficiency of heating systems and by reusing 'waste' heat from industry. Wherever possible, we should use renewable energy sources such as wind, water and solar power. Although many people object on safety grounds, making more use of nuclear energy would also help to reduce emissions of greenhouse gases.

In addition, we should limit the destruction of forests and use methods of agriculture that do not irreversibly damage the land. Of course, all of this is cancelled out if the human population continues to increase, so efforts should also be made to stabilise the world's population, to prevent further pressure on energy resources.

Many of these measures are being put into practice, and if people become more aware of this and other environmental problems, international cooperation might soon start to offset the effects of global warming.

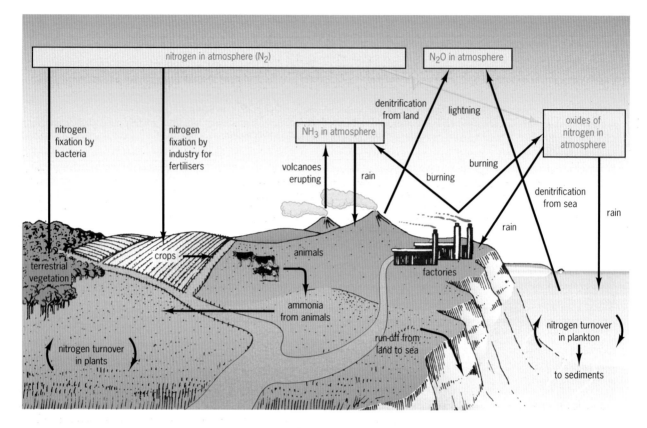

Fig 37.16 **The nitrogen cycle. Bacteria play an important role in the movement of nitrogen in the biosphere**

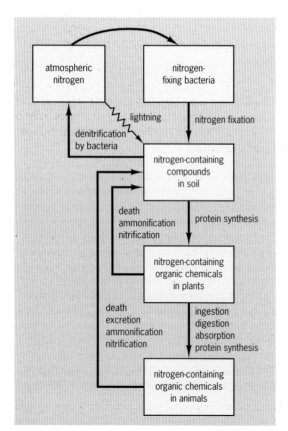

Fig 37.17 **Copiable diagram of the nitrogen cycle for exams**

living organisms cannot access this supply. Nitrogen gas is unreactive because the two atoms which make up the N_2 molecule are bound together by a strong triple bond. It can, however, be fixed by microorganisms, and then it cycles through ecosystems, as shown in Fig 37.16.

Nitrogen-rich waste from animals (compounds of ammonia, for example) is converted into nitrates by **nitrifying bacteria**. These microorganisms fix nitrogen by converting atmospheric nitrogen into useful products. In terrestrial ecosystems, nitrogen is fixed by free-living nitrifying bacteria in soil and by **symbiotic bacteria** that live in nodules, the swellings on the roots of legumes (plants like peas and beans). In aquatic ecosystems, some blue-green bacteria fix nitrogen. Nitrogen returns to the atmosphere through the action of **denitrifying bacteria**.

These processes, in more detail, are:

● **Nitrogen fixation**. This is carried out mainly by soil bacteria such as *Azotobacter*. Some bacteria, such as *Rhizobium* are symbiotic: they live in the root nodules of some plants such as peas and beans (Fig 37.18).

● **Ammonification**. This is the breakdown of organic nitrogen such as proteins and urea into ammonia. It is also known as **saprotrophic decay** (see Chapter 15) and is generally thought of as decomposition, or 'rotting'. Bacteria and fungi carry out most of the ammonification in the nitrogen cycle.

● **Nitrification**. Nitrifying bacteria such as *Nitrosomonas* oxidise ammonia to nitrite (NO_2^-). Other bacteria, such as *Nitrobacter* oxidise nitrite to nitrate (NO_3^-).

These oxidation reactions release energy which the bacteria use to make food (they are chemoautotrophs).

- **Denitrification**. Bacteria such as *Pseudomonas denitrificans* obtain energy by using the nitrite or nitrate ion as an electron acceptor for the oxidation of organic compounds. They live in conditions with low oxygen levels such as waterlogged soils, and can be important in reducing nitrate levels further.

Fig 37.18 **Legumes such as peas and beans have nodules on their roots. These house nitrogen-fixing species of the bacterium *Rhizobium*. The bacteria convert nitrogen gas from the atmosphere into ammonia. This passes out of the root nodules, dissolves in soil water and is then taken up by plant roots. The relationship between the bacteria and the plants is therefore symbiotic because both benefit: the plants get a steady supply of nitrogen and the bacteria get a home and a constant supply of sugars from the plant**

SUMMARY

■ Energy transfer through organisms in an ecosystem is one-way. Elements such as carbon and nitrogen cycle between living organisms and the abiotic environment.

■ In a **food chain**, energy is passed from **producers** to **primary consumers** to **secondary** and **tertiary consumers**. Each level of the food chain is called a **trophic level**. In most ecosystems, many food chains are linked together in a **food web**.

■ **Decomposers** break down the dead bodies of producers and consumers and play an important role in mineral cycling. Most decomposers are bacteria and fungi.

■ There are three main types of **ecological pyramid**: **pyramids of numbers**, **pyramids of biomass** and **pyramids of energy**. Pyramids of energy show that over 90 per cent of the energy transferred to the next trophic level is lost as heat.

■ The **carbon** and **nitrogen cycles** show how these elements circulate between the biotic and abiotic parts of an ecosystem. You should appreciate the overall cycle and be able to draw a clear summary diagram.

■ Carbon dioxide is a **greenhouse gas**: together with other gases in the atmosphere, it acts as a heat insulator. The presence of CO_2 in the atmosphere helps to maintain temperatures on Earth that are warm enough to sustain life.

■ Human activities have led to an increased concentration of CO_2 in the atmosphere. Many scientists believe that this is causing a general increase in average temperatures throughout the world, a phenomenon commonly called **global warming**.

QUESTIONS

1 In a study of an oak tree the following numbers of organisms were obtained at each trophic level.

Trophic level	Number of organisms
producer	1
primary consumer	260 000
secondary consumer	40
tertiary consumer	3

a) Sketch a pyramid of biomass to represent this food chain.

b) Suggest **two** reasons why there is such a large difference in the numbers of primary and secondary consumers.

c) The Venus flytrap is a green plant that catches and digests insects. It lives in wet, boggy places where mineral ions are scarce because they are washed out of the soil.
- **(i)** At which trophic level or levels would you put the Venus flytrap? Give a reason for your answer.
- **(ii)** Describe the difference between the way in which the Venus flytrap obtains its nitrogen and the way in which other plants normally obtain this element.

[NEAB June 1995 Biology: Module Test, Section A, q.5]

2 The graph of Fig 37.Q2 shows the changes in the numbers of microscopic floating organisms and physical factors in the surface waters of the North Sea during the course of a year.

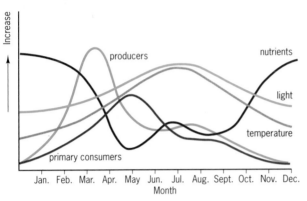

Fig 37.Q2

a) Use information from the graph to suggest **two** explanations for **each** of the following.
   (i) the rise in the number of producers during February and March;
   (ii) the decrease in the number of producers during May.

b) Explain how **one** abiotic factor may bring about the increase in nutrient levels of the surface waters from October to December.

[NEAB June 1995 Biology: Ecology Module Test, Section B q.5]

3 The graph of Fig 37.Q3 shows the relation between the productivity of phytoplankton (producers) and zooplankton (primary consumers) in a large lake.

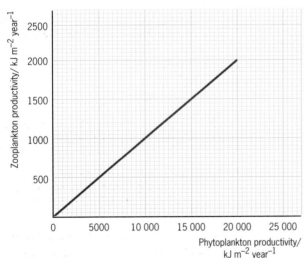

Fig 37.Q3

a) Use the graph to calculate the proportion of energy in phytoplankton which goes directly to the zooplankton. Show how you arrive at your answer.

b) Give **two** different fates of the energy that does not go directly to the zooplankton from the phytoplankton.

[AEB June 1994, Biology: Paper 1, q.14]

4 Fig 37.Q4 shows the global carbon cycle. The numbers represent the amount of carbon moving between each stage of the cycle, in 10^{15} g year^{-1}.

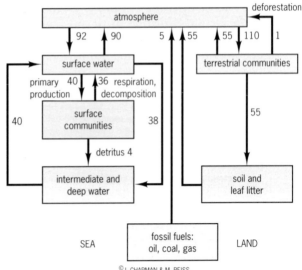

©J. CHAPMAN & M. REISS,
Ecology – Principles & Practice, adapted by permission of Cambridge University Press.

Fig 37.Q4

a) Using the information in Fig 37.Q4:
   (i) Calculate the net annual increase in the carbon concentration of the atmosphere. Show your working.
   (ii) State where, other than in the atmosphere, there is a net annual increase in the amount of carbon.

b) Suggest how deforestation can result in an increrase in the carbon dioxide concentration of the atmosphere.
   (i) Explain how leaf litter and soil contribute to the carbon dioxide in the atmosphere.
   (ii) Describe **two** possible consequences of a rise in atmosphereic carbon dioxide concentration.

[UCLES June 1995 Modular Biology: Ecology and Conservation Module, q.2]

5 The diagram below shows the conversion of ammonium ions to nitrate ions in the nitrogen cycle.

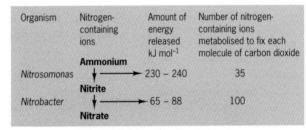

a) What is the name given to the conversion of ammonium ions to nitrate ions?

b) **(i)** What happens to the energy released in the rections shown in the diagram?
   (ii) Suggest why *Nitrobacter* metabolises more nitrogen-containing ions than *Nitrosomonas* in fixing each molecule of carbon dioxide.

[AEB June 1994 Biology: Paper 1, q.13]

WHY ARE BIG FIERCE ANIMALS SO RARE?

We normally think of large predators like lions or killer whales as being the most successful organism in their ecosystem because they can take the food that they need without worrying about predators themselves. But if you look at their numbers relative to the numbers of other animals further down the food chain, it is clear that these top carnivores are really quite rare. In this Assignment, we see why.

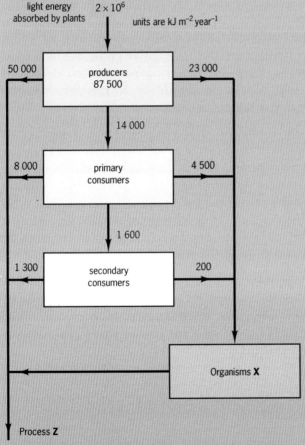

Fig 37.A1 **In this schematic diagram, the producers are the grass plants which make up most of the vegetation. The primary consumers are deer and the secondary consumers are lions**

1 Look at Fig 37.A1. This shows an energy flow diagram for a food chain in the Serengeti.
a) What is process **Z**?
b) What are organisms **X**?
c) How many trophic levels are there in the food chain?
d) Calculate the percentage efficiency with which light energy is converted into food energy in the producers.
e) What percentage of the light energy absorbed by plants is eventually available for secondary productivity in the lion?

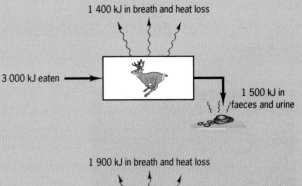

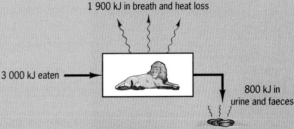

Fig 37.A2 **A comparison of the proportion of energy lost as heat and in faeces and urine in a deer and a lion which eat 3000 kJ of food**

2 Look at Fig 37.A2.
a) For every 3000 kJ of food eaten, what amount of energy is available to the deer for growth and repair?
b) What amount is available to the lion for this purpose?
c) Try to suggest why the deer loses a higher proportion of energy in faeces and urine than the lion.
d) Why should the lion use up more energy in cell respiration than the deer?
e) Suggest why the proportion of energy available for secondary productivity is about three times greater in the lion than in the deer.

3 From the information you have gained so far, why are large carnivores so rare?

Discussion questions

4 Do you think that the fact that humans are omnivores has anything to do with their evolutionary success?

5 Some people hold the view that, if humans were all vegetarian, the world food shortage would be largely solved. Do you agree or disagree, and why?

6 Several species of fish, such as trout and turbot, can be farmed. These fish are carnivores. Do you think this is an efficient way of producing food? Discuss the reasons for your answer.

38 Population ecology

The lemming: a victim of plant defences?

MOST PEOPLE KNOW the story about hordes of lemmings rushing off the edges of cliffs and riverbanks to commit suicide by drowning. Ecologists have studied lemmings and have found no evidence for this 'mad' behaviour. However, they have uncovered information about the life cycle of lemmings which may help to explain where the story originally came from.

Rather than remaining stable, a lemming population lurches from a concentration of about two animals per unit area to more than 40, in four-year cycles. Because lemmings are hunted by the fox and the lynx, the populations of these species also go up and down in a regular cycle, as their food supply fluctuates. Why the lemming population rises and then falls, or why the cycle should be so regular, is much more difficult to explain.

The most likely theory is that lemmings eat a species of plant which produces some sort of toxin on a four-year cycle. When the plant is low in toxin, the lemming population increases. When it is high in toxin, the lemming population crashes. This may help to give some credibility to the 'mass suicide' story: if lemmings die due to eating a poisonous plant, we would expect many more of them to be found dead at the time of the population crash.

Although no-one has yet managed to identify the plant, a similar phenomenon occurs in Canadian hares, and some of the plants they eat have been shown to increase toxin production in their new shoots for two or three years after a season of heavy grazing.

1 HOW POPULATIONS GROW

Populations are changing constantly. Imagine a new population of a few individuals starting in a particular area, with ideal environmental conditions – plenty of food, water etc. For example, a pair of healthy rabbits could be introduced onto a small island where there is plenty of edible vegetation and no predators. The population grows slowly at first: this is phase A in the graph of Fig 38.1.

After a while, the population begins to grow very rapidly and doubles at regular intervals. This type of growth is called **exponential growth**, and is illustrated as phase B in the same graph. As more rabbits are born, they breed and produce a huge rabbit population in a couple of years.

At some point, the population is prevented from increasing further by various **limiting factors** in the environment – phase C. When food, space and water supply are stretched to their limit, individuals

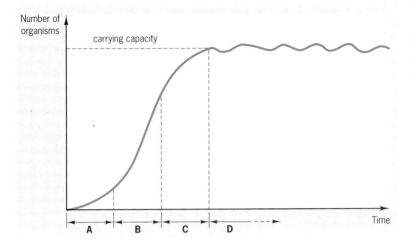

Fig 38.1 **This graph shows an idealised curve of population growth. As the curve A to C looks like a flattened S, it is called an** S-shaped **or** sigmoid curve. **The text describes phases A to D in detail**

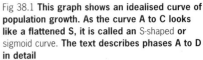

The **biotic potential** is the maximum rate at which a population can increase in size when there are no limits on its growth.

The **environmental resistance** which affects a population is made up from all the factors which act together to limit its size.

When the biotic potential and the environmental resistance reach a balance, the size of the population levels off: it reaches its **carrying capacity**.

die and breed less successfully. (If they can, they leave the environment.) Because there is less food to go around, they may be eating the plants faster than they can grow. Most populations now stabilise and the growth curve becomes flat. The number of individuals in the population at this point is the **carrying capacity** of the environment – phase D.

Sometimes there is a sudden increase in numbers and a population overshoots its carrying capacity (Fig 38.2). The environment cannot provide the resources for this number of organisms and a **population crash** usually follows. The population dies back below the level of the carrying capacity, and then takes time to stabilise. If the overshoot has damaged the environment, the population might stabilise at a different level.

The two growth curves described in Figs 38.1 and 38.2 occur in laboratory situations but they are not often found in real ecosystems. Fig 38.4 shows the three main types of growth curve that occur in natural conditions:

- **Stable**: the population size remains at roughly the same level over a long time. A study of a population of trees in a well-established woodland would produce this sort of curve.

- **Irruptive**: the population size is basically stable, with the occasional dramatic increase followed by a population crash. This population occurs in algal. A population is mostly stable but, in hot weather, when water is contaminated by phosphate or nitrate fertilisers, there is an **algal bloom** (Fig 38.3).

- **Cyclic**: the population increases and decreases in a regular cycle of growth and die-back. This describes how lemming populations change (see the Opener of this chapter).

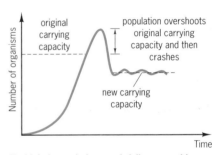

Fig 38.2 **A population crash follows a sudden increase in numbers which exceeds the carrying capacity. If the environment is damaged by the temporary overshoot, there may be a new carrying capacity**

Fig 38.3 **An algal bloom on a pond**

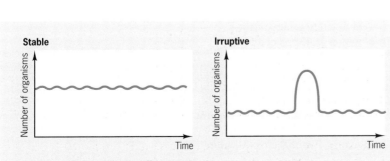

Fig 38.4 **Stable, irruptive and cyclic population curves**

FIELD STUDY NOTES: STUDYING POPULATIONS

When we study populations, it is useful to know how many organisms live in a particular area. We can estimate **population density** (the number of organisms per unit area) using the following techniques:

Direct counting. Useful for large organisms in an open habitat. Ecologists count large mammals, nesting birds or woodland trees in this way.

Sampling. By taking a sample of organisms from a small area, we can estimate the number of organisms that live in a much larger area. The sample must be representative of the whole habitat. Table 38.1 lists a range of techniques.

Technique	What you do	Organisms you can sample
Sweep netting	wave a net in the air	flying insects
Kick sampling	kick stones and gravel on a river bed and catch the disturbed organisms with a net placed downstream	aquatic invertebrates
Trapping	set sticky traps such as sheets of paper painted with honey	flying insects
	set light traps	moths and other nocturnal insects
	set pitfall traps, Fig 38.4(a)	insects that crawl on the ground
	set Longworth traps, Fig 38.4(b)	small mammals
Quadrat sampling	mark out a small area of ground with a wire grid	animal and plant species in an area of ground
Transect sampling	mark out a line or narrow belt and study along it	changes in landscape and living organism distribution up a slope, for example
Indirect techniques	use data from dead animals (beaver pelts, dead whales), count rabbit droppings, molehills or the exit holes of beetle larvae in wood	specific animal species

Table 38.1 **Sampling techniques for studying populations**

Fig 38.5(a) **Pitfall traps are set into the ground to catch crawling insects**

Capturing and marking. We can use this technique in field biology to give reliable estimates of large populations. It involves:

- taking a sample and marking the animals as they are caught,
- releasing the marked animals back into the general population,
- counting the marked animals in a second sample captured later.

Generally, the percentage of marked animals that are recaptured in the second sample tells us what percentage of the whole population we managed to mark in the first capture. So, if our second sample contains 5 per cent marked animals, we assume that our original sample represents 5 per cent of the total population of those animals in that area. This can be described by a formula called the **Lincoln index**:

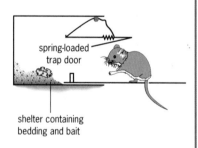

spring-loaded trap door

shelter containing bedding and bait

Fig 38.5(b) **A Longworth trap**

$$\text{Population size, } P = \frac{\left(\begin{array}{c}\text{number of organisms}\\ \text{in 1st sample}\end{array}\right) \times \left(\begin{array}{c}\text{number of organisms}\\ \text{in 2nd sample}\end{array}\right)}{\text{number of marked organisms recaptured}}$$

You must remember that this is a rough estimate, not an actual value and, to make sure that your results are as accurate as possible, you need to take the following precautions:

- Make sure that the population you intend to study is fairly stable: choosing a migratory bird at the start of winter is not a good idea.
- Capture animals randomly: do not select a particular group.
- Mix marked individuals evenly with unmarked individuals before releasing your sample back into the general population (this usually means leaving enough time for the animals to mix in with the population).
- Use marks which cannot be washed, rubbed or licked off between capture and recapture.
- Do not make the marks too obvious: the marked animals may be more easily seen and eaten by predators.

What limits population size?

In an ecosystem, abiotic (physical) factors in the environment can limit population size. When we look at the whole range of factors that affect population size, we can classify them into three general categories:

- Limiting factors that are *always present*. These include space and light. A cliff that supports a population of kittiwakes, for example, is a fixed size and can accommodate a fixed number of birds (Fig 38.6).

- Limiting factors that *vary predictably over time*. These include changes in temperature and weather patterns in different seasons which lead to population movement or a change in behaviour among different organisms. As winter approaches, birds which live in an ecosystem in the summer might migrate to a warmer place (Fig 38.7). Small mammals such as dormice hibernate.

- Limiting factors that are *unpredictable*. Some unpredictable limiting factors are **density dependent**: they cause a greater effect if there is a high population than if the population is quite small. Sudden changes in the availability of water and nutrients, or outbreaks of disease, for example, can devastate an overcrowded population but leave a widely dispersed population relatively untouched.

- Other factors are **density independent**: they can devastate a population, whatever its size. Fires, floods, severe frosts and other catastrophes, for example, might destroy part or the whole of an ecosystem without warning (Fig 38.8).

A Students on a field trip selected a small area of woodland by a stream and captured 75 snails by walking through the area and picking snails wherever they saw them. All were marked using a small dab of dark red nail varnish. 24 hours after the marked snails were released, 61 snails were captured in the same area: 15 of these had been marked. Use the Lincoln index to estimate the size of the snail population in the area.

B On field trips, small mammals such as shrews and voles caught in traps overnight are often found dead when the traps are checked next morning. Suggest why this happens, and what modification might solve the problem.

Fig 38.8 **Catastrophes, including those caused by human activity such as oil spills, have a density-independent effect on populations**

Fig 38.6 **The size of this ledge is an unchanging limiting factor. The kittiwakes can get nesting sites only by displacing other birds**

Fig 38.7 **Arctic terns are the world's champion travellers. In their search for food and the right breeding conditions, they range from deep inside the Arctic Circle, over the Atlantic to the pack ice of Antarctica**

HUMAN POPULATION GROWTH: IMPLICATIONS FOR THE NEXT CENTURY

HUMAN POPULATION GROWTH shows the classic J-shaped curve of exponential growth. If you compare the graph shown in Fig 38.9 with the one in Fig 38.1, you can see that the J-curve is the first part of the classic S-curve.

As in any exponential growth curve, the population starts off increasing slowly: from the early origins of human beings to the last 200 years or so, the curve remained fairly flat. However, we are now in the period where the curve is becoming extremely steep, with the human population of the world doubling about every 50 years or so. At some point, the curve will level off and the population will stabilise. It is difficult to know when this will happen and what forces will bring it about.

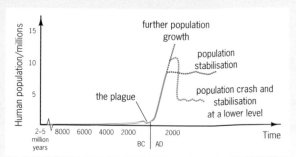

Fig 38.9 **Growth in the world human population. The dotted lines show three possible predictions of what will happen to the human population in the next two hundred years or so. It is inevitable that in some way the population will stabilise at some point in the future: the two main worries are when and how it will happen**

3 INTERACTIONS BETWEEN ORGANISMS

Every living organism in an ecosystem affects the others. Some of these interactions are **intraspecific**: they take place between organisms of the same species. Others are **interspecific**: they take place between organisms of different species.

Intraspecific interactions

Relationships between organisms of the same species in an ecosystem fall into three main categories:

- Reproduction and care of the young.
- Social behaviour.
- Competition for resources.

We consider these in more detail below.

Reproduction and care of the young

Organisms of the same species interact at many different stages of sexual reproduction. In some species the relationship is brief and minimal: a wind-pollinated plant simply releases pollen into the air and accepts pollen from another plant of the same species. In others, the association is complex and long term. For example, swans, like humans, find a mate and stay together for many years, sometimes for life (Fig 38.10). Both parents feed and care for the young and, to some extent, provide companionship and help for each other.

In both examples, the level of interaction achieves the aim of reproduction: producing enough surviving offspring to maintain the population level in the ecosystem.

In many animal species, reproduction involves interactions with more than just one other member of the same species. Many animals compete with each other for a suitable mate: for example, male red deer fight fiercely for control of groups of females. Animals that are more social, such as termites and meercats, provide joint care for each others' young. Some adults in the group, not necessarily just the parents, guard, transport and protect the young, while the others go off to search for food.

Fig 38.10 **The female swan carries her young, while the male displays to deter intruders**

Social behaviour

Many animals are social: they live in groups rather than as a collection of individuals that just happen to live in the same place. Group living has its advantages and disadvantages. Animals that live in groups are usually less likely to be eaten by predators than solitary animals. There seem to be several reasons for this:

- A large number of potential prey in one place can confuse a predator – the predator cannot decide which one to attack.
- Different individuals in a large group of animals can perform different tasks: some can look for food, others care for the young and some can act as look-outs for predators. Large groups therefore spot approaching predators much sooner than animals on their own and so can take more successful evading action.
- A large number of animals are more successful than just a few at defending against an attacking predator. Elephants group together, forming a circle to protect their young which are herded into the centre of the group if a predator threatens.

Animals that live in groups also stand more chance of hunting or finding food successfully (Fig 38.11, and see also page 395). Hyenas, for example, hunt in packs and can bring down a deer that is much larger than one hyena would be prepared to tackle. Other group associations can help to protect animals from the cold as they huddle together (see page 398).

Fig 38.11 **There is evidence that bees can communicate the position of good pollen sources by elaborate dancing displays to other bees in the hive. Here, a worker bee with full pollen sacs is dancing to indicate the amount, distance and direction of its source of pollen**

Competition for resources

Most of the disadvantages of group living arise because animals have to compete with each other for vital resources. Hyenas may make a bigger kill when they hunt in a pack, but the meat has to fill more stomachs. If food becomes scarce, rivalry within the pack becomes intense and some may need to find a new habitat to survive. Animals of the same species also compete for mates, shelter, water and social position.

Interspecific interactions

A full list of the interactions that can occur between individuals of different species in an ecosystem is given in Chapter 36, Table 36.1. In this chapter we will look in detail at four of these:

- Predator–prey relationships.
- Parasitism and disease.
- Mutualism, or symbiosis.
- Competition for resources.

Predator–prey relationships

A basic definition of a predator is an organisms which eats another organism to obtain energy and nutrients. This definition includes animals that feed on plants.

The relationship between predator and prey is based on a constant battle. The actual 'catching and eating' part of the battle is only a small part of the story. The more serious conflict involves evolution.

All species evolve by the processes of genetic mutation and natural selection (see Chapter 33): prey species are constantly evolving and becoming better able to deal with predators; predator

Fig 38.12 **Bats have evolved sophisticated sonar systems to detect flying insects with great accuracy. Some insects, however, have evolved the ability to detect bat sonar signals and respond by closing their wings and dropping out of the sky. This makes it much harder for the bat to catch them and they gain an important survival advantage over other insect species**

species are constantly evolving to out-manoeuvre their prey (Fig 38.12). Sometimes, either the predator or prey loses the long-term battle. If the prey evolves to completely avoid a predator, and that predator has no alternative food supply, the predator species can be driven to extinction. If a predator out-evolves its prey, it will do well in the short term, but, if the prey species becomes extinct, the predator then needs to find an alternative food supply to survive.

In general, predators thrive when they develop adaptations that allow them to eat well and make the most of their food. Herbivores, animals that eat plants, often have specialised mouthparts and digestive system to gain the most nourishment from the plants available to them. Many herbivores are ruminants: their digestive system has four stomachs and they chew their food twice, once after eating it, once after regurgitating it (see page 232).

Prey species continue to survive because they have evolved defence mechanisms against predators. In animals, these include camouflage, spines and barbs and chemical deterrents. Many, particularly insects, produce foul tasting or poisonous chemicals and usually advertise this by the colours and patterns on their bodies and wings (Fig 38.13).

Although plants cannot run away from grazing animals, they are not totally defenceless. Their defences include thorns, tough tissues and thick waxy cuticles. Many plants have also developed **secondary compounds**, chemicals that have no metabolic role but which put off potential grazers. Examples include tannins in oak leaves and cardiac glycosides in foxglove plants.

Symbiosis comes from a Greek word which means 'to live together'. We use it to describe a close physical interaction between two organisms from different species. There are three types of symbiosis:

Parasitism: one of the partners benefits, the other is harmed.

Commensalism: one partner benefits and the other is unaffected.

Mutualism: both partners benefit from the association.

The camouflage of this iguana matches the colours of the trunk and lichens in a Peruvian rainforest

The deadly sting of a scorpion deters predators as well as being useful for hunting food

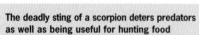

When under threat, a porcupine runs backwards at the enemy with quills raised. They can fall off, and an embedded quill can cause a septic wound that would kill a lion

Fig 38.13 **Animals use many different techniques to avoid capture by predators**

Parasitism and disease

Predators usually kill their prey: parasites have a different strategy because they depend on their prey, the **host**, staying alive. They live and reproduce using the resources of the host organism, a different species, and it is damaged in the process. This is why many parasites cause disease in humans and other animals (Table 38.2). The more successful parasites cause less damage to their host. The tapeworm, for example, can survive in the human digestive system for many years, causing relatively minor damage in the affected person.

Table 38.2 **Some major parasitic diseases and the parasites that cause them**

Disease	Parasite	Symptoms and effects
Malaria	*Plasmodium* (4 different species of this protozoan parasite cause malaria)	Regular cycles of fever: disease transmitted to people by the anopheles mosquito
Sleeping sickness	*Trypanosoma* species (also a protozoan parasite)	Transmitted to people by insects such as the tsetse fly which bite infected animals
Leishmaniasis (also called kala-azar and oriental sore)	*Leishmania* species (another protozoan parasite)	Transmitted to humans via sandflies
Schistosomiasis (also called bilharzia)	*Schistosoma* (blood fluke)	Flukes live in water (their intermediate host is a water snail) and infect humans who bathe
Tapeworm infection	*Taenia solium* (tapeworm)	Parasite lives in intestine: passed to humans in infected and undercooked meat
Toxoplasmosis	*Toxoplasma* species (protozoan parasite)	Causes weakness and fever – passed from mother to baby across placenta. Can be transmitted by domestic cats

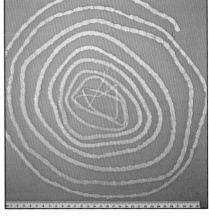

Fig 38.14 **Above: The beef tapeworm. The head is at the thin end, attached to many hundreds of segments. They break off and pass out of the host, to infect another victim.**

Below: At the head end (the scolex) are four suckers and a ring of hooks that attach the tapeworm to the gut wall of its host

Parasites show very specialised features which fit them to their lifestyle (Fig 38.14). All parasites show some general adaptations:

- They have ways of getting into the body of the host. Fungi which parasitise plants produce the enzyme **cellulase**, which breaks down plant cell walls. The schistosome (blood fluke) which causes bilharzia can bore through human skin.

- They have structures that enable them to attach to their host. Tapeworms have a ring of hooks called a **scolex** which they insert into the wall of the host's intestine.

- They have lost the organ systems that they no longer need. Tapeworms and other gut parasites have, for example, lost their own digestive system because they can absorb already digested food across their body surface. We call this loss of structure and function **parasitic degeneration**.

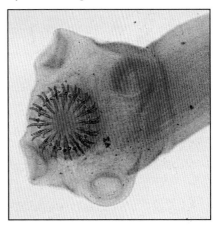

- They protect themselves against the internal defences of the host. Trypanosomes, which cause sleeping sickness in humans, produce an outer layer which varies constantly in its structure. Just as the host's body starts to make antibodies (see Chapter 26) to one type of coating, the trypanosome produces a different coat which the antibodies cannot recognise.

- They often have complex life cycles which allow them to infect new hosts. See the Feature box on malaria.

- They show great **fecundity**, or capacity for reproduction. Many tapeworm segments break off every day and pass out of the body in the faeces. Each segment can contain hundreds or even thousands of eggs.

C Organisms can interact intraspecifically or interspecifically. What do these two terms mean?

THE FIGHT AGAINST MALARIA

MALARIA IS A DISEASE caused by a parasite. Four species of protoctist of the genus *Plasmodium* are responsible for 200 to 300 million cases of malaria every year, worldwide. The disease is common in the tropical and subtropical regions of the world – home to 40 per cent of the world's population. The death rate is appalling, somewhere between one million and five million people each year, many of them children under five.

Malaria is extremely difficult to control for several reasons. Firstly, it has a complex life cycle (Fig 38.15). Any four of the *Plasmodium* species that cause it can be passed on to human hosts by any one of 60 species of *Anopheles* mosquito. Draining swamps and marshes and using insecticides to spray mosquito breeding areas was temporarily successful and reduced the spread of malaria during the 1950s and 1960s. However, not every species of carrier mosquito was wiped out. Many of them have developed resistance to the early insecticides and are now much more difficult to control.

Secondly, the parasite itself is a tricky customer. It produces symptoms which get worse and then better in cycles (Fig 38.16). Anti-malarial drugs target some but not all stages in the life cycle. The sufferer therefore needs to take the drugs over a long period of time, or the parasite 'hides' from their effects. Some of the common anti-malaria drugs have been used so widely that they have put *selection pressure* on the parasite and it has now developed resistance to many of them.

Although new drugs are being developed all the time (some of the most recent are based on extracts of rainforest plants), workers now concentrate on prevention. Any stagnant water that can be eliminated is removed and mosquito nets in houses are dipped in a safe insecticide such as permethrin. Biological methods of control of the mosquito larvae are being used – fish such as guppies are being bred to eat them. Vegetation around houses is cleared, and trees are being planted in marsh areas to destroy the wet breeding grounds of the mosquito.

Fig 38.15 **The life cycle of the malarial parasite, *Plasmodium vivax*. The anopheles mosquito is the first host, humans are the second host. People with malaria get regular cycles of illness because merozoites are continually released. When the parasite is 'hiding' in the red blood cells, symptoms are not too bad: it is the sudden rush of foreign proteins in the bloodstream that leads to the bouts of fever that are typical of malaria**

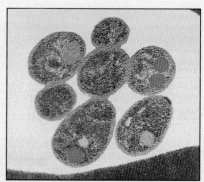
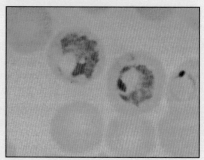

Fig 38.16 **The merozoite stage (above) and signet ring stage (below) of *Plasmodium vivax* in human red blood cells**

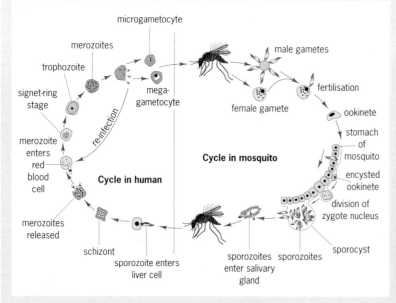

Mutualism and commensalism

The term **mutualism** is used to describe a relationship between two organisms from different species that receive *mutual* benefit from the association. Neither is harmed and both do better by being together than they would alone. The African bird, the oxpecker, for example, cleans the skin of the rhinoceros, removing all the parasites such as ticks and mites, which it eats. Legumes (see page 583), mychorrhizae and lichens (see page 184) are all examples of mutualistic associations.

Two organisms which are **commensal** live together and, while one of the partners benefits, the other does not. The relationship between clownfish and sea anemones is a good example of commensalism (Fig 38.17). The fish hide from predators in the mass of the anemone's stinging tentacles. Although the anemone is unharmed, it does not seem to benefit in any way.

Competing for a resource

Species can compete in different ways. Some organisms stop the organisms of other species from getting near a particular resource. Wrens, for example, defend food plants in their territory. In other situations, two species have access to the resource, but one is better at exploiting it.

Some species have evolved to coexist, sharing the resources that are available in an ecosystem. Some plants, for example, have a mesh of roots close to the surface of the soil to trap water from short showers of rain. Other neighbouring plants have long roots which take advantage of deeper groundwater. Both species are using the same resource, but they are sharing it in a way that allows both to benefit. This is called **resource partitioning** and it also occurs between animal species.

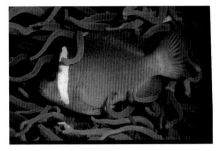

Fig 38.17 **There are 26 species of clownfish which all associate with sea anemones. This colourful fish hides from predators among the venom-producing tentacles of the sea anemone**

4 A WIDER LOOK AT COMMUNITIES IN AN ECOSYSTEM

We have seen that ecosystems are dynamic systems and that the physical parts of an ecosystem, such as the weather, affect and interact with the organisms that live there. In this section we look at how an ecosystem forms.

Imagine a completely physical environment such as the bare rock and ash left behind after a volcano has erupted and then become quiet. After the ground cools and water reenters the environment, some plants start to grow there. After a time, other species of plant establish themselves and eventually, probably after many years, woodland forms (Fig 38.18). This process is called **succession**.

Fig 38.18 **Succession in a bare site to woodland**

Boulders and rocks with mosses and lichens

Small plants and grasses grow as weathering produces finer pebbles. Plants with root nodules fix nitrogen. More plants grow as soil develops

Bushes and shrubs, such as hawthorn and bramble, grow. Most non-woody plants die out

Fast-growing trees, such as birch, grow up, forming dense, low forest

Larger, slower growing, but stronger oak trees grow above the birch and establish the climax community

Succession also describes the more minor changes in the vegetation of an ecosystem which occur after a less catastrophic environmental change. For example, after a small patch of trees in woodland die after a series of hot, dry summers, the land becomes repopulated by the same sorts of trees very quickly because their seeds are already present in fertile ground.

To distinguish between these two types of situation, we say that the process that results in new vegetation on a bare site is **primary succession.** This takes time and the final appearance of the community depends on which plant spores and seeds come into the area from neighbouring ecosystems. The process which repopulates a previously well-vegetated area after a minor environmental change is called **secondary succession**.

Strictly, succession is used only to describe the development of and changes in plant communities. In any ecosystem, however, the plant communities help to form the conditions which enable other classes of organism (animals, bacteria, fungi etc) to colonise the environment.

A developing ecosystem

In order to understand the stages that occur in succession, think of an environment consisting just of bare rock. Colonisation occurs in many different stages called **seres**. The rocks first provide a home for spiders which feed on insects in the area.

Succession begins as mosses and lichens start to grow on the bare rocks. These first plant species are called **pioneer species**. Over many years, perhaps thousands of years, the environment is battered by rain and wind and other climatic factors, and the surface of the rock is broken down into smaller fragments. The mosses and lichens help soil to form in cracks and crevices, and eventually a few small broad-leaved plants start to grow.

HOW OLD IS MY HEDGE?

AS YOU NEXT WALK DOWN a country lane, take a good look at the hedges that you pass. Many of these structures date from the Middle Ages. They were planted by medieval farmers as boundaries to mark off their land and to keep in livestock. They probably planted shrubs like hawthorn a few feet apart. The gaps at the base of these plants then allowed waves of colonisers.

First of all, fast growing, non-woody annual plants like typical garden weeds would have thrived: chickweed, groundsel and shepherd's purse for example. Then, non-woody perennials such as bluebells, primroses, stinging nettles, dandelions and thistles may have become established. These plants can produce food reserves to survive the winter. Eventually, woody perennials, shrubs and small trees such as elder, holly, and even some of the larger forest trees could have invaded the hedge.

Fig 38.19 **In theory, it is possible to date a hedge by looking at the range, or** diversity **of plants that are in it. The more diverse the community of the hedge, the older it is**

As the soil becomes enriched by organic matter from the action of decomposers on dead plants, more species of plant can grow, including grasses and then plants with root nodules that have nitrogen-fixing bacteria. As these plants start to thrive, a few larger plants, shrubs, begin to grow.

After this, things speed up because shrubs provide roosting and nesting sites for birds which then supply many more seeds in their droppings. Pioneer tree species which need high levels of light come into the area next. As woodland becomes more established, other, more shade-loving trees can colonise the area. Finally, the vegetation of the area becomes more or less stable: we describe the plant community that then exists as a **climax community**.

FIELD STUDY NOTES: QUADRATS AND TRANSECTS

Two of the sampling techniques described in Table 38.1 are used a lot by plant ecologists. They carry out **quadrat sampling** to survey a relatively small area, and **transect recording** to describe differences in vegetation that occur over a larger area of land.

Quadrat sampling

The basic technique of quadrat sampling is shown in Fig 38.20. When you are sampling using a quadrat you may come across problems: sometimes, for example, you may not be able to count how many individual plants there are in a dense clump. So you can use any of these methods:

- Estimate plant *density* by counting the number of individuals per unit area.
- Estimate plant *frequency* by looking at the distribution of individuals per unit area (are plants abundant or rare?).
- Assessing *plant cover* by measuring the proportion of the ground covered by a plant species.

Fig 38.20 **Quadrat sampling**

Transect recording

A transect is an imaginary line along which you make careful and systematic observations. There are two types of transect, the **belt transect** and the **line transect** (Fig 38.21). You can use transect recording to study how communities and ecosystems change along an environmental gradient, for example, through a woodland, along a rocky shore, or up the side of a hill

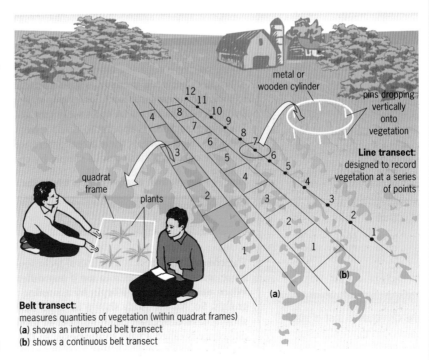

Belt transect:
measures quantities of vegetation (within quadrat frames)
(a) shows an interrupted belt transect
(b) shows a continuous belt transect

Fig 38.21 **Types of transect recording**

Line transect: designed to record vegetation at a series of points

SUMMARY

■ A population grows **exponentially** in ideal conditions. Growth slows down because of **limiting factors** in the environment. This produces an **S-shaped** or **sigmoid** growth curve.

■ When the size of the population becomes stable, we say that this is the **carrying capacity** of the environment. Sometimes, population growth exceeds the carrying capacity and a **population crash** follows. The population may then settle at a new carrying capacity.

■ Three main types of growth curve occur in nature: **stable**, **irruptive** and **cyclic**.

■ We can study populations using different techniques such as **direct counting**, **sampling** and **capturing and marking**. Statistical techniques, including standard deviation and chi-squared tests, are useful in analysing the results.

■ Some limiting factors which limit population growth are **density-dependent**: their effects are worse in a large, closely packed population. Others are **density-independent**: their effects do not depend on the population distribution or size.

■ Interactions between organisms occur between species: these are **interspecific interactions**. Within species, they are **intraspecific interactions**. You should know some examples of each type.

■ The living part of an ecosystem develops by the process of **succession**. If the area has never been colonised before, a newly formed volcanic island, for example, the process is called **primary succession**. If the area was previously covered with vegetation: a forest site that suffers a severe fire, for example, the process is called **secondary succession.**

QUESTIONS

1 Various types of trap are used to sample the animals which live in grassland. The animals are released later.

a) Fig 38.Q1(a) shows a *Longworth* trap which is used to catch small mammals. Explain why this type of trap should be checked at least twice each day.

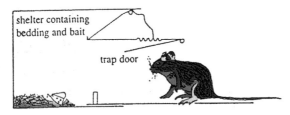

Fig 38.Q1(a)

b) Twenty *Longworth* traps were used over a period of ten days to sample the small mammals living in the grassland. Table 3.8Q1 shows the results which were obtained.

Table 3.8Q1

Mammal	Number trapped	% of total
Field vole	180	50
Woodmouse	108	
Bank vole	45	
Others	27	
Total	360	

   (i) Copy and complete Table 3.8Q1.
   (ii) Copy and complete the pie graph of Fig 38.Q2(b) to show the species composition of the small mammal population.

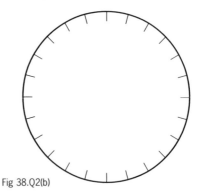

Fig 38.Q2(b)

c) Fig Fig 38.Q2(c) shows a pitfall trap which is used to catch invertebrate animals.

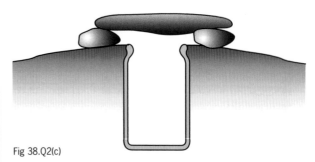

Fig 38.Q2(c)

Suggest **two** reasons why the results from this type of trap might give a false impression of the number and types of invertebrate animals which live in the grassland.

[NEAB: February 1996 AS Biology: Advanced Ecology, Section A, q.1]

2 Moose colonised *Isle Royale*, an island in Lake Michigan in Canada, in the early 1900s. They are large herbivores; no predators of the moose live on *Isle Royale*.

Table 38.Q2 shows the estimated population size of moose between 1915 and 1960.

Year	Estimated size of moose population
1915	200
1917	300
1921	1000
1925	2000
1928	2500
1930	3000
1934	400
1943	170
1947	600
1950	500
1960	600

Table 38.Q2

a) Describe the pattern shown in the results.

b) Give **two** density dependent factors which might account for the pattern shown in the results between 1915 and 1943. Explain how each factor may have had its effect.

[NEAB February 1996 AS Biology: Advanced Ecology, Section A, q.2]

3 A group of students investigated the distribution of plants bordering a small stream in a salt marsh. Fig 38.Q3 shows their results.

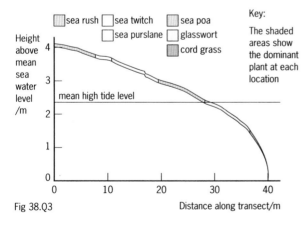

Key:
The shaded areas show the dominant plant at each location

Fig 38.Q3

a) Describe the techniques which the students could use to obtain the data shown in the diagram.

b) The distribution of plants in this habitat is governed mainly by abiotic factors.

Suggest **two** abiotic factors that could restrict the distribution of sea rush and cord grass in this habitat and explain how each factor would have its effect

[NEAB February 1996 AS Biology: Advanced Ecology, Section A, q.3]

4 Most cereal fields in Britain are sprayed with selective herbicides. In order to conserve wildlife, farmers are recommended to leave unsprayed a 6 metre strip, called a headland, around each cereal field.

a) Explain how spraying selective herbicides on a head-land might affect the number of insects living there.

Table 38.Q4 shows the results of an investigation to find the effect of leaving headlands unsprayed on populations of butterflies living there.

Table 38.Q4

Butterfly species	Number of each species recorded on headland sprayed with selective herbicide	Number of each species recorded on headland not sprayed with selective herbicide	χ^2	Significance (NS = not significant)
Small skipper	2	41	35.4	P 0.001
Large skipper	1	17	14.2	P 0.001
Large white	38	56	3.4	NS
Holly blue	13	29	6.1	P 0.05
Hedge brown	59	93	7.6	P 0.05
Small heath	0	11	11.0	P 0.01
Ringlet	23	52	11.2	P 0.01

b) Why was a χ^2 (chi-squared) test applied to the results?

c) What conclusions can be drawn from the results of the χ^2 test?

[NEAB February 1996 AS Biology: Advanced Ecology, Section A, q.4]

39 Human activity and the environment

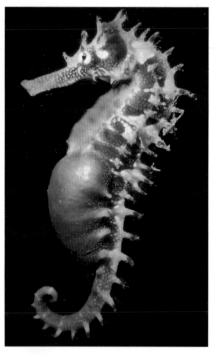

Sea-horses are unique in the animal kingdom: the male not the female gets pregnant and gives birth

MANY PROBLEMS IN the environment happen when people use a resource for today, hoping that everything will be okay tomorrow. In the Philippines sea-horse fishing is very important to the poor fishermen and their families: they can sell sea-horses to traders who then sell them on to the lucrative Chinese medicine market in Hong Kong and elsewhere. Without the sea-horses, many of the people would not be able to feed their families.

As fishing continues, the sea-horses have started to disappear. Fewer sea-horses means that the fishermen catch fewer and earn less and so start to catch smaller and younger sea-horses to supplement their income. The population is further depleted and a vicious circle is set up.

In a ground-breaking environmental project, scientists have studied the life-cycle and behaviour of sea-horses and are working with local fishermen to try to change fishing habits, so that the sea-horse population recovers. Rather than taking a standard environmentalist hard-line view and trying to persuade the villagers to stop fishing altogether so that all the sea-horses can be protected, the team recognises the conflict between human need and the needs of the environment. By showing the fishermen that an enforced 'safe area' will allow sea-horses to breed without threat, and teaching them to avoid catching pregnant male sea-horses, the scientists have proved to the local community that they can help to restore sea-horse numbers. Not only will the sea-horse population be saved for the future, the fishermen will get back their source of income and will be free from the worries of starvation.

1 WHY DO HUMANS AFFECT THEIR WORLD SO MUCH?

We humans are a unique species: we live in most of the ecosystems of the world, adapting our habitats to suit us, rather than having to adapt to the environment as most other species do (Fig 39.1). And, unlike other species, our behaviour affects the environment on a global as well as a local level.

The main human activity that affects the world environment is farming. Close behind is industry and mining for fuels such as coal, oil and gas. The more technologically advanced the society, the

Fig 39.1 We think of the British countryside as 'natural' but, in fact, it is almost entirely artificially made. Only a few areas in Scotland, Wales and Northern England remain in their 'original' form – all farmland and most forests are the result of human influence on the environment

greater the overall environmental impact of each individual person. So, for example, a child born in Britain into a fairly well-off family may live for 80 years or more. In that time the amount of resources they use is enormous: think of the furniture, electrical appliances, cars, fuel, food, clothing, etc used up by someone like this. At the other extreme, a child born in a poor family in India or South America might live only 20 years (Fig 39.2). They could live in poorly built housing, or be homeless and forced to live on the streets, probably never get quite enough to eat, and would certainly never use a car or any significant amount of fuel.

In this chapter, we take a brief look at some of the complex issues involved in controlling the effect people have on the ecology of the Earth. There is room only to give an overview of some of the problems and potential solutions.

Fig 39.2 **In many areas of the world extreme poverty, homelessness and malnutrition affect millions of people**

2 FARMING AND ITS EFFECTS ON THE ENVIRONMENT

It is very common to hear environmentalists complaining that farming is destroying hedgerows and wildlife habitats, that woodlands and forests are being cut down to make new farmland and that it should all be stopped. In addition, farmers are often portrayed as hard-line business men and women who put profit before anything else. There is a clear conflict here, but the issue is not as simple as these points of view suggest.

We must consider five basic facts:

● People need to have food to eat: staple foods need to be cheap.

● There are more people on Earth to feed than ever before.

● Farmers need to make a living.

● Farming needs land.

● Some farming techniques are damaging the environment so badly that the capacity of the land to produce food is being reduced.

The key to solving the problem of food production is to develop farming methods that are efficient enough to provide the amount of food that we need in the short term, but which protect the environment to ensure that food production levels can be maintained, or even increased, in the future.

?

A What is the different between a sustainable resource and an unsustainable one? Think of an example of each.

Food production on a global scale

On a global scale, food production has increased during the last 40 years (Fig 39.3). Until the early 1980's, this increase matched the increase in world population, so that, overall, more people had enough to eat than ever before. There were still, of course, countries and areas where food supply did not meet demand and many millions of people were malnourished or starving.

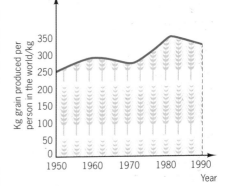

Fig 39.3 **The amount of grain produced per person in the world between 1950 and 1990**

Since 1984, population growth has increased faster than food production. Part of the reason for this has been the damaging effects of farming practices which produce food in the short term, but lead to long-term environmental damage, particularly soil erosion and desertification (see page 611). Very poor countries where food production has fallen since the mid-eighties have tried to compensate by importing food from Europe and the USA where more food is produced than is needed by the population. However, this only increases the debt of these countries, leading to more poverty and, in the long term, more malnutrition and starvation.

Many scientists believe that soil erosion, water shortages and pollution from excessive use of pesticides and fertilisers will lead to a crisis in food production during the twenty-first century. Only a change in farming techniques might prevent this.

The concept of sustainable agriculture

The problem of providing food for an increasing world population without reducing the capacity of the land to produce food is a difficult one, but it has some solutions. By using farming techniques which are sustainable, rather than successful in the short term only, many areas of the world should be able to continue food production at the right level for the foreseeable future.

Moves towards sustainable agriculture would include:

- Using new varieties of crop plants, perhaps produced by genetic engineering, which can survive and give good yields in poor soils, dry conditions and without the need for expensive pesticides and fertilisers (Fig 39.4; see also Chapter 34).

- Making use of non-traditional food sources – more plants that have nitrogen-fixing bacteria in their roots (these need no fertiliser), insects (many cultures of the world eat insects and their larvae as protein-rich delicacies) and unusual animals. In Europe, for example, ostrich farming is becoming more widespread.

Fig 39.4 **This scientist in the Philippines is testing the disease resistance of genetically modified varieties of rice**

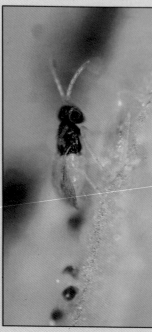

Fig 39.5 **Until recently the only biological control for whitefly, a common plague on greenhouse-grown tomatoes (left) was** Encarsia formosa **(centre), the parasitic wasp which eats only adult flies, leaving larvae to develop and so cause a recurring problem. Gardeners and commercial growers can now obtain the beetle** Delphastus pusillus **(above) which kills larvae as well as adult flies, so providing much more effective and long-lasting control**

- Reducing the amount of fish caught in the seas to levels where the sizes of the populations do not continue to decrease. At the same time, the number of fish farms could increase.
- Encouraging and rewarding farming methods which do not lead to soil erosion and other environmental problems.
- Using biological pest control, rather than artificial pesticides which can persist in the environment (Fig 39.5).
- Encouraging farmers in poor countries to grow a selection of crops that can feed local people, rather than cash crops such as tobacco and coffee for export.
- Increasing financial and technological aid to poor countries to allow them to set up more sustainable methods of farming and food production.

3 HUMAN ACTIVITIES AND POLLUTION

(a)

(b)

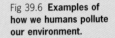
Fig 39.6 **Examples of how we humans pollute our environment.**

Fig 39.6(a) **The outlet pipe in the background is discharging raw sewage which is being washed up onto this beach**

Fig 39.6(b) **Effluent from the chemical works is polluting the river which supplies it with vital water**

(c)

(d)

Fig 39.6(c) **This landfill site will eventually be levelled and should look more pleasant but problems might still be caused by methane production under the surface**

Fig 39.6(d) **Pollutants in smoke from industrial plants add to the acid content of rain**

Pollution occurs when substances are released into the environment in harmful amounts, usually as a direct result of human activity (Fig 39.6). Most pollution results from excessively high concentrations of substances, but there are exceptions.

Some pollutants are substances which, although already found in the natural environment, are produced in very large quantities by human activity, such as CO_2 and organic waste (sewage). Other pollutants are toxic compounds produced by human activity, such as pesticides. These often remain in the environment for long periods of time, because there is no living organism to break them down. A third type of pollutant includes chemicals that seem to pass all the safety tests, but then have some totally unexpected effects. Pollutants can affect the air, the sea, and the land.

Fig 39.7 **Unpolluted rain has a pH of about 5.6 because of the CO_2 that occurs naturally in the atmosphere: rainwater in most industrialised countries has a pH of between 4 and 4.5. Some cities in the US are bathed in acid fog with a pH as low as 2.3 which is 1000 times as acidic as normal rain, and the same acidity as lemon juice**

Fig 39.8 **As they are carried in the air, oxides of sulphur and nitrogen form secondary pollutants such as nitric acid vapour, droplets of sulphuric acid and particles of nitrates and sulphates. These chemicals reach the surface of the Earth in two forms: *wet*, as acid rain, snow or cloud vapour; or *dry* as acidic particles**

Fig 39.9 **These trees in the Czech Republic have been killed by acid rain that has resulted from emissions of sulphur dioxide and oxides of nitrogen given off by heavy industry all over Europe**

Air pollution: acid rain

To many people, walking through a high Swiss meadow in summer during a shower of rain, sounds idyllic. Peace and quiet, fresh mountain air and refreshing rain. But what if the rain has the pH of vinegar or lemon juice (Fig 39.7)?

In Switzerland, Norway, Sweden and other parts of Scandinavia, acid rain falls on some of the most unspoilt natural landscape in the world. It starts off, as far as 1000 kilometres away, as emissions of sulphur and nitrogen oxides released from the more highly industrial areas of Europe. Although there are some natural sources of these gases, by far the largest amount is generated by burning fossil fuels: the petrol used to power road vehicles and the oil, coal and gas used in power stations (Fig 39.8).

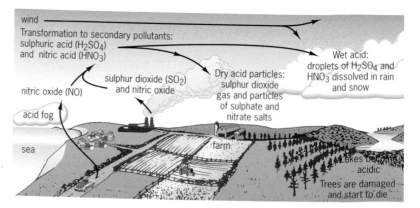

The effects of acid rain on the environment are still causing arguments. It is quite difficult to get good evidence to demonstrate a definite link between the occurrence of acid rain and the effects that ecologists and environmental scientists have noticed. For example, many European and North American forests have been damaged during the last 30 years, especially at high altitudes and at the edges of some forests which have large areas of dying trees (Fig 39.9). In Britain the pattern of loss of lichen species seems to follow the distribution of sulphur dioxide pollution. In fact, lichens are so sensitive to this sort of pollution that they are used as pollution indicators. Other possible effects of acid rain are listed in Table 39.1.

Many countries, including Britain, have now recognised acid rain as a serious problem and have reduced their emissions of sulphur dioxide in the last 25 years.

Table 39.1 **Possible effects of acid rain**

Possible effect of acid rain	How it might happen
Fish deaths in Scandinavian lakes and elsewhere	Indirect mercury poisoning: more acidic water is thought to convert inorganic mercury compounds in lake sediments into compounds which are soluble in the fatty tissues of fish
	Indirect aluminium poisoning: in acidic conditions, more aluminium ions are released from the soil and washed into lakes. Fish respond by making excess mucus which clogs their gills with fatal results
Damage to statues and buildings	Chemical reactions between components of acid rain and rock
Overstimulation of plant growth	Acid rain falling on the ground means excess nitrogen. The plants then use other nutrients faster, reducing soil fertility
Thought to make human respiratory problems, such as asthma, worse	Irritation of surface membranes in lungs and bronchi

Presenting data

The results of our investigations include pieces of information, or **data**. We can present raw data in several ways:

Line graphs: a quick and simple way of showing the relationship between two factors, or **variables**. A line graph has two axes. The variable that changes predictably (this includes time) is plotted on the horizontal axis. The variable that you measure is plotted on the vertical axis. See Fig 39.3 for an example of a line graph.

Bar charts: good for presenting data that is organised into sets or groups, such as the numbers of the same animal in different areas. See Fig 30.5 on page 486 for an example of a bar chart.

Histograms: these are similar to bar charts but are used to represent continuous data; measurements that occur along a continuous scale (heights of trees, numbers of leaves).

Kite diagrams: a type of bar chart that we use to present more detailed information. For example, a kite diagram can show the distribution and numbers of birds that live in a wood (Fig 39.10(a)).

Pie charts: used to display data in the form of a pie divided up into slices (Fig 39.10(b)).

Scatter plots: We use scatter plots when it is important to look at the relationship between two variables (Fig 39.10(c)).

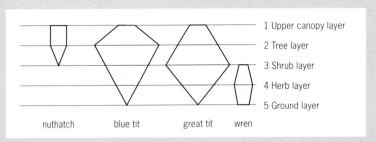

Fig 39.10(a) **This kite diagram shows the relative abundance of four different bird species that live in different zones in a wood**

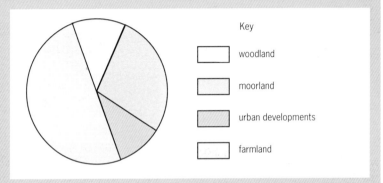

Fig 39.10(b) **This pie chart shows the proportion of vegetation in Britain that is woodland, moorland, urban development and farmland**

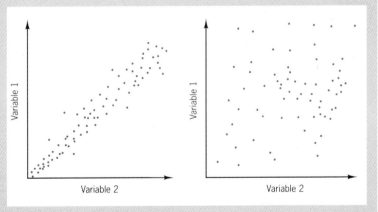

Fig 39.10(c) **A scatter plot like the one on the left shows that the two variables are linked. In this case there is a positive correlation between them. It actually shows a plot of age against length in fish: as you would expect, the plot shows that older fish are longer. The scatter plot on the right shows no correlation at all between the two variables**

Air pollution: the ozone layer

The effects of acid rain occur a few hundred miles away from the source of the pollutants which cause it. This seems bad enough, but other pollutants affect the whole planet. The effect of greenhouse gases on global warming is one example (see Chapter 36). Ozone depletion is another.

Ozone is a form of oxygen: its molecules contain 3 oxygen atoms, rather than the 2 that occur in molecules of atmospheric oxygen.

✔
Ozone (O_3) is present in the atmosphere as a whole in very small amounts, a few parts per million at the most. However, it is not evenly spread. It occurs as a definite layer in the stratosphere, 15–50 kilometres above the Earth's surface. Low level ozone occurs at ground level when bright sunlight acts on pollutants produced by heavy traffic – we see this as photochemical smog in large cities in high summer.

Both gases have been a major feature of the Earth's atmosphere for the last 450 million years, the oxygen as a result of the photosynthetic activities of blue-green bacteria and, later, of green plants (see Chapter 9). Oxygen is vital to life: it is the fuel for aerobic cell respiration. The role of ozone is less direct: it forms a sort of sunscreen high up in the atmosphere, shielding the biosphere from the harmful effects of ultraviolet light which beams down from the Sun.

Many scientists think that modern industrial pollutants are damaging the ozone layer, and that this may cause serious problems in the future. Chlorine- and bromine-containing compounds, mainly chlorofluorocarbons (CFCs), that we have released into the atmosphere in this century seem to be causing the measurable thinning of the ozone layer (Fig 39.11).

Fig 39.11 **In the mid-1980s, scientists discovered something unexpected: sunlight on the Antarctic in spring causing immense destruction of ozone and a huge hole appearing in that part of the ozone layer. This 'ozone hole' has been noticed every year since, and appears to be growing**

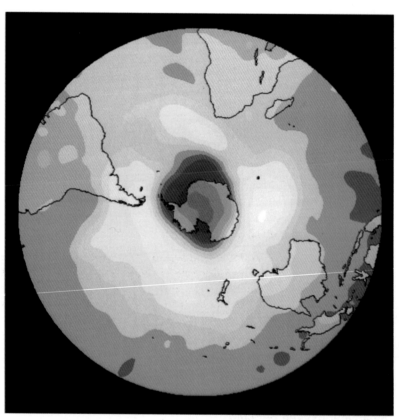

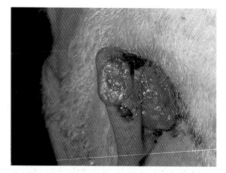

Fig 39.12 **In countries such as Australia, New Zealand, South Africa and Chile, which are protected only by a very thin layer of ozone for many months of the year, there have already been many more cases of skin cancer than ever before. Children in Australian schools are required by law to wear hats and blocking creams whenever they go outside**

Originally, CFCs were thought to be 'ideal' chemicals: they were non-flammable, non-toxic, non-corrosive, cheap to make and could be used as coolants in fridges, propellants in aerosol spray cans, and in the manufacture of the packing material Styrofoam. However, in the mid-1970s, scientists showed that CFCs rise slowly into the stratosphere and are then converted by the action of ultraviolet light into chlorine atoms. These accelerate the breakdown of ozone into O_2 and O. CFC molecules can stay in the stratosphere for about 100 years, and each one can destroy hundreds of thousands of molecules of ozone.

CFCs have now been phased out in most countries of the world, but their effects continue. Even if no more CFCs were put into the air from tomorrow, the ozone layer would still take about 100 years to recover from the worst of the damage. Less ozone in the stratosphere allows more of the harmful ultraviolet radiation from sunlight to reach the Earth's surface. Scientists predict that this will lead to a huge increase in skin cancers and cataracts (Fig 39.12).

✔
Brands of products which carry the phrase 'CFC-free' may do so as a marketing ploy to persuade the consumer that they are more 'environmentally worthy' than other products. In fact, by law, no aerosols sold in Britain contain CFCs.

Water pollution: getting into hot water

Fig 39.13 **As well as causing the death of fish directly, by depriving them of oxygen, warm water can also encourage the growth of parasites which would otherwise not be a problem**

When you think of water pollution, you probably think of chemical effluent, outflows of raw sewage or huge oil spills. While these undoubtedly are examples of pollution, probably the most over-looked cause of damage to inland aquatic ecosystems is the warm water released as a by-product of many industrial processes. Fish and water-living invertebrates are killed by warm water, not because they are scalded directly, but because warm water can carry much less dissolved oxygen than cool water and they 'suffocate' (Fig 39.13).

Water pollution and sewage

Contamination of rivers, lakes and seas by sewage causes two main problems. Firstly, it can introduce potentially dangerous pathogens (organisms that can cause disease) into the environment (see Fig 39.6). This is a particular problem if the water is drunk by animals or people. The second problem is that sewage and other organic wastes are decomposed by the action of aerobic bacteria. These use large quantities of oxygen, leaving less for other organisms living in the water. Sewage contamination can lead to **eutrophication** (see the Assignment).

Organic pollution consists of sewage, a mixture of faeces and urine. It contains salts, urea (a nitrogen-containing compound – see page 409), dead gut-lining cells, undigested food and living bacteria.

Biological oxygen demand

It is easy to find out the oxygen requirements of water from a particular site by taking a sample of the water and measuring its oxygen content. After keeping it sealed and in the dark for 5 days at 20 °C, the oxygen content is measured again. The rate at which the oxygen has been used up is the **biological oxygen demand (BOD)**. The BOD of unpolluted river water is about 3 mg O_2 per litre (dm^3) of water, per day. Raw sewage uses over *100 times* more oxygen in the same time (Table 39.2).

Although pollution by sewage is a major cause of eutrophication, there are others. If large amounts of nitrate- and phosphate-rich fertilisers are put onto farmland, the excess can run off and contaminate local rivers and lakes. Industrial processes which pollute local waterways with large amounts of concentrated sugars or other organic waste products can also cause the BOD of the water to increase significantly.

Find out more about eutrophication in the Assignment at the end of this chapter.

The term **eutrophic** means over-fertile. So, a eutrophic river or lake contains more nitrates and phosphates than normal, and abnormal growth, particularly of algae, can occur. The opposite of eutrophic is **oligotrophic**: this means under-fertile. So an oligotrophic river cannot sustain much life.

Table 39.2 **Typical BOD values per day of different types of water**

Water type	BOD value per day
Clean river water	3 mg dm^{-3}
Water from polluted stream	10 mg dm^{-3}
Domestic sewage, untreated	250–350 mg dm^{-3}

39.14 **This guillemot was caught in the oil spill which happened at Manobier beach, Wales in February 1996 after the tanker Sea Empress ran aground. Birds covered in oil cannot swim very easily and sometimes tire and drown. Those which end up on the beach try desperately to clean their feathers, ingesting and inhaling large amounts of oil which damages their digestive system and often causes pneumonia**

Contamination of water by oil and detergents

The names Exxon Valdez, Braer and Sea Empress may sound familiar, even if you can't quite remember that they are oil tankers that have famously run aground causing huge oil spills (Fig 39.14). These pollution disasters make headline news but, in 1993, a study by Friends of the Earth estimated that over 1000 times the amount of oil carried by the Exxon Valdez (which polluted the coast of Alaska in 1989) is lost in the US each year – not through national catastrophes, but through the normal operations of washing tankers and releasing the oily water, from pipeline and storage tank leaks and from accidental loss from offshore oil wells.

The effects of oil spills on ocean ecosystems depend on:

● The amount of oil released.

● How far away from the shore the spill happens.

● The weather conditions, water temperature and speed of ocean currents.

● The type of oil released: an area affected by a spill of refined oil takes over three times longer to recover than a similar area that has a crude oil spill.

In the clean-up operation following a large spill, oil is dispersed by detergents. The aim is to spread the oil molecules to limit damage to the local environment. Unfortunately, many detergents are extremely toxic and lead to many sea-bird deaths.

Spoiling the land

People can pollute areas of land directly by dumping toxic waste, by industrial processes such as mining and by simple neglect: leaving non-biodegradable rubbish behind after a picnic at a local beauty spot, for example. However, human activity, particularly farming, is indirectly causing two main problems: **deforestation** and **desertification**. Both have far-reaching global effects.

Deforestation

Deforestation has been practised by humans for centuries. As human population and technology have increased, so has the need to use wood. (The discovery of fire has had an enormous impact.) Many European forests were cleared by various different settlers from about the fourteenth century onwards. Today, the human population is larger than ever before and the process of deforestation has now spread to all parts of the world, particularly the tropical rain forests.

Deforestation is a process which results from the conflict between immediate human need and the necessity to protect the environment from serious and long-term damage. Tropical hardwoods are prized as building materials throughout the world, and other types of wood are used to make the thousands of tons of paper that we now use every day. Many people in the poor countries of the world depend on wood or charcoal for fuel, and wood, twigs, crop residues and grass are used for fuel in many of the 'better off' countries such as China, Brazil and Egypt. As well as cutting down forests to use the wood they provide, deforestation is also carried on to clear land for farming. People have cleared large areas of forest for plantations

PAPER FROM A NEW SOURCE

RECYCLING PAPER and using wood from sustainable sources can help us not to use so many valuable trees for paper-making. But what if there was a plant which could grow more than 100 times faster than the pine trees currently used for paper making? Supposing the paper made from this plant was just as good or perhaps even better than paper made from wood? And, to make it even more useful, the plant had a high level of disease resistance so that it needed little or no pesticides for a guaranteed crop?

It may surprise you to learn that such a plant actually exists. A woody annual plant called kenaf can provide fibres at a yield, per acre, that matches pine trees. And it can do it in 150 days, instead of 50 years. The plant has all the other properties of an ideal paper source but very few people have even heard of it. The difference between the real-life outcome of this find and the ideal, expected outcome comes down to money. Kenaf paper costs twice as much as paper from other sources. But there is hope for the future: as world-wide demand rises for paper from sources which do not destroy precious forest areas, the price of kenaf paper will probably drop. In ten years' time, most of the books that you will read may have a note in the front saying 'Printed on kenaf'.

of other trees (rubber and palm oil), and for cattle ranches and enormous wheat fields, in order to fulfil the needs of the human population for more raw materials and more food.

The consequences of deforestation are potentially serious. Locally, clearing the land of trees and plants leads to increased soil erosion (see below). Globally, reducing the biomass of productive trees may contribute to the greenhouse effect (see page 581) as less carbon dioxide is used in photosynthesis. Also, loss of rainforest, one of the most diverse ecosystems of the world, is likely to affect humans and other organisms in ways we cannot predict, although we already know that many of the plants found in rainforests are useful sources of powerful drugs (see page 202).

Balancing the needs of people with those of the environment is difficult. Some of the steps that might reduce deforestation are listed in the tick box opposite, in Chapter 36. But, as the Feature box on this page illustrates, apparently ideal solutions are often not put into practice because of cost.

Desertification

In Chapter 35 we saw how soil is an important abiotic factor that determines the distribution of organisms in an ecosystem: few plants can grow without the anchorage of soil or the water and minerals it supplies to them. Fig 36.3 shows what happens when the soil is lost, a process called **soil erosion** that occurs worldwide.

Soil erosion is a natural process: wind and water movements move surface debris and topsoil from one place to another. In ecosystems that are undisturbed by human activity, healthy plant roots 'bind' the soil, and soil is lost and made at about the same rate. Problems occur when the soil is removed faster than it is formed: this can happen because of farming, deforestation, building, over-grazing by animals, physical damage by off-road vehicles or fire and some kinds of mining.

Farming is probably the worst culprit. Topsoil is eroding on about a third of the world's cropland. Each year, the land must feed 90 million more people with about 25 billion tonnes less topsoil. In Africa, soil erosion is now 20 times faster than it was 30 years ago. The cattle-producing areas are now severely affected by desertification (Fig ___ are solutions to soil erosion and desertification, ___ to be put into practice (Table 39.3).

Many countries world-wide are now working together to try to reduce deforestation. To cover all of the different projects would need another whole book, but here are a few:

People are being encouraged to manage forests sustainably. This means using different techniques to make sure we have the wood we need without completely destroying large areas of forest.

People are being persuaded to use paper substitutes.

There is growing investment in rainforest communities to find new medicines from plants.

Eco-tourism is being promoted. If people pay to see wildlife in its natural surroundings, this contributes to the income of an area and makes it less likely that habitats are destroyed.

Assignment

JOURNEY ALONG A RIVER

As Fig 39.A1 shows, rivers tend to be relatively unpolluted near to their source. As they flow on towards the sea they are contaminated with organic pollution and industrial effluent, creating increased levels of pollution and eutrophication.

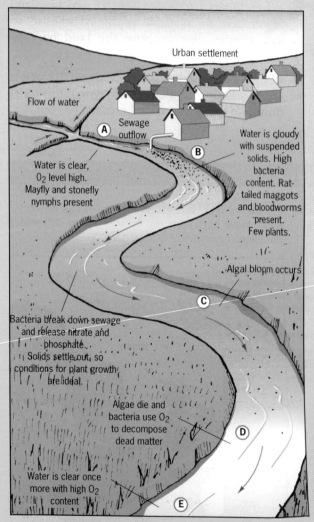

Fig 39.A1 **This diagram shows the different conditions found along a river**

It is often difficult to tell exactly how polluted the water is at any point, just by looking at it. We need to sample the water and sample the organisms that live there to find out whether the water is safe.

There are two easy methods which allow us to find out which organisms live in a particular part of the river. The first is to do a kick sample (see page 588). The number and diversity of organisms found are used to give the stretch of waterway a diversity index (see page 164).

1 What sort of diversity index would you expect to find in:
a) a stretch of unpolluted river, such as point **A** in Fig 39.A1?
b) a stretch of polluted river, such as point **B** in Fig 39.A1?

2
a) What factor is likely to be the most important in determining whether water can support a large or a small number of species?
b) What method would you use to measure this factor?

3 Initially, the water at point **B** is turbid (cloudy and murky). From your knowledge of the nature of sewage, explain why few animals or plants can grow in this zone of the river.

4 What do you think the BOD value may be for the water in this part of the river?

5 A student who was feeling particularly brave did a kick sample in this area and found very few species. They were the classic indicator species of heavy organic pollution: rat-tailed maggots, bloodworms and tubifex worms – see Fig 39.A2.

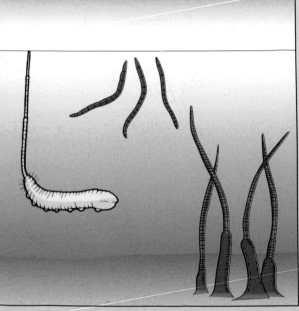

Fig 39.A2 **The larva of a rat-tailed maggot, some bloodworms and tubifex worms. The tails of the tubifex worms protrude from tubes that the worms make using mucus and debris**

a) What is an indicator species?
b) Explain why each of the three species shown in Fig 39.A2 is able to tolerate low oxygen levels.
c) When the water was tested in the laboratory, large numbers of *E. coli* bacteria were found. Using this and the other evidence, what would you conclude about the sewage outfall? (See page 5 for an extra...

More sophisticated investigations of the water further below the sewage outlet produced the data shown in Fig 39.A3.

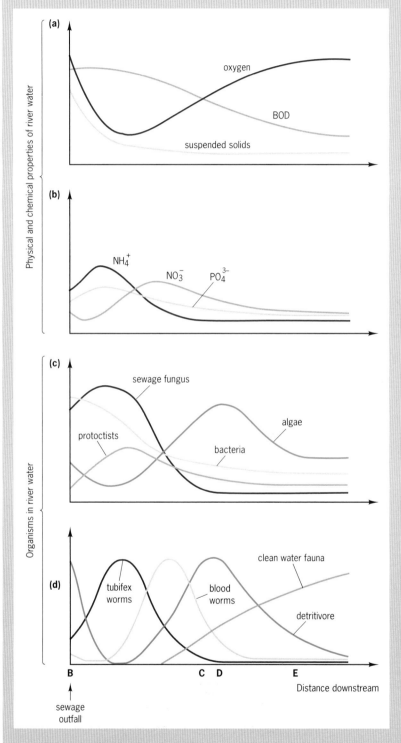

Fig 39.A3 **Changes in the river after a sewage outfall**

6 Look at Fig 39.A3(b).
a) Use your knowledge of the nitrogen cycle to explain why there is first a peak in the levels of dissolved NH_4^+, followed by a peak in NO_3^-.
b) The population of sewage fungus falls to zero, just before point **C**. What does this tell you about the metabolism of the sewage fungus?
c) Why does the level of bacteria not fall to zero?

7 At point **C**, the water begins to look like pea soup – conditions have become perfect for the growth of algae. Contrary to what you might first think, the problem here is that the water is too fertile – it has become eutrophic.
a) Use the graphs shown in Fig 39.A3 (a) and (b) to list three factors which are responsible for the algal bloom.
b) Explain why the algal bloom is harmful to the normal clean-water organisms.
c) Why do the populations of the different invertebrates present in the river water peak at different times?

8 Look at Fig 39.A3(d).
a) Find out the characteristics of tubifex worms and blood worms that allow them to flourish in the river between points **B** and **E**.
b) How do the detritivores help to change the water conditions between points **C** and **E**. Find out the names of some common detritivores that might be found in this sort of river.

Once the algae have absorbed the available nutrients, they start to die. Once again, bacterial activity becomes a problem, as decomposers rot the dead algae and use up large amounts of oxygen. After this, if there is no more organic discharge, the water further downstream begins to recover. Eventually, the water becomes clean once more.

Research and discussion questions

Find out about the basic principles behind sewage treatment. What major health problems does effective sewage treatment prevent?

Find out what the current laws are in the UK which control the discharge of sewage into rivers and into the sea. Do you think they are strict enough?

APPENDIX 1   THE CHEMISTRY OF WATER

There is no life without water: life almost certainly originated in water. Most biochemical reactions take place in solution and all organisms which live on land can do so only because they have their own internal aquatic environment. Some organisms can survive dehydration as spores or seeds, but these structures simply allow a state of dormancy which ends when water becomes available once again. The bodies of all organisms contain a high percentage of water. Even that most complex of organs, the human brain, is 85 per cent water.

The properties of water

All cells are tiny compartments of fluid. Living cells must exchange materials with the surroundings, and this can only take place in solution. So, as well as being full of water, cells are also surrounded by it. Water has unique physical and chemical properties, many of them significant in biology.

The water molecule: why is it sticky?

The water molecule consists of two atoms of hydrogen and one atom of oxygen. This gives water a relative molecular mass of 18. Many compounds of this size are gases: water is a liquid because its molecules are 'sticky' – they cling to each other.

Electrons are negatively charged. Since opposite charges attract and like charges repel, the lone pairs of electrons of the oxygen effectively repel the electrons of the hydrogen–oxygen bonds. This results in a 'v-shaped' molecule, as Fig A1.1 shows. The region around the oxygen atom has a slight negative charge, while the two hydrogen atoms have a slight positive charge. Molecules with different areas of positive and negative charge are said to be **polar**.

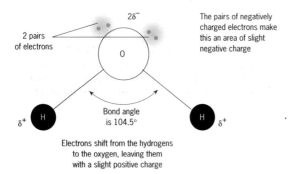

Fig A1.1 **Water is a polar molecule: it has distinct areas of positive and negative charge**

Attractive forces exist between water molecules: the positive charge on the hydrogen atoms of one molecule attracts the negative charge on the oxygen atoms of another molecule. These attractive forces are called **hydrogen bonds** (Fig A1.2). In effect, each water molecule acts as a 'mini magnet'. As well as 'sticking' to each other, water molecules also cling to other charged particles.

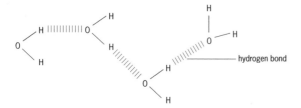

Fig A1.2 **Weak attractive forces exist between the positive areas of one water molecule and the negative areas of adjacent ones. Many of the unique properties of water are due to this hydrogen bonding**

Table A1.1 **Some properties of water**

Property	Significance in living systems
Cohesion	Hydrogen bonding causes water molecules to stick to each other. This is important, for example, when water evaporates from the leaves of a tall tree: water is so cohesive that a continuous column can be sucked up from roots which might be over 100 metres below. This cohesion causes **surface tension**. Water has a very high surface tension, and several organisms, such as pond skaters and rat-tailed maggots, take advantage of this property
The polar nature of water	This makes water a very good solvent – it has been called the 'universal solvent'. Though obviously not true since it doesn't dissolve everything, it does act as a solvent for all ionic and polar compounds
Melting and boiling point	Water freezes at 0 °C and boils at 100 °C and is therefore a liquid over most of the temperature range found on Earth
Specific heat capacity	A gram of water absorbs a lot (4.2 joules) of heat before its temperature rises by 1 °C. Organisms have a high water content and so can absorb a lot of energy (or lose a lot) before their temperature changes *significantly*. This makes water a useful **thermal buffer**. Most organisms cannot tolerate great changes in temperature, and the water in their bodies protects them from rapid changes in external temperature
Latent heat of vaporisation	The hydrogen bonding between water molecules makes it difficult for them to separate from each other and **vaporise** (evaporate) to form a gas. When water does evaporate, the escaping molecules take a lot of energy with them, and so vaporisation is a very efficient cooling process. Anyone who has ever stepped out of the shower and stood in a draught will know just how efficient! The processes of sweating, panting and transpiration all involve cooling by evaporation
Thermal conductivity	A naked person sitting in a room at 25 °C would find it comfortable. However, if immersed in water at 25 °C, the same person would feel cold. This is because the thermal conductivity of water allows allows heat to be transferred away from the body more quickly through the water than through air. Materials with a low thermal conductivity are good insulators. Fur is a good insulator because it traps air, but the insulating properties are lost if the fur gets wet

What happens when a substance dissolves in water?

Simple ionic substances such as sodium chloride dissolve readily in water. The sodium chloride completely **dissociates** (breaks up) into sodium ions and chloride ions. This is because the attractive forces between the water molecules and the ions are greater than between sodium ions and the chloride ions. The negatively charged area of water molecules and the positively charged sodium ions (Na^+) are attracted to each other. Water molecules surround each sodium ion, forming a sort of 'shell'. In a similar way, the positively charged parts of water molecules and negatively charged chloride ions (Cl^-) attract each other and Cl^- ions become surrounded by a 'shell' of water.

Many large molecules like proteins are too large to dissolve in water, but they do possess many polar groups on their surface. They attract a layer of water molecules which prevents them from settling out. We describe this as a **colloid**; each molecule in a colloid can be thought of as a small particle of solid suspended in a liquid. The water molecules which surround each large molecule are effectively 'tied up' and not free to move away from it. This is important in biological systems: these 'tied' water molecules exert what is known as a **colloid osmotic potential**: they affect the movement of water in or out of a system. In the mammalian circulation, for example, water drains back into blood due to the colloid osmotic potential exerted by the plasma proteins (see page 285).

APPENDIX 2 THE CHEMISTRY OF CARBON

Living things are composed mainly of organic molecules: molecules which have 'skeletons' of carbon atoms. This element can bond with itself repeatedly, to produce an infinite variety of molecules.

The carbon atom

Atoms have electrons in shells. Two electrons make the first shell full; eight electrons fill the second. Atoms with a full outer shell are stable. Atoms with a less than full outer shell are reactive and combine with others in order to fill the outer shell.

Carbon has an atomic number of 6: it has a nucleus containing 6 protons and 6 neutrons, orbited by 6 electrons, 2 in an inner shell, the other 4 in an outer shell (Fig A2.1). Carbon acquires a full, and therefore stable, outer shell of 8 electrons by sharing 4 more. So, carbon forms 4 covalent bonds - bonds with electrons shared between the atoms. We say it has a **valency** of 4.

Fig A2.1 **Dot and cross diagrams showing two simple carbon compounds.**

(a) Methane   **(b) Ethene**

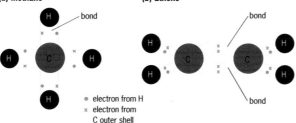

• electron from H
× electron from C outer shell

Notice how the carbon atom shares an electron with each of four hydrogen atoms to achieve a full outer shell of eight electrons

The two carbon atoms in this compound achieve a full outer shell of eight electrons by sharing electrons with hydrogen atoms and by forming a double bond

Carbon can form covalent bonds with other carbon atoms to form stable chains or rings. C–C bonds are the most common; compounds which contain single carbon bonds are said to be **saturated**. C=C bonds are also frequent and C≡C are possible. Compounds with double or triple bonds are said to be **unsaturated**. This means that they are not saturated with hydrogen (more hydrogen atoms could be added at their multiple bonds). This is particularly relevant with respect to the fatty acids – see page 46.

Carbon atoms commonly form covalent bonds with hydrogen, nitrogen, oxygen, phosphorus and sulphur. These elements make up the vast majority of organic molecules.

What is an organic molecule?

It is difficult to come up with a precise definition for an organic molecule but, for our purposes, we can say that it is one which contains carbon and hydrogen. Some organic molecules are enormous, with relative molecular masses in the millions. Table A2.1 shows some common examples of simple organic and inorganic compounds.

Table A2.1 **Some simple organic and inorganic compounds**

Inorganic	Organic
Water (H_2O)	Ethane (C_2H_6)
Nitrogen gas (N_2)	Ethanol (C_2H_5OH)
Ammonia (NH_3)	Ethanoic acid (CH_3COOH)
Carbon dioxide (CO_2)	Glucose ($C_6H_{12}O_6$)
Hydrogen sulphide (H_2S)	Amino acid ($CHRNH_2COOH$)
Nitrate ion (NO_3^-)	

In this book, mainly in Chapter 3, we look at four types of large organic molecules commonly found in living things: **carbohydrates**, **lipids**, **proteins** and **nucleic acids**.

Where do organic molecules come from?

'From thin air', could be the answer. The majority of organic molecules are originally made by plants which, through the remarkable process of photosynthesis, take carbon dioxide out of the air (or water) and use it to build up sugars, starch, cellulose, lipids, proteins and nucleic acids. In effect, light energy is 'trapped' in the chemical bonds which make up these organic molecules. Most other organisms obtain their organic molecules from plants, either directly or indirectly.

Organisms use organic molecules in two ways:

● They rearrange and incorporate them into their own bodies.

● They **respire** them – break them down to release energy by the process of cell respiration (see Chapter 8). This process is essentially the reverse of photosynthesis (see Chapter 9).

ACKNOWLEDGEMENTS

Grateful thanks are due to the following for their valued assistance during the preparation of this book: Bill Indge, Dr Peter Newham of Zeneca Parmaceuticals, the Biochemical Society, and Cellmark Diagnostics.

The publisher thanks examination boards for their permission to reproduce examination questions. Questions are acknowledged as follows:
AEB   The Associated Examining Board
NEAB   Northern Examinations and Assessment Board
UCLES University of Cambridge Local Examinations Syndicate
ULEAC University of London Examinations and Assessment Council

Every effort has been made to trace the holders of copyright. If any have been overlooked, the publisher will be pleased to make the necessary arrangements at the earliest opportunity.

Photographs
We are grateful to the following for permission to reproduce photographs:

A–Z Botanical Collection, 33.6 right (M Nimmo); ACE, 3.3d (Y Levy Phototake); 3.12 (A Hughes); Allsport, p.38 (J Gensheimer), 8.7, 8.A1 and 8.A2 (M Powell); Animal Physiology Unit, Babraham, Cambridge, 1.Q1; Heather Angel 1.25, 8.2, 8.19, 9.3, 10.4, 11.17a right, 11.18, 11.23, 12.9, 12.13b and c, 12.16, 12.24, 13.2 upper and lower, 13.6a, 13.6b 4th, 13.7 left and right, 13.9, 13.10, 13.12, 13.13, 15.15, 18.47, 28.1, 29.3, 29.15 centre, 35.3, 35.6d; Aquarius Library, 34.1; Ardea 8.1 lower left (DW Greenslade), 10.8 upper right (A Lindau), 11.24 right (JL Mason), 12.23 (M Ijima), 17.6c (J Clegg), 22.1 (J-P Ferrero), 27.14 (S Roberts), 29.9a (I Beanes), 29.22 (J Mason), 35.2 left, p.560 (J-P Ferrero), 36.2 left (P Morris), 36.3 (J-P Ferrero), 36.17 (C Haagner), 36.18 (Weisser), 36.19 lower (Kjaer), 38.7 (D & K Urry); Associated Press, p.188; John Barber, Cynthia Clarke, TD Allen, Paterson Institute 1.1; BBC Natural History Unit, 10.3 (J Foott), 18.37 (J Freund), 22.8 (LM Stone), 23.13 lower left (TD Mangelsen), 23.13 upper centre (P Coulter), 23.13 lower centre (A Shah), 24.15 (M Barton); BDH, 4.1; Beechams Pharmaceuticals, p.172 upper and lower; Biofotos, p.202 (B Rogers), 36.11 (J Venus); Biophoto Associates 1.3 left, 1.3 right, 1.9, 1.14.1–4, 1.15, 1.16, 1.18, 1.20, 1.29, 1.31a–d, 1.34, 1.35, 1.A1, 1.A2, 2.3 left, 2.7 upper, middle, lower, 3.3c, 3.13, 3.15d, 3.20, 3.22, 5.18 left, right, 5.19b, 5.20, 5.25, 5.A1, 6.1a, 6.6, 6.9a–e, 6.10, 6.13a–h, 6.16b, 7.6 lower, 7.10a and b, 8.1 lower right, 8.14 left, 8.26, 10.8 middle left and lower right, 11.11, 11.17a left, 11.21, 13.1b, 13.6b: 1st, 2nd, 3rd, 17.6d, 17.7d, 17.27, 17.29, 18.4b, 18.14, 8.19, 18.23a, 18.32, 18.33b, 18.39, 19.11b, 19.17, 19.Q2, 20.1, 20.4, 20.5, 20.7, 20.15 left, 21.20, 23.3c, 23.6, 23.13 upper left, 24.9, 25.3, 25.14e, 25.16, 26.1b, 26.3a, b, d, 26.6, 29.6, 29.9b, 29.10, 29.13c, 29.16, 29.19 left and right, 37.3, 38.6, 38.14, 38.16 lower; John Birdsall 17.3, 25.11 upper, 26.3c, 32.1; Anthony Blake 3.21 (G Buntrock), 27.1 (B Limage); Mike Boyle, 26.A2, 30.2 centre; Bubbles 22.1 left and lower, 22.A2; Cephas, 8.21 upper right (FB Higham), 8.21 upper left (M Rock), 8.21 lower right, 8.23 (M Rock), 8.24 (S Boreham); Collections/Shout, 14.2; Bruce Coleman 7.5, p.148 (J Taylor), 18.28a (K Taylor), p.362 (W Layer), 22.1; right (F Labhardt), 31.8 upper left (K Taylor), 31.8 upper right (A Purcell), 31.8 lower left (J Grayson), 31.8 lower right (F Labhardt), 31.13b (K Taylor), p.522, 33.8 (A Bacchella), 33.14a (J Foott), 33.14b (H Lange), 33.17 left (SJ Krasemann), 33.19 (K Taylor), 37.4 left (F Sauer), 37.4 right (D Davies), 38.11 and 38.12 (K Taylor); Edelman PR/Eli Lilly, 30.2 bottom left; Mary Evans Picture Library, 10.7 (A Roslin), 15.5, 32.14; Environmental Images, 39.6 upper right and 39.6 lower left (R Brook), 39.13 left (P Glendell), 39.13 right (R Brook), 39.14 (P Glendell); Vivien Fifield, 33.3; Geoscience Features 3.5, 3.15b, 9.A1, 11.6, 11.13a, 11.16, 12.26, 13.8, 15.8b, 15.8e (B Booth), 18.33a, 33.6 left; Peter Gould, 3.14 (3 photos); Prof D Grierson Nottingham Univ, 34.7 left and right; Green Moon, 25.6b; Prof. RJ Hay, Guy's & St Thomas's Med & Dent Sch, 11.9; Holt Studios International, p.iv upper right (N Cattlin), 18.A1 (N.Cattlin), 20.6c, 20.12 (B Gibbons), 20.13 (N Cattlin), 28.10 (P Peacock), 29.15a, 29.18 (N Cattlin), 29.21 (N Cattlin), 30.2 centre right (N Cattlin), 37.18 (N Cattlin), 39.1 (G Roberts), 39.4 (N Cattlin), 39.5 left (R Anthony), 39.5 centre and right (N.Cattlin); Geoff Howard, 21.1; Hulton Deutsch, p.28 upper and lower; Image Bank, 30.4; Kobal Collection, 31.13a; Andrew Lambert, 3.9, p.64, 8.3, 15.3; Frank Lane Picture Agency, 2.1 (S McCutcheon), p.161 (T Whittaker), 24.5a (R Wilmshurst), 24.14 (M Rose), 24.16 (R Wilmshurst), p.398 (W. Wisniewski), 25.5a (L Batten), 25.9 W Wisniewski, 28.4 (T Davidson), 33.4 (R Wilmshurst), 33.5 (S Hosking), 33.16 (E & D Hosking), 33.17 right (T Whittaker), 36.2 right (D Hosking), 36.5a (DP Wilson), 36.6 (B Borrell), 36.13 (A Wharton), 36.14 (M Gore), 38.3 (R Wilmshurst), 38.10 (P Perry), 38.17 (B Dembinsky); Life Science Images, 5.19a, 18.23b, 21.7; Meteorological Office, 36.15 left (JFP Galvin); Microscopix Photo Library, 15.8f , 17.10c, 17.15a, 17.23 (all 4: A Syred); Muscular Dystrophy Group, 32.15; NASA Johnsons Space Center, p.306; National Marine Mammal Laboratory, p.572; National Medical Slide Bank 5.5, 25.11 lower, 26.9, 31.11; Natural History Museum 21.33, 33.1; Natural History Photo Agency 12.20; Nature and Science, 5.17 (J Aribert), 11.24 left, 12.15b upper and lower, 13.15 (J Aribert), 17.31, 27.5; Oxford Scientific Films, 3.8 (GI Bernard), 3.15a (T Martin), 3.15c (S Osolinski), 3.15e (T & P Leeson), 3.17 (O Newman), 7.1 (GI Bernard), 7.3 (T Shepherd), 7.4 (R Blythe), 7.7 (K & D Dannen Photo Res Inc), 12.21 (M Deeble & V Stone), 12.22 (RA Tyrrell), 12.25 (A Root), 15.9 (JH Robinson), p.252 lower (T de Roy), 17.6b (P Kay), 17.6f (GI Bernard), 17.10b (T Shepherd), 18.26 (D Broomhall), 19.1a (P Parks), 19.2a (DJ Cox), 23.1 (S Stammers), 23.2 (JAL Cooke), 23.3a (G Gardener), 23.7 upper (LE Lauber), 23.7 lower (J Cooney), 24.1 (GI Bernard), 24.3 upper (A Ramage), 24.5b (E Neal), 24.8 (N Wu), 24.13 (W Shattil, B Rozinski), 35.1 (M Hill), 35.4 (Z Leszczynski), 35.6c (DB Fleetham), 35.9 (R Redfern); AM Page, 1.23, 1.32; Panos, p.124, 39.2 (B Klass); Papilio Photographic, 29.13a, 36.4; Dr C Paterson, Ninewells Hosp Dundee, 3.38; Planet Earth, 5.A2 (J Lee), 5.A3 (A Stevens), 7.6 centre (Purdy & Matthews), 7.6 upper right (T Dressler), 8.1 upper right (A Stevens), 8.A3 (J Scott), 9.9 (M Conlin), 12.14b (P David), 12.14c (J Seagrim), 12.17 (M Snyderman), 12.18 (P Oliveira), p.220 (G Bell), 15.7 (R Arnold), p.252 upper (G Douwma), 17.2 (P Chippendale), 17.5 (P Oxford), 17.6a and e (K Lucas), 17.7a (B Wood), 17.30 (J Lythgoe), 18.1 (M Conlin), 19.1b (A Kerstitch), p.320 (Space Frontiers), 21.16 (M Conlin), p.374 (P Rowlands), 23.13 right (S Bloom), 23.16 (JP Scott), p.388 24.3 lower (J Lythgoe), 24.12 (P Scoones), 25.2 (P Sayers), 25.A1 (R Salm), 28.3 (J Greenfield), 28.6 (G Douwma), 28.7 (R Matthews), 28.8 (D Perrube), p.466 (M Clay), 33.14c (M Snyderman), 33.14d (FJ Jackson), 35.6a (P Plailly), 35.6b (HC Heap), 36.5b (C Weston), 36.10a (WM Smithey), 36.10b (P Osford), 36.16 (H Charles), 36.19 upper (J Lythgoe), p.586 (I Menushina), 38.13: upper (A Bartchi), lower left (N Garbutt), lower right (M Gore), 38.19 (Mike Read), p.602 (J Mackinnon); PPL Laboratories, p.542; Professional Sport, 22.A1 (T Hindley); Redferns, p.504; Carol Redhead et al, 21.6; Roslyn Institute, Edinburgh, 6.1b, 6.1c, 34.6; Science & Society, 17.18; Science Photo Library, cover (P Dayanandan), p.iv left (B.P.Wolff), p.iv lower right (Penny Tweedie), p.1 (A B Dowsett), p.2 (Manfred Kage), 1.4 left (CNRI), 1.4 right (M.Wurtz/Biozentrum, Basel Univ), 1.5 (CNRI), 1.7 Moredul Animal Health Ltd, 1.11 (J Burgess), 1.13 (D Fawcett), 1.26, 2.3 right, 2.5 (P Plailly), 2.6 (B Nelson), 2.A1: upper (S Stammers), middle (M Devlin), lower (John Radcliffe Hosp), 2.A3, 2.A4 (BSIP Leca), p.38 (S Fraser RVI Newcastle upon Tyne), 3.3b, 3.3e, 3.28 (P Plailly), 3.39 (J King-Holmes), 3.40 (A Lesk), 4.2 (JC Revy), 4.8 W Baumeister, 4.21 (Saturn Stills), p.80 (E Nelson Custom Med Stock), p.100: upper (GJLP-CNRI), lower (P Marrazzi), 6.2 (PM Motta, J van Blerkom), 6.3 (D Scharf), 6.4 (M Sklar), 6.5, 6.7 (CNRI), 6.16a (G Parker), 6.17(b) (J Ashton), p.115 (S Fraser), p.116 (T van Sant Geosphere Proj Santa Monica), 7.2 (PM Motta, KR Porter, PM Andrews), 7.6 upper left (J Howard), 7.12 (St Bartholomew's Hosp), 8.1 upper left (M Dohrn Roy Coll Surgeons), 8.5 (JC Revy), 8.14 right (K Porter), 8.15 (AB Dowsett) 8.22 (M Kage), 8.25 (Sue Ford), 9.16, 10.2 (D Parker), 10.8: upper left (C Bjornberg), lower left (M Read), 11.5 (D Scharf), 11.7a and c (CNRI), 11.14: left (P Menzel), right (V Fleming), 12.7, 12.8, 12.10 and 12.11 (S Stammers), 12.12 (A Crump TDR WHO), 12.13a (A & H-F Michler), 12.19 (M Read), p.215 (Petit Format), 15.1 (M Abbey), 15.8: a (J Burgess), c (AB Dowsett), d (A Syred), p.236, 16.4, 16.7, 16.8: a and b (KFR Schiller), c (M Kage), d (CNRI), 16.13 (A Kage), 17.11a (M.Dohrn, R.C.Surgeons), 17.16 (P Tweedie), 17.21 (J Burgess), 17.A1 (D Scharf), 17.A2 (J Durham), 18.4b (Bioph Assoc), 18.4c, 18.6 (CNRI), 18.10 (D Lovegrove), 18.18 (RU Chandran TDR WHO), 18.28b (D Nunuk), 18.46, 19.1c (D Roberts), 19.2: b (J Burgess), c (M Walker), 19.8 (P Plailly Eurelios), 19.10: b (A Syred), c (PM Motta Dept Anat Univ la Sapienza Rome), 19.19 (P Menzel), 20.3 (R Kirby, D.Spears), 20.6a (JC Rew), 20.15 right (S Stammers), 20.16 (J Burgess), p.322 (A Hart-Davis), p.346 (J Selby), 21.34 (P Menzel). 21.35 (J Holmes), 22.3 (HC Robinson), 23.3b (A Syred), 23.9 (P Jude), 25.6a (R Clark, MR Goff), 25.13 (M Clarke), 25.14a (GCa CNRI), p.414, 26.1: a (BP Wolff), c (Omikron), d (V Deenen), 26.3d (Bioph Assoc), 26.4: a (NIBSC), b (PM Motta, G Macchiarell, SA Nottola), 26.7 (V Bradbury), 26.8 (A Brody), 26.11 (D Fawcett), 26.12b (JC Revy), 26.14 (Saturn Stills), 26.17 (G Jerrican), 26.18 (J King-Holmes), p.430 (SIU), 28.2 (C Nuridsany, M Perennou), 28.12b (PM Motta Dept Anat Univ la Sapienza Rome), 28.17 (S Terry), 28.21 (H Young), 28.25 (M Clarke), 28.A1 (H Morgan), 28.A2 (A Walker, Midland Fertility Services), 28.A3 (A Tsiaras), 29.11 and 29.15c (J Burgess), p.479 (K Seddon, T Evans, Queens Univ. Belfast), 30.1 (CNRI), 30.2 top and bottom right (P Menzel), 31.3(b) (Prof O Miller), 31.A1 (G Murti), 31.A2 (Omikron), 32.2, 32.7 (A & HF Michler), 32.12 (CNRI), 32.16 (A Hart-Davis), 33.2 (Nat Lib Med), 33.10 (T Craddock), 33.20 (P Marazzi), 34.5 (J Holmes Celltech), p.553 (M Read), 38.8 (S Fraser), 38.14 upper (E Grave), 38.16 upper (Moredun Animal Health), 39.6 upper left (S Fraser), 39.6 lower right (Y Hamel), 39.7 (T Craddock), 39.9 (S Fraser), 39.11 (NOAA), 39.12 (P Marazzi); South American Pictures, 8.21 lower left; Still Pictures, p.276; John Walmsley, 35.2 right, 36.15 right, 38.20; Zeneca Agrochemicals & Seeds, 34.7 upper right

ANSWERS TO SELF-TEST QUESTIONS

Chapter 1
A Mice, attracted by the smell, came to eat the wheat.

B **(a)** The full stop is about 0.3 mm = 300 μm. **(b)** 1 million.

C Dispersal of oil pollution, breakdown of waste plastic.

D 1000.

E They are all secretory.

F After (because the ER is a network of thin membranes which would not show up under a light microscope).

G All its tissues would be destroyed.

H We could destroy tumours.

I Because they could digest the organelle/cell which made them.

J **(a)** 500 **(b)** 200 **(c)** 5000.

K The root system, because it receives no light.

Chapter 2
A Ciliated epithelium.

B Smooth muscle because its contraction is slowest.

C It fatigues too quickly.

D The reproductive system.

Chapter 3
A Sucrose consists of glucose and fructose, both reducing sugars. In the sucrose molecule, the reactive groups of the two component sugars point inwards and so are 'hidden' in the three dimensional structure. They only become available to react with Benedict's solution when the sucrose has been split apart.

B The glycosidic link, which is between carbon 1 and carbon 4.

C Links: sucrose 1,2; lactose 1,4.

D Drinking artificial milk in which lactose had been replaced by glucose.

E It would cause an osmotic imbalance – the cytoplasm would become too concentrated.

F Because few organisms have the enzymes that can degrade cellulose.

G Oleic and linoleic acid.

H Oleic and linoleic. They are both unsaturated.

I 20 x 20 x 20 = 8000.

J 46 x 5 cm = 2.3 metres.

K Size, bases, lifetime, sugar.

L **(a)** TAGCAATGG. **(b)** UAGCAAUGG.

M Proline, serine, threonine.

N Stop.

Chapter 4
A Protein synthesis.

B Protein, lipid, nucleic acids.

C Because they all contain the same food type: starch.

D Tears, sweat, in the lysosomes of some white cells.

E It stimulates glycogen production.

F Peptide bonds.

G Proteins are denatured. This is irreversible.

H Because that is the optimum temperature for enzymes.

I It would be denatured by the acid.

J Because the enzymes they work with are present in tiny amounts.

K Vitamins are compounds made from complex molecules; many vitamins act as coenzymes. Trace elements are elements; many trace elements act as cofactors for enzymes.

L Competitive inhibitors compete with the normal substrate for the active site; non-competitive inhibitors bind away from the active site. Competitive inhibitors do not affect the ability

of the enzyme to bind to the normal substrate; non-competitive inhibitors modify the enzyme so that it cannot recognise the substrate.

M There would be no activity at all: the plot would go along the x-axis.

Chapter 5
A 1 mm.

B Water-loving, water-hating. In a phospholipid, the phosphorus end is attracted to water, the lipid end repels water.

C Because the atoms/molecules in solids aren't free to move.

D To maintain the diffusion gradient.

E Lung, intestine, leaf, kidney, placenta, root.

F Because there can never be a less concentrated solution, pure water will never absorb water by osmosis.

G Because water passes out of the bacteria by osmosis, dehydrating and killing them.

H You must have a solution to compare it to: Hypertonic to what?

I Its cytoplasm is isotonic with sea water, so it doesn't need one.

J Cyanide blocks aerobic respiration, so virtually no ATP is available.

K Solid particles cannot pass through the cell wall.

Chapter 6
A DNA contains the genetic code, RNA is built up on it, then RNA is used as the instructions for making proteins.

B $n = 23$.

C G_1, S and G_2.

D The amount of DNA would be halved at each division.

E **(a)** In ovaries and testes respectively. **(b)** Because it reduces the chromosome number (by half).

F In a human cell there are 23 bivalents.

G Older people's cells have undergone more cell divisions, so more chance for mistakes to be made which result in cancerous changes.

Chapter 7
A Small animals have a large surface area:volume ratio and so lose heat faster than larger organisms. They compensate by using most of the energy produced by respiration to maintain their body temperature.

B We still need to maintain body temperature, breathe, keep our hearts beating and keep cellular processes going. All this requires energy.

C False. Some heterotrophs eat other heterotrophs (eg lions eat gazelles).

D **(a)** It fixed carbon from carbon dioxide by photosynthesis. **(b)** From respiration. **(c)** Most lost as heat. The main compounds produced are carbon dioxide and water.

E When the temperature of the surroundings is the same as body temperature.

F **(a)** Their energy requirement goes down but they probably don't eat less, therefore more of their food energy is stored as fat. **(b)** About 800 kj per day.

G **(b)** Carbohydrates: 12 500/17 = 735 g. **(c)** Lipids: 12 500/38 = 329 g

Chapter 8
A More gas can be exchanged per unit time. Thin-walled, moist, good vascular network.

B Organic 'fuel' is oxidised to release energy and do useful work, and it produces heat as waste.

C It would take too long to break glucose down. ATP releases energy when one phosphate group is removed.

D The first uses a substrate as phosphate source, the second uses free phosphate and is powered by energy gained from redox reactions in the electron transport chain.

E An extra electron.

F It goes into the blood, is taken to the lungs and is breathed out.

G Because each glucose provides two pyruvates and therefore two acetyl-CoAs.

H These more metabolically active cells need more ATP to power their processes and therefore need more mitochondria (the sites of ATP synthesis in the cell).

I Active transport requires energy.

J **(a)** Each one of the 20 amino acids has a different number of carbon atoms. In respiration, the reactions which break each one down produce different volumes of CO_2 and use different volumes of O_2. **(b)** CO_2 is produced but no O_2 is used: any value divided by 0 is infinity.

K It prevents them being eaten.

M 50 units.

N The end-products are different (though the processes are similar).

Chapter 9
A Cyclic: the electrons flow in a loop. Photo: the process requires light. Phosphorylation: ATP is formed from ADP and P_i.

B In the stroma.

C Glyceraldehyde-3-phosphate + 9ADP + 8P_i + 6NADP.

D Glyceraldehyde 3-phosphate.

E X light, Y carbon dioxide, Z temperature.

F **(a)** Temperature **(b)** Between 20 ˚C and 25 ˚C.

G The amount of water needed for photosynthesis is tiny compared with the amount in the cell – by the time water limited photosynthesis, the cell would have dried up.

Chapter 10
A Yes: Fig 10.2 shows far fewer vertebrate species.

B The oak wood.

C **(a)** Feathers, shelled eggs, wings. **(b)** Live in water, scales, fins. **(c)** Inconspicuous flowers, parallel leaf veins, one cotyledon.

D The Carnivora.

E Breeds of dogs or other domestic animals; races of *Homo sapiens*.

F They take a characteristic that divides the first group of organisms into two or more subgroups, and continue to subdivide each group until a final characteristic identifies a species.

Chapter 11
A As the host cell divides it produces more copies of the virus as well as more copies of itself. All new cells contain the virus hidden in their DNA, so the virus gets into more cells without needing to actively reproduce.

B Blue-green bacteria produced large amounts of oxygen which was poisonous to many early bacteria.

C The bacteria die and *Pelomyxa* cannot survive without them.

D It has a membrane-bound nucleus.

E The fungus.

F The mycorrhizal network connects the trees.

G The food made by photosynthesis in the mature pine trees.

Chapter 12
A Animal cells do not have a cell wall, fungal cells do; animal cells have undulipodia, fungal cells do not; animals do not produce spores or hyphae, fungi do; animals have a nervous system, fungi do not.

B It produces medusae which are free-swimming.

C To avoid being moved by peristalsis and egested.

D Possession of a coelom, symmetry, three body layers.

E Food doesn't get mixed up with waste and the sense organs of the head are not fouled.

F Earthworms have some bristles (Oligochaetae = few bristles), but ragworms have many more (Polychaetae = many bristles).

G Insects: phylum Arthropoda, class Insecta.

Chapter 13

A Damp conditions.

B A tall plant can compete better for light. To grow tall, a plant must have supporting structures and an efficient transport system.

C Spores are haploid so must be formed by meiosis.

D **(a)** To prevent water loss. **(b)** Conifers can grow in regions that are colder because their wood is better able to withstand frost damage.

E Birds, mice, and rabbits eat seeds which pass through their bodies undigested and later drop to the ground in faeces. The animal may be several miles away from the original plant when this happens.

F Liliidae: daffodil, grass, yucca. Magnoliidae: fuchsia, oak, beech.

Chapter 14

A No. It has none of the characteristics of life (even though the tree was once alive).

B A Ferrari can move, burn fuel and 'excrete' waste, but it does not grow, reproduce, or respond independently to its surroundings and it is not made up from living cells.

Chapter 15

A They get fat.

B An animal that eats only meat: a carnivore.

C **(a)** Rice. **(b)** Inhabitants in the cold Arctic would have mainly a fish and meat diet (very poor in carbohydrates).

D The symptoms are those of vitamin A poisoning.

E Mussels don't burrow: they attach to rocks and don't need a long siphon.

F Sediments eaten by deposit feeders are mainly sand and mud – the small amount of organic material they do contain is often the indigestible parts left by other animals.

G Saprotrophs are mainly microscopic and digest food externally. Scavengers are macroscopic and digest food internally.

H Microphagous feeders: **(b)** baleen whale (the largest of all filter-feeders), **(c)** garden snail, **(g)** barnacle (filter-feeder), **(h)** earthworm (detritivore). Macrophagous feeders: **(e)** magpie, **(j)** locust. Fluid-feeders: **(a)** spider, **(d)** humming-bird, **(f)** cat flea, **(i)** leech.

I Meat has a much higher energy content than plants so carnivores need food less often; hunting takes a lot of effort and social cooperation, so large meals eaten less often are more efficient.

Chapter 16

A **(a)** Entrance blocked by soft palate. **(b)** Entrance blocked by epiglottis. **(c)** Way back out blocked by the curled tongue.

B **(a)** You would expect rennin production to decrease with age. It does. Young mammals on milk diets need it but older animals do not. **(b)** No, since only mammals produce milk.

C Almost 11 hours (6.35 m at a rate of 1cm min^{-1} = 635/60 = 10.58 h).

D So they do not self-digest the organs that produced them.

E An endopeptidase acts in the middle of the polypeptide, breaking it into smaller fragments; an exopeptidase works at the ends of a polypeptide, stripping off amino acids.

F The small intestine is a very absorptive region and can take in many of the nutrients produced by bacteria; only a small amount of absorption takes place in the colon, so nutrients produced here will just be lost in the faeces.

Chapter 17

A The gills collapse: their surfaces stick together like the bristles of a wet brush.

B 7.5 litres min^{-1}.

C One five thousandth.

D Dead space would be enormous and diver would breathe the same air in and out. Also, the chest muscles are not strong enough to keep expanding the lungs against the water pressure at that depth.

E The spongy cells are further from the illuminated surface of the leaf, so chloroplasts in them are less effective for photosynthesis.

F Large surface area; specialised cells form stomata to allow gases in and out; leaves are thin, so gases don't have to diffuse very far.

G Stomata close around noon to prevent water loss. Less water loss means less evaporation and less cooling effect at the leaf surface.

H Thick cuticle and several epidermal cell layers, leaf hairs, stomata in large pits on underside of leaf.

I Non-green tissues carry out respiration but not photosynthesis. They therefore take in oxygen and release carbon dioxide.

Chapter 18

A **(a)** 6 and **(b)** 15 litres per minute.

B **(a)** Increase. **(b)** Decrease.

C 120 mmHg = systolic pressure, 70 mmHg = diastolic pressure.

D **(a)** Increase. **(b)** Decrease.

E Hb 'steals' oxygen from an oxygen-rich environment and delivers it to tissues that are oxygen-poor.

F Ringing damages the phloem, cutting off food supply to the roots which die.

G Dead – they have no cytoplasm.

H The molecules would be pulled apart easily and the column of liquid would be broken.

I Xylem vessels containing air bubbles are useless because the water is no longer a continuous stream and cannot be pulled upwards.

J **(a)** Root pressure. **(b)** Very high humidity.

K **(a)** up **(b)** down **(c)** up **(d)** up **(e)** down.

L Metabolic energy - used in pumping sugars in and out.

M Aphids drip 'honeydew'. Excess sugar in the aphid's diet emerges from the anus undigested.

Chapter 19

A They have to support more weight.

B The spinal cord.

C Ligaments connect bones together at joints; ligaments need to be flexible to allow the joined bones to move.

D A tendon joins a muscle to a bone As the muscle contracts it exerts a force on the bone, moving it. If the tendon were elastic, the bone would move less than the change in muscle length.

E **(a)** smooth **(b)** smooth **(c)** skeletal **(d)** cardiac **(e)** skeletal.

Chapter 20

A Cells start with a primary cell wall. Later, more cellulose can form a strengthening secondary wall inside it. Secondary thickening is the formation of new xylem and phloem from cambium in the shoot and root.

B A branch increases in diameter each year, so last year's bark is outgrown.

C Auxin has diffused into it from the cut tip.

D **(a)** and **(b)**

E Inserting the mica sheet.

Chapter 21

A Stimulus: smell; receptor: receptor cells in nose; coordinator: brain; effector: muscles in arm and hand; response: food taken to mouth.

B **(a)** Zero voltage, because there is no potential difference. **(b)** To determine the difference in electrical charge inside and outside the cell.

C **(a)** To ensure that a neurone responds to a specific minimum level of stimulation. **(b)** Continuous depolarisation and fatigue of neurone.

D **(a)** Closed. **(b)** A closed ion channel prevents movement of ions. **(c)** Concentration gradient: high concentration of Na$^+$ ions outside the cell means that sodium flows into the cell down its concentration gradient when the ion channels open. Electrochemical gradient: the positively charged sodium ions are also attracted into the cell by the negatively charged protein molecules inside.

E Normally, as a nerve impulse travels, sodium channels close behind the wave of depolarisation. This sets up a refractory period and prevents the action potential moving back. If an action potential starts in the middle of an axon, it is free to travel in either direction.

F The larger the diameter of an axon, the faster it can conduct action potentials Squid giant axons conduct impulses very quickly: this enables the animal to have rapid escape responses.

G Myelinated nerves conduct impulses faster than unmyelinated ones.

H Substances need to diffuse across it from one neurone to the next: the shorter the distance they have to go, the faster the transmission of the impulse.

I The synaptic bulb has run out of neurotransmitter molecules.

J There will not be any new nerve impulses but the neurone is facilitated (ie more receptive to incoming signals).

K Acetylcholine is the neurotransmitter produced during 'fear, flight and fight' responses: nicotine binds to the same receptor sites.

L In the human brain there are 10^{10} neurones. If we assume that 10^7 make synapses with 10^4 neurones, the brain as a whole therefore makes 10^{12} connections per second (10^7 x 10^4 x 10^1 = 10^{12}).

M Sneezing, coughing, pupil constriction in bright light.

N Different segments of the body have different jobs. Those which do more work (eg thoracic ganglia) are larger. The cerebral ganglion is the largest because most of the animal's sense organs are in the head.

O **(a)** Hypothalamus. **(b)** Medulla oblongata. **(c)** Cerebral cortex.

P The sympathetic and parasympathetic divisions of the ANS both control breathing.

Q The physiological responses associated with emotional displays of anger, fear and embarrassment.

Chapter 22

A The digestive juices undiluted by food would attack the stomach lining. This could cause an ulcer.

Chapter 23

A **(a)** Photoreceptor; exteroreceptor. **(b)** Chemoreceptor; exteroreceptor. **(c)** Chemoreceptor; exteroreceptor.

B Exteroreceptors since they respond to stimuli from outside the body.

C The light sensitive film.

D Eyelids and lacrimal (tear) glands are adaptations for terrestrial life. They help protect the surface of the eye against dust and a dry atmosphere.

Frogs need eyelids and lacrimal glands as they live partly on land.

E **(a)** The ciliary muscles relax.
(b) The lens becomes thinner due to tension on the suspensory ligaments.
(c) The pupil dilates.

F Night sight (ie vision in dim light) needs rods that work properly. Vitamin A is necessary for the formation of retinal in rhodopsin.

G **(a)** Blue, green and red cones.
(b) Blue cones only.

Chapter 24

A The male is brightly coloured as a form of sexual advertisement. The female is drab as a form of camouflage.

B Negative phototaxis.

C It can obtain food immediately from the parent. It does not have to waste time by learning how to obtain food.

D We become used to the sound of a ticking clock in a room.

E Instrumental conditioning (using negative reinforcement).

F A large flock. The task of keeping watch is shared by more individuals, so there is less time watching and more time feeding.

Chapter 25

A Because they detect a change in internal conditions and then adjust it back to the normal level.

B In the tropics – they survive best in hot climates because they are ectotherms.

C Piloerection: we have no fur to make thicker.

D To replace the salt lost in the sweat.

E Warm: we have not much hair, relatively little fat, lots of sweat glands and we produce melanin.

F They are metabolically active and are involved in a rapid exchange of large amounts of materials.

G Any excess of the daily need would be deaminated and excreted.

Chapter 26

A Warm, moist, lots of food.

B Mucous secretions containing dust and dead cells, that should be swept by cilia to the exterior, remain trapped in the airways to cause congestion and infection.

C Anticoagulants stop blood clotting so that leech can feed freely. Uses: microsurgery, plastic surgery, dispersal of blood clots, treatment of chronic skin ulcers.

D Two basophils, a lymphocyte and a monocyte.

E Not likely to cause an allergic reaction.

F Some of the donors would have been blood group O, some of the recipients would have been blood group AB, and so no adverse reactions would have occurred.

G AB blood has A and B antigens on the surface of the red blood cells. These react with antibodies present in the blood of people from the other three groups.

Chapter 27

A 14.4 times per hour.

B Cortical, since an otter is not short of water.

C Blood pressure would drop, filtration would become ineffective, kidneys would suffer damage and fail.

D The loop of Henle is shaped like a hairpin, and it behaves in the same way as a countercurrent multiplier.

E Dark yellow urine indicates dehydration. Brown = trouble!

F Proximal and distal tubules, ascending limb of loop of Henle.

G Much more lost in faeces - dehydration.

H Makes the blood more concentrated, either by adding salt or removing water or both.

Chapter 28

A It defeats the object of sex – mixing genes from two individuals.

B Because those born by other means are certainly not dead.

C Heavily pregnant females cannot hunt.

D It would secrete more GnRF, thus stimulating more gonadotrophin release and therefore more oestrogen release.

E Temperature – not suitable because it is unreliable. Lunar cycles – actually quite good.

F Because they can easily find shelter and new food sources.

G Day 21.

Chapter 29

A They are pollinated by birds: their flapping would damage more delicate flowers.

B It maximises the chances of catching the pollen.

C No. They produce less pollen than wind pollinated plants; the grains are larger, sticky and are not carried by the air in significant amounts.

D **(a)** one set, **(b)** two sets, **(c)** three sets.

E Dicots.

F Bright colours attract animals, such as birds, which eat the fruit and disperse the seeds.

G Otherwise an insect visitor could move from thrum to pin on the same plant and the plant would then be self fertilised.

Chapter 30

A No: the species certainly goes on but individuals frequently perish.

B Tall.

Chapter 31

A DNA is double stranded, contains the sugar deoxyribose and the bases C, A, T, & G. RNA is single stranded, contains ribose and the bases C, A, U & G.

B **(a)** UAUGCGAUA. **(b)** 3. **(c)** tyrosine, alanine, isoleucine.

C They eat and digest protein, absorb amino acids, transport them in the circulation from the gut to cells.

D Anabolic.

E 61 different tRNAs carry only 20 different amino acids.

F Six: all ^{14}N strands, two hybrids.

G With no spindle, the chromosomes double but cannot separate, so a cell with twice the normal number of chromosomes results.

Chapter 32

A 50:50 (half Yy and half yy).

B If you use a plant that is TT, all the offspring of a cross with a Tt or a TT plant will be tall and you wouldn't gain any information.

C 700 000/2500 = 280.

D In previous centuries, villagers married people from the next village. Because we now have more transport, people move around the world more easily and so people meet and mate with people to whom they are unlikely to be related.

E X^hX^h. Female haemophiliacs must be a double recessive whose mother is a female carrier and whose father is a male sufferer. The chance of two such people meeting and marrying is unlikely and male haemophiliacs often don't live long enough to reproduce.

F One in two chance.

G No, If the mother is O and the father is A, the child could be only A or O.

H 9 agouti, 3 black, 4 albino.

Chapter 33

A Acquired characteristics don't affect an organism's DNA and so cannot be passed on to the next generation.

B Insects have short life-spans and high reproductive capacity; whales and rhinos have long life-spans and only produce a handful of offspring during their life time.

C Peppered moth, Biston betularia: selection of melanic form following industrial revolution. Hawks: selection for great visual acuity. Cheetah: selection for high speed running.

D **(a)** Analogous, **(b)** Homologous.

Chapter 34

A It performs transcription in reverse: it makes DNA from RNA.

B Because the smaller recognition site is more common in any DNA sequence.

C Optimum temperature (usually 37 ˚C); oxygen; nutrients (sugars, amino acids); sterile (no contaminating organisms).

Chapter 35

A Temperate deciduous forest.

B Abiotic: **(a)**, **(c)**, **(d)**. Biotic: **(b)**, **(e)**, **(f)**.

C Habitat is where an organism lives, the term ecological niche also describes what it does and how it relates to its physical environment and to other organisms.

Chapter 36

A Clay can lots of water and has very little air. Water heats up slowly, so a wet clay soil often feels cold. It is heavy: its particles pack together more tightly than more open sandy soil or loam.

B Loam. Sand lets water and minerals run through too easily. Clay is impervious and does not allow a build-up of nutrients.

C Water content would be different. Values measured depend on conditions, eg, on a hot day, soil would have less water in it at the end of the day because of evaporation.

D Bulbs have a large food supply which gives the young plant a head start so that it can complete its life cycle early, flowering before its light supply is cut off by taller plants coming into leaf.

E Keep out of direct sunlight, away from wind chill, ensure the thermometer does not get wet – evaporation cools.

F Intraspecific (within species) aggression.

Chapter 37

A The transfer of energy from one organism to another.

C 7 levels. Unusual: most food chains have only 4 or 5 trophic levels.

D **(a)** Normal pyramid, **(b)** inverted pyramid.

E Mammals are endotherms: they use much of their energy to maintain body temperature.

Chapter 38

A 305 snails.

B Small mammals have a high metabolic rate and need to feed often. Left without food and water, they die in a few hours. Put a small amount of food and some bedding in the traps.

C Intra: within a species. Inter: between different species.

Chapter 39

A Sustainable: used but continually replaced, eg solar power, hydroelectric power. Unsustainable: when used it is gone forever, eg coal deposits, oil, gas.

INDEX